INFORMATION THEORY OF MOLECULAR SYSTEMS

INFORMATION THEORY OF MOLECULAR SYSTEMS

ROMAN F. NALEWAJSKI

Department of Theoretical Chemistry
Jagiellonian University
Cracow, Poland

ELSEVIER

Amsterdam – Boston – Heidelberg – London – New York – Oxford
Paris – San Diego – San Francisco – Singapore – Sydney – Tokyo

Elsevier
Radarweg 29, PO Box 211, 1000 AE Amsterdam, The Netherlands
The Boulevard, Langford Lane, Kidlington, Oxford OX5 1GB, UK

First edition 2006

Library of Congress Cataloging-in-Publication Data
A catalog record for this book is available from the Library of Congress

British Library Cataloguing in Publication Data
A catalogue record for this book is available from the British Library

ISBN-13: 978-0-444-51966-5
ISBN-10: 0-444-51966-1

For information on all Elsevier publications
visit our website at books.elsevier.com

Printed and bound in the United Kingdom

Transferred to Digital Print 2011

PREFACE

In chemistry an understanding of the electronic structure of molecules comes from transforming the experimental or computational results into statements in terms of such chemical concepts as atoms-in-molecules (AIM), their collections, e.g., functional groups, and the chemical bonds representing molecular "connectivities" between bonded atoms. Atoms in molecules are known to be only slightly changed relative to their corresponding free-atom references, mostly in the outer (valence) shells. These displacements in atomic densities are represented by the familiar density difference (deformation) function of quantum chemistry. This book is an exposition of a novel perspective on the electronic structure of matter, which results from applying the entropy/information concepts of Fisher and Shannon to the electron probability distributions in molecular systems. It uses the standard principles and techniques of the *Information Theory* (IT) to extract a "chemical" interpretation of the molecular electron densities in terms of bonded atoms, reactants, chemical bonds, electron shells, lone electron-pairs, etc. Other classical issues in the theory of electronic structure can be tackled in a similar way, e.g., the information origin of the chemical bond, atomic valences and bond contributions, trends in chemical reactivity, and various aspects of molecular similarity. This book aims at introducing the information theory of molecular systems to graduate students and researchers from all areas of chemistry and physics, especially those with an interest in new ways of looking at the subject.

A general theme of this book is the electron density as a source and carrier of information about the molecular structure and reactivity. The classical structural and reactivity rules will be briefly reexamined, stressing the continuity of ideas in chemistry and exposing the interrelations between their original, mostly intuitive basis and more rigorous foundations in IT and *Density Functional Theory* (DFT) of Hohenberg, Kohn and Sham. The book is designed to introduce the new subject of the information theory of molecules and their constituent fragments to any scientist familiar with rudiments of the molecular quantum mechanics, DFT and basic elements of IT. For completeness, the background material will be briefly summarized in the first three chapters. It will be used in the remaining chapters, which contain the exposition of the book main theses.

Information theory provides measures of both the entropy/information contained in a single probability distribution and the information-distance (missing information, cross- or relative-entropy) quantities between different probability distributions, e.g., the entropy-deficiency concept of Kullback and Leibler or the conditional entropy and mutual information characteristics used in the theory of communication systems. Applying these concepts to the molecular realm introduces measures of the "disorder" (Shannon entropy) or "order" (Fisher information) contained in the electron density. The relevant variational principles, for assimilating in the optimized probability distribution the physical information contained in the constraints and references, in the most unbiased manner possible, provide convenient tools for making the informed-judgement decisions in the molecular structure/reactivity problems, and for extracting chemical information from the calculated distributions of electrons. In Chapter 3 the key elements of the theory for both a single and two probability schemes are introduced and the entropy descriptors of communication channels are summarized.

The basic concepts, principles and relations of an information-theoretic approach to the quantum theory of the electronic structure will be developed and applied to illustrative molecular and model systems. Such treatment gives rise to the *information-entropy representation* of the molecular states, which complements the familiar *energy-representation* of the density-functional and wave-function theories. Together these two levels of description provide a thermodynamic-like, unified perspective on molecules and their subsystems. The similarities of the combined DFT/IT description of molecular systems to ordinary phenomenological thermodynamics will be explored for both the equilibrium and non-equilibrium electron distributions.

Other major topics of the book will deal with the information and communication aspects of the chemical bond, molecular similarity and electron localization in molecules. A use of IT facilitates a formulation of new criteria of molecular similarity and generates novel indices of chemical reactivity. The reactivity indices developed in the combined DFT/IT approach include derivatives of both the system energy and its information entropy. In the theory of chemical reactivity, and particularly in the Charge Sensitivity Analysis (CSA), the second-order Taylor expansion of the system electronic energy in the reactant resolution plays an essential role in designing the adequate two-reactant indices. In Chapter 2 we shall provide an overview of such concepts. They include, e.g., the second-order derivatives of the system electronic energy, grand-potential, and other "thermodynamic" potentials, e.g., the chemical hardness and softness [or the Fukui Function (FF)] properties of the electron "gas" in molecules and their subsystems. The conceptual and computational advantages of DFT in this regard, particularly when supplemented by the entropic characteristics of subsystems from IT, will be emphasized throughout the book.

It follows from the basic theorems of DFT, which currently dominates the modern conceptual thinking about molecules and gives rise to an efficient and surprisingly accurate computational technique for very large systems, that the ground-state electron density carries the *complete* information about the molecular system in question, its electronic structure, trends in chemical reactivity, patterns of chemical bonds, etc. The equilibrium one-electron probability distribution uniquely determines the information content of the ground-state electron density, while the two-electron (conditional) probabilities in atomic resolution similarly determine the "communication" network between constituent atoms. Indeed, due to the electron delocalization, the information contained in atomic electron probabilities is transmitted via the network of chemical bonds throughout the whole molecule. In a sense the bonded-atoms, which continuously exchange electrons between themselves, "talk" to each other. The reorganization of the molecularly placed atomic distributions, from the densities of free atoms, which determine the "promolecular" reference of the familiar density-difference function, to the corresponding atomic pieces of the molecular electron density, can be viewed as a result of a flow of information throughout the *molecular communication channel*.

Molecules and their subsystems can be thus regarded as communication systems, which transmit "signals" of the electron allocations to AIM, from the free-atoms of the promolecular "input" to the bonded-atoms defining the molecular "output". Such information systems can be best described in terms of the entropic concepts developed in IT. The emerging communication theory of the chemical bond enables one to monitor the flow of information in the bond formation processes using a novel class of the entropy/information descriptors of both the overall chemical bond multiplicity and its covalent/ionic composition. In this approach the complementary "covalent" and "ionic" aspects of the chemical bond are reflected by the average uncertainty ("noise") of the channel and the amount of information flowing through the molecular communication network.

In the language of chemistry the concept of bonded atoms, the main building blocks of molecules, is paramount. The atomic fragments of the molecular electron density are known to retain most of the information contained in the corresponding free atoms of the periodic table of elements, exhibiting only subtle changes in their valence shells, due to the intra-atom "promotion" (hybridization, polarization) and the inter-atomic charge transfer to/from their respective molecular environments. Therefore, a substantial part of the book will be devoted to the information theory of atoms in a molecule.

The combined DFT/IT approach gives rise to a thermodynamic-like description of the equilibrium molecular systems and their fragments. The density fluctuations and the flows of electrons between subsystems can be also tackled using a related local description, which closely follows the ordinary irreversible thermodynamics. This development thus introduces an additional level of a thermodynamic-like causality into relations between perturbations and responses of molecular systems. Therefore, in the information-theoretic approach the whole experience of the ordinary thermodynamics can be employed in treating a variety of processes on a molecular level. It will be also demonstrated, that there is a wide range of problems in the theory of electronic structure and chemical reactivity, which can be successfully tackled by combining the concepts and techniques of IT and DFT.

This development emphasizes the importance of the complementary energy and entropy representations for gaining a more complete chemical interpretation of the molecular electronic structure in the subsystem resolution. Such dual variational principles have indeed been formulated in DFT: the minimum energy principle of Hohenberg and Kohn yields the ground-state density matching a given external potential due to the system nuclei, while the "entropic", fixed density (energy) search of Levy delivers the external potential matching a given v-representable density.

In fact the (information) entropy and energy representations of the molecular equilibrium states complement each other. Together they provide a versatile theoretical framework for describing a variety of displacements in the molecular electronic structure, which one encounters in chemistry. They include both the *"horizontal"* shifts, from one ground-state density to another, and the *"vertical"* changes, due to the flow of electrons between the constituent subsystems, for the fixed ground-state density of the system as a whole. An example of the "vertical" problem is represented by an "extraction" of the chemical interpretation of the fixed electron density of the molecule, e.g., in an exhaustive partitioning of the given molecular electron density into the AIM components. The generalized forces driving changes in the electronic structure, e.g., the so called *charge affinities*, can be defined using quantities defined in the complementary energy and entropy representations. For example, the entropic "forces" behind the flow of electrons between reactants in the donor-acceptor system can be defined in this way. Such descriptors combine the familiar FF response-quantities (derivatives of the system energy) with the information-distance densities (derivatives of the system entropy deficiency).

Also the effective external potentials of molecular subsystems can be in principle rigorously defined in the combined DFT/IT approach. Therefore, a given electron density of a molecular fragment can be viewed as representing the ground-state of an "embedded" part of the molecule. This perspective introduces an additional element of causality into the subsystem description, since each manipulation of the fragment density can then be interpreted as the ground-state response to the well defined perturbation of the embedding potential for the subsystem in question. Thus, a set of non-equilibrium densities of molecular fragments can be attributed an effective ground-state (equilibrium) representation. This is vital for the thermodynamic-like, phenomenological description of non-equilibrium, intermediate reconstructions of the electron distribution in molecular processes. In this way the equilibrium "thermodynamic" approach can be supplemented by the associated "instantaneous" description of the "vertical" processes involving molecular subsystems. The alternative representations of such a local treatment, corresponding to different measures of the missing information or alternative choices of independent state-parameters, have been examined and the corresponding "affinities" (*forces*) and "fluxes" (*responses*), which determine the associated "sources" of the local entropy deficiency, have been identified in close analogy to the phenomenological irreversible thermodynamics. For the linear dynamical processes in a (Markoffian) molecule they imply the local reciprocity rules, analogous to the familiar Onsager relations, which reflect basic symmetries between the linear effects of the subsystem affinities on fluxes.

This non-equilibrium development covers both the density fluctuations, relative to the (stationary) Hirshfeld distributions, which are always present in the open molecular subsystems, and the electron flows between constituent fragments of the molecule. The density fluctuations are the key ingredients of many chemical concepts, e.g., the chemical softness and Fukui function quantities. A freedom of choosing alternative state-parameters is also reminiscent of that present in the phenomenological thermodynamics. In this instantaneous IT-thermodynamic theory the stationary densities of the "stockholder" subsystems, previously regarded as static entities, appear as averages of instantaneous densities with the distribution of local fluctuations being related to the relevant missing-information density in a thermodynamic-like fashion.

Successful applications of IT to the electronic structure phenomena have demonstrated the theory potential in extracting the chemical interpretation from a given electron distribution in a molecule. For example, the theory was shown to provide the entropic justification of the "stockholder" principle, which was used by Hirshfeld to partition the molecular electron density into atomic pieces. The same approach applied to the joint *many*-electron distributions results in the generalized "stockholder" rule for dividing the molecular N-particle density. The resulting (overlapping) densities of bonded atoms exhibit proper asymptotic properties, equalize the subsystem chemical potentials, and satisfy unique equalization rules for several information-distance quantities, which make them attractive concepts for interpretations in chemistry. This inter-subsystem equalization of the entropy-deficiency densities of molecular fragments, satisfied only by the equilibrium (Hirshfeld) subsystems, provides a new information-theoretic perspective on the equilibria in the mutually open parts of a molecule. The stockholder AIM were shown to preserve as much as possible of the information contained in the electron densities of the free atoms. They generally exhibit a single cusp at the atomic nucleus and decay exponentially at large distances from the molecule. These bonded atoms reflect all typical, intuitively expected changes due to the formation of chemical bonds: an overall contraction of the AIM densities due to an increased attraction by remaining atoms, their hybridization/promotion and polarization towards the bonding partners, as well as the charge-transfer effects due to differences in atomic electronegativities.

The book will illustrate and emphasize the unifying role played by IT in physics and chemistry. It will show how important the "entropic" tools are for gaining a better understanding of the "chemistry" behind the calculated molecular electron distributions. In future such information-theoretic concepts should facilitate a more direct connection between the *ab initio* results of computational quantum chemistry and such concepts of the mostly intuitive language of chemistry as AIM, bond multiplicities, promotion energy, amount of charge-transfer, electronegativity, or the hardness/softness characteristics of the electron gas in a molecule.

I am greatly indebted to Professor Robert G. Parr, the chief protagonist of the Conceptual DFT, for many helpful discussions on various topics covered by the book, during Author's several visits to the University of North Carolina at Chapel Hill, where the subject began. His youthful, contagious enthusiasm for the conceptual developments in quantum chemistry, for still new ways of approaching classical issues in the quantum theory of electronic structure, have been both encouraging and stimulating over many years of our collaboration and friendship. I would also like to thank Prof. Artur Michalak, Dr. Sigfrido Escalante and Ms. Elżbieta Broniatowska for preparing the figures. Thanks are also due to Profs. P. W. Ayers, K. Jug, J. Korchowiec, A. M. Köster, A. Michalak and J. Mrozek, for helpful discussions on various parts of the manuscript.

Cracow, October 2005 ROMAN F. NALEWAJSKI

CONTENTS

ACRONYMS

A	Acidic Reactant, fragment
AB	Acid-Base, interaction, complex
ADL	Atom-Diatom Limit
AIM	Atoms-in-Molecules
AO	Atomic Orbitals
AP	Atomic Promolecule, reference state
B	Basic, reactant, fragment
BEBO	Bond-Energy–Bond-Order, surface
BO	Born-Oppenheimer, approximation
CA	Compliant Approach, to electronic and nuclear displacements
CBO	Charge-and-Bond-Order, matrix
CI	Configuration Interaction, theory, method
CM	Centre of Mass
CS	Charge Sensitivities
CSA	Charge Sensitivity Analysis
CT	Charge Transfer, stage, amount
CTA	Charge-Transfer Affinities
DA	Donor-Acceptor, interaction, complex, reactive system
DFT	Density Functional Theory
DFT/IT	DFT and IT, combined approach
DIM	Diatomics-In-Molecules, surface
DNM	Density Normal Modes
E	Electrophilic, site, reagent, reactivity
EE	Electronegativity Equalization
EEM	Electronegativity Equalization Method
EEP	Electronegativity Equalization Principle
EF	Electron-Following, perspective, transformations
ELF	Electron Localization Function
EP	Electron-Preceding, perspective, transformations
EPI	Extreme Physical Information, principle
F	Fisher information, intrinsic accuracy
FET	Frontier Electron Theory
FF	Fukui Function, electronic
FO	Frontier Orbital, theory, electron density
GGA	Generalized Gradient Approximation
GIP	Generalized Information Principle
H	Hirshfeld, stockholder partitioning of molecular electron density
HF	Hartree-Fock, theory, method
HK	Hohenberg-Kohn, theory, method
HOMO	Highest Occupied Molecular Orbital, frontier orbital
HR	Hardness Representation
HSAB	Hard-Soft-Acids-and-Bases, principle
IFA	Independent Fragment Approximation
IT	Information Theory
K	Kullback cross-entropy, entropy deficiency, missing information, information distance, (divergence)

KL	Kullback-Leibler cross-entropy, entropy deficiency, missing information, information distance (directed divergence)
KLI	Krieger-Li-Yafrate, method
KS	Kohn-Sham, theory, method, orbitals
LCAO	Linear Combinations of Atomic Orbitals
LDA	Local Density Approximation of DFT
LEPS	London-Eyring-Polanyi-Sato, surface
LR	Linear Response, function, approximation
LSDA	Local Spin-Density Approximation of DFT
LTR	Legendre-Transformed Representations
LUMO	Lowest Unoccupied Molecular Orbital, frontier orbital
MCP	Maximum Covalency Path
ME	Maximum Entropy, principle
MEC	Minimum Energy Coordinates, of nuclei
MED	Minimum Entropy Deficiency, principle
MEP	Minimum Energy Path
MGC	Maximum Global Covalency, structure
MO	Molecular Orbital, theory, method
MP	Møller-Plessett, theory, method
N	Nucleophilic, site, reagent, reactivity
NFF	Nuclear Fukui Function
OEP	Optimized Effective Potential, method
OPM	Optimized Potential Model
P	Polarization, stage
PES	Potential Energy Surface
PMO	Perturbational Molecular Orbital, theory
PT	Perturbation Theory
R	Radical, site, reagent, reactivity
S	Shanon, information theory, entropy
SAL	Separated-Atom Limit
SBC	Symmetric Binary Channel
SCF	Self-Consistent-Field, method
SFA	Separate Fragment Approximation
SO	Spin Orbitals
SP	Stockholder Principle
SR	Softness Representation
SRHF	Spin-Restricted Hartree-Fock, theory, method
SRL	Separated-Reactant Limit
SUHF	Spin-Unrestricted Hartree-Fock, theory, method
TS	Transition-State, theory, complex
UMP2	Spin-Unrestricted Møller-Plessett theory including double excitations
VB	Valence-Bond, theory, structures
ZMP	Zhao-Morrison-Parr, procedure

1. INTRODUCTION

A brief survey of modern concepts and principles of the electronic structure and chemical reactivity is presented with an emphasis on the importance of chemical concepts for understanding the molecular behavior. The specificity of the *chemical* interpretation of molecular processes, in terms of AIM, chemical bonds, functional groups, reactants, etc., is commented upon. The classical structural and reactivity rules are reviewed. The quadratic Taylor expansion of the electronic energy of molecular systems in powers of displacements (perturbations) of the system state-parameters is introduced. It is defined by the generalized response quantities: "potentials", the first partials of the energy, and charge sensitivities, the second partials of the energy with respect to the system parameters of state. This series constitutes an adequate framework for describing reactant subsystems in a bimolecular reactive system. The role of the electronic density as the source and carrier of the complete information about the system ground-state equilibrium and all its physical and chemical properties is stressed. Basic elements of the electron wave-function and density-functional theories of electronic structure are summarized and the conceptual advantages of DFT over the standard wave-function approach are emphasized. The Euler equation for the ground-state density, the DFT equivalent of the Schrödinger equation of the wave-function theory, is discussed in some detail. It embodies the crucial ground-state relation between the equilibrium distribution of electrons and the external potential due to the system nuclei. This equation is shown to imply the chemical potential (electronegativity) equalization throughout the physical space. A distinction is made between transitions from one ground-state density to another, called here the "*horizontal*" displacements of the system electronic structure, and those corresponding to flows of electrons between molecular subsystems, for the fixed density of the molecule as a whole, called the "*vertical*" displacements.

1.1. GENERAL OUTLOOK

The prediction of chemical reactivity presents a constant challenge to chemists, who desire to define the optimum conditions for performing specific reactions. The basic aim of the so called reactivity "theories" is to predict reactivity trends or to find an *explanation*, in chemical terms, of the experimentally or computationally determined course of a reaction. Such theories have to provide means of systematization, recognition of regularities and rationalization of the myriads of established experimental and computational facts, to disclose the fundamental causes governing the reactivity phenomena. The most general of them are formulated in terms of the appropriate variational principles or the most favorable "matching" rules for the crucial physical properties of reactants (global or regional), which uncover the decisive factors responsible for the preferred direction of a given chemical process.

Investigations into the primary sources of the observed chemical behaviour of molecules cover both the thermodynamic/statistical and quantum-mechanical laws of chemical change. For example, the concept of an activation energy in a bimolecular reaction is statistical in character, but the actual value of this critical energy of reactants, which is required for the reactive outcome of their collision, cannot be understood without the quantum-mechanical description of changes in the electronic structure of reacting species.

The ultimate goal of theoretical chemistry is to predict and understand the electronic structure of chemical compounds and their reactions using concepts and techniques of both the *static* and *dynamical* approaches. The basic objective of the dynamical treatment is to calculate the *rates* of chemical reactions from the first principles. Given the interaction potential for the nuclear motion in the specified system of reactants, one should in general be able to determine the probabilities, cross-sections, and rate constants for fundamental elementary reaction processes by solving the quantum-mechanical equation of motion for the system. This dynamical goal, however, has so far been realized only for very simple reactions involving only three or four atoms, due to the computationally immense task in the theoretic determination of the complete electronically adiabatic, Born–Oppenheimer (BO) Potential Energy Surface (PES), and in solving the Schrödinger equation for nuclear motions.

Therefore, much of the present understanding of the chemical reaction dynamics at the molecular level has come about by using limited information about the multidimensional PES. For example, the model (analytical) PES, reproducing a network of selected *ab initio* points, or approximate methods, e.g., the classical trajectories, have been used to probe the dynamics of elementary reactive collisions. Another familiar example is the statistical Transition-State (TS) theory, in which only the data on the geometry and frequencies of the separated reactants and the TS complex are required to convert this limited information about the interaction between reactants into measurable rate quantities. The DFT rooted molecular charge sensitivities (CS) constitute attractive (static) concepts, in terms of which the truly two-reactant reactivity criteria can be defined within CSA, for both the externally (or mutually) closed or open subsystems (see: Ayers and Parr, 2000, 2001; Baekelandt *et al.*, 1993; Cohen, 1996; Chattaraj and Parr, 1993; Gázquez, 1993; Gázquez *et al.*, Geerlings *et al.*, 2003; Nalewajski, 1993, 1995a,b, 1997a,b, 1999, 2000a, 2002d, 2003a; Nalewajski and Korchowiec, 1997; Nalewajski *et al.*, 1996).

Recently, the familiar HK variational principle of DFT, which determines the electron ground-state density, has been interpreted (Nalewajski, 2005d) as that for the extremum of the electronic energy subject to the information entropy constraint, in close analogy to the familiar criterion of the thermodynamical equilibrium of macroscopic systems in the *energy*-representation (Callen, 1962). The equivalent extremum rule for the molecular entropy deficiency, subject to the constraint of a constant electronic energy, has also been given, again paralleling the familiar *entropy*-representation principle of classical thermodynamics. In this development the electronic chemical potential (negative electronegativity) of DFT (see Parr and Yang, 1989) appears as the system global information "temperature". The associated local chemical potential gives rise to a similar "thermodynamic"-like description of the non-equilibrium electron densities of molecular systems in terms of the local information temperature. Of similar character is the application of the *Extreme Physical* (Fisher) *Information* (EPI) principle of Frieden (2000) to derive the Kohn-Sham (KS) (1965) equations of DFT (Nalewajski, 2003c), and to explore the entropic principles in Daudel's Loge Theory (Aslangul et al., 1972; Daudel, 1969, 1974) of the molecular electronic structure (Nalewajski, 2003d).

It will be demonstrated in the book that the combined DFT/IT approach allows one to treat objectively both the "horizontal" and "vertical" displacements of the molecular electronic structure in a thermodynamic-like fashion. The "vertical" problem is vital for extracting the chemical interpretation from the known molecular electron density, in terms of such chemical concepts as bonded atoms, functional groups, reactants, lone electron pairs, and bonds, which connect the constituent subsystems in the molecule. For example, it has recently been demonstrated (Nalewajski and Parr, 2000, 2001; Nalewajski, 2002a, 2003c) that IT can be successfully used to tackle the definition of AIM, by searching for atomic densities, which reproduce the density of the system as a whole and exhibit the least information distance relative to the corresponding free atoms of the "promolecule". These effective information-theoretic distributions of electrons in chemical atoms can be monitored at different stages of their reconstruction in a molecular environment, e.g., the optimum *polarization* (P) of the mutually closed atoms and after the *charge-transfer* (CT) between the system constituent atoms. Such information-theoretic AIM were shown to be identical with the familiar "*stockholder*" atoms of structural chemistry (Hirshfeld, 1977).

These information-theoretic atoms have been shown to be independent of the applied measure of the information distance, and they exhibit attractive, thermodynamic-like properties (Nalewajski and Parr, 2001; Nalewajski, 2002a-d, 2003a-c). In chemistry these infinite (overlapping) AIM, referenced to the corresponding free atoms of the promolecule and immersed in the molecular environment composed of the remaining atoms, constitute natural building units of molecules. Indeed, they conform to several classical ideas in chemistry, which strongly emphasize the atomic density/orbital overlap as the primary source of the chemical bond. The entropic definition of bonded atoms complements the famous Bader's (1990) concept of the non-overlapping topological atoms, defined by the partitioning the physical space into exclusive atomic "basins", separated by the so called "zero-flux" surfaces of the molecular electron density. These quantum-mechanically determined boundaries effectively partition the molecular electron densities into the exclusive atomic pieces, which are solely referenced to the *molecular* state. This is in contrast to the stockholder AIM, which are defined with respect to the free-atom (promolecular) reference.

The information-theoretic treatment of the sub-molecular reality of bonded molecular fragments gives rise to a "thermodynamic" description of molecules and their constituent fragments in terms of the entropy-equilibrium molecular subsystems, so important for the language of chemistry. Since molecular fragments do not constitute the quantum-mechanical "observables", they cannot be verified experimentally. The bonded atoms of chemistry ultimately represent the noumenons of Kant (Parr *et al.*, 2005). Nonetheless, they can be partially validated either by their ability to conform to the established chemical concepts or by the extra causality they offer in describing the molecular phenomena, e.g., via the demonstrated parallelism to the ordinary thermodynamics. It will be argued throughout the book that by using the IT approach to define molecular subsystems one indeed generates a "chemical" interpretation with thermodynamic-like causal relations between perturbations and responses of molecular subsystems. Therefore, within such an information-theoretic outlook on the molecular and sub-molecular electronic structure the whole experience of the ordinary thermodynamics can be employed in treating a variety of subtle processes in chemistry.

To summarize, besides providing the entropic justification of the "stockholder" AIM, IT has been shown to give rise to new criteria of molecular similarity (Nalewajski and Parr, 2000; Nalewajski and Broniatowska, 2003b), electron localization (Nalewajski *et al.*, 2005), the entropic treatment of the polarization (promotion) and CT stages of the reorganization of AIM (Nalewajski and Loska, 2001), when they form chemical bonds in a molecule, and a thermodynamic-like description of molecular systems and their fragments (Nalewajski and Parr, 2001; Nalewajski, 2002c, 2003a,b, 2004a). This development also includes new descriptors of the electron-transfer phenomena in reactive systems (Nalewajski and Świtka, 2002; Nalewajski, 2003a), entropic bond multiplicities and their ionic and covalent components (Nalewajski, 2000c, 2004b-e, 2005a-c). The density fluctuations and flows of electrons between subsystems have also been tackled in the local "thermodynamic" description (Nalewajski, 2002c, 2003a,b, 2004a), which closely follows the ordinary irreversible thermodynamics (Callen, 1962).

Finally, let us just mention other applications of IT in science (see, e.g., Brillouin, 1956), including spectacular applications in physics (Frieden, 2000), particularly in statistical thermodynamics, and in molecular biology (Yockey, 1992). IT plays the unifying role in physics by facilitating derivations of all its basic laws from the common EPI principle using the Fisher information measure (Frieden 2000). Illustrative applications in chemical physics also involve problems in chemical kinetics (Agmon and Levine, 1977; Bernstein, 1982; Levine, 1978), the definition of molecular "loges" (Aslangul *et al.*, 1972), the "surprisal" analysis and synthesis of the electron density (Gázquez and Parr, 1978; Politzer and Parr, 1976; Sears, 1980; Wang and Parr, 1977), the Compton profiles and momentum density (Gadre and Sears, 1979; Gadre 1984, 2002; Gadre, Bendale, *et al.*, 1985; Gadre *et al.*, 1985), density functionals, DFT interpreted as local thermodynamics, and the electron correlation problem (Acharya *et al.*, 1980; Hõ *et al.*, 1995; Esquivel *et al.*, 1996; Morrison *et al.*, 1990; Morrison and Parr, 1991; Nagy and Parr, 1994; 1996, 2000; Parr and Wang, 1997; Parr and Yang, 1989; Sears, 1980; Sears *et al.*, 1980; Yáñez *et al.*, 1995; Ziesche, 1995). Other examples include issues in the theory of transferability of molecular subsystems (Ayers, 2001) and some general topics in quantum mechanics (Mycielski and Białynicki-Birula, 1975; Białynicki-Birula and Mycielski, 1976; Frieden 2000; Gadre 2002).

1.2. A NEED FOR THE CONCEPTUAL APPROACH

The last decades have witnessed a dramatic growth of modern quantum chemistry, both in its conceptual ideas and computational techniques. The conceptual theory generates means for understanding the structure and chemical behavior of molecular systems, and for interpreting results of theoretical calculations. The *ab initio* data, often of an admirable accuracy, are now generated using both the wave-function and DFT methods, with a strong tendency of the latter to dominate calculations on very large systems. For recent reviews on conceptual developments in DFT the reader is referred to (Nalewajski, 2002d, Nalewajski et al., 1996; Nalewajski and Korchowiec, 1997; Mortier and Schoonheydt, 1997; Geerlings et al., 2003).

These qualitative and quantitative theoretical results are often synergetically combined with laboratory techniques, verifying experimental data and guiding the researchers in their planning of future experiments. However, the wavefunctions resulting from the modern high-level methods of computational quantum chemistry are so immensely complex that they cannot be immediately understood in simple and physically or chemically meaningful terms. The categorization and interpretation objectives in theoretical chemistry call for the well founded general principles and conceptual models, which are both transparent, intuitively appealing, and useful for both qualitative and semiquantitative applications to molecular systems of interest in chemistry.

The rates and mechanisms of chemical reactions can be predicted, in principle, by the standard methods of statistical thermodynamics, in terms of the partition functions of reactants and the transition-state complex. However, the range of applicability of the TS (absolute rate) theory is severely limited by the fact that an evaluation of the vibrational partition function for the TS complex of the elementary process of interest requires a detailed consideration of the whole PES for the reactive system. The calculation of the absolute rate constants is thus possible only for relatively simple systems. This indicates a need for a more approximate theoretical treatment of chemical reactions, i.e., the *conceptual reactivity theory*, which would allow chemists to go further in their predictions and understanding of properties of new compounds and outcomes of chemical interactions, particularly in large reactive systems of interest in the contemporary organic chemistry. Due to the diversity and ever increasing complexity of molecules and reactions, relatively crude assumptions have to be made in such simplified approaches to elementary molecular processes, and empirical factors are often introduced into theoretical expressions. Thus, from the purist point of view, such "theories" should be more appropriately classified as theoretical *models* of reactivity. An example of such a heuristic approach is the celebrated Hammond postulate of a relative similarity of the transition-state complex to reactants (products) in the exothermic (endothermic) reactions (Hammond, 1955; Johnson, 1975).

On one hand, such general conceptual tools *a posteriori* reduce the overwhelming amount of information embodied in the *ab initio* wavefunctions to a more manageable, qualitative level by extracting common roots of seemingly unrelated data. On the other hand, they provide a valuable means for the chemical understanding of the molecular structure and reactivity, enabling a subsequent informed "guess work" about the system behavior in a changed molecular environment and in planning a more precise characterization of future experiments. Such adequate theoretical models offer a rationale for *trends* within families of related compounds, and they bridge a gap between the rigorous quantum mechanics and empirical concepts of the intuitive, phenomenological chemistry.

The qualitative and quantitative theories/models of the electronic structure and chemical reactivity constitute inevitable and necessary ingredients of the scientific method of chemistry. Only a parallel advancement of both these branches marks the harmonious development of theoretical chemistry. The qualitative concepts determine the scientific vocabulary of interpretative chemistry, while the approximate model relations allow for a semi-quantitative prediction of trends implied by changing structural and experimental conditions.

A historical perspective on the molecular electronic structure, in terms of AIM bonds, electron-pairs, functional groups, etc., is the central and most fruitful theme in chemistry. A knowledge of the electronic and geometric structure parameters of isolated molecules already gives important clues for understanding the behavior of chemical compounds in different reactive environments. It constitutes a starting point for a subsequent, perturbative studies of molecular interactions. This *Separated Reactant Limit* (SRL) thus provides a natural and convenient reference state, at the early stage of the reactant mutual approach. The structure of separated reactants qualitatively reveals the expected main features of the preferred *Minimum Energy Path* (MEP), thus already determining gross features of the easiest ascent from reactants towards the transition-state complex, the exact location of which ultimately determines the activation barrier height.

Chemistry is concerned with properties and reactions of an enormous number of different compounds, which for the purpose of expediency are classified into similarity groups, e.g., those with the same functional group(s), so that the physical and chemical properties of a particular compound may be inferred from the behavior of any other member. A number of qualitative and quantitative relations have been formulated to relate properties of members belonging to the same and different similarity groups. Representative examples in the area of chemical reactivity are provided by the familiar directing influences of the electron-withdrawing and electron-donating substituents in benzene derivatives, as well as the related (experimental) correlations of Hammett (1935, 1937). These "*free-energy*" relationships (Marcus, 1969; Chapman and Shorter, 1972; Johnson, 1973) have been extremely valuable in helping chemists to predict the reactivity of chemical compounds and to understand a subtle inter-relationship between *reactivity* and *selectivity* in chemical processes.

The first task confronting the chemist is to identify the compound reactive sites as a function of the molecular structure, and to determine their relative reactivity trends. A complex organic molecule may contain several alternative *Nucleophilic* (N), *Electrophilic* (E), and/or *Radical* (R) centers, and hence the competition for these reaction sites is a very important general problem. To meet this challenge one has to understand how the molecular structure affects the reactivity at various active centers of the molecule. The relative reactivity of an active site may vary with the nature of the attacking agent (*ambident reactivity*). Ambidency may also be exhibited as a result of changing experimental conditions. Any *bona-fide* theory of chemical reactivity must provide a framework, which accounts for all these diverse reactivity phenomena. A distinction between the thermodynamic and kinetic controls of competing reactions is essential for a satisfactory explanation of such processes.

An understanding of reactivity trends in terms of the static reactivity criteria calls for the truly *two-reactant* theoretical treatment, which combines the molecule and attacking agent. Indeed, only such approaches provide an adequate basis for describing variations in reactivity of one reactant, and/or its particular site(s), with a changing character of the other reactant and its reactive sites. When two large species orient themselves relative to one another, at an early stage of a chemical reaction, an even more subtle challenge for the reactivity theory emerges. It is related to the fact that the very classification of chemical species as the *electrophilic* (electron-deficient, acceptor, acidic) or *nucleophilic* (electron rich, donor, basic) is only a relative one. More specifically, in such molecular interactions the relative acidic/basic properties of reactants or their respective active sites depend on the current state of the reaction partner, since reactants represent a strongly coupled parts of a single reactive system. Thus, a given molecular site may act as a base towards one (relatively acidic) site of the other reactant, while it can act as an acid towards another (relatively basic) site of the reaction partner.

The alternative functional groups in a molecular reactant are mutually coupled *via* the connecting atoms and bonds. Therefore, a chemical reaction taking place at one site is not without an influence on the current reactivity of the other site. The adequate reactivity theory must thus be both sufficiently rich in its conceptual basis and flexible in its theoretical framework, in order to fully account for all such *inductive* (coupled) reactivity effects.

A satisfactory reactivity theory must also be able to cover the issues of a subtle interplay between the electronic and geometrical coordinates of the reactive system (see, e.g., Nalewajski *et al.*, 1996; Nalewajski and Korchowiec, 1997; Nalewajski, 1999, 2000a; Cohen, 1996; Ayers and Parr, 2001). The so called *"mapping"* transformations between these two aspects of the molecular structure (Baekelandt *et al.*, 1995; Nalewajski, 1995b, 1999, 2000a; Nalewajski *et al.*, 1996; Nalewajski and Korchowiec, 1997; Mortier and Schoonheydt, 1997; Nalewajski and Sikora, 2000) provide unifying concepts for both the qualitative understanding and semi-quantitative characterization of such couplings in molecules and between reactants. Both the *Electron-Preceding* (EP) and *Electron-Following* (EF) transformations can be approached in this way. The former envisage a shift in the electron distribution as preceding, accelerating the associated motion of the system nuclei, while the latter, BO perspective, views the nuclear displacements as the driving force of changes in the molecular electron density (Nakatsuji, 1973, 1974).

The mapping relations involving both the local and normal coordinates of the system electronic and nuclear degrees-of-freedom have been formulated. Alternatively, the electronic atomic or collective structure parameters can be related to the *Minimum Energy Coordinates* (MEC) of nuclear displacements, which are formulated in the so called *Compliant Approach* (CA) to the electronic and nuclear displacements (Jones and Ryan, 1970; Decius, 1963; Swanson, 1976; Swanson and Satija, 1977; Nalewajski, 1995b; Nalewajski *et al.*, 1996; Nalewajski and Korchowiec, 1997). Through the mapping relations any shift in the nuclear position space can be "translated" into the conjugate displacement in the electron distribution, and *vice versa*. These transformations enrich a variety of diagnostic and interpretative tools of theoretical chemistry and provide a semi-quantitative characterization of the couplings between the nuclear and electronic molecular structures in chemical processes. For a more systematic review of physical quantities reflecting a coupling between the molecular electronic and nuclear structures the reader is referred to Nalewajski (1999, 2000a).

This development has decisively extended the range of applications of the theory of chemical reactivity, in comparison to the empirical structural rules of Gutmann (1978). The EF mapping relations, of the BO perspective, in principle allow one to diagnose trends in the electron redistribution, in response to a given (hypothetical or real) displacement of the system geometry. The "inverse" EP relations, of the Hellmann–Feynman-type perspective, are closer to an intuitive chemical thinking. They are required to solve another typical reactivity problem: how to manipulate the system electronic structure, e.g., the charge distribution of the *fine-grained*, local description or the effective oxidation states (net charges) of AIM in the *coarse-grained*, atomic resolution, to bring about a desired change in the system geometry and/or breaking the specific bond(s).

Other examples of the reactivity concepts related to the external potential are the *Nuclear Fukui Function* (NFF) and the linear density response function (see: Cohen, 1996; Cohen *et al.*, 1994, 1995, Parr and Yang, 1989; Nalewajski and Korchowiec, 1997; Nalewajski, 2000a; Ayers and Parr, 2001; Geerlings *et al.*, 2003).

1.3. CHEMICAL UNDERSTANDING OF MOLECULAR PROCESSES

Accurate theoretical calculations of the energy profile along the minimum-energy path on the molecular PES, which use the standard computational methods of quantum chemistry, cannot be itself regarded as a *theory*, but rather – as a computer *experiment*. In fact, most of the activity of theoretical chemists has been directed not towards an understanding the *rules* governing reactions, but towards a numerical determination of the physical properties of molecular systems. In the past the use of simplified models or principles, which permits a useful chemical information to be derived without such elaborate calculations, e.g., relative rates of admissible reaction channels, has provided a valuable insight into our understanding of molecular reactivity preferences. It also provided a way to analyze molecules and calculations on molecules in a theoretically consistent framework that allows chemists to *understand* the results of calculations in terms of intuitive concepts, which dominate the *language* of chemistry.

The *Molecular Orbital* (MO) theory has facilitated a deeper understanding of a wide range of physical properties of molecules, their relative reactivity trends, and the preferred pathways of chemical reactions. It provides a natural and standard framework to understand, at a qualitative or semi-quantitative level, what happens to the electronic structure, when a molecule is placed in a changed environment, e.g., in the presence of the catalyst or the reaction partner. This theory also warrants predictions of changes in the molecular geometry accompanying a given, real or hypothetical change in the system electronic structure. The MO reactivity theories have proven to be very useful to the experimentalist, who requires an understanding of *why* molecules react the way they do, *what* determines their electronic structure, and *how* this influences reactivity. They deliberately de-emphasize the computational aspect, aiming instead at a more qualitative understanding of electronic structure and chemical reactions. In other words, the MO reactivity ideas provide a basis for understanding the development of reactions along specific routes, even without an assistance of rather complex computer calculations. Such simple models invoke the classical concepts of the orbital symmetry, overlap of the electron distributions, as well as electronegativity and the hardness/softness descriptors of molecules and their constituent subsystems. The MO reactivity theories are not limited to the one-determinant description of the standard Hartree-Fock (SCF MO) method. Various *Configuration-Interaction* (CI) and *Valence-Bond* (VB) ideas are often invoked in qualitative models of chemical reactions (e.g., Epiotis, 1978; Shaik and Hiberty, 2004).

The concept of a conservation of the orbital symmetry in both organic and inorganic chemistry has proved to be a major advance in the theory of chemical reactivity (Woodward and Hoffmann, 1969, 1970). It has succeeded in bringing together and rationalizing diverse areas of the subject. This concept has provided a basis for the unified mechanistic approach to the cycloaddition reactions and various molecular rearrangements. Nowadays criteria of the orbital symmetry conservation and the related correlation diagrams constitute the standard part of the qualitative vocabulary of the modern organic chemistry (see, e.g., Gilchrist and Storr, 1972; Lehr and Marchand, 1972; Gill and Willis, 1974; Stone, 1978; Jones, 1979; Halevi, 1992). The celebrated *Woodward–Hoffmann Rules* have correlated a great number of existing chemical facts and stimulated further widespread experimentation. Similar ways of comprehending the geometry and reactivity of inorganic molecules have also been proposed (Albright *et al.*, 1985) thus demonstrating that simple concepts of symmetry and bonding are applicable to a chemical understanding of all molecules. General symmetry rules of chemical reactions have also been formulated by considering the symmetry restrictions on the excited-state contributions to the perturbed ground-state wave functions of reactants. Such terms describe the polarization and charge transfer due to the perturbation created by the normal-mode displacement of the nuclei along the reaction coordinate (Pearson, 1976; Bader, 1960; Bader and Bandrauk, 1968).

It should be stressed, however, that the characterization of reactions as *allowed* or *forbidden* by the symmetry criteria, carries no quantitative information. In many cases there are several allowed reaction paths, and it becomes necessary to distinguish between them, in order to determine which one is the most probable and, ideally, to estimate yield ratios and relative reaction rates. A number of perturbative methods have emerged for this purpose. Early work on organic reactivity concentrated on the conjugated π-electron systems, and as a result of this analysis various reactivity indices were proposed, e.g., the *π-electron density*, *free-valence* or the *self-polarizability* of atoms, with the high value of either index assumed to imply high reactivity. The localization energy method has assumed a model for the transition-state, the so called Wheland intermediate, in which both the attacking reagent and the substituted atom are bonded to a roughly tetrahedrally promoted carbon, unable to form π-bonds with the remainder of the original π-electron system. The associated change in the π-electron energy, the so called *localization energy*, is then taken as the reactivity index, by assuming that other contributions to the activation energy are likely to be approximately constant for a given type of the approaching agent. Dewar (1969) in his *Perturbational* MO (PMO) theory of organic chemistry has proposed an approximate way to estimate the localization energy, called the *reactivity number*.

The MO theory has been quite successful in interpreting and predicting the molecular orientations and stereoselections in a large variety of chemical reactions (Dewar, 1969; Fukui, 1975, 1987; Klopman, 1974; Dewar and Dougherty, 1975; McWeeny, 1979).The *Frontier Orbital* (FO) theory of Fukui and co-workers (e.g., Fukui, 1975, 1987; Fukui and Fujimoto, 1974) uses the FO density and the related *superdelocalizability* index to predict the reactivity preferences and most suitable orientations of molecular reactants. It has been found that the *electrophilic* (E) aromatic substitutions take place predominantly at the carbon position, where the π-electron density of the *Highest Occupied* MO (HOMO) reaches the maximum value. Accordingly, the atomic sites exhibiting the highest value of the π-electron density in the *Lowest Unoccupied* MO (LUMO) were confirmed as the preferred locations for the *nucleophilic* (N) aromatic substitution. These two crucial MO of each molecular system were termed the FO. They are expected to dominate the chemical interaction between reactants. Another reactivity index, called the *superdelocalizability*, has been derived taking into account the hyperconjugation in the transition-state complex of an aromatic substitution, between the aromatic π-electron system of the attacked molecule and the pseudo π-orbital of the subsystem consisting of the reagent and the hydrogen to be replaced in the reaction product.

The perturbational MO methods were also widely invoked after a formulation of the Woodward–Hoffmann rules, to treat chemical reactions more comprehensively (Klopman, 1974a). In these approaches the reactivity trends are not linked to a single term in the corresponding Taylor expansion of the interaction energy, but rather to the combined sum of contributions due to the steric interactions, electrostatic, polarization and electron-transfer effects, the solvation energy, etc. All these more elaborate treatments, using the semi-empirical formulation of the MO theory, take account of the charge distribution and the overlap between the reactant orbitals. In the familiar Klopman–Salem energy expression (Salem, 1968a,b, 1969; Klopman, 1968, 1974b) a domination of the electrostatic interaction between the substrates marks the so called Charge Control of the reaction, when reactants are both highly charged and relatively difficult to polarize. This is characteristic of the "*hard*" species in Pearson's (1973) terminology. One encounters the other extreme case of the dominating CT contribution to the interaction energy, due to the mixing of the filled orbitals on one molecule with the empty orbitals on the other molecule, when both reactants are uncharged and highly polarizable ("*soft*"). This category of chemical reactions is called to be the FO-controlled.

In this more general perturbational framework Klopman was able to rationalize the Pearson's (1973) *Hard and Soft Acids and Basis* (HSAB) principle that hard acids form stable complexes with hard bases, and soft acids with soft bases, respectively, whereas the complexes of hard acids with soft bases (or of soft acids with hard bases) remain relatively less stable. He was also able to take into account the nature of the attacking reagent in the electrophilic or nucleophilic aromatic substitution, and to show that the ratios of yields of the *ortho*, *meta*, and *para* products of the substitution of the benzene derivatives depend on the competition between the charge and frontier orbital controls.

This perturbation theory of chemical reactivity focuses on an early stage of the reactant approach, when the molecules are still distinct though close enough for the molecular orbital description of the combined reactive system to be valid, say separated by a distance of the order of 5–10 a.u. The implicit assumption is that the reaction profiles for the compared reaction paths are of similar shape, so that the trends of the predicted energy differences at an early point on the reaction coordinate reflect the differences in the activation energies.

The frontier-orbital approximation recognizes the interaction between the HOMO and LUMO on both reactants as the crucial effect controlling the course of a chemical process. In many cases an additional approximation is introduced by considering only a single HOMO–LUMO pair for the bimolecular system, for which the orbital energy separation is the smallest, e.g., the HOMO of the *donor* (basic) reactant and the LUMO of the *acceptor* (acidic) reactant. The argument against such a drastic approximation is that it neglects many contributions to the Klopman–Salem equation from other molecular orbitals, the combined effect of which may outweigh the selected frontier-orbital interaction. Further uncertainties arise in the unique determination of the orbital energies. To remedy this shortcoming the orbital energies have been substituted by the ionization potentials and electron affinities, by virtue of the familiar Koopmans' and Janak's theorems. Nevertheless, the frontier orbital theory undoubtedly works in most cases, though it may not be as universally successful as are the Woodward–Hoffmann rules.

In order to further justify the frontier-electron model Fukui has formulated the three supplementary principles (Fukui 1975): of the "positional parallelism between the Charge-Transfer and Bond-Interchange", of "narrowing the frontier-orbital separation", and of "growing the frontier-electron density" along the reaction path. Although these principles have been subject to some criticism by other theoreticians (see, e.g., Stone, 1978), they have correctly recognized the need to include the relaxation effects, of both the electronic and geometrical structures of reactants, with the progress along the reaction path. Indeed, a chemical reaction always involves a subtle coupling between the equilibrium electron (chemical bond) distribution on one side and the nuclear configuration on the other side, with the latter determining the external potential for the fast movements of electrons within the familiar Born–Oppenheimer (BO) approximation.

As we have already remarked in the preceding section, changes in the distribution of electrons due to the substrate interaction create extra forces acting on the nuclei. This Hellmann–Feynman, "*electron-preceding*" perspective (Nakatsuji, 1973, 1974a,b) is close to the intuitive chemical thinking, in which manipulations of the electronic structure are considered as preceding and ultimately accelerating the subsequent changes in the molecular geometry. The BO, "*electron-following*" perspective provides the complementary description, in which displacements in the nuclear positions precede the concomitant electronic relaxation (Nalewajski, 1999; 2000a; Nalewajski and Korchowiec, 1997; Nalewajski and Sikora, 2000). Clearly, the complete understanding of molecular mechanisms of chemical reactions must ultimately involve reactivity criteria relating to both these representations. In the former, the displacements in the reactant *electronic* degrees-of-freedom, e.g., electron densities or the condensed electron populations (or net charges) of AIM, are considered as the independent state-parameters of the reactive system, with *nuclear/geometric* parameters responding to this electron *perturbation.* In the latter, the displacements in the nuclear coordinates and the associated shifts in the reactant external potentials are viewed as the system independent state-variables, with electronic parameters responding to this perturbation in the system geometry.

The geometrical relaxations, in response to displacements in the electronic structure in the acid–base (acceptor–donor) reactive system, are also the subject of the intuitive bond-variation rules of Gutmann (1978). They also follow the Hellmann–Feynman (electron-preceding) perspective of Nakatsuji (1973, 1974a,b), who obtained interesting interrelations between changes in the electron density and nuclear configuration in a variety of contexts associated with chemical reactions. For example, it was observed that the centroid of a change in the electron density tends to lag the change in the nuclear coordinates in a movement away from a stable configuration, and it tends to lead the geometrical change in a movement away from an unstable geometrical structure towards the equilibrium one, thus creating the equilibrium geometry restoring force acting on the nuclei.

The interaction between the frontier orbitals is the strongest, when the mutual orientation of both reactants, at a given inter-molecular separation, gives the maximum overlap. The frontier electron theory identifies such a maximum overlap direction as the preferred one. The FO theory thus represents the *overlap* matching rule of chemical reactivity, valid for the frontier-controlled reactions in the Klopman classification. The corresponding rule for the charge-controlled reactions follows from the corresponding matching of the electrostatic potentials of both reactants. The largest charge stabilization should result, when the electron-deficient, positive regions of the electrostatic potential of one reactant overlap the most with the electron rich, negative regions of the electrostatic potential of the other reactant. The electrostatic potential analysis has now become firmly established as an effective guide to molecular interactions (see, e.g., Politzer and Truhlar, 1981; Murray and Sen, 1996). It is being applied to a variety of important chemical and biological systems, covering the preferred sites for the N and E attacks, solvent effects, catalysis, as well as the molecular cluster and crystal behavior.

The electrostatic potential is the unique functional of the molecular electron density and it exhibits interesting critical points, which reflect the opposing contributions from the nuclei and electrons. Its topological analysis (Gadre and Shirsat, 2000) supplements the related analysis of the electron density (Bader, 1990). Investigations of the local features of the electron density distribution and the associated Laplace field (Bader 1990) leads to the unique topological definition of AIM, chemical bonds, molecular structure, and structural changes. It can also be used to gain a valuable insight into the bonding mechanism, to identify the sites that are prone to the electrophilic or nucleophilic attacks, and the bonds that can be easily broken in a molecule (Bader, 1990; Kraka and Cremer, 1990). The maps of local features of the Laplasjan of the molecular density can be also used to predict the best matching of the electron-depleted regions of the acidic reactant with the electron-rich regions of the basic reactant, which can qualitatively determine their preferred mutual orientation in the course of a chemical reaction.

To conclude this short outline of main ideas of modern theories of the molecular electronic structure and chemical reactivity one should also recognize the important insights from the Valence-Bond (VB) theory (Heitler and London, 1927; London, 1928). This historic rival of the MO approach, has dominated chemistry until the mid-1950s and made a strong comeback from 1980s onward (see, e.g., Cooper, 2002; Shaik, 1989; Shaik and Hiberty, 1991, 1995, 2004). Its roots can be traced to the classical paper of Lewis (1916). The VB theory introduces into chemistry the concept of the bonding electron pair and the octet rule. The qualitative VB theory gives rise to a lucid insight into elementary chemical processes and produces key paradigms of chemical bonding and reactivity. It allows one to successfully tackle various issues in the molecular theory, including the aromaticity-antiaromaticity and the VB diagrams conceptualizing the chemical reactivity and the barrier formation by avoided crossing (resonance mixing) of the VB states that describe reactants and products. This qualitative theory also offers complementary insights into the factors that control the barrier heights and the competition between the σ and π electrons in determining the regular structure of the benzene ring. The quantitative variants, the *ab initio* VB methods, e.g., Generalized VB (GVB) scheme (Goddard and Harding, 1978), provide efficient computational tools for determining the outcomes of chemical reactions. One should also mention the use of the VB ideas in modeling the PES of elementary chemical reactions, e.g., the familiar LEPS (London-Eyring-Polanyi-Sato), DIM (Diatomics-in-Molecules), and BEBO (Bond-Energy–Bond-Order) surfaces used in numerical studies of the reactive scattering (see, e.g.: Murrell *et al.*, 1984).

1.4. TAYLOR EXPANSIONS OF THE ELECTRONIC ENERGY FOR MOLECULES AND REACTANTS

When considering a behavior of a single molecule or a family of chemically similar molecules in a given type of a chemical reaction, e.g., during the electrophilic, nucleophilic, or radical attack by a small agent, various *single-reactant* reactivity concepts have proven their utility in predicting the most reactive site (Dewar, 1969; Fukui, 1975, 1987; McWeeny, 1979). Such criteria are based upon the underlying notion of an inherent chemical reactivity of a molecule, or a hierarchy of relative reactivities of its parts, for the fixed reaction *stimulus* at each of the compared locations due to the perturbation created by the same attacking agent. This notion implies that the way the molecule reacts is somehow predetermined by its own structure. Clearly, this point of view is a very approximate one, since it neglects the mutual influence of one reactant upon the other.

A more subtle *two-reactant* description of chemical reactivity is required to probe alternative arrangements of two large molecules, e.g., in cyclization reactions, when several chemical bonds are being formed or broken. In order to account for the mutual influence of both molecular subsystems in a given bimolecular reactive system, the adequate reactivity criteria have to include the relevant *embedding* energy terms. For each part of the reactive system they involve the appropriate *reaction "stimulus"*, i.e., the subsystem perturbation created by a presence of the complementary subsystem at a given, say, early stage of the reactant mutual approach, and the conjugate *response* of the perturbed reactant. The normalized response quantities, per unit displacement in the system state-parameters, determine generalized "polarizabilities" (charge sensitivities) of reactants.

In general, the electronic ground-state energy E of a molecule can be regarded as functional of alternative sets of the system state-"variables", both global and local in character, which uniquely specify the system equilibrium state. The *global* state-parameters $\boldsymbol{X} = \{X_a\}$, refer to the system as a whole, while the *local* state-parameters, represented by functions $\boldsymbol{x}(\boldsymbol{r}) = \{x_\alpha(\boldsymbol{r})\}$, describe an infinitesimal local volume element at a given location in space. In what follows we shall denote the whole set of the system state parameters by the row vector $\boldsymbol{P} = (\boldsymbol{X}, \boldsymbol{x})$. Chemistry is fundamentally the science about transformations (reactions) and responses of molecules due to displacements in their environment, which modify the initial state-variables $\boldsymbol{P}^0$ by $\Delta\boldsymbol{P} = \boldsymbol{P} - \boldsymbol{P}^0 = \{\Delta\boldsymbol{X}, \Delta\boldsymbol{x}\}$. The chemical understanding of the molecular electronic structure is not limited to properties of isolated species, but it also covers a behavior of molecular reactants, when they are in contact with other agents, which create perturbation in their environment, reflected by the effective shifts $\Delta\boldsymbol{P}$ in the system state-parameters. These displacements can ultimately lead to a change in the pattern of bonds between the system constituent atoms, i.e. to an elementary chemical reaction. Such an understanding of the molecular behavior calls for CS, which measure the system responses to such hypothetical or real changes in the system physical degrees of freedom, both electronic and nuclear.

This scenario naturally connects to the quadratic Taylor expansion of the system electronic energy $E[\boldsymbol{P}]$, in powers of displacements in the system state-parameters, in which the partial derivatives, calculated for the system initial electronic/geometric structure, reflect the generalized potentials (first partials), grouped in the row vector $\boldsymbol{p} = \partial E(\boldsymbol{P})/\partial\boldsymbol{P}$, and polarizabilities (CS, second partials), grouped in the square matrix $\boldsymbol{\pi} = \partial^2 E(\boldsymbol{P})/\partial\boldsymbol{P}\partial\boldsymbol{P}' = \partial\boldsymbol{p}/\partial\boldsymbol{P}'$.

The 1st-order energy change, linear in displacements $\Delta\boldsymbol{P} = \{\Delta\boldsymbol{X}, \Delta\boldsymbol{x}\}$, is determined by the conjugate potentials $\boldsymbol{p} = (\boldsymbol{W}, \boldsymbol{w})$, global $\boldsymbol{W} = \{W_a = \partial E[\boldsymbol{X}, \boldsymbol{x}]/\partial X_a\}$, and local $\boldsymbol{w}$, defined by the *functional* partial derivatives (see Appendix A) with respect to the system local state-parameters, $\boldsymbol{w}(\boldsymbol{r}) = \{w_\alpha(\boldsymbol{r}) = \partial E[\boldsymbol{X}, \boldsymbol{x}]/\partial x_\alpha(\boldsymbol{r})\}$,

$$\Delta^{(1)}E[\boldsymbol{P}] \equiv \Delta^{(1)}E[\Delta\boldsymbol{P}] = \Sigma_a W_a \Delta X_a + \Sigma_\alpha \int w_\alpha(\boldsymbol{r})\, \Delta x_\alpha(\boldsymbol{r})\, d\boldsymbol{r} \equiv \boldsymbol{p}\, \Delta\boldsymbol{P}^{\mathrm{T}}. \tag{1.4.1}$$

In a bimolecular system $M = A\text{---}B$, consisting of reactants $Z = A, B$, the list of state-parameters includes those describing both subsystems: $\boldsymbol{P} = (\boldsymbol{P}_A, \boldsymbol{P}_B) = \{\boldsymbol{P}_Z\}$. The first-order energy change in M is thus determined by the additive 1st-order contributions (1.4.1) for each reactant:

$$\Delta^{(1)}E[\boldsymbol{P}_Z] \equiv \Delta^{(1)}E[\Delta\boldsymbol{P}_Z] = [\partial E(\boldsymbol{P})/\partial\boldsymbol{P}_Z]\, \Delta\boldsymbol{P}_Z^{\mathrm{T}} \equiv \boldsymbol{p}_Z\, \Delta\boldsymbol{P}_Z^{\mathrm{T}}, \qquad Z = A, B; \tag{1.4.2}$$

$$\Delta^{(1)}E[\mathrm{A\text{---}B}] = \Sigma_{Z=A,B}\, \Delta^{(1)}E[\Delta\boldsymbol{P}_Z]. \tag{1.4.3}$$

It should be realized, however, that in the Taylor expansion for the reactive system these energy terms couple the potentials $\boldsymbol{p}_Z = \{\boldsymbol{W}_Z, \boldsymbol{w}_Z\}$ of one reactant with the perturbations $\Delta\boldsymbol{P}_{Z'} = \{\Delta\boldsymbol{X}_{Z'}, \Delta\boldsymbol{x}_{Z'}\}$ due to the other reactant $Z' \neq Z$. Obviously, the influence of one reactant on the state-variables of the other reactant depends on the current electronic/geometric structure of both these subsystems in M. Therefore, already at the lowest, first-order expansion level the two-reactant nature of the interaction energy can be clearly recognized.

The quadratic form of displacements $\Delta\boldsymbol{P} = \{\Delta\boldsymbol{X}, \Delta\boldsymbol{x}\}$ in the system state-parameters, with coefficients representing the system generalized polarizabilities (CS) $\boldsymbol{\pi}$, defines the second-order energy change (see, e.g., Cohen, 1996; Coulson and Longuet-Higgins, 1947a,b; Geerlings *et al.*, 2003; Nalewajski, 1997, 1999, 2000a, 2002a,d; Nalewajski and Korchowiec, 1997; Nalewajski *et al.*, 1996; Parr and Yang, 1989; Sen, 1993):

$$\Delta^{(2)}E[\boldsymbol{P}] \equiv \Delta^{(2)}E[\Delta\boldsymbol{P}] \equiv \tfrac{1}{2}\int \Delta\boldsymbol{P}\,\boldsymbol{\pi}\,\Delta\boldsymbol{P}^{\mathrm{T}}\,d\tau$$

$$= \tfrac{1}{2}\,\{\textstyle\sum_a \sum_{a'} G_{a,a'}\,\Delta X_a\,\Delta X_{a'} + 2\sum_a \sum_\alpha \Delta X_a \int g_{a,\alpha}(\boldsymbol{r})\Delta x_\alpha(\boldsymbol{r})\,d\boldsymbol{r}$$

$$+ \textstyle\sum_\alpha \sum_{\alpha'} \int\int \Delta x_\alpha(\boldsymbol{r})\,\gamma_{\alpha,\alpha'}(\boldsymbol{r},\boldsymbol{r}')\,\Delta x_{\alpha'}(\boldsymbol{r}')\,d\boldsymbol{r}\,d\boldsymbol{r}'\}, \qquad (1.4.4)$$

where the symbol $\int d\tau$ denotes the integration over the relevant locations in space. It contains the global response quantities

$$\mathbf{G} = \{G_{a,a'} = \partial^2 E[\boldsymbol{X}, \boldsymbol{x}]/\partial X_a\,\partial X_{a'} = \partial W_a/\partial X_{a'} = \partial W_{a'}/\partial X_a\}, \qquad (1.4.5)$$

the local CS

$$\boldsymbol{\Gamma}(\boldsymbol{r}) = \{\Gamma_{a,\alpha}(\boldsymbol{r}) = \partial^2 E[\boldsymbol{X}, \boldsymbol{x}]/\partial X_a\,\partial x_\alpha(\boldsymbol{r}) = \partial W_a/\partial x_\alpha(\boldsymbol{r}) = \partial w_\alpha(\boldsymbol{r})/\partial X_a\}, \qquad (1.4.6)$$

and the non-local responses

$$\boldsymbol{\gamma}(\boldsymbol{r}, \boldsymbol{r}') = \{\gamma_{\alpha,\alpha'}(\boldsymbol{r}, \boldsymbol{r}') = \partial^2 E[\boldsymbol{X}, \boldsymbol{x}]/\partial x_\alpha(\boldsymbol{r})\,\partial x_{\alpha'}(\boldsymbol{r}')$$

$$= \partial w_{\alpha'}(\boldsymbol{r}')/\partial x_\alpha(\boldsymbol{r}) = \partial w_\alpha(\boldsymbol{r})/\partial x_{\alpha'}(\boldsymbol{r}')\}. \qquad (1.4.7)$$

Above, we have also listed the relevant Maxwell cross-differentiation identities.

The corresponding second-order energy for the bimolecular reactive system $M = A{-}{-}{-}B$, due to perturbations $\Delta\boldsymbol{P} = (\Delta\boldsymbol{P}_A, \Delta\boldsymbol{P}_B)$,

$$\Delta^{(2)}E[A{-}{-}{-}B] \equiv \Delta^{(2)}E[\Delta\boldsymbol{P}_A, \Delta\boldsymbol{P}_B] = \tfrac{1}{2}\textstyle\sum_Z \sum_{Z'} \int \Delta\boldsymbol{P}_Z\,\boldsymbol{\pi}_{Z,Z'}\,\Delta\boldsymbol{P}_{Z'}^{\mathrm{T}}\,d\tau \equiv \sum_Z \sum_{Z'} \Delta^{(2)}E_{Z,Z'}, \qquad (1.4.8)$$

contains the diagonal contributions [see Eq. (1.4.4)], for $Z = Z'$, $\Delta^{(2)}E_{Z,Z} = \Delta^{(2)}E[\Delta\boldsymbol{P}_Z]$, defined by the intra-reactant CS, $\boldsymbol{\pi}_{Z,Z} = \partial^2 E(\boldsymbol{P})/\partial\boldsymbol{P}_Z\partial\boldsymbol{P}_Z = \partial\boldsymbol{p}_Z/\partial\boldsymbol{P}_Z$, and the off-diagonal (inter-reactant) coupling terms, for $Z \neq Z'$, determined by the inter-reactant CS, $\{\boldsymbol{\pi}_{Z,Z'} = \partial^2 E(\boldsymbol{P})/\partial\boldsymbol{P}_Z\partial\boldsymbol{P}_{Z'} = \partial\boldsymbol{p}_{Z'}/\partial\boldsymbol{P}_Z = \partial\boldsymbol{p}_Z/\partial\boldsymbol{P}_{Z'}\}$:

$$\Delta^{(2)}E_{Z,Z'} = \tfrac{1}{2}\int \Delta\boldsymbol{P}_Z\, \boldsymbol{\pi}_{Z,Z'}\, \Delta\boldsymbol{P}_{Z'}^{\mathrm{T}}\, d\tau = \tfrac{1}{2}\{\textstyle\sum_{a,Z}\sum_{a',Z'} G_{a,Z;\,a',Z'}\, \Delta X_{a,Z}\Delta X_{a',Z'}$$
$$+ \textstyle\sum_{a,Z}\sum_{\alpha',Z'} \Delta X_{a,Z}\int g_{a,\alpha'}(\boldsymbol{r})\, \Delta x_{\alpha',Z'}(\boldsymbol{r})\, d\boldsymbol{r} + \sum_{a',Z'} \Delta X_{a',Z'} \sum_{\alpha,Z}\int g_{a',\alpha}(\boldsymbol{r})\, \Delta x_{\alpha}(\boldsymbol{r})\, d\boldsymbol{r}$$
$$+ \textstyle\sum_{\alpha,Z}\sum_{\alpha',Z'}\int\int \Delta x_{\alpha,Z}(\boldsymbol{r})\, \gamma_{\alpha,Z;\,\alpha',Z'}(\boldsymbol{r},\boldsymbol{r}')\, \Delta x_{\alpha',Z'}(\boldsymbol{r}')\, d\boldsymbol{r}\, d\boldsymbol{r}'\}, \qquad (1.4.9)$$

Such embedding energy contributions are in principle included in all 2nd-order perturbational approaches to reactive systems (Coulson and Longuet-Higgins, 1947; Klopman, 1974b; Fujimoto and Fukui, 1974; Fukui, 1975; McWeeny, 1979; Parr and Yang, 1989; Alonso and Balbás, 1993; Baekelandt *et al.*,1993; Chattaraj and Parr, 1993; Ciosłowski and Mixon, 1993; Gázquez, 1993; Nalewajski, 1993, 1995, 1997a, 1999, 2000a, 2002a,d; Nalewajski *et al.*, 1996; Mortier and Schoonheydt, 1997; Nalewajski and Korchowiec, 1997). As shown in Eqs. (1.4.8) and (1.4.9), the responses of reactants can be classified as "diagonal" (intra-reactant), when both displacements in the defining second derivative of the system electronic energy refer to the same subsystem, or "off-diagonal" (inter-reactant), when the two perturbations in the energy derivative correspond to different subsystems. In particular, the diagonal CS representing the normalized non-local responses reflect the influence of an attack (perturbation) on one site of the subsystem under consideration on its reactivity at the other location. Accordingly, the off-diagonal two-point CS account for the influence of an attack at the specified location in one reactant on the reactivity at the other location in the reaction partner.

We conclude that the second-order Taylor expansion for the bimolecular reactive system, summarized in Eqs. (1.4.1)-(1.4.3), (1.4.8), and (1.4.9), provides a consistent two-reactant framework, in which all couplings between the reactant descriptors can be adequately accounted for. It should be realized, however, that in this approximation one probes the reactant behavior only through their leading responses, which are linear functions of the applied perturbations. In other words, the quadratic Taylor expansion approach amounts to the *Linear-Response* (LR) approximation. To include the non-linear effects an expansion to higher-orders, e.g., cubic terms, should be used. However, the extra energy contributions in such more elaborate Taylor series should generally be small compared to those already present in the quadratic, LR approximation. These higher-order terms can be thought of as only slightly modifying the charge-sensitivities of the quadratic approach, into those representing the "dressed" molecular fragments in reactive system.

This coupling between reactants indicates that for the fixed reaction stimuli (perturbations) in the reactive system the trends in the electronic energy (reactivity) have to be indexed by vectors of generalized potentials and matrices of CS, since no single response quantity can fully reflect the net effect of a complicated pattern of all couplings present in the system. Moreover, changes in the electronic energy have to be supplemented in the full (second-order) interaction potential of the BO approximation, $\Delta W^{(1+2)}$, relative to the SRL, with the nuclear repulsion between reactants, V_{nn},

$$\Delta W^{(1+2)} \equiv \Delta E^{(1+2)} + V_{nn}, \qquad (1.4.10)$$

for the current geometry of the whole reactive system. The V_{nn} term has to be taken into account also in approximate, semiquantitative treatments, which use the separated reactant responses to approximate the Taylor expansion of the electronic energy at the finite inter-reactant separation.

For a series of similar reactions, e.g., attacks at alternative sites of the same molecule by atomic agents of a similar chemical character, one can assume that at comparable stages of the reactant approach the sum of the 1st-order electronic energy and V_{nn} remains approximately constant for all compared locations, so that the 2nd-order energy determines the preferred reaction path. For comparable values of the reaction perturbations (stimuli) at alternative sites the trends in the quadratic terms of the electronic energy are then reflected by CS themselves (Coulson and Longuet-Higgins, 1947a,b; Parr and Yang, 1984, 1989).

Notice, however, that such a simplified treatment cannot be used in a general reactive system involving two large reactants, the size and reactivity of which changes in a series of compared mutual orientations, often leading to different reaction paths. This is because at each site alternative orientations of reactants imply a different matching between the perturbation/potential and responses quantities. In such a general case the whole set of molecular potentials and CS has to be combined with the appropriate reaction stimuli to produce the overall interaction energy, to be eventually compared for a series of the probed geometries of the reactive system. This is important for predicting a direction of the energetically preferred approach of reactants at the crucial, early stage of the reaction, which sets the least activation course of the process and thus selects the preferred reaction event.

A remarkable progress in DFT of Hohenberg, Kohn and Sham ((1964, 1965) (see, e.g., Parr and Yang, 1989; Dreizler and Gross, 1990; Nalewajski, 1996a) besides offering efficient schemes for the electronic structure computations has provided an attractive framework for formulating novel concepts and rules describing behavior of molecular systems in different chemical environments (Parr and Yang, 1984, 1989; Nalewajski and Parr, 1982; Chattaraj and Parr, 1993; Ciosłowski and Mixon, 1993; Gázquez, 1993; Berkowitz and Parr, 1998; Nalewajski *et al.*, 1996; Cohen, 1996; Nalewajski and Korchowiec, 1987; Nalewajski, 1984, 1985, 1993, 1997b, 1999, 2000a,d, 2002a,d; Ayers and Parr, 2000; Geerlings *et al.*, 2003). In chemistry, and particularly in reactivity theory, this conceptual development has had a distinctly unifying character. For example, some of the originally intuitive, but remarkably successful tools of chemistry, such as the electronegativity (Mulliken, 1934; Sanderson, 1951, 1976; Iczkowski and Margrave, 1961; Gyftopoulos and Hatsopoulos, 1965; Sen and Jørgensen, 1987) and hardness (Pearson, 1973; Sen, 1993), which have long been part of the chemical vocabulary, have been shown to be fundamental an well defined (Parr *et al.*, 1978; Parr and Pearson, 1983).

According to the famous theorem by Hohenberg and Kohn (1964) and its shape-function reformulation for Coulomb systems by Ayers (2000), the ground-state electron density $\rho(\boldsymbol{r})$ or the density per electron $\sigma(\boldsymbol{r}) = \rho(\boldsymbol{r})/N$ (one-electron probability distribution called the *shape*-factor) carries the complete "information" about the non-degenerate quantum-mechanical state of the molecule. It uniquely identifies the shape of the system external potential due to the nuclei and the overall number of electrons and hence also the Coulombic molecular Hamiltonian. This exact result has given a new impetus towards using DFT and the IT of Fisher (1925), Shannon (1948, 1949), and Kullback and Leibler (1951, 1959) (see also: Brillouin, 1956; Jaynes, 1957; Abramson, 1963; Ash, 1965; Mathai and Rathie, 1975, Frieden, 2000) in the density-based chemical interpretation of the electronic structure of molecular systems in terms of AIM and bond multiplicities (Cedillo *et al.*, 2000; Nalewajski and Parr, 2000, 2001; Nalewajski, 2000c, 2002b-e; Nalewajski and Jug, 2001; Nalewajski and Loska 2001; Ayers, 2001). In DFT this approach leads to the IT "thermodynamics" of molecular systems and their fragments (Nalewajski and Parr, 2001; Nalewajski, 2002c, 2003b, 2004a, 2005d), a description in the spirit of the earlier DFT approaches (Ghosh *et al.*, 1984; Ghosh and Berkowitz, 1985; Nagy and Parr, 1994).

1.5. ELECTRON WAVE-FUNCTION AND DENSITY THEORIES

The molecular wave-function in the BO approximation for N electrons moving in the external field due to m nuclei in their fixed positions $\mathbf{R} = (\boldsymbol{R}_1, \boldsymbol{R}_2, \ldots, \boldsymbol{R}_m)$,

$$\Psi(N) = \Psi(\boldsymbol{x}_1, \boldsymbol{x}_2, \ldots, \boldsymbol{x}_N; \boldsymbol{X}_1, \boldsymbol{X}_2, \ldots, \boldsymbol{X}_m) \equiv \Psi(\mathbf{x}; \mathbf{X}) \equiv \Psi[(\mathbf{r}, \boldsymbol{\sigma}); (\mathbf{R}, \boldsymbol{\Sigma})] \equiv \Psi(\mathbf{x}), \quad (1.5.1)$$

of the Schrödinger quantum mechanics carries the complete information that can be known about a molecular system for its assumed geometry specified by nuclear positions (parameters). It depends on the spin $\boldsymbol{\sigma} = (\sigma_1, \sigma_2, \ldots, \sigma_N)$ and position $\mathbf{r} = (\boldsymbol{r}_1, \boldsymbol{r}_2, \ldots, \boldsymbol{r}_N)$ coordinates of electrons and, parametrically, on the set of the nuclear spins $\boldsymbol{\Sigma} = (\Sigma_1, \Sigma_2, \ldots, \Sigma_m)$ and their position coordinates $\mathbf{R} = \{\boldsymbol{R}_\alpha\}$: $\mathbf{X} = (\mathbf{R}, \boldsymbol{\Sigma})$. The fixed identity of the nuclei and their assumed positions determine the external potential of ith electron at position $\boldsymbol{r}_i$, generated by the nuclear charges $\boldsymbol{Z} = (Z_1, Z_2, \ldots, Z_m)$ (a.u.):

$$v(\boldsymbol{r}_i) = -\sum_{\alpha=1}^{m} \frac{Z_\alpha}{r_{i,\alpha}}, \qquad r_{i,\alpha} = |\boldsymbol{r}_i - \boldsymbol{R}_\alpha|. \quad (1.5.2)$$

The atomic units will be used throughout the book, unless explicitly specified otherwise.

The stationary states of electrons in molecular systems satisfy the time-independent Schrödinger equation:

$$\hat{\mathrm{H}}\Psi = E\Psi, \quad (1.5.3)$$

where $\hat{H} = \hat{H}(N, v)$ is the electronic Coulomb Hamiltonian of the molecule:

$$\hat{H}(N, v) = \sum_{i=1}^{N} v(\boldsymbol{r}_i) - \frac{1}{2}\sum_{i=1}^{N} \nabla_i^2 + \sum_{i=1}^{N-1}\sum_{j=i+1}^{N} \frac{1}{|\boldsymbol{r}_i - \boldsymbol{r}_j|} \equiv \hat{H}(\mathbf{r})$$
$$\equiv \hat{V}_{ne}(N, v) + [\hat{T}(N) + \hat{V}_{ee}(N)] \equiv \hat{V}_{ne}(N, v) + \hat{F}(N). \qquad (1.5.4)$$

Here $\Delta_i = \nabla_i^2$ stands for the Laplacian acting on coordinates of *i*th electron, $\hat{T}(N)$ is the quantum-mechanical operator of the electronic kinetic energy, $\hat{V}_{ne}(N, v)$ denotes the multiplicative operator of the electron-nucleus attraction energy, $\hat{V}_{ee}(N)$ stands for the operator of the repulsion energy between electrons, and $\hat{F}(N)$ denotes for the universal (*v*-independent) part of the molecular electronic Hamiltonian.

The variational principle determining the ground-state (*gs*) wave-function for the given number of electrons *N* and external (BO) potential *v* due to the nuclei in their "frozen" positions, $\Psi_{gs} = \Psi[N, v]$, which yields the electronic Schrödinger equation (1.5.3) as the associated Euler equation, involves the minimization of the expectation value of the system energy,

$$E_v[\Psi] = \int \Psi^*(\mathbf{x})\, \hat{H}(\mathbf{r})\, \Psi(\mathbf{x})\, d\mathbf{x} \equiv \langle \Psi | \hat{H} | \Psi \rangle, \qquad (1.5.5)$$

subject to the subsidiary condition of the wave-function normalization:

$$\int \Psi^*(\mathbf{x})\, \Psi(\mathbf{x})\, d\mathbf{x} \equiv \langle \Psi | \Psi \rangle = 1. \qquad (1.5.6)$$

At the minimum of the electronic energy, for the ground-state $\Psi_{gs} = \Psi[N, v]$,

$$\delta\{E_v[\Psi] - E[N, v](\langle \Psi | \Psi \rangle - 1)\}\big|_{\Psi[N,v]} = 0, \qquad (1.5.7)$$

when the expectation value $E_v[\Psi_{g.s.}] = E[N, v] = E_{g.s} = \min_\Psi E_v[\Psi]\,\big|_{<\Psi|\Psi> = 1}$ reaches the ground-state electronic energy $E = E_{gs}$ of Eq. (1.5.3), the Schrödinger equation for the stationary ground-state is satisfied:

$$\hat{H}\, \Psi_{gs} = E_{gs}\, \Psi_{gs}. \qquad (1.5.8)$$

In the energy minimum principle of Eq. (1.5.7) the ground-state energy plays the role of the Lagrange multiplier, which enforces the wave-function normalization constraint. The stationary-state Schrödinger equation also implies for the exact eigenfunction of the electronic Hamiltonian that the *local* energy,

$$E[\mathbf{x}] \equiv \hat{H}(\mathbf{r})\, \Psi(\mathbf{x}) / \Psi(\mathbf{x}), \qquad (1.5.9a)$$

is equalized throughout the whole configuration space of electrons at the common level of the exact eigenvalue, e.g.,

$$E_{gs}[\mathbf{x}] = \hat{\mathrm{H}}(\mathbf{r})\Psi_{gs}(\mathbf{x})/\Psi_{gs}(\mathbf{x}) = E[N, v]. \qquad (1.5.9b)$$

As demonstrated by Hohenberg and Kohn (HK) (1964), the external potential is uniquely determined, up to an additive constant, by the ground-state electron density:

$$v(\boldsymbol{r}) = v[\rho_{gs}; \boldsymbol{r}], \qquad (1.5.10)$$

$$\rho_{gs}(\boldsymbol{r}) = \langle\Psi_{gs}|\textstyle\sum_i \delta(\boldsymbol{r}_i - \boldsymbol{r})\,|\Psi_{gs}\rangle \equiv \langle\Psi_{gs}|\,\hat{\rho}(\boldsymbol{r})\,|\Psi_{gs}\rangle \equiv \rho[\Psi_{gs}; \boldsymbol{r}] \equiv \rho_{gs}[N, v; \boldsymbol{r}], \qquad (1.5.11)$$

where the density operator $\hat{\rho}(\boldsymbol{r})$ is given by the sum of the Dirac deltas $\delta(\boldsymbol{r}_i - \boldsymbol{r})$ (see Appendix A).

Also, by the Hellmann-Feynman theorem:

$$\rho_{gs}(\boldsymbol{r}) = \left(\frac{\partial E[N,v]}{\partial v(\boldsymbol{r})}\right)_N = \langle\Psi_{gs}|\frac{\partial \hat{\mathrm{H}}(N,v)}{\partial v(\boldsymbol{r})}|\Psi_{gs}\rangle \equiv \rho[N, v; \boldsymbol{r}] \equiv \rho(\boldsymbol{r}). \qquad (1.5.12)$$

As indicated in the last equation, to simplify notation from now on we drop the *gs* subscript of the electron density. The electron density $\rho(\boldsymbol{r})$ thus implies the equilibrium, ground-state distribution of electrons in a molecule, unless explicitly specified otherwise.

Therefore, since $N = \int\rho(\boldsymbol{r})\, d\boldsymbol{r} \equiv N[\rho]$ is also uniquely determined by the density, the system Hamiltonian is the unique functional of the ground-state density, $\hat{\mathrm{H}}(N[\rho], v[\rho]) = \hat{\mathrm{H}}[\rho]$, and so is the non-degenerate ground-state,

$$\Psi_{gs} = \Psi[N, v] = \Psi[\rho[N, v]], \qquad (1.5.13)$$

and all physical properties of the system. For example, the ground-state energy is the unique functional of electron density:

$$E[N, v] = E_v[\rho] \equiv \int v(\boldsymbol{r})\rho(\boldsymbol{r})\, d\boldsymbol{r} + \langle\Psi_{gs}|\,\hat{\mathrm{F}}(N)\,|\Psi_{gs}]\,\rangle \equiv V_{ne}[\rho] + F[\rho], \qquad (1.5.14)$$

where $V_{ne}[\rho] = \langle\Psi_{gs}|\,\hat{\mathrm{V}}_{ne}(N,v)|\Psi_{gs}\rangle = \int v(\boldsymbol{r})\, \rho(\boldsymbol{r})\, d\boldsymbol{r}$ is the expectation value of the electron nuclear attraction energy and the universal density functional $F[\rho]$ generates the sum of the electronic kinetic and repulsion energies:

$$F[\rho] = T[\rho] + V_{ee}[\rho] \equiv F[N, v], \qquad (1.5.15)$$

where:

$$T[\rho] = \langle\Psi_{gs}\,|\,\hat{\mathrm{T}}(N)\,|\,\Psi_{gs}\rangle \equiv T[N, v], \qquad (1.5.16)$$

$$V_{ee}[\rho] = \langle\Psi_{gs}\,|\,\hat{\mathrm{V}}_{ee}(N)|\,\Psi_{gs}\rangle \equiv V_{ee}[N, v]. \qquad (1.5.17)$$

Thus, the ground-state electron density ρ represents the alternative, exact specification of the molecular non-degenerate ground-state. In other words, there is a unique mapping between $\Psi_{gs.}$ and ρ., $\Psi_{gs}\leftrightarrow\rho$, so that both these functions carry the *complete* information about the system quantum-mechanical state. That ρ uniquely identifies the electronic Coulomb Hamiltonian should not come as a surprise. Indeed, the nuclear cusps of the electron density in an atom, molecule or solid, identify both positions **R** of the nuclei and their charges $\boldsymbol{Z}$, and this constitute sufficient information to generate the external potential due to the nuclei.

More specifically, the nuclear cusps of the molecular electron density are required for the kinetic and nuclear-attraction terms in the Hamiltonian not to cause divergences in the term $\hat{\mathrm{H}}\Psi$ of the Schrödinger equation (Kato, 1957). The cusp relation for nucleus α reads:

$$\left.\frac{\partial\rho_{av}(r_\alpha)}{\partial r_\alpha}\right|_{r_\alpha=0} = -2Z_\alpha\rho_{av}(0), \qquad r_\alpha = |\boldsymbol{r} - \boldsymbol{R}_\alpha| \equiv |\boldsymbol{r}_\alpha|, \tag{1.5.18}$$

where the derivative of the spherical average $\rho_{av}(r_\alpha)$ of $\rho(\boldsymbol{r}_\alpha) \equiv \rho(r_\alpha, \theta_\alpha, \varphi_\alpha)$,

$$\rho_{av}(r_\alpha) \equiv \frac{1}{4\pi}\int_0^\pi \sin\theta_\alpha d\theta_\alpha \int_0^{2\pi} d\varphi_\alpha\ \rho(r_\alpha, \theta_\alpha, \varphi_\alpha), \tag{1.5.19}$$

is calculated for $r_\alpha = 0$.

The second theorem of Hohenberg and Kohn (1964) states the DFT variational principle:

$$\delta\{E_v[\rho] - \mu[N, v]\,(N[\rho] - N)\}\big|_{\rho[N,v]} = 0, \tag{1.5.20}$$

where the Lagrange multiplier μ associated with the density normalization constraint $N[\rho] = \int \rho(\boldsymbol{r})\, d\boldsymbol{r} = N$ is the *chemical potential* of electrons (Parr, Donnelly, Levy, and Palke, 1978; Parr and Yang 1989; Nalewajski and Korchowiec, 1997):

$$\begin{aligned}\mu[N, v] = \mu[\rho] &\equiv \left(\frac{\partial E[N,v]}{\partial N}\right)_v = \left.\left(\frac{\delta E_v[\rho]}{\delta\rho(\boldsymbol{r})}\right)_v\right|_{\rho[N,v]} \\ &= v(\boldsymbol{r}) + \frac{\delta F[\rho]}{\delta\rho(\boldsymbol{r})} \equiv \mu[N, v; \boldsymbol{r}] \equiv \mu(\boldsymbol{r}).\end{aligned} \tag{1.5.21}$$

This equation identifies the equalized local chemical potential $\mu(\boldsymbol{r}) = \mu$ as the local energy "intensity" representing the energy conjugate of the ground-state density $\rho(\boldsymbol{r})$.

This Euler equation for the electron density defines the ground-state functional relation between the equilibrium density and the external potential relative to the global chemical potential level, $u(\boldsymbol{r}) = v(\boldsymbol{r}) - \mu$,

$$u(\boldsymbol{r}) = -\frac{\delta F[\rho]}{\delta \rho(\boldsymbol{r})} \equiv u[\rho; \boldsymbol{r}]. \qquad (1.5.22)$$

Therefore, the ground-state electron density minimizes the density functional for the electronic energy subject to the constraint of the density normalization. The last equation also demonstrates that the local chemical potential $\mu(\boldsymbol{r})$ remains equalized throughout the space at the global chemical potential level $\mu[N, v]$.

The preceding equation represents the HK Euler equation for the optimum (ground-state) density, expressing the fact that the electron distribution matches the relative external potential: $u = u[\rho]$ and $\rho = \rho[u]$. It is satisfied for each *horizontal* displacement of the molecular electronic and geometrical structures, along the ground-state energy surface of the system in question. The density for which there exists a local $u = u[\rho]$ is called *v*-representable.

Moreover, since $dN = -dQ$, where the net electric charge of the system as a whole, $Q = (\sum_\alpha Z_\alpha) - N = \int [\sum_\alpha Z_\alpha \delta(\boldsymbol{r} - \boldsymbol{R}_\alpha) - \rho(\boldsymbol{r})]\, d\boldsymbol{r} \equiv \int q(\boldsymbol{r}) d\boldsymbol{r}$, the chemical potential equalization principle (1.5.21) also implies the equalization of the local *electronegativity* $\chi[N, v; \boldsymbol{r}] = \chi(\boldsymbol{r})$ (Sanderson, 1951, 1976) at the global level $\chi[N, v]$:

$$\chi[N, v] = \chi[\rho] \equiv \left(\frac{\partial \bar{E}[Q, v]}{\partial Q}\right)_v = -\mu[N, v] = \left(\frac{\delta \bar{E}_v[q]}{\delta q(\boldsymbol{r})}\right)_v\Bigg|_{\rho[N,v]}$$
$$\equiv \chi(\boldsymbol{r}) = -\mu(\boldsymbol{r}), \qquad (1.5.23)$$

where $\bar{E}[Q, v] = E[N, v]$ and the differential of a local charge density $dq(\boldsymbol{r}) = -d\rho(\boldsymbol{r})$.

In the EF perspective of the BO approximation the distribution of electrons depends on the current positions of the system nuclei. Therefore, in this approach the local *electronic* state-function $\rho(\boldsymbol{r})$ represents the dependent state-variable, the functional of the independent *nuclear* state-parameter $u(\boldsymbol{r})$. In what follows we shall call such a description the *softness representation* of molecular electronic structure, since this theoretical framework defines one of the principal integral kernels of the DFT theory, the *softness kernel*, which measures the negative density response at one point, per unit displacement in the relative external potential at another point.

In the "inverse" EP perspective, in the spirit of the Hellmann-Feynman theorem, the electronic distribution $\rho(\boldsymbol{r})$ plays the role of the independent state-parameter, to which the nuclear state-variable $u(\boldsymbol{r})$ responds. We shall call this description level the *hardness representation* of molecular systems, since it defines the other principal kernel of the DFT treatment of molecules and reactive systems, called the *hardness kernel*. The latter is defined by the inverse of the softness kernel, and measures the negative response of the relative external potential at one point, to unit displacement of the ground-state electron density at another point.

Both these approaches constitute the complete theoretical framework for describing and ultimately understanding diverse molecular processes in terms of the structure-reactivity relationships, for diagnosing the reactivity preferences of molecules and their fragments, and for the speculative chemical thinking on how to manipulate the electronic structure of molecular systems, in order to bring about the desired change in their bonding pattern, e.g., forming or breaking the specified bond.

To summarize, the ground-state wavefunction, defined in the $4N$-dimensional *configurational space*, and the electron density, defined in the 3-dimensional *physical space*, constitute exactly equivalent definitions of the non-degenerate quantum-mechanical state of the N-electron molecular system. They both contain the complete physical information about the system. The theory of electronic structure of matter can thus be rigorously based on the electronic distribution, which offers both computational and conceptual/interpretative advantages. It should be emphasized that the ground-state density ρ determines *all* molecular properties, including those characterizing the excited states, since it uniquely identifies the system Hamiltonian.

1.6. HORIZONTAL AND VERTICAL DISPLACEMENTS OF MOLECULAR ELECTRONIC STRUCTURE

In the "thermodynamic" treatment of molecular systems (Ghosh, *et al.*, 1984; Ghosh and Berkowitz, 1985; Nagy and Parr, 1994; Nalewajski and Parr, 2001; Nalewajski 2002c, 2003a,b, 2004a, 2005d) one aims at determining changes in the electronic structure accompanying the displacement from one equilibrium (ground-state) density, $\rho_1 = \rho[N_1, v_1]$, to another, $\rho_2 = \rho[N_2, v_2]$. These electron distributions are uniquely determined by the two state-parameters determining the corresponding electronic Hamiltonians: the overall number of electrons N_i and the external potential due to the nuclei, $v_i(\boldsymbol{r})$, $i = 1, 2$. We call such shifts in the system electronic structure the "*horizontal*" displacements (Nalewajski, 1999; Nalewajski and Parr, 2001) on the ground-state density $\rho[N, v] \equiv \rho$ and energy $E[N, v] = E_v[\rho] \equiv E_{gs}[\rho]$ "surfaces".

The horizontal displacement, from one equilibrium (v-representable) electron density to another, gives rise to the associated change in the *generalized* density functional for the ground-state energy,

$$E[\rho] \equiv \int v[\rho; \boldsymbol{r}]\, \rho(\boldsymbol{r})\, d\boldsymbol{r} + F[\rho] \equiv E[N[\rho], v[\rho]] = E_{v[\rho]}[\rho], \qquad (1.6.1)$$

in which the external potential changes with the electron density in such a way that the current electron distribution ρ matches $v = v[\rho]$ as its ground-state (equilibrium) density.

It should be emphasized that this density functional for the *ground-state* energy differs from the Hohenberg–Kohn functional of Eq. (1.5.14), for the *variational* energy $E_v[\rho]$, in which the external potential is fixed (not related to the trial density ρ). Only for the exact ground-state density, satisfying the HK minimum principle of Eq. (1.5.20),

$$E_v[\rho[N, v]] = \bar{E}_u[\rho[u]] \equiv \bar{E}[u] = E_\rho[u[\rho]] \equiv E[\rho]. \qquad (1.6.2)$$

We have indicated in the preceding equation that the energy functional $E[\rho]$ can be alternatively interpreted as the functional of the relative external potential $u(\boldsymbol{r})$, $\overline{E}[u]$. This is because the ground-state Euler equation (1.5.22) shows that the density and relative external potential are unique functionals of each other.

In what follows we shall use the term *Softness Representation* (SR) to denote the description of molecular equilibrium states using the (nuclear) external potential as the *independent* local state-parameter. In this approach, to which both the variational functional $E_v[\rho]$, for a trial ρ, and the ground-state functional $\overline{E}_u[\rho[u]] = \overline{E}[u]$ belong, the electron density represents the *dependent* state-variable. Therefore, such a *softness* "picture" adopts the EF perspective of Section 1.2. Selecting the (electron) density as an *independent* state-variable, e.g., in the variational functional $E_\rho[u]$, for a trial u, and the ground-state functional $E_\rho[u[\rho]] = E[\rho]$, gives the complementary *Hardness Representation* (HR), which adopts the EP perspective of Section 1.2.

The Euler equation (1.5.22) implies that the external potential $v(\boldsymbol{r})$ in open molecular systems is determined by ρ only to a constant μ, which can be related to the chemical potential of an external *electron reservoir*($\mathfrak{R}$): $\mu = \mu_{\mathfrak{R}}$. Therefore, the unique specification of the external potential as the functional of ρ additionally requires the knowledge of this global state-parameter, the "*intensive*" conjugate of N:

$$v(\boldsymbol{r}) = u(\boldsymbol{r}) + \mu = \mu_{\mathfrak{R}} - \delta F/\delta\rho(\boldsymbol{r}) \equiv v[\mu, \rho; \boldsymbol{r}]. \tag{1.6.3}$$

The "horizontal" character of changes in the ground-state electronic structure is in contrast to a search for the equilibrium (exhaustive) partition of the molecular ground-state density $\rho(\boldsymbol{r})$ into densities $\boldsymbol{\rho}(\boldsymbol{r}) \equiv \{\rho_\alpha(\boldsymbol{r}) = \rho_\alpha[\rho; \boldsymbol{r}]\}$ (a row vector) of the constituent subsystems, e.g., the density pieces of m constituent AIM, which at each point sum up to this given molecular density: $\rho(\boldsymbol{r}) = \sum_\alpha \rho_\alpha(\boldsymbol{r})$. This density division problem is "*vertical*" in character (Nalewajski and Parr, 2000, 2001) being performed for the fixed molecular density.

This is also the case in the Levy (1979) constrained search construction of the universal functional $F[\rho]$ of Eqs. (1.5.14) and (1.5.15):

$$F[\rho] = \inf_{\Psi\to\rho}\langle\Psi|\,\hat{\mathrm{F}}\,|\Psi\rangle, \tag{1.6.4}$$

in which one searches over all wave-functions (or density operators) yielding a given electron density, and calculates the density functional $F[\rho]$ as the lowest value (infimum) of the expectation (or ensemble-average) values of the sum of the electron kinetic and repulsion energy operators. Since this search is performed for the fixed (ground-state) density, it also implies the fixed value of the system electronic energy. Therefore, by analogy to the maximum principle of the thermodynamic entropy for constant internal energy in the ordinary phenomenological thermodynamics, this DFT minimum principle can be regarded as being also "entropic" in character.

Since $u = u[\rho]$ and $\rho = \rho[u]$, by Eq. (1.5.22), the universal density functional $F[\rho]$ of the HR (in the EP perspective), can be alternatively regarded as the associated functional of the relative external potential $\overline{F}[u]$ of the SR (in the EF perspective):

$$F[\rho] = F[\rho[u]] \equiv \overline{F}[u]. \tag{1.6.5}$$

The corresponding density constrained search for the external potential matching the given (v-representable) density ρ reads (see, e.g., Lieb, 1982; Nalewajski and Parr, 1982; Lieb, 1983; Nalewajski and Korchowiec, 1997; Colonna and Savin, 1999):

$$F[\rho] = \sup_{v'\to\rho} \{E[N, v'] - \int v'(\boldsymbol{r})\, \rho(\boldsymbol{r})\, d\boldsymbol{r}\}. \tag{1.6.6a}$$

In this extremum principle one searches over external potentials $v'\to\rho$, which give rise to for the specified ground-state density ρ, and determines the maximum/ suprimum of the Legendre transform (see the next chapter) of the system energy, which replaces the external potential v' by the ground-state density ρ, in the list of the system independent state-parameters. At the solution point this variational principle yields the optimum external potential $v = v[\rho]$, which identifies the specified (v-representable) density as its ground-state, equilibrium distribution. It should be recalled that ρ fixes the shape of $v' = v'[\mu', \rho]$, where μ' denotes the system chemical potential attributed to an external reservoir. Therefore, the trial potentials in this constrained search can only differ by a constant $\mu = \mu_{\mathfrak{R}}$. This constrained-search construction can be generalized to any trial density ρ' (Lieb, 1983; Colonna and Savin, 1999):

$$F[\rho'] = \sup_{v'} \{E_{v'}[\rho'] - \int v'(\boldsymbol{r})\, \rho'(\boldsymbol{r})\, d\boldsymbol{r}\}, \tag{1.6.6b}$$

in which $v' \neq v'[\mu', \rho]$.

As argued elsewhere (Nalewajski and Parr, 2001), the AIM division problem of the fixed molecular density ρ also represents a search for the optimum *effective* external potentials of atomic subsystems $\boldsymbol{v}^{eff} = \{v_\alpha^{eff}\}$:

$$v_\alpha^{eff}(\boldsymbol{r}) = v_\alpha^{eff}[\boldsymbol{\rho}[\rho]; \boldsymbol{r}] = v(\boldsymbol{r}) + \left(\frac{\partial \widetilde{F}^n[\boldsymbol{\rho}]}{\partial \rho_\alpha(\boldsymbol{r})}\right)_{\beta\neq\alpha}, \qquad \alpha = 1, 2, \ldots, m. \tag{1.6.7}$$

Here, the partial differentiation with respect to $\rho_\alpha(\boldsymbol{r})$ of the *non-additive* part $\widetilde{F}^n[\boldsymbol{\rho}]$ of the *total* Hohenberg–Kohn–Levy functional in the AIM resolution, $F[\rho] \equiv \widetilde{F}[\boldsymbol{\rho}]$,

$$\widetilde{F}^n[\boldsymbol{\rho}] \equiv \widetilde{F}[\boldsymbol{\rho}] - \Sigma_\gamma F[\rho_\gamma] \equiv \widetilde{F}[\boldsymbol{\rho}] - \widetilde{F}^a[\boldsymbol{\rho}], \tag{1.6.8}$$

where $\widetilde{F}^a[\boldsymbol{\rho}]$ denotes the *additive* part of $\widetilde{F}[\boldsymbol{\rho}]$, is carried out for the fixed densities of the remaining subsystems $\{\rho_{\beta\neq\alpha}\}$.

These effective external potentials of the *embedded* bonded-atoms in a molecule are then related to their respective densities through the global-like ground-state Euler equation of DFT [Eqs. (1.5.21,22]:

$$v_\alpha^{eff}(\boldsymbol{r}) - \mu_\alpha = u_\alpha[\boldsymbol{\rho}[\rho]; \boldsymbol{r}] \equiv -\frac{\delta F[\rho_\alpha]}{\delta \rho_\alpha(\boldsymbol{r})}, \tag{1.6.9}$$

where the equalized subsystem chemical potential $\mu_\alpha = \mu_\alpha(\boldsymbol{r})$ is given by the partial derivatives,

$$\mu_\alpha(\boldsymbol{r}) = \frac{\partial \tilde{E}_v[\boldsymbol{\rho}]}{\partial \rho_\alpha(\boldsymbol{r})} = \left(\frac{\partial \tilde{E}[\boldsymbol{N}, v]}{\partial N_\alpha}\right)_v, \qquad \boldsymbol{N} = (N_1, N_2, \ldots, N_m) \tag{1.6.10}$$

of the system electronic energy in the AIM resolution:

$$E_v[\rho] \equiv \tilde{E}_v[\boldsymbol{\rho}[\rho]] = E[N, v] \equiv \tilde{E}[\boldsymbol{N}, v]. \tag{1.6.11}$$

Clearly, for the mutually open atomic subsystems, with no barriers preventing the flow of electrons between the bonded atoms, the AIM chemical potentials are equalized at the global chemical potential level of Eq. (1.5.21), which characterizes the molecular system as a whole:

$$\boldsymbol{\mu}(\boldsymbol{r}) \equiv \{\mu_\alpha(\boldsymbol{r})\} = \boldsymbol{\mu} \equiv \{\mu_\alpha\} = \mu \boldsymbol{1}, \tag{1.6.12}$$

where the unit row vector $\boldsymbol{1} = (1, 1, \ldots)$ and the vector $\boldsymbol{N} \equiv \{N_\alpha = \int \rho_\alpha(\boldsymbol{r})\, d\boldsymbol{r} \equiv N[\rho_\alpha]\}$ groups the average numbers of electrons in atomic subsystems.

The same, equalized chemical potentials of AIM derived from the molecular ground-state density follow from the density functional for the electronic energy of the embedded atom α:

$$\mathcal{E}_{v\alpha}[\boldsymbol{\rho}] = \{\int v(\boldsymbol{r})\, \rho_\alpha(\boldsymbol{r})\, d\boldsymbol{r} + F[\rho_\alpha]\} + F^n[\boldsymbol{\rho}] \equiv E_v[\rho_\alpha] + F^n[\boldsymbol{\rho}], \tag{1.6.13}$$

where $E_v[\rho_\alpha]$ stands for the electronic energy of ρ_α alone, and $F^n[\boldsymbol{\rho}]$ represents the *embedding energy* due to the presence of electrons of the remaining subsystems (Nalewajski and Parr, 2001):

$$\mu_\alpha(\boldsymbol{r}) = \partial \mathcal{E}_{v\alpha}[\boldsymbol{\rho}] / \partial \rho_\alpha(\boldsymbol{r}) = \mu_\alpha = \mu, \qquad \alpha = 1, 2, \ldots, m. \tag{1.6.14}$$

We therefore conclude that the DFT description of the equilibrium states in molecules and their mutually-open subsystems are isomorphic, since the subsystem effective (relative) external potentials are related to their electron densities through the same "horizontal"-type Euler equation linking the complementary softness (EF) and hardness (EP) representations of the molecular electronic structure.

2. ALTERNATIVE PERSPECTIVES AND REPRESENTATIONS

The alternative *maximum entropy* and *minimum energy* principles of ordinary thermodynamics are recalled and their equivalent Legendre transformed formulations, corresponding to alternative sets of the system independent state-parameters, are briefly summarized. The DFT descriptions of equilibria in both the externally *closed* (N-controlled) and *open* (μ-controlled) molecular systems are explored in the complementary *Electron Following* and *Electron Preceding* perspectives on molecular processes. In the chemical *Softness* (EF) *Representation* the external potential due to the system nuclei constitutes the independent state-parameter, while in the chemical *Hardness* (EP) *Representation* the ground-state electron density determines the equilibrium state of the molecule. The equivalence of these alternative formulations is demonstrated and the associated quadratic Taylor expansions of the corresponding "thermodynamic" potentials for each representation are derived and discussed. Their generalized potentials (first partials) and charge sensitivities (second partials) are derived and interpreted. The basic concepts of the DFT reactivity theory include the global and local softness (hardness) descriptors of the electron "gas", the electronic and nuclear Fukui functions, and the theory basic kernels: the density linear-response function of the closed molecular system and the softness and hardness kernels of the externally open system. The fundamental relations between these basic quantities are derived and interpreted. In each representation the transformations between the system *perturbations*, i.e., displacements of the system independent parameters, and the equilibrium linear *responses* of the conjugate (dependent, unconstrained) state-variables are formulated in terms of the relevant charge sensitivities.

2.1. ENERGY AND ENTROPY PRINCIPLES IN THERMODYNAMICS

The formal structure of the phenomenological thermodynamics (see, e.g., Callen, 1960; Tisza, 1977) can be formulated in the equivalent *energy* or *entropy representations*. The former is based upon the *energy* principal relation expressing the system internal energy U in terms of the system *entropy*, S, and remaining *extensive* parameters $\{X_1, X_2, \ldots\}$, e.g., the volume, V, mole numbers $\{N_i\}$ of the system components, etc.:

$$U = U(S, X_1, X_2, \ldots). \tag{2.1.1}$$

The partial derivatives of this function with respect to the extensive parameters of state determine the corresponding energetic *intensive* parameters $\{P_1, P_2, \ldots.)$, where $P_k = \partial U/\partial X_k$, e.g., the system temperature,

$$T = T(S, X_1, X_2, \ldots) = \left(\frac{\partial U}{\partial S}\right)_{X_1, X_2 \ldots} \equiv \frac{\partial U}{\partial S}, \tag{2.1.2}$$

the negative pressure, $-P = \partial U/\partial V$, the chemical potentials of the system components, $\{\mu_i = \partial U/\partial N_i\}$, etc.

The equivalent fundamental relation in the *entropy* representation,

$$S = S(U, X_1, X_2, \ldots), \tag{2.1.3}$$

similarly defines the entropic *intensive* parameters, $F_k = \partial S/\partial X_k$, e.g., the inverse temperature

$$T^{-1} = T^{-1}(U, X_1, X_2, \ldots) = \left(\frac{\partial S}{\partial U}\right)_{X_1, X_2 \ldots} \equiv \frac{\partial S}{\partial U}, \tag{2.1.4}$$

and the remaining entropic intensities $F_k = -P_k/T$.

This double representation of the theory is also reflected by the equivalent principles determining the equilibrium state of macroscopic systems:

> **Energy Minimum Principle.** *The equilibrium values of unconstrained internal parameters minimize the system internal energy for the given value of the system total entropy,* $S = S^0$.

> **Entropy Maximum Principle.** *The equilibrium values of unconstrained internal parameters maximize the system entropy for the given value of the system total internal energy,* $U = U^0$.

The underlying conditions for the equivalence of both these principles follow from the analytic postulates determining the geometrical form of the fundamental surface of Eqs. (2.1.1) and (2.1.3), that $\partial S/\partial U > 0$ [Eq. (2.1.4)] and that U is a single-valued continuous function of S [Eq. (2.1.1)].

2.2. LEGENDRE TRANSFORMATIONS

The formalism of thermodynamics can be recast in such a way that intensive parameters (slopes of the fundamental surfaces) replace the conjugate extensive parameters as mathematically independent variables in the theory fundamental relations. The systematic procedure for achieving this objective is the Legendre transformation.

In the simplest case of the fundamental relation involving a single extensive parameter X, i.e., for the fundamental surface being defined by a function $Y = Y(X)$, the intensive parameter measures the slope of the fundamental curve, $P = dY/dX$. In the conventional *point-geometry* this curve represents the locus of points satisfying the fundamental relation $Y = Y(X)$. It can be alternatively considered within the so called *line-geometry* as the envelope of a family of tangent lines defined by their intercepts y along the Y-axis, expressed as a function of slopes P: $y = y(P)$. Indeed, just as every point in the (X,Y) plane is uniquely described by the two numbers, the point coordinates, so every straight line in this plane is uniquely specified by the line slope P and the intercept y defined by the equation $P = (Y - y)/(X - 0)$ or

$$y = Y - PX = Y - \frac{dY}{dX}\,X. \tag{2.2.1}$$

The preceding equation represents the Legendre transformation of the fundamental relation in the "Y-representation", $Y = Y(X)$, into the equivalent fundamental relation in the intercept, "y-representation". Eliminating the extensive variables X and Y in Eq. (2.2.1), using the equations $P = P(X, Y)$ and $y = y(X, Y)$, finally defines the intercept function expressed solely in terms of the slope:

$$y = y(P) \equiv Y[P] \equiv \mathcal{L}_Y[P]\,. \tag{2.2.2}$$

This function is referred to as the *Legendre transform* $Y[P] \equiv \mathcal{L}_Y[P]$ of $Y(X)$, which corresponds to replacing the extensive parameter X by its conjugate intensive parameter P. In thermodynamics this elimination is assured by the criterion of stability expressing a dependence of P on X: $dP/dX = d^2Y/dX^2 \neq 0$.

The generalization of this transformation to functions of many variables is straightforward. In ordinary thermodynamics this transformation gives rise to *thermodynamic potentials*. Consider the simplest case of the single component system, which is not exposed to external electric and/or magnetic fields. The corresponding lists of extensive parameters in the energy and entropy representations, respectively, then read: $\{S, V, N\}$ and $\{U, V, N\}$. The *Helmholtz free energy* is the partial Legendre transform of the internal energy, which replaces S by T in the list of independent parameters:

$$F(T, V, N) = U[T] = \mathcal{L}_U[T] = U - \frac{\partial U}{\partial S} S = U - TS, \tag{2.2.3}$$

while the *enthalpy* represents the partial Legendre transform of the internal energy replacing V by $-P$:

$$H(S, P, N) = U[P] = \mathcal{L}_U[P] = U - \frac{\partial U}{\partial V} V = U + PV, \tag{2.2.4}$$

Other examples include the *Gibbs free energy* (*free enthalpy*),

$$G(T, P, N) = U[T, P] = \mathcal{L}_U[T, P] = U - \frac{\partial U}{\partial S} S - \frac{\partial U}{\partial V} V = U - TS + PV, \tag{2.2.5}$$

and the *grand-canonical potential* (*grand potential*):

$$\Omega(T, V, \mu) = U[T, \mu] = \mathcal{L}_U[T, \mu] = U - \frac{\partial U}{\partial S} S - \frac{\partial U}{\partial N} N = U - TS - \mu N. \tag{2.2.6}$$

Clearly, the complete transformation $U[T, P, \mu] = \mathcal{L}_U[T, P, \mu] = 0$, by the Euler equation for homogeneous functions

$$U = \frac{\partial U}{\partial S} S + \sum_i \frac{\partial U}{\partial X_i} X_i = TS + \sum_i P_i X_i, \tag{2.2.7}$$

applied to the internal energy exhibiting the homogeneous first-order property (see, e.g., Callen, 1960):

$$U(\lambda S, \lambda X_1, \lambda X_2, \ldots) = \lambda\, U(S, X_1, X_2, \ldots). \tag{2.2.8}$$

This homogeneity equation also implies the differential form of the relationship among intensive parameters, called the Gibbs-Duhem equation:

$$S\, dT + \sum_i X_i\, dP_i = 0. \tag{2.2.9}$$

It should be observed that the *equations of state*, expressing intensive parameters in terms of the independent extensive parameters,

$$P_i = P_i(S, X_1, X_2, \ldots) \qquad \text{and} \qquad F_i = F_i(U, X_1, X_2, \ldots), \tag{2.2.10}$$

are homogeneous zero-order, since the fundamental relations (2.1.1) and (2.1.3) are homogeneous first-order.

Similar Legendre transformations can be performed on the entropy function of Eq. (2.1.3). Again, they correspond to replacing the specified entropy extensive parameters by the conjugate entropic "intensities" and give rise to the *Massieu functions* $S[F_1, F_2, \ldots] = \mathcal{L}_S[F_1, F_2, \ldots]$ of thermodynamics. The latter represent the generalized "potentials" in the entropy representation (Callen, 1960). The Gibbs-Duhem relation in the entropy representation again states that the sum of products of the extensive parameters and the differentials of the corresponding intensive parameters identically vanishes:

$$U\,d(1/T) + \sum_i X_i\, dF_i = 0. \tag{2.2.11}$$

Other examples of the Legendre transformation can be found in the Lagrangian and Hamiltonian mechanics. More specifically, the system Lagrangian, $L = L(\{w_i\}, \{q_i\})$, expressed as a function of the generalized coordinates $\{q_i\}$ and velocities $\{w_i\}$, completely determines the dynamics of a mechanical system and thus expresses the fundamental relation in mechanics. The generalized momenta are the conjugate "intensities" of the system velocities, representing the velocity "slopes" of the Lagrangian: $\{p_i = \partial L/\partial w_i\}$. The negative Hamiltonian, defined as the Legendre transform of the Lagrangian which replaces velocities by momenta $\{p_i\}$ as independent variables, $-H(\{p_i\}, \{q_i\}) = L - \sum_i p_i\, v_i$, constitutes the fundamental equation of mechanics in the Hamiltonian representation.

In the energy representation the internal energy in the system equilibrium state is minimum for constant entropy, and hence each Legendre transform of the energy is minimum for constant values of the transformed intensive variables. Accordingly, in the entropy representation the entropy of the system equilibrium state is maximum for constant energy, and from this it follows that each Legendre transform of the entropy is maximum for constant values of the transformed entropic intensive variables.

The extremum principles in the Legendre transformed representations, which allow one to select the most convenient independent variables for a given problem, are thus expressed as the minimum principles for the corresponding thermodynamic potentials in the energy representation of phenomenological thermodynamics, or as the maximum principles for the associated Massieu functions in the entropy representation. It should be observed that fixing the intensive parameters of the given Legendre-transformed representation selects the manifold of states, which are consistent with this requirement. Therefore, the minimum principle for the relevant potential determines the equilibrium values of the unconstrained internal parameters by searching for the potential minimum over this partial manifold. This can be generally expressed in the following way:

> **Thermodynamic Potential Minimum Principle.** *The equilibrium values of unconstrained internal parameters of the system in contact with a set of reservoirs* $\mathcal{R} = (\mathcal{R}_1, \mathcal{R}_2, \ldots)$ *characterized by their energy intensive parameters* $\mathbf{P}(\mathcal{R}) = (P_1^{\mathcal{R}}, P_2^{\mathcal{R}}, \ldots)$, *minimize the thermodynamic potential* $U[P_1, P_2, \ldots]$ *at constant* $\mathbf{P} = \mathbf{P}(\mathcal{R})$, i.e., $(P_1 = P_1^{\mathcal{R}},\ P_2 = P_2^{\mathcal{R}}, \ldots)$.

The corresponding general result for the entropy Legendre transforms reads:

> **Massieu Function Maximum Principle.** *The equilibrium values of unconstrained internal parameters of the system in contact with a set of reservoirs* $\mathfrak{R} = (\mathfrak{R}_1, \mathfrak{R}_2, \ldots)$ *characterized by their entropy intensive parameters* $\boldsymbol{F}(\mathfrak{R}) = (F_1^{\mathfrak{R}}, F_2^{\mathfrak{R}}, \ldots)$, *maximize the Massieu function* $S[F_1, F_2, \ldots]$ *at constant* $\boldsymbol{F} = \boldsymbol{F}(\mathfrak{R})$, i.e., $(F_1 = F_1^{\mathfrak{R}},\ F_2 = F_2^{\mathfrak{R}}, \ldots)$.

This brief reminder of the formal structure, general principles, and equivalent formulations of the fundamental relations in thermodynamics demonstrates the richness and flexibility of the thermodynamic description of equilibrium states of macroscopic systems. Now, after this short detour into the textbook thermodynamics, we go back to the molecular equilibrium states and their alternative representations in DFT. The main goal of the following analysis of the alternative Legendre transformed representations of equilibrium states in molecules and their constituent fragments is to extract formal similarities and to identify main differences between the DFT of atoms and molecules and phenomenological thermodynamics.

We first observe that in the molecular world the distinction between the "intensive" and "extensive" state-parameters is blurred, since the system energy is not the homogeneous function of the independent parameters of state. Nevertheless, we shall often refer to the molecular system number of electrons N as the global "extensive" parameter of state, since the overall number of electrons in the composite system is the sum of their values in each of the subsystems. The same additive property characterizes the thermodynamic extensive parameters. Accordingly, the energy conjugate parameter, the electronic chemical potential μ, will be referred to as the global "intensive" parameter.

In the electron-following (BO) perspective the local independent parameter has a *nuclear*, character. In this chemical softness representation the external potential due to the nuclei, $v(\boldsymbol{r})$, will be regarded as the system local "extensive" parameter. Indeed, the external potential of the composite system is the sum of the external potentials of molecular subsystems. Its energy conjugate, the electron density $\rho(\boldsymbol{r})$, represents the unconstrained, dependent variable in the EF formulation, which is in the spirit of the BO approximation. As such it will be referred to as the *local "intensive"*, dependent quantity of the chemical softness representation.

These roles of the local state-parameters are reversed in the electron-preceding perspective, i.e., the chemical hardness representation, in which the electron density is regarded as the controlling, independent parameter of state, while the external potential responds to the specified redistribution of electrons, thus representing a dependent (unconstrained) parameter of state. Therefore, since again the density of the composite system is the sum of densities of the constituent subsystems, we shall classify $\rho(\boldsymbol{r})$ as the local "extensive" parameter of the EP representation. In this formulation the external potential $v(\boldsymbol{r})$ will be regarded as the system local "intensive" parameter of this chemical hardness formulation of the theory.

2.3. THE CHEMICAL SOFTNESS REPRESENTATION

The emergence of the modern DFT (Hohenberg and Kohn, 1964; Kohn and Sham, 1965) has generated a new impetus and a convenient theoretical framework for formulating new, thermodynamic-like approaches to classical problems in chemistry (see, e.g.: Par and Yang, 1989; Nalewajski and Korchowiec, 1997; Nalewajski, 1996a). For example, the origin of chemical bonding, identity of AIM, factors determining the nature and relative importance of alternative reaction sites and pathways in large reactive and catalytic systems, stability of molecular charge distribution, similarity of molecules, electron localization, etc., have all been approached afresh (see, e.g., Parr and Yang, 1989; Sen, 1993; Nalewajski, 1996b, Nalewajski and Korchowiec, 1997; Geerlings *et al.*, 2003).

The theory provides an alternative point of view, from which one can approach all physical/chemical properties and processes involving atomic, molecular and reactive systems. This novel perspective is in the spirit of the old *Electronegativity Equalization* (EE) approach of Sanderson (1951, 1976) for the equilibrium (ground-state) distribution of electrons. As we have demonstrated in the preceding chapter, in DFT the electron cloud is regarded as a "fluid", which is fully characterized by its density distribution $\rho(\boldsymbol{r})$. This outlook has enriched the theory of electronic structure and chemical reactivity by both helping to rationalize and quantify classical concepts and rules of chemistry, e.g., the HSAB principle of Pearson (1973) (Parr and Pearson, 1983; Nalewajski, 1984), the EE rule of Sanderson (Parr *et al.*, 1978), and provided definitions of the hardness/softness characteristics of the electron distribution in atoms and molecules (Parr and Pearson, 1983; Berkowitz and Parr, 1988; Chattaraj and Parr, 1993; Gazquez, 1993; Nalewajski, 1993; Nalewajski *et al.*, 1996; Nalewajski and Korchowiec, 1997). It has brought about a deeper understanding of the nature of chemical bond, various factors determining its strength and composition, and of subtle reactivity preferences (see, e.g.: Parr and Yang, 1989, Nalewajski and Korchowiec, 1997; Geerlings *et al.*, 2003).

For example, in DFT the *Frontier Electron Theory* (FET) of Fukui (1975, 1987) has been given a more rigorous foundation in terms of the related density response index called the electronic *Fukui Function* (FF) (Parr and Yang, 1984; 1989; see also: Nalewajski *et al.*, 1996; Nalewajski and Korchowiec, 1997; Korchowiec and Uchimaru, 1998; Michalak *et al.*, 1999; Ayers and Levy, 2000). It represents the normalized response of the electron density per unit displacement in the system global number of electrons N:

$$f(\boldsymbol{r}) = \left(\frac{\partial \rho(\boldsymbol{r})}{\partial N}\right)_v = \frac{\partial^2 E[N, v]}{\partial N\, \partial v(\boldsymbol{r})}, \qquad \int f(\boldsymbol{r})\, d\boldsymbol{r} = 1 . \qquad (2.3.1)$$

The corresponding *nuclear* FF (NFF) (Cohen, 1996; Nalewajski, 1999, 2000a) has also been introduced:

$$\varphi_\alpha = \left(\frac{\partial \boldsymbol{F}_\alpha}{\partial N}\right)_v = -\frac{\partial^2 W[N, v(\mathbf{R})]}{\partial N\, \partial \boldsymbol{R}_\alpha} = \left(\frac{\partial \mu}{\partial \boldsymbol{R}_\alpha}\right)_N, \tag{2.3.2}$$

where [see Eq. (1.4.10)] the total BO potential for nuclear motions

$$W[N, v(\mathbf{R})] = E[N, v(\mathbf{R})] + V_{nn}(\mathbf{R}) \equiv W(\mathbf{R}), \tag{2.3.3}$$

$$V_{nn}(\mathbf{R}) = \sum_{\alpha=1}^{m-1} \sum_{\beta=\alpha+1}^{m} Z_\alpha Z_\beta \,/\left|\boldsymbol{R}_\alpha - \boldsymbol{R}_\beta\right| \tag{2.3.4}$$

denotes the nuclear repulsion energy, and $\partial/\partial\boldsymbol{R}_\alpha = \nabla_\alpha$ stands for the gradient with respect to the position of αth nucleus. A reference to Eq. (2.3.2) shows that NFF measures the normalized response in the force acting on nucleus α, $\boldsymbol{F}_\alpha = -\nabla_\alpha W(\mathbf{R})$, per unit displacement in the global number of electrons. This cross-differentiation identity (Maxwell relation) also shows that this index can be alternatively interpreted as the response in the system electronic chemical potential per unit displacement of $\boldsymbol{R}_\alpha$. The NFF represents an example of the reactivity index, which measures the coupling between the geometric and electronic structure parameters of a given molecular system (Nalewajski, 1999, 2000a, 2002f).

2.3.1. Closed (N-Controlled) Systems

It follows from Eqs. (1.5.12) and (1.5.21) that the *partial* ("intensive") energy conjugates $\boldsymbol{b} = \partial E[\boldsymbol{a}]/\partial \boldsymbol{a}$ of the ground-state electronic energy $E[\boldsymbol{a}] = E[N, v]$, with respect to the "extensive" state-parameters, $\boldsymbol{a} = (N, v)$, include the ground-state chemical potential and the electronic density:

$$\boldsymbol{b} = \left\{\left(\frac{\partial E[N,v]}{\partial N}\right)_v = \mu,\ \left(\frac{\partial E[N,v]}{\partial v(\boldsymbol{r})}\right)_N = \rho(\boldsymbol{r})\right\}. \tag{2.3.5}$$

In the preceding equation, the local partial derivative, identifying the energy conjugate of the external potential, states the familiar Hellmann-Feynman theorem.

The corresponding matrix of the second partial derivatives,

$$\mathbf{c} = \frac{\partial^2 E[\boldsymbol{a}]}{\partial \boldsymbol{a}'\partial \boldsymbol{a}} = \frac{\partial \boldsymbol{b}}{\partial \boldsymbol{a}'}, \tag{2.3.6}$$

groups the so called *principal* CS of the molecular system under consideration. They include the *global hardness* (Parr and Pearson, 1983)

$$\eta = \left(\frac{\partial^2 E[N,v]}{\partial N^2}\right)_v = \left(\frac{\partial \mu}{\partial N}\right)_v, \tag{2.3.7}$$

the inverse of the *global softness* of the SR in the EP perspective,

$$S = \left(\frac{\partial N}{\partial \mu}\right)_v = 1/\eta, \tag{2.3.8}$$

the electronic FF of Eq. (2.3.1) (Parr and Yang, 1984, 1989),

$$f(\boldsymbol{r}) = \frac{\partial^2 E[N,v]}{\partial N\, \partial v(\boldsymbol{r})} = \left(\frac{\partial \rho(\boldsymbol{r})}{\partial N}\right)_v = \left(\frac{\partial \mu}{\partial v(\boldsymbol{r})}\right)_N, \tag{2.3.9}$$

and the density *Linear Response* (LR) function (Nalewajski and Parr, 1982, Berkowitz and Parr, 1988):

$$\beta(\boldsymbol{r},\boldsymbol{r}') = \left(\frac{\partial^2 E[N,v]}{\partial v(\boldsymbol{r})\, \partial v(\boldsymbol{r}')}\right)_N = \left(\frac{\partial \rho(\boldsymbol{r}')}{\partial v(\boldsymbol{r})}\right)_N = \left(\frac{\partial \rho(\boldsymbol{r})}{\partial v(\boldsymbol{r}')}\right)_N = \beta(\boldsymbol{r}',\boldsymbol{r}). \tag{2.3.10}$$

The cross-differentiation identities in the last two equations are the corresponding Maxwell relations (Callen, 1962; Nalewajski and Parr, 1982; Nalewajski, 1983). It follows from Eq. (2.3.9) that, the electronic FF has a double physical interpretation. On one hand, it reflects the local density response to a unit shift in the system global number of electrons. On the other hand, it measures the response in the system global chemical potential of electrons per unit shift in the local value of the external potential due to the nuclei. The latter interpretation shows that also FF represents the index measuring a coupling between the molecular electronic and nuclear (geometric) structures, in the spirit of the EF perspective. The linear response kernel of Eq. (2.3.10) represents the softness kernel of the externally closed molecular systems. It also couples the local electronic and nuclear (geometric) parameters of a molecule.

The derivative quantities of Eqs. (2.3.5) and (2.3.7)-(2.3.10) determine the associated quadratic Taylor expansion of the system electronic energy, of the molecular system as a whole, in terms of powers of $\Delta \boldsymbol{a}$:

$$\Delta^{(1+2)} E[N,v] = \mu\,\Delta N + \int \rho(\boldsymbol{r})\,\Delta v(\boldsymbol{r})\,d\boldsymbol{r}$$
$$+ \frac{1}{2}\Big[\eta(\Delta N)^2 + 2\Delta N \int f(\boldsymbol{r})\,\Delta v(\boldsymbol{r})\,d\boldsymbol{r} + \iint \Delta v(\boldsymbol{r})\,\beta(\boldsymbol{r},\boldsymbol{r}')\,\Delta v(\boldsymbol{r}')\,d\boldsymbol{r}\,d\boldsymbol{r}'\Big]. \tag{2.3.11}$$

This expansion involves hypothetical (or real) displacements (perturbations) in the system external potential, due to the probing shifts in the molecular geometry. It thus represents an example of the Taylor expansion within the EF description, in which the nuclear shifts or the presence of other external potential sources in the molecular environment, e.g., due to the reaction partner, create new conditions for the movements of electrons and ultimately induce adjustments in their equilibrium distribution. It should be observed that the displacements in Eq. (2.3.11) are along the true ground-state energy surface, $E = E[N, v]$, with the equilibrium electron density $\rho = \rho[N, v]$ responding to current displacements in the system external potential. Therefore, this expansion is "horizontal" in character.

The related quantities for molecular subsystems have also been proposed (Nalewajski *et al.*, 1996; Nalewajski and Korchowiec, 1997; Nalewajski, 1997a,b, 1998, 2000a, 1999; Korchowiec and Uchimaru, 1998), e.g., those describing reactants in the bimolecular reactive system $M_R = A\text{---}B$, where for definiteness A and B respectively denote the *acidic* (electron acceptor) and *basic* (electron donor) subsystems. The CS of the externally-open reactants, in contact with the hypothetical electron reservoir, have also been determined (Nalewajski, 1999, 2000b), e.g., for chemical species adsorbed on the catalyst surface. In such an approach each reactant $X = (A, B)$ is attributed an average number of electrons, N_X, and the effective external potential v_X^{eff}, which includes the embedding contribution due to the presence of the other reactant [see Eq. (1.6.7)]. For a given, fixed geometry of subsystems in M_R each reactant undergoes displacements $\Delta \boldsymbol{a}_X = (\Delta N_X, \Delta v_X^{eff}) = \boldsymbol{a}_X - \boldsymbol{a}_X^0$ in its effective state-parameters $\boldsymbol{a}_X = (N_X, v_X^{eff})$ in M_R, relative to the corresponding values in the SRL: $\boldsymbol{a}_X^0 = (N_X^0, v_X^0)$. The corresponding second-order change in the system electronic energy can be expressed using the reactant-resolved Taylor expansions discussed in Section 1.4.

The (N,v)-perturbation expansion of Eq. (2.3.11) and the related expansions for reactants (molecular subsystems) belong to the chemical softness representation, of the N-controlled molecular or reactive systems, in which the overall (integer) number of electrons in M_R as a whole is fixed. Such N-controlled systems are regarded as being closed with respect to the external electron reservoirs. Obviously, in the global internal equilibrium its constituent fragments are regarded as being mutually open, free to polarize and exchange electrons. The related Taylor series in the subsystem resolution provides a formal basis for most of the DFT treatments of molecular reactants perturbed by the presence of the reaction partner or the catalyst (see, e.g.: Alonso and Balbás, 1993; Ayers and Levy, 2000; Ayers and Parr, 2000, 2001; Baekelandt, 1993; Chattaraj and Parr, 1993; Gázquez, 1993; Mortier, and Schoonheydt, 1997; Nalewajski, 1984, 1993, 1995a,b, 1997b, 1998b, 1999, 2000a, 2002d,e; Nalewajski and Korchowiec, 1997; Nalewajski *et al.*, 1996).

In the N-controlled reactive systems, the shifts in the subsystem average number of electrons are due to the internal CT between reactants, while $\Delta v_X^{eff}(\boldsymbol{r})$ originates from the presence, changes in the geometry and the electronic structure of the complementary subsystem. These perturbations of the reactive system generate the reorganization of electronic structure on both coupled (interacting) reactants, which in turn determines the forces acting on their nuclei towards the new (displaced) equilibrium configuration. The normalized responses in forces $\mathbf{F} = \{\boldsymbol{F}_X = \{F_{X,\alpha}\}\}$ acting on the system nuclei $\alpha \in X = (A, B)$, per unit shifts in the reactant overall numbers of electrons $\Delta \boldsymbol{N} = \{\Delta N_X\}$, are reflected by the *matrix* of subsystem NFF [see Eq. (2.3.2)]:

$$\boldsymbol{\varphi} = \partial \mathbf{F}/\partial \boldsymbol{N} = \{\boldsymbol{\varphi}_X = \partial \boldsymbol{F}_X/\partial \boldsymbol{N} = \{\partial F_{X\alpha}/\partial N_Y,\ (X, Y) = A, B\}\}. \tag{2.3.12}$$

One can alternatively probe the couplings between the electronic and geometric degrees-of-freedom in M_R using the explicit EP mapping transformations formulated in CSA (Nalewajski and Korchowiec, 1997) or in the *Electronegativity Equalization Method* (EEM) of Baekelandt *et al.* (1993) (see e.g., Baekelandt *et al.*, 1995; Mortier and Schoonheydt, 1997; Nalewajski, 1995, 1999, 2000a; Nalewajski and Korchowiec, 1997; Nalewajski *et al.*, 1996; Nalewajski and Sikora, 2000).

The alternative, *variational* form of the quadratic energy expansion for the *fixed* external potential results from expanding the HK density functional $E_v[\rho']$ [Eq. (1.5.14)] in powers of displacements $\Delta\rho'(\boldsymbol{r}) = \rho'(\boldsymbol{r}) - \rho(\boldsymbol{r})$ of the *trial* density $\rho'(\boldsymbol{r})$, representing the system unconstrained local variable, from to the ground-state density $\rho(\boldsymbol{r}) = \rho[N, v; \boldsymbol{r}]$:

$$\Delta^{(1+2)}E_v[\rho'] = \mu \int \Delta\rho'(\boldsymbol{r})\, d\boldsymbol{r} + \frac{1}{2}\iint \Delta\rho'(\boldsymbol{r})\, \eta(\boldsymbol{r},\boldsymbol{r}')\, \Delta\rho'(\boldsymbol{r}')\, d\boldsymbol{r}\, d\boldsymbol{r}' . \qquad (2.3.13)$$

Here the *hardness kernel* of the EP perspective,

$$\eta(\boldsymbol{r},\boldsymbol{r}') = \left(\frac{\partial^2 E_v[\rho]}{\partial\rho(\boldsymbol{r})\,\partial\rho(\boldsymbol{r}')}\right)_v = \frac{\delta^2 F[\rho]}{\delta\rho(\boldsymbol{r})\,\delta\rho(\boldsymbol{r}')} = \left(\frac{\partial\mu(\boldsymbol{r}')}{\partial\rho(\boldsymbol{r})}\right)_v = \left(\frac{\partial\mu(\boldsymbol{r})}{\partial\rho(\boldsymbol{r}')}\right)_v$$

$$= -\frac{\delta u(\boldsymbol{r}')}{\delta\rho(\boldsymbol{r})} = -\frac{\delta u(\boldsymbol{r})}{\delta\rho(\boldsymbol{r}')}. \qquad (2.3.14)$$

In the Taylor expansion (2.3.13) the displaced electron density $\rho' = \rho + \Delta\rho'$ ceases to be the ground-state density, when the external potential remains "frozen". In other words, trial densities are not the equilibrium ones for the fixed external potential, so that the expansion is not "horizontal" in character. Neither is it "vertical", since it involves finite displacements in the electronic density from the initial, equilibrium distribution. Therefore, in this power series of the density-displacements the variational shifts in the distribution of electrons are not related to the underlying, matching shifts in the system external potential, as the equilibrium responses to this perturbation. Therefore, it cannot be classified as belonging to either the EF or EP approaches. Moreover, since Eq. (2.3.13) involves the hardness kernel, in which the shifts in the local density are regarded as perturbations, of the independent *electronic* variable, it should be more appropriately classified as belonging to the chemical hardness representation.

2.3.2. Open (μ-Controlled) Systems

The externally-*open* molecular systems and reactants, in equilibrium with the electron reservoir, $\mu = \mu_{\mathfrak{R}}$, is characterized by the *grand potential* [Perdew *et al.*, 1982; Perdew, 1985; Parr and Yang, 1989]:

$$\Omega[\mu,v] = E - \left(\frac{\partial E[N,v]}{\partial N}\right)_v N = E - N\mu = \Omega[u,\rho[u]] \equiv \Omega[u]$$
$$= \int u(\boldsymbol{r})\rho(\boldsymbol{r})\,d\boldsymbol{r} + F[\rho] \equiv \Omega_u[\rho], \qquad (2.3.15)$$

where the relative external potential $u(\boldsymbol{r}) = v(\boldsymbol{r}) - \mu$ [see Eq. (1.5.22)]. It corresponds to the Legendre transformed representation (see Section 2.2) $\mathcal{E}[\mu, v] \equiv \mathcal{L}_E[\mu, v] = \mathcal{E}[u] \equiv \mathcal{L}_E[u]$ of the system electronic energy E, in which the chemical potential of the external electron reservoir, $\mu = \mu_{\mathfrak{R}}$, replaces the number of electrons in the list of the open system state-parameters.

It follows from the preceding equation that the first differential of the grand potential reads:

$$d\Omega[\mu,v] = -N\,d\mu + \int \rho(\boldsymbol{r})\,dv(\boldsymbol{r})\,d\boldsymbol{r} = \int \rho(\boldsymbol{r})\,du(\boldsymbol{r})\,d\boldsymbol{r} = d\Omega_u[\rho]. \qquad (2.3.16)$$

Actually, we have already encountered this quantity in the HK variational principle (for externally-open systems) [Eq. (1.5.20)], in which the extra $\mu N[\rho]$ term supplementing the electronic energy was added to enforce the correct normalization of the optimum density. This minimum principle gives rise to the Euler equation (1.5.22) and allows one to express the grand potential as the minimum of searches over the trial (ensemble average) numbers of electrons N' or variational densities ρ' (see, e.g., Nalewajski and Parr, 1982; Nalewajski and Korchowiec, 1997), which constitute the unconstrained electronic global and local parameters of state of this representation:

$$\Omega[\mu,v] = \min_{N'}\{E[N',v] - \mu N'\}$$
$$= \min_{\rho'}\{E_v[\rho'] - \mu N[\rho']\} = \min_{\rho'} \Omega_u[\rho'] = \Omega_u[\rho]. \qquad (2.3.17)$$

These u-constrained searches deliver the optimum *electronic* structure parameters, the grand-ensemble average number of electrons $N = \int \rho(\boldsymbol{r})d\boldsymbol{r}$ and the ensemble ground-state density ρ, which match the fixed (independent) state-parameter of the open molecular system, $u(\boldsymbol{r})$, determined by to the system *nuclei* and the external reservoir of electrons. As already indicated in Eq. (2.3.15), it is the relative external potential $u(\boldsymbol{r})$ [Eq. (1.5.22)] which defines the equilibrium state in this representation. The functional derivative with respect to this independent state parameter, defining the grand potential conjugate of the external potential [see Eq. (2.3.16)],

$$\frac{\delta\Omega_u[\rho]}{\delta u(\boldsymbol{r})} = \rho(\boldsymbol{r}), \qquad (2.3.18)$$

expresses the Hellmann-Feynman theorem for open molecular systems [compare Eq.(1.5.12)].

The *partial* conjugates of $\Omega[\mu, v]$ define the grand-potential conjugates of the open system global and local state-variables, respectively,

$$\boldsymbol{e} = \frac{\partial \Omega[\boldsymbol{d}]}{\partial \boldsymbol{d}} = \left\{ \left(\frac{\partial \Omega[\mu, v]}{\partial \mu} \right)_v = -N, \ \left(\frac{\partial \Omega[\mu, v]}{\partial v(\boldsymbol{r})} \right)_\mu = \rho(\boldsymbol{r}) \right\}. \qquad (2.3.19)$$

The corresponding matrix of the open-system CS is defined by the second partial derivatives of $\Omega[\boldsymbol{d}]$:

$$\mathbf{g} = \frac{\partial^2 \Omega[\boldsymbol{d}]}{\partial \boldsymbol{d}' \partial \boldsymbol{d}} = \frac{\partial \boldsymbol{e}}{\partial \boldsymbol{d}'}. \qquad (2.3.20)$$

It consists of the *global softness* of Eq. (2.3.8),

$$S = \left(\frac{\partial N}{\partial \mu} \right)_v = -\left(\frac{\partial^2 \Omega[\mu, v]}{\partial \mu^2} \right)_v, \qquad (2.3.21)$$

the mixed derivative defining the *local softness* of the electron gas,

$$s(\boldsymbol{r}) = \frac{\partial^2 \Omega[\mu, v]}{\partial \mu\, \partial v(\boldsymbol{r})} = \left(\frac{\partial \rho(\boldsymbol{r})}{\partial \mu} \right)_v = -\left(\frac{\partial N}{\partial v(\boldsymbol{r})} \right)_\mu = -\frac{\delta N}{\delta u(\boldsymbol{r})}, \qquad (2.3.22)$$

and the softness kernel,

$$\sigma(\boldsymbol{r}, \boldsymbol{r}') = -\left(\frac{\partial^2 \Omega[\mu, v]}{\partial v(\boldsymbol{r})\, \partial v(\boldsymbol{r}')} \right)_\mu = -\left(\frac{\partial \rho(\boldsymbol{r}')}{\partial v(\boldsymbol{r})} \right)_\mu = -\left(\frac{\partial \rho(\boldsymbol{r})}{\partial v(\boldsymbol{r}')} \right)_\mu = \sigma(\boldsymbol{r}', \boldsymbol{r})$$

$$= -\frac{\delta^2 \Omega[u]}{\delta u(\boldsymbol{r})\, \delta u(\boldsymbol{r}')} = -\frac{\delta \rho(\boldsymbol{r})}{\delta u(\boldsymbol{r}')} = -\frac{\delta \rho(\boldsymbol{r}')}{\delta u(\boldsymbol{r})}, \qquad (2.3.23)$$

the inverse of the hardness kernel of Eq. (2.3.14) (see Appendix A):

$$\int \eta(\boldsymbol{r}, \boldsymbol{r}'') \sigma(\boldsymbol{r}'', \boldsymbol{r}')\, d\boldsymbol{r}'' = \delta(\boldsymbol{r} - \boldsymbol{r}'). \qquad (2.3.24)$$

It should be observed that the grand potential of Eq. (2.3.15) can be also regarded as the Legendre transform of the density functional for the electronic energy [Eq. (1.5.14)], $E_v[\rho] \equiv E[\rho; v]$, where the semicolon separates the system dependent (electronic) variable $\rho = \rho[N, v]$ from the fixed (nuclear) parameter $v(\boldsymbol{r})$:

$$\Omega_u[\rho] = E_v[\rho] - \int \frac{\delta E_v[\rho]}{\delta \rho(\boldsymbol{r})} \rho(\boldsymbol{r})\, d\boldsymbol{r} = E_v[\rho] - \int \mu(\boldsymbol{r}) \rho(\boldsymbol{r})\, d\boldsymbol{r}$$

$$= E_v[\rho] - N[\rho]\mu = \int u(\boldsymbol{r}) \rho(\boldsymbol{r})\, d\boldsymbol{r} + F[\rho] = \mathcal{L}_E[\mu; v]. \qquad (2.3.25)$$

In this local Legendre transform $\mathcal{E}[\mu; v] \equiv \mathcal{L}_E[\mu; v]$ of the energy density functional $E[\rho; v]$ the local value of the ground-state density $\rho(\boldsymbol{r})$ has been replaced by its energy conjugate, the local (equalized) chemical potential $\mu(\boldsymbol{r}) = \mu$ [Eq. (1.5.21)].

It follows from Eqs. (1.5.22), (2.3.14) and (2.3.16) that

$$\frac{\delta\Omega_u[\rho]}{\delta\rho(\boldsymbol{r})} = u(\boldsymbol{r}) + \frac{\delta F[\rho]}{\delta\rho(\boldsymbol{r})} = 0 \quad \text{and} \quad \frac{\delta^2\Omega_u[\rho]}{\delta\rho(\boldsymbol{r})\,\delta\rho(\boldsymbol{r}')} = \frac{\delta^2 F[\rho]}{\delta\rho(\boldsymbol{r})\,\delta\rho(\boldsymbol{r}')} = \eta(\boldsymbol{r},\boldsymbol{r}'). \quad (2.3.26)$$

The first of the preceding equations is just the Euler equation for the ground-state density [Eq. (1.5.22)], which implies that $\rho = \rho[u]$ and thus $\Omega_u[\rho[u]] \equiv \Omega_u[\rho[u]]$, as indeed confirmed by the differential of Eq. (2.3.16).

The electron density argument in these expressions is the *average* density of the grand-canonical ensemble (see, e.g., Parr and Yang, 1989). The vanishing first functional derivative in the preceding equation shows that the grand potential is stationary with respect to displacements from the ground-state density, in accordance with the HK variational principle of Eq. (1.5.20).

The derivatives of Eqs. (2.3.18)-(2.3.23) define the quadratic Taylor expansion of the grand potential in terms of powers of displacements $\Delta\boldsymbol{d} = (\Delta\mu, \Delta v) = \boldsymbol{d} - \boldsymbol{d}^0$ in its state-parameters $\boldsymbol{d} = (\mu, v)$ of the externally-open molecular system, relative to the reference values $\boldsymbol{d}^0 = (\mu^0, v^0)$:

$$\begin{aligned}\Delta^{(1+2)}\Omega[\mu,v] &= -N\,\Delta\mu + \int\rho(\boldsymbol{r})\,\Delta v(\boldsymbol{r})\,d\boldsymbol{r} \\ &+ \frac{1}{2}\Big[-S(\Delta\mu)^2 + 2\,\Delta\mu\int s(\boldsymbol{r})\,\Delta v(\boldsymbol{r})\,d\boldsymbol{r} - \iint\Delta v(\boldsymbol{r})\,\sigma(\boldsymbol{r},\boldsymbol{r}')\,\Delta v(\boldsymbol{r}')\,d\boldsymbol{r}\,d\boldsymbol{r}'\Big] \\ &= \int\Delta u(\boldsymbol{r})\,\rho(\boldsymbol{r})\,d\boldsymbol{r} - \frac{1}{2}\iint\Delta u(\boldsymbol{r})\,\sigma(\boldsymbol{r},\boldsymbol{r}')\,\Delta u(\boldsymbol{r}')\,d\boldsymbol{r}\,d\boldsymbol{r}'. \quad (2.3.27)\end{aligned}$$

This series approximates to the second-order horizontal displacements along the grand potential surface $\Omega_u[\rho]$, with the electron density satisfying the HK Euler equation (1.5.22) for the displaced relative external potential $u'(\boldsymbol{r}) = u(\boldsymbol{r}) + \Delta u(\boldsymbol{r})$: $\rho' = \rho[u']$.

Alternatively, using derivatives of Eq. (2.3.26) one arrives at the corresponding *variational* expansion of the density functional $\Omega_u[\rho']$, for the fixed relative potential u and trial densities $\rho' = \rho + \Delta\rho'$, around the ground-state density ρ:

$$\Delta^{(1+2)}\Omega_u[\rho'] = \Delta^{(2)}\Omega_u[\rho'] = \frac{1}{2}\iint\Delta\rho'(\boldsymbol{r})\,\eta(\boldsymbol{r},\boldsymbol{r}')\,\Delta\rho'(\boldsymbol{r}')\,d\boldsymbol{r}\,d\boldsymbol{r}'\,. \quad (2.3.28)$$

Again, there is a fundamental difference between the expansions (2.3.27) and (2.3.28) of the grand potential. The former is horizontal in character, approximating changes $\Delta^{(1+2)}\Omega[\mu,v]$ in the *equilibrium* grand-potential, to the second-order in displacements of the relative external potential, while the latter reflects displacements $\Delta^{(1+2)}\Omega_u[\rho']$ in the grand potential, due to the *trial* grand-ensemble densities ρ', which are not the equilibrium ones for the fixed relative external potential u, thus deviating from the ensemble equilibrium density $\rho[u]$.

The expansion (2.3.27) also defines the formal basis for the DFT treatment of reactants in the open reactive system $M_R = A$---B, which can exchange particles with the external electron reservoir (see, e.g.: Nalewajski, 1984, 1993, 1995a,b, 1997b, 1998b, 1999, 2000a, 2002d,e; Nalewajski and Korchowiec, 1997; Nalewajski *et al.*, 1996). In the reactant resolution each subsystem $X = (A, B)$ exhibits displacements in its effective independent state-parameters, $\boldsymbol{d}_X = (\mu_X, v_X)$, which define the relative potential $u_X = v_X - \mu_X$ of X in the externally open M_R, $\Delta\boldsymbol{d}_X = \boldsymbol{d}_X - \boldsymbol{d}_X^0 = (\Delta\mu_X, \Delta v_X)$, or $\Delta u_X = \Delta v_X - \Delta\mu_X$, relative to the corresponding values for the externally open separated reactants: $\boldsymbol{d}_X^0 = (\mu_X^0, v_X^0)$ or $u_X^0 = v_X^0 - \mu_X^0$. This is due to the presence of the reaction partner. The horizontal reactant-resolved (μ_X, v_X)- or u_X-perturbational expansions of the coupled, interacting reactants in M_R define the *chemical softness representation* of the open reacting species coupled to an external electron reservoir (hypothetical or real).

The basic derivative properties characterizing the closed and open molecular systems (or reactants) are mutually related. For example, in the closed molecular system, in which density displacements conserve the fixed (integer) number of electrons $N = N^0$, the *internal hardness kernel* of the closed molecular system

$$\eta^{int}(\boldsymbol{r}, \boldsymbol{r}') = -\left(\frac{\partial v(\boldsymbol{r}')}{\partial \rho(\boldsymbol{r})}\right)_N = -\beta^{-1}(\boldsymbol{r}, \boldsymbol{r}'), \tag{2.3.29}$$

reflects the equilibrium responses in the external potential in the *polarizational* displacements of the system electron density. It differs from the open system hardness kernel $\eta(\boldsymbol{r}, \boldsymbol{r}')$ [Eqs. (2.3.14) and (2.3.26)], which also includes the external CT contribution. These two normalized potential responses to local density displacements represent the basic kernels of the "inverse" *chemical softness representation*, which will be the subject of the next section. Indeed, in both these charge sensitivities the primary perturbation $\delta\rho(\boldsymbol{r})$ is electronic in character while the monitored responses $\delta v(\boldsymbol{r}')$ or $\delta u(\boldsymbol{r}')$ are of the nuclear (geometric) origin. As such this kernel indeed belongs to the EP perspective.

As we have already indicated in Eq. (2.3.24) the softness kernel is the inverse of the hardness kernel. Indeed, by the functional chain-rule (see Appendix A)

$$\int \sigma(\boldsymbol{r}, \boldsymbol{r}'')\, \eta(\boldsymbol{r}'', \boldsymbol{r}')\, d\boldsymbol{r}'' = \int \frac{\delta\rho(\boldsymbol{r}'')}{\delta u(\boldsymbol{r})} \frac{\delta u(\boldsymbol{r}')}{\delta\rho(\boldsymbol{r}'')}\, d\boldsymbol{r}'' = \frac{\delta u(\boldsymbol{r}')}{\delta u(\boldsymbol{r})} = \delta(\boldsymbol{r}' - \boldsymbol{r}). \tag{2.3.30}$$

It can be expressed using the relevant chain-rule, or the functional Jacobian transformation of the defining derivative, in terms of the density LR [Eq. (2.3.10)], measuring the *closed* system density response, and the extra term for *open* molecular systems involving the electronic FF [Eqs. (2.3.1) and (2.3.9)] and the system global softness [Eqs. (2.3.8)-(1.31)]. For this purpose we express the ground-state density as functional of the independent state-parameters in the energy representation [see Eq. (1.5.12)], $\rho = \rho[N, v]$:

$$\sigma(\boldsymbol{r},\boldsymbol{r}') = \eta(\boldsymbol{r}',\boldsymbol{r})^{-1} = -\frac{\delta\rho(\boldsymbol{r}')}{\delta u(\boldsymbol{r})} = -\left(\frac{\partial\rho(\boldsymbol{r}')}{\partial v(\boldsymbol{r})}\right)_{\mu} = -\left(\frac{\partial\rho(\boldsymbol{r}')}{\partial v(\boldsymbol{r})}\right)_{N} - \left(\frac{\partial N}{\partial v(\boldsymbol{r})}\right)_{\mu}\left(\frac{\partial\rho(\boldsymbol{r}')}{\partial N}\right)_{v}$$
$$= -\beta(\boldsymbol{r},\boldsymbol{r}') + s(\boldsymbol{r})f(\boldsymbol{r}') = -\beta(\boldsymbol{r},\boldsymbol{r}') + f(\boldsymbol{r})\,S\,f(\boldsymbol{r}'). \quad (2.3.31)$$

The first contribution in the preceding equation measures the density response due to the system internal polarizational (for the fixed $N = N^0$), while the second component reflects the effects due to the external *CT* (Berkowitz and Parr, 1988).

In the preceding equation we have used the relevant definitions of Eqs. (2.3.1), (2.3.10), (2.3.22), and the alternative expression for the local softness

$$s(\boldsymbol{r}) = \left(\frac{\partial\rho(\boldsymbol{r})}{\partial\mu}\right)_{v} = \left(\frac{\partial\rho(\boldsymbol{r})}{\partial N}\right)_{v}\left(\frac{\partial N}{\partial\mu}\right)_{v} = f(\boldsymbol{r})\,S. \quad (2.3.32)$$

The global softness combines the additive softness-kernel and local-softness contributions:

$$S = \left(\frac{\partial N}{\partial\mu}\right)_{v} = \iint\frac{\delta\rho(\boldsymbol{r}')}{\delta u(\boldsymbol{r})}\left(\frac{\delta u(\boldsymbol{r})}{\delta\mu}\right)_{v} d\boldsymbol{r}\, d\boldsymbol{r}' = \iint\sigma(\boldsymbol{r},\boldsymbol{r}')\, d\boldsymbol{r}\, d\boldsymbol{r}'$$
$$= \int\left(\frac{\partial\rho(\boldsymbol{r})}{\partial\mu}\right)_{v} d\boldsymbol{r} = \int s(\boldsymbol{r})\, d\boldsymbol{r}\,, \quad (2.3.33)$$

so that

$$s(\boldsymbol{r}) = \int\sigma(\boldsymbol{r},\boldsymbol{r}')\, d\boldsymbol{r}'\,. \quad (2.3.34)$$

It also follows from Eq. (2.32) that the electronic FF represents the renormalized local softness:

$$f(\boldsymbol{r}) = s(\boldsymbol{r})/S = s(\boldsymbol{r})\eta. \quad (2.3.35)$$

The system global hardness of Eq. (2.3.7) can be similarly expressed as the FF-weighed average of the hardness kernel:

$$\eta = \left(\frac{\partial\mu}{\partial N}\right)_{v} = \iint\left(\frac{\partial\rho(\boldsymbol{r})}{\partial N}\right)_{v}\left(\frac{\partial^2 E_v[\rho]}{\partial\rho(\boldsymbol{r})\partial\rho(\boldsymbol{r}')}\right)_{v}\left(\frac{\partial\rho(\boldsymbol{r}')}{\partial N}\right)_{v} d\boldsymbol{r}\, d\boldsymbol{r}'$$
$$= \iint f(\boldsymbol{r})\,\eta(\boldsymbol{r},\boldsymbol{r}')\,f(\boldsymbol{r}')\, d\boldsymbol{r}\, d\boldsymbol{r}'\,. \quad (2.3.36)$$

The same result is obtained for the *local hardness* of the ground-state distribution of electrons, defined by the derivative

$$\eta(\boldsymbol{r}) = \left(\frac{\partial \mu(\boldsymbol{r})}{\partial N}\right)_v = \int \left(\frac{\partial^2 E_v[\rho]}{\partial \rho(\boldsymbol{r})\partial \rho(\boldsymbol{r}')}\right)_v \left(\frac{\partial \rho(\boldsymbol{r}')}{\partial N}\right)_v d\boldsymbol{r}' = \int \eta(\boldsymbol{r},\boldsymbol{r}')\, f(\boldsymbol{r}')\, d\boldsymbol{r}' = \left(\frac{\partial \mu}{\partial N}\right)_v = \eta\,. \tag{2.3.37}$$

The last equation expresses the *local hardness equalization* rule, which is a direct consequence of the chemical potential equalization principle of Eq. (1.5.21).

As we have seen, all relations between the ground-state charge sensitivities of the open and closed systems can be derived using the Euler equation (1.5.22) and the relevant functional chain-rule transformations or the equivalent Jacobian manipulations of the defining derivatives (see, e.g.: Berkowitz and Parr, 1988; Nalewajski, 2002d,e, 2003a; Nalewajski *et al.*, 1996; Nalewajski and Korchowiec, 1997; Parr and Yang, 1989).

The subsystem-resolved analogs [see Eqs. (1.4.3), (1.4.8) and (1.4.9)] of the quadratic Taylor expansions of Eqs. (2.3.11) and (2.3.27) provide a flexible theoretical framework for the *two-reactant* treatment of diverse phenomena of chemical reactivity, covering both displacements in the global electronic $\{N$ or $\mu\}$ and the local nuclear $\{v(\mathbf{R})$, $\mathbf{F}(\mathbf{R})$, or $\mathbf{R}\}$ degrees-of-freedom of the molecular subsystems in $M_R = A\text{---}B$. This description can adequately treat both the *promoted* state of the closed reactants at the intermediate *polarization* stage of a chemical reaction, when the overall number of electrons in each subsystem is fixed, and at the *equilibrium* state of the mutually-open reactants at the final *charge-transfer* stage of the reaction, when electrons can freely flow between reactants. Therefore, this perturbative DFT approach naturally connects to the conventional stages in which the reaction mechanism is described in chemistry: the initial v-driven polarization of the mutually closed reactants being followed by the N-driven CT between the already polarized species.

This chemical softness representation of the closed or open molecular systems, in which the external potentials v or u play the role of the constrained local state-parameters, provide the theoretical framework for the *electron-following* perspective on molecular displacements. It emphasizes the primary role attributed in this BO-type approach to shifts in the external potential due to the system nuclei, which induce the subsequent electron redistribution in the molecule under consideration. In this description the probing shifts in nuclear positions, i.e., the system geometrical parameters, define the *perturbations*, which induce subsequent *responses* in the molecular electronic structure. In other words, the molecular state is specified by the local external potential, which represents the independent state-parameter, while the electronic density plays the role of the dependent (unconstrained) state variable, which adjusts itself to the current form of the displaced effective external potential, e.g., due to changed positions of nuclei or electron flows in the system chemical environment.

However, for the *complete* theoretical framework, capable of tackling *all* issues in the chemical reactivity, one additionally requires a related development within the *chemical hardness representation*, in which the roles of the nuclear and electronic local state-variables are reversed. In such a description one is interested in the Taylor expansions in powers of displacements in the ground-state electron density, which represent independent perturbations of the molecular system, and are regarded as preceding the subsequent movement of the nuclei. In this *electron preceding* perspective the local values of the electronic density $\rho(\boldsymbol{r})$ constitute the independent state-parameters, to which the nuclear (geometrical) factors respond. Such an attitude is close to the chemical thinking about how to manipulate molecules in order to induce desirable changes in the molecular or reactive system. Indeed, chemists often successfully modify the pattern of the chemical bonds by judiciously designing a crucial electronic perturbation of the system, which in turn induces the coupled movement of the nuclei. We shall examine the elements of such an approach in the next section.

2.4. THE CHEMICAL HARDNESS REPRESENTATION

Displacements of the electronic local state-variables define the system perturbations in the chemical hardness representation of the EP perspective on molecular processes. They generate forces driving responses in nuclear positions, giving rise to the adjusted external potential v (or u). This is in the spirit of the Hellmann-Feynman theorem, since in the EP description the electronic structure is seen as the primary cause of the observed geometrical structures of molecular systems. This way of viewing and – ultimately – manipulating molecular changes dominates the speculative chemical thinking about molecules and reactants.

The chemical hardness representation of the closed or open molecular systems is thus obtained, when the ground-state electron density replaces the external potential in the list of state-parameters (see, e.g., Nalewajski, 1999, 2002d, 2003a; Nalewajski and Korchowiec, 1997). This is accomplished through the relevant $v \to \rho$ Legendre transformation (see Section 2.2). The relevant "thermodynamic" potentials of the hardness representation, which determine the molecular equilibria in the externally closed and open systems, respectively, are thus given by the following Legendre transforms of $E[N, v]$ (for the N-controlled equilibria) or $\Omega[\mu, v]$ (for the μ-controlled equilibria):

$$\mathsf{F}[N,\rho] = E - \int \left(\frac{\partial E[N,v]}{\partial v(\boldsymbol{r})}\right)_N v(\boldsymbol{r})\,d\boldsymbol{r} = E - \int \rho(\boldsymbol{r}) v(\boldsymbol{r})\,d\boldsymbol{r} = F[\rho]$$
$$\equiv \mathcal{E}[N, \rho] \equiv \mathcal{L}_E[N, \rho], \tag{2.4.1}$$

$$\mathsf{R}[\mu,\rho] = \Omega - \int \left(\frac{\partial \Omega[\mu,v]}{\partial v(\boldsymbol{r})}\right)_\mu v(\boldsymbol{r})\,d\boldsymbol{r} = \Omega - \int \rho(\boldsymbol{r}) v(\boldsymbol{r})\,d\boldsymbol{r} = F[\rho] - \mu N[\rho] = R_\mu[\rho]$$
$$\equiv \mathcal{E}[\mu, \rho] \equiv \mathcal{L}_E[\mu, \rho]. \tag{2.4.2}$$

These Legendre transforms of the system energy give rise to the corresponding first differentials:

$$d\mathsf{F}[N,\rho] = \mu dN - \int v(\boldsymbol{r})\, d\rho(\boldsymbol{r})\, d\boldsymbol{r} = -\int u(\boldsymbol{r})\, d\rho(\boldsymbol{r})\, d\boldsymbol{r}, \tag{2.4.3}$$

$$d\mathsf{R}[\mu,\rho] = -Nd\mu - \int v(\boldsymbol{r})\, d\rho(\boldsymbol{r})\, d\boldsymbol{r}. \tag{2.4.4}$$

Therefore, in these two Legendre-transformed representations of the EP perspective (hardness representation) the electronic energy $E_v[\rho] = E[N, v]$ of the EF perspective (softness representation) is replaced by the universal functional $F[\rho] = \mathsf{F}[N,\rho]$ for the expectation value of the sum of the electron repulsion and kinetic energies. As seen in the two preceding equations, these additional "thermodynamic" potentials have been also expressed as corresponding density functionals: $\mathsf{F}[N,\rho] = F[\rho]$ and $\mathsf{R}[\mu,\rho] = R_\mu[\rho]$. For the equilibrium (ground-state) density the thermodynamic potential $\mathsf{F}[N,\rho] = F[\rho]$ defines the repulsive, universal part of the density functional for the system electronic energy, while the *free F*-potential $\mathsf{R}[\mu,\rho] = R_\mu[\rho]$ represents its open system analog.

It should be observed that the $\mathsf{R}[\mu,\rho]$ potential represents the *complete* Legendre transform of the system electronic energy, $\mathcal{E}[\mu, \rho] \equiv \mathcal{L}_E[\mu, \rho]$, in which all parameters of $E[N, v]$ have been replaced by their respective energy conjugates (μ, ρ). Contrary to the ordinary thermodynamics, where such complete transformation gives rise to the identically vanishing potential, due to the Euler equation (2.2.7) and the homogeneous first-order property of the internal energy, the molecular potential $\mathcal{L}_E[\mu, \rho] = \mathsf{R}[\mu,\rho]$ does not vanish. This is because the molecular electronic energy is not the homogeneous function of degree 1 of its natural (principal) state parameters $[N, v]$. Another manifestation of this fact is the non-vanishing character of the molecular analog of the Gibbs-Duhem equation (2.2.9):

$$-d\mathsf{R}[\mu,\rho] = N\, d\mu + \int v(\boldsymbol{r})\, d\rho(\boldsymbol{r})\, d\boldsymbol{r} \neq 0. \tag{2.4.5}$$

2.4.1. Closed (N-Controlled) Systems

In the chemical hardness representation $\mathsf{F}[N,\rho]$ the ground-state density constitutes the independent parameter of state. It uniquely identifies the system Hamiltonian and thus completely specifies the molecular ground-state. The conjugate variable of the Euler equation (1.5.22), the relative external potential $u = u[\rho]$, plays the role of the dependent (nuclear) variable. Therefore, the potential (2.4.1) can be also interpreted as the functional of the optimum external potential, which "matches" the specified electron density:

$$\mathsf{F}[N,\rho] \equiv F_\rho[u[\rho]] \equiv F[u;\, \rho] = F[\rho]. \tag{2.4.6}$$

The notation $F_\rho[u] = F[u;\, \rho]$ emphasizes the dependent role of the unconstrained external potential u and the independent character of the electronic state-parameter ρ. Clearly, for a trial (variational) external potential u', which does not match through Eq. (1.5.22) the specified v-representable density ρ, $F_\rho[u'] \neq F[\rho]$. Therefore, the density-functional notation $F[\rho]$ is less general, covering only the equilibrium case, when the external potential identifies the specified density as its ground-state density: $F[\rho] = F[u[\rho]; \rho]$. It cannot account for the variational, non-equilibrium situation $F[u'; \rho]$, when u' does not match the given ρ.

In fact, the equilibrium potential $\mathsf{F}[N,\rho] = F[\rho]$ can be constructed via the density constrained search for the external potential (see, e.g., Nalewajski and Parr, 1982; Nalewajski and Korchowiec, 1997; Colonna and Savin, 1999),

$$\mathsf{F}[N,\rho] = \max_{v'} \{E[N, v'] - \int v'(\boldsymbol{r})\, \rho(\boldsymbol{r})\, d\boldsymbol{r}\} \equiv \max_{v'} \mathsf{F}[v'; N,\rho] = F[\rho], \qquad (2.4.7)$$

which yields the external potential $v = v[\rho]$ matching the specified (v-representable) density ρ as its ground-state distribution of electrons. It follow from this maximum principle of the auxiliary (non-equilibrium) Legendre transform functional $\mathsf{F}[v';N,\rho]$, with respect to trial external potentials $\{v'\}$, that at the solution point $v' = v[\rho] \equiv v$ the functional derivative

$$\left(\frac{\partial \mathbf{F}[v; N,\rho]}{\partial v(\boldsymbol{r})}\right)_{v=v[\rho]} = 0. \qquad (2.4.8)$$

Moreover, since the ground-state density is uniquely determined by the relative external potential u, $\rho = \rho[u]$, from the HK Euler equation (1.5.22) for the optimum density, the equilibrium functional $F[\rho]$ can be alternatively regarded as the composite functional of u:

$$F[\rho] = F[\rho[u]] = \widetilde{F}[u]. \qquad (2.4.9)$$

The Euler Eq. (1.5.22) identifies the F-conjugate of the ground-state density ρ, i.e., the local "intensity" defined by the functional derivative of $F[\rho]$, as the negative relative external potential, $-u$, while the corresponding kernel of the second functional derivatives is given by the hardness kernel $\eta(\boldsymbol{r}, \boldsymbol{r}')$ [Eq. (2.3.14)]. Together these functional derivatives define the second-order Taylor expansion of $F[\rho]$ in powers of displacements of the electronic state-parameter ρ in this EP perspective:

$$\Delta^{(1+2)} F[\rho] = -\int u(\boldsymbol{r})\, \Delta\rho(\boldsymbol{r})\, d\boldsymbol{r} + \frac{1}{2}\iint \Delta\rho(\boldsymbol{r})\, \eta(\boldsymbol{r},\boldsymbol{r}')\, \Delta\rho(\boldsymbol{r}')\, d\boldsymbol{r}\, d\boldsymbol{r}' . \qquad (2.4.10)$$

Let us now examine the equivalent expansion of the composite functional $\widetilde{F}[u]$of Eq.(2.4.9), in powers of displacements $\Delta u(\boldsymbol{r})$, around the relative potential $u = u[\rho]$, the equilibrium one for the specified electron density $\rho(\boldsymbol{r})$. Using the functional chain-rule transformations of Appendix A one determines the following functional derivatives of $\widetilde{F}[u]$ [see Eqs. (2.3.14) and (2.3.23)]:

$$\frac{\delta \widetilde{F}[u]}{\delta u(\boldsymbol{r})} = \int \frac{\delta F[\rho]}{\delta \rho(\boldsymbol{r}')} \frac{\delta \rho(\boldsymbol{r}')}{\delta u(\boldsymbol{r})} d\boldsymbol{r}' \equiv \int \sigma(\boldsymbol{r},\boldsymbol{r}') u(\boldsymbol{r}') d\boldsymbol{r}', \tag{2.4.11}$$

$$\frac{\delta^2 \widetilde{F}[u]}{\delta u(\boldsymbol{r})\, \delta u(\boldsymbol{r}')} = \iint \frac{\delta \rho(\boldsymbol{r}''')}{\delta u(\boldsymbol{r})} \frac{\delta^2 F[\rho]}{\delta \rho(\boldsymbol{r}''')\, \delta \rho(\boldsymbol{r}'')} \frac{\delta \rho(\boldsymbol{r}'')}{\delta u(\boldsymbol{r}')} d\boldsymbol{r}''\, d\boldsymbol{r}'''$$

$$= \iint \sigma(\boldsymbol{r},\boldsymbol{r}''')\eta(\boldsymbol{r}''',\boldsymbol{r}'') d\boldsymbol{r}''' \sigma(\boldsymbol{r}',\boldsymbol{r}'')\, d\boldsymbol{r}'' = \int \delta(\boldsymbol{r}-\boldsymbol{r}'')\sigma(\boldsymbol{r}',\boldsymbol{r}'') d\boldsymbol{r}'' = \sigma(\boldsymbol{r}',\boldsymbol{r}). \tag{2.4.12}$$

In the second line of the last equation we have used the inverse property of the softness and hardness kernels [Eq.(2.3.24)].

Hence the quadratic Taylor expansion of the composite functional $\widetilde{F}[u]$, in the spirit of the EF perspective of the chemical softness representation, becomes

$$\Delta^{(1+2)} \widetilde{F}[u] = [\int \Delta u(\boldsymbol{r}) \int \sigma(\boldsymbol{r},\boldsymbol{r}') d\boldsymbol{r}]\, u(\boldsymbol{r}') d\boldsymbol{r}' + \frac{1}{2} \iint \Delta u(\boldsymbol{r}) \sigma(\boldsymbol{r},\boldsymbol{r}') \Delta u(\boldsymbol{r}') d\boldsymbol{r}\, d\boldsymbol{r}'. \tag{2.4.13}$$

One can easily verify the equivalence of the expansions (2.4.10) and (2.4.13). For example, the square bracket of the first-order contribution in the preceding equation generates, via the chain-rule, $-\Delta\rho[\Delta u]$, so that the resulting first-order change amounts to the first differential of Eqs. (2.4.3) and (2.4.10). It should be stressed that both these Taylor series are "horizontal" in character, since a given electronic perturbation $\Delta\rho$ implies the equilibrium displacement in the conjugate nuclear state-quantity, $\Delta u = \Delta u[\Delta\rho]$, and vice versa, a given shift in the relative potential due to the nuclei, Δu, results in the concomitant change in the electron density, $\Delta\rho = \Delta\rho[\Delta u]$. The LR approximations of these functional relations read:

$$\Delta u(\boldsymbol{r}) = -\int \Delta\rho(\boldsymbol{r}')\eta(\boldsymbol{r}',\boldsymbol{r})\, d\boldsymbol{r}' = \Delta u[\Delta\rho;\, \boldsymbol{r}],$$

$$\Delta\rho(\boldsymbol{r}) = -\int \Delta u(\boldsymbol{r}')\sigma(\boldsymbol{r}',\boldsymbol{r})\, d\boldsymbol{r}' = \Delta\rho[\Delta u;\, \boldsymbol{r}]. \tag{2.4.14}$$

In terms of these coupled shifts in the complementary state-variables the second-order terms in Eqs. (2.4.10) and (2.4.13) have a common physical interpretation:

$$\Delta^{(2)} F[\rho] = -\frac{1}{2} \int \Delta\rho(\boldsymbol{r})\, \Delta u[\Delta\rho;\boldsymbol{r}]\, d\boldsymbol{r} = \Delta^{(2)} \widetilde{F}[u] = -\frac{1}{2} \int \Delta u(\boldsymbol{r})\, \Delta\rho[\Delta u;\boldsymbol{r}]\, d\boldsymbol{r}. \tag{2.4.15}$$

Next, let us separate the *internal* system polarization, for the fixed overall number of electrons N (i.e., $\Delta N = 0$), from changes in the electron distribution induced by the external CT, due to a finite inflow/outflow of electrons ($\Delta N \neq 0$) to/from the molecular system under consideration (see also Appendix B). This is accomplished by expressing the electron density as the product of its overall normalization, N, and the unity-normalized *shape* (probability) factor $p(\boldsymbol{r})$:

$$\rho(\boldsymbol{r}) = N\, p(\boldsymbol{r}), \qquad \int p(\boldsymbol{r})\, d\boldsymbol{r} = 1 \quad \text{or} \quad \int \delta p(\boldsymbol{r})\, d\boldsymbol{r} = 0 \,. \tag{2.4.16}$$

Hence,

$$\left(\frac{\partial \rho(\boldsymbol{r})}{\partial N}\right)_p = p(\boldsymbol{r}) \quad \text{and} \quad \left(\frac{\partial \rho(\boldsymbol{r}')}{\partial p(\boldsymbol{r})}\right)_N = N\delta(\boldsymbol{r}-\boldsymbol{r}'). \tag{2.4.17}$$

This factorization indeed distinguishes between the two contributions to displacements of the electron density:

$$\Delta\rho(\boldsymbol{r}) = N\Delta p(\boldsymbol{r}) + p(\boldsymbol{r})\Delta N = [\Delta\rho(\boldsymbol{r})]_N + [\Delta\rho(\boldsymbol{r})]_p \equiv \Delta\rho_P(\boldsymbol{r}) + \Delta\rho_{CT}(\boldsymbol{r}), \tag{2.4.18}$$

where the first term, $\Delta\rho_P(\boldsymbol{r})$, originates from the polarizational (P) changes in the density *shape* factor, while the second contribution, $\Delta\rho_{CT}(\boldsymbol{r})$, combines all effects due to the external CT.

Consider, as an illustrative example, the KS (spin-resolved) ansatz for the molecular electron density, expressed in terms of the occupied orbitals $\{\varphi_{i\sigma}(\boldsymbol{r})\}$ for the spin orientation $\sigma = \{\uparrow, \downarrow\}$ and their occupations $\{0 \le n_{i\sigma} \le 1\}$, $\sum_i n_{i\sigma} = N_\sigma$ and $\sum_\sigma N_\sigma = N$:

$$\rho(\boldsymbol{r}) = \sum_\sigma \sum_i n_{i\sigma} |\varphi_{i\sigma}(\boldsymbol{r})|^2 \equiv \sum_\sigma \sum_i n_{i\sigma}\, p_{i\sigma}(\boldsymbol{r}) = \sum_\sigma \sum_i \rho_{i\sigma}(\boldsymbol{r}) = \sum_\sigma \rho_\sigma(\boldsymbol{r}). \tag{2.4.19}$$

Here $\rho_{i\sigma}(\boldsymbol{r})$ and $p_{i\sigma}(\boldsymbol{r})$ denote the orbital density and its shape factor, respectively, while $\rho_\sigma(\boldsymbol{r})$ stands for the molecular spin-density. The first, P-term in Eq. (2.4.18) involves changing the KS orbitals, i.e., the shape factors of their densities, for the fixed MO occupations, while the second, CT-term, deals with displacements of the KS orbital occupations for the fixed shapes of their orbital densities.

It follows from the partition of Eq. (2.4.18) that the fixed density constraint introduces an implicit dependence between the infinitesimal displacements of the local shape factor $p(\boldsymbol{r})$ and the overall density normalization N:

$$d\rho(\boldsymbol{r}) = N\, dp(\boldsymbol{r}) + p(\boldsymbol{r})\, dN \equiv [d\rho(\boldsymbol{r})]_N + [d\rho(\boldsymbol{r})]_p = 0. \tag{2.4.20}$$

It is reflected by the derivatives:

$$\left(\frac{\partial p(\boldsymbol{r})}{\partial N}\right)_\rho = -\frac{p(\boldsymbol{r})}{N} = \left(\frac{\partial N}{\partial p(\boldsymbol{r})}\right)_\rho^{-1}. \tag{2.4.21}$$

Consider now the first *partial* derivatives of $F[Np] \equiv F[\rho]$. From the Euler equation (1.5.22) one obtains:

$$\left(\frac{\partial F[Np]}{\partial N}\right)_p = \int\left(\frac{\delta F}{\delta\rho(\boldsymbol{r})}\right)\left(\frac{\partial\rho(\boldsymbol{r})}{\partial N}\right)_p d\boldsymbol{r} = -\int u(\boldsymbol{r})\, p(\boldsymbol{r})\, d\boldsymbol{r} \equiv -\bar{u}$$

$$= \mu - \int v(\boldsymbol{r})\, p(\boldsymbol{r})\, d\boldsymbol{r} \equiv \mu - \bar{v},$$

$$\left(\frac{\partial F[Np]}{\partial p(\boldsymbol{r})}\right)_N = \left(\frac{\delta F}{\delta\rho(\boldsymbol{r})}\right)\left(\frac{\partial\rho(\boldsymbol{r})}{\partial p(\boldsymbol{r})}\right)_N = -Nu(\boldsymbol{r}), \tag{2.4.22}$$

It can be straightforwardly verified that these partial derivatives reproduce the first differential of Eq. (2.4.3):

$$dF[\rho] = -\int u(\boldsymbol{r})\, d\rho(\boldsymbol{r})\, d\boldsymbol{r} = -\bar{u}\, dN - N\int u(\boldsymbol{r})\, dp(\boldsymbol{r})\, d\boldsymbol{r} = dF[Np]. \tag{2.4.23}$$

Therefore, one can formally define the first partial derivatives of $\mathsf{F}[N,\rho] \equiv F[Np]$:

$$\left(\frac{\partial \mathsf{F}[N,\rho]}{\partial\rho(\boldsymbol{r})}\right)_N = \int\left(\frac{\partial F[Np]}{\partial p(\boldsymbol{r}')}\right)_N\left(\frac{\partial p(\boldsymbol{r}')}{\partial\rho(\boldsymbol{r})}\right)_N d\boldsymbol{r}' \equiv -\int u(\boldsymbol{r}')\,\delta(\boldsymbol{r}'-\boldsymbol{r})\, d\boldsymbol{r}' = -u(\boldsymbol{r}),$$

$$\left(\frac{\partial \mathsf{F}[N,\rho]}{\partial N}\right)_\rho = \int\left(\frac{\partial F[Np]}{\partial p(\boldsymbol{r})}\right)_N\left(\frac{\partial p(\boldsymbol{r})}{\partial N}\right)_\rho d\boldsymbol{r} + \left(\frac{\partial F[Np]}{\partial N}\right)_p = \bar{u} - \bar{u} = 0. \tag{2.4.24}$$

Again, the above derivatives give rise to the first differential of Eq. (2.4.3): $d\mathsf{F}[N,\rho] = dF[\rho] = dF[Np]$.

Let us now examine the second partials of $F[Np]$. The "diagonal" derivative with respect to the overall density normalization, N, results from the following chain-rule transformation of $F[\rho]$:

$$\left(\frac{\partial^2 F[Np]}{\partial N^2}\right)_p = -\left(\frac{\partial\bar{u}}{\partial N}\right)_p = \iint\left(\frac{\partial\rho(\boldsymbol{r})}{\partial N}\right)_p\left(\frac{\partial^2 F[\rho]}{\partial\rho(\boldsymbol{r})\,\partial\rho(\boldsymbol{r}')}\right)\left(\frac{\partial\rho(\boldsymbol{r}')}{\partial N}\right)_p d\boldsymbol{r}\, d\boldsymbol{r}'$$

$$= \iint p(\boldsymbol{r})\,\eta(\boldsymbol{r},\boldsymbol{r}')\,p(\boldsymbol{r}')\,d\boldsymbol{r}\,d\boldsymbol{r}' \equiv \bar{\eta}. \tag{2.4.25}$$

This derivative represents the shape-factor averaged hardness kernel, $\bar{\eta}$, defining the *mean global hardness*. A similar transformation of the other "diagonal" derivative, with respect to the local shape factor, gives the *internal hardness kernel*, of the closed molecular system (for constant N):

$$\left(\frac{\partial^2 F[Np]}{\partial p(\boldsymbol{r})\,\partial p(\boldsymbol{r}')}\right)_N = -N\left(\frac{\delta u(\boldsymbol{r}')}{\delta\rho(\boldsymbol{r})}\right)\left(\frac{\partial\rho(\boldsymbol{r})}{\partial p(\boldsymbol{r})}\right)_N = N^2\eta(\boldsymbol{r},\boldsymbol{r}'). \tag{2.4.26}$$

It represents the renormalized hardness kernel $\eta(\boldsymbol{r},\boldsymbol{r}')$ of Eq. (2.3.14).

Finally, for the mixed derivative of $F[Np]$ one obtains:

$$\left(\frac{\partial}{\partial N}\left(\frac{\partial F[Np]}{\partial p(\boldsymbol{r})}\right)_N\right)_p = -\int\left(\frac{\partial u(\boldsymbol{r}')}{\partial N}\right)_p\left(\frac{\partial \rho(\boldsymbol{r}')}{\partial p(\boldsymbol{r})}\right)_N d\boldsymbol{r}' = -N\left(\frac{\partial u(\boldsymbol{r})}{\partial N}\right)_p \tag{2.4.27}$$

$$=\left(\frac{\partial}{\partial p(\boldsymbol{r})}\left(\frac{\partial F[Np]}{\partial N}\right)_p\right)_N = -\int\left(\frac{\partial u(\boldsymbol{r}')}{\partial p(\boldsymbol{r})}\right)_N\left(\frac{\partial \rho(\boldsymbol{r}')}{\partial N}\right)_p d\boldsymbol{r}' = -\int p(\boldsymbol{r}')\left(\frac{\partial u(\boldsymbol{r}')}{\partial p(\boldsymbol{r})}\right)_N d\boldsymbol{r}' \tag{2.4.28}$$

$$=\iint\left(\frac{\partial \rho(\boldsymbol{r}'')}{\partial N}\right)_p\left(\frac{\partial^2 F[\rho]}{\partial \rho(\boldsymbol{r}'')\,\partial \rho(\boldsymbol{r}')}\right)\left(\frac{\partial \rho(\boldsymbol{r}')}{\partial p(\boldsymbol{r})}\right)_N d\boldsymbol{r}''\,d\boldsymbol{r}'$$

$$=N\iint p(\boldsymbol{r}'')\eta(\boldsymbol{r}'',\boldsymbol{r}')\delta(\boldsymbol{r}'-\boldsymbol{r})\,d\boldsymbol{r}''d\boldsymbol{r}' = N\,\overline{\eta}(\boldsymbol{r}), \tag{2.4.29}$$

in which the *mean* (p-averaged) *local hardness* $\overline{\eta}(\boldsymbol{r}) = \int p(\boldsymbol{r}')\eta(\boldsymbol{r}',\boldsymbol{r})\,d\boldsymbol{r}'$.

These alternative expressions imply the following Maxwell relation:

$$N\left(\frac{\partial u(\boldsymbol{r})}{\partial N}\right)_p = \int p(\boldsymbol{r}')\left(\frac{\partial u(\boldsymbol{r}')}{\partial p(\boldsymbol{r})}\right)_N d\boldsymbol{r}'. \tag{2.4.30}$$

It can be easily verified by calculating the two partial derivatives involved:

$$\left(\frac{\partial u(\boldsymbol{r})}{\partial N}\right)_p = \int\frac{\delta u(\boldsymbol{r})}{\delta \rho(\boldsymbol{r}')}\left(\frac{\partial \rho(\boldsymbol{r}')}{\partial N}\right)_p d\boldsymbol{r}' = -\overline{\eta}(\boldsymbol{r}), \tag{2.4.31}$$

$$\left(\frac{\partial u(\boldsymbol{r}')}{\partial p(\boldsymbol{r})}\right)_N = \int\frac{\delta u(\boldsymbol{r})}{\delta \rho(\boldsymbol{r}'')}\left(\frac{\partial \rho(\boldsymbol{r}'')}{\partial p(\boldsymbol{r}')}\right)_N d\boldsymbol{r}'' = -N\eta(\boldsymbol{r}',\boldsymbol{r}). \tag{2.4.32}$$

The derivatives (2.4.25), (2.4.26) and (2.4.29) give rise to the following second differential of $F[Np]$:

$$d^2F[Np] = \frac{1}{2}[\overline{\eta}(dN)^2 + 2N\,dN\int\overline{\eta}(\boldsymbol{r})dp(\boldsymbol{r})\,d\boldsymbol{r} + N^2\iint dp(\boldsymbol{r})\eta(\boldsymbol{r},\boldsymbol{r}')dp(\boldsymbol{r}')\,d\boldsymbol{r}\,d\boldsymbol{r}'] = d^2F[\rho]. \tag{2.4.33}$$

Indeed, it can be easily verified that this expansion is equivalent to the second differential implied by the quadratic forms in Eqs. (2.3.13), (2.3.28) and (2.4.10).

It follows from Eq. (2.4.24) that the corresponding second partial derivatives of $\mathsf{F}[N,\rho]$ are:

$$\left(\frac{\partial^2 \mathsf{F}[N,\rho]}{\partial \rho(\boldsymbol{r})\,\partial \rho(\boldsymbol{r}')}\right)_N = -\left(\frac{\partial u(\boldsymbol{r}')}{\partial \rho(\boldsymbol{r})}\right)_N = \iint\left(\frac{\partial p(\boldsymbol{r})}{\partial \rho(\boldsymbol{r})}\right)_N\left(\frac{\partial^2 \boldsymbol{F}[Np]}{\partial p(\boldsymbol{r})\,\partial p(\boldsymbol{r}')}\right)_N\left(\frac{\partial p(\boldsymbol{r}')}{\partial \rho(\boldsymbol{r})}\right)_N d\boldsymbol{r}\,d\boldsymbol{r}'$$

$$= \eta(\boldsymbol{r},\boldsymbol{r}'),$$

$$\left(\frac{\partial^2 \mathsf{F}[N,\rho]}{\partial N^2}\right)_\rho = \left(\frac{\partial}{\partial N}\left(\frac{\partial \mathsf{F}[N,\rho]}{\partial \rho(\boldsymbol{r})}\right)_N\right)_\rho = \left(\frac{\partial}{\partial \rho(\boldsymbol{r})}\left(\frac{\partial \mathsf{F}[N,\rho]}{\partial N}\right)_\rho\right)_N = 0. \qquad (2.4.34)$$

The *CT*-contribution in Eq. (2.4.18), also has a transparent interpretation in the LDA variant of the KS method, in which the exchange-correlation energy density at a given location $\boldsymbol{r}$ in space is that for the homogeneous electron gas of the same density: $\rho(\boldsymbol{r}) = \rho_{hom.}$. The Wigner-Seitz radius, of the sphere containing one electron,

$$r_0(\boldsymbol{r}) = \left(\frac{3}{4\pi\rho_{hom.}}\right)^{1/3}, \qquad (2.4.35)$$

or its dimensionless measure $r_s = r_0/a_0$, where a_0 is the Bohr radius, define the sphere volume $V_s(1) = 1/\rho(\boldsymbol{r})$. One observes that the global probability of any single electron in the molecule is $P(1) = 1/N$. The *CT*-term in Eq. (2.4.18) amounts to a change in the local density, which results from adding $P(1)\Delta N$ electrons to $V_s(1)$:

$$\Delta\rho_{hom.} = P(1)\Delta N/V_s(1) = [\rho(\boldsymbol{r})/N]\,\Delta N = p(\boldsymbol{r})\,\Delta N. \qquad (2.4.36)$$

It should be observed that the *CT*-induced density polarization, reflected by a change in the equilibrium polarization of the ($N+\Delta N$)-electron system relative to that of the N-electron system, is the *second*-order effect and as such it is not present in the first-differential of Eq. (2.4.18). This effect will be reflected, however, by the CS of the open molecular systems (see, e.g., Nalewajski, 2002e).

2.4.2. Open (μ-Controlled) Systems

The thermodynamic potential, which determines the equilibrium state of the externally *open* molecular system identified by its chemical potential μ and the grand-ensemble average electron density $\rho(\boldsymbol{r})$, is given by the Legendre transform $\mathsf{R}[\mu,\rho] = \mathcal{E}[\mu, \rho] = \mathcal{L}_E[\mu, \rho]$ [Eq. (2.4.2)]. Its first differential of Eq. (2.4.4) identifies the corresponding partial derivatives with respect to the system state-parameters:

$$\left(\frac{\partial \mathsf{R}}{\partial \mu}\right)_\rho = -N, \quad \left(\frac{\partial \mathsf{R}}{\partial \rho(\boldsymbol{r})}\right)_\mu = -v(\boldsymbol{r}). \qquad (2.4.37)$$

The external potential due to the nuclei, $v(\boldsymbol{r})$, is the representation only dependent (unconstrained) variable, since N is uniquely determined by the density.

Therefore, this chemical hardness representation provides the EP perspective $\mathsf{R}[\mu,\rho] \equiv R_\mu[v[\rho]] \equiv R_\rho[v]$, which can be regarded as the "inverse" description relative to the EF perspective of the system electronic energy $E[N, v] = E_v[\rho[v]]$, of the chemical softness representation. In the same spirit one could regard the $\mathsf{F}[N,\rho] = F_\rho[u[\rho]]$ and $\Omega[\mu,v] = \Omega_u[\rho[u]]$ representations as mutually "inverse" descriptions of the molecular equilibrium states.

The notation $R_\rho[v]$ stresses the dependent role of v in determining the equilibrium state defined by μ and ρ. Indeed, the equilibrium value of the Legendre transform (2.4.2) follows from the following v-maximum principle:

$$\mathsf{R}[\mu,\rho] = \max_{v'}\{\Omega[\mu, v'] - \int v'(\boldsymbol{r})\,\rho(\boldsymbol{r})\,d\boldsymbol{r}\} \equiv \max_{v'}\mathsf{R}[v';\mu,\rho] = R_\rho[v[\mu,\rho]]. \quad (2.4.38)$$

Next, let us examine the second partials of $\mathsf{R}[\mu,\rho]$:

$$\left(\frac{\partial^2\mathsf{R}}{\partial\mu^2}\right)_\rho = -\left(\frac{\partial N}{\partial\mu}\right)_\rho = 0, \quad \left(\frac{\partial^2\mathsf{R}}{\partial\rho(\boldsymbol{r}')\partial\rho(\boldsymbol{r})}\right)_\mu = -\left(\frac{\partial v(\boldsymbol{r})}{\partial\rho(\boldsymbol{r}')}\right)_\mu = -\left(\frac{\partial u(\boldsymbol{r})}{\partial\rho(\boldsymbol{r}')}\right)_\mu = \eta(\boldsymbol{r}',\boldsymbol{r}),$$

$$\frac{\partial^2\mathsf{R}}{\partial\rho(\boldsymbol{r}')\partial\mu} = -\left(\frac{\partial N}{\partial\rho(\boldsymbol{r})}\right)_\mu = -\left(\frac{\partial v(\boldsymbol{r})}{\partial\mu}\right)_\rho = -1. \quad (2.4.39)$$

The last equation expresses the fact that the ground-state density determines the external potential only to the additive constant μ. The derivatives (2.4.37) and (2.4.39) determine the following (horizontal) quadratic expansion of $\mathsf{R}[\mu,\rho]$ in terms of displacements in the representation state-parameters:

$$\Delta^{(1+2)}\mathsf{R}[\mu,\rho] = -N\Delta\mu - \int v(\boldsymbol{r})\Delta\rho(\boldsymbol{r})\,d\boldsymbol{r} - \Delta\mu\,\Delta N + \frac{1}{2}\iint\Delta\rho(\boldsymbol{r})\eta(\boldsymbol{r},\boldsymbol{r}')\Delta\rho(\boldsymbol{r}')\,d\boldsymbol{r}\,d\boldsymbol{r}', \quad (2.4.40)$$

where $\Delta N = \int \Delta\rho\, d\boldsymbol{r}$.

Since $\rho = \rho[u]$ [see Eq. (1.5.22)] the "thermodynamic" potential $\mathsf{R}[\mu,\rho] \equiv R_\mu[v[\rho]] \equiv R_\mu[\rho]$ can be also regarded as the composite functional of the relative external potential u: $R_\mu[\rho[u]] \equiv \tilde{R}_\mu[u]$. Its functional derivatives for the fixed chemical potential of the reservoir read:

$$\frac{\delta\tilde{R}_\mu}{\delta u(\boldsymbol{r})} = \int\frac{\delta\rho(\boldsymbol{r}')}{\delta u(\boldsymbol{r})}\frac{\delta R_\mu}{\delta\rho(\boldsymbol{r}')}d\boldsymbol{r}' = \int\sigma(\boldsymbol{r},\boldsymbol{r}')v(\boldsymbol{r}')\,d\boldsymbol{r}', \quad (2.4.41)$$

$$\frac{\delta^2\tilde{R}_\mu}{\delta u(\boldsymbol{r})\delta u(\boldsymbol{r}')} = \int\frac{\delta\rho(\boldsymbol{r}'')}{\delta u(\boldsymbol{r})}\frac{\delta^2 R_\mu}{\delta\rho(\boldsymbol{r}'')\delta\rho(\boldsymbol{r}''')}\frac{\delta\rho(\boldsymbol{r}''')}{\delta u(\boldsymbol{r}')}d\boldsymbol{r}''\,d\boldsymbol{r}''' = \sigma(\boldsymbol{r},\boldsymbol{r}'). \quad (2.4.42)$$

They give rise to the associated second-order Taylor expansion:

$$\Delta^{(1+2)}\tilde{R}_\mu[u] = \int\int\sigma(\boldsymbol{r},\boldsymbol{r}')v(\boldsymbol{r}')\,d\boldsymbol{r}'\,\Delta u(\boldsymbol{r})\,d\boldsymbol{r} + \frac{1}{2}\iint\Delta u(\boldsymbol{r})\sigma(\boldsymbol{r},\boldsymbol{r}')\Delta u(\boldsymbol{r}')\,d\boldsymbol{r}\,d\boldsymbol{r}', \quad (2.4.43)$$

which represents a transcription of the $\Delta\mu = 0$ cut of Eq. (2.4.55), within the chemical hardness representation, into the corresponding expression in the chemical softness representation.

The two horizontal Taylor expansions (2.4.40) and (2.4.43) provide a basis for the equilibrium reactivity problems, within the EP and EF perspectives, respectively, of the open molecular and/or reactive systems in contact with the external electron reservoir, when they are specified by the ground-state distribution of electrons.

2.5. TRANSFORMATIONS BETWEEN PERTURBATIONS AND RESPONSES

In the global equilibrium state of the given molecular system, when all constituent subsystems are mutually open, free to exchange electrons between themselves, the linear responses of the unconstrained (dependent) state-variables can be expressed as transformations of the corresponding perturbations, i.e., displacements in the representation independent state-parameters (Nalewajski, 2002d, 2003a). In this section we shall briefly summarize these relations.

In the $E[N,v]$-representation of the EF perspective (see Section 2.3.1) the displacements $[\Delta N, \Delta v]$ of the independent state-parameters generate the following equilibrium linear responses of the dependent, energy-conjugate variables [see Eqs. (2.3.5), (2.3.7)-(2.3.11)]:

$$\Delta\mu = \left(\frac{\partial\mu}{\partial N}\right)_v \Delta N + \int\left(\frac{\partial\mu}{\partial v(\boldsymbol{r})}\right)_v \Delta v(\boldsymbol{r})d\boldsymbol{r} = \eta\,\Delta N + \int f(\boldsymbol{r})\,\Delta v(\boldsymbol{r})\,d\boldsymbol{r}, \tag{2.5.1}$$

$$\Delta\rho(\boldsymbol{r}) = \left(\frac{\partial\rho(\boldsymbol{r})}{\partial N}\right)_v \Delta N + \int\left(\frac{\partial\rho(\boldsymbol{r})}{\partial v(\boldsymbol{r}')}\right)_N \Delta v(\boldsymbol{r}')d\boldsymbol{r}' = f(\boldsymbol{r})\Delta N + \int \Delta v(\boldsymbol{r}')\,\beta(\boldsymbol{r}',\boldsymbol{r})\,d\boldsymbol{r}'. \tag{2.5.2}$$

These two relations can be jointly expressed in the following matrix (integral) transformation:

$$[\Delta\mu, \Delta\rho(\boldsymbol{r})] = [\Delta N, \quad \int d\boldsymbol{r}'\,\Delta v(\boldsymbol{r}')\,]\begin{bmatrix} \eta & f(\boldsymbol{r}) \\ f(\boldsymbol{r}') & \beta(\boldsymbol{r}',\boldsymbol{r}) \end{bmatrix}. \tag{2.5.3}$$

Consider next the inverse representation $\mathrm{R}[\mu,\rho]$ of the EP perspective, discussed in the preceding section, in which the independent perturbations $[\Delta\mu, \Delta\rho]$ determine the linear responses in the conjugate (unconstrained variables of state [see derivatives of Eqs. (2.4.52) and (2.4.54)]: $\Delta N = \int \Delta\rho(\boldsymbol{r})\,d\boldsymbol{r}$ and

$$\Delta v(\boldsymbol{r}) = \left(\frac{\partial v(\boldsymbol{r})}{\partial\mu}\right)_\rho \Delta\mu + \int\left(\frac{\partial v(\boldsymbol{r})}{\partial\rho(\boldsymbol{r}')}\right)_\mu \Delta\rho(\boldsymbol{r}')d\boldsymbol{r}' = \Delta\mu - \int \Delta\rho(\boldsymbol{r}')\eta(\boldsymbol{r}',\boldsymbol{r})\,d\boldsymbol{r}'. \tag{2.5.4}$$

These linear transformations can be combined into the single matrix (integral) relation:

$$[\Delta N,\ \Delta v(\boldsymbol{r})] = [\Delta\mu,\ \ \int d\boldsymbol{r}'\,\Delta\rho(\boldsymbol{r}')\,]\begin{bmatrix} 0 & 1 \\ 1 & -\eta(\boldsymbol{r}',\boldsymbol{r}) \end{bmatrix}. \tag{2.5.5}$$

The corresponding overall transformations in the $F[\rho]$ and $\Omega[u]$ representations are summarized in Eq. (2.4.14). These transformations can also give a more resolved form in terms of the corresponding partial derivatives (see Sections 2.3.2 and 2.4.1). Consider for example the transformations of perturbations $[\Delta\mu,\ \Delta v]$ of the $\Omega[\mu,\ v]$-representation into the conjugate responses $[\Delta N,\ \Delta\rho]$:

$$\Delta N = \left(\frac{\partial N}{\partial \mu}\right)_v \Delta\mu + \int \left(\frac{\partial N}{\partial v(\boldsymbol{r})}\right)_\mu \Delta v(\boldsymbol{r})\,d\boldsymbol{r} = S\,\Delta\mu - \int s(\boldsymbol{r})\,\Delta v(\boldsymbol{r})\,d\boldsymbol{r}\,, \tag{2.5.6}$$

$$\Delta\rho(\boldsymbol{r}) = \left(\frac{\partial \rho(\boldsymbol{r})}{\partial \mu}\right)_v \Delta\mu + \int \left(\frac{\partial \rho(\boldsymbol{r})}{\partial v(\boldsymbol{r}')}\right)_\mu \Delta v(\boldsymbol{r}')\,d\boldsymbol{r}' = s(\boldsymbol{r})\Delta\mu - \int \Delta v(\boldsymbol{r}')\,\sigma(\boldsymbol{r}',\boldsymbol{r})\,d\boldsymbol{r}'\,, \tag{2.5.7}$$

where we have used Eqs. (2.3.21)-(2.3.23). Combining these transformation into the joint matrix form gives the following integral transformation:

$$[\Delta N,\ \Delta\rho(\boldsymbol{r})] = [\Delta\mu,\ \ \int d\boldsymbol{r}'\,\Delta v(\boldsymbol{r}')\,]\begin{bmatrix} S & s(\boldsymbol{r}) \\ -s(\boldsymbol{r}') & -\sigma(\boldsymbol{r}',\boldsymbol{r}) \end{bmatrix}. \tag{2.5.8}$$

In Appendix B the *"geometric"* interpretation of charge rearrangements in molecules is given. It separates the *polarizational* (*P*) and *charge-transfer* (CT) components of general changes of the electron density, for $\Delta N = 0$ and $\Delta N \neq 0$, respectively. This approach uses the projector operators acting in the vector (Hilbert) space spanned by the independent modes of density displacements, and gives rise to the (*P*, *CT*)-resolution of the hardness and softness kernels. The geometric partition allows one to view the overall relations of Eq (2.4.14) in terms of the separate *P*- and *CT*-effects, as well as the contributions reflecting the coupling between these two familiar stages of the density reconstruction in molecular systems.

3. ENTROPY, INFORMATION AND COMMUNICATION CHANNELS

The *Information Theory* (IT) of Fisher and Shannon provides convenient tools for the systematic, unbiased extraction of the *chemical* interpretation of the known (experimental or calculated) electron distribution in a molecule, in terms of the overlapping bonded atoms, functional groups, reactants, chemical bonds, etc. A short overview is presented of the basic concepts, relations and techniques of IT, which will be used in this book. The Shannon entropy and Fisher intrinsic-accuracy measures of the amount of the uncertainty (disorder) contained in the given probability distribution are introduced and their main properties are briefly examined. The grouping rules for combining the subsystem entropy data into the corresponding global information measures will be derived. The relative (*"cross"*)-entropy (entropy deficiency, missing information, directed-divergence) concept of Kullback and Leibler, measuring the information distance between the compared probability distributions, is introduced. The rudiments of the entropy/information characteristics of the communication channels are outlined and applied to specific, illustrative signal-transmitting networks. The average conditional-entropy and mutual-information quantities of communication systems are discussed in a more detail in view of their importance for interpreting the covalent and ionic bond components within the "communication" theory of the chemical bond. The rules for combining the sub-channel information data into the corresponding descriptors of the channel as a whole will be established. The channel reductions, carried out by combining selected groups of inputs and/or outputs into a single unit, allow one to extract the intra- and inter-group entropy/information characteristics, which later will be used to estimate the corresponding IT bond indices of molecular fragments. The variational principle for the constrained extremum of an admissible measure of information is advocated as a powerful tool for an unbiased assimilation in the optimum probability distribution the information contained in the physical constraints and/or references. Finally, the Extreme Physical Information (EPI) principle is summarized and its close relation to the theory of measurement is emphasized. As illustrations the Schrödinger and Kohn-Sham equations are interpreted as information principles.

3.1. ENTROPY AND INFORMATION

In this book several applications of IT to selected problems in the theory of molecular electronic structure and chemical reactivity will be presented. This chapter is intended as an introduction to alternative measures of the information contained in probability distributions, and a summary of the key elements of the theory, both concepts and techniques, which will be used to extract the chemical interpretation from the known (experimental or calculated) electron distributions in molecules and reactive systems, in terms of AIM, reactants, functional groups, chemical bonds, etc.

The IT (Fisher, 1922, 1925, 1959; Shannon, 1948, 1949; Kullback and Leibler, 1951; Brillouin, 1956; Khinchin, 1957; Kullback, 1959; Jaynes, 1957, 1963; Abramson, 1963; Ash, 1965; Pfeifer, 1978; Frieden, 2000) is one of the youngest branches of the applied probability theory, in which the probability ideas have been introduced into the field of communication, control, and data processing. Its foundations have been laid in 1920s by Sir R. A. Fisher (1922, 1925) in his classical measurement theory, and in 1940s by C. E. Shannon (1948, 1949) in his mathematical theory of communication.

We begin this short overview with the *entropy* concept of Shannon for a finite set of the observation events, e.g., the outcomes of a measurement of the physical quantity or a detection of a signal in the communication channel. The entropy provides a measure of the average *uncertainty* (disorder) contained in the probability distribution for such events. It has originated from the needs of practice, to create a theoretical model for the transmission of information, and evolved into an important chapter of the general theory of probability.

The central idea of Shannon's theory of communication is the concept of a measure of the amount of information acquired as a result of carrying out specific observations. The possible outcomes $\boldsymbol{a} = (a_1, a_2, \ldots, a_n)$ of such "experiments", which define the complete set of exclusive observation "events", can be considered as the random variables observed with probabilities

$$\boldsymbol{P}(\boldsymbol{a}) = [P(a_1), P(a_2), \ldots, P(a_n)] = (p_1, p_2, \ldots, p_n) \equiv \boldsymbol{p}, \qquad p_i \geq 0, \quad \sum_i p_i = 1. \qquad (3.1.1)$$

This implies that there is an *uncertainty*, which event of the measurement process or the signal detection will actually occur. In communication systems this uncertainty originates from both the message *source* (*input*) and the *transmission channel*, since it is not known with certainty, which of the possible input messages will be sent or which signal was transmitted to produce the observed signal in the channel *receiver* (*output*). An appropriate measure of this uncertainty must be related in some manner to the set of probabilities for the observation events under consideration.

A number of intuitive and theoretical considerations led to the introduction of the event uncertainty measure, which depends upon the logarithm of the outcome probability. The so called *self-information* in the event a_i, i.e., the amount of information received after the outcome a_i has been observed, is defined as

$$I(a_i) = -\log p_i = I(p_i), \tag{3.1.2}$$

where the logarithm is taken to an arbitrary but fixed base. In keeping with the custom in works on IT the logarithm, when taken to base 2, corresponds to the information measured in *bits* (a contraction of the term *binary digit*). If any other base had been chosen, the result would be to multiply this measure by an appropriate constant, which is equivalent to a scale change. For example, selecting log = ln expresses the amount of information in *nats* (*natural units*), while another choice log = $\log_{10}$ counts the information content in *hartleys*, to commemorate R. V. L. Hartley (1928), who first proposed the logarithmic information measure:

1 nat = 1.44 bits and 1 hartley = 3.32 bits.

The *average* uncertainty of a given probability distribution (3.1.1) must depend on the whole probability vector $\boldsymbol{p}$ or the probability density function $p(x)$ of the continuous random variable x, satisfying the normalization relation: $\int p(x)\, dx = 1$. The properties of the quantity discovered by Shannon (and anticipated by others), called the *entropy*, admit of an interpretation in terms of the average uncertainty measure. The Shannon entropy plays also a central role in the statistical mechanics (Jaynes, 1957, 1985) being proportional to the thermodynamic entropy of Boltzmann.

The average entropy in the probability distribution $\boldsymbol{p}$ is the mean (expectation) value of the self-information values for individual events,

$$S(\boldsymbol{p}) = \Sigma_i\, p_i\, I(p_i) = -\, \Sigma_i\, p_i \log p_i \equiv S(\boldsymbol{P}(\boldsymbol{a})), \tag{3.1.3}$$

in which the outcome probability provides the relevant "weighting" factor, and $p_i \log p_i = 0$ for $p_i = 0$. This definition can be naturally generalized into the random variable assuming the continuous series of values with the normalized probability density $p(x)$:

$$S[p] = -\int p(x) \log p(x)\, dx \equiv \int p(x)\, I(p(x))\, dx. \tag{3.1.4}$$

where the definite integration is over the whole range of the random variable x, and $p(x) \log p(x) = 0$ for those values of x for which $p(x) = 0$.

When one carries out the experiment, the possible outcomes of which are the events $\boldsymbol{a}$ or $\{x\}$, one obtains some *information*, by finding out which of the events has actually occurred. As a result the uncertainty of Eqs. (3.1.3) or (3.1.4) has been completely eliminated. This implies that the information given us by carrying out the experiment consists in removing the uncertainty, which existed before the experiment. In other words, the larger the average uncertainty in $\boldsymbol{p}$ or $p(x)$, the larger the average amount of information obtained by eliminating it. It is therefore natural to express the information by an increasing function of $S(\boldsymbol{p})$ or $S[p]$. A given choice of this function implies the corresponding choice of the unit for the quantity of information and is therefore fundamentally unimportant. The properties of the information entropy show that it is especially convenient to take the quantity of information to be proportional to the entropy. However, the proportionality constant can be taken as unity, by an appropriate choice of units. Therefore, we consider the *amount of information* given by the experiment establishing the probability distribution to be equal to the average uncertainty in $\boldsymbol{p}$ or $p(x)$, i.e., to its *overall entropy*.

Thus, the average entropy has a double interpretation. On one hand it represents the average uncertainty in the discrete or continuous probability distribution of the complete set of the mutually exclusive events. On the other hand, it measures the average quantity of information received by observing these events in the experiment. The uncertainty for the individual event, $I(a_i)$ or $I(x)$, has a similar double meaning: it measures the uncertainty of observing a particular event in a certain trial, and the amount of information gained after actually observing this very outcome in the experiment.

3.2. PROPERTIES OF SHANNON ENTROPY

Let us summarize the main properties of the average entropy, which might be expected of a reasonable measure of uncertainty/information. Again, for reasons of simplicity, we first consider the case of a finite, discrete probability distribution $\boldsymbol{p}$. First of all, we see immediately that $S(\boldsymbol{p}) = 0$ if and only if one of the probabilities in $\boldsymbol{p}$ is one and all the others are zero. Indeed, this is the case, where the outcome of the experiment carries no uncertainty, since the result can be predicted beforehand with complete certainty. In all other cases $S(\boldsymbol{p}) > 0$. For the fixed number of n outcomes the probability scheme exhibiting the most uncertainty, i.e., the largest value of the entropy, is the one with equally likely outcomes: $p_i = 1/n$.

Example 3.1. This can be demonstrated using the constrained principle of the *Maximum Entropy* (ME) of Shannon and Jaynes,

$$\delta\,[S(\boldsymbol{p}) - \lambda \textstyle\sum_i p_i] \equiv \delta\Phi(\boldsymbol{p}, \lambda) = 0, \qquad (3.2.1)$$

where λ is the Lagrange multiplier enforcing the probability normalization, to be determined from the value of this constraint. Assuming log = ln gives the Euler equation

$$\partial\Phi(\boldsymbol{p}, \lambda)/\partial p_k = \ln p_k + 1 + \lambda \equiv \ln\,[C(\lambda)\, p_k] = 0$$

$$\text{or} \quad p_k = 1/C(\lambda), \quad k = 1, 2, \ldots, \qquad (3.2.2)$$

where $C(\lambda) = \exp(1 + \lambda)$, predicting the equal probabilities for all events. Finally, from the probability normalization condition, $C(\lambda) = n$, which completes this short demonstration.

The ME rule represents a powerful method for determining the equilibrium probability distribution of a physical system or process, given some information about it. It has been successfully used to reconstruct the equilibrium thermodynamics (Jaynes, 1957, 1985; Brillouin, 1956; Fisher, 1959; Frieden, 2000), asserting at the same time the principle applicability to far more general problems. It can be stated in the following way:

> ***given data* d *for a certain experiment, the probability distribution* p*, which describes* d *most objectively, must maximize the entropy* S(p) *with respect to all* p*'s satisfying* d.**

This principle represents a device allowing one to assimilate in the optimum probability distribution the physical information contained in the constraints in the most unbiased manner possible. Equivalently, one might say that the entropy maximization results in the most "evenly spread" of all distributions consistent with the imposed constraints.

The ME rule thus involves the maximization of Shannon's entropy subject to the imposed constraints and provides a unifying principle in statistical physics, allowing a construction of thermodynamic laws based upon statistical inference and the unbiased assimilation of available data. In other words, the entropy becomes the starting point in the construction of the statistical mechanics, instead of being identified in the end as a byproduct of the energy centered arguments. The Shannon-Jaynes principle is the uniquely correct method for inductive inference, when a new information is given in the form of the statistical expectation values.

Example 3.2. Let us now consider another illustration of how the principle of ME is applied. Using the continuous analog of the variational principle (3.2.1), for the constrained maximum of the entropy functional of Eq. (3.1.4), one can straightforwardly demonstrate that among all normalized probability distribution laws $p(x)$ of a continuous random variable x with the fixed variance D (the square of the standard mean deviation σ) with respect to the given average value $\bar{x}$,

$$D \equiv \sigma^2 = \int (x - \bar{x})^2 p(x)\, dx, \tag{3.2.3}$$

the normal distribution is the one which maximizes the Shannon entropy.

The relevant extremum principle for the auxiliary entropy functional, which incorporates these constraints,

$$\delta \{S[p] - \lambda_1 \int p(x)\, dx - \lambda_2 \int (x - \bar{x})^2 p(x)\, dx\} \equiv \delta\Omega\, [p, \lambda_1, \lambda_2] = 0, \tag{3.2.4}$$

gives the Euler equation for the optimum distribution $p(x)$ that realizes the two constraints, marking the vanishing partial functional derivative of the auxiliary functional $\Omega\, [p, \lambda_1, \lambda_2]$:

$$-\frac{\delta\Omega[p, \lambda_1, \lambda_2]}{\delta p(x)} = \ln p(x) + 1 + \lambda_1 + \lambda_2 (x - \bar{x})^2 = 0. \tag{3.2.5}$$

Hence, $p(x) = C(\lambda_1) \exp[-\lambda_2(x - \bar{x})^2]$, where $C(\lambda_1) = \exp(1 + \lambda_1)$. From the constraints one finally finds: $C(\lambda_1) = (2\pi D)^{-1/2}$ and $\lambda_2 = (2D)^{-1}$. Therefore, for the fixed variance D the ME rule indeed gives the normal distribution law as the optimum solution:

$$p^n(x) = (2\pi D)^{-1/2} \exp[-(x - \bar{x})^2/(2D)]. \qquad (3.2.6)$$

Example 3.3. Clearly the same solution should follow from the modified set of constraints, in which the average value $\bar{x}$ of the preceding example is not given from the start, but fixed by the additional constraint $\int x\, p(x)\, dx = \bar{x}$ in the modified variational principle:

$$\delta\,\{S[p] - \lambda_1 \int p(x)\, dx - \lambda_2 \int x\, p(x)\, dx - \lambda_3 \int (x - \bar{x})^2\, p(x)\, dx\}$$
$$\equiv \delta\Omega\,[p, \lambda_1, \lambda_2, \lambda_3] = 0. \qquad (3.2.7)$$

The associated Euler equation for this problem,

$$\ln p(x) + 1 + \lambda_1 + \lambda_2 x + \lambda_3(x - \bar{x})^2 = \ln p(x) + (1 + \lambda_1 + \lambda_3 \bar{x}^2) + (\lambda_2 - 2\lambda_3 \bar{x})x + \lambda_3 x^2 = 0$$

$$\text{or} \qquad \ln\,[p(x)/A(\lambda_1, \lambda_3)] = B(\lambda_2, \lambda_3)x - \lambda_3 x^2 = 0,$$

gives

$$p(x) = A \exp(Bx - \lambda_3 x^2) = A' \exp\{-\lambda_3[x - B/(2\lambda_3)]^2\}. \qquad (3.2.8)$$

The three constraints involved then identify the coefficient A' and the two exponents of the above Gaussian function, giving rise to: $A' = \lambda_3 = (2D)^{-1}$, $B/(2\lambda_3) = \bar{x}$, and $= (2\pi D)^{-1/2}$, and thus to the normal distribution law of Eq. (3.2.6).

As we shall demonstrate in the next example, this exponential, closed-form solution is a general feature of variational principles involving the Shannon measure of the distribution disorder.

Example 3.4. Specific cases of the constrained Shannon entropy problems can be summarized by the following general form of the ME variational principle:

$$\delta\,\{S[p] - \lambda \int p(x)\, dx - \omega \int f(x)\, p(x)\, dx\} \equiv \delta\Omega\,[p, \lambda, \omega] = 0, \qquad (3.2.9)$$

where λ and ω are the Lagrange multipliers (constants) and the *known* function $f(x)$ is the resultant "kernel" generated by all the other physical constraints imposed upon the system under consideration. The associated Euler-Lagrange equation,

$$-\frac{\delta\Omega[p,\lambda,\omega]}{\delta p(x)} = \ln p(x) + 1 + \lambda + \omega f(x) = 0,$$

gives the exponential solution

$$p(x) = A(\lambda)\exp[-\omega f(x)]. \tag{3.2.10}$$

It can be straightforwardly demonstrated that this solution corresponds to the maximum of $\Omega[p]$, in accord with the ME principle. Indeed, the second functional derivative of this functional (see Appendix A) reads:

$$\frac{\delta^2\Omega[p,\lambda,\omega]}{\delta p(x)\,\delta p(x')} = -\frac{\delta \ln p(x)}{\delta p(x')} = -\delta(x-x')/p(x).$$

It thus gives rise to the negative second differential for a finite variation $dp(x)$, independent of the specific form of the optimum probability distribution and the constraints imposed,

$$d^{(2)}\Omega[p] = -\frac{1}{2}\iint dp(x)\frac{\delta(x-x')}{p(x)}dp(x')\,dx\,dx' = -\frac{1}{2}\int\frac{[dp(x)]^2}{p(x)}dx < 0, \tag{3.2.11}$$

since the integrand is non-negative everywhere.

Suppose now that we have two finite (normalized) probability distributions: for the events $\boldsymbol{a} = (a_1, a_2, \ldots, a_n)$, occurring with the probabilities $\boldsymbol{P}(\boldsymbol{a}) = (p_1, p_2, \ldots, p_n) \equiv \boldsymbol{p}$, and for the outcomes $\boldsymbol{b} = (b_1, b_2, \ldots, b_m)$ being observed with probabilities $\boldsymbol{P}(\boldsymbol{b}) = (q_1, q_2, \ldots, q_m) \equiv \boldsymbol{q}$. Consider next the probabilities of the Cartesian product of the two sets of the observation events,

$$\mathbf{P}(\boldsymbol{a}\otimes\boldsymbol{b}) = \{P(a_i b_j) \equiv \pi_{i,j},\ i = 1, 2, \ldots, n,\ j = 1, 2, \ldots, m\} \equiv \boldsymbol{\pi}, \tag{3.2.12}$$

i.e., of the joint (simultaneous) occurrence of the events defining the two probability schemes. The product events define yet another finite set consisting of mn joint-events $\boldsymbol{a}\otimes\boldsymbol{b} = \{a_i \cap b_j \equiv a_i b_j \equiv (i,j)\}$.

For the two mutually *independent* probability distributions $\mathbf{P}^0(\boldsymbol{a}\otimes\boldsymbol{b}) = \{\pi_{i,j}^{\,0} = p_i q_j\} = \boldsymbol{p}\otimes\boldsymbol{q} \equiv \boldsymbol{\pi}^0$, and hence the entropy of these joint-event probabilities is the sum of entropies of two separate probability distributions:

$$\begin{aligned} S(\boldsymbol{\pi}^0) &= -\textstyle\sum_i\sum_j \pi_{i,j}^{\,0}\log \pi_{i,j}^{\,0} = -\sum_i\sum_j p_i q_j(\log p_i + \log q_j) \\ &= -(\textstyle\sum_j q_j)\sum_i p_i \log p_i - (\sum_i p_i)\sum_j q_j \log q_j = S(\boldsymbol{p}) + S(\boldsymbol{q}). \end{aligned} \tag{3.2.13}$$

Independence of events a_i and b_j implies that the knowledge of the occurrence of one of them gives no knowledge of the occurrence of the other. A general case of the mutually dependent events will be considered in Section 3.5.

Consider next the distribution $\boldsymbol{q}$, which is not substantially different from $\boldsymbol{p}$, by including an additional event of zero probability: $\boldsymbol{q} = (\boldsymbol{p}, 0)$. Then $S(\boldsymbol{p}) = S(\boldsymbol{q})$, since adding the impossible event (or any number of impossible events) to the scheme cannot change its average entropy (uncertainty) content. This equality expresses the so called "zero indifference" of the Shannon entropy.

The *uniqueness theorem* of the information theory (Khinchin, 1957) states that a measure of the uncertainty of a finite probability scheme, the *continuous* function with respect to all its probability arguments and satisfying the conditions of *uniformity* of probability distribution at the maximum of entropy, *additivity* of the amount of information, and *equality* of the information content of the substantially identical schemes, must be proportional to the Shannon entropy of Eq. (3.1.3). For completeness, we also observe that the Shannon uncertainty measure is independent of the order in which the probabilities appear in the probability vector. In other words, it is invariant with respect to the permutation of the order of the observation events.

3.3. ENTROPY DEFICIENCY

Kullback and Leibler (*KL*) have proposed an important generalization of Shannon's entropy, the so called *relative ("cross") entropy*, also known as the entropy deficiency, missing information, or the directed divergence (Kullback and Leibler, 1951). It measures the *information "distance"* between the two (normalized) probability distributions for the same set of events. For example, in the discrete scheme case, this discrimination information of the distribution $\boldsymbol{p} = \{P(a_i) = p_i\}$ with respect to the *reference* distribution $\boldsymbol{p}^0 = \{P^0(a_i) = p_i^0\}$ is defined as

$$\Delta S(\boldsymbol{p}\,|\,\boldsymbol{p}^0) \equiv I^{KL}(\boldsymbol{p}\,|\,\boldsymbol{p}^0) = \textstyle\sum_i p_i \log (p_i/p_i^0)$$
$$\equiv \textstyle\sum_i p_i\, I(p_i/p_i^0) \equiv \langle I(p_i/p_i^0)\rangle_{\boldsymbol{p}} \equiv \sum_i \Delta s(p_i\,|\,p_i^0) \geq 0, \qquad (3.3.1)$$

where the symbol $\langle\rangle_{\boldsymbol{p}}$ stands for the $\boldsymbol{p}$-weighted average value. In the continuous case the directed divergence measure of the entropy deficiency of the probability density $p(x)$, relative to the prior distribution $p^0(x)$, is defined by the functional:

$$\Delta S[p\,|\,p^0] \equiv I^{KL}[p\,|\,p^0] = \textstyle\int p(x) \log[p(x)/p^0(x)]\, dx$$
$$\equiv \textstyle\int p(x)\, I[p(x)/p^0(x)]\, dx \equiv \langle I[p/p^0]\rangle_p \equiv \int \Delta s[p(x)\,|\,p^0(x)]\, dx \geq 0. \qquad (3.3.2)$$

In Eqs. (3.3.1) and (3.3.2) we have explicitly indicated that the average entropy deficiency is the mean measure over the current probability distribution of the logarithm of the probability ratio $I[p_i/p_i^0]$ or $I[p(x)/p^0(x)]$, called the *surprisal*, which provides for the individual events a measure of the information in the current probability with reference to the prior probability value. The equality in the last two equations takes place, when the two compared probability distributions are identical, i.e., when the surprisal vanishes for all discrete and continuous events.

Therefore, the entropy deficiency provides a measure of an information resemblance between the two compared probability distributions. When they are identical the surprisal vanishes for all events giving rise to the zero value of the average missing information. The more the two probability schemes differ from one another, the larger the information distance. The proof of its non-negative character is based on the observation that the line $y = z - 1$ lies above the curve $y = \log z$, with the two functions having equal (zero) value only for $z = 1$. Consider, e.g., the continuous case. Taking $z(x) = p^0(x)/p(x)$ and using the probability normalization then directly gives: $\Delta S[p|p^0] = -\int p(x) \log z(x)\, dx \geq \int p(x)\,[z(x) - 1]\, dx = \int [p^0(x) - p(x)]\, dx = 0$. Notice, however, that the surprisal becomes negative, when the current probability is lower than the reference value.

Kullback (1959) has proposed the principle of the *Minimum Entropy Deficiency* (MED), a generalization of the ME rule, as another unbiased procedure for assimilating in the optimum probability distribution the constraints of available experimental data, and thus its highest similarity to the reference distribution afforded by the constraints. In fact both principles are equivalent when the reference distribution is uniform. Suppose that an experiment has been performed, which yields the expectation values of several functions of the discrete probability distribution $\boldsymbol{p}$, $\{F_i(\boldsymbol{p}) = F_i^0\}$, or functionals of the probability density $p(x)$ in the continuous case, $\{F_i[p] = F_i^0\}$. The Kullback principle asserts that the task of assimilating this information in the optimum probability distribution, which is to resemble the reference distribution as much as possible, can be accomplished by minimizing the entropy deficiency subject to these constraints:

$$\delta\{\Delta S[\boldsymbol{p}|\boldsymbol{p}^0] - \Sigma_i \lambda_i[F_i(\boldsymbol{p}) - F_i^0]\} = 0 \quad \text{or} \quad \delta\{\Delta S[p|p^0] - \Sigma_i \lambda_i(F_i[p] - F_i^0)\} = 0, \quad (3.3.3)$$

where $\{\lambda_i\}$ are the Lagrange multipliers to be ultimately determined from the values of these constraints. This principle also provides a method for a non-parametric probability estimation.

Example 3.5. Consider the following trivial problem: find the optimum, normalized probability density $p(x)$ most resembling the reference distribution $p^0(x)$. Clearly, the answer $p = p^0$ is known beforehand, since this solution gives the lowest possible value of the missing information $\Delta S[p^0|p^0] = 0$. The corresponding MED principle with incorporated normalization constraint,

$$\delta\,\{\Delta S[p|p^0] - \lambda \int p(x)\, dx\} = 0, \quad (3.3.4)$$

gives the Euler equation ($\log \equiv \ln$),

$$\ln[p(x)/p^0(x)] + 1 - \lambda = \ln\{p(x)/[C(\lambda)p^0(x)]\} = 0 \quad \text{or} \quad p(x) = C(\lambda)\, p^0(x) = p^0(x), \quad (3.3.5)$$

since $C(\lambda) = \exp(1 + \lambda) = 1$, by the assumed normalization of both distributions: $C(\lambda) = 1/[\int p^0(x)\, dx] = 1$. Thus, as expected, the optimum distribution is found to be identical with the reference probability density.

It should be observed that the entropy deficiency is not symmetrical with respect to the two probability distributions and exhibits negative values of the integrand. To avoid this Kullback (K) has proposed an alternative measure, called the *divergence*, given by the symmetrized combination of two directed-divergences (Kullback,1959):

$$\Delta S(\boldsymbol{p}, \boldsymbol{p}^0) \equiv I^K(\boldsymbol{p}, \boldsymbol{p}^0) = \Delta S(\boldsymbol{p} \,|\, \boldsymbol{p}^0) + \Delta S(\boldsymbol{p}^0 \,|\, \boldsymbol{p}) = \textstyle\sum_i (p_i - p_i^0)\, I[p_i/p_i^0]$$
$$\equiv \textstyle\sum_i \Delta p_i\, I[p_i/p_i^0] \equiv \sum_i \Delta s(p_i, p_i^0) \ge 0, \qquad \Delta s(p_i, p_i^0) \ge 0, \qquad (3.3.6)$$

$$\Delta S[p, p^0] \equiv I^K[p, p^0] = \Delta S[p \,|\, p^0] + \Delta S[p^0 \,|\, p] = \int [p(x) - p^0(x)]\, I[p(x)/p^0(x)]\, dx$$
$$\equiv \int \Delta p(x) I[p(x)/p^0(x)]\, dx \equiv \int \Delta s[p(x), p^0(x)]\, dx \ge 0, \quad \Delta s[p(x), p^0(x)] \ge 0. \quad (3.3.7)$$

The Shannon entropy is a measure of the "*disorder*" (uncertainty, "smoothness") of the probability distribution. On a finite interval, the distribution possessing the highest entropy is the uniform distribution. Any deviation from uniformity thus indicates the "perturbing" presence of "*order*". The Kullback-Leibler measure, i.e., the referenced Shannon's entropy, provides a similar description, but with reference to some other distribution. It exhibits important properties to be satisfied by any admissible measure of information: *non-negativity*, with the zero value corresponding to $\boldsymbol{p} = \boldsymbol{p}^0$ (or $p = p^0$); *additivity* for independent distributions: $I^{KL}[p \,|\, p^0] = I^{KL}[q \,|\, q^0] + I^{KL}[r \,|\, r^0]$, when $p(x, y) = q(x)r(y)$ and $p^0(x, y) = q^0(x)\, r^0(y)$; *invariance* to variable transformation: $I^{KL}[p(x) \,|\, p^0(x)] = I^{KL}[p(y) \,|\, p^0(y)]$, for $y = f(x)$.

In what follows we shall briefly examine the complementary, Fisher (1922, 1925, 1959) information-measure, which reflects the overall "*order*" ("sharpness") of the probability distribution. The normalized molecular one-electron probability density $p(\boldsymbol{r})$, $\int p(\boldsymbol{r})\, d\boldsymbol{r} = 1$, called the *shape-factor* of the electron density $\rho(\boldsymbol{r}) = Np(\boldsymbol{r})$, where N is the overall number of electrons, $N = \int \rho(\boldsymbol{r})\, d\boldsymbol{r}$, involves continuous events of all possible locations of an electron in the physical space. In the next section we shall examine the Fisher information measure for such probabilities depending upon these vector, locality-event variables $\boldsymbol{r}$, $\boldsymbol{r}'$, etc.

3.4. FISHER INFORMATION

The parametric information measure for *locality* (Fisher, 1922, 1925, 1959; see also: Frieden, 2000), called the *intrinsic accuracy*, historically predates the familiar Shannon form by about 25 years, being proposed in about the same time when the final form of the quantum mechanics was formulated. For a single-component probability distribution $p(\boldsymbol{r})$ defined in the physical space $\boldsymbol{r} = (x, y, z)$ it is defined by the functional

$$I^F[p] = \int p(\boldsymbol{r})\, [\nabla \ln p(\boldsymbol{r})]^2\, d\boldsymbol{r} = \int [\nabla p(\boldsymbol{r})]^2/p(\boldsymbol{r})\, d\boldsymbol{r}, \qquad (3.4.1)$$

reminiscent of the von Weizsäcker (1935) inhomogeneity correction to the Thomas-Fermi electronic kinetic energy. It characterizes the *"narrowness"* of the probability density $p(\boldsymbol{r})$.

Example 3.6. In the normal-distribution case of $p^n(x)$ [Eq. (3.2.6)] one finds:

$$I^F[p^n] = \int [p^n(x)]^{-1}\left(\frac{dp^n(x)}{dx}\right)^2 dx = \int p^n(x)\left(\frac{(x-\bar{x})}{D}\right)^2 dx. \qquad (3.4.2)$$

The substitution $z = (x-\bar{x})/\sqrt{D}$ and a use of the typical integral

$$\int_0^\infty z^2 \exp(-az^2)\,dz = \frac{1}{4}\sqrt{\frac{\pi}{a^3}}, \qquad (3.4.3)$$

then give:

$$I^F[p^n] = \int_{-\infty}^{+\infty} p^n(x)\left(\frac{(x-\bar{x})}{D}\right)^2 dx = \frac{1}{D}\sqrt{\frac{2}{\pi}}\int_0^\infty z^2 \exp(-\frac{z^2}{2})\,dz = \frac{1}{D}. \qquad (3.4.4)$$

In a similar way one calculates the Shannon entropy of this normal distribution:

$$S[p^n] = -\int_{-\infty}^{+\infty} p^n(x)\ln p^n(x)\,dx = \ln(2\pi D) + \frac{1}{2}, \quad \text{in bits.} \qquad (3.4.5)$$

Therefore, the Fisher information of the normal distribution measures the inverse of its variance, called the *invariance*, while the complementary Shannon entropy is proportional to the logarithm of variance, thus monotonically increasing with the spread of the Gaussian distribution.

In fact the Shannon entropy $S[p] \equiv I^S[p]$ and the Fisher information $I^F[p]$ describe the complementary facets of the probability distribution: the former reflects distribution's *"spread"* (a measure of uncertainty, "disorder"), while the latter provides a measure of its *"narrowness"* ("order"). Important properties of this functional, non-negativity, additivity for independent events, and invariance to coordinate transformation, confirm it as an alternative admissible measure of information.

The intrinsic-accuracy functional (3.4.1) can be simplified by expressing it as a functional of the associated (real) "amplitude" function $\psi(\boldsymbol{r})$ of the probability distribution, $p(\boldsymbol{r}) = \psi^2(\boldsymbol{r})$,

$$I^F[p] = 4\int [\nabla\psi(\boldsymbol{r})]^2\,d\boldsymbol{r} = I^F[\psi]. \qquad (3.4.6)$$

Therefore, $I^F[\psi]$ measures the gradient content in the distribution amplitude.

The extension of the Fisher information to the multi-component spatial probabilities $\boldsymbol{p}(\boldsymbol{r}) \equiv \{p_n(\boldsymbol{r}) \equiv \psi_n(\boldsymbol{r})^2\}$, in terms of the component probability amplitudes $\boldsymbol{\psi}(\boldsymbol{r}) \equiv \{\psi_n(\boldsymbol{r})\}$, reads (Frieden, 2000):

$$I^F[\boldsymbol{p}] = 4\sum_n \int \sum_{v=x,y,z} \left(\frac{\partial \psi_n}{\partial x_v}\right)^2 d\boldsymbol{r} = 4\sum_n \int \nabla\psi_n \bullet \nabla\psi_n \, d\boldsymbol{r} = I^F[\boldsymbol{\psi}]\cdot \qquad (3.4.7)$$

The Fisher information in the probability density $p(x)$ is proportional to the Kullback-Leibler entropy deficiency (see, e.g., Frieden, 2000),

$$I^{KL}[p|p'] = \int p(x) \log[p(x)/p'(x)]\, dx, \qquad (3.4.8)$$

between the probability density $p(x)$ itself and its infinitesimally shifted version $p'(x) = p(x + \Delta x)$ (see Example 3.10):

$$I^F[p] = \lim_{\Delta x\to 0} \{[2(\Delta x)^{-2}]\, I^{KL}[p|p']\}. \qquad (3.4.9)$$

This Kullback-Leibler limit of the Fisher information stresses the local character of this information measure, which probes through the gradient of the amplitude the local inhomogeneity of the probability distribution.

The analytical properties of the Shannon and Fisher information functionals are quite different. The former probes an average measure of the distribution smoothness, while the latter provides a local measure of the distribution sharpness (inhomogeneity). Hence, when extremized through variation of the probability density the Shannon entropy gives directly the same exponential form of the optimum solution of Eq. (3.2.9), while the Fisher information gives rises to a differential equation, and hence to the multiple solutions specified by appropriate boundary conditions. As demonstrated by Frieden (2000), only the use of the Fisher measure of information allows one to derive the differential equations of physics from the principle of the Extreme Physical Information (see Section 3.8), which is closely related to the theory of measurement.

Example 3.7. Let us examine the following general problem of the extreme (minimum) Fisher information (Huber, 1981; Frieden, 2000) contained in the probability amplitude in one dimension, $\psi(x) = p(x)^{1/2}$, $I^F[\psi] = 4\int \psi'(x)\, dx$, where $\psi'(x) = d\psi(x)/dx$:

$$\underset{\psi}{\text{Min}}\, \{I^F[\psi] - \lambda_1 \int \psi(x)\, g(x)\, dx - \lambda_2 \int \psi^2(x)\, h(x)\, dx\} \equiv \underset{\psi}{\text{Min}}\, K^F[\psi, \lambda_2, \lambda_1]. \qquad (3.4.10)$$

Here λ_1 and λ_2 are the Lagrange multipliers enforcing the constraints defined in terms of the probability amplitude and the probability distribution, respectively, defined by the known "kernel" functions $g(x)$ and $h(x)$. Calculating the functional derivative (see Appendix A) gives the following Euler equation for the optimum amplitude:

$$\frac{\delta K^F[\psi]}{\delta\psi(x)} = -\lambda_1 g(x) - 2\lambda_2 h(x)\psi(x) - 8\frac{d\psi'(x)}{dx} = 0$$

$$\text{or} \quad \psi''(x) + \left(\frac{\lambda_2}{4}\right)h(x)\psi(x) + \left(\frac{\lambda_1}{8}\right)g(x) = 0. \tag{3.4.11}$$

Therefore, the multiple solutions to the constrained Fisher information problem (3.4.10) are determined by the second-order differential equation. This basic difference between the Shannon and Fisher constrained variational principles is due to the lack of the derivative (gradient) in the integrand of the Shannon functional.

The maximum Shannon entropy and the minimum Fisher information principles, for a given set of constraints, may have coincident solutions (see, e.g., Frieden, 2000). The Maxwell-Boltzmann velocity dispersion law of the equilibrium statistical thermodynamics results from both information principles, with the Fisher information principle generating additional, non-equilibrium solutions as subsidiary minima, with the absolute minimum being attained by the Maxwell-Boltzmann solution. The coincidence is observed at the equilibrium level of statistical mechanics, with the Shannon ME rule being unable to cover the non-equilibrium phenomena.

In the Fisher information representation one can also define the functionals providing measures of the information distance between two probability distributions (Nalewajski, 2004a), which closely follow the corresponding Kullback-Leibler (directed divergence) or Kullback (divergence) entropy-deficiencies derived from the Shannon entropy. In the single-component case, of two spatial probability densities $p(\boldsymbol{r})$ and $p^0(\boldsymbol{r})$ defining the associated enhancement ratio $w(\boldsymbol{r}) = p(\boldsymbol{r})/p^0(\boldsymbol{r})$, the Fisher analog of the directed divergence of Eq. (3.3.2) reads:

$$I^F[p\,|\,p^0] = \int p(\boldsymbol{r})\{\nabla\ln[p(\boldsymbol{r})/p^0(\boldsymbol{r})]\}^2 d\boldsymbol{r} = \int p^0(\boldsymbol{r})\,[\nabla w(\boldsymbol{r})]^2/w(\boldsymbol{r})\, d\boldsymbol{r}$$
$$\equiv \int p(\boldsymbol{r})\{\nabla I[w(\boldsymbol{r})]\}^2 d\boldsymbol{r} \equiv \left\langle [\nabla I(w)]^2 \right\rangle_p \geq 0. \tag{3.4.12}$$

Reversing the role of the current and reference densities gives the other Fisher directed divergence:

$$I^F[p^0\,|\,p] = \int p^0(\boldsymbol{r})\{\nabla\ln[p^0(\boldsymbol{r})/p(\boldsymbol{r})]\}^2 d\boldsymbol{r} = \int p^0(\boldsymbol{r})\{\nabla\ln[p(\boldsymbol{r})/p^0(\boldsymbol{r})]\}^2 d\boldsymbol{r}$$
$$= \int p^0(\boldsymbol{r})\,[\nabla w(\boldsymbol{r})/w(\boldsymbol{r})]^2\, d\boldsymbol{r} = \int p^0(\boldsymbol{r})\{\nabla I[w(\boldsymbol{r})]\}^2 d\boldsymbol{r} \equiv \left\langle [\nabla I(w)]^2 \right\rangle_{p^0} \geq 0. \tag{3.4.13}$$

The symmetrized Fisher information distance, the analog of Kullback's divergence of Eq. (3.3.7), is thus defined by the arithmetic average distribution, $p^{av}(\boldsymbol{r}) = ½[p(\boldsymbol{r}) + p^0(\boldsymbol{r})]$,

$$I^F[p, p^0] \equiv ½\,(I^F[p\,|\,p^0] + I^F[p^0\,|\,p]) = \left\langle [\nabla I(w)]^2 \right\rangle_{p^{av}} \geq 0. \tag{3.4.14}$$

The Fisher information originates from the classical measurement theory as a measure of the expected error in a "smart" measurement, in the context of efficient estimators of a parameter (see, e.g., Frieden, 2000). Fisher (1922, 1925) was the first to suggest that data samples in an experiment together with a given parametric distribution model contain the statistical information about the parameter(s). In fact, the intrinsic accuracy of Eq. (3.4.1) is a special case of the parametric Fisher measure, for the *locality parameter*. Indeed, when the parameter is one of locality, the parametric Fisher information provides the information about the probability distribution itself.

Let $p(x|\theta)$ be the probability distribution function of x depending upon the parameter θ. The parametric Fisher measure is defined as

$$I_F(\theta) \equiv \int p(x|\theta)\left(\frac{\partial \ln p(x|\theta)}{\partial\theta}\right)^2 dx = \int \frac{[p'(x|\theta)]^2}{p(x|\theta)}dx, \quad p'(x|\theta) = \frac{\partial p(x|\theta)}{\partial\theta}. \tag{3.4.15}$$

We shall now illustrate this information quantity for two specific choices of the parameter, when it shifts the locality and scales the probability distribution, respectively.

Example 3.8. It can be easily verified that by taking θ to be a parameter of locality,

$$p(x|\theta) = p(x+\theta) = p(x'), \tag{3.4.16}$$

one recovers the intrinsic accuracy measure of Eq. (3.4.1). Indeed, since

$$\frac{\partial p(x|\theta)}{\partial\theta} = \frac{\partial p(x')}{\partial x'}\frac{\partial x'}{\partial\theta} = \frac{\partial p(x')}{\partial x'}, \tag{3.4.17}$$

$$I_F(\theta) = \int \frac{[p'(x')]^2}{p(x')}dx' = I^F[p]. \tag{3.4.18}$$

Example 3.9. Consider now the parametric family of probability distributions defined by the *scale* parameter:

$$p(x|\theta) = \theta^{-1}p(x/\theta) = \theta^{-1}p(x'). \tag{3.4.19}$$

Substituting the chain-rule transformed derivative with respect to the scale parameter,

$$\frac{\partial p(x|\theta)}{\partial\theta} = -\frac{p(x')}{\theta^2} + \frac{1}{\theta}\frac{\partial p(x')}{\partial x'}\frac{\partial x'}{\partial\theta} = -\frac{1}{\theta^2}\left[p(x') + x'\frac{\partial p(x')}{\partial x'}\right], \tag{3.4.20}$$

into Eq. (3.4.15) gives the Fisher information for scaling:

$$I_F(\theta) = \int \frac{[p'(x|\theta)]^2}{p(x|\theta)} dx = \frac{1}{\theta^2} \int \left(p(x') + 2x' \frac{dp(x')}{dx'} + x'^2 \, p^{-1}(x') \left[\frac{dp(x')}{dx'} \right]^2 \right) dx'$$

$$= \frac{1}{\theta^2} [\int x'^2 \, p^{-1}(x') \left[\frac{dp(x')}{dx'} \right]^2 dx' - 1]. \qquad (3.4.21)$$

In the preceding expression we have transformed by parts one of the integrals:

$$\int x' \frac{dp(x')}{dx'} dx' = -\int p(x')\, dx' = -1, \qquad (3.4.22)$$

since $p(x) \to 0$ for $|x| \to \infty$.

3.5. DEPENDENT PROBABILITY DISTRIBUTIONS

In a general case of two mutually *dependent*, discrete probability distributions (see Section 3.2), $\boldsymbol{p} = \{P(a_i) \equiv p_i\}$ and $\boldsymbol{q} = \{P(b_j) \equiv q_j\}$, we decompose the joint-probability $\pi_{i,j} = P(a_i b_j)$ as the product $\pi_{i,j} = p_i\, P(j|i)$ of the *marginal* probability of *i*th event in one set of events, $p_i = p(a_i)$, and the *conditional* probability $P(j|i) = P(a_i b_j)/P(a_i)$, of the event b_j of the other set of outcomes, given that the event a_i has already occurred. The relevant normalization conditions for the joint probabilities and conditional probabilities $\mathbf{P}(\boldsymbol{b}\,|\,\boldsymbol{a}) = \{P(j|i)\}$ read: $\sum_j \pi_{i,j} = p_i$, $\sum_i \pi_{i,j} = q_j$, $\sum_i \sum_j \pi_{i,j} = 1$, and $\sum_j P(j|i) = 1$, $i = 1, 2, ..., n$. In this case the Shannon entropy of the product distribution $\boldsymbol{\pi} = \{\pi_{i,j}\}$,

$$\begin{aligned} S(\boldsymbol{\pi}) &= -\textstyle\sum_i \sum_j \pi_{i,j} \log \pi_{i,j} = -\sum_i \sum_j p_i\, P(j|i)\, [\log p_i + \log P(j|i)] \\ &= -[\textstyle\sum_j P(j|i)] \sum_i p_i \log p_i - \sum_i p_i\, [\sum_j P(j|i) \log P(j|i)] \\ &\equiv S(\boldsymbol{p}) + \textstyle\sum_i p_i\, S(\boldsymbol{q}|i) \equiv S(\boldsymbol{p}) + S(\boldsymbol{q}|\boldsymbol{p}), \end{aligned} \qquad (3.5.1)$$

is the sum of the average entropy of the marginal probability distribution, $S(\boldsymbol{p})$, and the *average conditional entropy*

$$S(\boldsymbol{q}\,|\,\boldsymbol{p}) = -\textstyle\sum_i p_i\, [\sum_j P(j|i) \log P(j|i)] \equiv \sum_i p_i\, S(\boldsymbol{q}|i) \equiv \sum_i \sum_j \pi_{i,j}\, I(j|i). \qquad (3.5.2)$$

The latter represents the extra amount of uncertainty about the occurrence of events $\boldsymbol{b} = \{b_j\}$, given that the events $\boldsymbol{a} = \{a_i\}$ are known to have occurred. In other words: the amount of information obtained as a result of simultaneously observing the events $\boldsymbol{a}$ and $\boldsymbol{b}$ of the two discrete probability distributions $\boldsymbol{p} = \boldsymbol{P}(\boldsymbol{a})$ and $\boldsymbol{q} = \boldsymbol{P}(\boldsymbol{b})$, respectively, equals to the amount of information observed in one set, say $\boldsymbol{a}$, supplemented by the extra information provided by the occurrence of events in the other set $\boldsymbol{b}$, when $\boldsymbol{a}$ are known to have occurred already. This is schematically illustrated in Fig. 3.1.

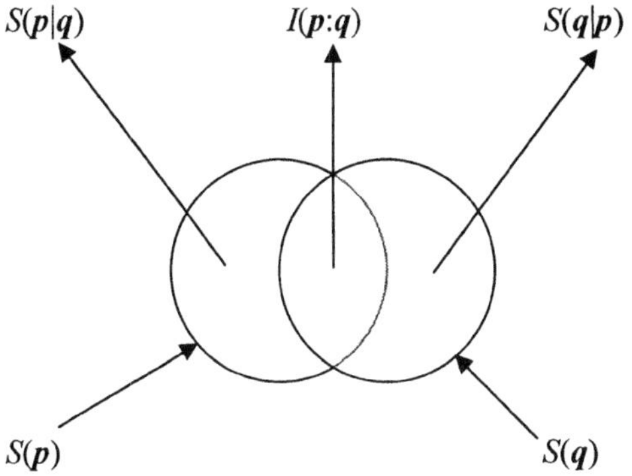

Figure 3.1. A qualitative diagram of the conditional entropy and mutual information quantities of two probability distributions $\boldsymbol{p}$ and $\boldsymbol{q}$. Two circles enclose the areas representing the entropies $S(\boldsymbol{p})$ and $S(\boldsymbol{q})$ of the two *separate* distributions. The common (overlap) area of the two circles corresponds to the mutual information $I(\boldsymbol{p}:\boldsymbol{q})$ in both distributions. The remaining parts of two circles represent the corresponding conditional entropies $S(\boldsymbol{p}|\boldsymbol{q})$ and $S(\boldsymbol{q}|\boldsymbol{p})$, measuring the residual uncertainty about events in a given set, when one has the full knowledge of the occurrence of the events in the other set of outcomes. The area enclosed by the envelope of the two overlapping circles then represents the entropy of the "product" (joint) distribution: $S(\boldsymbol{\pi}) = S(\mathbf{P}(\boldsymbol{a}\otimes\boldsymbol{b})) = S(\boldsymbol{p}) + S(\boldsymbol{q}) - I(\boldsymbol{p}:\boldsymbol{q}) = S(\boldsymbol{p}) + S(\boldsymbol{q}|\boldsymbol{p}) = S(\boldsymbol{q}) + S(\boldsymbol{p}|\boldsymbol{q})$.

Clearly, by using the other probability distribution $\boldsymbol{q} = \{P(b_i)\}$ as the marginal one, one arrives at the alternative expression for $S(\boldsymbol{\pi})$:

$$\begin{aligned} S(\boldsymbol{\pi}) &= -\textstyle\sum_i \sum_j \pi_{i,j} \log\pi_{i,j} = -\sum_i \sum_j q_j P(i|j) \,[\log q_j + \log P(i|j)] \\ &= -[\textstyle\sum_i P(i|j)] \sum_j q_j \log q_j - \sum_j q_j [\sum_i P(i|j) \log P(i|j)] \\ &\equiv S(\boldsymbol{q}) + \textstyle\sum_j q_j S(\boldsymbol{p}|j) \equiv S(\boldsymbol{q}) + S(\boldsymbol{p}|\boldsymbol{q}), \end{aligned} \tag{3.5.3}$$

where the average conditional entropy

$$S(\boldsymbol{p}|\boldsymbol{q}) = -\textstyle\sum_j q_j [\sum_i P(i|j) \log P(i|j)] \equiv \sum_j q_j S(\boldsymbol{p}|j) \equiv \sum_i \sum_j \pi_{i,j} I(i|j) \tag{3.5.4}$$

and the conditional probabilities $P(\boldsymbol{a}|\boldsymbol{b}) = \{P(i|j)\}$ are normalized to satisfy the equations: $\sum_i P(i|j) = 1$, $j = 1, 2, ..., m$. Equation (3.5.3) is also illustrated in Fig. 3.1. The average conditional entropy $S(\boldsymbol{p}|\boldsymbol{q})$ represents the residual uncertainty about the events $\boldsymbol{a}$, when the events $\boldsymbol{b}$ are known to have been observed already.

Properties of the conditional entropy are similar to those of the entropy itself, since the conditional probability is a probability measure. It follows from Eq. (3.5.2) that the average conditional entropy $S(\boldsymbol{q}|\boldsymbol{p})$ represents the mean value taken over the marginal probability distribution p_i of the information $S(\boldsymbol{q}|i) = -\sum_j P(j|i) \log P(j|i) =$

$S[\boldsymbol{P}(\boldsymbol{q}|i)]$ contained in the conditional probability vector $\boldsymbol{P}(\boldsymbol{q}|i) = \{P(b_j|a_i), j = 1, 2, \ldots, m\}$, which measures the residual amount of information in $\boldsymbol{q}$, given that the event a_i has already occurred.

In the same equation the conditional entropy has been further resolved in terms of the *conditional self-information* $\mathbf{I}(\boldsymbol{q}|\boldsymbol{p}) \equiv \{I(j|i) = -\log P(j|i) \geq 0\}$, where equality takes place if and only if $P(j|i) = 1$, i.e., when there is no uncertainty about the occurrence of b_j, given that a_i has already been observed. A finite value of $I(j|i)$ is the residual uncertainty about b_j, when one has the knowledge of the occurrence of a_i. The average conditional entropies of Eqs. (3.5.2) and (3.5.4) have been expressed as the joint-probability–weighted averages of the relevant conditional self-information.

The common amount of information in two events a_i and b_j, $I(i{:}j)$, measuring the information about a_i provided by the occurrence of b_j, or the information about b_j provided by the occurrence of a_i, is called the *mutual information* in these two events:

$$I(i{:}j) = \log[P(a_i b_j)/P(a_i)P(b_j)] = \log[\pi_{i,j}/(p_i\, q_j)] \equiv \log[P(i|j)/p_i] \equiv \log\,[P(j|i)/q_j] = I(j{:}i). \qquad (3.5.5)$$

The mutual information in two events may take on any real value, positive, negative, or zero. It vanishes, when both events are independent, i.e., when the occurrence of one event does not influence (or condition) the probability of the occurrence of the other event, and it is negative under the condition that the occurrence of one event makes the nonoccurrence of the other event more likely.

The mutual information of an event with itself defines its *self-information*: $I(i{:}i) \equiv I(i) \geq 0$, since $P(i|i) = 1$. It vanishes when $p_i = 1$, i.e., when there is no uncertainty about the occurrence of a_i, so that the occurrence of this event removes no uncertainty, hence conveys no information. This quantity provides a measure of the uncertainty about the occurrence of the event, i.e., the information received when the event occurs. It also follows from Eq. (3.5.5) that

$$I(i{:}j) = I(i) - I(i|j) = I(j) - I(j|i) = I(i) + I(j) - I(i, j)$$
$$\text{or} \qquad I(i, j) = I(i) + I(j) - I(i{:}j), \qquad (3.5.6)$$

where the self-information of the joint event $I(i, j) = -\log\pi_{i,j}$. Thus, the information in the joint occurrence of the events a_i and b_j is the information in the occurrence of a_i plus that in the occurrence of b_j minus the mutual information. Clearly, for the independent events $I(i, j) = I(i) + I(j)$, so that $I(i{:}j) = 0$.

One defines the *average mutual information* in two discrete probability distributions, $I(\boldsymbol{p}{:}\boldsymbol{q})$, as the $\boldsymbol{\pi}$-weighted expectation value of the mutual information quantities for the individual joint events:

$$I(\boldsymbol{p}{:}\boldsymbol{q}) = \sum_i \sum_j \pi_{i,j}\, I(i{:}j) = \sum_i \sum_j \pi_{i,j} \log\,(\pi_{i,j}/\pi_{i,j}^{\,0})$$
$$= S(\boldsymbol{p}) + S(\boldsymbol{q}) - S(\boldsymbol{\pi}) = S(\boldsymbol{p}) - S(\boldsymbol{p}\,|\,\boldsymbol{q}) = S(\boldsymbol{q}) - S(\boldsymbol{q}\,|\,\boldsymbol{p}) \geq 0. \qquad (3.5.7)$$

where the equality holds only for the independent distributions, when $\pi_{i,j} = \pi_{i,j}^0 = p_i q_j$. These average entropy/information relations are also illustrated in Fig. 3.1.

Again, the non-negative character of this average information measure can be demonstrated using the inequality $\log z \le z - 1$ or $\log (1/z) \ge 1 - z$. Defining $z = \pi_{i,j}^0/\pi_{i,j}$ and taking into account the normalization conditions of $\boldsymbol{\pi}$ and $\boldsymbol{\pi}^0$ then gives:

$$I(\boldsymbol{p}:\boldsymbol{q}) = \sum_i \sum_j \pi_{i,j} \log (\pi_{i,j}/\pi_{i,j}^0) \ge \sum_i \sum_j (\pi_{i,j} - \pi_{i,j}^0) = 0. \quad (3.5.8)$$

The above average conditional entropy and mutual information quantities can be extended into the continuous probability densities $p(x)$ and $q(x')$, normalized to satisfy the relation $\int p(x)\, dx = \int q(x')\, dx' = 1$:

i) entropy of the joint probability density for the *independent* events: $\pi^0(x, x') = p(x)\, q(x')$,

$$\begin{aligned} S[\pi^0] &= -\iint \pi^0(x, x') \log \pi^0(x, x')\, dx\, dx' = -\iint p(x)\, q(x') [\log p(x) + \log q(x')]\, dx\, dx' \\ &= -[\int q(x')\, dx'] \int p(x) \log p(x)\, dx - [\int p(x)\, dx] \int q(x') \log q(x')\, dx' \\ &= S[p] + S[q]; \end{aligned} \quad (3.5.9)$$

ii) entropy of the marginally decomposed joint probability density of *dependent* events: $\pi(x, x') = p(x)\, p(x'|x) = q(x')\, p(x|x')$, normalized in accordance with the sum rules: $\iint \pi(x, x')\, dx\, dx' = 1$, $\int \pi(x, x')\, dx = q(x')$, $\int \pi(x, x')\, dx' = p(x)$, with the conditional probability density $p(x'|x) = \pi(x, x')/p(x)$ or $p(x|x') = \pi(x, x')/q(x')$, normalized to satisfy the conditions, $\int p(x'|x)\, dx' = \int p(x|x')\, dx = 1$,

$$\begin{aligned} S[\pi] &= -\iint \pi(x, x') \log \pi(x, x')\, dx\, dx' \\ &= -\iint p(x) p(x'|x) [\log p(x) + \log p(x'|x)]\, dx\, dx' \\ &= -[\int p(x'|x)\, dx'] \int p(x) \log p(x)\, dx - \iint \pi(x, x') \log p(x'|x)]\, dx\, dx' \\ &\equiv S[p] + \iint \pi(x, x')\, I(x'|x)\, dx\, dx' \equiv S[p] + S[q|p] \\ &= -\iint q(x') p(x|x') [\log q(x') + \log p(x|x')]\, dx\, dx' \\ &= S[q] + S[p|q] = S[p] + S[q] - I[p:q]; \end{aligned} \quad (3.5.10)$$

iii) the average conditional entropy:

$$S[q\,|\,p] = -\iint \pi(x, x') \log p(x'|x)\, dx\, dx'; \quad (3.5.11)$$

iv) the average mutual information:

$$\begin{aligned} I[p:q] &= \iint \pi(x, x') \log[\pi(x, x')/\pi^0(x, x')]\, dx\, dx' \equiv \iint \pi(x, x')\, I(x{:}x')\, dx\, dx' \\ &= S[p] + S[q] - S[\pi] = S[p] - S[p|q] = S[q] - S[q|p] \ge 0. \end{aligned} \quad (3.5.12)$$

The amount of uncertainty in $\boldsymbol{q}$ (or q) can only decrease, when $\boldsymbol{p}$ (or p) has been known beforehand, $S(\boldsymbol{q}) \geq S(\boldsymbol{q}|\boldsymbol{p})$ (or $S[q] \geq S[q|p]$), as indeed seen in Fig. 3.1, with equality being observed only when the two sets of events are independent (non-overlapping).

It should be observed that the average mutual information quantities of Eqs. (3.5.7) and (3.5.12) are examples of the entropy deficiency functional of Kullback and Leibler, which measure the missing information between joint probability $\boldsymbol{\pi}$ [or $\pi(x, x')$] of the *dependent* events, and the corresponding joint probability distributions $\boldsymbol{\pi}^0 = \boldsymbol{p}\otimes\boldsymbol{q}$ [or $\pi^0(x, x') = p(x)\, q(x')$] for the independent events:

$$\Delta S(\boldsymbol{\pi}|\boldsymbol{\pi}^0) = I(\boldsymbol{p}:\boldsymbol{q}) \qquad \text{and} \qquad \Delta S[\pi|\pi^0] = I[p:q]. \tag{3.5.13}$$

These average mutual informations thus measure a degree of dependence between the events of the two compared probability distributions.

A similar information-distance interpretation can be given to the conditional entropies of Eqs. (3.5.4) and (3.5.11):

$$S(\boldsymbol{p}|\boldsymbol{q}) = -\sum_i \sum_j \pi_{i,j} \log(\pi_{i,j}/q_j) = -\sum_i \sum_j \pi_{i,j} \log\{\pi_{i,j}\, p_i/[p_i\, q_j]\} = S(\boldsymbol{p}) - \Delta S(\boldsymbol{\pi}|\boldsymbol{\pi}^0),$$

$$S[p|q] = -\iint \pi(x, x') \log [\pi(x, x')/q(x')]\, dx\, dx'$$
$$= -\iint \pi(x, x') \log [\pi(x, x')\, p(x)/\pi^0(x, x')]\, dx\, dx' = S[p] - \Delta S[\pi\,|\pi^0]. \tag{3.5.14}$$

3.6. GROUPING/COMBINATION RULES

We now turn to the *grouping principles* of information quantities, which reflect their dependence upon the stages, in which the probability distributions are generated. Alternatively, the terms in these entropy/information combination rules can be associated with the way, in which the uncertainty contained in the distribution is removed in the process of acquiring the information (removing the uncertainty).

Consider an arbitrary distribution of N outcomes $\boldsymbol{a} = (a_1, a_2,\ldots, a_N)$ defining the overall (normalized) probability vector $\boldsymbol{p} = \{P(a_i) = p_i\}$, which is divided into exhaustive and exclusive groups $\boldsymbol{G} = \{G_\alpha(n_\alpha)\}$ of outcomes $\boldsymbol{a} = \{\boldsymbol{a}_\alpha\}$, where n_α is the number of outcomes in group α, $\boldsymbol{a}_\alpha = (\alpha_1, \alpha_2, \ldots, \alpha_{n_\alpha})$, and $\sum_\alpha n_\alpha = N$. This division partitions the probability vector into the group probability vectors, $\boldsymbol{p}_G = \{\boldsymbol{p}_\alpha\}$, which determine the group *condensed* probabilities $\boldsymbol{P}_G = \{P_\alpha = \sum_{m\in\alpha} p_m\}$. The ratios $\boldsymbol{p}_\alpha/P_\alpha = \{p_m/P_\alpha = \pi(m|\alpha)\} \equiv \boldsymbol{\pi}(\boldsymbol{a}|\alpha)$ then define the conditional probabilities of an event α_m in group α, when it is known for sure that the outcome is in the group α (a parameter). They are normalized to satisfy the condition $\sum_{m\in\alpha} \pi(m|\alpha) = 1$.

This division can reflect the way the outcomes are organized in the process of establishing the overall probability itself. First, each subset of events is considered separately, to determine the *intra*-group (conditional) probabilities, and then by examining the relative frequencies of outcomes in different groups, one determines the group condensed probabilities, and hence the absolute probabilities of outcomes in the whole set of events.

Let us transform the Shannon entropy $S(\boldsymbol{p})$ [Eq. (3.1.3)] into the equivalent group-resolved expression. A straightforward manipulation of probabilities gives:

$$S(\boldsymbol{p}) = -\sum_i p_i \log p_i = -\sum_{\alpha\in \boldsymbol{G}} \sum_{m\in\alpha} p_m \log p_m$$

$$= -\sum_{\alpha\in \boldsymbol{G}} \sum_{m\in\alpha} P_\alpha\, \pi(m|\alpha)[\log P_\alpha + \log \pi(m|\alpha)]$$

$$= -[\sum_{m\in\alpha} \pi(m|\alpha)] \sum_{\alpha\in \boldsymbol{G}} P_\alpha \log P_\alpha - \sum_{\alpha\in \boldsymbol{G}} P_\alpha \sum_{m\in\alpha} \pi(m|\alpha) \log \pi(m|\alpha)$$

$$\equiv -\sum_{\alpha\in \boldsymbol{G}} P_\alpha \log P_\alpha + \sum_{\alpha\in \boldsymbol{G}} P_\alpha S(\boldsymbol{\pi}(\boldsymbol{\alpha}|\alpha)) \equiv S(\boldsymbol{P}_G) + S(\boldsymbol{p}|\boldsymbol{G}). \qquad (3.6.1)$$

Therefore, the uncertainty in the overall probability distribution $S(\boldsymbol{p})$ equals the group uncertainty $S(\boldsymbol{P}_G)$ plus the conditional entropy $S(\boldsymbol{p}|\boldsymbol{G})$ given by the $\boldsymbol{P}_G$-weighted sum of the intra-group uncertainties $\{S(\boldsymbol{\pi}(\boldsymbol{\alpha}|\alpha))\}$. These contributions can be interpreted in terms of an associated experiment, which is intended to remove the uncertainty in $\boldsymbol{p}$ by selecting the "correct" outcome out of all N possible outcomes, by taking informative steps towards this end. First one identifies which group the outcome is in. The uncertainty removed, after the group origin of the searched outcome has finally been established, is given by the first term $S(\boldsymbol{P}_G)$ of the preceding equation. At this stage of acquiring the information the intra-group uncertainty still remains; it equals $S(\boldsymbol{\pi}(\boldsymbol{\alpha}|\alpha))$, when the outcome is in the group α. However, in the average intra-group uncertainty, for all alternative groups in the partitioning, these intra-group uncertainties, each characterizing the *separate* group of outcomes, have to be weighted in accordance with the condensed group probabilities in the whole set of events, thus giving rise to the second, conditional entropy term $S(\boldsymbol{p}_G|\boldsymbol{G})$ of the last equation.

In order to derive the associated grouping rule for the Fisher information contained in the discrete, histogramic representation of the continuous probability distribution function $p(x) \approx \boldsymbol{p}_n = [p(x_1), p(x_2), \ldots, p(x_n)]$, where $x_{i+1} - x_i = \Delta x$, $i = 1, 2, \ldots, n-1$, we express the functional of Eq. (3.4.1) as the associated histogramic limit, for $\Delta x \to 0$ ($n \to \infty$):

$$I^F[p] = \lim_{\Delta x\to 0} \frac{1}{\Delta x} \sum_i \frac{[p(x_{i+1}) - p(x_i)]^2}{p(x_i)} = \lim_{\Delta x\to 0} \frac{1}{\Delta x} \sum_i p(x_i) \left[\frac{p(x_{i+1}) - p(x_i)}{p(x_i)}\right]^2$$

$$= \lim_{\Delta x\to 0} \frac{1}{\Delta x} \sum_i p(x_i) \left[\frac{p(x_i + \Delta x)}{p(x_i)} - 1\right]^2. \qquad (3.6.2)$$

Hence, for the above division of the finite, discrete representation $\boldsymbol{p}_n$ of the continuous probability distribution $p(x)$ one obtains the following grouping formula corresponding to the ordered (sequential) groups of histogramic points,

$$\ldots < (x_{1,\alpha-1} < x_{2,\alpha-1}, \ldots) < (x_{1,\alpha} < x_{2,\alpha}, \ldots) < (x_{1,\alpha+1} < x_{2,\alpha+1}, \ldots) < \ldots, \tag{3.6.3}$$

$$I^F(\boldsymbol{p}) \equiv \frac{1}{\Delta x}\sum_i \frac{[p(x_{i+1}) - p(x_i)]^2}{p(x_i)} = \sum_{\alpha\in G} P_\alpha \left(\frac{1}{\Delta x}\sum_{m\in\alpha}\frac{[\pi(m+1|\alpha) - \pi(m|\alpha)]^2}{\pi(m|\alpha)}\right)$$
$$\equiv \sum_{\alpha\in G} P_\alpha I^F(\boldsymbol{\pi}_n(\boldsymbol{\alpha}|\alpha)) \equiv I^F(\boldsymbol{\pi}_n|\boldsymbol{G}). \tag{3.6.4a}$$

The emphasis on the *ordered* set of localization events is crucial in the case of the Fisher measure of information, since changing the order of points in $\boldsymbol{p}_n$ would have dramatic effect upon the probability derivatives (see the preceding equation), nd thus the average information content. We recall that the global Shannon entropy was invariant with respect to changing the order of events in the distribution.

The same result is obtained, when one partitions the whole range of the continuous random variable x into the mutually exclusive domains, as is the case in the topological partitioning of the molecular electron density (Bader, 1990),

$$\boldsymbol{d} \equiv \{d_\alpha\} = \{x_\alpha \le x < x_\alpha + D_\alpha \equiv x_{\alpha+1}\}, \quad D_\alpha > 0,$$

of monotonically increasing coordinate values: $\ldots < x_{\alpha-1} < x_\alpha < x_{\alpha+1} < \ldots$:

$$I^F[p] = \int p(x)^{-1}\left(\frac{dp(x)}{dx}\right)^2 dx = \sum_\alpha \int_{x_\alpha}^{x_\alpha+D_\alpha} p(x)^{-1}\left(\frac{dp(x)}{dx}\right)^2 dx$$
$$= \sum_\alpha P_\alpha \int_{x_\alpha}^{x_\alpha+D_\alpha} \tilde{\pi}(x|\alpha)^{-1}\left(\frac{d\tilde{\pi}(x|\alpha)}{dx}\right)^2 dx = \sum_\alpha P_\alpha I^F[\tilde{\pi}(x|\alpha)], \tag{3.6.4b}$$

where

$$P_\alpha = \int_{x_\alpha}^{x_\alpha+D_\alpha} p(x)\,dx \quad \text{and} \quad \tilde{\pi}(x|\alpha) = p(x)/P_\alpha, \ \text{for } x \in d_\alpha. \tag{3.6.4c}$$

Therefore, the Fisher information of the finite, discrete probability distribution is the $\boldsymbol{P}_G$-weighted mean value of the intra-group Fisher information quantities. A comparison with the corresponding combination principle for the *global* Shannon entropy [Eq. (3.6.1)] shows that the above Fisher rule misses the inter-group term, which depends solely on the condensed probabilities $\boldsymbol{P}_G$. This is due to the local character of the Fisher measure of the information content. We recall that the Fisher information is related to the relative entropy (entropy deficiency) of Kullback and Leibler [Eq. (3.4.9)] between the probability distribution and its locally shifted modification. We shall demonstrate this in the next example.

Example 3.10. We shall justify this relation using the discrete representation of Eq. (3.6.2). We first observe that in the final expression of this equation the ratio $p(x_i + \Delta x)/p(x_i) \cong 1$, so that the quantity in the square brackets of the final expression of this equation is small: $p(x_i + \Delta x)/p(x_i) - 1 \equiv r \cong 0$. Then, using the second-order Taylor expansion $\ln(1 + r) = r - r^2/2$, and hence $r^2 = 2[r - \ln(1 + r)]$, allows one transform the discrete expression for the Fisher information into the equivalent entropy-deficiency term between the continuous distribution $p(x)$ and its shifted version $p(x + \Delta x) = \tilde{p}(x)$:

$$
\begin{aligned}
I^F[p] &= \lim_{\Delta x \to 0}\left\{\frac{1}{\Delta x}\sum_i p(x_i)\left[\frac{p(x_i+\Delta x)}{p(x_i)}-1\right]^2\right\} \\
&= \lim_{\Delta x \to 0}\left\{\frac{-2}{\Delta x}\sum_i p(x_i)\ln\left[\frac{p(x_i+\Delta x)}{p(x_i)}\right]+\frac{2}{\Delta x}\sum_i p(x_i)\left[\frac{p(x_i+\Delta x)}{p(x_i)}-1\right]\right\} \\
&= \lim_{\Delta x \to 0}\left\{\frac{-2}{\Delta x}\sum_i p(x_i)\ln\left[\frac{p(x_i+\Delta x)}{p(x_i)}\right]\right\} = \lim_{\Delta x \to 0}\left\{\frac{-2}{(\Delta x)^2}\int p(x)\ln\frac{p(x+\Delta x)}{p(x)}\,dx\right\} \\
&= \lim_{\Delta x \to 0}\frac{2}{(\Delta x)^2}\,\Delta S[p(x)|p(x+\Delta x)]. \qquad (3.6.5)
\end{aligned}
$$

The grouping rule for the entropy deficiency follows that for the Shannon entropy [Eq. (3.6.1)]. We assume the same grouping pattern $\boldsymbol{G} = \{G_\alpha\}$ for the events in the two discrete distributions: $\boldsymbol{p} = (p_1, p_2, \ldots, p_n)$ and $\boldsymbol{p}^0 = (p_1^0, p_2^0, \ldots, p_n^0)$, which defines the corresponding group probabilities: condensed, $\boldsymbol{P_G} = \{P_\alpha = \sum_{m\in\alpha} p_m\}$, $\boldsymbol{P_G}^0 = \{P_\alpha^0 = \sum_{m\in\alpha} p_m^0\}$, and conditional, $\boldsymbol{\pi}(\boldsymbol{\alpha}|\alpha) \equiv \boldsymbol{\pi}_\alpha = \{\pi(m|\alpha) = p_m/P_\alpha\}$, $\boldsymbol{\pi}^0(\boldsymbol{\alpha}|\alpha) \equiv \boldsymbol{\pi}_\alpha^0 = \{\pi^0(m|\alpha) = p_m^0/P_\alpha^0\}$, where $m\in\alpha$. Taking into account the corresponding normalization conditions,

$$\textstyle\sum_{\alpha\in \boldsymbol{G}} P_\alpha = \sum_{\alpha\in \boldsymbol{G}} P_\alpha^0 = \sum_{m\in\alpha} \pi(m|\alpha) = \sum_{m\in\alpha} \pi^0(m|\alpha) = 1, \qquad (3.6.6)$$

then gives:

$$
\begin{aligned}
\Delta S(\boldsymbol{p}|\boldsymbol{p}^0) &= \sum_i p_i \log\frac{p_i}{p_i^0} = \sum_{\alpha\in \boldsymbol{G}}\sum_{m\in\alpha} p_m \log\frac{p_m}{p_m^0} = \sum_{\alpha\in \boldsymbol{G}}\sum_{m\in\alpha} P_\alpha \pi(m|\alpha)\log\frac{P_\alpha \pi(m|\alpha)}{P_\alpha^0 \pi^0(m|\alpha)} \\
&= \sum_{m\in\alpha} P_\alpha \log\frac{P_\alpha}{P_\alpha^0} + \sum_{\alpha\in \boldsymbol{G}} P_\alpha \sum_{m\in\alpha}\pi(m|\alpha)\log\frac{\pi(m|\alpha)}{\pi^0(m|\alpha)} \equiv \Delta S(\boldsymbol{P_G}|\boldsymbol{P_G^0}) + \sum_{\alpha\in \boldsymbol{G}} P_\alpha\, \Delta S(\boldsymbol{\pi}_\alpha|\boldsymbol{\pi}_\alpha^0). \qquad (3.6.7)
\end{aligned}
$$

Therefore, the entropy deficiency of the overall distribution is the sum of the group information distance and the weighted average of the intra-group entropy deficiencies.

Turning again to the Fisher information representation of Eq. (3.6.5), we observe that in the limit $\Delta x \to 0$ $p(x) = \tilde{p}(x)$ thus giving rise to the identical sets of condensed probabilities, $\boldsymbol{P}_G = \widetilde{\boldsymbol{P}}_G$, and hence the vanishing group missing information. This additionally explains the form of the grouping rule of Eq. (3.6.4).

Let us next examine a slightly different *combination* scenario for the continuous probability distributions. Consider the *composite* (normalized) probability distribution function, which for reasons of simplicity is taken to be defined in one dimension x: $p = p(x)$. It consists of the *component* contributions, $\boldsymbol{p}_G(x) = \{p_\alpha(x)\}$,

$$p(x) = \Sigma_{\alpha\in G}\, p_\alpha(x), \qquad \Sigma_{\alpha\in G} \int p_\alpha(x)\, dx = \int p(x)\, dx = 1, \tag{3.6.8}$$

which separately satisfy the subsystem normalization to the fragment condensed probability, $\int p_\alpha(x)\, dx = P_\alpha$, defining the associated (normalized) probability vector $\boldsymbol{P}_G = \{P_\alpha\}$, $\Sigma_{\alpha\in G} P_\alpha = 1$. In this partition of $p(x)$ each component covers the same, infinite set of continuous locality events, which are identified by the coordinate x. For example, in molecular systems the component distributions may describe molecular fragments: MO, AIM, reactants, etc.

There are two sets of parametric, conditional probabilities of such fragments. On one hand, one defines the *local* conditional probabilities for each component:

$$\boldsymbol{\pi}(x|\boldsymbol{G}) = \{\pi(x|\alpha) = p_\alpha(x)/P_\alpha\}, \qquad \int \pi(x|\alpha)\, dx = 1, \tag{3.6.9}$$

in which x is the random locality variable, while the component label α is the subsystem parameter. Here $\pi(x|\alpha)$ is the probability of observing the outcome x in the *isolated* (separated) component α. Only the condensed probability P_α reflects the fact that this component is a part of the overall, composite distribution.

On the other hand, one can define the other set of conditional probabilities of probability components, for the specified, parametric value of the continuous (locality) event:

$$\boldsymbol{\pi}(\boldsymbol{G}|x) = \{\pi(\alpha|x) = p_\alpha(x)/p(x)\}, \qquad \Sigma_{\alpha\in G}\, \pi(\alpha|x) = 1. \tag{3.6.10}$$

In these parametrically dependent probabilities the role of variables and parameters has been reversed relative to probabilities of Eq. (3.6.9). Here $\pi(\alpha|x)$ stands for the probability that the specified local event x in the system as a whole, refers to the αth component of the overall distribution. In fact it represents the local *share* of αth subsystem in the overall probability density. This alternative set of the component conditional probabilities thus characterizes each component as a part of the composite distribution, e.g., the constituent fragments in a molecule.

The last two equations imply the following relations between these two sets of the component conditional probabilities:

$$p_\alpha(x) = \pi(\alpha|x)\, p(x) = \pi(x|\alpha)\, P_\alpha \quad \text{or} \quad \pi(x|\alpha) = \pi(\alpha|x)[p(x)/P_\alpha]. \tag{3.6.11}$$

Expressing the Shannon entropy $S[p]$ of Eq. (3.1.4) in terms of probabilities $p(x)$ and $\boldsymbol{\pi}(\boldsymbol{G}|x)$ gives:

$$S[p] = -\int p(x) \log p(x)\, dx = -\sum_{\alpha\in \boldsymbol{G}} \int p_\alpha(x) \log [p_\alpha(x)/\pi(\alpha|x)]\, dx$$
$$= -\sum_{\alpha\in \boldsymbol{G}} \int p_\alpha(x) \log p_\alpha(x)\, dx - \int p(x)\, [-\sum_{\alpha\in \boldsymbol{G}} \pi(\alpha|x) \log \pi(\alpha|x)]\, dx$$
$$= \sum_{\alpha\in \boldsymbol{G}} S[p_\alpha] - \int p(x)\, S[\boldsymbol{\pi}(\boldsymbol{G}|x)]\, dx \equiv S^{add}[\boldsymbol{p}_G] + S^{nadd}[\boldsymbol{p}_G, \boldsymbol{\pi}(\boldsymbol{G}|x)]. \qquad (3.6.12)$$

The first, *additive* term, $S^{add}[\boldsymbol{p}_G]$, is the sum of entropies $\{S[p_\alpha]\}$ of the component probabilities $\boldsymbol{p}(x)$, while the second, *non-additive* contribution $S^{nadd}[\boldsymbol{p}_G, \boldsymbol{\pi}(\boldsymbol{G}|x)]$ is the negative $p(x)$-weighted average value of the local conditional entropy $S[\boldsymbol{\pi}(\boldsymbol{G}|x)]$ contained in the subsystem conditional probabilities $\boldsymbol{\pi}(\boldsymbol{G}|x)$.

Clearly, the additive contribution can be also expressed in terms of the entropies of the condensed and conditional probabilities of the probability distribution [compare Eq. (3.6.1)] by using the other set $\boldsymbol{\pi}(x|\boldsymbol{G})$ of parametric probabilities:

$$S^{add}[\boldsymbol{p}_G] = -\sum_{\alpha\in \boldsymbol{G}} P_\alpha \int \pi(x|\alpha)\, [\log P_\alpha + \log \pi(x|\alpha)]\, dx$$
$$= -\sum_{\alpha\in \boldsymbol{G}} P_\alpha \log P_\alpha\; [\int \pi(x|\alpha)\, dx] - \sum_{\alpha\in \boldsymbol{G}} P_\alpha \int \pi(x|\alpha) \log \pi(x|\alpha)\, dx$$
$$= S(\boldsymbol{P}_G) + \sum_{\alpha\in \boldsymbol{G}} P_\alpha S[\pi(x|\alpha)] \equiv S(\boldsymbol{P}_G) + S[p|\boldsymbol{G}]. \qquad (3.6.13)$$

Therefore, the additive Shannon entropy in the component resolution satisfies the familiar grouping theorem of Eq. (3.6.1).

The non-additive entropy $S^{nadd}[\boldsymbol{p}_G, \boldsymbol{\pi}(\boldsymbol{G}|x)]$ can be alternatively expressed as the $\boldsymbol{P}_G$–weighted average of the component entropy-deficiency terms. First, expressing $p_\alpha(x)$ in terms of $\pi(x|\alpha)$ gives:

$$S^{nadd}[\boldsymbol{p}_G, \boldsymbol{\pi}(\boldsymbol{G}|x)] = \sum_{\alpha\in \boldsymbol{G}} \int p_\alpha(x) \log \pi(\alpha|x)\, dx$$
$$= \sum_{\alpha\in \boldsymbol{G}} P_\alpha \int \pi(x|\alpha) \log \pi(\alpha|x)\, dx. \qquad (3.6.14)$$

Then, using Eq. (3.6.11) one can express $\pi(\alpha|x)$ in terms of $\pi(x|\alpha)$:

$$\sum_{\alpha\in G} P_\alpha \int \pi(x|\alpha) \log \pi(\alpha|x)\, dx = \sum_{\alpha\in G} P_\alpha \int \pi(x|\alpha) \log\left(\frac{P_\alpha\, \pi(x|\alpha)}{p(x)}\right) dx$$
$$\equiv \sum_{\alpha\in G} P_\alpha \log P_\alpha + \sum_{\alpha\in G} P_\alpha \int \pi(x|\alpha) \log\left(\frac{\pi(x|\alpha)}{p(x)}\right) dx$$
$$= -S(\boldsymbol{P}_G) + \sum_{\alpha\in G} P_\alpha\, \Delta S[\pi(x|\alpha)|p(x)]. \qquad (3.6.15)$$

This final expression shows that the non-additive entropy term in the component resolution is the sum of the group entropy $S(\boldsymbol{P}_G)$ and the $\boldsymbol{P}_G$–weighted average of the subsystem entropy deficiencies between the conditional probability $\pi(x|\alpha)$ and the overall distribution $p(x)$.

Later in the book, when presenting the subsystem development within the communication theory of the chemical bond, we shall also examine the combination formulas for the conditional entropy and mutual information quantities defined in Section 3.5.

Example 3.11. Consider as an illustration of the above subsystem combination rules the exhaustive Hirshfeld (H) (1977) partition of the molecular electron density $\rho(\boldsymbol{r}) = Np(\boldsymbol{r})$ into densities $\boldsymbol{\rho}^H(\boldsymbol{r}) = \{\rho_\alpha{}^H(\boldsymbol{r})\} \equiv N\boldsymbol{p}^H(\boldsymbol{r})$ of bonded atoms, which we shall examine in a more detail in Chapters 5 and 6:

$$\rho(\boldsymbol{r}) = \Sigma_\alpha \rho_\alpha{}^H(\boldsymbol{r}), \qquad \rho_\alpha{}^H(\boldsymbol{r}) = \rho_\alpha{}^0(\boldsymbol{r})[\rho(\boldsymbol{r})/\rho^0(\boldsymbol{r})], \qquad \rho^0(\boldsymbol{r}) = \Sigma_\alpha \rho_\alpha{}^0(\boldsymbol{r}), \tag{3.6.16}$$

where $\rho^0(\boldsymbol{r})$ denotes the electron density of the isoelectronic promolecule, resulting from the "frozen", free-atom densities $\boldsymbol{\rho}^0(\boldsymbol{r}) = \{\rho_\alpha{}^0(\boldsymbol{r})\}$ shifted to the actual positions of AIM.

It follows from the preceding equation that the Hirshfeld AIM components of the molecular electron density satisfy the following *"stockholder"* rule:

$$\rho_\alpha{}^H(\boldsymbol{r})/\rho(\boldsymbol{r}) = \rho_\alpha{}^0(\boldsymbol{r})/\rho^0(\boldsymbol{r}). \tag{3.6.17}$$

Indeed, by regarding $\rho^0(\boldsymbol{r})$ as the overall "investment" and $\rho(\boldsymbol{r})$ as the overall "profit" the AIM participation in the molecular profit is seen to be equal to its share in the overall investment. One finds the atomic average numbers of electrons by integrating the subsystem densities:

$$N_\alpha{}^H = \int \rho_\alpha{}^H(\boldsymbol{r})\, d\boldsymbol{r}, \qquad N_\alpha{}^0 = \int \rho_\alpha{}^0(\boldsymbol{r})\, d\boldsymbol{r}; \qquad N = \Sigma_\alpha N_\alpha{}^H = N^0 = \Sigma_\alpha N_\alpha{}^0, \tag{3.6.18}$$

which determine the condensed probabilities of constituent atoms in the molecular or promolecular system as a whole:

$$P_\alpha{}^H = N_\alpha{}^H / N \qquad \text{and} \qquad P_\alpha{}^0 = N_\alpha{}^0 / N. \tag{3.6.19}$$

The molecularly normalized atomic probability distributions $\boldsymbol{p}^H(\boldsymbol{r}) = \{p_\alpha{}^H(\boldsymbol{r})\} \equiv \boldsymbol{\rho}^H(\boldsymbol{r})/N$ and $\boldsymbol{p}^0(\boldsymbol{r}) = \{p_\alpha{}^0(\boldsymbol{r})\} \equiv \boldsymbol{\rho}^0(\boldsymbol{r})/N$, of the atomic fragments in molecule and promolecule, respectively, are the system renormalized shape-factors of the subsystem densities:

$$p_\alpha{}^H(\boldsymbol{r}) = \rho_\alpha{}^H(\boldsymbol{r})/N, \qquad \int p_\alpha{}^H(\boldsymbol{r})\, d\boldsymbol{r} = P_\alpha{}^H;$$
$$p_\alpha{}^0(\boldsymbol{r}) = \rho_\alpha{}^0(\boldsymbol{r})/N^0, \qquad \int p_\alpha{}^0(\boldsymbol{r})\, d\boldsymbol{r} = P_\alpha{}^0. \tag{3.6.20}$$

The two sets of the Hirshfeld conditional probabilities of constituent atoms in molecule and promolecule, as well as the associated normalization conditions are:

$$\pi^H(\boldsymbol{r}|\alpha) = p_\alpha{}^H(\boldsymbol{r})/P_\alpha{}^H = \rho_\alpha{}^H(\boldsymbol{r})/N_\alpha{}^H, \qquad \pi^0(\boldsymbol{r}|\alpha) = p_\alpha{}^0(\boldsymbol{r})/P_\alpha{}^0 = \rho_\alpha{}^0(\boldsymbol{r})/N_\alpha{}^0,$$
$$\int \pi^H(\boldsymbol{r}|\alpha)\, d\boldsymbol{r} = \int \pi^0(\boldsymbol{r}|\alpha)\, d\boldsymbol{r} = 1;$$

$$\pi^H(\alpha|\boldsymbol{r}) = p_\alpha^{\ H}(\boldsymbol{r})/p(\boldsymbol{r}) = \rho_\alpha^{\ H}(\boldsymbol{r})/\rho(\boldsymbol{r}); \qquad \pi^0(\alpha|\boldsymbol{r}) = p_\alpha^{\ 0}(\boldsymbol{r})/p^0(\boldsymbol{r}) = \rho_\alpha^{\ 0}(\boldsymbol{r})/\rho^0(\boldsymbol{r});$$
$$\Sigma_\alpha \pi^H(\alpha|\boldsymbol{r}) = \Sigma_\alpha \pi^0(\alpha|\boldsymbol{r}) = 1. \qquad (3.6.21)$$

The additive entropy of the Hirshfeld probability components, expressed in terms of the AIM electron densities, becomes [see Eqs. (3.6.12) and (3.6.13)]

$$S^{add}[\boldsymbol{p}^H] = \Sigma_\alpha S^{add}[p_\alpha^{\ H}] = N^{-1}\, \Sigma_\alpha S[\rho_\alpha^{\ H}] + \log N \equiv N^{-1} S^{add}[\boldsymbol{\rho}^H] + \log N, \qquad (3.6.22)$$

while the non-additive entropy of Eq. (3.6.14) gives:

$$S^{nadd}[\boldsymbol{p}^H] = S[p] - S^{add}[\boldsymbol{p}^H]. \qquad (3.6.23)$$

Therefore, the sum of these two contributions recovers the Shannon entropy in the molecular distribution, which represents the total average uncertainty in AIM resolution:

$$S^{add}[\boldsymbol{p}^H] + S^{nadd}[\boldsymbol{p}^H] = S[p] \equiv S^{total}[\boldsymbol{p}^H]. \qquad (3.6.24)$$

Example 3.12. We shall now examine the non-additivity of the entropy deficiency of the Hirshfeld probability components. The *total* entropy in the bonded AIM resolution, with respect to the free-atom electron distributions of the promolecular prototype, is given by the Kullback-Leibler functional

$$\Delta S[p|p^0] = \int p(\boldsymbol{r}) \log[p(\boldsymbol{r})/p^0(\boldsymbol{r})]\, d\boldsymbol{r} \equiv \Delta S^{total}[\boldsymbol{p}^H|\boldsymbol{p}^0] = N^{-1}\int \rho(\boldsymbol{r}) \log[\rho(\boldsymbol{r})/\rho^0(\boldsymbol{r})]\, d\boldsymbol{r}. \qquad (3.6.25)$$

Consider next its additive part in the Hirshfeld resolution [see Eq. (3.6.16)]:

$$\Delta S^{add}[\boldsymbol{p}^H|\boldsymbol{p}^0] = \Sigma_\alpha \int p_\alpha^{\ H}(\boldsymbol{r}) \log[p_\alpha^{\ H}(\boldsymbol{r})/p_\alpha^{\ 0}(\boldsymbol{r})]\, d\boldsymbol{r} = N^{-1}\Sigma_\alpha \int \rho_\alpha^{\ H}(\boldsymbol{r}) \log[\rho(\boldsymbol{r})/\rho^0(\boldsymbol{r})]\, d\boldsymbol{r}$$
$$= \Delta S[p|p^0] = \Delta S^{total}[\boldsymbol{p}^H|\boldsymbol{p}^0]. \qquad (3.6.25)$$

Therefore, for this particular partitioning of the molecular electron density the information-distance non-additive term exactly vanishes:

$$\Delta S^{nadd}[\boldsymbol{p}^H|\boldsymbol{p}^0] \equiv \Delta S^{total}[\boldsymbol{p}^H|\boldsymbol{p}^0] - \Delta S^{add}[\boldsymbol{p}^H|\boldsymbol{p}^0] = 0. \qquad (3.6.26)$$

This attractive feature of the stockholder atoms makes them unique information-theoretic concepts, which are particularly useful in a thermodynamic-like interpretation of electron distributions in molecular systems.

3.7. COMMUNICATION CHANNELS

We continue this short overview of the IT concepts and techniques with the key entropy/information quantities describing the transmission of signals in the communication systems (Shannon, 1948, 1949; Abramson, 1963; Pfeifer, 1978). The basic elements of such a "device" are shown in Scheme 3.1. The *input signal* emitted from n "inputs" $\boldsymbol{a} = (a_1, a_2, \ldots, a_n)$ of the channel *source* (**A**) is characterized by the *input probability distribution* $\boldsymbol{P}(\boldsymbol{a}) = \boldsymbol{p} = (p_1, p_2, \ldots, p_n) \equiv \boldsymbol{P}(\mathbf{A})$. It can be received at m "outputs" $\boldsymbol{b} = (b_1, b_2, \ldots, b_m)$ of the system *receiver* (**B**).

Input (Source): **A** **P(B|A)** *Output* (Receiver): **B**

a_1 b_1

a_2 b_2

....

$p_i \rightarrow a_i$ ——— $P(b_j|a_i) \equiv P(j|i)$ ———→ $b_j \rightarrow q_j$

....

a_n b_m

Scheme 3.1. A schematic diagram of the communication system, characterized by the probability vectors: $\boldsymbol{P}(\boldsymbol{a}) = \{p(a_i)\} = \boldsymbol{p} = (p_1, \ldots, p_n) \equiv \boldsymbol{P}(\mathbf{A})$, of the channel *"input"* events $\boldsymbol{a} = (a_1, \ldots, a_n)$ in the system *source* **A**, and $\boldsymbol{P}(\boldsymbol{b}) = \{p(b_j)\} = \boldsymbol{q} = (q_1, \ldots, q_m) \equiv \boldsymbol{P}(\mathbf{B})$, of the *"output"* events $\boldsymbol{b} = (b_1, \ldots, b_m)$ in the system *receiver* **B**. The transmission of signals via a network of communication channels is described by the $(n \times m)$-matrix of the conditional probabilities $\mathbf{P}(\mathbf{B}|\mathbf{A}) = \{P(b_j|a_i) \equiv P(j|i)\}$, of observing different "outputs" (*columns*, $j = 1, 2, \ldots, m$), given the specified "inputs" (*rows*, $i = 1, 2, \ldots, n$).

The distribution of the output signal among $\boldsymbol{b}$ gives rise to the *output probability distribution* $\boldsymbol{P}(\boldsymbol{b}) = \boldsymbol{q} = (q_1, q_2, \ldots, q_m) \equiv \boldsymbol{P}(\mathbf{B})$. The transmission of signals is randomly disturbed within the communication system, thus exhibiting a communication *noise*, so that in general a signal sent at a given input can be received with a non-zero probability at several outputs. This feature of communication systems is described by the conditional probabilities of the outputs given inputs, $\mathbf{P}(\mathbf{B}|\mathbf{A}) = \{P(b_j|a_i) = P(a_i b_j)/p(a_i) \equiv P(j|i)\}$, or the conditional probabilities of the inputs, given outputs, $\mathbf{P}(\mathbf{A}|\mathbf{B}) = \{P(a_i|b_j) = P(a_i b_j)/p(b_j) \equiv P(i|j)\}$, where $P(a_i b_j) \equiv \pi_{i,j}$ stands for the probability of the joint occurrence of the specified pair of the output and input events.

The Shannon input entropy, of the source probability vector $\boldsymbol{p}$, $S(\mathbf{A}) \equiv S(\boldsymbol{p})$, determines the so called *a priori* entropy. The average *conditional entropy* $S(\mathbf{A}|\mathbf{B}) \equiv S(\boldsymbol{p}|\boldsymbol{q})$ [see Section 3.5] and its output resolved contributions $\{S(\boldsymbol{p}|j)\}$ provide the *a posteriori* entropy of the input, given the known outcomes in the output. The latter depend on the probabilities about the inputs $\mathbf{P}(\mathbf{A}|\mathbf{B}) = \{P(a_i|b_j) = P(i|j)\}$, when the signal at the output has already been received.

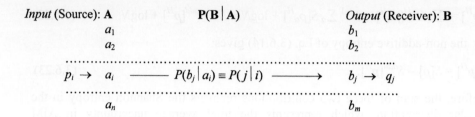

The average conditional entropy $S(\mathbf{A}|\mathbf{B})$ measures the indeterminacy of the source with respect to the receiver. Similarly, the conditional entropy $S(\mathbf{B}|\mathbf{A}) \equiv S(\boldsymbol{q}|\boldsymbol{p})$ provides a measure of the uncertainty of the receiver relative to the source. Thus, an observation of a single output signal provides on average the amount of information given by the difference between the *a priori* and *a posteriori*

uncertainties, $S(\boldsymbol{p}) - S(\boldsymbol{p}|\boldsymbol{q}) = I(\boldsymbol{p}:\boldsymbol{q}) \equiv I(\mathbf{A}:\mathbf{B})$ (see Section 3.5), which defines the *mutual information* in the source and receiver probability vectors. In other words, the mutual information measures the amount of information transmitted through the communication channel.

Example 3.13. As an illustration let us consider the *Symmetric Binary Channel* (SBC) shown in Scheme 3.2. It consists of two inputs and two outputs. We assume the input probabilities $\boldsymbol{p} = (x, 1 - x)$, and the symmetric conditional probability matrix $\mathbf{P}(\mathbf{B}|\mathbf{A}) = \{P(1|2) = P(2|1) \equiv \omega$, and $P(1|1) = P(2|2) = 1 - \omega\}$. Note that by the normalization of conditional probabilities the sum of all conditional probabilities in each row of this matrix must give 1. These assumptions also imply the following joint probability matrix: $\mathbf{P}(\mathbf{AB}) = \boldsymbol{\pi} = \{\pi_{1,1} = p_1 P(1|1) = x(1 - \omega),\ \pi_{1,2} = p_1 P(2|1) = x\omega,$ $\pi_{2,1} = p_2 P(1|2) = (1 - x)\omega,\ \pi_{2,2} = p_2 P(2|2) = (1 - x)(1 - \omega)\}$, and the output probability vector $\boldsymbol{q} = (q_1 = \pi_{1,1} + \pi_{2,1},\ q_2 = 1 - q_1 = \pi_{1,2} + \pi_{2,2})$.

The SBC entropies of the source and receiver probability vectors are determined by the binary entropy function $H(x)$ shown in Fig. 3.2,

$$S(\mathbf{A}) = -x \log x - (1 - x) \log (1 - x) \equiv H(x), \tag{3.7.1}$$

$$S(\mathbf{B}) = H(z), \quad z(x, \omega) \equiv q_2 = x\omega + (1 - x)(1 - \omega) \quad \text{(see Scheme 3.2)}, \tag{3.7.2}$$

and so is the conditional entropy $S(\mathbf{B}|\mathbf{A})$:

$$S(\mathbf{B}|\mathbf{A}) = x\, H(\omega) + (1 - x)\, H(\omega) = H(\omega). \tag{3.7.3}$$

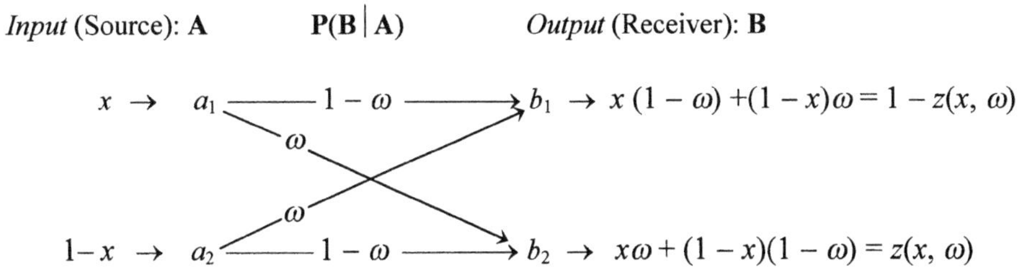

Scheme 3.2. The symmetric binary channel (SBC).

Hence, the entropy of the joint input-output events

$$S(\mathbf{AB}) \equiv S(\boldsymbol{\pi}) = S(\mathbf{A}) + S(\mathbf{B}) - S(\mathbf{A}:\mathbf{B}) = S(\mathbf{A}) + S(\mathbf{B}|\mathbf{A}) = S(\mathbf{B}) + S(\mathbf{A}|\mathbf{B})$$
$$= H(x) + H(\omega), \tag{3.7.4}$$

and the mutual information between the system inputs and outputs [Eq. (A.2.11)]:

$$I(\mathbf{A{:}B}) = S(\mathbf{A}) - S(\mathbf{A}|\mathbf{B}) = S(\mathbf{B}) - S(\mathbf{B}|\mathbf{A}) = H[z(x, \omega)] - H(\omega). \quad (3.7.5)$$

These relations are illustrated in Fig. 3.2. It should be observed that z always lies between ω and $(1 - \omega)$, so that $H(z) = H(1 - z) \geq H(\omega) = H(1 - \omega)$. This demonstrates the non-negative character of the mutual information.

The amount of information flowing through the SBC, $I(\mathbf{A{:}B})$, thus depends on both the conditional probability parameter ω, characterizing the communication system itself, and on the input probability parameter x, which determines the probabilities of the input signals, i.e., the way the channel is used. Let us qualitatively examine the x-dependence of this quantity. For $x = 0$ (or 1) $H(z) = H(\omega)$ and thus $I(\mathbf{A{:}B}) = 0$, i.e., there is no flow of information from the source to the receiver whatever. For $x = ½$, when the two input signals are equally probable, one finds that $H(z) = 1$ bit, thus giving rise to the maximum value of the channel mutual information, determining the so called system transmission *capacity*:

$$C(\omega) \equiv \max_p I(\mathbf{A{:}B}) = \max_x \{H[z(x, \omega)] - H(\omega)\} = 1 - H(\omega) \text{ (bits)}. \quad (3.7.6)$$

This definition implies that the capacity of a communication system is determined solely by the conditional probabilities of the communication channel itself, being independent of the input probabilities, which are fixed by the condition of the mutual information maximum. It follows from the preceding equation that for $\omega = ½$, when both output signals are equally probable for a given input signal, the information capacity of SBC identically vanishes.

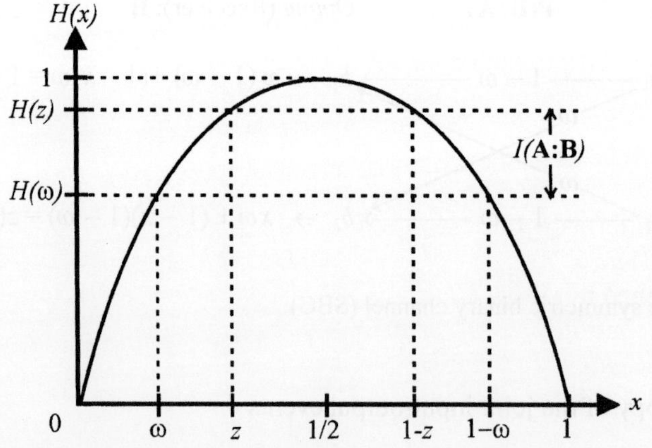

Figure 3.2. The binary entropy function (in bits), $H(x) = -x \log_2 x - (1 - x) \log_2 (1 - x)$, and the geometric interpretation of the conditional entropy $S(\mathbf{B}|\mathbf{A}) = H(\omega)$ and mutual information $I(\mathbf{A{:}B}) = H(z) - H(\omega)$.

Next, let us briefly summarize selected examples of communication systems, determined by the specific structure of their conditional probability matrices $\mathbf{P}(\mathbf{B}|\mathbf{A}) = \{P(j|i)\}$, of observing a signal in the output b_j given the input a_i, or $\mathbf{P}(\mathbf{A}|\mathbf{B}) = \{P(i|j)\}$, of observing the input a_i given the output b_j. The so called *noiseless* (*n*) system includes only one non-zero element in each *column* of $\mathbf{P}(\mathbf{B}|\mathbf{A})$, so that by observing the signal at a given output one knows for sure, from which input it has originated. Hence, the conditional probabilities $\mathbf{P}_n(\mathbf{A}|\mathbf{B}) = \{P_n(i|j)\}$ are either 1 or 0 and

$$S_n(\mathbf{A}|\mathbf{B}) = -\sum_j q_j \sum_i P_n(i|j) \log P_n(i|j) = 0 \quad \text{and} \quad I_n(\mathbf{A}{:}\mathbf{B}) = S(\mathbf{A}). \tag{3.7.7}$$

Indeed, in this channel the knowledge of the output signals gives no additional information about the input signals, so that amount of information transmitted through the noiseless system is equal to the uncertainty of the input probability scheme, as measured by the *a priori* entropy.

In the *deterministic* (*d*) system only a single element in each *row* of $\mathbf{P}_d(\mathbf{B}|\mathbf{A})$ is different from zero. Since, by the normalization of conditional probabilities, the sum of elements in a given row has to be equal to 1, each row then contains 1 and $(m-1)$ zeros. In other words, it is know with certainty, to which output a given input signal is transmitted, although several inputs may contribute to a single output. Thus, the conditional probabilities $\mathbf{P}_d(\mathbf{B}|\mathbf{A})$ are either 1 or 0, which gives $S_d(\mathbf{B}|\mathbf{A}) = 0$ and hence $I_d(\mathbf{A}{:}\mathbf{B}) = S(\mathbf{B})$. Therefore, the amount of information flowing through the deterministic channel is equal to the output entropy.

The conditional probability matrix $\mathbf{P}_u(\mathbf{B}|\mathbf{A})$, in which the elements of each row are permutations of the elements of the first row, describes the so called *uniform* (*u*) communication system. The simplest example of such a communication network is provided by the SBC of Scheme 3.2. In uniform channels the equi-probable input signals give rise to the equi-probable output signals. Moreover, the amount of information flowing through the uniform channel can be expressed as:

$$\begin{aligned} I_u(\mathbf{A}{:}\mathbf{B}) &= S(\mathbf{B}) - S_u(\mathbf{B}|\mathbf{A}) = S(\mathbf{B}) + \sum_i p_i \left[\sum_j P_u(j|i) \log P_u(j|i)\right] \\ &= S(\mathbf{B}) + \sum_j P_u(j|i) \log P_u(j|i) \equiv S(\mathbf{B}) - S_u(\mathbf{B}|i), \end{aligned} \tag{3.7.8}$$

since for the uniform system the internal sum, over columns j, is the same for all rows and $\sum_i p_i = 1$. Hence, $S_u(\mathbf{B}|i) = S_u(\boldsymbol{q}|i)$ is independent of i. The maximum of $I_u(\mathbf{A}{:}\mathbf{B})$ is thus determined by the maximum of $S(\mathbf{B})$. For the equi-probable input signals $\{p_i = 1/n\}$ the output entropy $S(\mathbf{B})$ of the uniform system reaches the maximum value, $S(\mathbf{B}) = \log m$, where m denotes the number of outputs in $\mathbf{B}$, so that the capacity of the uniform channel is given by the expression:

$$C_u = \log m - S_u(\mathbf{B}|i). \tag{3.7.9}$$

Communication channels can be simplified by appropriate *reductions*, carried out by combining several inputs and/or outputs into a single unit. This manipulation diminishes or, at best, leaves unchanged the amount of information for the original input and output signals of the primary channel (before reduction). The particular reductions, which preserve in the reduced channel the amount of information of the primary channel, are called the *satisfactory reductions.* For the explicit conditions, which must be satisfied by the satisfactory *elementary* reductions of two signals, when they become a single signal, the reader is referred to Abramson (1963). Later in the book (Section 8.9) we shall apply the reduced communication channels of molecular fragments to extract the internal and external descriptors of the chemical bonds of molecular subsystems (Nalewajski, 2005a).

3.8. PRINCIPLE OF THE EXTREME PHYSICAL INFORMATION

The ME and MED variational principles for the extreme entropy/information content in the optimized probability distribution p can be generalized in terms of the Fisher information. In fact in the *Generalized Information Principle* (GIP) (Brillouin, 1956; Frieden, 2000) one optimizes any admissible, statistically acceptable information functional $I[p;\, \boldsymbol{q}^0]$ defined with respect to references $\boldsymbol{q}^0 = \{q_\alpha{}^0\}$ and subject to the relevant normalization and physical constraints $\{F_i[p] = F_i^0\}$:

$$\delta\{I[p;\, \boldsymbol{q}^0] - \sum_i \lambda_i F_i[p]\} \equiv \delta K[p;\, \boldsymbol{q}^0] \equiv \delta I[p;\, \boldsymbol{q}^0] - \delta J[p] = 0. \qquad (3.8.1a)$$

This principle may involve several optimized probability distributions $\boldsymbol{p} = \{p_\beta\}$ in multi-component systems:

$$\delta\{I[\boldsymbol{p};\, \boldsymbol{q}^0] - \sum_i \lambda_i F_i[\boldsymbol{p}]\} \equiv \delta K[\boldsymbol{p};\, \boldsymbol{q}^0] \equiv \delta I[\boldsymbol{p};\, \boldsymbol{q}^0] - \delta J[\boldsymbol{p}] = 0. \qquad (3.8.1b)$$

This general principle of the *Extreme Physical Information* (EPI) $K[p;\, \boldsymbol{q}^0]$ (or $K[\boldsymbol{p};\, \boldsymbol{q}^0]$) consists of two additive components: the *intrinsic* information $I[p;\, \boldsymbol{q}^0]$ (or $I[\boldsymbol{p};\, \boldsymbol{q}^0]$) and the *bound* (constrained) information $J[p]$ (or $J[\boldsymbol{p}]$). The latter usually represents the information associated with the specific physical parameters under measurement, thus characterizing the effect of the measurement process, whereas the former always has the same form, regardless of the physical parameter that is being measured. Actually, to derive the fifferential equations of physics one requires $I[p] = I^F[p]$ or $I[\boldsymbol{p}] = I^F[\boldsymbol{p}]$ (Frieden, 2000). The EPI roots of the fundamental physical laws demonstrate the unifying character of the Fisher information concept in physics. The EPI approach allows one to view the physical laws within a unified framework of the measurement theory. In fact, the Fisher information, representing the limiting form of many different measures of information, can be regarded as a kind of "mother" information. The information contained in the acquired data originates from the physical phenomenon or system, which is probed by the measurement. Each measurement effects transfer of the Fisher information from the phenomenon to the instrument intrinsic data. As demonstrated by Brillouin (1956) the Second Law of thermodynamics requires that the time averages of the two information contributions of the physical information in Eq. (3.8.1) satisfy the inequality $J \geq I$. Prior to a measurement of a given physical quantity the system has a bound information measure J corresponding to the specified physical data defining the constraints (input). Then a measurement takes place, which assimilates the bound information in the intrinsic (data) information, initiating an information "transition" $J \rightarrow I$, from the phenomenon ("input") to the measuring device ("output"), with the Brillouin inequality indicating some loss, in time, of information in this transformation. The perturbation of the system due to the measurement process perturbs J and thus also I by amounts δJ and δI, respectively, which at the EPI solution point satisfy the conservation law of the information change: $\delta I = \delta J$.

Example 3.14. The stationary Schrödinger equation (1.5.3), the basic relation of the molecular quantum mechanics, provides an example of the constrained information principle (Sears, 1980). In the Born-Oppenheimer approximation, for the fixed positions of the nuclei, it represents the eigenvalue problem of the electronic energy operator $\hat{\mathrm{H}}(N,v) = \hat{\mathrm{T}}(N) + \hat{\mathrm{V}}(N,v)$ [Eq. (1.5.4)}, for N particles moving in the external potential $v(\boldsymbol{r})$ due to the "frozen" nuclear framework [Eq. (1.5.2)]. As we have already demonstrated in Section 1.5, it results from the wave-function variational principle for the trial wave function $\Psi(\mathbf{x})$ subject to the normalization constraint $\langle\Psi|\Psi\rangle = 1$. This constraint can be alternatively interpreted as the normalization condition of the N-electron probability distribution:

$$D(\mathbf{x}) = \Psi^*(\mathbf{x})\Psi(\mathbf{x}), \qquad \langle\Psi|\Psi\rangle = \int D(\mathbf{x})\, d\mathbf{x} = 1. \tag{3.8.2}$$

On partial integration the expectation value of the kinetic energy in state $\Psi(\mathbf{x})$,

$$T = T[\Psi] \equiv \langle\Psi|\hat{\mathrm{T}}|\Psi\rangle = -\tfrac{1}{2}\textstyle\sum_i \int \Psi^*(\mathbf{x})\nabla_i^2\Psi(\mathbf{x})\, d\mathbf{x}, \tag{3.8.3}$$

can be transformed into the equivalent, Fisher-type functional of the many-body probability density:

$$T = T[D] = (N/8) \int [\nabla_1 D(\mathbf{x})]^2/D(\mathbf{x})\, d\mathbf{x} \equiv (N/8)\, I^F[D], \tag{3.8.4}$$

called the multivariate density functional for the kinetic energy (Sears *et al.*, 1980).

By the marginal decomposition of this probability distribution, in terms of the one-electron probability density $p(\boldsymbol{x}_1)$ and the associated many-electron conditional probability density $P(\boldsymbol{x}_2, \ldots, \boldsymbol{x}_N|\boldsymbol{x}_1) = D(\mathbf{x})/p(\boldsymbol{x}_1)$ of the remaining $(N-1)$ electrons,

$$D(\mathbf{x}) = p(\boldsymbol{x}_1)\, [D(\mathbf{x})/p(\boldsymbol{x}_1)] \equiv p(\boldsymbol{x}_1)\, P(\boldsymbol{x}_2, \ldots, \boldsymbol{x}_N|\boldsymbol{x}_1),$$
$$\int\ldots\int P(\boldsymbol{x}_2, \ldots, \boldsymbol{x}_N|\boldsymbol{x}_1)\, d\boldsymbol{x}_2 \ldots d\boldsymbol{x}_N \equiv \int P(\mathbf{x}'|\boldsymbol{x}_1)\, d\mathbf{x}' = 1. \tag{3.8.5}$$

The one-electron probability distribution is the shape-factor of the electron spin-density $\rho(\boldsymbol{x}_1)$, $p(\boldsymbol{x}_1) = \rho(\boldsymbol{x}_1)/N = \int D(\mathbf{x})\, d\mathbf{x}'$. In the conditional probability distribution $P(\boldsymbol{x}_2, \ldots, \boldsymbol{x}_N|\boldsymbol{x}_1) \equiv P(\mathbf{x}'|\boldsymbol{x}_1)$, the coordinates $\mathbf{x}'$ are variables, while those of electron "1", $\boldsymbol{x}_1$, have been down-graded to parameters. Using the last equation one can extract the part of $T[D]$ which depends on the one-electron distribution:

$$T[D] = (N/8)\, \{\int [\nabla_1 p(\boldsymbol{x}_1)]^2/p(\boldsymbol{x}_1)\, d\boldsymbol{x}_1 + \int p(\boldsymbol{x}_1)(\int[\nabla_1 P(\mathbf{x}'|\boldsymbol{x}_1)]^2/P(\mathbf{x}'|\boldsymbol{x}_1)\, d\mathbf{x}')\, d\boldsymbol{x}_1\}$$
$$\equiv (N/8)\{I^F[p] + \int\ldots\int p(\boldsymbol{x}_1)\, I^F[P; \boldsymbol{x}_1]\, d\boldsymbol{x}_1\}. \tag{3.8.6}$$

Therefore, the kinetic energy is given by the Fisher information contained in the many-body probability density, which can be rigorously partitioned into the one-electron density functional, proportional to von Weizsäcker's (1935) non-homogeneity correction, and the average of Fisher information in the conditional probability density of remaining electrons taken over the marginal distribution $p(\boldsymbol{x}_1)$.

Clearly, the average total potential energy of N electrons, $V[\Psi] = \langle\Psi|\hat{V}(N,v)|\Psi\rangle$, the expectation value of the N-electron multiplicative operator $\hat{V}(N,v) = \hat{V}_{ne}(N,v) + \hat{V}_{ee}(N)$ [see Eq. (1.5.4)], can be also expressed as a straightforward functional of the N-electron probability density D,

$$V[D] = \int \hat{V}(\mathbf{x}, v)\, D(\mathbf{x})\, d\mathbf{x}, \tag{3.8.7}$$

thus giving the associated functional for the expectation value of the total electronic energy:

$$E[D] = T[D] + V[D]\,. \tag{3.8.8}$$

Therefore, the Schrödinger variational principle of Eq. (1.5.7) can be also interpreted as the information principle for the many-electron probability density:

$$\delta\,\{(N/8)I^F[D\,] + V_e[D] - E[N, v] \int D(\mathbf{x})\, d\mathbf{x}\,\} = 0, \tag{3.8.9a}$$

with the exact electronic energy $E[N, v]$ playing the role of the Lagrange multiplier associated with the probability density normalization [Eq.(3.8.2)] and the multiplicative operator $\hat{V}(\mathbf{x}, v) = V(N, v)$ representing the N-electron Lagrange multiplier function, including all one- and two-body terms, which enforces the potential energy constraint: $V[D] = V[N, v]$. In this variational principle the Fisher measure of the information contained in the N-electron distribution is optimized, subject to the potential energy and normalization constraints. The preceding information principle can be alternatively interpreted as the Schrödinger(S) EPI principle of Eq. (3.8.1) with the bound information being related to the "experiment" measuring the N-electron density:

$$\delta I^F[D\,] \equiv \delta I^S[\Psi(D)] = (8/N)\delta\{E[N, v] \int D(\mathbf{x})\, d\mathbf{x} - V_e[D]\} \equiv \delta J^S[\Psi(D)]. \tag{3.8.9b}$$

Example 3.15. We shall now demonstrate that the Kohn-Sham (1965) equations of computational DFT (see Appendix C) also result from the EPI principle (Nalewajski, 2003c). Following Frieden (2000) we assume the intrinsic data information functional in the multi-component Fisher form $I^F[\psi]$ [Eq. (3.4.7)], where $\psi = \{\psi_n\}$ groups the singly occupied spin-orbitals, since in KS theory each orbital probability density constitutes a distinct component of the overall one–electron probability distribution, in the spirit of the combination scenario of Eq. (3.6.8). The constraint (phenomenon) part $J[\psi]$ of the physical information functional, $K[\psi] = I_F[\psi] - J[\psi]$, then uniquely defines the EPI problem for the molecular system in question.

Consider the key one-body problem of KS theory (Appendix C) of N *non-interacting* electrons moving in an effective (local) external potential $v_{KS}(\boldsymbol{r})$, which also determines the ground-state density of N *interacting* electrons moving in the external potential $v(\boldsymbol{r})$ due to the system nuclei. It describes electrons by a single KS determinant $\Psi^{KS} \equiv |\psi| = \det\{\varphi_n \chi_n\}$ constructed from N orthonormal, singly-occupied molecular spin-orbitals defined by the (real) spatial functions [the molecular orbitals (MO)] $\boldsymbol{\varphi} = \{\varphi_n\}$ and the corresponding spin functions $\boldsymbol{\chi} = \{\chi_n\}$.

By the Hohenberg-Kohn (1964) theorem $v_{KS}(\boldsymbol{r})$ is the unique functional of the ground–state electron density of the non-interacting system, $v_{KS}(\boldsymbol{r}) = v_{KS}[\boldsymbol{r}; \rho]$, where

$$\rho(\boldsymbol{r}) = \sum_n \varphi_n(\boldsymbol{r})^2 \equiv \sum_n \rho_n(\boldsymbol{r}), \tag{3.8.10}$$

which by hypothesis is equal to that of the real system of N interacting electrons.

The density functional for the electronic energy, $E_v[\rho]$ [Eq. (1.5.14)], consists of the universal Hohenberg-Kohn-Levy functional $F[\rho] = T[\rho] + V_{ee}[\rho]$ [Eqs. (1.5.14), (1.5.15)] and the external potential energy $V_{ne}[\rho]$. The total kinetic energy of electrons includes both the non-interacting (s) and correlation (c) parts:

$$T[\rho] = T_s[\rho] + T_c[\rho]. \tag{3.8.11}$$

The electron repulsion energy of the interacting system,

$$V_{ee}[\rho] = V_{ee}^{class}[\rho] + (E_{xc}[\rho] - T_c[\rho]), \tag{3.8.12}$$

is expressed in the KS theory as the sum of the classical (Hartree) repulsion energy,

$$V_{ee}^{class}[\rho] = \frac{1}{2}\iint \frac{\rho(\boldsymbol{r})\rho(\boldsymbol{r}')}{|\boldsymbol{r}-\boldsymbol{r}'|}\, d\boldsymbol{r}\, d\boldsymbol{r}', \tag{3.8.13}$$

and the corresponding electron-correlation contribution, $E_{xc}[\rho] - T_c[\rho]$, where $E_{xc}[\rho]$ stands for the KS *exchange-correlation energy*, which determines the correlation part $v_{xc}(\boldsymbol{r})$ of the effective *KS* potential

$$v_{KS}(\boldsymbol{r}) = v(\boldsymbol{r}) + \int \rho(\boldsymbol{r}')\, |\boldsymbol{r} - \boldsymbol{r}'|^{-1}\, d\boldsymbol{r}' + \delta E_{xc}[\rho]/\delta\rho(\boldsymbol{r}) \equiv v(\boldsymbol{r}) + v_H(\boldsymbol{r}) + v_{xc}(\boldsymbol{r}) \tag{3.8.14}$$

in the effective *KS*-Hamiltonian

$$\hat{\mathrm{H}}_{KS}(\boldsymbol{r}) = -\tfrac{1}{2}\nabla^2 + v_{KS}(\boldsymbol{r}). \tag{3.8.15}$$

To summarize, in the *KS* theory the electronic energy of the interacting system is given by the following density functional:

$$E_v[\rho] = T_s[\rho] + V_{ne}[\rho] + V_{ee}^{class}[\rho] + E_{xc}[\rho] \equiv T_s[\boldsymbol{\varphi}[\rho]] + V_e^{KS}[\rho], \tag{3.8.16}$$

where the *KS* functional for the electronic potential energy $V_e^{KS}[\rho]$ contains the correlation part of the kinetic energy of interacting electrons:

$$V_e^{KS}[\rho] = V_e[\rho] + T_c[\rho] = V_{ne}[\rho] + V_{ee}[\rho] + T_c[\rho]. \tag{3.8.17}$$

The original derivation of the KS equations (see Appendix C), which determine the optimum orbitals of the hypothetical non-interacting system and thus the electron density and energy of the interacting system,

$$\hat{H}_{KS}(\boldsymbol{r})\,\varphi_n(\boldsymbol{r}) = \varepsilon_n\,\varphi_n(\boldsymbol{r}), \qquad n = 1, 2, \ldots, N, \tag{3.8.18}$$

where $\{\varepsilon_n\}$ denote the KS eigenvalues (orbital energies), follow from the Hohenberg-Kohn (1964) variational principle for the system electronic energy of Eq. (3.8.16),

$$\delta\,\{E_v[\rho] - \Sigma_n \Sigma_m\, \Theta_{mn}\langle\varphi_m|\varphi_n\rangle\} \equiv \delta K^{KS}[\boldsymbol{\varphi}[\rho]] = 0 \quad \text{or}$$

$$8\delta T_s[\boldsymbol{\varphi}] \equiv \delta I^{KS}[\boldsymbol{\varphi}[\rho]] = 8\Sigma_n \Sigma_m\, \Theta_{mn}\, \delta\langle\varphi_m|\varphi_n\rangle - 8\,\delta V_e^{KS}[\rho] \equiv \delta J^{KS}[\boldsymbol{\varphi}[\rho]], \tag{3.8.19}$$

where in the canonical representation the Lagrange multipliers, which enforce the MO normalization and orthogonality constraints, become $\Theta_{mn} = \varepsilon_n\delta_{mn}$. The preceding equation identifies the intinsic ($J^{KS}[\boldsymbol{\varphi}[\rho]]$) and bound ($J^{KS}[\rho[\boldsymbol{\varphi}]]$) information terms for the *KS* problem, by a comparison with the EPI relation (3.8.1). Indeed, a straightforward integration by parts shows that the expectation value of the kinetic energy of non-interacting electrons of the *separable* KS problem,

$$T_s[\boldsymbol{\varphi}] = -\tfrac{1}{2}\,\Sigma_n \int \varphi_n(\boldsymbol{r})\nabla^2\varphi_n(\boldsymbol{r})\, d\boldsymbol{r} = \tfrac{1}{2}\,\Sigma_n \int [\nabla\varphi_n(\boldsymbol{r})]^2\, d\boldsymbol{r}, \tag{3.8.20}$$

is proportional to the multi-component Fisher information [Eq. (3.4.7)]:

$$I^F[\boldsymbol{\varphi}] = 4\,\Sigma_n \int [\nabla\varphi_n(\boldsymbol{r})]^2\, d\boldsymbol{r} = \Sigma_n \int [\nabla\rho_n(\boldsymbol{r})]^2/\rho_n(\boldsymbol{r})\, d\boldsymbol{r} = N\,\Sigma_n \int [\nabla p_n(\boldsymbol{r})]^2/p_n(\boldsymbol{r})\, d\boldsymbol{r}$$

$$= 8T_s[\boldsymbol{\varphi}] = I^{KS}[\boldsymbol{\varphi}[\rho]]. \tag{3.8.21}$$

In fact, the KS MO can be considered as the probability amplitudes which generate the *conditional* probability distributions of Eq. (3.6.9), $\boldsymbol{\pi}(\boldsymbol{r}) = \{\pi(\boldsymbol{r}|n) = \varphi_n(\boldsymbol{r})^2 = \rho_n(\boldsymbol{r}) \equiv N\,p_n(\boldsymbol{r})\}$ and the system overall electron density $\rho(\boldsymbol{r}) = \Sigma_n\,\rho_n(\boldsymbol{r})$ [Eq. (3.8.10)].

The density "measurement" determines the "phenomenon" of the KS-EPI principle, which effects a transfer of the Fisher information from the *input*, specified by *bound information* $J[\boldsymbol{\varphi}] = J^{KS}[\boldsymbol{\varphi}[\rho]]$, to the *intrinsic information* $I^F[\boldsymbol{\varphi}] = I^{KS}[\boldsymbol{\varphi}[\rho]]$, which represents the kinetic energy of thew non-interacting electrons. By the Hohenberg-Kohn variational principle, the optimum KS orbitals, which mark the solution point of the underlying KS-EPI principle, are unique functionals of the system electron density: $\boldsymbol{\varphi} = \boldsymbol{\varphi}[\rho]$.

Therefore, the *KS* variational principle (3.8.19) can indeed be interpreted as the EPI principle of Eq. (3.8.1):

$$\delta\{I^{KS}[\boldsymbol{\varphi}[\rho]] - J^{KS}[\boldsymbol{\varphi}[\rho]]\} = \delta K^{KS}[\boldsymbol{\varphi}[\rho]] = 0. \tag{3.8.22}$$

4. PROBING THE MOLECULAR ELECTRON DISTRIBUTIONS

The molecular ground-state density ρ and the promolecular electron distribution ρ^0, representing the "frozen" free-atom density components in positions of AIM, and the associated one-electron probability distributions $p = \rho/N$ and $p^0 = \rho^0/N^0$, are used to diagnose the effects due to the chemical bonds in molecular systems. The difference function $\Delta\rho = \rho - \rho^0$, representing the density displacement in a molecule relative to the promolecular reference, which is commonly used as diagnostic tool to extract the effects due to the chemical bonds, is related to the information distance quantities resulting from the Shannon entropy. The approximate relations between the alternative entropy-deficiency densities and $\Delta\rho$ are derived using the first-order Taylor expansion of the surprisal in the missing information functional, thus ascribing to the molecular density reconstruction the complementary information-theoretic interpretation. The molecular entropy displacement and cross-entropy quantities are used as probes of changes in the electron uncertainty distribution during the bond formation process. These quantities are compared for selected linear molecules exhibiting both single and multiple chemical bonds and varying covalent/ionic composition. The global results testify to the smallness of the reconstruction of the electronic structure, which accompanies the formation of chemical bonds in molecules, while the information-distance plots, strongly resembling those of the density displacement themselves, are shown to reflect all familiar aspects of the "promotion" of the free atoms to their corresponding valence states in a molecule. The entropy-displacement and missing-information diagrams provide complementary, information-theoretic interpretation of the familiar density difference plots, and are advocated as sensitive diagnostic tools for detection of specific aspects of the chemical bonds. Illustrative application of such an information-theoretic analysis to problem of the central bond in propellanes is also reported. Finally, the key ingredient of the Electron Localization Function (ELF) of DFT, which has been designed to visualize the electronic shells and the distribution of bonding and nonbonding pairs of the valence electrons in molecules, is shown to measure the non-additive Fisher information in the MO resolution. Alternative, information-theoretic ELF is proposed and tested against the original ELF for typical molecules.

4.1. ENTROPY–DEFICIENCY DESCRIPTORS OF MOLECULAR ELECTRON DENSITIES

The densities $\{\rho_i^0\}$ of the separated atoms define the molecular (isoelectronic) prototype called the atomic *"promolecule"* (Hirshfeld, 1977), given by the sum of the free-atom electron densities shifted to the actual locations of atoms in the molecule (see Example 3.11). The resulting electron density $\rho^0 = \sum_I \rho_i^0$ of this collection of the "frozen" atomic electron distributions defining the initial stage in the bond-formation process, determines a natural reference for extracting changes due to the chemical bonds. Indeed, the familiar *density difference function,* $\Delta\rho = \rho - \rho^0$, has been widely used to probe changes in the electronic structure, which are responsible for the chemical bonds in the molecule. That atomic ground-states, and/or small perturbation thereof, are uniquely appropriate reference states for a detailed description of AIM, was well understood by the pioneers, for instance Pauling (1960) and Mulliken (1935, 1955) in their use of the concepts of hybridization, promotion, polarization and ionic character.

In this chapter we shall use the information-theoretic concepts to explore the molecular ground-state electron distributions $\rho(\boldsymbol{r})$ or their shape (probability)-factors $p(\boldsymbol{r})=\rho(\boldsymbol{r})/N$ [see Eqs. (1.5.12) and (2.4.16)], generated by the Kohn-Sham DFT calculations. Consider the density $\Delta s(\boldsymbol{r})$ of the entropy-deficiency (directed-divergence) functional of Kullback and Leibler (1951) [Eq. (3.3.2)], between the molecular electron distribution $\rho(\boldsymbol{r})$ and the promolecular density $\rho^0(\boldsymbol{r})$,

$$\Delta S[\rho\,|\,\rho^0] = \int \rho(\boldsymbol{r}) \ln [\rho(\boldsymbol{r})/\rho^0(\boldsymbol{r})]\, d\boldsymbol{r} = \int \rho(\boldsymbol{r})\, I[w(\boldsymbol{r})]\, d\boldsymbol{r} \equiv \int \Delta s(\boldsymbol{r})\, d\boldsymbol{r}. \tag{4.1.1}$$

It represents the renormalized missing information between the shape-factors $p(\boldsymbol{r})$ and $p^0(\boldsymbol{r})=\rho^0(\boldsymbol{r})/N^0$, $N=N^0$, of the two compared electron densities:

$$\Delta s(\boldsymbol{r}) \equiv \Delta s[\rho(\boldsymbol{r})|\rho^0(\boldsymbol{r})] = N\, p(\boldsymbol{r}) \log [p(\boldsymbol{r})/p^0(\boldsymbol{r})] \equiv N\Delta s[p(\boldsymbol{r})|p^0(\boldsymbol{r})], \tag{4.1.2}$$

where $w(\boldsymbol{r})$ denotes the local density/probability *enhancement factor* and $I[w(\boldsymbol{r})]$ is the corresponding *"surprisal"*.

We shall demonstrate that this local information distance or the related Kullback measure [Eq.(3.3.7)],

$$\Delta S[\rho, \rho^0] = \int[\rho(\boldsymbol{r}) - \rho^0(\boldsymbol{r})] \ln[\rho(\boldsymbol{r})/\rho^0(\boldsymbol{r})]\, d\boldsymbol{r} = \int\Delta\rho(\boldsymbol{r})\, I[w(\boldsymbol{r})]\, d\boldsymbol{r} \equiv \int\Delta D(\boldsymbol{r})\, d\boldsymbol{r}, \tag{4.1.3}$$

where the divergence density

$$\Delta D(\boldsymbol{r}) \equiv \Delta s[\rho(\boldsymbol{r}), \rho^0(\boldsymbol{r})] = \Delta\rho(\boldsymbol{r}) I[w(\boldsymbol{r})] = Np(\boldsymbol{r})\, I[w(\boldsymbol{r})] = N\Delta s[p(\boldsymbol{r}), p^0(\boldsymbol{r})], \tag{4.1.4}$$

are related to the molecular density difference function, which is used to interpret the electronic origins of the chemical bond.

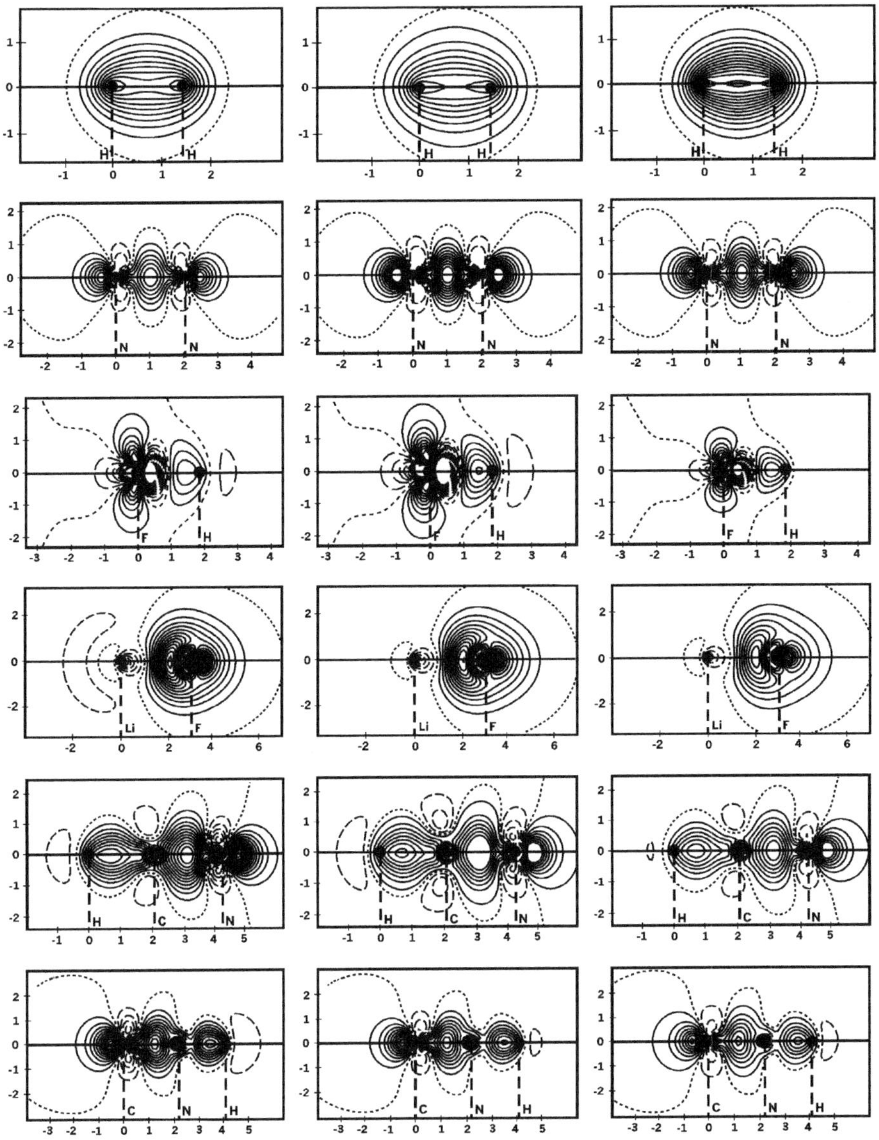

Figure 4.1. The contour diagrams of the molecular density difference function, $\Delta\rho(\boldsymbol{r}) = \rho(\boldsymbol{r}) - \rho^0(\boldsymbol{r})$ (first column), the information-distance density, $\Delta s(\boldsymbol{r}) = \rho(\boldsymbol{r})\, I[w(\boldsymbol{r})]$ (second column) and its approximate, first-order expansion, $\Delta s(\boldsymbol{r}) \cong \Delta\rho(\boldsymbol{r})\, w(\boldsymbol{r})$ (third column), for selected diatomic and linear triatomic molecules: H_2, HF, LiF, HCN and HNC. The solid, pointed and broken lines denote the positive, zero and negative values, respectively, of the equally spaced contours. The same convention is applied in Figs. 4.2-6 (Nalewajski *et al.*, 2002).

We shall also examine the molecular displacements of the Shannon entropy [see Eq. (3.1.4)] relative to the free-atom, promolecular reference,

$$\mathcal{H}[\rho] \equiv S[\rho] - S[\rho^0] = -\int \rho(\boldsymbol{r}) \ln\rho(\boldsymbol{r})\, d\boldsymbol{r} + \int \rho^0(\boldsymbol{r}) \ln\rho^0(\boldsymbol{r})\, d\boldsymbol{r} \equiv \int h_\rho(\boldsymbol{r})\, d\boldsymbol{r}, \qquad (4.1.5a)$$

and its density $h_\rho(\boldsymbol{r})$ as alternative probes of the chemical bond effects in molecules. The corresponding entropy shifts in terms of the normalized probability distributions

$$\mathcal{H}[p] \equiv S[p] - S[p^0] = -\int p(\boldsymbol{r}) \ln p(\boldsymbol{r})\, d\boldsymbol{r} + \int p^0(\boldsymbol{r}) \ln p^0(\boldsymbol{r})\, d\boldsymbol{r} \equiv \int h_p(\boldsymbol{r})\, d\boldsymbol{r}, \qquad (4.1.5b)$$

can also be used to explore the electron uncertainty relaxation, which accompanies the bond formation in molecules. In fact all these entropy/information and electron density displacement quantities can be regarded as complementary tools for diagnosing the chemical bonds and effective valence states of the bonded atoms in the molecular environment (Nalewajski *et al.*, 2002; Nalewajski and Broniatowska, 2003a).

It follows from Eq. (4.1.1) that the molecular surprisal $I[w(\boldsymbol{r})]$ measures the density-per-electron of the system entropy-deficiency:

$$I[w(\boldsymbol{r})] = \Delta s(\boldsymbol{r})/\rho(\boldsymbol{r}). \qquad (4.1.6)$$

It can be also interpreted [see Eq. (4.1.4)] as the density-per-electron of the molecular divergence:

$$I[w(\boldsymbol{r})] = \Delta D(\boldsymbol{r})/\Delta\rho(\boldsymbol{r}). \qquad (4.1.7)$$

In the first column of Fig. 4.1 we have displayed contour maps of the density difference function $\Delta\rho(\boldsymbol{r})$ for typical linear diatomic and triatomic molecules. They exhibit all typical aspects of the equilibrium reconstructions of free-atoms during formation of the single and multiple chemical bonds with varying degree of bond covalency and ionicity. Consider first the purely covalent bond in the homonuclear diatomics. The single bond in H_2 gives rise to a relative accumulation of electrons in the bond region, between the two nuclei. The triple-bond pattern for N_2 is more complex, reflecting the density accumulation in the bonding region due to both σ and π bonds, and the accompanying increase in the density of the lone pairs on both nitrogen atoms, due to their expected *sp*-hybridization in the valence state. One also observes the electron density decrease in the vicinity of the nuclei and an outflow of electrons from the $2p_\pi$ atomic orbitals, a clear sign of their involvement in the π bond system. The remaining two diatomics, HF and LiF, represent partially-ionic bonds, between the two atoms exhibiting small and large differences, respectively, in their electronegativity (chemical hardness) descriptors. One observes a pattern of density displacement in HF, which corresponds to a weakly ionic (strongly covalent) bond, while in LiF the two AIM are seen to be connected by the strongly ionic (weakly covalent) bond. Indeed, in HF one detects a relatively high degree of a common possession of the valence electrons by the two atoms, which significantly contribute to the shared bond-charge located between them, and a comparatively weak H→F polarization. In LiF a substantial Li→F electron transfer can be seen so that the ion-pair picture provides an adequate description of the chemical bond in this system. These observations are further supported by the effective charges $\{q_X^H\}$ of the stockholder AIM which are reported in Table 4.1 together with the AIM entropy deficiencies $\{\Delta S_X[p_X^H|p_X^0]\}$ and the global missing information $\Delta S[p|p^0]$ comparing the molecular and promolecular probability distributions. Their rather low magnitudes confirm a strong similarity between molecular and promolecular electron densities.

Table 4.1. Representative net charges of the stockholder AIM (see Example 3.11), $\{q_X^H = N_X^H - Z_X\}$ (a.u.), where Z_X is the charge of the nucleus X, the AIM entropy deficiencies $\Delta S_X[p_X^H|p_X^0]$, and global entropy deficiency $\Delta S[p|p^0]$ (in bits), for the linear molecules of Fig. 4.1. (Nalewajski *et al.*, 2002).

Molecule	X	q_X^H	$\Delta S_X[p_X^H\|p_X^0]$	$\Delta S[p\|p^0]$
H_2	H	0.00	0.056	0.056
N_2	N	0.00	0.006	0.006
HF	H	0.24	0.144	0.020
	F	−0.24	0.005	
LiH	Li	0.35	0.157	0.136
	H	−0.35	0.012	
LiF	Li	0.58	0.244	0.063
	F	−0.58	0.007	
LiCl	Li	0.53	0.212	0.033
	Cl	−0.53	0.003	
HCN	H	0.14	0.104	0.017
	C	0.03	0.015	
	N	−0.17	0.005	
HNC	H	0.20	0.110	0.018
	N	−0.10	0.008	
	C	−0.10	0.011	
HNCS	H	0.19	0.114	0.008
	N	−0.13	0.007	
	C	0.05	0.008	
	S	−0.11	0.002	
HSCN	H	0.22	0.088	0.008
	S	−0.07	0.002	
	C	0.04	0.008	
	N	−0.19	0.004	

In the two triatomic molecules shown in Fig. 4.1 one similarly finds a strongly covalent pattern of the electron density displacements in the regions of the single N–H and C–H atoms. A typical buildup of the bond charge due to the multiple CN bonds in the two isomers HCN and HNC can be also observed. The increase in the lone-pair electron density on the terminal heavy atom, N in HCN and or C in HNC, can be also detected thus confirming the expected *sp*-hybridization of these bonded atoms in their promoted, valence state.

After this short survey of the main relaxation trends in the overall electron density, which accompany the formation of chemical bonds in these illustrative molecular systems, let us examine the associated maps showing adjustments in the information-distance density relative to the promolecular reference. They are displayed in the second column of the Fig. 4.1. These entropy-deficiency diagrams, relative to the collection of the free-atoms of the promolecule, provide the electron uncertainty representation of the density-difference function. A comparison between the corresponding panels of the first two columns in the figure shows that the two displacement maps so strongly resemble one another that they are hardly distinguishable. This confirms a close relation between the local density and entropy-deficiency relaxation patters, thus attributing to the former a complementary information-theoretic interpretation. A strong resemblance between these two types of molecular diagrams indicates that the local inflow of electrons increases the relative entropy, while the outflow of electrons gives rise to a diminished level of the relative-uncertainty content of the electron distribution in the molecule. The density displacement and the missing-information distribution can be thus viewed as complementary probes of the system chemical bonds.

4.2. APPROXIMATE RELATIONS IN TERMS OF THE DENSITY DIFFERENCE FUNCTION

We have seen in the preceding section that molecular electron density $\rho(\boldsymbol{r})$ is only slightly modified relative to the promolecular distribution $\rho^0(\boldsymbol{r})$, $\rho(\boldsymbol{r}) \approx \rho^0(\boldsymbol{r})$ or $w(\boldsymbol{r}) \approx 1$. Indeed, the formation of chemical bonds involves only a minor reconstruction of the electronic structure, mainly in the valence-shells of the constituent AIM:

$$|\Delta\rho(\boldsymbol{r})| \equiv |\rho(\boldsymbol{r}) - \rho^0(\boldsymbol{r})| << \rho(\boldsymbol{r}) \approx \rho^0(\boldsymbol{r}). \qquad (4.2.1)$$

Therefore, on average the ratio $\Delta\rho(\boldsymbol{r})/\rho(\boldsymbol{r}) \approx \Delta\rho(\boldsymbol{r})/\rho^0(\boldsymbol{r})$ may be expected to be small in the energetically important regions of large density values.

As explicitly shown in the first column of Fig. 4.1, the largest density differences $\Delta\rho(\boldsymbol{r})$ are observed mainly in the bond region, between the nuclei of bonded atoms. However, the reconstruction of atomic lone pairs can also lead to an appreciable displacement in the molecular electron densitiy. This is clearly seen in the contour maps of the relative displacement in the overall molecular density,

$\Delta\rho(\boldsymbol{r})/\rho^0(\boldsymbol{r})$, which has been plotted in the second column of Fig. 4.2. It should be observed, that relatively large magnitudes of this ratio at large distances from the molecule are projected out by the density/probability values at these locations in the functionals generating the *average* entropy deficiency measures.

By expanding the logarithms of the molecular surprisal $I[w(\boldsymbol{r})]$, around $w(\boldsymbol{r}) = 1$ to the first-order in the relative displacement of the electron density, one obtains the following approximate relations between the local value of the molecular surprisal and that of the density-difference function:

$$I[w(\boldsymbol{r})] = \ln[\rho(\boldsymbol{r})/\rho^0(\boldsymbol{r})] \cong \Delta\rho(\boldsymbol{r})/\rho^0(\boldsymbol{r}) \approx \Delta\rho(\boldsymbol{r})/\rho(\boldsymbol{r}). \qquad (4.2.2)$$

The performance of this approximation has been tested in a qualitative comparison of Fig. 4.2. A good overall similarity between the contours of the molecular surprisal (first column) and those of the relative density displacements (second column), demonstrates a semi-quantitative character of this approximate relationship. It thus provides a semi-quantitative information-theoretic interpretation of the relative density difference diagrams and links the local surprisal of IT to the density difference function of quantum chemistry.

Equation (4.2.2) also relates the integrands of the alternative information-distance functionals with the corresponding displacements in the electron density:

$$\Delta s(\boldsymbol{r}) = \rho(\boldsymbol{r})I[w(\boldsymbol{r})] \cong \Delta\rho(\boldsymbol{r})\, w(\boldsymbol{r}) \approx \Delta\rho(\boldsymbol{r}), \qquad (4.2.3)$$

$$\Delta D(\boldsymbol{r}) = \Delta\rho(\boldsymbol{r})\, I[w(\boldsymbol{r})] \cong [\Delta\rho(\boldsymbol{r})]^2/\rho^0(\boldsymbol{r}) \geq 0. \qquad (4.2.4)$$

Equation (4.2.3) is numerically verified in Fig. 4.1, where the contour diagrams of the directed divergence density $\Delta s(\boldsymbol{r})$ (second column) are compared with the corresponding maps of its first-order approximation, $\Delta\rho(\boldsymbol{r})w(\boldsymbol{r})$ (third column), and the density difference itself (first column). A general similarity between the three diagrams in each row of the figure confirms a semi-quantitative character of the first-order expansions of the directed divergence and divergence functionals. The corresponding numerical validation of Eq. (4.2.4) is shown in Fig. 4.3, where the contour maps of Kullback's divergence density $\Delta D(\boldsymbol{r})$ (first column) are compared with the corresponding diagrams of its first-order approximation $[\Delta\rho(\boldsymbol{r})]^2/\rho^0(\boldsymbol{r})$ (second column). One again observes a semi-quantitative similarity between the diagrams in each row, which demonstrates the validity of this approximate relation.

By expanding the logarithms of the subsystem and molecular surprisals one can thus relate alternative measures of the information-distance density to the density-difference function $\Delta\rho$ of quantum chemistry. We have also found that the surprisal function $I(\boldsymbol{r})$ is semi-quantuitatively represented by the relative displacement $\Delta\rho/\rho^0$. We have also demonstrated numerically that the local information-distance $\Delta s(\boldsymbol{r})$ is adequately approximated by the density displacement itself, while the density of Kullback's divergence ΔD approximately equals $(\Delta\rho)^2/\rho^0$.

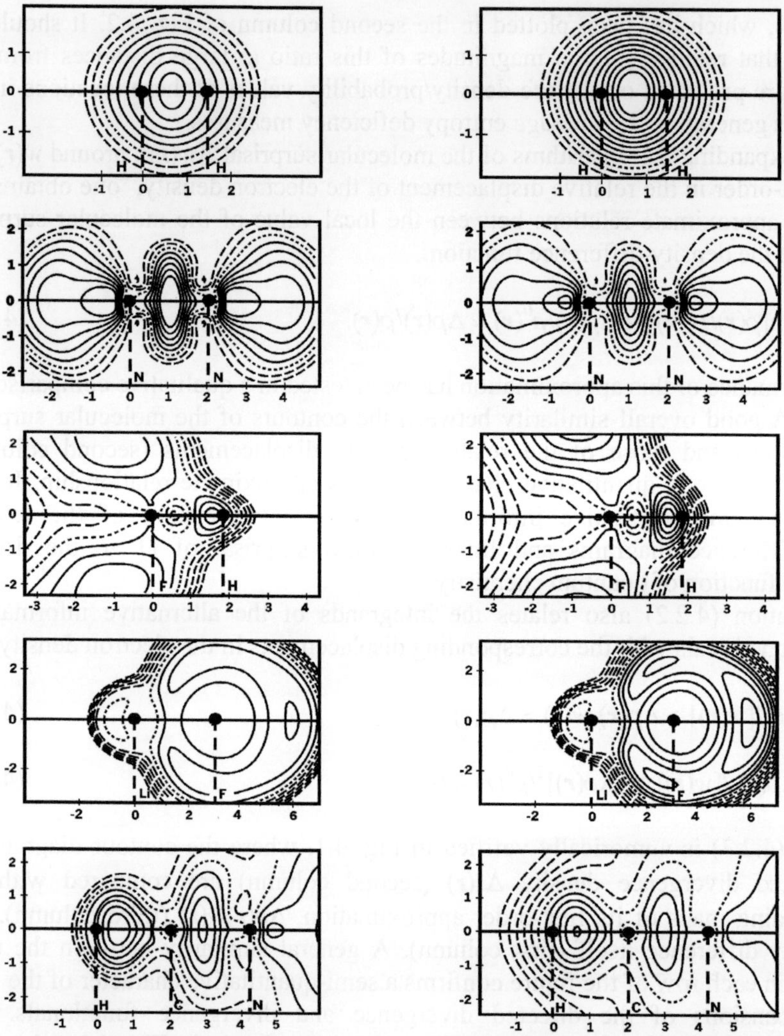

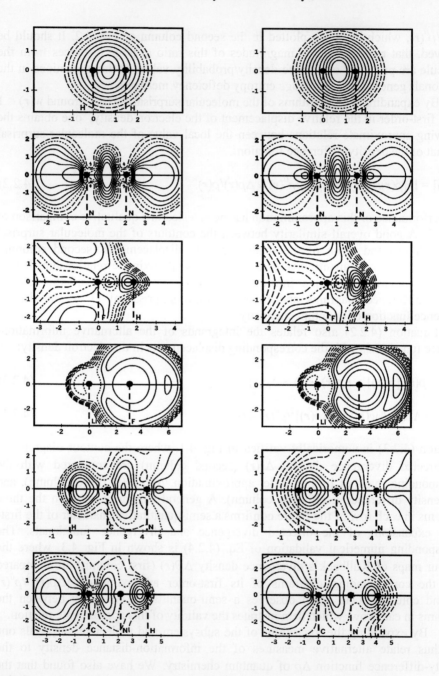

Figure 4.2. The contour maps of the molecular surprisal, $I[w(\boldsymbol{r})]$ (first column), and the relative displacements of the molecular density, $\Delta\rho(\boldsymbol{r})/\rho^0(\boldsymbol{r})$ (second column), for the linear molecules of Fig. 4.1. This comparison validates the approximate relation of Eq. (4.2.2) (Nalewajski *et al.*, 2002).

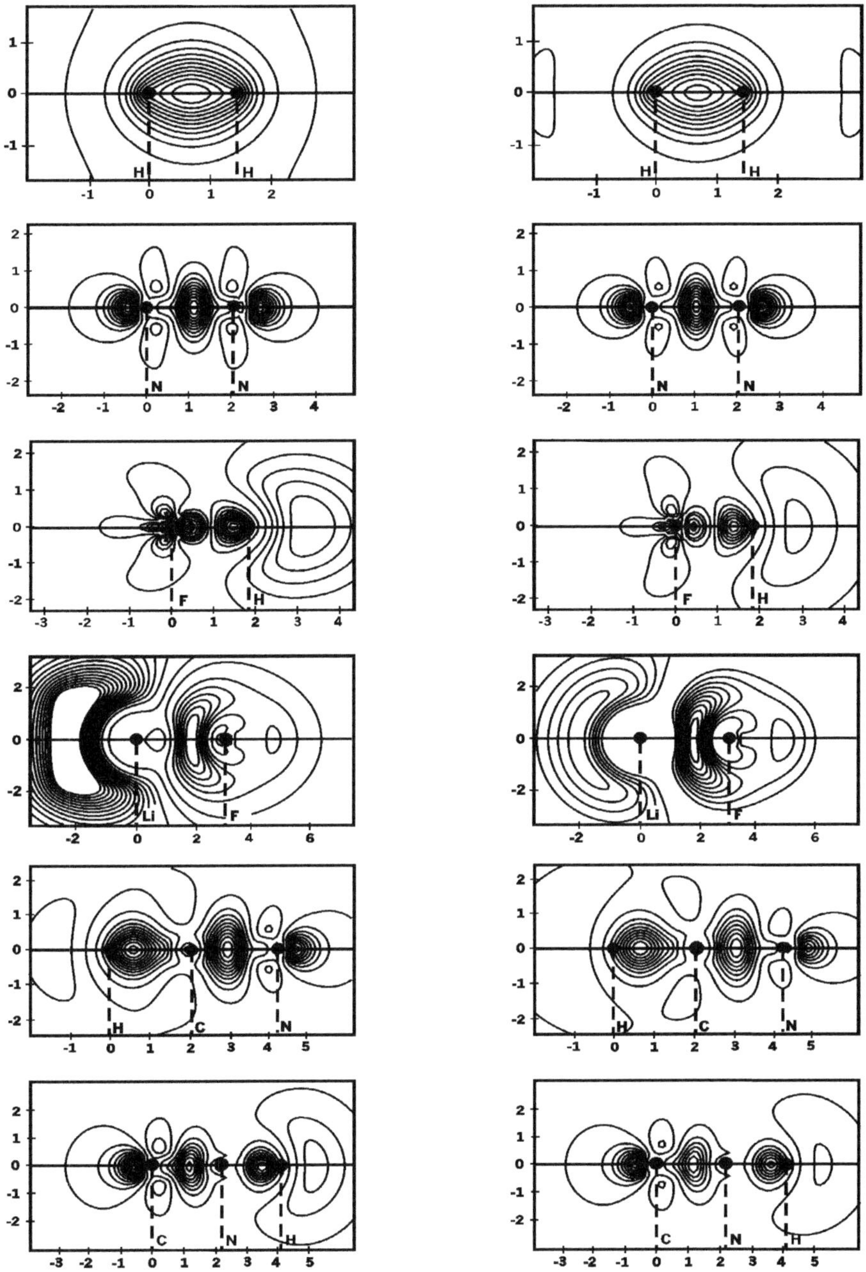

Figure 4.3. A comparison between the contour maps of the molecular divergence density, $\Delta D(\boldsymbol{r}) = \Delta\rho(\boldsymbol{r}) I[w(\boldsymbol{r})]$ (first column), and the $\Delta\rho(\boldsymbol{r})$-"weighted" relative density displacement, $[\Delta\rho(\boldsymbol{r})]^2/\rho^0(\boldsymbol{r})$ (second column), for the molecules of Fig. 4.1, which validates the approximate relation of Eq. (4.2.4) {Nalewajski *et al.*, 2002}.

4.3. DISPLACEMENTS OF MOLECULAR SHANNON ENTROPY

Of interest also are displacements of the overall Shannon entropy and its density relative to the corresponding promolecular reference values. They provide additional tools for examining the local entropy/information redistribution due to the formation of chemical bonds (Nalewajski and Broniatowska, 2003a). The global displacement of the Shannon entropy of the molecular electron density ρ, relative to that contained in the promolecular density ρ^0 [Eq. (4.1.5a)] also defines the entropy displacement density $h_\rho(\boldsymbol{r})$ which reflects a local contribution to the overall displacement in the electron uncertainty in the molecule, relative to the promolecule. This function can be also approximately related to the density-difference function:

$$\begin{aligned} h_\rho(\boldsymbol{r}) &= -\rho(\boldsymbol{r})\ln\rho(\boldsymbol{r}) + \rho^0(\boldsymbol{r})\ln\rho^0(\boldsymbol{r}) \\ &= -[\rho^0(\boldsymbol{r}) + \Delta\rho(\boldsymbol{r})]\{\ln[1 + \Delta\rho(\boldsymbol{r})/\rho^0(\boldsymbol{r})] + \ln\rho^0(\boldsymbol{r})\} + \rho^0(\boldsymbol{r})\ln\rho^0(\boldsymbol{r}) \\ &\cong -\Delta\rho(\boldsymbol{r})[\ln\rho^0(\boldsymbol{r}) + 1 + \Delta\rho(\boldsymbol{r})/\rho^0(\boldsymbol{r})]. \end{aligned} \tag{4.3.1}$$

The entropy displacement of Eq. (4.1.5a) can be also expressed in terms of the directed divergences between ρ and ρ^0, and the corresponding functional of the density difference:

$$\begin{aligned} \mathcal{H}[\rho] &= -\Delta S[\rho\,|\rho^0] - \int\Delta\rho(\boldsymbol{r})\ln\rho^0(\boldsymbol{r})\,d\boldsymbol{r} = \Delta S[\rho^0|\rho] - \int\Delta\rho(\boldsymbol{r})\ln\rho(\boldsymbol{r})\,d\boldsymbol{r} \\ &= \tfrac{1}{2}\{\Delta S[\rho^0|\rho] - \Delta S[\rho\,|\rho^0]\} - \int\Delta\rho(\boldsymbol{r})\ln[\rho^0(\boldsymbol{r})\rho(\boldsymbol{r})]^{1/2}\,d\boldsymbol{r}. \end{aligned} \tag{4.3.2}$$

In the last equation the geometric mean of the two compared electron densities, $\rho^g(\boldsymbol{r}) \equiv [\rho^0(\boldsymbol{r})\rho(\boldsymbol{r})]^{1/2}$, represents the "transition"-density between the initial (promolecular) and the final (molecular) states in the bond-formation process.

The density functional for the Shannon entropy difference represents the renormalized functional of the molecular shape-factor [Eq. (4.1.5b)]:

$$\mathcal{H}[p] = \mathcal{H}[\rho]/N, \tag{4.3.3}$$

where the functional density $h_p(\boldsymbol{r})$ can be again expressed in terms of $h_\rho(\boldsymbol{r})$ and $\Delta\rho(\boldsymbol{r})$:

$$h_p(\boldsymbol{r}) = (1/N)\{h_\rho(\boldsymbol{r}) + \Delta\rho(\boldsymbol{r})\ln N\}. \tag{4.3.4}$$

In Fig. 4.4 the contour maps of the entropy-displacement density $h_\rho(\boldsymbol{r})$ are compared with the corresponding density-difference diagrams for representative linear molecules of Fig. 4.1. The non-equidistant contour values have been selected to facilitate a qualitative comparison between the topographies of these two scalar fields. Therefore, only the profile of $h_\rho(\boldsymbol{r})$, shown in the third column of the figure, reflects the relative importance of each feature.

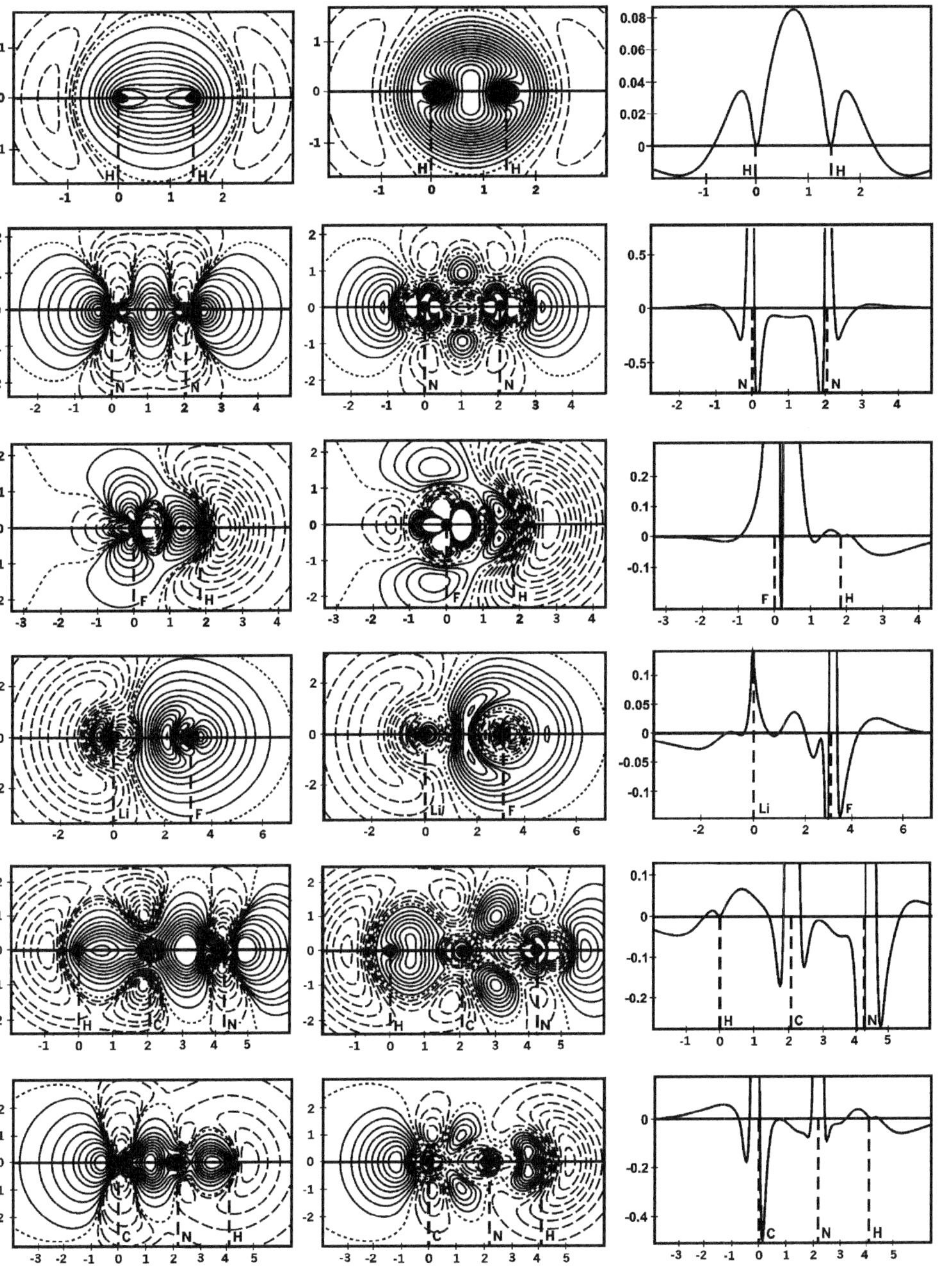

Figure 4.4. A comparison between the (non-equidistant) contours of the density-difference $\Delta\rho(\boldsymbol{r})$ (first column) and entropy-difference $h_\rho(\boldsymbol{r})$ (second column) functions for the linear molecules of Fig. 4.1. The corresponding profiles of $h_\rho(\boldsymbol{r})$ for the cuts along the bond axis are shown in the third column of the figure (Nalewajski and Broniatowska, 2003a).

When interpreting these plots one should realize that a negative (positive) value of $h_\rho(\boldsymbol{r})$ [or $h_p(\boldsymbol{r})$] signifies a decrease (increase) in the local electron *uncertainty* in the molecule relative to the promolecular reference value. The $\Delta\rho$ and h_ρ diagrams for H_2, displayed in the first row of Fig. 4.4, qualitatively resemble one another and the corresponding map of the related entropy-deficiency density Δs shown in Fig. 4.1. The main feature of the h_ρ-diagram, an increase in the electron uncertainty in the bonding region between the two nuclei, is due to the inflow of electrons to this region. This manifests the bond-covalency, which can be attributed to the electron sharing effect and a delocalization of the bonding electrons in the molecule, effectively moving in the field of both nuclei. One detects in all these maps a similar nodal structure. One also finds that the nonbonding regions exhibit a decreased uncertainty, due to the transfer of the electron density from this area to the vicinity of the two nuclei and the bond-region between them.

Consider next the h_ρ diagram for N_2. It generates a much richer nodal structure in comparison to the corresponding contour maps of $\Delta\rho$ and Δs for this homonuclear diatomic (Fig. 4.1). The σ and π electron regions are now separated by the nodal surface, with additional nodes dividing the inner (in the vicinity of the nuclei) and outer (valence) parts of the entropy-difference distribution. In contrast to the corresponding $\Delta\rho$ and Δs plots, the molecular entropy difference function reveals a negative feature in the σ component of the triple N≡N bond, thus marking a decrease in the electron uncertainty in this σ bond-region of the molecular electron distribution, around the bond axis, relative to the promolecule level of the local entropy density.

This pattern represents a resultant effect of changes in the shapes of atomic orbitals in the molecule, due to their contraction and hybridization, and of displacements in the orbital electron occupations resulting from an effective excitation (promotion) of the atomic valence electrons in the molecule. Indeed, the bonded atoms are promoted to the valence-state configuration, which is effectively excited in comparison to that of the free atoms of the promolecule. More specifically, the ground-state configuration of the valence electrons in free nitrogen, $N^0 = [2s^2 2p_\sigma^{\ 1} 2p_x^{\ 1} 2p_y^{\ 1}]$, can be compared with that characterizing the AO of the bonded nitrogen in N_2: $N[N_2] = [2s^{3/2} 2p_\sigma^{\ 3/2} 2p_x^{\ 1} 2p_y^{\ 1}]$. The latter configuration results from the molecular symmetry and an elementary MO diagram for the minimum basis set of the valence AO. It gives rise to the following MO electron occupations: $N_2 = [\sigma^2 \pi_x^{\ 2} \pi_y^{\ 2} n_1^{\ 2} n_2^{\ 2}]$, where the σ (bonding) MO represents a symmetric combination of the two (bonding) $(2s, 2p_\sigma)$-hybrids directed towards the bonding partner, π_x is the symmetric combination of two $2p_x$ orbitals on both centres, and n_i stands for the nonbonding $(2s_i, 2p_{i\sigma})$ hybrid on *i*th atom, directed away from the bonding partner. A comparison between these two electron configurations reveals that the bonded nitrogen exhibits an effective $2s \rightarrow 2p_\sigma$ excitation to the amount of a half of an electron. This AO occupation transfer signifies an effective lowering of the symmetry of the atomic σ electrons, in comparison to their state in the promolecule, which implies less uncertainty (more order) in their effective distributions in a molecule. Fig. 4.4 also shows that the π bonds in N_2 increase the electron uncertainty in the bond-region, due to the inflow of electrons from the atomic regions of the maxima of the electron distributions of $2p_\pi$ orbitals of the free nitrogens, where indeed a negative feature is detected in the entropy difference density. One also recognizes the buildup of the electron uncertainty in the outer regions of the two lone pairs, a direct manifestation of the $(2s,2p_\sigma)$-hybridization which accompanies formation of the σ bond.

Next, we turn to the entropy difference maps for the two heteronuclear diatomics, HF (third row) and LiF (fourth row). In HF the free atoms exhibit lower hardness and electronegativity differences, compared to those in LiF, for which a higher bond-ionicity (lower bond-covalency) should thus be expected. The contour maps of the entropy-difference function again exhibit the valence shell patterns similar to those observed in the corresponding density-difference (left) panels. They indeed reflect the expected differences in these polarized bonds. In HF one observes a partial electron transfer from the nonbonding part of the hydrogen electron distribution to the bonding charge shared by the two atoms and to the lone pair AO's on fluorine.

An elementary MO diagram involving only the valence shell AO suggests that the two nonbonding $2p_\pi$ AO of fluorine, with occupancy of 5/3 in F^0 and 2 in F[HF], accept in the molecular valence-state 1/3 of an electron each from the σ electrons of both atoms. This $\sigma \rightarrow \pi$ promotion in the molecule is indeed seen in the entropy-difference diagram, where an increase in the local electron uncertainty is observed. An outflow of the outer part of the hydrogen electron distribution is seen to result in lowering of the one-electron entropy in this region. A similar though weaker trend is detected on Li in LiF. In this molecule the inflow of electrons to the fluorine (acceptor) atom raises the local entropy of the outer (valence electrons) relative to the F^0 value, while the opposite effect is detected on the Li (donor) atom.

The remaining contour maps in Fig. 4.4 are devoted to the triatomic isomers: HCN (fifth row) and HNC (sixth row). In the regions of the triple C≡N bond they exhibit patterns similar to those observed in N_2 (second row), while the strongly covalent C–H and N–H bonds give rise to the entropy difference maps which are reminiscent of those previously observed for the single H–H and F–H bonds. The effects of the (s, p)-hybridization are again clearly seen on the peripheral heavy atom, and the negative feature of the σ component of the triple bond is observed. These results provide an additional confirmation of the applicability of the entropy-difference maps in diagnosing the information origins of the chemical bonds.

Therefore, the molecular entropy-difference function displays all typical changes in the electron distribution in a molecule relative to the corresponding free atoms. Its diagrams provide an a new tool for diagnosing the chemical bonds in terms of displacements in the uncertainty content of the molecular electron densities. The comparison of Fig. 4.4 demonstrates that in many respects the entropy difference plot provides a more detailed account of the reorganization of the electronic structure in the bond-formation process than does the corresponding density-difference diagram. The entropy-displacement function can be thus regarded as a complementary tool for probing the electronic structure of AIM and to detect changes they undergo in the bond-forming–bond-breaking processes. These diagrams generate a representation of all major effects due to the chemical bonds. They represent complementary probes to the density-difference and information-distance maps reported in Figs. 4.1-3. We have demonstrated that using these information-theoretic quantities to monitor changes in the information content of the molecular electron distribution allows one to identify typical displacements reflecting the bonded-atom hybridization and promotion relative to the promolecular (free-atom) reference, and to recognize the main entropy/information flows accompanying formation of the chemical bonds between atomic subsystems.

In Table 4.2 we have listed representative values of the molecular entropy differences of Eq. (4.1.5a) together with the Shannon entropies for the molecular and promolecular electron densities. These results show that the molecular distributions give rise to a lower level of the information-entropy (less uncertainty) compared to the respective promolecules.

Table 4.2. Displacements of the molecular Shannon entropies (in bits) for representative molecules of Fig. 4.1.*

Molecule	$\mathcal{H}[\rho] = S[\rho] - S[\rho^0]$	$S[\rho]$	$S[\rho^0]$
H_2	– 0.84	6.61	7.45
N_2	– 0.68	8.95	9.63
HF	– 1.00	3.00	4.00
LiF	– 3.16	5.12	8.28
HCN	– 1.44	12.99	14.45
HNC	– 1.39	13.06	14.45

*Nalewajski and Broniatowska, 2003a.

Thus, on average, the degree of uncertainty contained in the electron distribution decreases, when the constituent free-atoms form chemicxal bonds in the molecule. Indeed, the dominating overall contraction of atomic electron distributions in the field of all nuclear attractors in the molecule should imply a higher degree of "order" (less uncertainty) in the molecular electron density in comparison to that present in the promolecular distribution. The largest magnitude of this relative decrease in the entropy content of the molecular electron density is observed for LiF, which exhibits the most ionic bond (largest amount of charge transfer) among all molecules included in the table.

There is no apparent correlation in Table 4.2 between the global entropy displacement and the bond multiplicity. For example, a triple covalent bond in N_2 generates less overall entropy loss than does a single bond in H_2. The reason for a low magnitude of the entropy displacement in N_2 is the result of a mutual cancellation of the negative and positive contributions due to valence electrons. Indeed, the orbital hybridization, AO contraction, and the charge transfer should lower the entropy of the atomic electron distribution, since they increase charge inhomogeneity in the molecule, relative to the promolecule. By the same criterion, the effective AIM promotion and the electron delocalization via the system chemical bonds should have the opposite effect, of increasing the uncertainty content of the electron distribution in the molecule. Should one assume a similar entropy displacement of about –0.7 bits for all triple bonds in a series of isoelectronic molecules, N_2, HCN and CNH, one obtains a contribution due to a single C–H or N–H bond of about –0.7, a result close to that found for the H–H bond.

4.4. ILLUSTRATIVE APPLICATION TO PROPELLANES

As an additional illustration we shall present next the combined density difference, entropy-displacement, and information-distance analysis of the central bond in the propellane systems shown in Fig. 4.5 (Nalewajski and Broniatowska, 2003a). The main purpose of this study was to examine the effect of an increase in the bridge size, in the series of the [1.1.1], [2.1.1], [2.2.1], and [2.2.2] propellanes, on the central C'–C' bond, between the (primed) "bridgehead" carbon atoms.

Figure 4.6 reports the contour maps of the molecular density difference function $\Delta\rho(\boldsymbol{r})$, the Kullback-Leibler integrand $\Delta s(\boldsymbol{r})$, and the entropy displacement function $h_\rho(\boldsymbol{r})$, for the planes of sections shown in Fig. 4.5. The corresponding central-bond profiles of the density and entropy difference functions are shown in Fig. 4.7. The optimized geometries of propellanes have been determined from the UHF calculations (GAMESS program) using the 3-21G basis set. The contour maps have been generated using the DFT calculations in the extended (DZVP) basis set containing the split-valence and polarization functions.

The density difference plots of Figs. 4.6 and 4.7 show that there is on average a depletion of the electron density, relative to the promolecule, between the bridgehead carbon atoms in the [1.1.1] and [2.1.1] propellanes, while the [2.2.1] and [2.2.2] systems exhibit a net density buildup in this region. A similar conclusion follows from the entropy-displacement and entropy-deficiency plots of these figures. The two entropic diagrams are again seen to be qualitatively similar to the corresponding density-difference plots. This resemblance is particularly strong between the diagrams of the first two columns in Fig. 4.6, where the density displacement $\Delta\rho(\boldsymbol{r})$ and missing information density $\Delta s(\boldsymbol{r})$ are shown, respectively.

The numerical bond-orders from the two-electron difference approach (Nalewajski *et al.*, 1996), which are reported in Fig. 4.7, and the corresponding profiles shown in this figure reveal a changing nature of the central bond in the four propellanes. The central "bonds" in the smallest propellanes, lacking the accumulation of the electron density or the entropy (entropy-deficiency) density between the bridgehead atoms, are seen to be mostly of the intermediate character, being realized "through-bridges" rather than "through-space". A gradual emergence of the direct, "through-space" component, due to accumulation of the electron density and the entropy (entropy-deficiency) density between the bridgehead carbons, is observed when the bridges are enlarged in the two largest propellanes of Fig. 4.5.

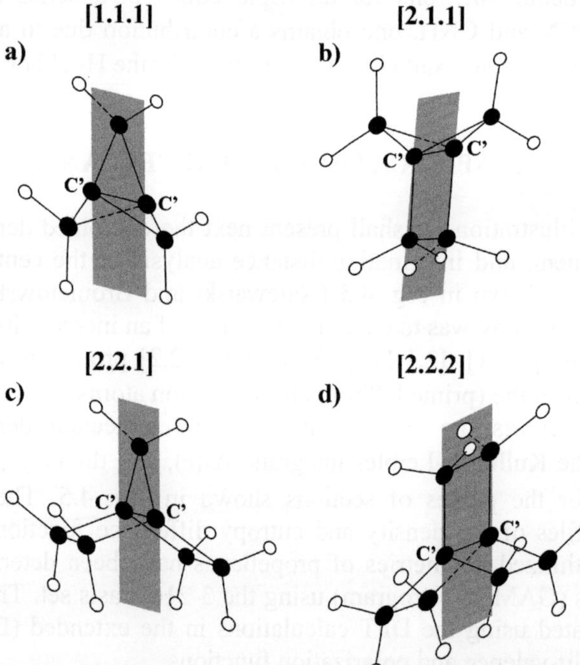

Figure 4.5. The propellane structures and the planes of sections containing the bridge and bridgehead (C') atoms for the diagrams of Figs. 4.6 and 4.7.

Using the two-electron difference approach (Nalewajski and Mrozek, 1994, 1996; Nalewajski *et al.*, 1993, 1996, 1997; Mrozek *et al.*, 1999) one roughly estimates (Nalewajski *et al.*, 1996) a full single bond in the [2.2.1] and [2.2.2] propellanes and approximately 0.8 bond-order in the [1.1.1] propellane. Using the latter estimate as a measure of the "through-bridges" component in the largest propellane one predicts about 0.2 bond-order measure for the "through-space" component of the central bond in the largest [2.2.2] propellane.

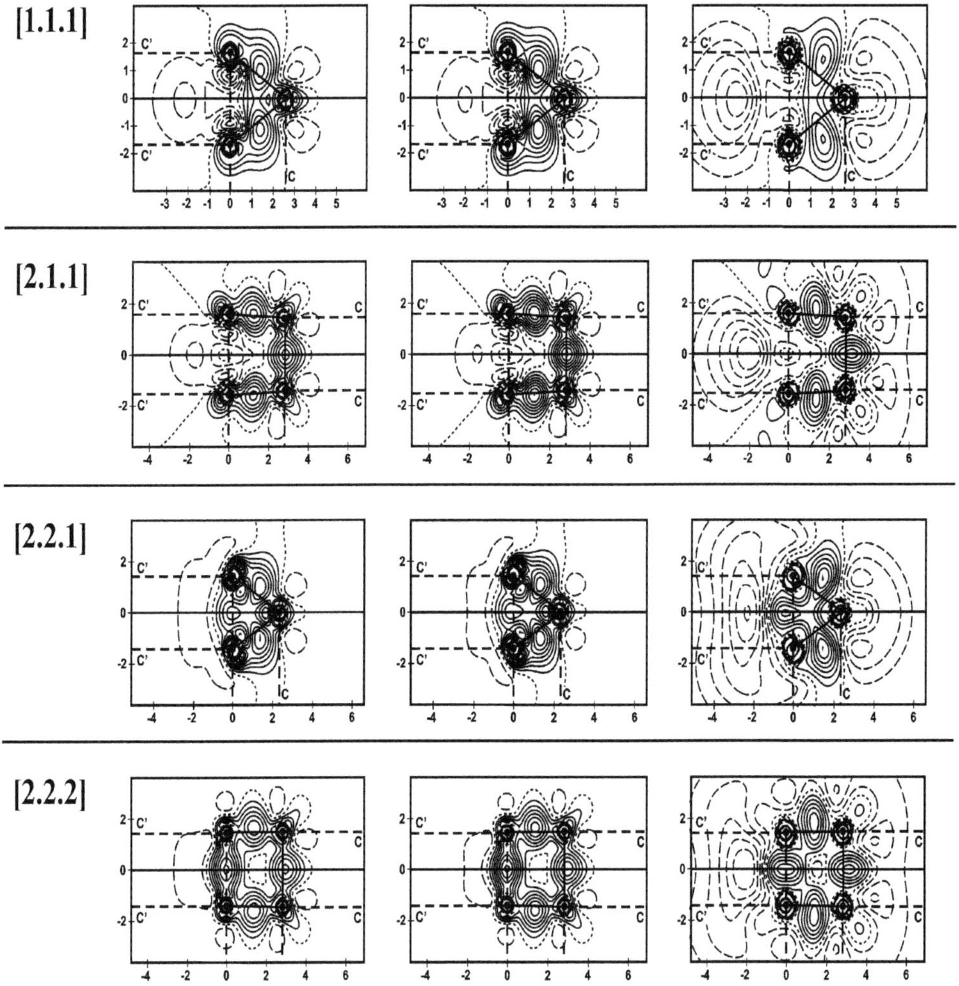

Figure 4.6. A comparison between the equidistant-contour maps of the density-difference function $\Delta\rho(\boldsymbol{r})$ (first column), the information-distance density $\Delta s(\boldsymbol{r})$ (second column), and the entropy-displacement density $h_\rho(\boldsymbol{r})$ (third column), for the four propellanes of Fig. 4.5 (Nalewajski and Broniatowska, 2003a).

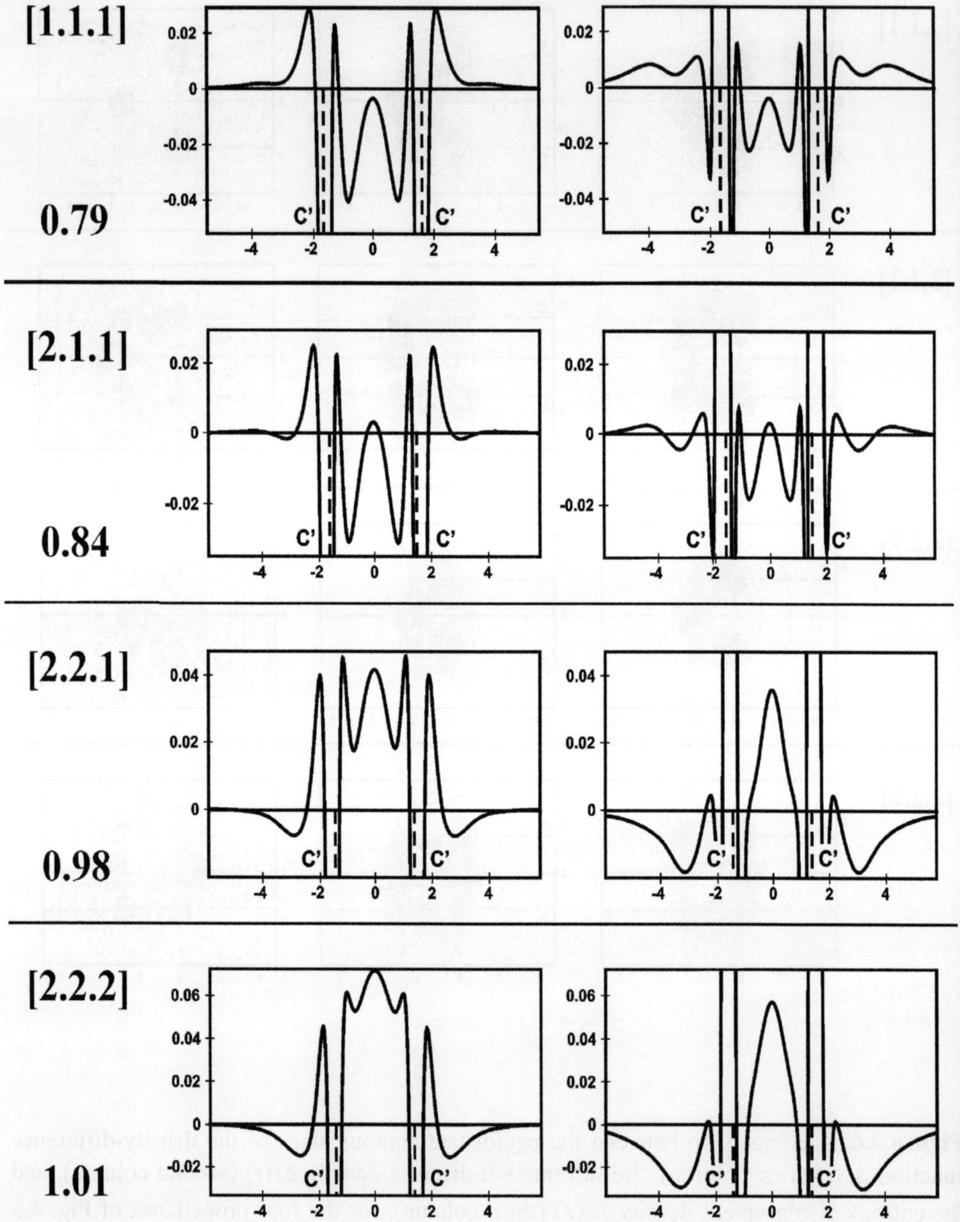

Figure 4.7. The bridgehead bond profiles of the density difference function (left panel) and molecular entropy displacement (right panel) for the four propellanes of Fig. 4.5. For comparison the numerical values of the bond multiplicities from the difference approach (Nalewajski *et al.*, 1996) are also reported.

4.5. ELECTRON LOCALIZATION FUNCTION AS INFORMATION MEASURE

The *Electron Localization Function* (ELF) for atomic and molecular systems has been proposed by Becke and Edgecombe (1990) to visualize both the atomic shell structure and the distribution of the bonding and lone-pair electrons in molecules, in order to classify and distinguish between the single and multiple chemical bonds, and to monitor changes in the electron distribution in the bond-forming–bond-breaking processes (Savin *et al.*, 1997; Silvi and Savin, 1994). An extension of ELF to the time-dependent systems has also been reported (Marques and Gross, 2004).

In this section we shall provide the information-theoretic interpretation to the key ingredient of ELF in terms of the Fisher (locality) measure of the information content (see Section 3.4). It has been demonstrated (Nalewajski *et al.*, 2005) that this analysis leads to an alternative information measure of the electron localization in molecular systems, which has been successfully tested against the original ELF in representative atoms and molecules.

The key ingredient of ELF is the leading term of the Taylor expansion of the spherically-averaged Hartree-Fock *conditional* (c) pair-probability $P_c^{\sigma\sigma}(s|\boldsymbol{r})$, which measures the probability of finding in the distance s from a given (*reference*) electron at $\boldsymbol{r}$ the other (*dependent*) electron of the same spin σ,

$$P_c^{\sigma\sigma}(s|\boldsymbol{r}) = \tfrac{1}{3}\, D_\sigma(\boldsymbol{r})\, s^2 + \dots, \qquad (4.5.1)$$

$$\begin{aligned} D_\sigma(\boldsymbol{r}) \equiv D_\sigma[\rho_\sigma; \boldsymbol{r}] &= \textstyle\sum_i^\sigma |\nabla \varphi_{i\sigma}(\boldsymbol{r})|^2 - \tfrac{1}{4}\, |\nabla \rho_\sigma(\boldsymbol{r})|^2/\rho_\sigma(\boldsymbol{r}) \\ &\equiv \textstyle\sum_i^\sigma \tau_{i\sigma}(\boldsymbol{r}) - \tfrac{1}{4}|\nabla \rho_\sigma(\boldsymbol{r})|^2/\rho_\sigma(\boldsymbol{r}) \geq 0, \end{aligned} \qquad (4.5.2)$$

where both the kinetic energy density $\tau_\sigma(\boldsymbol{r}) = \sum_i^\sigma \tau_{i\sigma}(\boldsymbol{r})$ and the spin density

$$\rho_\sigma(\boldsymbol{r}) = \textstyle\sum_i^\sigma [\varphi_{i\sigma}(\boldsymbol{r})]^2 \equiv \sum_i^\sigma \rho_{i\sigma}(\boldsymbol{r}) \qquad (4.5.3)$$

are given by the sums of additive contributions due to the *molecular orbitals* (MO) $\boldsymbol{\varphi}_\sigma = \{\varphi_{i\sigma}\}$ representing the spatial parts of the σ–type spin-orbitals. In the related one-determinant approximation of the computational DFT, the Kohn Sham (1965) theory (see Example 3.15 and Appendix C), the orbital densities $\boldsymbol{\rho}_\sigma = \{\rho_{i\sigma}\}$ are the mutually-closed pieces of the overall spin density ρ_σ^s of the *separable* (s), non-interacting system, which by hypothesis equals to that of the real, interacting system of electrons in a molecule, $\rho_\sigma = \rho_\sigma^s$.

The probability function $D_\sigma(\boldsymbol{r})$ provides a measure of the localization of the reference electron, reaching the small values for the highly localized distribution of the reference electron. Indeed, it vanishes in the limiting case of the one-electron system and in the regions of the multi-electron systems dominated by a single, localized σ spin-orbital of the reference electron, which effectively excludes by the Pauli principle another spin-like electron from its vicinity.

The two terms of Eq. (4.5.2) have precise information-theoretic interpretations in terms of the the density of the Fisher information $I^F[p]$ [Eqs. (3.4.2) and (3.4.6)], of the one-electron probability density $p(\boldsymbol{r}) = \psi^2(\boldsymbol{r})$, $\int p(\boldsymbol{r})\, d\boldsymbol{r} = 1$, where $\psi(\boldsymbol{r})$ stands for the distribution *amplitude*:

$$I^F[p] = \int p(\boldsymbol{r})\, [\nabla \ln p(\boldsymbol{r})]^2\, d\boldsymbol{r} = \int |\nabla p(\boldsymbol{r})|^2 / p(\boldsymbol{r})\, d\boldsymbol{r} \equiv \int f[p; \boldsymbol{r}]\, d\boldsymbol{r}$$
$$= I^F[\psi] = 4 \int |\nabla \psi(\boldsymbol{r})|^2\, d\boldsymbol{r} \equiv \int f[\psi; \boldsymbol{r}]\, d\boldsymbol{r}. \qquad (4.5.4)$$

As we have already observed in Section 3.4, this information measure characterizes the distribution "sharpness" (localization, "order") and provides a complementary description of the probability distribution to the global Shannon entropy of the molecular probability density, which reflects the distribution "smoothness" (spread), thus indexing the uncertainty contained in *p*.

The kinetic energy term in Eq. (4.5.2) is seen to represent the sum of the additive orbital Fisher information densities in the amplitude representation (see Eq. (3.4.7):

$$\tau_\sigma(\boldsymbol{r}) = \textstyle\sum_i^\sigma \tau_{i\sigma}(\boldsymbol{r}) = \frac{1}{4} \sum_i^\sigma f[\varphi_{i\sigma}; \boldsymbol{r}] \equiv \frac{1}{4} f_\sigma^a[\boldsymbol{\varphi}_\sigma; \boldsymbol{r}]$$
$$\equiv \tfrac{1}{4} \textstyle\sum_i^\sigma f_{i\sigma}[\rho_{i\sigma}; \boldsymbol{r}] \equiv \frac{1}{4} f_\sigma^a[\boldsymbol{\rho}_\sigma; \boldsymbol{r}]. \qquad (4.5.5)$$

It represents a quarter of the *additive* (*a*) Fisher information density $f_\sigma^a[\boldsymbol{\varphi}_\sigma; \boldsymbol{r}] = f_\sigma^a[\boldsymbol{\rho}_\sigma; \boldsymbol{r}]$ contained in the MO probability amplitudes $\boldsymbol{\varphi}_\sigma$ or the associated components $\boldsymbol{\rho}_\sigma = \{\rho_{i\sigma}\}$ of the overall spin-density ρ_σ.

A reference to Eq. (4.5.4) also shows that the second term in Eq. (4.5.2) can be similarly identified as providing a quarter of the *total* (*t*) Fisher information density in ρ_σ, $f_\sigma[\rho_\sigma; \boldsymbol{r}] \equiv f_\sigma^t[\rho_\sigma; \boldsymbol{r}] \equiv f_\sigma^t(\boldsymbol{r})$, combining the *additive* (*a*) and *non-additive* (*n*) parts of the Fisher information density in MO resolution which is contained in ρ_σ:

$$\tfrac{1}{4} f_\sigma^t(\boldsymbol{r}) = \tfrac{1}{4} |\nabla \rho_\sigma(\boldsymbol{r})|^2 / \rho_\sigma(\boldsymbol{r}) = \tfrac{1}{4} f_\sigma^t[\boldsymbol{\rho}_\sigma; \boldsymbol{r}] \equiv \tfrac{1}{4}\{f_\sigma^a[\boldsymbol{\rho}_\sigma; \boldsymbol{r}] + f_\sigma^n[\boldsymbol{\rho}_\sigma; \boldsymbol{r}]\}. \qquad (4.5.6)$$

Therefore, the $D_\sigma(\boldsymbol{r})$ function of Eq. (4.5.2) is proportional to the negative non-additive contribution to $f_\sigma^t(\boldsymbol{r})$, defined in terms of the MO electron densities,

$$D_\sigma(\boldsymbol{r}) = -\tfrac{1}{4} f_\sigma^n[\boldsymbol{\rho}_\sigma; \boldsymbol{r}]. \qquad (4.5.7)$$

This key ingredient of the ELF thus has a direct information-theoretic interpretation in addition to its conditional-probability meaning in the original definition of Eq. (4.5.1).

As we have already remarked above, in the KS theory the MO densities $\boldsymbol{\rho}_\sigma$, or their probability amplitudes (KS orbitals) $\boldsymbol{\varphi}_\sigma$, refer to the hypothetical non-interacting system, while the overall density ρ_σ corresponds to both the interacting and non-interacting systems. Hence, the source of the non-additivity of Eq. (4.5.7) is the electron interaction in the real system. Therefore, the KS MO *partitioning* non-additivity $f_\sigma^n[\rho_\sigma; \boldsymbol{r}]$ of the total Fisher information density $f_\sigma^t[\rho_\sigma; \boldsymbol{r}]$ in fact represents the electron *interaction* non-additivity present in the interacting molecular system. Moreover, since the additive Fisher information density $f_\sigma^a(\boldsymbol{r})$ combines all the *intra*-orbital contributions, $f_\sigma^a[\rho_\sigma; \boldsymbol{r}] = f_\sigma^{intra}[\rho_\sigma; \boldsymbol{r}]$, the non-additive part can be also interpreted as measuring the *inter*-orbital Fisher information density: $f_\sigma^n[\rho_\sigma; \boldsymbol{r}] = f_\sigma^{inter}[\rho_\sigma; \boldsymbol{r}]$.

As indicated in Eq. (4.5.4), the Fisher information functional is properly defined in terms of the unity normalized probability distributions, i.e., the *shape* factors of the corresponding electron densities, e.g.,

$$p_\sigma(\boldsymbol{r}) = \rho_\sigma(\boldsymbol{r})/N_\sigma, \qquad \int p_\sigma(\boldsymbol{r})\, d\boldsymbol{r} = 1, \tag{4.5.8}$$

where $N_\sigma = \int \rho_\sigma(\boldsymbol{r})\, d\boldsymbol{r}$ is the total number of electrons of the spin variety σ, i.e., the number of the (singly occupied) KS spin-orbitals $\{\varphi_{i\sigma}\}$. Then, the total Fisher information density of the electron shape-function $p_\sigma(\boldsymbol{r})$ reads

$$f_\sigma[p_\sigma; \boldsymbol{r}] = f_\sigma[\rho_\sigma; \boldsymbol{r}]/N_\sigma. \tag{4.5.9}$$

It should be realized that each orbital density represents the unity normalized *conditional* probability distribution, $\pi_{i\sigma}(\boldsymbol{r}) \equiv \pi_\sigma(\boldsymbol{r}|i)$, of finding an electron with spin σ at $\boldsymbol{r}$, when it is known beforehand that this electron occupies the specified, ith MO (a parameter) (see Section 3.6 and Example 3.11):

$$\begin{aligned} &\rho_{i\sigma}(\boldsymbol{r}) = p_{i\sigma}(\boldsymbol{r})/P_{i\sigma} \equiv \pi_{i\sigma}(\boldsymbol{r}), \quad P_{i\sigma} = \int p_{i\sigma}(\boldsymbol{r})\, d\boldsymbol{r} = 1/N_\sigma \equiv P_\sigma, \\ &\quad \textstyle\sum_i^\sigma p_{i\sigma}(\boldsymbol{r}) = p_\sigma(\boldsymbol{r}), \qquad \sum_i^\sigma \int p_{i\sigma}(\boldsymbol{r})\, d\boldsymbol{r} = \int \pi_{i\sigma}(\boldsymbol{r})\, d\boldsymbol{r} = \sum_i^\sigma P_{i\sigma} = 1. \end{aligned} \tag{4.5.10}$$

Here $p_{i\sigma}(\boldsymbol{r}) \equiv p_\sigma(i, \boldsymbol{r})$ is the *joint* probability of simultaneous events that an electron of spin σ is found at $\boldsymbol{r}$ and that it originates from spin orbital $\varphi_{i\sigma}$, while the orbital probability vector $\boldsymbol{P}_\sigma = \{P_{i\sigma} = 1/N_\sigma\}$ groups the *condensed* probabilities of finding an electron of spin σ on specified MO.

It should be emphasized that the molecular spin probability distribution $p_\sigma(\boldsymbol{r})$ should be compared in terms of the Fisher information descriptors only with the *molecularly* normalized MO *joint* distributions $\boldsymbol{p}_\sigma(\boldsymbol{r}) = \{p_{i\sigma}(\boldsymbol{r})\}$, since only these orbital probability densities characterize the orbital distributions of electrons as *subsystems-in-the-molecule*. This is in contrast to the conditional probabilities $\boldsymbol{\pi}_\sigma(\boldsymbol{r}) = \{\pi_\sigma(\boldsymbol{r}|i)\} = \boldsymbol{\rho}_\sigma(\boldsymbol{r})$, equal to the orbital electron densities, which describe the *separate* probability distributions, as indeed reflected by the unity normalization of each of them. Since the gradient of the logarithm of the joint probability distributions $\boldsymbol{p}_\sigma(\boldsymbol{r}) = \{p_{i\sigma}(\boldsymbol{r})\}$ is solely determined by the local, conditional probability factor $\pi_{i\sigma}(\boldsymbol{r})$,

$$\nabla \ln p_{i\sigma}(\boldsymbol{r}) = \nabla \ln[P_{i\sigma}\, \pi_\sigma(\boldsymbol{r}|i)] = \nabla \ln \pi_\sigma(\boldsymbol{r}|i), \tag{4.5.11}$$

the additive Fisher information density of the joint probability distribution satisfies the *grouping rule* of Eq. (3.6.4b) for combining the intra-orbital contributions into the molecular information content,

$$f_\sigma^a[\mathbf{p}_\sigma; \mathbf{r}] = \Sigma_i^\sigma P_{i\sigma} f_{i\sigma}[\pi_{i\sigma}; \mathbf{r}] = P_\sigma f_\sigma^a[\boldsymbol{\pi}_\sigma; \mathbf{r}] \equiv d_\sigma[\mathbf{p}_\sigma; \mathbf{r}], \qquad (4.5.12)$$

and hence

$$I^{Fa}[\mathbf{p}_\sigma] = \int f_\sigma^a[\mathbf{p}_\sigma; \mathbf{r}]\, d\mathbf{r} = \Sigma_i^\sigma P_{i\sigma} \int f_{i\sigma}[\pi_{i\sigma}; \mathbf{r}]\, d\mathbf{r} = \Sigma_i^\sigma P_{i\sigma} I^F[\pi_{i\sigma}]. \qquad (4.5.13)$$

In other words, the additive Fisher information in the joint MO probabilities $\mathbf{p}_\sigma(\mathbf{r}) = \{p_{i\sigma}(\mathbf{r})\}$ is the mean value of the Fisher information contained in the orbital densities $\boldsymbol{\rho}_\sigma$, i.e., the MO conditional probabilities $\boldsymbol{\pi}_\sigma$, with the unbiased weighting factors for each orbital, $P_{i\sigma} = P_\sigma = 1/N_\sigma$, determining the relevant weighting factor for each MO "group" of local events.

Therefore, the total and additive Fisher information densities, which determine the non-additive component in the MO resolution, when expressed in terms of the molecularly normalized probabilities, have the same, linear scaling with the overall number of electrons:

$$f_\sigma[\rho_\sigma; \mathbf{r}] = N_\sigma f_\sigma[p_\sigma; \mathbf{r}] \qquad \text{and} \qquad f_\sigma^a[\boldsymbol{\rho}_\sigma; \mathbf{r}] = f_\sigma^a[\boldsymbol{\pi}_\sigma; \mathbf{r}] = N_\sigma f_\sigma^a[\mathbf{p}_\sigma; \mathbf{r}], \quad (4.5.14)$$

and hence

$$D_\sigma(\mathbf{r}) = -\tfrac{1}{4} N_\sigma (f_\sigma[p_\sigma; \mathbf{r}] - f_\sigma^a[\mathbf{p}_\sigma; \mathbf{r}]) \equiv -\tfrac{1}{4} N_\sigma f_\sigma^n[\mathbf{p}_\sigma; \mathbf{r}]. \qquad (4.5.15)$$

The conditional probability densities $\boldsymbol{\pi}_\sigma(\mathbf{r}) = \{\pi_\sigma(\mathbf{r}|i)\} = \boldsymbol{\rho}_\sigma(\mathbf{r})$ provide the convenient framework for an explicit expression for the non-additive Fisher information density:

$$f_\sigma^n[\mathbf{p}_\sigma; \mathbf{r}] = |\nabla p_\sigma(\mathbf{r})|^2 / p_\sigma(\mathbf{r}) - \Sigma_i^\sigma |\nabla p_{i\sigma}(\mathbf{r})|^2 / p_{i\sigma}(\mathbf{r}) = N_\sigma^{-1}\{f_\sigma[\rho_\sigma; \mathbf{r}] - f_\sigma^a[\boldsymbol{\rho}_\sigma; \mathbf{r}]\}$$
$$= N_\sigma^{-1} \Sigma_i^\sigma \Sigma_j^\sigma \nabla\pi_{i\sigma}(\mathbf{r}) \bullet \nabla\pi_{j\sigma}(\mathbf{r})\, [\rho_\sigma(\mathbf{r})^{-1} - \pi_{i\sigma}(\mathbf{r})^{-1}\delta_{ij}]. \qquad (4.5.16)$$

It follows from the original, two-electron conditional probability meaning of $D_\sigma(\mathbf{r})$ that the smaller the probability of finding a second spin-like electron near the reference point at $\mathbf{r}$, the more highly localized is the reference electron. Therefore, there is an overall "inverse" relationship between this conditional probability, proportional to the negative non-additive component of the Fisher information in MO resolution, and a realistic measure of the electron localization.

The original ELF (Becke and Edgecombe, 1990) has been constructed using the following "squared" reciprocity relation:

$$ELF_\sigma(\boldsymbol{r}) = [1 + \chi_\sigma(\rho_\sigma; \boldsymbol{r})^2]^{-1}, \qquad \chi_\sigma[\rho_\sigma; \boldsymbol{r}] = D_\sigma[\rho_\sigma; \boldsymbol{r}]/D_\sigma^{\,0}(\boldsymbol{r}), \tag{4.5.17}$$

with respect to the *Local Density Approximation* (LDA) value:

$$D_\sigma^{\,0}(\boldsymbol{r}) = D_\sigma^{\,0}[\rho_\sigma; \boldsymbol{r}] = [3(6\pi)^{2/3}/5]\, \rho_\sigma(\boldsymbol{r})^{5/3}. \tag{4.5.18}$$

This definition, dimensionless and invariant with respect to the unitary transformation of orbitals, has been designed to directly reflect the electron localization relative to this LDA reference. It assumes the values between 0 and 1 and exhibits the desirable features of reaching the upper limit $ELF = 1$ for the perfect localization and $ELF = ½$ for a delocalized (homogeneous) electron gas.

It should be realized, however, that this expression has been "tailored" somewhat arbitrarily, by selecting the square of $\chi_\sigma = D_\sigma/D_\sigma^{\,0}$ in the denominator and adopting the uniform-density electron gas reference to give $ELF = ½$. This particular version of ELF was shown to realistically reveal the location of atomic shells as well as the core and valence (binding and lone) electron pairs in molecules. Clearly, any alternative choice of ELF should also deliver all these features of the electron configuration with comparable accuracy and clarity of the graphical visualization.

As we have already argued above, the overall and MO information-theoretic quantities should be expressed in terms of the relevant (molecularly normalized) shape factors, of the system as a whole and probabilities of the orbital subsystems in molecules. The LDA reference function expressed in terms of the molecular probability distribution reads:

$$d_\sigma^{\,0}[p_\sigma(\boldsymbol{r})] = [3(6\pi)^{2/3}/5][\rho_\sigma(\boldsymbol{r})/N_\sigma]^{5/3} = D_\sigma^{\,0}[\rho_\sigma; \boldsymbol{r}]/N_\sigma^{\,5/3}. \tag{4.5.19}$$

Hence the modified ELF ratio,

$$\chi_\sigma[\boldsymbol{p}_\sigma; \boldsymbol{r}] = d_\sigma[\boldsymbol{p}_\sigma; \boldsymbol{r}]/d_\sigma^{\,0}[p_\sigma(\boldsymbol{r})] = N_\sigma^{\,2/3} D_\sigma(\boldsymbol{r})/D_\sigma^{\,0}(\boldsymbol{r}) = N_\sigma^{\,2/3}\chi_\sigma[\boldsymbol{\rho}_\sigma; \boldsymbol{r}], \tag{4.5.20}$$

which can be used to construct the IT-ELF. The simplest option is to use the ordinary "inverse" relationship in the spirit of the original expression (4.5.17):

$$elf_\sigma(\boldsymbol{r}) = N_\sigma^{\,2/3}/(N_\sigma^{\,2/3} + \chi_\sigma[\boldsymbol{p}_\sigma; \boldsymbol{r}]) = (1 + \chi_\sigma[\boldsymbol{\rho}_\sigma; \boldsymbol{r}])^{-1}. \tag{4.5.21}$$

Above, we have modified the original construction to regain the assumed normalization of Becke and Edgecombe: $elf_\sigma(\boldsymbol{r}) = ½$, for the perfectly delocalized, homogeneous electron gas, when $\chi_\sigma[\boldsymbol{p}_\sigma; \boldsymbol{r}] = 1$, and $elf_{i,\sigma}(\boldsymbol{r}) = 1$, for the perfectly localized case, when $\chi_\sigma(\boldsymbol{p}_\sigma; \boldsymbol{r}) = 0$.

It should be observed that the original ELF of Eq. (4.5.17) is recovered through the associated "squared" inverse relationship:

$$ELF_\sigma(\boldsymbol{r}) = N_\sigma^{\,4/3}/(N_\sigma^{\,4/3} + \chi_\sigma[\boldsymbol{p}_\sigma; \boldsymbol{r}]^2) = [1 + \chi_\sigma(\rho_\sigma; \boldsymbol{r})^2]^{-1}. \tag{4.5.22}$$

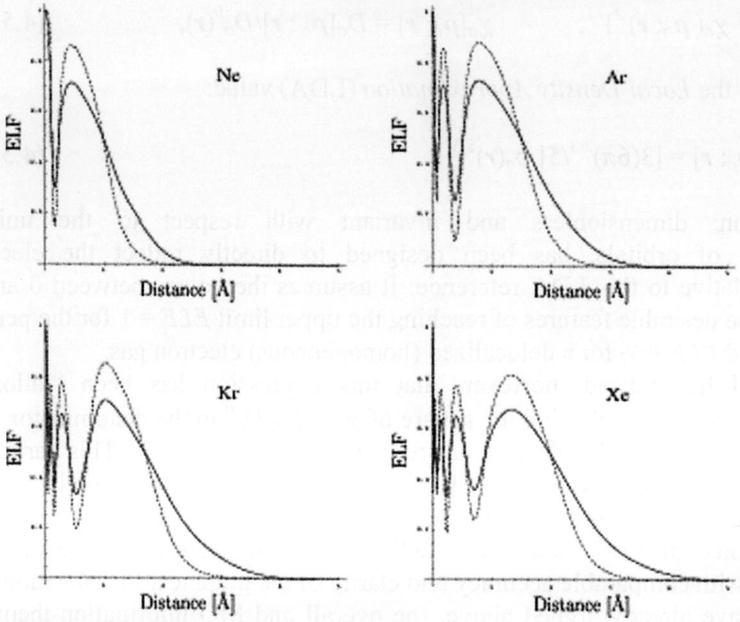

Figure 4.8. Plots of ELF (dashed line) and IT-ELF (solid line) for Ne, Ar, Kr and Xe.

In Fig. 4.8 representative graphs are presented of the IT-ELF function $elf_\sigma(\boldsymbol{r})$ for the rare-gas atoms Ne, Ar, Kr and Xe. For comparison the dashed curves represent the original function $ELF_\sigma(\boldsymbol{r})$. The qualitative behavior of the two curves is seen to be very similar. In general the IT-ELF exhibits smaller outer amplitudes and thus a larger spatial extension than the original ELF. Figures 4.9 and 4.10 report illustrative comparisons of the molecular ELF and IT-ELF plots. Again, the topology of the two functions is qualitatively the same. The atomic shell structures as well as lone pairs are clearly displayed by both functions. As already discussed for the rare-gas atoms, the main difference is a decay of the outer amplitudes, being faster in the case of the squared inverse relationship of the original ELF. As a consequence the IT-ELF generates a chemically "softer" distribution of the localized electrons in comparison to a relatively "harder" distribution resulting from the original ELF definition. A reference to the NH_3 and PH_3 plots in Figs 4.9 and 4.10 also reveals that the IT-ELF distinguishes lone-pairs and hydrogen atoms somewhat more clearly, than the original localization function.

We have thus demonstrated that the conditional two-electron probability function, which defines the ELF, in fact measures in the MO resolution the non-additive part of the density of the Fisher information for locality (intrinsic accuracy). This interpretation gives rise to the modified IT-ELF, based upon the first-power inverse relationship, which compares favorably with the original ELF.

Figure 4.9. Plots of ELF (left column) and IT-ELF(right column) for N_2(first row), H_2O (second row), and, NH_3 (third row) on selected planes. The color scale for the ELF values is given in the bottom of the figure.

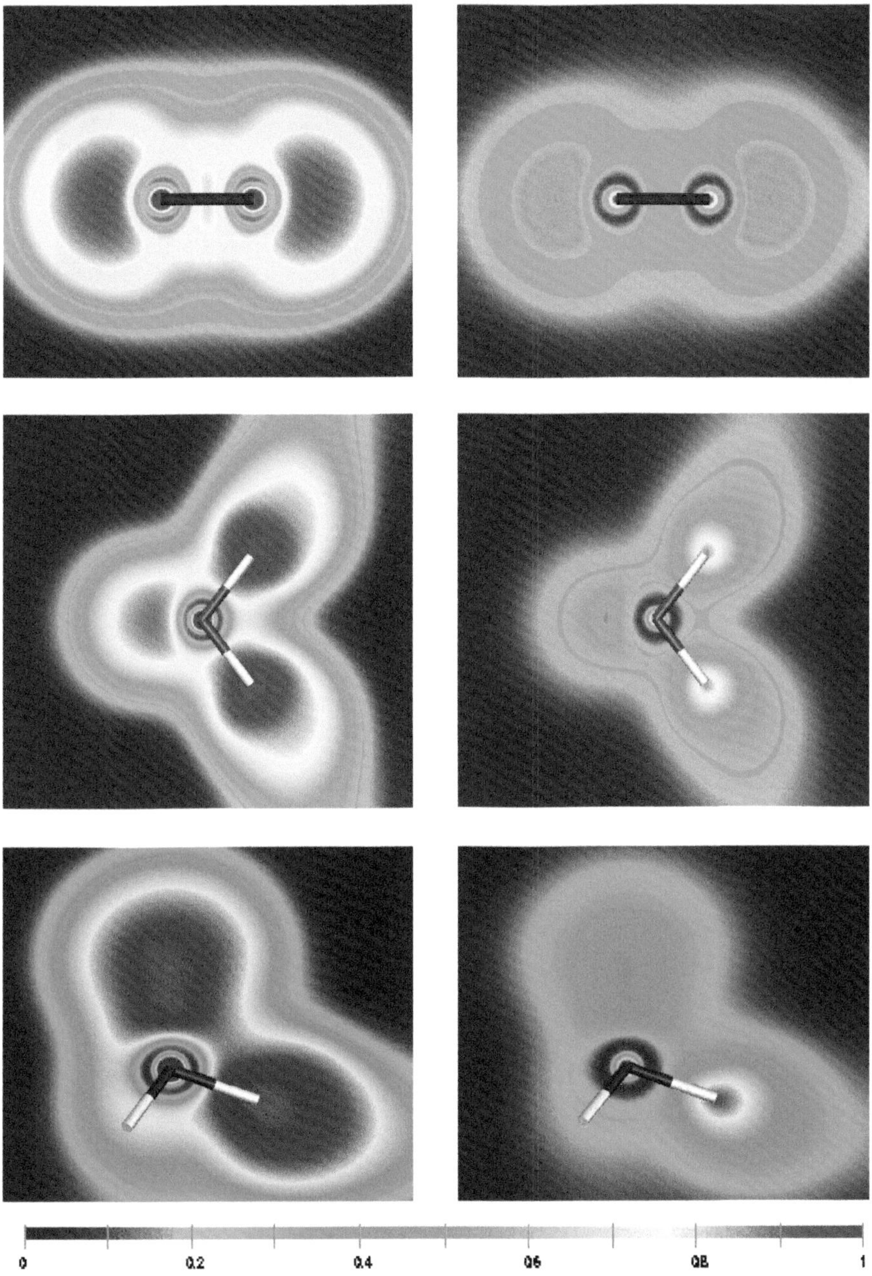

Figure 4.10. Plots of ELF(first column) and IT-ELF (second column) for PH_3 (first row) and B_2H_6 (second and third rows) on selected planes. The color scale for the ELF values is given in the bottom of the figure.

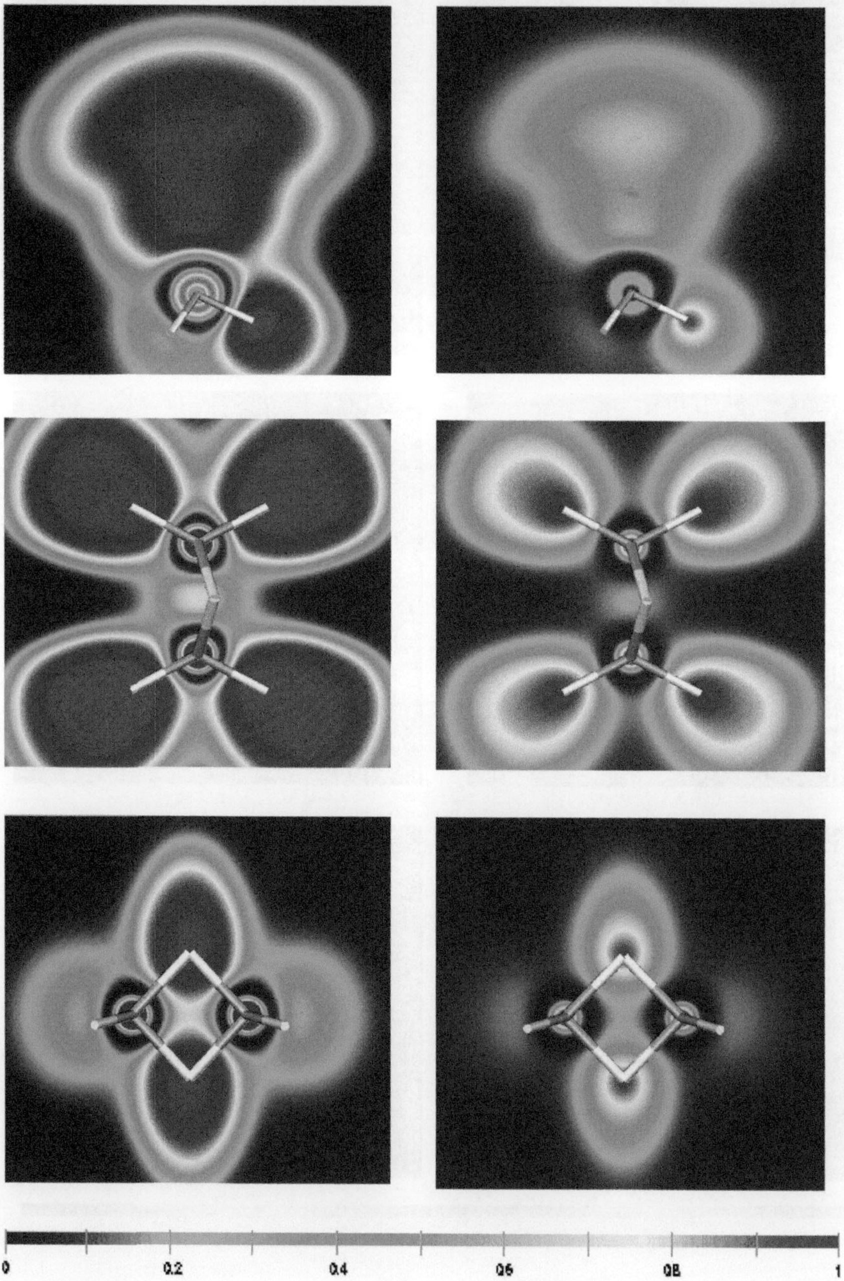

Figure 4.11. Plots of ELF(first column) and IT-ELF (second column) for the [1,1,1] (top row) and (2,2,2) (bottom row) propellanes of Fig. 4.5 on selected planes of sections. The color scale for the ELF values is given in the bottom of the figure.

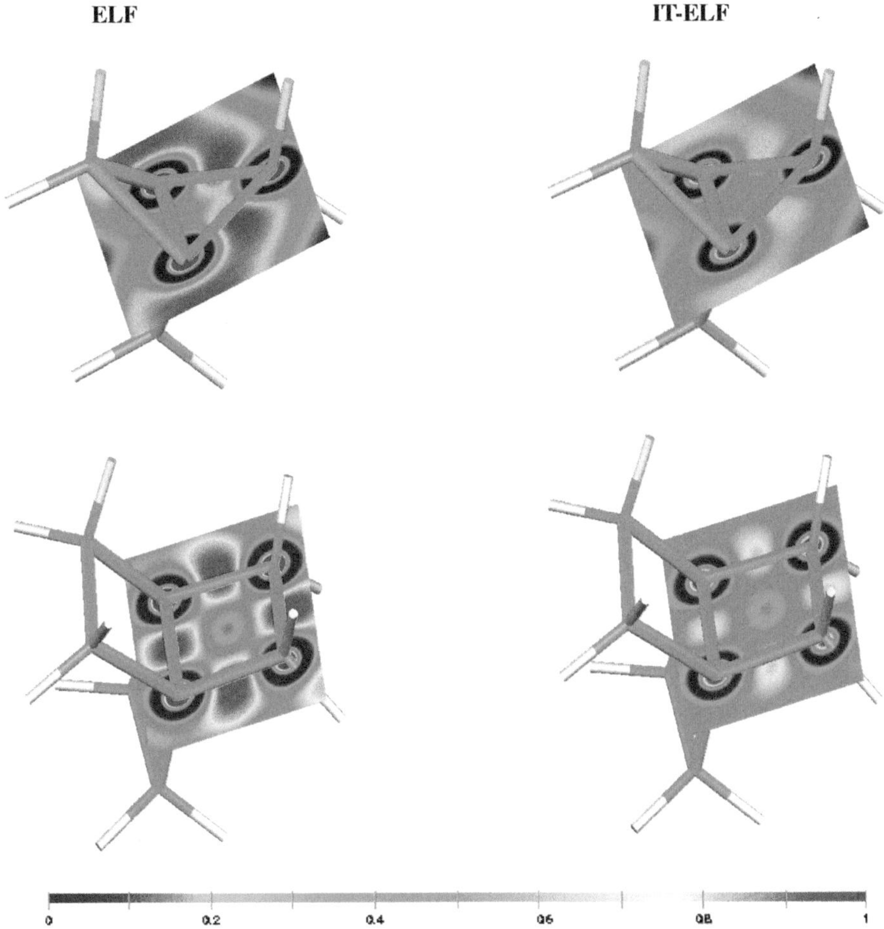

5. ATOMS-IN-MOLECULES FROM THE INFORMATION THEORY

The stockholder rule of Hirshfeld, for the exhaustive partition of the molecular electron density into the AIM components, is derived using the information variational principles. The densities of these bonded atoms are reported for illustrative diatomics and several integral descriptors of the Hirshfeld AIM in selected linear molecules of the preceding chapter are discussed, including the atomic net charges and the entropy deficiencies relative to the free-atom reference. The asymptotic properties of these bonded atoms are briefly examined. The stockholder principle of Hirshfeld is justified using both the local and global entropy deficiency principles of Kullback and Leibler in the Information Theory. The illustrative two-reference problems, e.g., those encountered in the AIM polarization (promotion) and molecular similarity problems, are also treated using the relevant constrained variational principles of the information-distance. The minimum missing-information principle is then used to generalize the *one*-electron division scheme of Hirshfeld into the corresponding *many*-electron stockholder rule, for an unbiased partitioning of the molecular joint probabilities involving several electrons. The two-electron case of this generalized stockholder partition is illustrated for the diatomic case. It is shown to provide an attractive perspective on chemical interactions between the *hard* and *soft* Lewis *acids* and *bases*. The effective one-electron densities of the two-electron stockholder AIM are compared with the corresponding Hirshfeld distributions. It is demonstrated for illustrative diatomics including hydrogen that the bonded atoms, which originate from the stockholder partitioning of the two-electron densities, emphasize more strongly the bonding region in comparison to their one-electron (Hirshfeld) analogs. For heavier constituent atoms the two stockholder partitions give rise to practically identical one-electron distributions. The variations of the information-theoretic AIM accompanying the bond dissociation are investigated. The influence of the *vertical* and *horizontal* effects of the electron correlation on the effective electron distributions of the two-electron stockholder atoms and their "covalent" (off-diagonal) and "ionic" (diagonal) components is examined in a more detail.

5.1. INTRODUCTION

In chemistry an understanding of the electronic structure of molecules and their preferences in reactions comes from transforming the computed (or experimental) electron distributions into statements in terms of chemical concepts, such as bonded atoms - building blocks of molecules, their collections defining larger fragments of the whole system, and the chemical bonds (e.g.: Mulliken, 1934, 1935, 1955; Pauling, 1960; Parr and Yang, 1989; Maksić, 1991; Bader, 1994; Nalewajski *et al.*, 1996; Nalewajski and Korchowiec, 1997; Nalewajski, 1993, 1995a,b, 1997a,b, 1998, 2003a). The latter represent the molecular "connectivities" between atoms and molecular fragments, on which an important part of the chemical science is based.

Indeed, chemistry deals with the mutually open subsystems, capable of exchanging electrons, which to a large extent preserve their identity in different molecular environments, being only slightly perturbed in their valence shells relative to the corresponding free-subsystem reference. The Information Theory (IT) (Shannon, 1948; Shannon and Weaver, 1949; Kullback and Leibler, 1951; Brillouin, 1956; Fisher, 1922, 1925, 1959; Kullback, 1959; Abramson, 1963; Ash, 1965; Frieden, 2000; see also Chapter 3) has recently been demonstrated to provide a framework for an unbiased extraction of such a chemical interpretation from the known molecular distributions of electrons (Nalewajski and Parr, 2000, 2001; Nalewajski, 2000c, 2002a-c, 2003a-d, 2004a-h; Nalewajski and Loska, 2001; Nalewajski and Broniatowska, 2003 a,b; Nalewajski *et al.*, 2002; Nalewajski and Świtka, 2002). It also allows one to formulate a thermodynamic-like description of molecules and their constituent parts (Nalewajski and Parr, 2001; Nalewajski, 2002c, 2003a,b, 2004a, 2005d) and - as we have shown in the preceding chapter - provides tools for probing the chemical bonds (Nalewajski, 2000c, 2004b-e, 2005a-c; Nalewajski and Jug, 2002; Nalewajski *et al.*, 2002; Nalewajski and Broniatowska, 2003a).

Fundamental to chemistry is an understanding of molecules as combinations of atoms. It is not surprising, then, that the concept of *Atoms-in-Molecules* (AIM) has been much discussed in the scientific literature. Let us recall that chemistry deals with mainly small changes of bonded atoms and larger molecular fragments, with reasonably well understood and *transferable* molecular invariants, such as AIM, functional groups, molecular subsystems, e.g., reactants and products of an elementary chemical reaction, etc., which tend to maintain their identity in different molecular environments. Most molecular systems may be thought of as consisting of only slightly perturbed atoms (or atomic ions), deformed by the presence of the molecular remainder and exhibiting modified net charges. These displacements in the atomic electronic structure are due to the coupled processes of the intra-atomic *polarization* (P) and the inter-atomic *charge transfer* (CT), which accompany the formation of chemical bonds.

The bonded fragments of a given molecular system represent the mutually-*open* subsystems capable of exchanging electrons with their respective molecular reminders. One would hope to find that a given AIM, like its free (non-bonded) analog, would possess a single cusp at the nucleus in its electron density, linked to the effective atomic number of the nucleus. The bonded atoms are pieces of the *molecule*, so that their electron densities $\{\rho_i\}$ must sum up to the molecular electron density $\rho = \sum_i \rho_i$. However, since the bonded atom preserves to a remarkably high degree the free-atom identity, the AIM distributions should be also closely related to densities $\{\rho_i^0\}$ of the separated atoms, which define the molecular prototype, called the *Atomic "Promolecule"* (AP) (Hirshfeld, 1977), given by the sum of the free-atom electron densities shifted to the actual AIM locations in the molecule (see Example 3.11). The resulting electron density, $\rho^0 = \sum_i \rho_i^0$, of this collection of the "frozen" atomic electron distributions, which defines the initial stage in the bond formation process, determines a natural reference for extracting changes due to the chemical bonds. Indeed, the familiar density difference function, $\Delta\rho = \rho - \rho^0$, has been widely used to probe changes in the electronic structure, which are ultimately responsible for the chemical bonds in the molecule (see the preceding chapter).

One would also expect some overlap between the densities of bonded atoms and a degree of their polarization towards the bonding partner, to reflect the presence of chemical bonds. That atomic ground-states, and/or small perturbation thereof, are uniquely appropriate reference states for a detailed description of AIM, was well understood by the pioneers, for instance Pauling (1960) and Mulliken (1935, 1955), in their use of such concepts as the orbital *hybridization*, atomic *promotion*, density polarization and bond *ionic/covalent* character. Ideally, the AIM definition should preserve as much *information* as possible about the separated atoms, since this implies their least promotion to the *valence-state* in the molecule.

The electronic structure of molecular systems is characterized by the *one-*, *two-* and *many-*electron probability distributions of the continuous ("fine-grained") description. To obtain its *chemical* interpretation, e.g., in terms of AIM, functional groups, reactants or other type of chemically significant subsystems, e.g., the σ or π electrons, these overall distributions have to be "discretized" in terms of the relevant pieces of the overall density attributed to the constituent parts of the molecular system under consideration, e.g., the bonded atoms or molecular fragments (AIM clusters). The densities of molecular subsystems constitute their *fine-grained* description. By an appropriate integration of the seelectron/probability densities one can then obtain the corresponding *condensed* descriptors of the electronic structure of molecules and their fragments, providing the discrete (*coarse-grained*) indices of the associated continuous electron distributions.

Additional resolution level resulting from theoretical calculations is provided by the *Atomic Orbital* (AO) or *Molecular Orbital* (MO) representations. For example, such a subatomic division scheme provides a basis for the familiar Mulliken *population analysis* of electron distributions in molecular systems and its numerous modifications and extensions.

Clearly, an exhaustive partitioning of a given molecular electron density between constituent (bonded) atoms, which determines the AIM effective net charges (oxidation states) in a given molecular environment, is not unique, since it depends on the adopted criterion for such a division (see, e.g., Parr *et al.*, 2005). The imposing variety of published theoretical methods for partitioning the molecular density into *"best"* AIM contributions testifies to the importance of this theme in chemistry. Different methods are based on different principles, some to a degree arbitrary or heuristic, which can produce conflicting trends in the associated AIM charges. Methods differ in the theoretical techniques used, e.g., topological analysis of the density, wave-function description, or the density-functional approach. They also differ in the physical/heuristic principles invoked, e.g., electronegativity equalization, zero flux, the minimum promotion-energy rules, and the minimum entropy-deficiency (information-distance, missing-information).

The historically first scheme of the Mulliken/Löwdin population analysis have used the *functional-space* partitioning, in which one distributes electrons between AO, which form the basis set for expanding the molecular orbitals of the Hartree-Fock (SCF LCAO MO) theory, non-orthogonal in the Mulliken approach and the symmetrically orthogonalized in the Löwdin variant. Another popular approach of Bader (1990), with a solid topological and quantum-mechanical basis, uses the *physical space* partitioning, i.e., a division of space into the exclusive atomic basins, with the boundaries determined by the zero-flux surfaces, on which the flow of electrons between subsystems vanishes. In the latter approach the spherical, spatially non-confined bonded atoms of the population analysis are replaced by the topological, non-spherical pieces of the molecular density, obtained as cuts along the zero-flux surfaces. As a result the topological AIM represent the spatially-confined (non-overlapping) and strongly non-symmetrical atoms.

Yet another exhaustive division scheme of Hirshfeld (1977) (see Example 3.11), which is widely used in crystallography, uses the "common-sense" *local* partitioning principle. It parallels the stock-market rule that in forming a molecule each atom locally partakes of a density gain or loss in the molecule ("profit") in proportion to its share in the promolecule density ("investment"). The overlapping stockholder AIM are, by construction, infinitely extending (not spatially confined) and known to be only slightly polarized relative to the free-atom reference. The topological and stockholder atoms are derived from the molecular electron density. They both preserve a "memory" of the original overall distribution of electrons in a molecule.

This *stockholder* division scheme has recently been shown to have a strong basis in IT (Nalewajski and Parr, 2000; Nalewajski, 2002a, 2003a,c). However, this partition is by no means unique, since the *two*-electron generalization of the stockholder principle (Nalewajski, 2002a) generates in molecules involving hydrogen slightly different effective one-electron distributions of chemical atoms, which in some cases emphasize more strongly the bonding (overlap) regions between the bond partners, relative to the corresponding one-electron stockholder atoms of Hirshfeld (Nalewajski and Broniatowska, 2005; Broniatowska, 2005).

Each division scheme has its own merits and specific disadvantages. For example, the net charges of the population analysis are known to suffer from a strong basis-set dependence. The spatially-confined, strongly non-symmetric atoms obtained from Bader's analysis, cannot be linked as ground-state densities to any effective external potential, due to their step-like behavior at the atomic basin boundary. Only their collection, i.e., the molecular ground-state density is in one-to-one correspondence with the molecular external potential (the Hohenberg-Kohn theorem). Using the familiar DFT terminology, one thus concludes that these confined density pieces are not the *subsystem* effective-external potential representable [see Eq. (1.6.7)] or v_i-representable for short. However, they all result from the spatial division of the molecular ground-state density, which itself is the *system* v-representable.

The v_i-representability of ith component ρ_i of $\rho = \sum_i \rho_i$ implies that it can be associated as the ground-state density with the subsystem *effective* external potential $v_i = v + v_i^e$ [Eq. (1.6.7)], including the molecular bare-nuclei contribution v and the embedding correction v_i^e due to the presence of electrons in the molecular reminder (Nalewajski and Parr, 2001). The fulfillment of the v_i-representability requirement introduces an attractive element of *causality* into such an AIM description, since then adjustments in the AIM densities relative to the isolated atoms could be viewed as the ground-state responses to displacements in the atomic effective external potentials.

In defining AIM, how can one preserve, to the extent possible, the information content of the ground-state free-atoms? In order to do it objectively, in an unbiased manner, it is natural to use some information-theoretic principle. And here is where DFT (Hohenberg and Kohn, 1964; Kohn and Sham, 1965; Parr and Yang, 1989) helps, since DFT states that the electron density itself carries all the information about the ground-state (see Section 1.5). So, we may define AIM in a way that makes the atomic densities resemble as much as possible the isolated atom densities, and thereby achieve the "best" atoms we can have in a molecule in an information-theoretic sense. As we shall see later in this chapter such an approach applied to a division of the electron density, or the associated one-electron probability distribution, leads to the Hirshfeld atoms, which, until recently, had suffered from a lack of a deeper theoretical justification. We shall also demonstrate that the stockholder atoms are solutions of the minimum entropy-deficiency principle of IT, which involves the *cross-entropy* similarity criterion of Kullback and Leibler (1951) between the atomic pieces of the molecular electron density and the corresponding densities of the free atoms defining the promolecule. As such they are the best, fully objective (unbiased) atomic one-electron distributions, which resemble the most the non-bonded electron densities of constituent atoms. In the next chapter we shall examine some general properties exhibited by the molecular Hirshfeld subsystems.

Alternative information quantities measuring the entropy deficiency between the two compared probability distributions, as well as the conditional entropy and mutual information concepts of the communication theory (see Chapter 3) have proven their utility in diverse molecular applications. They have been shown to be capable of directly addressing typical questions that chemists formulate in their investigations, ranging from the very definition of molecular subsystems and their similarity in different chemical environments, through the nature and origins of the chemical bond, to general rules of chemistry. The entropy/information descriptors of the chemical bond-orders will be covered in Chapters 7 and 8, which are devoted to the Communication Theory of the chemical bond. We shall also demonstrate in Chapter 10 that IT facilitates thermodynamic-like description of electronic gas in molecular systems, within the local approach to both the molecular fragment *equilibria* and *irreversible processes* in the subsystem resolution. The elements of a general thermodynamic description will be also covered, including the variational principles in the energy and information entropy representations.

In this chapter we shall explore the basic elements of the information-theoretic approach to bonded atoms. This development introduces the (information) *entropy representation*, complementary to the familiar *energy representation* of quantum mechanics, which facilitates the chemical interpretation of molecular properties and processes. The IT approach complements the familiar energetic characteristics of molecules and their fragments with the additional entropy descriptors, which – as we shall argue later in the book – are crucial for a more complete chemical interpretation of computed results.

5.2. ONE-ELECTRON STOCKHOLDER PRINCIPLE

It has been shown by Hirshfeld (1977) that the electron density $\rho(\boldsymbol{r})$ of the molecular system $M = (A^H \mid B^H \mid \ldots)$, consisting of the mutually-open atoms $X^H = (A^H, B^H, \ldots)$, as marked by the perpendicular broken lines separating the AIM symbols in M, is exhaustively partitioned $\rho(\boldsymbol{r}) = \sum_X \rho_X^H(\boldsymbol{r})$ into the "stockholder" AIM densities $\{\rho_X^H(\boldsymbol{r})\} \equiv \boldsymbol{\rho}^H(\boldsymbol{r})$ (see Example 3.11):

$$\rho_X^H(\boldsymbol{r}) = \rho_X^0(\boldsymbol{r})\,[\rho(\boldsymbol{r})/\rho^0(\boldsymbol{r})] \equiv \rho_X^0(\boldsymbol{r})\,w(\boldsymbol{r}) = \rho(\boldsymbol{r})\,[\rho_X^0(\boldsymbol{r})/\rho^0(\boldsymbol{r})] \equiv \rho(\boldsymbol{r})\,d_X^H(\boldsymbol{r}),$$
$$\sum_X d_X^H(\boldsymbol{r}) = 1. \qquad (5.2.1)$$

Here $\boldsymbol{\rho}^0(\boldsymbol{r}) = \{\rho_X^0(\boldsymbol{r})\}$ groups the densities of the free atoms, giving rise to the reference electron density $\rho^0(\boldsymbol{r}) = \sum_X \rho_X^0(\boldsymbol{r})$ of the isoelectronic promolecule $M^0 = (A^0 | B^0 | \ldots)$, consisting of the non-bonded (mutually closed) atoms $X^0 = A^0, B^0, \ldots$,, as marked by the perpendicular *solid* lines separating the AIM symbols in M^0:

$$N^0 = \int \rho^0(\boldsymbol{r})\,d\boldsymbol{r} = \sum_X \int \rho_X^0(\boldsymbol{r})\,d\boldsymbol{r} = \sum_X N_X^0 = N = \int \rho(\boldsymbol{r})d\boldsymbol{r} = \sum_X \int \rho_X^H(\boldsymbol{r})d\boldsymbol{r} = \sum_X N_X^H. \quad (5.2.2)$$

The free-atom densities $\boldsymbol{\rho}^0$ in AP are shifted to the respective atomic positions in the molecule and the vectors $\boldsymbol{N}^H = \{N_X^H\}$ and $\boldsymbol{N}^0 = \{N_X^0\}$ group the atomic average numbers of electrons of the bonded and free atoms, respectively. As we have already observed in the preceding chapter, the same promolecular reference is used to determine the familiar density difference function,

$$\Delta\rho(\boldsymbol{r}) = \rho(\boldsymbol{r}) - \rho^0(\boldsymbol{r}), \qquad (5.2.3)$$

which extracts changes in the electron distribution due to chemical bonds.

A reference to Eq. (5.2.1) shows that the Hirshfeld AIM densities satisfy the local principle of the *one*-electron stockholder division (see Example 3.11), which can be stated as the following equality between the local molecular and promolecular *conditional* probabilities [see Eq. (3.6.10) and Example 3.11]:

$$d_X^H(\boldsymbol{r}) = \rho_X^H(\boldsymbol{r})/\rho(\boldsymbol{r}) \equiv \pi^H(X|\boldsymbol{r}) = d_X^0(\boldsymbol{r}) = \rho_X^0(\boldsymbol{r})/\rho^0(\boldsymbol{r}) \equiv \pi^0(X|\boldsymbol{r}),$$
$$\Sigma_X\, \pi^H(X|\boldsymbol{r}) = \Sigma_X\, \pi^0(X|\boldsymbol{r}) = 1. \qquad (5.2.4)$$

It has been interpreted by Hirshfeld (1977) using the stock-market analogy: each atom participates locally in the molecular "profit" $\rho(\boldsymbol{r})$ in proportion to its share $d_X^0(\boldsymbol{r}) = \pi^0(X|\boldsymbol{r})$ in the promolecular "investment" $\rho^0(\boldsymbol{r})$. In the next section we shall demonstrate that this common-sense division rule has a solid basis in IT.

Let us extract the overall number of electrons N from the molecular and subsystem densities,

$$\rho(\boldsymbol{r}) = N\, p(\boldsymbol{r}) = N\, \Sigma_X\, p_X^H(\boldsymbol{r}) \qquad \text{and} \qquad \boldsymbol{\rho}^H(\boldsymbol{r}) = N\, \boldsymbol{p}^H(\boldsymbol{r}) = N\, \{p_X^H(\boldsymbol{r})\},$$
$$\Sigma_X \int p_X^H(\boldsymbol{r})\, d\boldsymbol{r} = \Sigma_X\, (N_X^H/N) \equiv \Sigma_X\, P_X^H = 1, \qquad (5.2.5)$$

where $p(\boldsymbol{r})$ and $\boldsymbol{p}^H(\boldsymbol{r})$ stand for the *molecularly* normalized shape-factors of the system as a whole and of its Hirshfeld atoms, respectively, while the vector $\boldsymbol{P}^H = \{P_X^H\}$ groups the condensed probabilities of finding an electron of M on the specified stockholder AIM.

The normalization of Eq. (5.2.5) reflects the important fact that bonded atoms are constituent parts of the molecule, so that the full normalization condition has to involve the summation/integration over the complete set of one-electron events, consisting of all possible "values" of the discrete argument X (atomic label) and all spatial locations of an electron, identified by continuous coordinates $\boldsymbol{r}$ in the subsystem probability distributions: $\boldsymbol{p}^H(\boldsymbol{r}) = \{p_X^H(\boldsymbol{r}) \equiv p^H(X, \boldsymbol{r})\}$.

The same normalization has to be adopted for the free-atom pieces of the one-electron probability distribution in the isoelectronic promolecule and for its free atom components, respectively,

$$p^0(\boldsymbol{r}) = \rho^0(\boldsymbol{r})/N = \Sigma_X\, p_X^0(\boldsymbol{r}), \qquad \text{and}$$
$$\boldsymbol{p}^0(\boldsymbol{r}) = \boldsymbol{\rho}^0(\boldsymbol{r})/N = \{p_X^H(\boldsymbol{r})\}, \qquad \Sigma_X \int p_X^0(\boldsymbol{r})\, d\boldsymbol{r} = \Sigma_X\, (N_X^0/N) \equiv \Sigma_X\, P_X^0 = 1, \qquad (5.2.6)$$

where $\boldsymbol{P}^0 = \{P_X^0\}$ collects the condensed probabilities of observing an electron of M^0 on the specified free atom. Again the *full* normalization of the shape (probability) factors $\boldsymbol{p}^0(\boldsymbol{r}) = \{p_X^0(\boldsymbol{r}) \equiv p^0(X, \boldsymbol{r})\}$ of the non-bonded atoms in the promolecular system involves summation over the discrete atomic "variable" X and integration over all positions $\boldsymbol{r}$ of an electron, the latter representing a continuous event arguments of the probability distributions of atomic fragments in M^0.

Therefore, the probability distributions of molecular/promolecular subsystems have to be spatially (partially) normalized to the fragment condensed probability in the system as a whole. In fact, the unity(atom)-normalized densities,

$$\boldsymbol{\pi}^H(\boldsymbol{r}) = \{\pi_X{}^H(\boldsymbol{r}) \equiv p^H(X, \boldsymbol{r})/P_X{}^H \equiv \pi^H(\boldsymbol{r}\,|X),\ \int \pi^H(\boldsymbol{r}\,|X)\, d\boldsymbol{r} = 1\} \qquad \text{and}$$
$$\boldsymbol{\pi}^0(\boldsymbol{r}) = \{\pi_X{}^0(\boldsymbol{r}) \equiv p^0(X, \boldsymbol{r})/P_X{}^0 \equiv \pi^0(\boldsymbol{r}\,|X),\ \int \pi^0(\boldsymbol{r}\,|X)\, d\boldsymbol{r} = 1\}, \qquad (5.2.7)$$

represent the complementary set of conditional probabilities (see Eq. (3.6.9) and Example 3.11), of finding an electron of X at position $\boldsymbol{r}$, in which the atomic label is not a variable but the *parameter*. Indeed, the normalization of such conditional distributions involves the integration over the position variable only. They represent the *separate* components of the molecular electron density, without any reference to the molecular or promolecular system.

As we have already observed in Eq. (5.2.4), the share factors represent the other set of the conditional probabilities of atomic subsystems in the molecule, which we have discussed in Section 3.6, of attributing an electron of M (or M^0), already found at $\boldsymbol{r}$, to a given atomic fragment X^H (or X^0). They represent the true AIM components in the molecule. In these local conditional probabilities the atomic label represents the discrete probability argument, while the electron space position plays the role of a continuous vector parameter.

It also follows from Eq. (5.2.1) that in the *one*-electron stockholder division each free subsystem density (or shape factor) is locally modified in accordance with the molecular (subsystem independent) density-enhancement factor $w(\boldsymbol{r})$:

$$w_X{}^H(\boldsymbol{r}) \equiv \rho_X{}^H(\boldsymbol{r})/\rho_X{}^0(\boldsymbol{r}) = p_X{}^H(\boldsymbol{r})/p_X{}^0(\boldsymbol{r}) = \rho(\boldsymbol{r})/\rho^0(\boldsymbol{r}) = p(\boldsymbol{r})/p^0(\boldsymbol{r}) \equiv w(\boldsymbol{r}). \qquad (5.2.8)$$

Therefore, this procedure is devoid of any subsystem bias and as such appears to be fully objective.

Representative plots of the overlapping electron densities of bonded hydrogen atoms in H_2 are shown in Fig. 5.1. They are seen to be distributed all over the physical space, decaying exponentially at large distances from the molecule and exhibiting in the bond density profile a single cusp at the atomic nucleus. They also display the expected polarization towards the bonding partner. These subsystem densities are highly transferable (Hirshfeld, 1977) and their overlap in the molecule accords with the classical interpretation of the origin of the chemical bond. One also observes a higher AIM density at the atomic nucleus, in comparison to the free-hydrogen density, i.e., a contraction of the AIM distribution at the expense of the nonbonding part of the free-atom density. This is due to the presence of the other atom causing an effective lowering of the molecular external potential relative to the atomic external potential. The stockholder pieces of the molecular electron density also exhibit several additional properties which make them attractive tools for chemical interpretations. Below we shall investigate some of these features in a more detail.

Let us first examine the asymptotic properties of the stockholder atomic densities. For simplicity we consider a diatomic system $M = (A^H|B^H)$, $\rho = \rho_A^H + \rho_B^H$, consisting of two Hirshfeld atoms A^H and B^H, the free analogs of which exhibit the relative electron *acceptor* (acidic) and *donor* (basic) properties respectively. This implies $\mu_A^0 = -I_A^0 < \mu_B^0 = -I_B^0$, where μ_X^0 and I_X^0 denote the chemical potential (negative electronegativity) and the ionization potential of X^0. Rewriting Eq. (5.2.1) in terms of the local density ratio $x = \rho_B^0/\rho_A^0$ gives:

$$\rho_A^H = (1 + x)^{-1}\rho \quad \text{and} \quad \rho_B^H = (1 + x^{-1})^{-1}\rho. \tag{5.2.9}$$

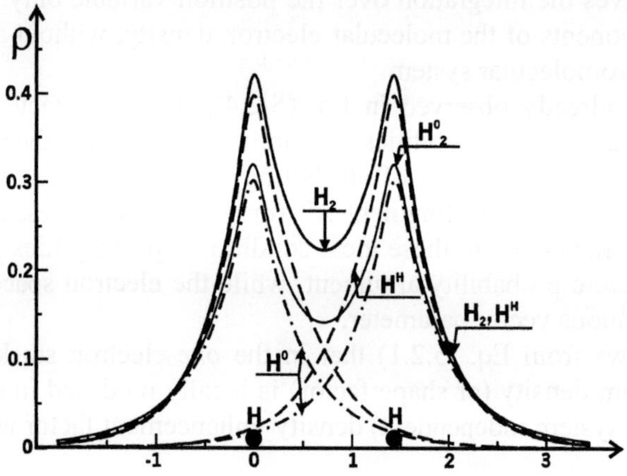

Figure 5.1. The Hirshfeld electron densities (H^H) of bonded hydrogen atoms obtained from the molecular density (H_2). The free-hydrogen densities (H^0) and the resulting electron density of the promolecule (H_2^0) are also shown for comparison. The density and inter-nuclear distance are in a.u. The zero cusps at nuclear positions are the artifacts of the Gaussian basis set used in DFT calculations.

Hence, for $r \to \infty$, when the distances from both nuclei become large compared to inter-atomic distance R_{AB}, $x \to \infty$ since the asymptotic behavior of the free subsystems of the promolecule are determined by the their ionization potentials, i.e., negative energies of the highest occupied Kohn-Sham orbitals (atomic chemical potentials):

$$\rho_X^0 \to \exp[-2(2I_X^0)\, r]\ (r \to \infty), \qquad X = A, B. \tag{5.2.10}$$

Thus, $r \to \infty$ implies $\rho_A^H \to 0$ and $\rho_B^H \to \rho$, so that the density of the softer (donor) atom B has a dominant contribution to the molecular density at distances from the molecule large compared to R_{AB}. This result is in full agreement with the subsystem Kohn-Sham analysis by van Leeuwen *et al.* (1996).

In Table 4.1 we have reported the net charges and entropy-deficiencies of the Hirshfeld AIM for illustrative linear molecules discussed in the preceding chapter (Nalewajski *et al.*, 2002). It is seen in the table that the Hirshfeld charges represent the chemical intuition quite well. For example, in the series of heteronuclear diatomics: HF, LiH, LiCl, and LiF, of increasing bond ionicity due to a growing electronegativity difference, the amount of charge transfer monotonically increases, as intuitively expected. The atomic missing information

$$\Delta S_X[p_X{}^H | p_X{}^0] = \int p_X{}^H(\boldsymbol{r}) \log[p_X{}^H(\boldsymbol{r})/p_X{}^0(\boldsymbol{r})] \, d\boldsymbol{r} , \qquad (5.2.11)$$

reflects the information distance between the atomic shape-factors $p_X{}^H(\boldsymbol{r})$ and $p_X{}^0(\boldsymbol{r})$. The reported values of these quantities are quite small, thus showing how strongly the bonded atoms resemble their free-atom analogs. The same general conclusion follows from examining the reported global entropy deficiencies (Example 3.12):

$$\Delta S[p | p^0] = \sum_X \Delta S_X[p_X{}^H | p_X{}^0] = \int p(\boldsymbol{r}) \log[p(\boldsymbol{r})/p^0(\boldsymbol{r})] \, d\boldsymbol{r} \equiv \int p(\boldsymbol{r}) \, I[w(\boldsymbol{r})] \, d\boldsymbol{r}. \qquad (5.2.12)$$

This strong similarity between the molecular and promolecular electron distributions is also seen in Fig. 5.1: the appreciable changes of the free-atom densities in the molecule are only observed around the nuclei (a contraction of the free-atom density) and in the bond region between the two nuclei (a polarization of the free atoms towards the bonding partner). In heavier atoms only a slight distortion of the valence (external) electrons is observed in the stockholder (bonded) atoms, with the inner shell structure left practically intact.

5.3. INFORMATION-THEORETIC JUSTIFICATION

As we have argued in the preceding section, the optimum local partition of the molecular density can be best formulated in terms of the unknown *conditional* probabilities $\boldsymbol{d}(\boldsymbol{r}) = \{d_X(\boldsymbol{r}) \equiv \pi(X|\boldsymbol{r})\} \equiv \boldsymbol{\pi}(\boldsymbol{r})$, which uniquely determine the AIM pieces of $p(\boldsymbol{r})$, $\{p_X(\boldsymbol{r}) = d_X(\boldsymbol{r})\, p(\boldsymbol{r})\}$ (Nalewajski, 2002a, 2003c). The local Kullback-Leibler function of the unknown share factors $\boldsymbol{d}(\boldsymbol{r})$, which measures the overall local information distance relative to the promolecule reference values $\boldsymbol{d}^0(\boldsymbol{r}) = \{\pi^0(X|\boldsymbol{r})\} \equiv \boldsymbol{\pi}^0(\boldsymbol{r})$, is given by the sum of subsystem contributions:

$$\Delta S[\boldsymbol{\pi}(\boldsymbol{r}) | \boldsymbol{\pi}^0(\boldsymbol{r})] = \sum_X \pi(X|\boldsymbol{r}) \ln[\pi(X|\boldsymbol{r})/\pi^0(X|\boldsymbol{r})]. \qquad (5.3.1)$$

The best (unbiased) share factors of subsystems minimize this missing information function subject to the normalization of the local conditional probabilities,

$$\delta \{\Delta S[\boldsymbol{\pi}(\boldsymbol{r}) | \boldsymbol{\pi}^0(\boldsymbol{r})] - \lambda \sum_X \pi (X | \boldsymbol{r})\} = 0, \qquad (5.3.2)$$

where λ stands for the appropriate Lagrange multiplier. The resulting Euler equation for the optimum local conditional probabilities,

$$\ln [\pi^{opt}(X|\boldsymbol{r})/\pi^{0}(X|\boldsymbol{r})] + (1-\lambda) \equiv \ln \{\pi^{opt}(X|\boldsymbol{r})/[C\pi^{0}(X|\boldsymbol{r})]\} = 0, \qquad (5.3.3)$$

or $\pi^{opt}(X|\boldsymbol{r}) = C\pi^{0}(X|\boldsymbol{r})$, when combined with the normalization constraint, $1 = \sum_X \pi^{opt}(X|\boldsymbol{r}) = C \sum_X \pi^{0}(X|\boldsymbol{r}) = C$, gives the Hirshfeld solution of Eqs. (5.2.1) and (5.2.4): $\boldsymbol{d}^{opt}(\boldsymbol{r}) = \boldsymbol{d}^{H}(\boldsymbol{r}) = \boldsymbol{d}^{0}(\boldsymbol{r})$. Therefore, the Hirshfeld choice of the local share factors minimizes the local information distance of Eq. (5.3.1) to the lowest value possible: $\Delta S[\boldsymbol{d}^{H}(\boldsymbol{r}) \,|\, \boldsymbol{d}^{0}(\boldsymbol{r})] = 0$.

The same answer follows from the relevant *global* information principles formulated in terms of either the electron densities or their shape factors (Nalewajski and Parr, 2000), in which one seeks the optimum atomic (or fragment) distributions exhibiting the strongest resemblance to the corresponding non-bonded, reference distributions, i.e., the minimum information distance between the corresponding free and bonded subsystems. We define the Kullback-Leibler (1951) information distance functional between the trial one-electron densities of atomic fragments $\{\rho_X\} \equiv \boldsymbol{\rho}$ (or the associated probability distributions $\{p_X\} \equiv \boldsymbol{p} = \boldsymbol{\rho}/N$) of the bonded atoms and the corresponding reference densities $\{\rho_X^{0}\} \equiv \boldsymbol{\rho}^{0}$ (or $\{p_X^{0}\} \equiv \boldsymbol{p}^{0} = \boldsymbol{\rho}^{0}/N$) of the free-atoms:

$$\begin{aligned}\Delta S[\boldsymbol{\rho}\,|\,\boldsymbol{\rho}^{0}] &= \textstyle\sum_X \int \rho_X(\boldsymbol{r}) \ln[\rho_X(\boldsymbol{r})/\rho_X^{0}(\boldsymbol{r})]\, d\boldsymbol{r} \equiv \sum_X \int \rho_X(\boldsymbol{r})\, I_X[w_X(\boldsymbol{r})]\, d\boldsymbol{r} \equiv \sum_X \Delta S_X[\rho_X \,|\, \rho_X^{0}] \\ &= N\,\Delta S[\boldsymbol{p}\,|\,\boldsymbol{p}^{0}] = N \textstyle\sum_X \int p_X(\boldsymbol{r}) \ln[p_X(\boldsymbol{r})/p_X^{0}(\boldsymbol{r})]\, d\boldsymbol{r} \\ &\equiv N \textstyle\sum_X \int p_X(\boldsymbol{r})\, I_X[w_X(\boldsymbol{r})]\, d\boldsymbol{r} \equiv N \sum_X \Delta S_X[p_X \,|\, p_X^{0}]. \end{aligned} \qquad (5.3.4)$$

Here, $I_X[w_X(\boldsymbol{r})] = \ln[p_X(\boldsymbol{r})/p_X^{0}(\boldsymbol{r})] = \ln[\rho_X(\boldsymbol{r})/\rho_X^{0}(\boldsymbol{r})]$, stands for the atomic surprisal for the current value of the AIM local enhancement factor relative to the free-subsystem reference: $w_X(\boldsymbol{r}) = \rho_X(\boldsymbol{r})/\rho_X^{0}(\boldsymbol{r}) = p_X(\boldsymbol{r})/p_X^{0}(\boldsymbol{r})$.

We have also indicated in the preceding equation that the directed divergence of $\boldsymbol{\rho}$ relative to $\boldsymbol{\rho}^{0}$, $\Delta S[\boldsymbol{\rho}\,|\,\boldsymbol{\rho}^{0}]$ is just N times the entropy deficiency $\Delta S[\boldsymbol{p}\,|\,\boldsymbol{p}^{0}]$. The same relation holds between the subsystem missing-information quantities: $\Delta S_X[\rho_X \,|\, \rho_X^{0}] = N\, \Delta S_X[p_X \,|\, p_X^{0}]$. Therefore, for the isoelectronic molecule and promolecule systems, the problem of normalization of the compared electronic densities does not influence the corresponding constrained variational principle for determining the optimum densities:

$$\delta\{\Delta S[\boldsymbol{\rho}\,|\,\boldsymbol{\rho}^{0}] - \textstyle\int \lambda(\boldsymbol{r}) \sum_X \rho_X(\boldsymbol{r})\, d\boldsymbol{r}\} = N\delta\{\Delta S[\boldsymbol{p}\,|\,\boldsymbol{p}^{0}] - \int \lambda(\boldsymbol{r}) \sum_X p_X(\boldsymbol{r})\, d\boldsymbol{r}\} = 0, \qquad (5.3.5)$$

where the local Lagrange multiplier $\lambda(\boldsymbol{r})$ enforces the exhaustive division condition at point $\boldsymbol{r}$: $\sum_X \rho_X(\boldsymbol{r}) = \rho(\boldsymbol{r})$ or $\sum_X p_X(\boldsymbol{r}) = p(\boldsymbol{r})$. Indeed, the above variational problem in terms of electron densities is then equivalent to the associated principle in terms of the probability distributions (shape-factors):

$$\delta\{\Delta S[\boldsymbol{p}\,|\,\boldsymbol{p}^0] - \int \lambda(\boldsymbol{r})\, \Sigma_X p_X(\boldsymbol{r})\, d\boldsymbol{r}\} = 0, \qquad (5.3.6)$$

with $\lambda(\boldsymbol{r})$ now multiplying the local exhaustive division constraint $\Sigma_X p_X(\boldsymbol{r}) = p(\boldsymbol{r})$.

It can be easily verified that both these variational principles give the same answer of the local stockholder division: $p_X^{opt}(\boldsymbol{r}) = p_X^H(\boldsymbol{r}) = \rho_X^H(\boldsymbol{r})/N$. Thus, the issue of the subsystem density normalization is not really a problem, provided that the shape factors of molecular fragments are properly normalized, as probability distributions of the constituent parts of the molecular system (see Parr *et al.*, 2005).

The Hirshfeld solution is independent of the adopted entropy/information measure. For example, the complementary directed divergence $\Delta S[\boldsymbol{\rho}^0\,|\,\boldsymbol{\rho}] = N\, \Delta S[\boldsymbol{p}^0\,|\,\boldsymbol{p}]$ or the symmetrized *divergence* of Kullback (1959) [see Eq. (3.3.7)],

$$\begin{aligned}\Delta S[\boldsymbol{\rho}^0, \boldsymbol{\rho}] &= \Delta S[\boldsymbol{\rho}\,|\,\boldsymbol{\rho}^0] + \Delta S[\boldsymbol{\rho}^0\,|\,\boldsymbol{\rho}] \\ &= \Sigma_X \int \Delta\rho_X(\boldsymbol{r}) \ln[\rho_X(\boldsymbol{r})/\rho_X^0(\boldsymbol{r})]\, d\boldsymbol{r} = N\, \Delta S[\boldsymbol{p}^0, \boldsymbol{p}],\end{aligned} \qquad (5.3.7)$$

where $\Delta\rho_X(\boldsymbol{r}) = \rho_X(\boldsymbol{r}) - \rho_X^0(\boldsymbol{r}) = N\, \Delta p_X(\boldsymbol{r})$, can be shown to give rise to the same solution (Nalewajski *et al.*, 2002; Nalewajski, 2004a).

The Fisher information measure for locality (intrinsic accuracy) gives rise to the differential equation for the optimum subsystem distributions. One of its particular solutions is the Hirshfeld prescription for the bonded atoms (Nalewajski, 2004a).

The bonding character of the stockholder AIM originates from the maximum similarity criterion of the densities of bonded atoms to the corresponding densities of free atoms, under the constraint of the fixed molecular density. In the global minimum entropy deficiency principle the missing information term provides an entropy/information "penalty" for the AIM densities deviating from the free atom densities.

5.4. ILLUSTRATIVE TWO-REFERENCE PROBLEMS

It is also instructive to examine the related information problem (Nalewajski and Parr, 2000), with the densities of AIM subject to a less stringent condition of the required overall normalization of the molecular electron density $\rho'(\boldsymbol{r}) = \Sigma_X \rho_X(\boldsymbol{r})$ resulting from the current AIM densities $\boldsymbol{\rho}$, $N[\rho'] = \int \rho'(\boldsymbol{r})\, d\boldsymbol{r} = N$, instead of the "rigid" condition of the exhaustive division at each point in space. We now assume the two-term penalty function, $\Delta S[\boldsymbol{\rho}\,|\,\boldsymbol{\rho}^0] + \Delta S[\rho'\,|\,\rho]$, to be minimized in the corresponding constrained variational principle. The first term reaches the minimum value, when the subsystem densities deviate as little as possible from the free-atom densities, while the second, competing penalty contribution calls for the maximum overall similarity of the trial molecular density $\rho'(\boldsymbol{r}) \equiv \Sigma_X \rho_X(\boldsymbol{r})$ to the true molecular density $\rho(\boldsymbol{r})$:

$$\delta\{\Delta S[\boldsymbol{\rho}\,|\,\boldsymbol{\rho}^0] + \Delta S[\rho'\,|\,\rho] - \zeta \textstyle\sum_X \int \rho_X(\boldsymbol{r})\, d\boldsymbol{r}\} = 0. \tag{5.4.1}$$

The auxiliary entropy functional now includes two sets of references: $\boldsymbol{\rho}^0$ (giving rise to ρ^0) and ρ (resulting from $\boldsymbol{\rho}^H$), with the equal (unbiased) "weight" attributed to each penalty term. Therefore, the variational principle (5.4.1) should produce only the partially "promoted" AIM densities, slightly different from those of bonded atoms, which mark a "transition" between the "promolecule" and molecule. The resulting Euler equation reads:

$$\ln[\rho_X^{opt}(\boldsymbol{r})/\rho_X^0(\boldsymbol{r})] + \ln[\rho^{opt}(\boldsymbol{r})/\rho(\boldsymbol{r})] + 2 - \zeta \equiv \ln\{\rho_X^{opt}(\boldsymbol{r})\rho^{opt}(\boldsymbol{r})/[\rho_X^0(\boldsymbol{r})\rho(\boldsymbol{r})\, c(\zeta)]\} = 0$$

$$\text{or} \qquad \rho_X^{opt}(\boldsymbol{r}) = c(\zeta)\, \rho_X^0(\boldsymbol{r})\, [\rho(\boldsymbol{r})/\rho^{opt}(\boldsymbol{r})], \tag{5.4.2}$$

where $\ln c(\zeta) = \zeta - 2$. Hence, $\rho^{opt}(\boldsymbol{r}) = \sum_X \rho_X^{opt}(\boldsymbol{r}) = c(\zeta)\rho(\boldsymbol{r})\rho^0(\boldsymbol{r})/\rho^{opt}(\boldsymbol{r})$ or

$$\rho^{opt}(\boldsymbol{r}) = [c(\zeta)]^{1/2}\, [\rho(\boldsymbol{r})\rho^0(\boldsymbol{r})]^{1/2}. \tag{5.4.3}$$

The optimum overall density is thus proportional to the geometric average of the initial (promolecular) and final (molecular) electron densities. The proportionality constant $c(\zeta)$ then follows from the value of the global constraint:

$$N = \int \rho^{opt}(\boldsymbol{r})\, d\boldsymbol{r} = c(\zeta)^{1/2} \int [\rho^0(\boldsymbol{r})\rho(\boldsymbol{r})]^{1/2}\, d\boldsymbol{r}. \tag{5.4.4}$$

Finally, combining Eqs. (5.4.2-4) gives

$$\rho_X^{opt}(\boldsymbol{r}) = \{N/\int (\rho^0\rho)^{1/2} d\boldsymbol{r}'\}\, \rho_X^0(\boldsymbol{r})[\rho(\boldsymbol{r})/\rho^0(\boldsymbol{r})]^{1/2} \equiv C\rho_X^0(\boldsymbol{r})w(\boldsymbol{r})^{1/2} = C[\rho_X^0(\boldsymbol{r})\, \rho_X^H(\boldsymbol{r})]^{1/2}. \tag{5.4.5}$$

Since the geometric mean $\rho^g(\boldsymbol{r}) \equiv [\rho^0(\boldsymbol{r})\rho(\boldsymbol{r})]^{1/2}$ of the isoelectronic molecular and promolecular electron densities, both integrating to N electrons, should also integrate to $N[\rho^g] \cong N^0$, $C \cong 1$. Thus, the optimum densities of this modified, two-reference entropy deficiency principle exhibit a "half-way" local enhancement, equal to the square root of the corresponding Hirshfeld value of Eq. (5.2.8).

Therefore, when the equally-weighted (unbiased) entropy deficiency (information penalty) terms in the information principle demand a simultaneous similarity of the optimized densities to two overall densities, $\rho[\boldsymbol{\rho}^0] = \rho^0$ and $\rho[\boldsymbol{\rho}^H] = \rho$, the optimum solution involves the geometric average of the two references. Indeed, the reader can also verify that the solution of Eq. (5.4.5) also follows from the explicit variational principle for the overall density $\rho' = \sum_X \rho_X(\boldsymbol{r})$:

$$\delta\{\Delta S[\rho'\,|\,\rho^0] + \Delta S[\rho'\,|\,\rho] - \zeta \int \rho'(\boldsymbol{r})\, d\boldsymbol{r}\} = 0. \tag{5.4.6}$$

A similar feature can be observed in the optimum subsystem densities of Eq. (5.4.5), where the free-atom densities $\{\rho_X^0(\boldsymbol{r})\}$ are proportional to the geometric mean of the two sets of subsystem references:

$$\rho_X^{opt}(N; \boldsymbol{r}) = C\rho_X^0(\boldsymbol{r})\,[\rho(\boldsymbol{r})/\rho^0(\boldsymbol{r})]^{1/2} = C\rho_X^0(\boldsymbol{r})[\rho_X^H(\boldsymbol{r})/\rho_X^0(\boldsymbol{r})]^{1/2}$$
$$\cong [\rho_X^H(\boldsymbol{r})\rho_X^0(\boldsymbol{r})]^{1/2} \equiv \rho_X^g(\boldsymbol{r}). \qquad (5.4.7)$$

The same solution also follows from the explicit principle for the optimum *"transition"* densities of promoted atoms, the information content of which is to be as close as possible to the corresponding densities of the free (initial) and bonded (final, Hirshfeld) AIM:

$$\delta\{\Delta S[\boldsymbol{\rho}\,|\,\boldsymbol{\rho}^0] + \Delta S[\boldsymbol{\rho}\,|\,\boldsymbol{\rho}^H] - \zeta \textstyle\sum_X \int \rho_X(\boldsymbol{r})\, d\boldsymbol{r}\} = 0. \qquad (5.4.8)$$

As an additional illustration of the versatility of the IT-approach let us modify the preceding equation by imposing the *closed* subsystem constraints, $\int \rho_X(\boldsymbol{r})\, d\boldsymbol{r} = N_X^0$, $X = A, B, \ldots$, instead of the global closure relation of Eq. (5.4.8):

$$\delta\{\Delta S[\boldsymbol{\rho}\,|\,\boldsymbol{\rho}^0] + \Delta S[\boldsymbol{\rho}\,|\,\boldsymbol{\rho}^H] - \textstyle\sum_X \lambda_X \int \rho_X(\boldsymbol{r})\, d\boldsymbol{r}\} = 0. \qquad (5.4.9)$$

The optimum densities of such *polarized*, mutually-closed and molecularly-"promoted" AIM in $M^c \equiv (A\,|\,B\,|\ldots)$ again involve the geometric mean of Eqs. (5.4.5) and (5.4.7), with slightly modified normalization constant:

$$\rho_X^{opt}(N_X^0; \boldsymbol{r}) = (N_X^0/\textstyle\int \rho_X^g d\boldsymbol{r}')\, \rho_X^g(\boldsymbol{r}) \equiv C_X\, \rho_X^g(\boldsymbol{r}) \cong \rho_X^g(\boldsymbol{r}), \qquad (5.4.10)$$

since again $\rho_X^g(\boldsymbol{r})$, the geometric mean of two subsystem densities containing N_X^0 electrons, should also integrate to $N[\rho_X^g] \cong N_X^0$, thus giving $C_X \cong 1$.

A similar two-reference problem arises in designing the information-theoretic measures of *molecular similarity*, a concept invoked to characterize structural resemblance of different molecules or their fragments (Nalewajski and Parr, 2000). Other measures have been successfully used in the past for this purpose, including the overlap integrals between the electronic densities, electrostatic potentials, and Fukui functions of Parr and Yang (1984, 1989).

In the IT approach the similarity of electronic structure implies a closeness of the information content of the compared electronic distributions, e.g., one-electron densities. Obviously, the most natural and direct indices of such an *information similarity* are the alternative measures of the information distance (entropy deficiency, missing information) or the mutual information (see Section 3.5). For example, in a typical screening search through the candidate species with densities $\{\rho_i\}$, which are being tested for chemical activity similar to that exhibited by the reference system with density ρ_0, one would select the system giving rise to the minimum value of $\Delta S[\rho_i|\rho_0]$ as the most promising molecule.

The simplest, *single-reference* variational criterion for the minimum entropy deficiency between the test (variational) molecular/subsystem shape factor $p(\boldsymbol{r}) = \rho(\boldsymbol{r})/N[\rho]$ and the reference distribution $p_0(\boldsymbol{r}) = \rho_0(\boldsymbol{r})/N[\rho_0]$, subject to the usual normalization constraint $\int p(\boldsymbol{r})\, d\boldsymbol{r} = 1$, gives the optimum solution $p = p_0$ (see Example

3.5). The most favorable matching is achieved, when the two shape factors are identical. The *two-reference* similarity problem arises, when one searches for the optimum matching between the trial electron probability distribution p and the two reference distributions, $p_1^{\,0}$ and $p_2^{\,0}$, subject to the usual probability normalization constraint:

$$\delta\{\Delta S[p|p_1^{\,0}] + \Delta S[p|p_2^{\,0}] - \omega \int p(\boldsymbol{r})\, d\boldsymbol{r}\} = 0. \qquad (5.4.11)$$

Solution of this entropy-deficiency principle is again proportional to the geometric mean of the two reference distributions:

$$p(\boldsymbol{r}) = [\int (p_1^{\,0} p_2^{\,0})^{1/2}\, d\boldsymbol{r}']^{-1}\, [p_1^{\,0}(\boldsymbol{r})\, p_2^{\,0}(\boldsymbol{r})]^{1/2} \cong [p_1^{\,0}(\boldsymbol{r})\, p_2^{\,0}(\boldsymbol{r})]^{1/2} \equiv p^{g}(\boldsymbol{r}). \qquad (5.4.12)$$

The highest value of the geometric mean of the two spatially separated reference probability distributions is expected in the region of the maximum "overlap" between them. Hence, the IT criterion of Eq. (5.4.12) calls for p exhibiting the maximum similarity to the overlap between the two reference probability densities.

One also encounters such similarity problems in the reactivity problems of the heterogeneous catalysis, when matching the adsorbate shape factor (p) with those of the two active sites of a catalyst, represented by their regional electron shape factors ($p_1^{\,0}$, $p_2^{\,0}$), which are involved in bonding a large adsorbate. Another illustrative case is a molecule binding to the two sites of another molecule, e.g., in cyclization reactions.

5.5. MANY-ELECTRON STOCKHOLDER PRINCIPLE

We shall now apply the entropy-deficiency principle to a division of the molecular k-electron joint distribution into the corresponding pieces corresponding to the AIM k-clusters (Nalewajski, 2002a, 2003a; Nalewajski and Broniatowska, 2005b). We shall use the same free-atom/promolecule reference as in the one-electron case. This information-theoretic approach gives rise to a generalization of Hirshfeld's one-electron division scheme into an appropriate "stockholder" rule for dividing the molecular joint probabilities of simultaneously finding $1 < k \leq N$ electrons at the specified locations in space, into the corresponding contributions in the AIM resolution (Nalewajski, 2002a, 2003a).

Consider an exhaustive partition of the molecular k-electron probability density (see Section 1.5),

$$p_k(\boldsymbol{r}, \boldsymbol{r}', \ldots, \boldsymbol{r}^{(k)}) = \rho_k(\boldsymbol{r}, \boldsymbol{r}', \ldots, \boldsymbol{r}^{(k)})/[N(N-1)\ \ldots\ (N-k+1)]$$
$$= \langle\Psi\,|\,\delta(\boldsymbol{r}_1 - \boldsymbol{r})\delta(\boldsymbol{r}_2 - \boldsymbol{r}')\ \ldots\ \delta(\boldsymbol{r}_k - \boldsymbol{r}^{(k)})\,|\,\Psi\rangle, \qquad 1 < k \leq N,$$
$$\int \ldots \int p_k(\boldsymbol{r}, \boldsymbol{r}', \ldots, \boldsymbol{r}^{(k)})\, d\boldsymbol{r}\, d\boldsymbol{r}' \ldots d\boldsymbol{r}^{(k)} = \langle\Psi\,|\,\Psi\rangle = 1, \qquad (5.5.1)$$

into contributions due to the AIM k-clusters $(X_1, X_2, ..., X_k) \equiv (\alpha, \beta, ..., \gamma) \equiv G$:

$$p_k(\boldsymbol{r}, \boldsymbol{r}', ..., \boldsymbol{r}^{(k)}) = \Sigma_\alpha \Sigma_\beta ... \Sigma_\gamma\, p_{\alpha\beta...\gamma}(\boldsymbol{r}, \boldsymbol{r}', ..., \boldsymbol{r}^{(k)}) \equiv \Sigma_G\, p_G(\boldsymbol{r}, \boldsymbol{r}', ..., \boldsymbol{r}^{(k)}). \qquad (5.5.2a)$$

Here $p_G(\boldsymbol{r}, \boldsymbol{r}', ..., \boldsymbol{r}^{(k)})$ is the probability of the joint k-electron event that an electron of the molecule, found at $\boldsymbol{r}$, originates from atom α, and another electron, located at $\boldsymbol{r}'$, is from atom β, etc. It satisfies the normalization condition:

$$\Sigma_\alpha \Sigma_\beta ... \Sigma_\gamma \int ... \int p_{\alpha\beta...\gamma}(\boldsymbol{r}, \boldsymbol{r}', ..., \boldsymbol{r}^{(k)})\, d\boldsymbol{r}\, d\boldsymbol{r}' ... d\boldsymbol{r}^{(k)} \equiv \Sigma_G P_G = 1, \qquad (5.5.3)$$

where $P_{\alpha\beta...\gamma} \equiv P_G$ stands for the condensed probability of the cluster G of AIM.

The isoelectronic ($N = N^0$) reference state of the (distinguishable) free-atoms of the promolecule is given by the product of the atomic (antisymmetric) states:

$$\Psi^0(\mathbf{x}) = \Pi_X \Psi_X^0(\boldsymbol{x}_X). \qquad (5.5.4)$$

Hence the reference k-electron distribution for such a collection of non-bonded atoms

$$\begin{aligned} p_k^0(\boldsymbol{r}, \boldsymbol{r}', ..., \boldsymbol{r}^{(k)}) &= \rho_k^0(\boldsymbol{r}, \boldsymbol{r}', ..., \boldsymbol{r}^{(k)})/[N(N-1) ... (N-k+1)] \\ &= \langle \Psi^0 | \delta(\boldsymbol{r}_1 - \boldsymbol{r})\delta(\boldsymbol{r}_2 - \boldsymbol{r}') ... \delta(\boldsymbol{r}_k - \boldsymbol{r}^{(k)}) | \Psi^0 \rangle, \\ \int ... \int p_k^0(\boldsymbol{r}, \boldsymbol{r}', ..., \boldsymbol{r}^{(k)})\, d\boldsymbol{r}\, d\boldsymbol{r}' ... d\boldsymbol{r}^{(k)} &= \langle \Psi^0 | \Psi^0 \rangle = 1, \end{aligned} \qquad (5.5.5)$$

and its AIM k-cluster components:

$$p_k^0(\boldsymbol{r}, \boldsymbol{r}', ..., \boldsymbol{r}^{(k)}) = \Sigma_\alpha \Sigma_\beta ... \Sigma_\gamma\, p_{\alpha\beta...\gamma}(\boldsymbol{r}, \boldsymbol{r}', ..., \boldsymbol{r}^{(k)}) \equiv \Sigma_G\, p_G^0(\boldsymbol{r}, \boldsymbol{r}', ..., \boldsymbol{r}^{(k)}), \qquad (5.5.2b)$$

$$\begin{aligned} p_{\alpha\beta...\gamma}^0(\boldsymbol{r}, \boldsymbol{r}', ..., \boldsymbol{r}^{(k)}) &= \langle \Psi^0 | \delta(\boldsymbol{r}_{i\in\alpha} - \boldsymbol{r})\delta(\boldsymbol{r}_{j(\neq i)\in\beta} - \boldsymbol{r}') ... \delta(\boldsymbol{r}_{k[\neq(i,j,...)]\in\gamma} - \boldsymbol{r}^{(k)}) | \Psi^0 \rangle \\ &= \rho_{\alpha,\beta,...,\gamma}^0(\boldsymbol{r}, \boldsymbol{r}', ..., \boldsymbol{r}^{(k)})/[N(N-1) ... (N-k+1)]. \end{aligned} \qquad (5.5.6)$$

The latter satisfy the normalization condition:

$$\Sigma_\alpha \Sigma_\beta ... \Sigma_\gamma \int ... \int p_{\alpha\beta...\gamma}^0(\boldsymbol{r}, \boldsymbol{r}', ..., \boldsymbol{r}^{(k)})\, d\boldsymbol{r}\, d\boldsymbol{r}' ... d\boldsymbol{r}^{(k)} \equiv \Sigma_G P_G^0 = 1, \qquad (5.5.7)$$

where P_G^0 denotes the condensed probability of cluster G of free atoms in the promolecule.

Following the Hirshfeld development the k-electron generalization of the *stockholder* (S) principle should call for the participation of the AIM k-cluster G in the overall k-electron molecular "profit", $p_k(\boldsymbol{r}, \boldsymbol{r}', ..., \boldsymbol{r}^k)$, in accordance with this cluster "share" (conditional probability)

$$d_G^0(\boldsymbol{r}, \boldsymbol{r}', ..., \boldsymbol{r}^{(k)}) = p_G^0(\boldsymbol{r}, \boldsymbol{r}', ..., \boldsymbol{r}^{(k)})/p_k^0(\boldsymbol{r}, \boldsymbol{r}', ..., \boldsymbol{r}^{(k)}) \equiv \pi^0(G | \boldsymbol{r}, \boldsymbol{r}', ..., \boldsymbol{r}^{(k)}) \qquad (5.5.8)$$

in the overall k-electron "investment" $p_k^0(\mathbf{r}, \mathbf{r}', ..., \mathbf{r}^{(k)})$ of the promolecule.

The k-electron share factors $\{d_G^0(\mathbf{r}, \mathbf{r}', ..., \mathbf{r}^{(k)})\}$ represent the conditional probabilities $\{\pi^0(\alpha, \beta, ..., \gamma | \mathbf{r}, \mathbf{r}', ..., \mathbf{r}^{(k)}) \equiv \pi^0(G | \mathbf{r}, \mathbf{r}', ..., \mathbf{r}^{(k)})\}$ that the k electrons, already simultaneously located at indicated electron positions $(\mathbf{r}, \mathbf{r}', ..., \mathbf{r}^{(k)})$ (parameters), are attributed to the specified AIM $(\alpha, \beta, ..., \gamma)$ (variables). They must satisfy the appropriate normalization condition,

$$\Sigma_\alpha \Sigma_\beta ... \Sigma_\gamma \pi^0(\alpha, \beta, ..., \gamma | \mathbf{r}, \mathbf{r}', ..., \mathbf{r}^{(k)}) = \Sigma_G \pi^0(G | \mathbf{r}, \mathbf{r}', ..., \mathbf{r}^{(k)}) = 1. \quad (5.5.9)$$

which assures the exhaustive stockholder partitioning of Eq. (5.5.2a).

Therefore, the unbiased k-electron stockholder principle should read:

$$d_G^S(\mathbf{r}, \mathbf{r}', ..., \mathbf{r}^{(k)}) = p_G^S(\mathbf{r}, \mathbf{r}', ..., \mathbf{r}^{(k)})/p_k(\mathbf{r}, \mathbf{r}', ..., \mathbf{r}^{(k)}) \equiv \pi^S(G|\mathbf{r}, \mathbf{r}', ..., \mathbf{r}^{(k)})$$
$$= d_G^0(\mathbf{r}, \mathbf{r}', ..., \mathbf{r}^{(k)}) \equiv \pi^0(G|\mathbf{r}, \mathbf{r}', ..., \mathbf{r}^{(k)}), \quad (5.5.10)$$

where the normalization of the molecular conditional probabilities also obeys the relevant normalization:

$$\Sigma_G \pi^S(G | \mathbf{r}, \mathbf{r}', ..., \mathbf{r}^{(k)}) = 1. \quad (5.5.11)$$

This generalized stockholder-division scheme can be also interpreted as the unbiased (AIM-cluster independent) "enhancement" of the k-cluster distributions of the free (non-bonded) atoms defining the promolecule:

$$p_G^S(\mathbf{r}, \mathbf{r}', ..., \mathbf{r}^{(k)}) = d_G^0(\mathbf{r}, \mathbf{r}', ..., \mathbf{r}^{(k)})\, p_k(\mathbf{r}, \mathbf{r}', ..., \mathbf{r}^{(k)})$$
$$= p_G^0(\mathbf{r}, \mathbf{r}', ..., \mathbf{r}^{(k)})\, [p_k(\mathbf{r}, \mathbf{r}', ..., \mathbf{r}^{(k)})/p_k^0(\mathbf{r}, \mathbf{r}', ..., \mathbf{r}^{(k)})]$$
$$\equiv p_G^0(\mathbf{r}, \mathbf{r}', ..., \mathbf{r}^{(k)})\, w_k(\mathbf{r}, \mathbf{r}', ..., \mathbf{r}^{(k)}). \quad (5.5.12)$$

Clearly, the same k-electron shares or enhancements apply to the division of the corresponding electron densities, for which the above probability distributions provide the associated shape factors:

$$\rho_G^S(\mathbf{r}, \mathbf{r}', ..., \mathbf{r}^{(k)}) = N(N-1) ... (N-k+1)\, p_G^S(\mathbf{r}, \mathbf{r}', ..., \mathbf{r}^{(k)})$$
$$= d_G^0(\mathbf{r}, \mathbf{r}', ..., \mathbf{r}^{(k)})\, \rho_k(\mathbf{r}, \mathbf{r}', ..., \mathbf{r}^{(k)})$$
$$= \rho_G^0(\mathbf{r}, \mathbf{r}', ..., \mathbf{r}^{(k)})\, w_k(\mathbf{r}, \mathbf{r}', ..., \mathbf{r}^{(k)}). \quad (5.5.13)$$

This unbiased division rule for the molecular k-electron distributions can be straightforwardly justified using the relevant missing-information principle. For example, one can formulate the Kullback-Leibler entropy deficiency principle in terms of the unknown k-electron share-factors $\mathbf{d}_k(\mathbf{r}, \mathbf{r}', ..., \mathbf{r}^{(k)}) = \{d_G(\mathbf{r}, \mathbf{r}', ..., \mathbf{r}^{(k)}) = \pi(G | \mathbf{r}, \mathbf{r}', ..., \mathbf{r}^{(k)})\}$, referenced to their known free-atom analogs $\mathbf{d}_k^0(\mathbf{r}, \mathbf{r}', ..., \mathbf{r}^{(k)}) =$

$\{d_G^0(\boldsymbol{r}, \boldsymbol{r}', ..., \boldsymbol{r}^{(k)}) = \pi^0(G|\boldsymbol{r}, \boldsymbol{r}', ..., \boldsymbol{r}^{(k)})\}$. As we have already indicated in Eqs. (5.5.8) and (5.5.10), these probability ratios represent the conditional probabilities that k monitored electrons, found simultaneously in positions $(\boldsymbol{r}, \boldsymbol{r}', ..., \boldsymbol{r}^{(k)})$, originate from the current AIM cluster $G = (\alpha, \beta, ..., \gamma)$. Thus, for a given set of the simultaneous positions of k electrons, the relevant information principle for the optimum values of $\mathbf{d}_k(\boldsymbol{r},..., \boldsymbol{r}^{(k)})$ must now involve the minimum of the k-electron entropy deficiency function,

$$\Delta S[\mathbf{d}_k|\mathbf{d}_k^0] = \Sigma_G\, d_G(\boldsymbol{r}, ..., \boldsymbol{r}^{(k)}) \ln[d_G(\boldsymbol{r}, ..., \boldsymbol{r}^{(k)})/d_G^0(\boldsymbol{r}, ..., \boldsymbol{r}^{(k)})], \tag{5.5.14}$$

subject to the normalization of Eq. (5.5.11):

$$\delta\{\Delta S[\mathbf{d}_k|\mathbf{d}_k^0] - \lambda\, \Sigma_G\, d_G(\boldsymbol{r}, ..., \boldsymbol{r}^{(k)})\} = 0. \tag{5.5.15}$$

The resulting Euler Equation,

$$\ln\{d_G(\boldsymbol{r}, ..., \boldsymbol{r}^{(k)})/[C\, d_G^0(\boldsymbol{r}, ..., \boldsymbol{r}^{(k)})]\} = 0, \qquad \ln C = \lambda - 1,$$
$$\text{or} \qquad d_G(\boldsymbol{r}, ..., \boldsymbol{r}^{(k)}) = C\, d_G^0(\boldsymbol{r}, ..., \boldsymbol{r}^{(k)}), \tag{5.5.16}$$

with the constant $C = 1$ determined from the normalization constraint, then gives the k-electron stockholder principle of Eq. (5.5.10): $\mathbf{d}_k^{opt} = \mathbf{d}_k^S = \mathbf{d}_k^0$.

The same optimum partitioning follows from the minimum principle for the overall (integral) entropy deficiency between the unknown k-electron probability pieces $\mathbf{p}_k(\boldsymbol{r}, ..., \boldsymbol{r}^{(k)}) = \{p_G(\boldsymbol{r}, ..., \boldsymbol{r}^{(k)})\}$ and their free-atom/promolecule reference analogs, $\mathbf{p}_k^0(\boldsymbol{r}, ..., \boldsymbol{r}^{(k)}) = \{p_G^0(\boldsymbol{r}, ..., \boldsymbol{r}^{(k)})\}$, subject to the subsidiary condition of the exhaustive division [Eq. (5.5.2a)] (Nalewajski, 2002a):

$$\delta\{\Delta S[\mathbf{p}_k|\mathbf{p}_k^0] - \int \lambda(\boldsymbol{r}, ..., \boldsymbol{r}^{(k)})\, \Sigma_G\, p_G(\boldsymbol{r}, ..., \boldsymbol{r}^{(k)})\, d\boldsymbol{r} \ldots d\boldsymbol{r}^{(k)}\} = 0, \tag{5.5.17}$$

where

$$\Delta S[\mathbf{p}_k|\mathbf{p}_k^0] = \Sigma_G \int \ldots \int p_G(\boldsymbol{r}, ..., \boldsymbol{r}^{(k)})\, I_G(\boldsymbol{r}, ..., \boldsymbol{r}^{(k)})\, d\boldsymbol{r} \ldots d\boldsymbol{r}^{(k)}, \tag{5.5.18}$$

and the k-electron surprisal of cluster G

$$I_G(\boldsymbol{r}, ..., \boldsymbol{r}^{(k)}) = \ln[p_G(\boldsymbol{r}, ..., \boldsymbol{r}^{(k)})/p_G^0(\boldsymbol{r}, ..., \boldsymbol{r}^{(k)})] \equiv I_G[w_G(\boldsymbol{r}, ..., \boldsymbol{r}^{(k)})]. \tag{5.5.19}$$

In Eq. (5.5.17) $\lambda(\boldsymbol{r},..., \boldsymbol{r}^{(k)})$ stands for the Lagrange multiplier function enforcing the exhaustive division constraint for the specified positions of k electrons. It can be easily verified, using the corresponding Euler equation,

$$I_G(\boldsymbol{r}, ..., \boldsymbol{r}^{(k)}) + 1 - \lambda(\boldsymbol{r}, ..., \boldsymbol{r}^{(k)}) \equiv \ln\{p_G(\boldsymbol{r}, ..., \boldsymbol{r}^{(k)})/[D(\boldsymbol{r}, ..., \boldsymbol{r}^{(k)})p_G^0(\boldsymbol{r}, ..., \boldsymbol{r}^{(k)})]\} = 0$$
$$\text{or} \qquad p_G(\boldsymbol{r}, ..., \boldsymbol{r}^{(k)}) = D(\boldsymbol{r}, ..., \boldsymbol{r}^{(k)})\, p_G^0(\boldsymbol{r}, ..., \boldsymbol{r}^{(k)}), \tag{5.5.20}$$

and the local constraint,

$$p_k(\boldsymbol{r}, \ldots, \boldsymbol{r}^{(k)}) = D(\boldsymbol{r}, \ldots, \boldsymbol{r}^{(k)})\, p_k^{\,0}(\boldsymbol{r}, \ldots, \boldsymbol{r}^{(k)}), \tag{5.5.21}$$

that this information principle also gives the stockholder solution of Eq. (5.5.10):

$$\begin{aligned} p_G^{\,opt}(\boldsymbol{r}, \ldots, \boldsymbol{r}^{(k)}) &= p_G^{\,S}(\boldsymbol{r}, \ldots, \boldsymbol{r}^{(k)}) = d_G^{\,0}(\boldsymbol{r}, \ldots, \boldsymbol{r}^{(k)})\, p_k(\boldsymbol{r}, \ldots, \boldsymbol{r}^{(k)}) \\ &= p_G^{\,0}(\boldsymbol{r}, \ldots, \boldsymbol{r}^{(k)})\, w_k(\boldsymbol{r}, \ldots, \boldsymbol{r}^{(k)}). \end{aligned} \tag{5.5.22}$$

The optimum AIM probability distributions are thus characterized by the cluster independent, universal k-electron enhancement factor:

$$w_G^{\,opt}(\boldsymbol{r}, \ldots, \boldsymbol{r}^{(k)}) = w_G^{\,S}(\boldsymbol{r}, \ldots, \boldsymbol{r}^{(k)}) = w_k(\boldsymbol{r}, \ldots, \boldsymbol{r}^{(k)}) = p_k(\boldsymbol{r}, \ldots, \boldsymbol{r}^{(k)})/p_k^{\,0}(\boldsymbol{r}, \ldots, \boldsymbol{r}^{(k)}). \tag{5.5.23}$$

Moreover, since the same, unbiased local enhancement is applied to all atomic clusters, the preceding equation also implies the equalization of the local cluster surprisals at the corresponding value for the system as a whole:

$$I_G[w_G^{\,S}(\boldsymbol{r}, \ldots, \boldsymbol{r}^{(k)})] = I_{G'}[w_{G'}^{\,S}(\boldsymbol{r}, \ldots, \boldsymbol{r}^{(k)})] = \ldots = I[w_k(\boldsymbol{r}, \ldots, \boldsymbol{r}^{(k)})]. \tag{5.5.24}$$

5.6. STOCKHOLDER PARTITION OF TWO-ELECTRON DISTRIBUTIONS IN DIATOMICS

As an illustrative example let us consider a diatomic molecule $M = A\text{–}B$, for which the stockholder division of its joint two-electron probability distribution $p_2(\boldsymbol{r}, \boldsymbol{r}')$ [or the corresponding electron pair-density $\rho_2(\boldsymbol{r}, \boldsymbol{r}')$] involves four AIM 2-cluster components:

$$\begin{aligned} &\mathbf{p}^S(\boldsymbol{r}, \boldsymbol{r}') = \{p_{\alpha\beta}^{\,S}(\boldsymbol{r}, \boldsymbol{r}')\}, \qquad \boldsymbol{\rho}_2^{\,S}(\boldsymbol{r}, \boldsymbol{r}') = \{\rho_{\alpha\beta}^{\,S}(\boldsymbol{r}, \boldsymbol{r}')\}, \qquad \alpha, \beta = A, B; \\ &p_2(\boldsymbol{r}, \boldsymbol{r}') = \textstyle\sum_{\alpha=A,B} \sum_{\beta=A,B} p_{\alpha\beta}^{\,S}(\boldsymbol{r}, \boldsymbol{r}'), \\ &\rho_2(\boldsymbol{r}, \boldsymbol{r}') = N(N-1)\, p_2(\boldsymbol{r}, \boldsymbol{r}') = \textstyle\sum_{\alpha=A,B} \sum_{\beta=A,B} \rho_{\alpha\beta}^{\,S}(\boldsymbol{r}, \boldsymbol{r}'), \\ &\rho_{\alpha\beta}^{\,S}(\boldsymbol{r}, \boldsymbol{r}') = N(N-1)\, p_{\alpha\beta}^{\,S}(\boldsymbol{r}, \boldsymbol{r}'). \end{aligned} \tag{5.6.1}$$

These four pieces of the molecular two-electron probability distribution are defined by Eqs. (5.5.22) and (5.5.23):

$$\begin{aligned} p_{\alpha\beta}^{\,S}(\boldsymbol{r}, \boldsymbol{r}') &= [p_{\alpha\beta}^{\,0}(\boldsymbol{r}, \boldsymbol{r}')/p_2^{\,0}(\boldsymbol{r}, \boldsymbol{r}')]\, p_2(\boldsymbol{r}, \boldsymbol{r}') \equiv d_{\alpha\beta}(\boldsymbol{r}, \boldsymbol{r}')\, p_2(\boldsymbol{r}, \boldsymbol{r}') \\ &= [p_2(\boldsymbol{r}, \boldsymbol{r}')/p_2^{\,0}(\boldsymbol{r}, \boldsymbol{r}')]\, p_{\alpha\beta}^{\,0}(\boldsymbol{r}, \boldsymbol{r}') \equiv w_2(\boldsymbol{r}, \boldsymbol{r}')\, p_{\alpha\beta}^{\,0}(\boldsymbol{r}, \boldsymbol{r}'), \end{aligned} \tag{5.6.2}$$

where the reference two-electron probability distribution of the isoelectronic promolecule $p_2^{\,0}(\boldsymbol{r}, \boldsymbol{r}') = \sum_{\alpha=A,B} \sum_{\beta=A,B} p_{\alpha\beta}^{\,0}(\boldsymbol{r}, \boldsymbol{r}')$, $d_{\alpha\beta}(\boldsymbol{r}, \boldsymbol{r}')$ stands for the two-electron *share factor* of the ($\alpha\beta$)-cluster and $w_2(\boldsymbol{r}, \boldsymbol{r}')$ denotes the universal (molecular, cluster-independent) *enhancement factor* for the specified positions of two electrons in space. There are the two *atomic* (diagonal, $\alpha = \beta$) and two *diatomic* (off-diagonal, $\alpha \neq \beta$) contributions in Eq. (5.6.1). Their share-factors are proportional to the corresponding reference distributions $\mathbf{p}^0(\boldsymbol{r}, \boldsymbol{r}') = \{p_{\alpha\beta}{}^0(\boldsymbol{r}, \boldsymbol{r}')\}$ of the free atoms in AP:

$$p_{\alpha\beta}{}^0(\boldsymbol{r}, \boldsymbol{r}') = \{\, p_{\alpha\alpha}{}^0(\boldsymbol{r}, \boldsymbol{r}') = P_{\alpha,\alpha}{}^0\ \pi_{2,\alpha}{}^0(\boldsymbol{r}, \boldsymbol{r}'), \qquad \alpha = \beta;$$
$$p_{\alpha\beta}{}^0(\boldsymbol{r}, \boldsymbol{r}') = P_{\alpha,\beta}{}^0\ \pi_\alpha{}^0(\boldsymbol{r})\pi_\beta{}^0(\boldsymbol{r}'), \qquad \alpha \neq \beta\}, \tag{5.6.3}$$

where $\pi_\alpha{}^0(\boldsymbol{r})$ and $\pi_{2,\alpha}{}^0(\boldsymbol{r}, \boldsymbol{r}')$ denote the (unity-normalized) one- and two-electron probability densities, respectively, of the *separate* free-atom α. Here $\{P_{\alpha,\beta}{}^0\} \equiv \mathbf{P}^0$ are the condensed atomic two-electron probabilities in the promolecule: $P_{\alpha,\alpha}{}^0 = \iint p_{\alpha\beta}{}^0(\boldsymbol{r}, \boldsymbol{r}')\, d\boldsymbol{r}\, d\boldsymbol{r}' = N_\alpha{}^0(N_\alpha{}^0 - 1)/[N^0(N^0 - 1)]$ and $P_{\alpha,\beta}{}^0 = \iint p_{\alpha\beta}{}^0(\boldsymbol{r}, \boldsymbol{r}')\, d\boldsymbol{r}\, d\boldsymbol{r}' = N_\alpha{}^0\, N_\beta{}^0/[N^0(N^0 - 1)]$ ($\alpha \neq \beta$). The diagonal components $\{p_{\alpha\alpha}{}^0(\boldsymbol{r}, \boldsymbol{r}')\}$ measure the joint probability of finding two electrons on a given AIM, thus corresponding to the *ionic* valence structures of the VB theory of Heitler and London (1927). Similarly, the off-diagonal terms $\{p_{\alpha\beta}{}^0(\boldsymbol{r}, \boldsymbol{r}')\}$, measuring the joint probabilities of finding two electrons on different atoms, can be naturally associated with the *covalent* valence structures of the VB theory. Therefore, the stockholder partitioning of the molecular two-electron densities provides a convenient theoretical framework for separating these two VB components of the chemical bond.

To summarize, the four components of $\mathbf{p}^S(\boldsymbol{r}, \boldsymbol{r}')$ or $\boldsymbol{\rho}_2{}^S(\boldsymbol{r}, \boldsymbol{r}')$ in a diatomic molecule represent natural channels for the electron redistribution among the two atoms forming the chemical bond. The diagonal kernels are associated with the *ionic* bond component, and the off-diagonal kernels describe the *covalent* part of the molecular adjustment of the two-electron density relative to the promolecular reference.

It directly follows from the preceding equation that the diagonal share of the free hydrogen or any other one-electron system vanishes identically, since in this case $p_{2,\alpha}{}^0(\boldsymbol{r}, \boldsymbol{r}') = 0$. Therefore, the two hydrogen atoms in H_2 (see the next section) divide between themselves the molecular two-electron density only through the equal off-diagonal components of the AIM 2-clusters, which are proportional to the product of one-electron distributions of free atoms. Similarly, in the LiH case, which we shall examine in the next section, only the diagonal Li component and the two off-diagonal contributions determine the *effective* one-electron distributions of the bonded stockholder atoms, which originates from the molecular two-electron density (Nalewajski and Broniatowska, 2005b).

These stockholder parts of the molecular two-electron distributions in atomic resolution also provide a convenient perspective for describing the chemical interactions between the *Hard*(*H*) and *Soft*(*S*) *Acids*(*A*) and *Bases*(*B*) (HSAB). The bonding preferences in such *Donor-Acceptor* (DA) systems are summarized by the HSAB Principle of chemistry (Pearson, 1963, 1966, 1973) that the most stable DA complexes are formed between a pair of the Lewis acid and base, which exhibit comparable chemical potential (electronegativity) and hardness/softness descriptors.

Thus, the *like* categories of hardness of DA reactants determine the optimum conditions for the stable chemical bonds. However, the bond character differs in the two stable matching cases, being predominantly covalent in the *SS* complex and strongly ionic for the *HH* system (Parr and Pearson, 1983; Nalewajski, 1984).

In a diatomic molecule a relative "weight" of the ionic (diagonal) and covalent (off-diagonal) contributions of $\mathbf{p}^S(\boldsymbol{r}, \boldsymbol{r}')$ or $\boldsymbol{\rho}_2^S(\boldsymbol{r}, \boldsymbol{r}')$ in the associated molecular distributions is determined by the corresponding share factors of Eq. (5.6.2). They represent the two-electron *conditional* probabilities,

$$d_{\alpha\beta}(\boldsymbol{r}, \boldsymbol{r}') = p_{\alpha\beta}{}^S(\boldsymbol{r}, \boldsymbol{r}')/p_2(\boldsymbol{r}, \boldsymbol{r}') \equiv \pi^S(\alpha, \beta \,|\, \boldsymbol{r}, \boldsymbol{r}'),$$

$$\textstyle\sum_{\alpha=A,B} \sum_{\beta=A,B} \pi^S(\alpha, \beta \,|\, \boldsymbol{r}, \boldsymbol{r}') = 1, \qquad (5.6.4)$$

that the two electrons, already found at the specified locations in space, originate from a given pair of bonded atoms. The above normalization condition implies a *competition* between the two bond components. Indeed, an increase in the (off-diagonal) bond *covalency* is at the expense of the (diagonal) bond *ionicity* and *vice versa*. This effect also transpires from other two-electron descriptors of the chemical bond multiplicities (Wiberg, 1968; Gopinathan and Jug, 1983; Mayer, 1983; Jug and Gopinathan, 1990; Ponec and Strnad, 1994; Ponec and Uhlik, 1997; Bochicchio *et al.*, 1998}, in the difference approach of the MO theory (Nalewajski *et al.*, 1993, 1994, 1996, 1997; Nalewajski and Mrozek, 1994, 1996; Mrozek *et al.*, 1998) and in the communication theory of the chemical bond (Nalewajski, 2000c, 2004b-e, 2005a-c; Nalewajski and Jug, 2002; Nalewajski and Broniatowska, 2005a), which will be the subject of Chapters 7 and 8.

The partially integrated components of the two-electron stockholder AIM,

$$\mathbf{p}_S{}^{eff}(\boldsymbol{r}) = \int \mathbf{p}^S(\boldsymbol{r}, \boldsymbol{r}')\, d\boldsymbol{r}' = \{p_{\alpha\beta}{}^{eff}(\boldsymbol{r})\}, \qquad (5.6.5)$$

generate contributions to the associated *effective* one-electron distributions of bonded atoms:

$$p_\alpha{}^{eff}(\boldsymbol{r}) = \textstyle\sum_{\beta=A,B} p_{\alpha\beta}{}^{eff}(\boldsymbol{r}), \qquad \alpha = A, B. \qquad (5.6.6)$$

The diagonal part $p_{\alpha\alpha}{}^{eff}(\boldsymbol{r})$ of $p_\alpha{}^{eff}(\boldsymbol{r})$ accounts for the *intra*-atomic (ionic) rearrangement of the one-electron probability density, while the off-diagonal ($\beta \neq \alpha$) part $p_{\alpha\beta}{}^{eff}(\boldsymbol{r})$ of $p_\alpha{}^{eff}(\boldsymbol{r})$ represents the associated *inter*-atomic (covalent) displacement.

In qualitative considerations the crucial regions for the covalent and ionic bond components are the bond and atomic regions, respectively. The former extends between the two nuclei, while the latter refers to the vicinity of the atomic nuclei. The off-diagonal (covalent) two-electron components of Eq. (5.6.3) are proportional to the overlap between one-electron distributions of two free-atoms, for the actual interatomic distance in the molecule. Therefore, when two electrons are located in the bonding region it assumes a relatively high values for the *SS* complex and relatively small values for both the *HH* and the mixed hardness pairs *SH* or *HS* of the

Basic (*B*, donor) and *Acidic* (*A*, acceptor) atoms. Accordingly, the diagonal (ionic) conditional probability $\pi^S(\alpha,\alpha|\boldsymbol{r}, \boldsymbol{r}')$ is proportional to the two-electron densities of free atom α. In the atomic region this quantity should exhibit a relatively high value for a relatively compact valence distribution of the hard atom, the electrons of which are spatially confined by the strong atomic attractor of a weakly screened nucleus, and relatively low value for the diffused distribution of valence electrons of the soft atom, which move in the field of the effective external potential due to a more completely screened nucleus.

It should be further realized that an efficient chemical bonding of the DA (coordination) type, a characteristic feature of the *Acid-Base* (AB) interactions, can be formed only when the electrons donated by *B* can be efficiently accommodated by *A*. This is emphasized by the *frontier-electron* approach (Fukui, 1975, 1987), emphasizing the chemical interactions between the HOMO of the donor reactant and the LUMO of the acceptor reagent. The frontier electrons, the densities of which approximate the electronic Fukui function (Parr and Yang, 1984, 1989), may be expected to dominate the $\mathbf{p}^S(\boldsymbol{r}, \boldsymbol{r}')$ or $\boldsymbol{\rho}_2{}^S(\boldsymbol{r}, \boldsymbol{r}')$ kernels in the *SS* and *HH* interactions, when the chemical hardness descriptors of interacting atoms are comparable. This is not the case, when only one diagonal component $\pi^S(\alpha,\alpha|\boldsymbol{r}, \boldsymbol{r}')$ is large, e.g., when the hard atom forms a bond with the soft partner. As a result of such non-comparable hardness characteristics of two atoms, the ionicity of chemical interactions in such mixed *H-S* or *S-H* atomic has predominantly *intra*-atomic, polarizational character. Only when the chemical hardness descriptors of both atoms are comparable, there is an appreciable *inter*-atomic ionicity, which can be linked to the charge transfer between the DA reactants.

We can thus conclude that the four components of the stockholder partitioning of the two-electron distributions in diatomic molecules provide a new, convenient framework for separating the covalent and ionic bond components. It also gives rise to a transparent interpretation of the main bond preferences summarized by the HSAB principle. More specifically, it directly follows from this two-electron perspective that the most favorable *HH* or *SS* pairs of atoms form stable chemical bonds mainly due to the ionic and covalent interactions, respectively, with the mixed hardness pairs giving rise to a smaller chemical stabilization of the DA system.

The four kernels $\mathbf{p}^S(\boldsymbol{r}, \boldsymbol{r}')$, or their partially integrated, effective one-electron components $\mathbf{p}^{eff}(\boldsymbol{r})$ also define an attractive representation of the main promotional displacements, which the bonded atoms exhibit relative to their free analogs. In the next section we shall examine specific examples of the effective one-electron densities of bonded atoms in H_2, LiH, HF, LiF and N_2, which originate from the stockholder division of the molecular two-electron density:

$$\sum_{\beta=A,B} \int \rho_{\alpha\beta}{}^S(\boldsymbol{r}, \boldsymbol{r}')\, d\boldsymbol{r}' \equiv \sum_{\beta=A,B} \rho_{\alpha\beta}{}^{eff}(\boldsymbol{r}) = \rho_\alpha{}^{eff}(\boldsymbol{r}), \quad \alpha = A, B. \tag{5.6.7}$$

We shall compare them with the corresponding Hirshfeld electron densities obtained from the stockholder partitioning of the molecular one-electron density. In what follows we call these two sets of AIM the *two-* and *one*-electron *stockholder* atoms, respectively, or 2-*S* and 1-*S* bonded atoms, for short.

5.7. ELECTRON DISTRIBUTIONS OF ONE- AND TWO-ELECTRON STOCKHOLDER ATOMS

The effective one-electron densities $\rho_\alpha^{eff}(\boldsymbol{r})$ of the 2-*S* AIM and the electron densities $\rho_\alpha^{H}(\boldsymbol{r})$ of the Hirshfeld 1-*S* atoms (α^H), will be compared for H_2, LiH, HF, LiF and N_2, These molecules cover the purely covalent bonds, single in H_2 and triple in N_2, as well as the partly ionic bonds in the remaining heteronuclear diatomics. This selection allows one to examine the effect of a growing number of the open two-electron channels (AIM 2-clusters) in the 2-*S* partitioning procedure: 2 off-diagonal channels in H_2, 1 diagonal and 2 off-diagonal channels in LiH and HF, and all 4 channels (2 diagonal and 2 off-diagonal) in N_2 and LiF.

The numerical results will be generated using the standard UHF, UHF+MP2 ≡ UMP2 and Kohn-Sham (KS) DFT (KS/LDA and KS/B3LYP) calculations in the extended (DZV or DZVP) basis sets, which involve the split-valence atomic orbitals (DZV) in some calculations supplemented by the polarization functions (DZVP). This analysis has been performed for several internuclear separations ranging from the equilibrium bond length to the near-dissociation distances. The AIM densities will be also compared with the corresponding reference densities $\rho^0(\boldsymbol{r})$ of the separated (free) constituent atoms. The equilibrium geometries in the adopted basis sets have been used. In order to precisely separate the Coulomb correlation effects on the effective AIM distributions, the common UHF equilibrium geometry has been used in both the 1-*S* and 2-*S* partitions.

In the present analysis we are interested in differences observed in the effective *one*-electron distributions of the *two*-electron stockholder (2-*S*) AIM, relative to those characterizing the Hirshfeld (1-*S*) bonded atoms, which give rise to the *same* molecular electron density ρ. In order to facilitate such a comparison, the constraint of the fixed density $\rho = \rho^{\mathrm{UHF}}$ was imposed in the UMP2 numerical calculations. It should be realized, however, that the same electron density and external potential in both these partition schemes in DFT imply the same molecular electronic energies corresponding to the two sets of constituent atoms. Therefore, only the *"vertical"* (for the identical electron densities) influence of the Coulomb correlation holes on the resultant electron distributions of 2-*S* AIM, estimated using the MP2 theory for the fixed UHF (Coulomb-uncorrelated) molecular electron density, was examined.

Clearly, the Coulomb-correlated calculations ultimately give rise to the optimum electron density, which is slightly modified relative to the UHF result, $\rho^{\mathrm{UMP2}} \neq \rho^{\mathrm{UHF}}$, which we call the *"horizontal"* displacement in the system electronic structure. This change in the equilibrium distribution of electrons lowers the system electronic energy by the Coulomb-correlation energy: $E^{\mathrm{UMP2}} - E^{\mathrm{UHF}} = \Delta E_c < 0$. The corresponding expression for the electron pair-density in terms of the exchange-correlation (*xc*) hole $h_{xc}^{\mathrm{UMP2}}(\boldsymbol{r}'|\boldsymbol{r})$ gives the following partition into the independent-particle (Hartree, H) contribution and the correlation corrections due to the Fermi (x) and Coulomb (c) holes, respectively:

$$\begin{aligned}\rho_2^{\mathrm{UMP2}}(\boldsymbol{r}, \boldsymbol{r}') &= \rho^{\mathrm{UMP2}}(\boldsymbol{r})[\rho^{\mathrm{UMP2}}(\boldsymbol{r}') + h_{xc}^{\mathrm{UMP2}}(\boldsymbol{r}'|\boldsymbol{r})] \equiv \rho_{2,H}^{\mathrm{UMP2}}(\boldsymbol{r}, \boldsymbol{r}') + \rho_{2,xc}^{\mathrm{UMP2}}(\boldsymbol{r}, \boldsymbol{r}') \\ &= \rho^{\mathrm{UMP2}}(\boldsymbol{r})\{[\rho^{\mathrm{UMP2}}(\boldsymbol{r}') + h_x^{\mathrm{UMP2}}(\boldsymbol{r}'|\boldsymbol{r})] + h_c^{\mathrm{UMP2}}(\boldsymbol{r}'|\boldsymbol{r})\} \\ &\equiv \rho_{2,Hx}^{\mathrm{UMP2}}(\boldsymbol{r}, \boldsymbol{r}') + \rho_{2,c}^{\mathrm{UMP2}}(\boldsymbol{r}, \boldsymbol{r}'). \end{aligned} \tag{5.7.1}$$

Above we have used the familiar partition of the resultant hole into the separate exchange (x) and (Coulomb) correlation (c) contributions (see Appendix C). Here, the subscript Hx symbolizes the sum of the Hartree and exchange contributions.

The present analysis of the *"vertical"* (fixed-density) correlation effects in the 2-S partition of the UHF density is obtained by approximating in the foregoing equation $\rho^{\mathrm{UMP2}}(\boldsymbol{r}) \cong \rho^{\mathrm{UHF}}(\boldsymbol{r})$ and hence also

$$\rho_{2,H}^{\mathrm{UMP2}}(\boldsymbol{r}, \boldsymbol{r}') \cong \rho_{2,H}^{\mathrm{UHF}}(\boldsymbol{r}, \boldsymbol{r}') \quad \text{and} \quad \rho_{2,Hx}^{\mathrm{UMP2}}(\boldsymbol{r}, \boldsymbol{r}') \cong \rho_2^{\mathrm{UHF}}(\boldsymbol{r}, \boldsymbol{r}'), \tag{5.7.2}$$

since the UHF MO have been used in the subsequent MP2 procedure for estimating the Coulomb-hole. These approximations give the following final expression for the electron pair-density, which constitutes the basis of the present "vertical" analysis:

$$\rho_2^{\mathrm{UMP2}}(\boldsymbol{r}, \boldsymbol{r}') \approx \rho_2^{\mathrm{UHF}}(\boldsymbol{r}, \boldsymbol{r}') + \rho^{\mathrm{UHF}}(\boldsymbol{r})\, h_c^{\mathrm{UMP2}}(\boldsymbol{r}'|\boldsymbol{r}). \tag{5.7.3}$$

It can be straightforwardly verified using the familiar sum-rules satisfied by the correlation holes that the integration over $\boldsymbol{r}'$ indeed recovers $\rho^{\mathrm{UHF}}(\boldsymbol{r}) = (N-1)^{-1} \int \rho_2^{\mathrm{UHF}}(\boldsymbol{r}, \boldsymbol{r}')\, d\boldsymbol{r}'$ since the second, Coulomb-correlation contribution exactly vanishes: $\int h_c^{\mathrm{UMP2}}(\boldsymbol{r}'|\boldsymbol{r})\, d\boldsymbol{r}' = 0$.

The familiar antisymmetry requirement imposed on the N-electron wave-function implies the statistical, *exchange* (x) correlation between the movements of like-spin electrons. In our approximation, the two-electron conditional probabilities, which take into account this Pauli *exclusion-principle*, describe the *non-interacting* system of the *same* electron density, as that of the *real* system, consisting of fully *interacting* electrons. These joint two-electron probabilities, which carry the exchange correlation, are fully determined by the KS orbitals. Since in the KS limit, for the vanishing coupling-constant in the adiabatic connection of DFT (Appendix C), the Coulomb correlation due to finite charges of electrons is completely turned off, the Fermi correlation accounts for the whole correlation of the non-interacting system.

The UHF approximation, which in principle can be regarded as roughly equivalent to the exact exchange–only DFT, has been selected to generate the reference data in the near-dissociation limit. Since the correlation effects in typical DFT methods are only correct on the level of the *average* correlation holes, we have limited the DFT analysis to the KS (non-interacting) limit only, which exclusively covers the exchange effects embodied in the two-electron probabilities generated by the KS molecular orbitals. We denote such UHF-like approach as the UHF-KS scheme. Indeed the results obtained for molecules in UHF and UHF-KS approximations will be shown to be practically indistinguishable. Only at internuclear distances close to the bond dissociation one detects appreciable differences due to a general inadequacy of the UHF-KS inter-atomic exchange holes in such strongly bond-elongated systems, both in the LDA and GGA/B3LYP variants. The effects of the Coulomb correlation will be determined using the spin-unrestricted Møller-Plesset (MP) theory (UMP2), which takes into account the double excitations from the ground-state UHF electron configuration.

5.7.1. *Vertical* Effects of Electron Correlation

Figure 5.2 compares the bond profiles of the molecular and atomic electron densities in H_2 and LiH. The reported UHF plots are also representative of the UHF-KS (LDA and B3LYP) results in the same basis sets, which are practically indistinguishable for the equilibrium bond lengths assumed in the figure. It follows from Panel *a* of the figure that in H_2 both the H^S (2-*S*) and H^H (1-*S*) atoms assume a distinctly molecular character, being polarized towards the bonding partner and exhibiting the extra contraction near the nucleus, in comparison to the spherical probability distribution of the free hydrogen. However, the two profiles of the bonded hydrogen atoms are quite different: the effective probability plot from the two-electron stockholder distribution extends further towards the other atom, thus penetrating more effectively the bonding region of the molecule at the expense of the area around the atomic nucleus, where a weaker contraction can be detected. In H_2, due to the one-electron reference of the neutral hydrogen, only the two off-diagonal contributions, which we call "covalent" by analogy to the Valence-Bond structures of the Heitler-London theory, participate in the two-electron stockholder division. The exactly vanishing diagonal ("ionic") channels imply that the observed deviations between 1-*S* and 2-*S* densities observed in Fig. 1*a* for such a *Hard-Hard* diatomic are solely of the "covalent" origin.

These observations hold true in LiH as well (see Fig. 5.2*b*), in which the extra diagonal ("ionic") channel on Li opens in the 2-*S* partitioning. In this molecule the largest deviations between the 1-*S* and 2-*S* electron distributions can be detected on Li. Indeed, in this *Soft*(Li)–*Hard*(H) molecule the effective density of the heavy atom contains both the diagonal (*ionic*) and a weak off-diagonal (*covalent*) contributions, while the hydrogen electron density contains only a small *covalent* part. We shall demonstrate in Section 5.8 that the bonding "shoulder" of the Li effective distribution in Fig. 5.2b is due to the off-diagonal (covalent), [Li,H]-contribution. It is seen to extend well into the region around the hydrogen nucleus.

The contour maps reported in Fig. 5.3 further illustrate the extra-bonding character of the 2-*S* AIM, compared to the Hirshfeld (1-*S*) bonded atoms in these two molecules. Indeed, in the H_2 case the 2-*S* hydrogen exhibits a stronger cylindrical polarization around the bond-axis, compared to the weakly polarized, almost spherical 1-*S* (Hirshfeld) atom. In LiH the *acceptor* (hydrogen) atom is seen to remain almost spherical in both one- and two-electron stockholder division schemes, while the *donor* (Li) atom is seen to be more strongly polarized towards the hydrogen in the 2-*S* partition. In this molecule the density contours of the heavy atom Li^S exhibit a distinct directional character. This feature is lacking in Li^H, in the one-electron (Hirshfeld) division, where the directions of the maximum polarization are seen to lie on the surface of the cone, at a roughly 45 degree angle relative to the Li–H bond axis.

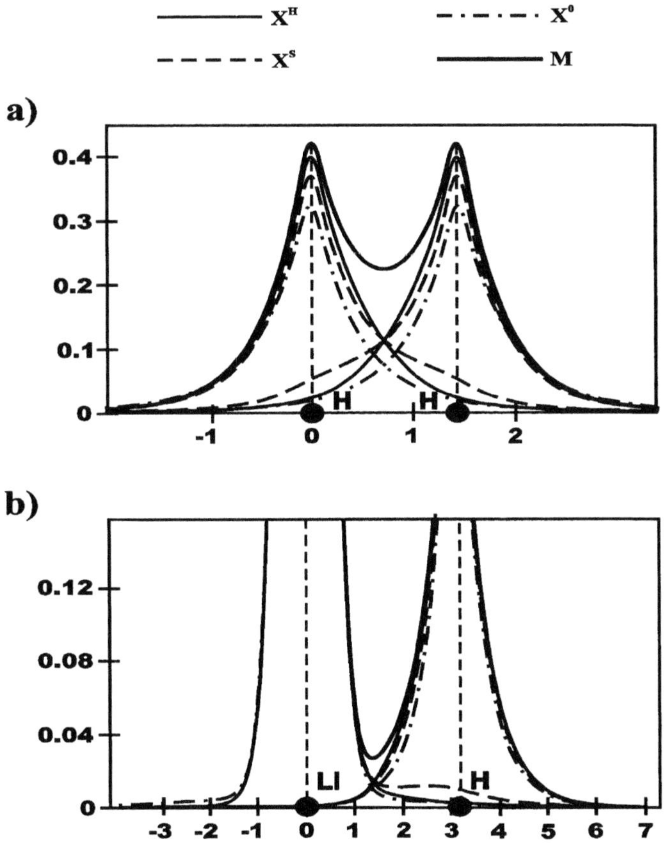

Figure 5.2. A comparison of the molecular $\rho(\boldsymbol{r})$ [M] and atomic, $\rho_\alpha{}^0(\boldsymbol{r})$ [α^0], $\rho_\alpha{}^{eff}(\boldsymbol{r})$ [α^S], and $\rho_\alpha{}^H(\boldsymbol{r})$ [α^H], electron densities in H_2 (Panel *a*) and LiH (Panel *b*) obtained from the UHF calculations. The associated plots generated by the DFT calculations in the LDA and B3LYP approximations (UHF-KS scheme) are practically indistinguishable from the reported UHF results (DZVP basis set). The equilibrium bond lengths have been assumed and the atomic units are used throughout this chapter.

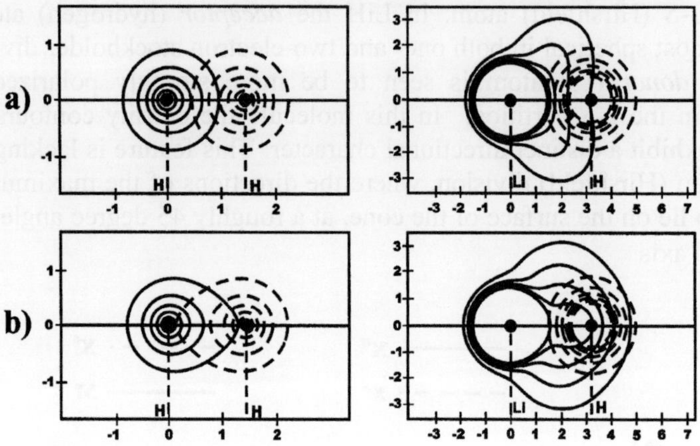

Figure 5.3. A comparison of the contour diagrams of the $\rho_\alpha^{H}(\boldsymbol{r})$ (Panel *a*) and $\rho_\alpha^{eff}(\boldsymbol{r})$ (Panel *b*) distributions of bonded atoms in H_2 (left column) and LiH (right column) from the UHF calculations. The associated DFT (UHF-KS) plots in the LDA and B3LYP approximations are practically indistinguishable from the reported UHF diagrams. The equilibrium bond lengths have been assumed (DZVP basis set).

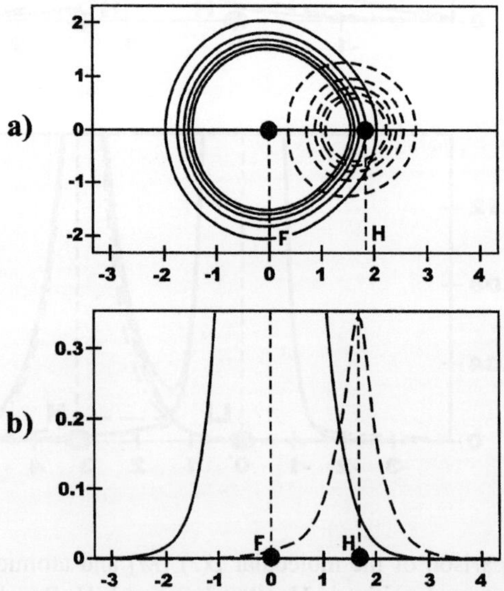

Figure 5.4. Contour diagrams (Panel *a*) of the effective electron density of the 2-*S* AIM and the electron densities of the 1-*S* (Hirshfeld) atoms (both plots are indistinguishable in the scale of the figure) in HF (UMP2 approximation), together with the corresponding bond-axis profiles (Panel *b*) (DZV basis set).

a)

b)

c)

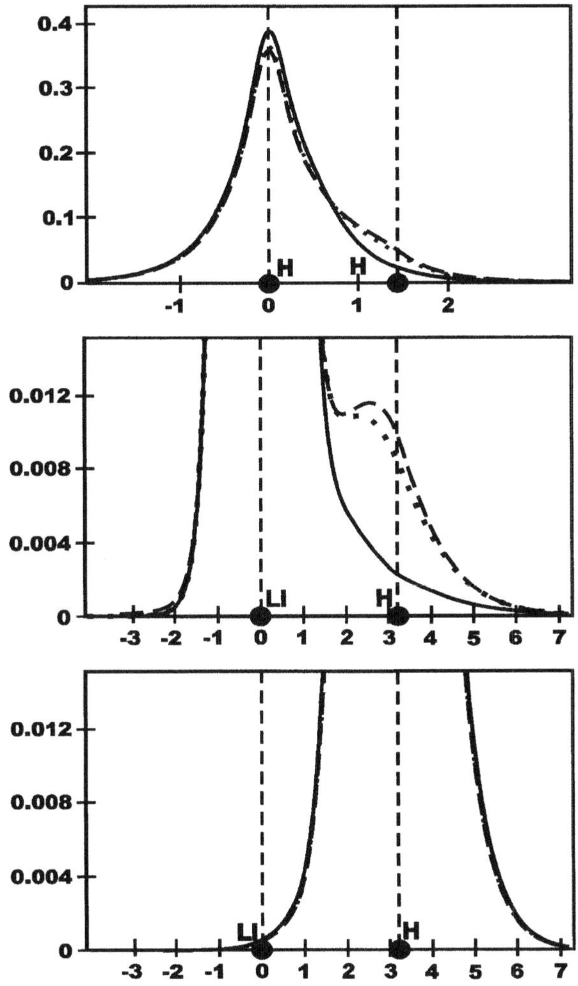

Figure 5.5. The effect of the *vertical* exchange (UHF, broken line) and Coulomb (UMP2, dotted line) electron correlation on the bond-axis profiles of the effective electron densities of 2-*S* AIM in H_2 (Panel *a*) and LiH (Panels *b* and *c*). The 1-*S* (Hirshfeld, solid line) plots are also shown for comparison (DZV basis set).

A similar comparison of the contour maps and density profiles for constituent atoms in HF (Fig. 5.4) reveals that there are no appreciable differences between the 1-*S* and 2-*S* stockholder AIM in this molecule. Therefore, with a growing number of the valence electrons the subtle differences in the shapes of the stockholder AIM due to the electron correlation gradually disappear. The same general conclusion follows from examining the logarithmic AIM density plots in N_2 and LiF, which we shall report in Section 5.8. Thus, in heavier molecules the 1-*S* and 2-*S* stockholder AIM become for all practical purposes identical.

In Fig. 5.5 we have examined in a more detail the effect of the vertical electron correlation on the effective density profiles of 2-*S* bonded atoms in H_2 and LiH. A reference to Panel *a* of the figure shows that the bonding shoulder of the 2-*S* hydrogen atom in H_2 is only slightly lowered and there is a bit more density contraction in the vicinity of the nucleus, when one turns on the Coulomb correlation on top of the exact exchange. A more bonding character of the 2-*S* AIM relative to the 1-S bonded atom is again clearly seen in this density profile. The same general conclusion follows from examining Panel *b* of this figure: the inclusion of the Coulomb correlation slightly lowers the bonding part of the density profile of the bonded 2-*S* Li atom, in comparison to the exchange-only plot. This blow-up picture emphasizes the bonding region, where the differences between the 2-*S* and 1-*S* profiles are most apparent. However, as the corresponding hydrogen plots of Panel *c* and Fig. 2*b* clearly show, the overall distributions of the 2-*S* and 1-*S* AIM are not that very different. With the exception of the bonding region the two profiles are almost identical. This closeness of the two sets of the bonded stockholder atoms becomes even stronger in diatomics consisting two heavy atoms, e.g., LiF and N_2, in which all four two-electron channels (AIM 2-clusters) become available for the stockholder partitioning of the molecular pair-density.

The details of generally minor local differences between the effective electron distributions of the 2-*S* AIM, relative to the electron densities of the 1-*S* (Hirshfeld) atoms, in all five representative diatomics are displayed in the *left column* of Fig. 5.6, where the relevant *difference* profiles are displayed. As expected, these deviations decay fast with the growing number of electrons from H_2 ($N = 2$) to N_2 ($N = 14$). The largest differences, observed in Panel *a*, indicate that the 2-*S* bonded hydrogens in H_2 shift the electron density from the non-bonding and nuclear regions towards the other nucleus, relative to the corresponding Hirshfeld atomic pieces of the molecular electron density.

A different pattern of such AIM density relocations due to the vertical electron correlation is observed in heteronuclear diatomics of Panels *b-d* of the figure left column. In LiH (Panel *b*) the lithium atom gains electrons at the expense of hydrogen atom, with the maximum of these deviations being observed near the nucleus of a more electronegative (harder) hydrogen atom. This polarization shows that the 2-*S* AIM slightly lower the charge separation exhibited by the net charges of the Hirshfeld AIM, which have been reported in Table 5.1.

Similar deviations are exhibited by the difference plots of Panel *c*, for HF, which are limited basically to the bonding region between the two nuclei. This time the positively charged 2-*S* hydrogen atom gains electrons, relative to the corresponding Hirshfeld density, while the negatively charged fluorine atom looses part of its excess electron density. Thus, the 2-*S* partition partly moderates the AIM charge separation generated by the 1-*S* partition. A reference to Panel *d* shows the same effect in LiF, although the magnitude of the maximum density difference is now ten times lower.

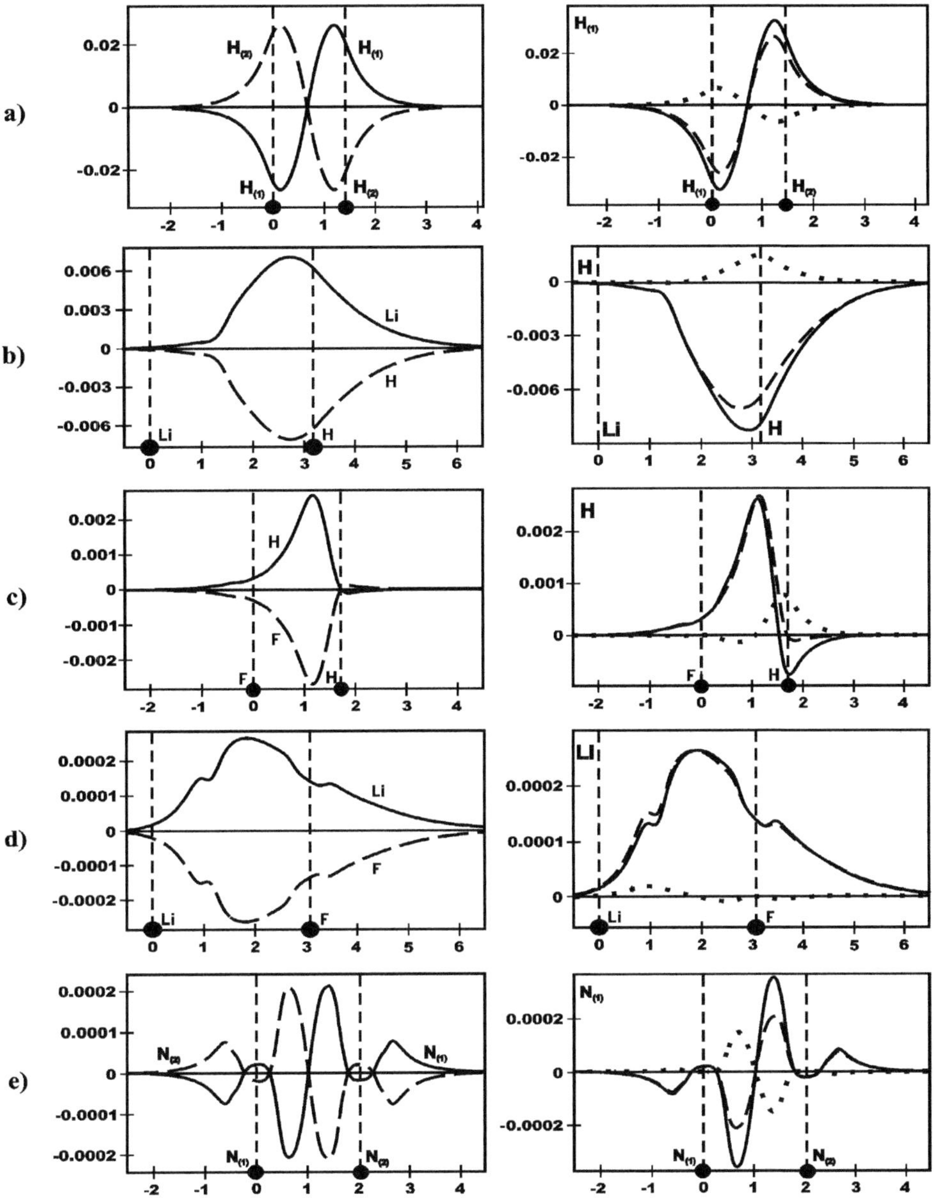

Figure 5.6. Left column: deviations between the densities $\{\rho_\alpha^{eff}(\boldsymbol{r})\}$ of the 2-*S* AIM and $\{\rho_\alpha^{H}(\boldsymbol{r})\}$ of the 1-*S* bonded atoms along the bond axis (DZV basis set, UMP2 approximation). Right column plots provide a resolution of the overall displacements (broken line) into the corresponding shifts due to the Fermi (exchange, solid line) and Coulomb (dotted line) vertical correlation between electrons. Molecules: H_2 (Panels *a*), LiH (Panels *b*), HF (Panels *c*), LiF (Panels *d*), and N_2 (Panels e).

Table 5.1. A comparison of the net charges (a.u.) of the bonded AIM in LiH, HF and LiF, obtained from the *one*-electron (*H*) and *two*-electron (*S*) stockholder division of the molecular electron distributions (DFT LDA calculations, DZVP basis set). The 2-*S* AIM have been generated using the two-electron joint probabilities generated by the KS orbitals (UHF-KS scheme).

Molecule	AIM (α)	q_α^H	q_α^S
LiH	Li	0.35	0.24
HF	H	0.24	0.21
LiF	Li	0.58	0.56

A more complex redistribution pattern is found for N_2 (Panel *e*). The 2-*S* nitrogen atom, when compared with its 1-*S* analog, is seen to slightly shift its valence electrons, from both the non-bonding and bonding regions, to the valence shell of the other nitrogen. Therefore, the electron correlation delicately increases the bonding character of the two-electron stockholder nitrogens, in comparison to the Hirshfeld (1-*S*) atomic subsystems. However, as witnessed by the magnitude of the density oscillations, comparable to those observed in LiF, this change is so minute that it is practically invisible in the absolute density plots.

Finally, let us examine the resolution of these overall (vertical) correlation-induced differences between the 1-*S* and 2-*S* AIM, repeated as broken-line profiles in the right column in Fig. 5.6, into contributions due to the exchange (solid line) and Coulomb (dotted line) correlation effects. This partition is shown in the *right* column of Fig. 5.6. A general conclusion following from these density-difference profiles is that the Coulomb correlation slightly moderates the displacements due to the Pauli exclusion principle, with the density oscillations due to Coulomb correlation exhibiting opposite phase in comparison to the corresponding exchange curves. The Fermi correlation contribution dominates the overall displacements and gives rise to a slightly stronger bonding character of the effective electron densities of the 2-*S* atoms, while the Coulomb correlation makes the shape of the 2-*S* AIM closer to that of the Hirshfeld density pieces.

5.7.2. *Horizontal* Density Displacements due to Coulomb Correlation

It should be emphasized that the above development has taken into account only the "vertical" correlation influence on the resulting densities of 2-*S* AIM,

$$\Delta\rho^{vertical} = \rho^S(\rho)^{\mathrm{UMP2}} - \rho^S(\rho)^{\mathrm{UHF}}, \tag{5.7.4}$$

for the fixed molecular one-electron distribution ρ, common to both the 1-S and 2-S partitions. However, an inclusion of the Coulomb correlation changes the system electron density relative to that in the UHF treatment. This "horizontal" correlation influence further affects both the Hirshfeld and two-electron stockholder AIM.

In Fig. 5.7 we have examined the horizontal shifts of the *molecular* electron density due to the Coulomb correlation,

$$\Delta\rho^{horizontal} = \rho^{\mathrm{UMP2}} - \rho^{\mathrm{UHF}}, \tag{5.7.5}$$

for all five diatomics considered in this analysis. Their Hirshfeld components are displayed in Fig. 5.8.

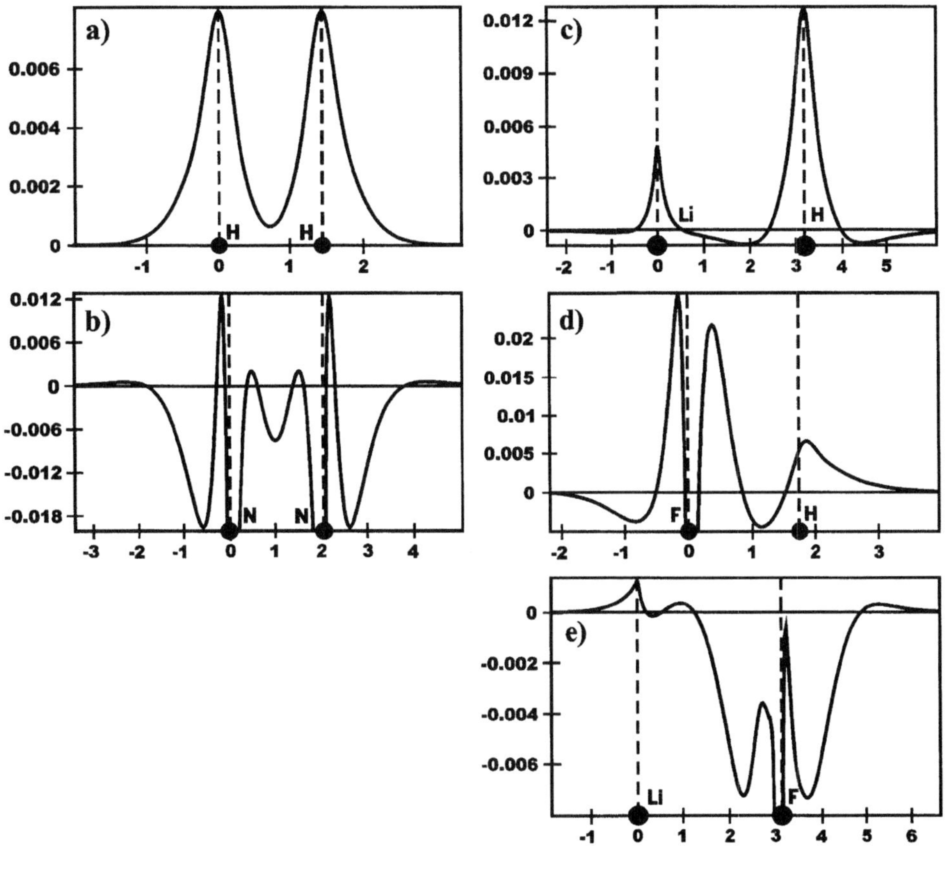

Figure 5.7. The "horizontal" shifts in the molecular electron density due to the Coulomb correlation, $\Delta\rho^{horizontal} = \rho^{\mathrm{UMP2}} - \rho^{\mathrm{UHF}}$, for selected homonuclear diatomics (left column), H_2 (Panel *a*), N_2 (Panel *b*), and heteronuclear diatomic molecules (right column), LiH (Panel *c*), HF (Panel *d*) and LiF (Panel *e*).

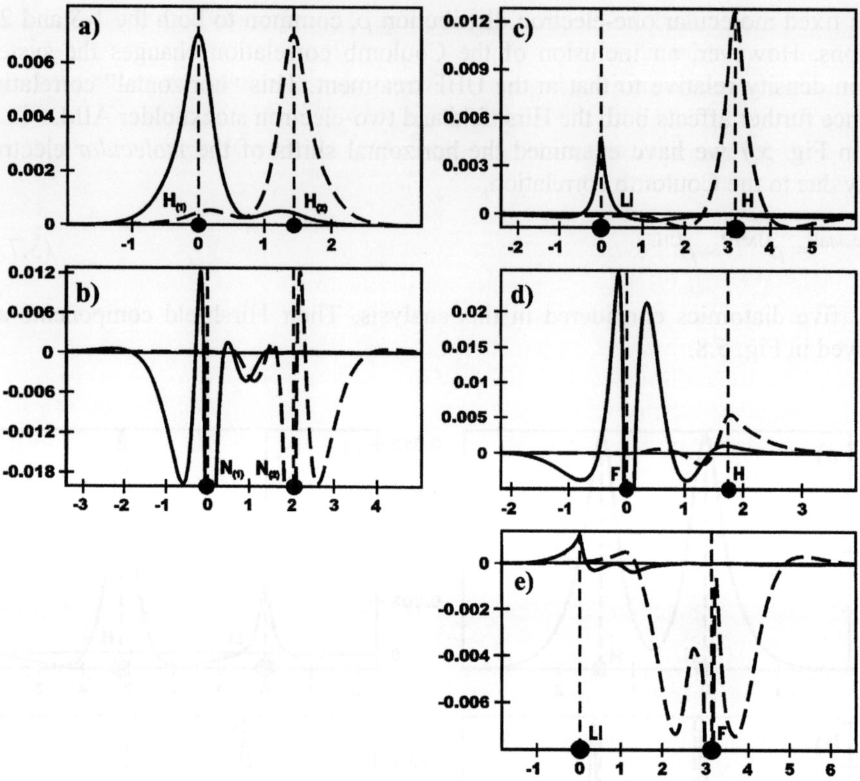

Figure 5.8. The "horizontal" shifts in the electron densities of the Hirshfeld AIM in diatomic molecules of Fig. 5.7 due to the Coulomb correlation: $\Delta\rho_\alpha^{H,horizontal} = \rho_\alpha^{H,\mathrm{UMP2}} - \rho_X^{H,\mathrm{UHF}}$. These diagrams provide the 1-*S* atomic resolution of the molecular diagrams of Fig. 5.7.

A reference to Fig. 5.7*a* shows that the Coulombic part of the electron correlation shifts the electron density from far-outside regions of the UHF molecular distribution towards the nuclei and, to a lesser degree, to the bonding region between them. This pattern should indeed be intuitively expected, since the two spin-paired electrons occupying the bonding MO are not Fermi-correlated, so that their close, near-coalescence encounters are not excluded by the Pauli principle. Therefore, the inclusion of the Coulomb correlation between the two electrons will be felt most strongly in the regions of the highest electron concentration, i.e., near the nuclei and in the bonding region. Since the correlation avoidance of two electrons effectively increases the average distance between them, thus lowering their repulsion, one should expect a slight electron-density inflow to these regions of the highest accumulation of the UHF electron density in the molecule.

A more complicated density redistribution pattern is observed in Fig. 5.7*b*. The strongest relocations are again observed in the atomic *core*-regions, around the two nuclei, dominated by the spin-paired 1*s* electrons, where the turning-on the Coulomb correlation should indeed be felt most. It should be observed that these horizontal correlation effects polarize the atomic cores away from the triple-bond electrons. The lone-electron pair regions are seen to gain electrons at large distances from the nuclei, at the expense of the valence-regions, less distant from the bonding electron pairs, where a decrease in the electron density of the predominantly (2*s*, 2*p*)-hybrid is observed. The average effect in the region of the bond-charge accumulation due to both σ and π bonds is also observed to be a slight lowering the electron density. Turning-on the Coulomb correlation affects not only the interactions within the lone-pair but also the interactions between mainly the opposite-spin electrons of the lone- and bonded-pairs of electrons, respectively. In the triple-bond molecule the latter effect dominates, thus giving rise to the horizontal-correlation redistrubution pattern in N_2, which is opposite that observed in H_2: on average electrons are expelled from the bonding and lone-pair regions of the molecule, thus partially reversing the major bond-formation effects observed at the UHF level of theory. We also observe in the density-displacement profile an "inductive" effect, of the regions of positive horizontal density changes being followed be regions of the negative shift in, which less distant the electron density.

Let us now turn to the heteronuclear diagrams shown in the right column of Fig. 5.7. The LiH horizontal profile again exhibits an extra concentration of the electron density around the nuclei, at the expense of the bonding and nonbonding regions, as in H_2. Notice, however, that the negatively charged bonded hydrogen is predicted to receive relatively more of the electron density than its positively charged bond partner, so that the horizontal Coulomb-correlation displacement *increases* the UHF charge separation. This is contrary to the vertical Coulomb correlation influence, seen in Fig. 5.6*b*, which acted in the opposite direction, towards a slight moderation of the UHF net charges on AIM. A reference to Fig. 5.7*d* indicates that the bonded fluorine atom in HF undergoes a similar charge redistribution as that observed for the bonded nitrogen in N_2, particularly in the core region, while the area in the vicinity of the proton gains electrons, as in H_2 and LiH. Around the H–F bond this horizontal Coulomb correction is seen to shift electrons from the hydrogen to fluorine, thus again strengthening the UHF charge separation, contrary to the vertical moderating influence observed in Fig. 5.6*c*. An opposite overall effect is detected in the most ionic LiF bond profile of Fig. 5.7*e*, with the fluorine atom on average loosing electrons and the bonded lithium atom gaining electrons. This horizontal Coulomb correlation influence partly moderates the magnitude of the UHF atomic charges thus acting in-phase with the vertical pattern of Fig. 5.6*d*.

Finally, in Fig. 5.8 these overall Coulomb-correlation ("horizontal") influences on the molecular electron densities of diatomic molecules have been resolved into the corresponding Hirshfeld AIM-components. The figure shows changes in the electron distribution of the 1-*S* bonded atoms obtained from the ground-state density in the UMP2 approximation, relative to their analogs derived from the molecular

UHF electron density, for the same (UHF optimum) internuclear distance. It follows from Fig. 5.8*a* that the dominant horizontal effect of the Coulomb correlation on the density of the Hirshfeld hydrogen in H_2 is located in the region around the nucleus, giving rise to a more compact (contracted) electron distribution. Another distinct, but relatively smaller effect is observed in the bonding region and around the nucleus of the other atom. This effect confirms our earlier conjecture from the 2-*S* diagrams for H_2 that an inclusion of the Coulomb correlation makes the stockholder hydrogens a bit more bonding in character. The N_2 plots (Fig. 5.8*b*) exhibit a strong core polarization. The Hirshfeld nitrogen atoms are seen to slightly diminish their electron density in the lone-pair region, occupied by the non-bonding $2s$-$2p_\sigma$ hybrid, and – to a lower degree – in the bonding region between the two nuclei. This pattern shows that the horizontal correlation effects slightly moderate some excess bonding character determined in the UHF approximation. A similar trend is observed in the fluorine plot of Fig. 5.8*d*. The electron distribution of the bonded F in HF is seen to become a bit less diffused, when the Coulomb correlation is turned-on. Both atoms are seen to withdraw part of their bond charge, thus effectively lowering the covalent bond component, relative to the UHF reference. However, each atom is also seen to slightly increase its density in the valence region of its bond partner. Finally, in the LiF diagrams of Fig. 5.8*e* one detects the dominant reduction of the excess electron population on the bonded fluorine atom, which effectively lowers the ionic bond component. Finally, a reference to both AIM plots of Fig. 5.8*c* shows that the horizontal Coulomb correlation slightly increases the Li→H CT, relative to the UHF charge separation. Again, the AIM electron distributions become more contracted around the nuclei at the expense of both the bonding and nonbonding regions of the molecule.

5.7.3. Near-Dissociation Bond Elongation

It is also of interest to examine the evolution of the electron distributions of the information-theoretic AIM when the chemical bond is elongated from the equilibrium value to the near dissociation limit. First, we compare the reference UHF results, which can be regarded as equivalent to the exact-exchange-only UHF-KS DFT calculations.

The representative plots for H_2 and LiH are reported in Figs. 5.9 and 5.10. These diagrams imply a smooth transition of the electron densities of the bonded stockholder atoms into those of the corresponding free-atoms, when the chemical bond is elongated to the practically near-dissociation inter-nuclear distance. At each stage the two-electron stockholder atoms are seen to exhibit a stronger and longer-lasting bond polarization towards the other atom, in comparison to the one-electron (Hirshfeld) AIM. This bonded character can indeed be detected in the shapes of 2-*S* atoms at relatively higher bond-length values in comparison to 1-*S* atoms, which faster assume the free-atom distribution with bond elongation.

Next, let us investigate the effect of approximations introduced in the DFT treatment of the Fermi holes, which characterize the exchange correlation in molecules, on the effective distributions of the bonded atoms obtained from the 2-*S* partition of the molecular electron densities. The two-electron probabilities from the (LDA, B3LYP) DFT calculations approximate the exact exchange holes resulting from the UHF theory. We shall now examine how these approximations affect the evolution of shapes of AIM during the bond dissociation. Clearly, the adequate representation of the "tails" of the *x*-holes, which ultimately determine the inter-atomic exchange-correlation energy at large distances between dissociating atoms, may be expected to be crucial for a smooth transition from the bonded (*promoted*) atoms into the isolated (*separated*) atoms of the dissociation limit.

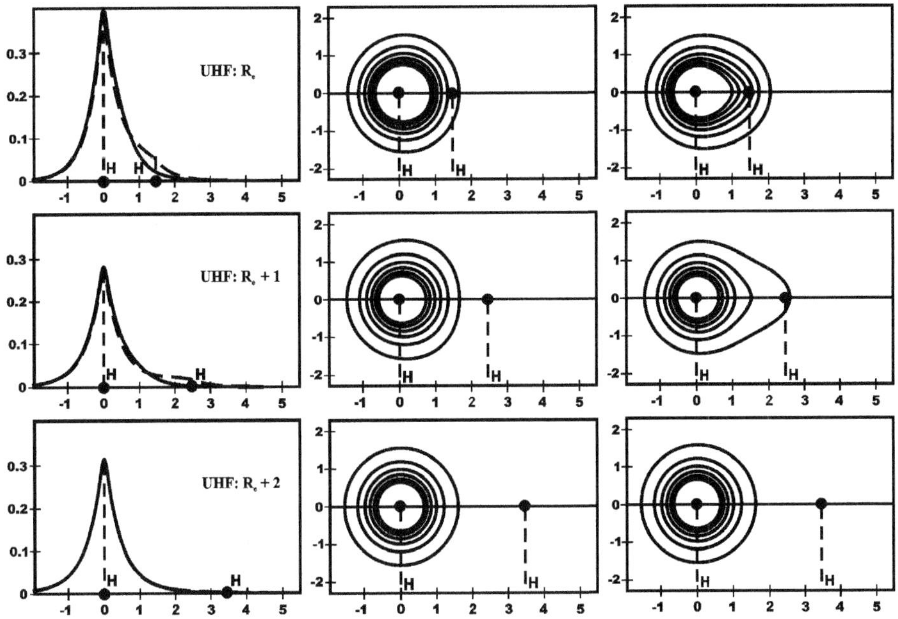

Figure 5.9. Variations with increasing internuclear distance (a.u.) of the UHF electron densities of the one- and two-electron stockholder AIM in H_2, as reflected by the bond density profiles (first column) and the contour maps (remaining columns). The Hirshfeld density pieces are shown in Columns 1 (solid lines) and 2, while two-electron stockholder AIM are displayed in Columns 1 (broken lines) and 3. The same convention is used in Figs. 5.10-14 (DZVP basis set).

In Figs. 5.11 and 5.12 the evolution of the one- and two-electron stockholder AIM in the H-H bond-elongation process is examined, as predicted by the DFT calculations using the LDA and B3LYP functionals approximating the exchange-correlation energy. These plots should be compared with the corresponding reference (UHF) diagrams shown in Fig. 5.9. It should be emphasized that in the heteronuclear diatomics the adopted variants of DFT calculations give rise to the ion pair in the dissociation limit, instead of neutral atoms. This is because the LDA and B3LYP functionals do not reproduce the so called N-discontinuities of the effective and chemical potentials, which are properly taken into account by the orbital-density functionals, e.g., the familiar *Optimized Effective Potential* (OEP) model, *Krieger-Li-Iafrate* (KLI) approximation, or the related exchange-only realization known as the *Optimized Potential Model* (OPM) (Krieger *et al.*, 1995; Grabo *et al.*, 1998), which in principle represents the DFT analog of the UHF theory.

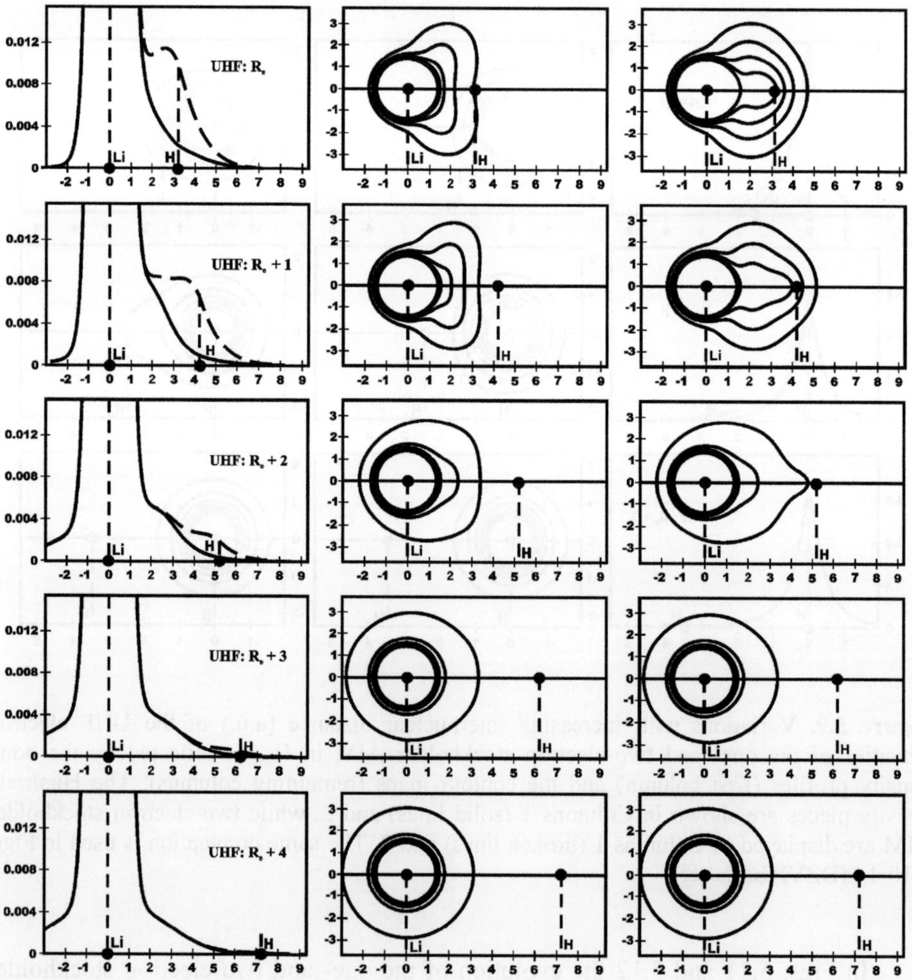

Figure 5.10. Same as in Fig. 5.9 for Li in LiH.

This comparison shows that in the two reported DFT calculations the Hirshfeld division scheme leads to the *single cusp* in the atomic density pieces, which eventually become identical with the free-atom/ion densities at large internuclear distances.

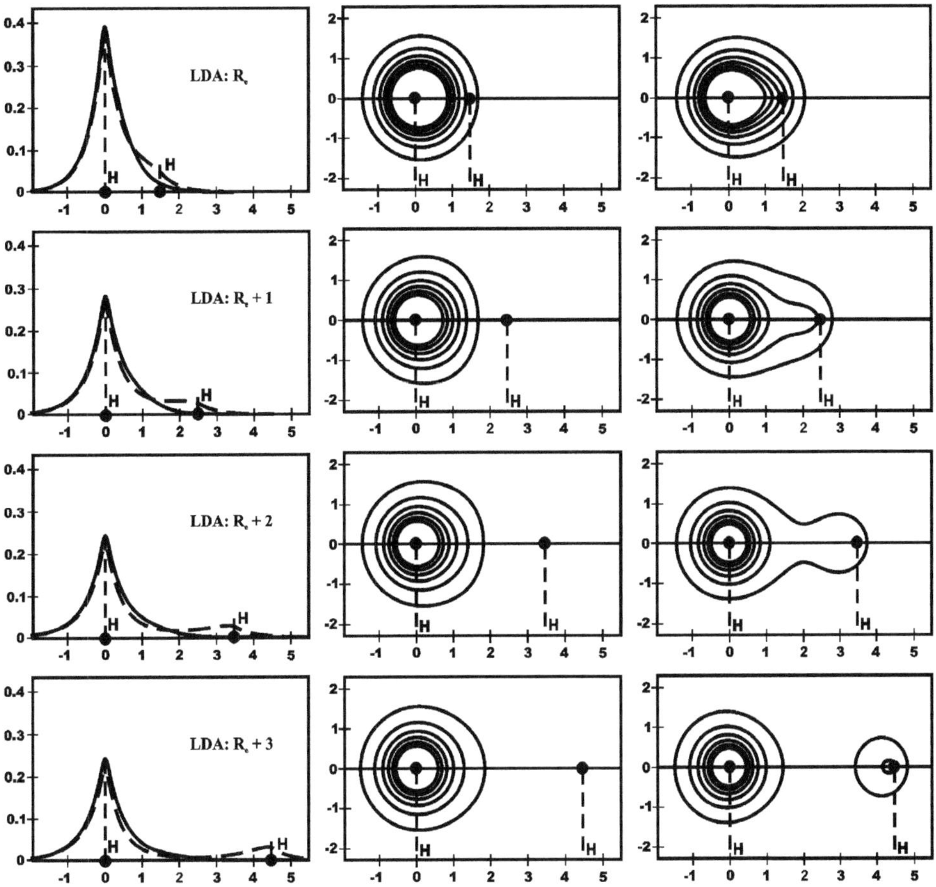

Figure 5.11. Same as in Fig. 5.9 for the UHF-KS scheme based upon the LDA approximation of DFT.

This is no longer the case in the 2-*S* partition scheme, where the most approximate LDA variant is seen in Fig. 5.11 to give rise to the two-cusp feature in the bonded hydrogens at the near-dissociation bond lengths. The more exact B3LYP functional is seen in Fig. 5.12 to remedy this incorrect prediction of the KS-LDA theory at large distances between atoms although a near-cusp shoulder can still be observed at the intermediate internuclear separation in the second row of the figure.

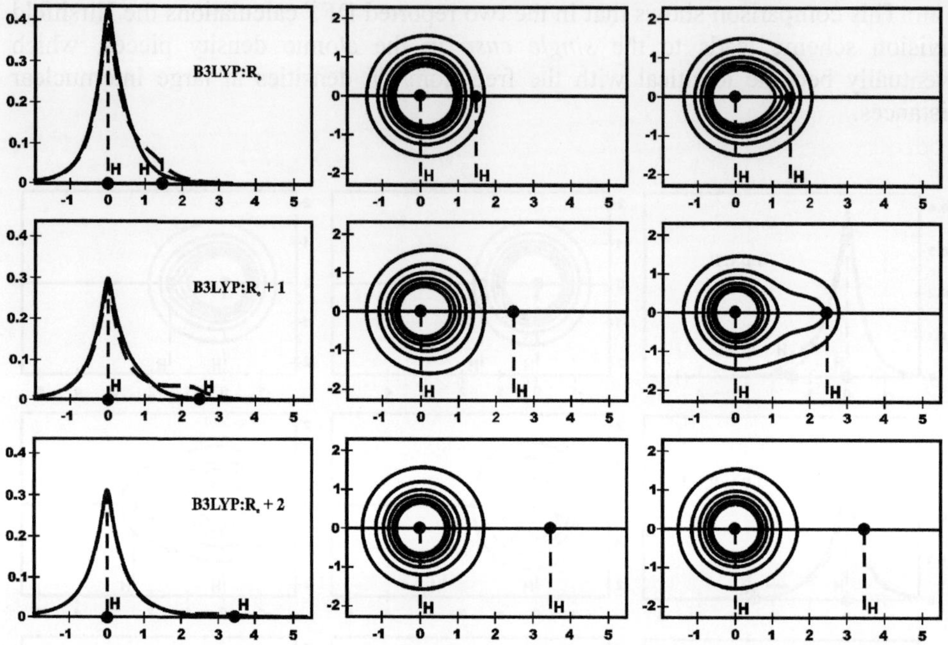

Figure 5.12. Same as in Fig.5.11 for the B3LYP exchange-correlation energy functional.

This sensitivity of the two-electron stockholder partition with respect to the quality of the molecular two-electron density is more strongly demonstrated in the corresponding plots for LiH (Figs. 5.13 and 5.14). These DFT results should be compared with the reference UHF electron densities of Fig. 5.10. Only Li plots are shown in Figs. 5.13 and 5.14, since the bonded hydrogen atoms have been found to be relatively insensitive to the quality of the approximate representations of the exchange-effects, with both the one- and two-electron stockholder divisions predicting practically identical hydrogen AIM.

In the LDA approximation the two cusp feature of the 2-*S* Li atom can be clearly seen already at the equilibrium bond length. This qualitative feature of the atomic electron density is seen to be relatively enhanced at larger distances between the two atoms. This unphysical artifact of the LDA approximation is remedied in Fig. 5.14 at large internuclear separations by the less approximate B3LYP-DFT calculations using the gradient dependent exchange-correlation functional, although it can still be detected at intermediate bond elongations.

One can conclude from these illustrative results that a meaningful description of the two-electron stockholder atoms close to the bond dissociation can be obtained only from calculations, in which the tails of correlation holes are realistically represented. The above illustrative results show that the LDA of DFT does not satisfy this requirement, leading to the unphysical two-cusp densities of bonded atoms at large internuclear distances. In this regard the UHF results or the orbital-dependent exchange-correlation energy functional schemes of DFT, e.g., the OEP/KLI approximation, or the related exchange-only case known as the OPM, which in principle represents the exact UHF-KS scheme, should all provide the adequate description of the evolution of the effective *one*-electron densities of the 2-*S* AIM in the bond dissociation processes.

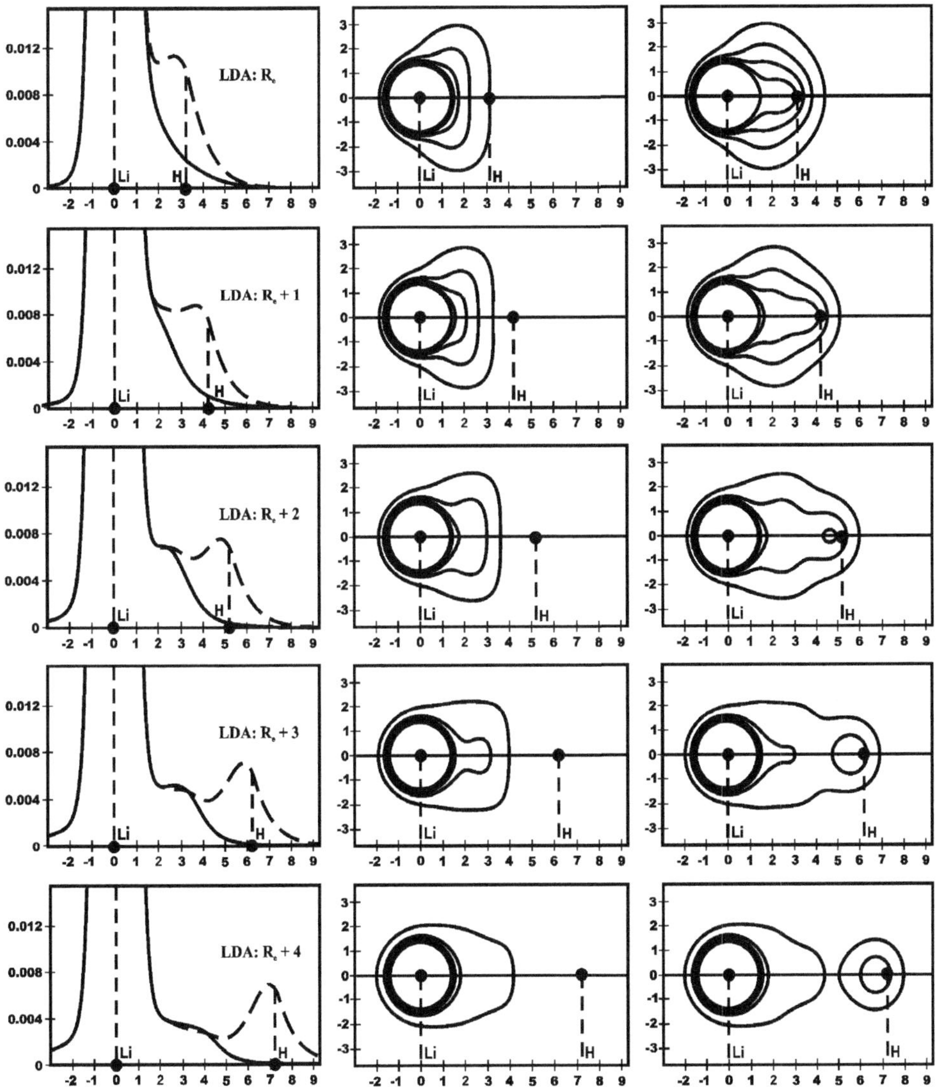

Figure 5.13. Same as in Fig. 5.10 for Li in LiH and the LDA approximation of DFT.

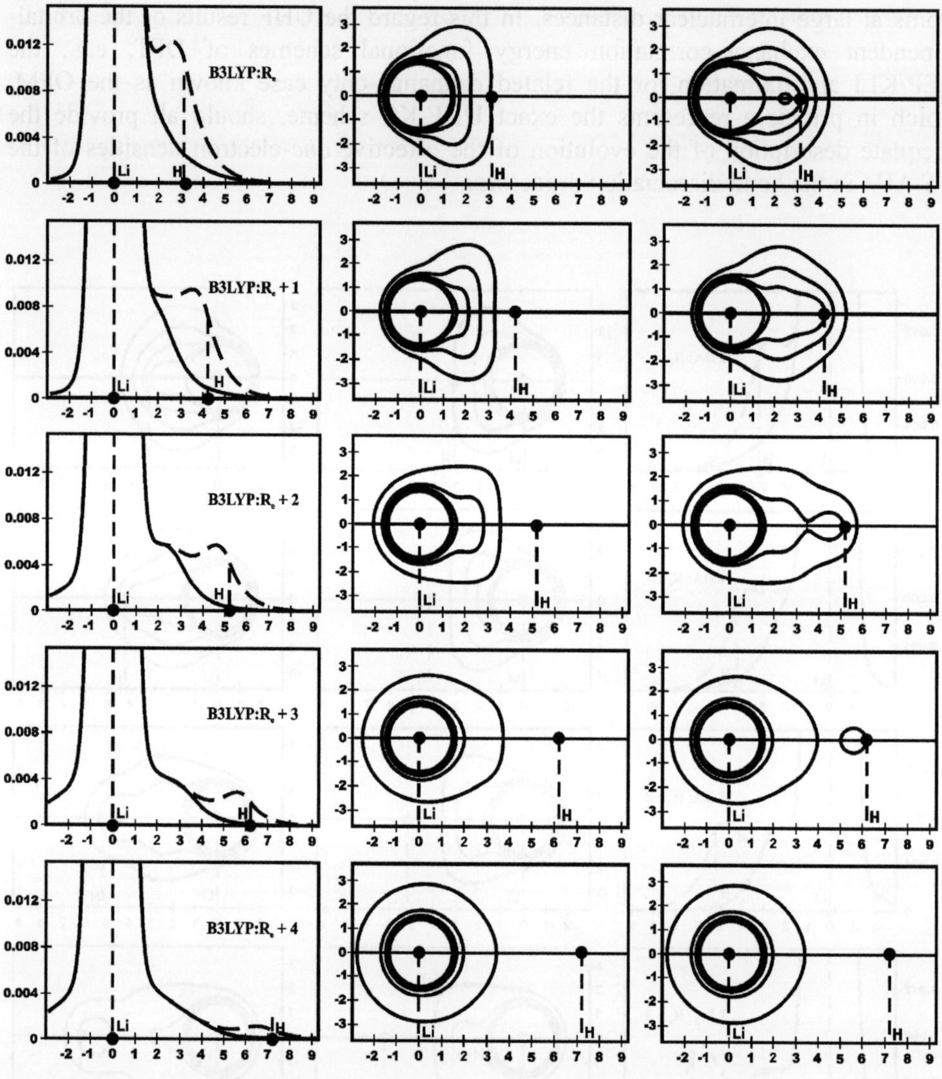

Figure 5.14. Same as in Fig. 5.10 for Li in LiH and the B3LYP approximation of DFT.

5.8. CLUSTER COMPONENTS OF TWO-ELECTRON STOCKHOLDER AIM IN DIATOMICS

Let us next examine for representative diatomics $M = AB$ the "diagonal" (one-centre) and "off-diagonal" (two-centre) components of the AIM 2-clusters, $\{\rho_{\alpha\beta}^{S}(\boldsymbol{r}, \boldsymbol{r}') = N(N-1)\ p_{\alpha\beta}^{S}(\boldsymbol{r}, \boldsymbol{r}')\}$ of the *two*-electron stockholder (2-*S*) AIM, which we have introduced in the preceding section. In order to visualize their contributions to the effective *one*-electron distribution of bonded atoms, we shall generate the bond axis profiles of the partly integrated components:

$$\rho_{\alpha\beta}^{\,eff}(\boldsymbol{r}) = \int \rho_{\alpha\beta}^{\,S}(\boldsymbol{r}, \boldsymbol{r}')\, d\boldsymbol{r}', \qquad \alpha = A, B. \tag{5.8.1}$$

These cluster contributions are defined by the stockholder fractions (shares) $\{d_{\alpha\beta}^{\,S}(\boldsymbol{r}, \boldsymbol{r}')\}$ of the molecular pair-density $\rho_2(\boldsymbol{r}, \boldsymbol{r}')$:

$$\rho_{\alpha\beta}^{\,S}(\boldsymbol{r}, \boldsymbol{r}') = d_{\alpha\beta}^{\,S}(\boldsymbol{r}, \boldsymbol{r}')\, \rho_2(\boldsymbol{r}, \boldsymbol{r}'), \tag{5.8.2}$$

which are determined by the free-atom distributions describing the promolecular reference. They represent the two-electron conditional probabilities $\{\pi^S(\alpha, \beta|\boldsymbol{r}, \boldsymbol{r}')\}$, the normalization of which involves the summation over all pairs of AIM labels $\{\alpha, \beta\}$ (variables) for the fixed electron positions (parameters): $\Sigma_\alpha\Sigma_\beta\, \pi^S(\alpha, \beta\,|\boldsymbol{r}, \boldsymbol{r}') = 1$.

As we have also observed before, the 2-cluster density components can be alternatively viewed as the (cluster-independent, unbiased) molecular *enhancement* of the corresponding promolecular 2-cluster distributions $\{\rho_{\alpha\beta}^{\,0}(\boldsymbol{r}, \boldsymbol{r}') = N(N-1)\, p_{\alpha\beta}^{\,0}(\boldsymbol{r}, \boldsymbol{r}')\}$,

$$w_2(\boldsymbol{r}, \boldsymbol{r}') = \rho_2(\boldsymbol{r}, \boldsymbol{r}')/\rho_2^{\,0}(\boldsymbol{r}, \boldsymbol{r}') \cong 1, \tag{5.8.3}$$

which give rise to the promolecular pair-density $\rho_2^{\,0}(\boldsymbol{r}, \boldsymbol{r}') = \Sigma_\alpha\Sigma_\beta\, \rho_{\alpha\beta}^{\,0}(\boldsymbol{r}, \boldsymbol{r}')$. In the preceding equation we have indicated that the molecular pair-density strongly resembles that of the promolecule.

The associated two-electron probability distributions $\{p_{\alpha\beta}^{\,0}(\boldsymbol{r}, \boldsymbol{r}')\}$ satisfy the overall promolecular normalization:

$$\Sigma_\alpha\Sigma_\beta \int\!\!\int p_{\alpha\beta}^{\,0}(\boldsymbol{r}, \boldsymbol{r}')\, d\boldsymbol{r}\, d\boldsymbol{r}' \equiv \Sigma_\alpha\Sigma_\beta\; P_{\alpha\beta}^{\,0} = 1; \tag{5.8.4}$$

here the *condensed* cluster probabilities in the promolecule, $\mathbf{P}^0 = \{P_{\alpha\beta}^{\,0}\}$ are given by the following expressions in terms of the overall numbers of electrons in the isolated constituent atoms $\{N_\alpha^{\,0}\}$ and that of the (isoelectronic) promolecule, $N^0 = \Sigma_\alpha N_\alpha^{\,0} = N$:

$$P_{\alpha\alpha}^{\,0} = N_\alpha^{\,0}\,(N_\alpha^{\,0} - 1)/[N^0(N^0 - 1)]; \quad P_{\alpha\beta}^{\,0} = N_\alpha^{\,0}\, N_\beta^{\,0}/[N^0(N^0 - 1)], \quad \alpha \neq \beta. \tag{5.8.5}$$

Of interest also are the partial normalizations of the cluster two-electron probabilities:

$$\int p_{\alpha\alpha}^{\,0}(\boldsymbol{r}, \boldsymbol{r}')\, d\boldsymbol{r}' = [(N_\alpha^{\,0} - 1)/(N^0 - 1)]\, p_\alpha^{\,0}(\boldsymbol{r}), \tag{5.8.6}$$

$$\int p_{\alpha\beta}^{\,0}(\boldsymbol{r}, \boldsymbol{r}')\, d\boldsymbol{r}' = [N_\beta^{\,0}/(N^0 - 1)]\, p_\alpha^{\,0}(\boldsymbol{r}), \qquad \alpha \neq \beta, \tag{5.8.7}$$

where the atomic one-electron distribution $p_\alpha^{\,0}(\boldsymbol{r})$ gives rise to the condensed atomic probability in the promolecule as a whole:

$$\int p_\alpha{}^0(\boldsymbol{r})\, d\boldsymbol{r} = N_\alpha{}^0/N^0 \equiv P_\alpha{}^0, \qquad \Sigma_\alpha P_\alpha{}^0 = 1. \tag{5.8.8}$$

Therefore, the off-diagonal ($\alpha \neq \beta$) 2-cluster component is roughly separable:

$$p_{\alpha\beta}{}^S(\boldsymbol{r}, \boldsymbol{r}') \cong p_\alpha{}^0(\boldsymbol{r})\, p_\beta{}^0(\boldsymbol{r}'), \tag{5.8.9}$$

thus giving rise to the α-cusped effective off-diagonal contribution due to the other atom β, to the density distribution of αth AIM:

$$p_{\alpha\beta}{}^{eff}(\boldsymbol{r}) \cong p_\alpha{}^0(\boldsymbol{r}) \int p_\beta{}^0(\boldsymbol{r}')\, d\boldsymbol{r}' = [N_\beta{}^0/(N^0 - 1)]\, p_\alpha{}^0(\boldsymbol{r}). \tag{5.8.10}$$

Clearly, the promolecular dominance of the molecular diagonal component,

$$p_{\alpha\alpha}{}^S(\boldsymbol{r}, \boldsymbol{r}') \cong p_{2,\alpha}{}^0(\boldsymbol{r}, \boldsymbol{r}'), \tag{5.8.11}$$

will also generate the effective contribution proportional to $p_\alpha{}^0(\boldsymbol{r})$:

$$p_{\alpha\alpha}{}^{eff}(\boldsymbol{r}) \cong \int p_{2,\alpha}{}^0(\boldsymbol{r}, \boldsymbol{r}')\, d\boldsymbol{r}' = [(N_\alpha{}^0 - 1)/(N^0 - 1)]\, p_\alpha{}^0(\boldsymbol{r}). \tag{5.8.12}$$

Hence, both $p_{\alpha\beta}{}^{eff}(\boldsymbol{r})$ and $p_{\alpha\alpha}{}^{eff}(\boldsymbol{r})$ components can be expected to strongly resemble the indicated fractions of the promolecularly normalized free-atom distribution $p_\alpha{}^0(\boldsymbol{r})$:

$$p_{\alpha\alpha}{}^{eff}(\boldsymbol{r}) + p_{\alpha\beta}{}^{eff}(\boldsymbol{r}) \cong p_\alpha{}^0(\boldsymbol{r})\, (N_\alpha{}^0 - 1 + N_\beta{}^0)/(N^0 - 1) = p_\alpha{}^0(\boldsymbol{r}). \tag{5.8.13}$$

These expectations are supported by the numerical results for the representative diatomics, H_2, LiH, HF, LiF, and N_2, which are reported in Figs. 5.15-17. The figures summarize the 2-*S* partitioning of the molecular two-electron densities obtained from the KS-LSDA approximation (DZVP basis set). These calculations can be regarded as being practically equivalent to the UHF approximation, which exactly takes into account the exchange (Fermi) correlation between electrons and completely neglects the many-body effects due to the Coulomb interaction. In Fig. 5.18 the influence of the Coulomb correlation, estimated using the UMP2 method (DZV basis set), is additionally examined.

Since in the isolated hydrogen atom (one-electron system) $p_{2,\mathrm{H}}{}^0(\boldsymbol{r}, \boldsymbol{r}') = 0$, the diagonal component of the 2-*S* division identically vanishes. Therefore, the effective one-electron distribution of the 2-*S* hydrogens in a diatomic molecule, which are referenced to the isolated (neutral) atom, has only the off-diagonal cluster contribution, shown already in Figs. 5.2-5, as well as in the logarithmic plot of the left panel of Fig. 5.16. It indeed follows from the approximate relation (5.8.10) that this term gives rise to $\rho_\mathrm{H}{}^S(\boldsymbol{r}) \cong \rho_\mathrm{H}{}^0(\boldsymbol{r})$.

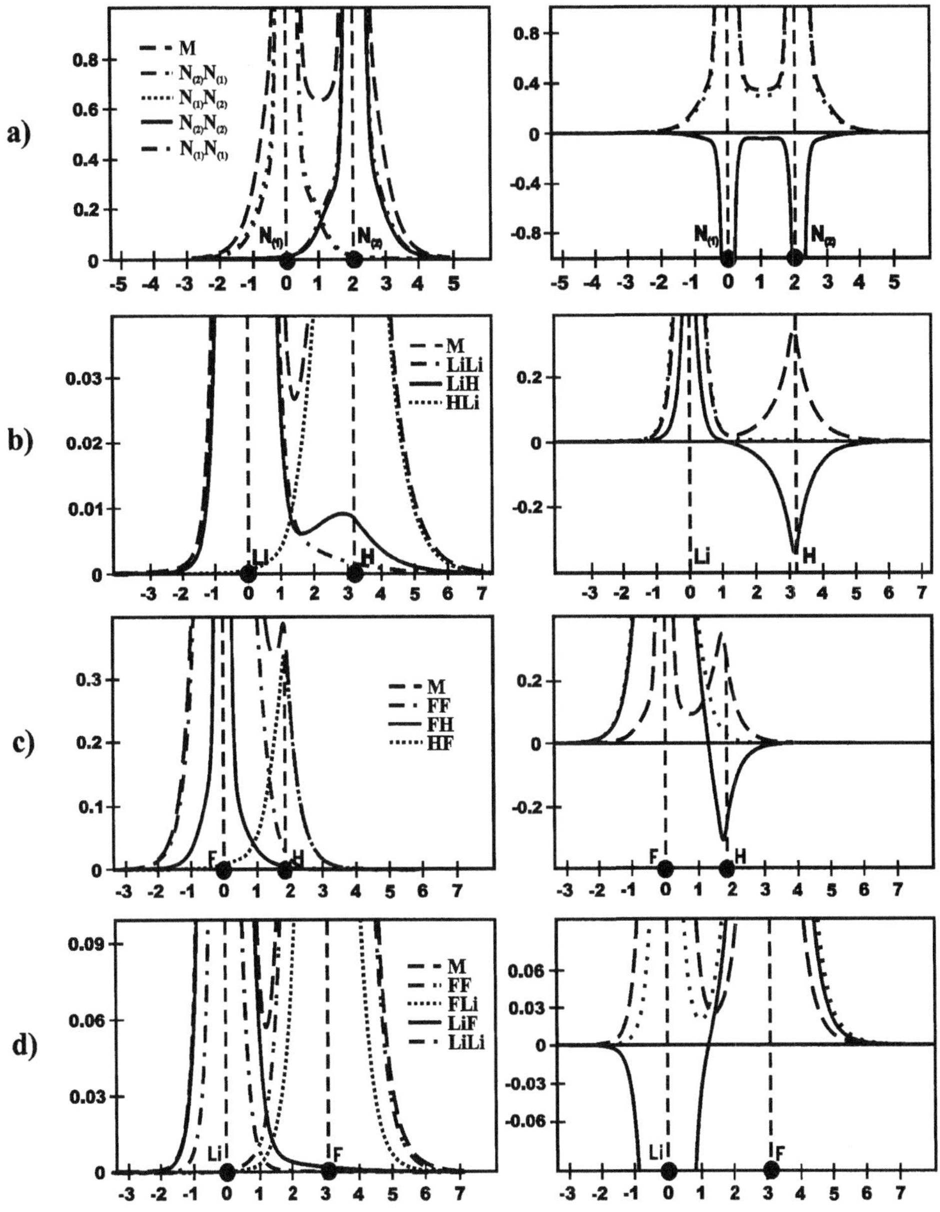

Figure 5.15. Left column: the bond axis profiles of the 2-cluster components $\{\rho_{\alpha\beta}^{eff}(\boldsymbol{r})\}$ of the 2-*S* AIM in: N_2 (*a*), LiH (*b*), HF(*c*), and LiF (d). Right column shows the corresponding sums of the diagonal (dotted line) and off diagonal (broken line) components, $C(\boldsymbol{r}) \equiv \rho_{\alpha\alpha}^{eff}(\boldsymbol{r}) + \rho_{\beta\beta}^{eff}(\boldsymbol{r})$ and $D(\boldsymbol{r}) \equiv \rho_{\alpha\beta}^{eff}(\boldsymbol{r}) + \rho_{\beta\alpha}^{eff}(\boldsymbol{r})$, respectively, while the solid line represents the difference $C(\boldsymbol{r}) - D(\boldsymbol{r})$ [UHF-KS(LDA) scheme, DZVP basis set].

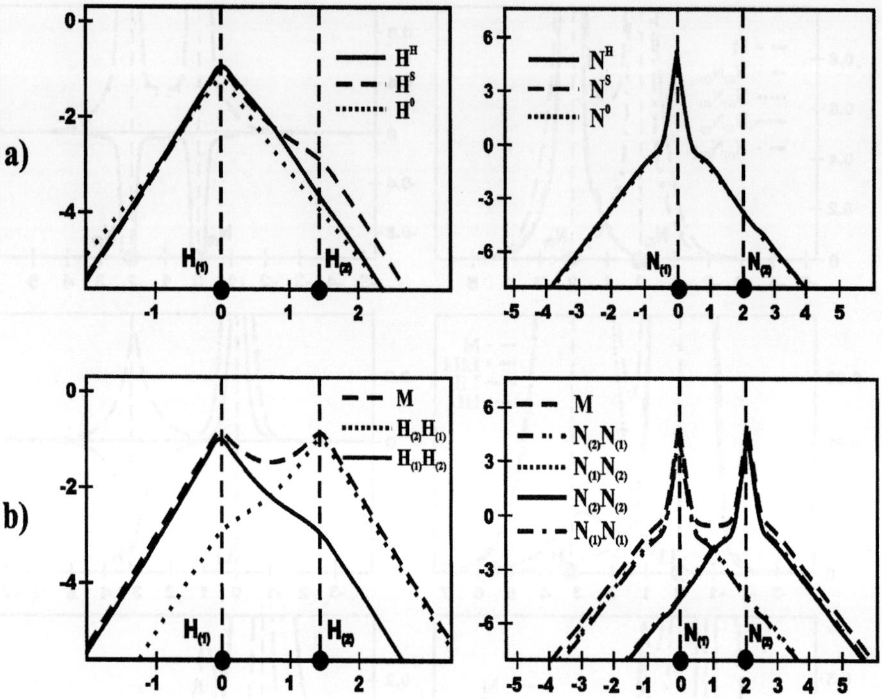

Figure 5.16. A comparison of the logarithmic plots of the electron density profiles along the bond axis of the isolated (^{0}), 1-S (^{H}), and 2-S (^{S}) atoms (Part a), and of the molecular (M) and the AIM 2-cluster components of the effective atomic densities (Part b), in H_2 (left column) and N_2 (right column) [UHF-KS(LDA) scheme, DZVP basis set].

Thus, changes in the electron distribution of the bonded hydrogen, due to the polarization and CT, have to be reflected by the off-diagonal component linking it with the bond partner. This is particularly constraining in the LiH case [see Figs 5.2*b*, right panel of 5.3*b*, 5.5(*b*,*c*), 5.15*b*, and 5.17 (left column)], where the negatively charged hydrogen has to accommodate the extra electron transfer from Li. The resulting CT "shoulder" is clearly visible in the $p_{\text{LiH}}{}^{eff}(\boldsymbol{r})$ profiles shown in Figs. 5.15*b* and 5.17*c* (left panel). It should be realized, however, that this artificial feature practically disappears in HF [Figs. 5.15*c* and 5.17*c* (medium column)], i.e., for the positively charged 2-*S* hydrogen, when it acts as the net electron donor. This is because the inflowing charge on F can be freely accommodated by the open diagonal (ionic) channel on fluorine atom, $p_{\text{FF}}{}^{eff}(\boldsymbol{r})$. The other reason for this difference is three times as high number of electrons in F compared to Li. Nevertheless, in the logarithmic plot of panel *c* in the middle column of Fig. 5.17, one still detects a minor "shoulder" feature in the $p_{\text{FH}}{}^{eff}(\boldsymbol{r})$ profile at the proton position.

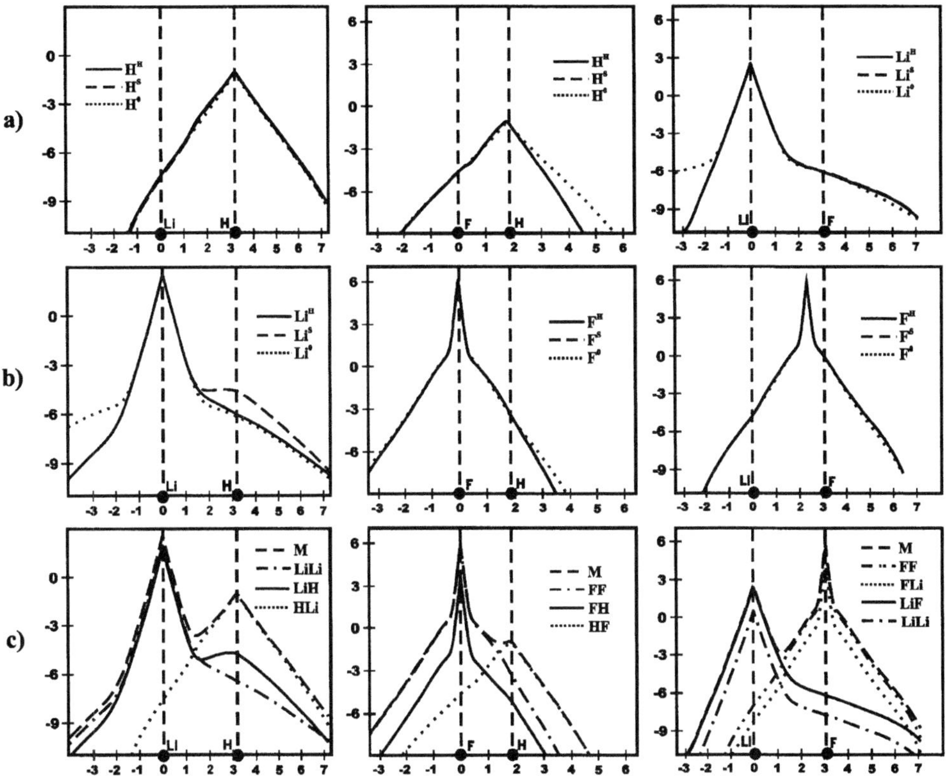

Figure 5.17. A comparison of the logarithmic plots of the electron density profiles along the bond axis of the isolated (^{0}), 1-S (^{H}), and 2-S (^{S}) atoms (Parts *a* and *b*), and of the molecular (*M*) and the AIM 2-cluster components of the effective atomic densities (Part *c*), in LiH (left column), HF (middle column), and LiF (right column) [UHF-KS(LDA) scheme, DZVP basis set].

In the homonuclear N_2 system [Fig. 5.15*a* and 5.16*b* (right panel)] the diagonal ("ionic") and off-diagonal ("covalent") contributions to the bonded nitrogen atom are seen to be almost equal, with the exception of the 1*s* core regions around nuclei, where the covalent term dominates. This feature is clearly reflected in the right panel of Fig. 5.15*a*, where the balance between the overall diagonal and off-diagonal components is investigated.

This overall equality of both these contributions in the chemically most important valence-shell, identifies the purely covalent matching between the diagonal and off-diagonal pair-density components of the 2-S AIM. It is reminiscent of the equal share of the *ionic* and *covalent* VB-structures in the minimum basis set SCF MO description of the purely covalent chemical bond in H_2 (see Section 8.2.3).

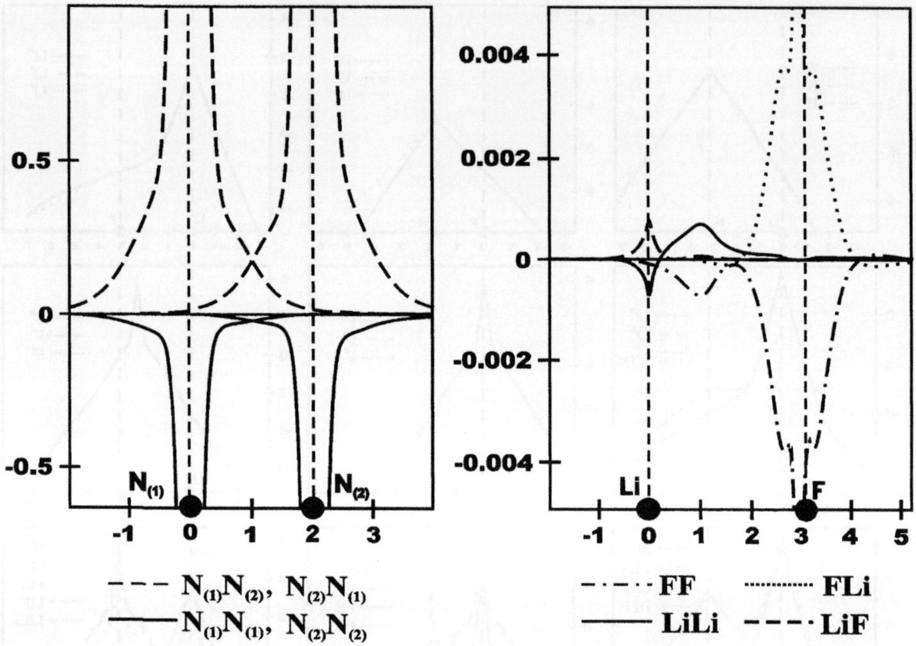

Figure 5.18. The effect of the extra Coulomb correlation on the exchange-only correlated 2-cluster components of the effective AIM densities in N_2 (left panel) and LiF (right panel).

The difference between the overall diagonal and off-diagonal density contributions in heteronuclear diatomics reflects the bond ionicity and the CT between the bonded 2-*S* AIM. This is most transparently seen in LiF, in the right panel of Fig. 5.15*d*, where both covalent and ionic channels of both atoms are open. The overall off-diagonal contribution is found to dominate the (donor) Li atom region, while the opposite trend is detected in the (acceptor) F atom. The difference (solid line) plot directly reflects the Li→F CT. A similar H→F CT pattern can be detected in HF density profile (see the right panel of Fig. 5.15*c*), for which the open diagonal channel on fluorine can accommodate the inflowing charge donated by the hydrogen.

The opposite trend observed in LiH is not representative of this general balance, since this is the artefact of the closed ionic (diagonal) channel on the acceptor hydrogen atom. Indeed, the net inflow of electrons to the bonded, acceptor 2-*S* hydrogen, cannot be accommodated through the off-diagonal (H,Li) component, which strongly resembles the isolated hydrogen density. It can be absorbed only in the other (Li,H) off-diagonal component, which is attributed in the 2-*S* division scheme to the bonded lithium atom. This artificially creates the nonphysical dominance of the overall "covalent" density component in the hydrogen region and the net "acceptor" difference plot in the lithium region.

In Table 5.1 we have listed the net AIM charges for the constituent 1-*S* (Hirshfeld) and 2-*S* stockholder atoms in these three heteronuclear diatomics. It follows from the table that with the increasing number of electrons the differences between the net charges of the 1-*S* and 2-*S* AIM gradually disappear. The largest deviation is observed for LiH for the reasons discussed above.

The two sets of the AIM charges reflect qualitatively the chemically expected bond polarization due to the electronegativity difference exhibited by the isolated atoms. A general rule emerging from this comparison is that the 2-*S* AIM, which effectively contain the Fermi correlation effects, exhibit less charge separation in comparison to the 1-*S* (Hirshfeld) atoms. For example, the 2-*S* Li in LiH exhibits distinctly less positive net charge, $q_{\mathrm{Li}}{}^{S} = 0.24$, compared to the 1-*S* Li^{H} in LiH, $q_{\mathrm{Li}}{}^{H} = 0.35$, which accords with the closed diagonal (acceptor) channel on the bonded hydrogen. In the two remaining diatomic molecules this difference is seen to be substantially reduced.

In the logarithmic plots of Figs. 5.16 and 5.17 we have examined in a more detail the valence-shell decays of the effective electron densities of the 2-*S* bonded atoms in the homonuclear diatomics, and their AIM 2-cluster components. These plots are compared against the corresponding free- and Hirshfeld-atom densities and the molecular electron density.

The diagrams for the homonuclear diatomics (Fig. 5.16) confirm the molecular character of the bonded hydrogen atoms resulting from both divisions, with the constant decay rate at large distances determined by the molecular ionization potential. In Fig. 5.16*a* the extra-bonding character of the 2-*S* hydrogen atoms in H_2, relative to the Hirshfeld analogs, is now pronounced. It should be observed that the differences between the 1-*S* and 2-*S* nitrogen atoms in N_2 practically disappear also for the low values of the electron densities, which are emphasized in these logarithmic plots. The approximate equality of the ionic (diagonal) and covalent (off-diagonal) density contributions of the 2-*S* nitrogens, marking the purely covalent, multiple bond in N_2, is seen to be satisfied also at these valence regions of low electron density.

Similar conclusions follow from examining the logarithmic density profiles of Fig. 5.17, for the three representative heteronuclear diatomics: LiH (left column), HF (medium column), and LiF (right column). The H in LiH plots of the Hirshfeld and 2-*S* AIM, shown in the upper panel, exhibit only minor differences at the low density region in comparison to free-hydrogen, while the extra density polarization of the Li valence 2*s* electron from the non-bonded region towards the hydrogen can be detected in the medium panel of the left column in the figure. It is more pronounced in the two-electron stockholder lithium, due to the off-diagonal (covalent) LiH component seen in the lowest panel. The density decay of the polarized LiH molecule is seen to be governed by the two components due to the softer Li atom: $\{\rho_{Li\alpha}{}^{eff}(\boldsymbol{r}),\ \alpha = \mathrm{H, Li}\}$. The CT polarization of the bonded hydrogen in HF is clearly seen in the upper panel of the middle column. No substantial differences between the 1-*S*(*H*) and 2-*S*(*S*) constituent atoms are being observed for this diatomic consisting of two *hard* atoms, which generate the partly ionic bond with a relatively strong covalent component. A typical behavior of the logarithmic profiles for the electron donor (Li) and acceptor (F) is also seen in the third column of the figure, with the bonded lithium exhibiting a strong polarization of its valence electron density towards fluorine. Both *one-* and *two-*electron divisions give rise to practically identical sets of density profiles at the low density regions.

In Fig. 5.18 we have additinally examined the effect of the Coulomb correlation on the 2-cluster components of the effective one-electron distributions of constituent AIM in N_2 and LiF. As intuitively expected, in the homonuclear diatomic (left panel) the diagonal (ionic) contributions, corresponding to events of finding two electrons on the same atom, are lowered in the immediate vicinity of the nuclei, where the accumulation of electrons is the highest. This is accompanied by a strong increase of the off-diagonal (covalent, delocalization) components in these two nuclear regions, due to an increased sharing of the bonding electrons by the two AIM. In other words, the Coulomb correlation lowers the one-centre ionic terms in favour of the two-centre covalent contributions to the effective electron density of the 2-*S* bonded AIM. A reference to the right panel in the figure shows that this electron-correlation effect also moderates the charge separation resulting from the exchange-only approximation. Indeed, the FF profile implies a slight decrease of the effective electron density on F, resulting from the fluorine one-centre two-electron contribution. This diagonal component also implies a lowering of the density in the bond-region, near Li atom, which is compensated by the lithium one-center contribution. We also observe a complementary increase in the two-centre FLi profile on the fluorine due to the Coulomb correlation, a direct manifestation of an increased covalency of the Li—F bond, i.e., a more extensive sharing of the valence electrons between the two bonding partners.

This analysis shows that an inclusion of the "vertical" Coulomb correlation moderates a somewhat inflated charge separation of the (Hirshfeld) stockholder division of the molecular electron density. The dominating correlation effect seen in the 2-*S* AIM distributions results from the exchange (Fermi) correlation, with the remaining Coulomb correlation introducing only a minor modification of the exchange-only two-electron joint probabilities.

5.9. CONCLUSION

In this chapter we have summarized the "stockholder" partitioning of the molecular electron distributions into fragments representing the bonded atoms. In the *one-*electron case it gives the Hirshfeld (1977) division scheme, which was shown to have a solid information-theoretic basis. It has been demonstrated that this division minimizes the *global* information distance between the AIM electron distributions and the corresponding densities of the constituent free-atoms defining the isoelectronic (atomic) promolecule, subject to the local constraints of the exhaustive partition. Alternatively, the Hirshfeld AIM discretization of the molecular electron density can be interpreted as the *local* division of the AIM conditional probabilities, for the fixed locality of an electron. In such a local perspective the stockholder partition minimizes the local entropy deficiency between the AIM conditional probabilities and the known free-atom conditional probabilities of the promolecule, subject to the relevant probability normalization constraint. The stockholder rule was shown to be independent of the particular measure of the information distance used in the promolecule-referenced, constrained variational principle for the minimum of the entropy deficiency penalty term.

It has also been demonstrated that in a consistent treatment of the bonded- or free-atoms as parts of a larger molecular or promolecular systems, respectively, the problem of the subsystem density normalization does not influence the entropy/information variational principles. In other words, the extremum principles in terms of subsystem electron densities and their probability distributions are exactly equivalent.

The usefulness of the information variational problems, formulated in terms of the missing-information functional, has been also demonstrated in applications to the illustrative maximum similarity issues involving two-references. In these principles the cross (relative) entropy terms also determine the information "penalty" for the optimum probability densities of molecules or their fragments deviating from the specified reference distributions. Such problems emerge in the molecular similarity considerations and in defining the polarizational-promotion of the mutually closed atoms in molecular systems.

Applying the constrained entropy-deficiency principle to a more general problem of dividing the molecular joint *many*-electron distributions gives rise to an extension of the *one*-electron stockholder principle of Hirshfeld to the related schemes for the optimum (exhaustive) partition of the molecular joint k-electron probabilities, $1 < k \leq N$, into contributions from the corresponding k-clusters of AIM. At different $k > 1$, in the k-S AIM, slight changes in the effective one-electron distributions of bonded atoms can be expected, compared to the $k = 1$ (Hirshfeld) distributions of the 1-S AIM.

The $k = 2$ case, which takes into account the dominant two-electron correlation effects, has been examined in a more detail for illustrative diatomics. The influence of both the vertical and horizontal correlation effects on atomic densities has been investigated. It has been argued that the two-electron stockholder terms corresponding to the diagonal and off-diagonal AIM 2-clusters provide a natural framework for separating the ionic and covalent bond components, respectively. When applied to the classical problems in chemistry, e.g., the interactions between hard and soft acids and bases, the two-electron stockholder division has been shown to generate a transparent interpretation of the main stability preferences and bond compositions of the DA interaction, as summarized by the familiar HSAB principle of Pearson.

Since the two-particle correlations account for a major part of the overall correlation energy, it was of particular interest to compare the one-electron distributions of the 1-S and 2-S AIM. The electron densities of bonded atoms in selected diatomics, H_2, LiH, HF, LiF, and N_2, show that in diatomics containing hydrogen the two-electron stockholder treatment emphasizes more strongly the bonding region of the atomic distribution, in comparison to its one-electron (Hirshfeld) analog, giving rise to a slightly more pronounced polarization of AIM atom towards their bond partners. However, this subtle difference was found to fast disappear with the increasing number of electrons in molecular system. For heavier atoms the distinction between the electron densities and probability distributions of the 1-*S* and 2-*S* AIM practically disappears.

It has been shown that the adequate two-electron density components can be obtained only from calculations using a correct description of inter-atom electron correlation effects. The most approximate (LDA) variant of DFT has been shown to give rise to a qualitatively incorrect, two-cusp AIM densities in the bond dissociation limit. This shortcoming is partly remedied in the GGA-type B3LYP functional which in the 2-*S* division leads to a dissociation into the separated free atoms/ions.

Therefore, the Hirshfeld atomic subsystems, representing the optimum (equilibrium) molecular fragments resulting from the minimum entropy-deficiency principle, are not exactly unique in the Information Theory, since each *k*-electron stockholder division defines slightly different effective one-electron distributions of bonded atoms in a given molecular system. However, with these differences being so minute for most of the constituent atoms, one can safely conclude that the promolecule-referenced *k*-electron division problems gives rise to practically identical sets of *k*-*S* AIM. This "*invariance*" property of the stockholder bonded atoms from IT partly explains the wide use of the Hirshfeld atoms in crystallography and their several unique properties, which make them important concepts for interpretations in chemistry. We shall examine some of these characteristic features of the molecular stockholder-fragments in the next chapter.

For illustrative diatomic molecules we have explored in some detail the *vertical* and *horizontal* correlation influences on the atomic and molecular electron densities, as well as their one- and two-centre components in the 2-*S* division scheme. The diagonal (one-center) AIM-components of this partition measure the joint probability of finding two electrons on the same AIM, thus corresponding to the *ionic* valence structures of the VB theory of Heitler and London. Similarly, the two-centre off-diagonal contributions, measuring the joint probabilities of finding two electrons on different bonded atoms, can be naturally associated with the *covalent* valence structures of the VB theory. In this perspective, the stockholder partitioning of the molecular *two*-electron densities provides a convenient theoretical framework for describing these two components of the chemical bond. We have also observed in Section 4 that the purely covalent triple bond in N_2, in which all division channels are accessible, corresponds to almost-equalized one- and two-centre components of the 2-*S* partition, again in perfect analogy to the VB description. Therefore, one should associate a deviation from this balance as a reflection of the bond ionicity. A notable exception to this rule is provided by the 2-*S* hydrogen atoms, originating from the *one*-electron free-atom reference, which participate in the molecular density only through the off-diagonal (H,α)-components.

6. OTHER PROPERTIES OF STOCKHOLDER SUBSYSTEMS

Several attractive features of the stockholder pieces of the molecular electron distribution, which define the entropy-deficiency equilibrium molecular fragments in the information-theoretic approach, are examined and illustrated for selected molecules discussed in Chapter 4. The local equalization of the entropy/information densities of the Hirshfeld AIM is emphasized. It incorporates the unbiased character of the stockholder division principle and is shown to give rise to the exact additivity of the subsystem entropy deficiencies. The cross-entropy terms in the variational principles, which give rise to the stockholder partition of the molecular electron distributions, are interpreted as effective entropy "penalty" terms, which generate the *localized* (single-cusped) electron distributions in bonded atoms. The additive Shannon entropy of atomic distributions yields the perfectly *delocalized* (multi-cusped) electron densities of molecular fragments. The entropy/information quantities describing changes in the stockholder AIM relative to the free-atoms can be also used to diagnose the AIM promotion in the molecular valence-state and – ultimately – to understand the entropy/information origins of the chemical bond. An illustrative application of such an analysis to linear diatomics and propellanes is reported. In the theory of chemical reactivity the subsystem-resolved, second-order Taylor expansions of the molecular "thermodynamic" potentials, the Legendre transforms of the system electronic energy, which are relevant for alternative sets of the system independent state-parameters, play an essential role in designing the adequate reactivity criteria. In this chapter an overview of such *chemical softness* (density response) and *hardness* (potential response) properties of the Hirshfeld atoms will be given. The relation between such descriptors of molecules and their stockholder fragments will be examined. As mutually open fragments of the molecule the Hirshfeld fragments of the molecule satisfy the chemical potential (electronegativity) equalization in the molecular ground-state. They are also shown to be in principle effective external potential representable. The *partial* and *total* energy-conjugates of the stockholder-AIM pieces of the molecular electron density are identified. The charge sensitivities (CS) of the stockholder AIM, measuring their generalized "polarizabilities", are examined in both the chemical softness and hardness representations, within the *electron-following* (EF) and *electron-preceding* (EP) perspectives, respectively. These subsystem-resolved CS correspond to the global equilibria in the *mutually* open fragments of the molecular or reactive system, when they are either *externally* open or closed relative to an external reservoir.

6.1. LOCAL ENTROPY/INFORMATION EQUALIZATION RULES

In the preceding chapter the common-sense, unbiased scheme of Hirshfeld (1977) for the (local) exhaustive partitioning of the molecular electron density into the infinitely extending AIM pieces has been given a solid information-theoretic basis. We have demonstrated that this scheme results from the localy-constrained minimum principle for the missing information in densities of bonded atoms relative to the corresponding free-atom (promolecular) distributions, which provide the reference for the density-difference function. In this chapter several additional properties of these entropy-deficiency equilibrium (stable) 1-*S* atoms will be established.

These densities of overlapping atoms preserve as much as possible the information contained in the electron distributions of the free atoms. The Hirshfeld atom was found to exhibit a cusp at the atomic nucleus, which reflects its effective nuclear charge in a molecule. Its density decays exponentially at large distances from the molecule. As we shall argue in this chapter, the stockholder atom also manifests the intuitively expected changes due to a formation of the chemical bond. The Hirshfeld atoms are reflecting their “promotion” to the valence-state, orbital hybridization, and *polarization* towards the bonding partners, as well as the charge transfer due to electronegativity differences of the constituent free atoms. As mutually open fragments of the molecule these 1-*S* subsystems satisfy the chemical potential equalization principle in the molecular ground-state. The unbiased (molecular) character of the local enhancement factor of the stockholder AIM, giving rise to the molecular (fragment-independent) local surprisals, also implies the equalization of the fragment local information-distance densities at the corresponding molecular level which characterizes the system as a whole.

The missing-information density of the bonded atom and the related entropy displacement can be approximately related to atomic density-difference function. In fact the entropy/information and electron density displacement quantities of AIM can be regarded as complementary tools for probing the effective states of the bonded atoms in their molecular environments. In this chapter several illustrative examples will be given of using these alternative tools for diagnosing the AIM promotion in the molecular systems discussed in Chapter 4.

The uniqueness of the Hirshfeld subsystems in the entropy-deficiency representation makes them attractive concepts for chemical interpretations of the calculated or experimental electron densities of molecular and reactive systems (Nalewajski and Parr, 2000, 2001; Nalewajski, 2002a-c, 2003a-d, 2004a,c). In this chapter we shall examine the charge sensitivities of the stockholder AIM, given by the second derivatives of the electronic energy in subsystem resolution. They measure the atom responses to typical perturbations in the molecule (see Chapter 2 and Appendix D). Specific expressions for these generalized polarizabilities in terms of the corresponding descriptors of the system as a whole will be examined. The local character of the stockholder division principle will be shown to greatly simplify the chain-rule transformations which determine these relations.

In the preceding chapter we have already discussed several unique features of the electron densities of the stockholder subsystems (Nalewajski and Parr, 2001; Nalewajski, 2002a; Nalewajski *at al.*, 2002; Nalewajski and Świtka, 2002}, e.g., the inter-subsystem *equalization* of the local values of the *enhancement factors* of molecular fragments,

$$\begin{aligned} w_X^H(\boldsymbol{r}) &\equiv \rho_X^H(\boldsymbol{r})/\rho_X^0(\boldsymbol{r}) = p_X^H(\boldsymbol{r})/p_X^0(\boldsymbol{r}) \\ &= \rho(\boldsymbol{r})/\rho^0(\boldsymbol{r}) = p(\boldsymbol{r})/p^0(\boldsymbol{r}) \equiv w(\boldsymbol{r}), \qquad X = A, B, \ldots, \end{aligned} \tag{6.1.1}$$

which takes place only for this special, unbiased division scheme. Thus, for a given location in space such quantities are subsystem independent, all being equal to the corresponding value of the global enhancement $w(\boldsymbol{r})$ of the molecular density relative to the promolecular reference value.

The preceding equation also implies in the stockholder partition the equalization of any function of the subsystem enhancement factors, e.g., of the density/probability surprisals of AIM:

$$I_X[w_X^H(\boldsymbol{r})] \equiv I_X^H(\boldsymbol{r}) = \log w_X^H(\boldsymbol{r}) = \log w(\boldsymbol{r}) = I[w(\boldsymbol{r})] \equiv I(\boldsymbol{r}), \qquad X = A, B, \ldots \tag{6.1.2}$$

The subsystem surprisal $I_X[w_X^H(\boldsymbol{r})]$ measures the density per electron of the subsystem entropy deficiency:

$$I_X^H(\boldsymbol{r}) = \Delta s_X^H(\boldsymbol{r})/\rho_X^H(\boldsymbol{r}), \tag{6.1.3}$$

where $\Delta s_X^H(\boldsymbol{r})$ stands for the density of the Hirshfeld-fragment cross-entropy:

$$\begin{aligned} \Delta S[\rho_X^H \,|\, \rho_X^0] &= \int \rho_X^H(\boldsymbol{r}) \log[\rho_X^H(\boldsymbol{r})/\rho_X^0(\boldsymbol{r})]\, d\boldsymbol{r} = \int \pi^H(X\,|\boldsymbol{r})\rho(\boldsymbol{r})\, I(\boldsymbol{r})\, d\boldsymbol{r} \\ &\equiv \int \pi^H(X\,|\boldsymbol{r})\, \Delta s(\boldsymbol{r})\, d\boldsymbol{r} \equiv \int \Delta s_X^H(\boldsymbol{r})\, d\boldsymbol{r}, \qquad \textstyle\sum_X \pi^H(X|\boldsymbol{r}) = 1, \end{aligned} \tag{6.1.4}$$

where the conditional probability $\pi^H(X|\boldsymbol{r})$ is defined by stockholder share-factor $d_X^H(\boldsymbol{r})$ of Eq. (5.2.4). The atomic surprisal can be also interpreted as the density per electron of the subsystem divergence,

$$I_X^H(\boldsymbol{r}) = \Delta D_X^H(\boldsymbol{r})/\rho_X^H(\boldsymbol{r}), \tag{6.1.5}$$

where $\Delta D_X^H(\boldsymbol{r})$ is the fragment density of the Kullback missing information measure:

$$\begin{aligned} \Delta S[\rho_X^H, \rho_X^0] &= \int [\rho_X^H(\boldsymbol{r}) - \rho_X^0(\boldsymbol{r})] \log[\rho_X^H(\boldsymbol{r})/\rho_X^0(\boldsymbol{r})]\, d\boldsymbol{r} \\ &= \int \pi^H(X\,|\boldsymbol{r})\, \Delta\rho(\boldsymbol{r})\, I[w(\boldsymbol{r})]\, d\boldsymbol{r} \\ &\equiv \int \pi^H(X\,|\boldsymbol{r})\, \Delta D(\boldsymbol{r})\, d\boldsymbol{r} \equiv \int \Delta D_X^H(\boldsymbol{r})\, d\boldsymbol{r}. \end{aligned} \tag{6.1.6}$$

As we have observed in Eq. (6.1.2), for the Hirshfeld partition the equalization takes place of the local subsystem surprisals at the global surprisal value:

$$I_X^H(\boldsymbol{r}) = \Delta s_X^H(\boldsymbol{r})/\rho_X^H(\boldsymbol{r}) = \Delta D_X(\boldsymbol{r})/\Delta\rho_X(\boldsymbol{r}) = I[w(\boldsymbol{r})] = \Delta s(\boldsymbol{r})/\rho(\boldsymbol{r}) = \Delta D(\boldsymbol{r})/\Delta\rho(\boldsymbol{r}). \quad (6.1.7)$$

We have indicated in the preceding equation that the subsystem surprisals have analogous interpretations in terms of the subsystem missing information or divergence densities as do the corresponding information densities for the whole molecule. It measures the fragment entropy deficiency per electron, or – alternatively - the subsystem divergence density $\Delta D_X(\boldsymbol{r}) = [\rho_X(\boldsymbol{r}) - \rho_X^0(\boldsymbol{r})]I_X^H(\boldsymbol{r}) = \Delta\rho_X(\boldsymbol{r})\, I_X^H(\boldsymbol{r})$ per unit shift in the fragment local electron density.

These local inter-subsystem entropy/information equalization rules, e.g., those for the enhancement factors and surprisals of molecular fragments, provide the following *entropic perspective* on the equilibrium state of bonded molecular fragments in the molecule: among all admissible local divisions of the molecular electron density the entropy equilibrium (stockholder) subsystems are uniquely identified as those, for which each subsystem reaches locally the same information-distance density relative to its free-analog, as does the molecule as a whole relative to its promolecular prototype.

Clearly, any function of the subsystem surprisal and/or the enhancement factor is also equalized for the stockholder partitioning at the corresponding global value of that function. For example, the subsystem information-distances representing the alternative divergence densities of the Hirshfeld subsystems per electron of the promolecule,

$$\check{s}_X^H(\boldsymbol{r}) \equiv \Delta s_X^H(\boldsymbol{r})/\rho_X^0(\boldsymbol{r}) = w_X^H(\boldsymbol{r})\, I_X^H(\boldsymbol{r}), \quad (6.1.8)$$

$$\check{D}_X^H(\boldsymbol{r}) \equiv \Delta D_X^H(\boldsymbol{r})/\rho_X^0(\boldsymbol{r}) = [w_X^H(\boldsymbol{r}) - 1]\, I_X^H(\boldsymbol{r}), \quad (6.1.9)$$

are equalized for the stockholder pieces of the molecular density at the corresponding global values:

$$\check{s}_X^H(\boldsymbol{r}) = \Delta s(\boldsymbol{r})/\rho^0(\boldsymbol{r}) \equiv \check{s}(\boldsymbol{r}) = w(\boldsymbol{r})I(\boldsymbol{r}), \quad (6.1.10)$$

$$\check{D}_X^H(\boldsymbol{r}) = \Delta D(\boldsymbol{r})/\rho^0(\boldsymbol{r}) \equiv \check{D}(\boldsymbol{r}) = [w(\boldsymbol{r}) - 1]\, I(\boldsymbol{r}). \quad (6.1.11)$$

Other important examples of the locally equalized subsystem quantities are the entropy-deficiency conjugates of the subsystem densities, defined by the corresponding partial functional derivatives:

$$\left(\frac{\partial \Delta S[\boldsymbol{\rho}^H|\boldsymbol{\rho}^0]}{\partial \rho_X^H(\boldsymbol{r})}\right)_{Y\neq X} = \frac{\delta \Delta S_X[\rho_X^H|\rho_X^0]}{\delta \rho_X^H(\boldsymbol{r})} = I_X^H(\boldsymbol{r}) + 1 \equiv F_X^H(\boldsymbol{r}), \quad (6.1.12)$$

$$\left(\frac{\partial \Delta S[\boldsymbol{\rho}^H,\boldsymbol{\rho}^0]}{\partial \rho_X^H(\boldsymbol{r})}\right)_{Y\neq X} = \frac{\delta \Delta S_X[\rho_X^H,\rho_X^0]}{\delta \rho_X^H(\boldsymbol{r})} = I_X^H(\boldsymbol{r}) + 1 - \frac{1}{w_X^H(\boldsymbol{r})} \equiv G_X^H(\boldsymbol{r}), \quad (6.1.13)$$

where the subscript $Y \neq X$ implies that electron densities of the remaining molecular fragments are fixed. For the stockholder division of the molecular density these alternative measures of the local subsystem information distance from the free-fragment reference are equalized at the corresponding global values:

$$F_X^H(\boldsymbol{r}) = I_X^H(\boldsymbol{r}) + 1 = \frac{\delta \Delta S[\rho|\rho^0]}{\delta \rho(\boldsymbol{r})} = I(\boldsymbol{r}) + 1 \equiv F(\boldsymbol{r}), \tag{6.1.14}$$

$$G_X^H(\boldsymbol{r}) = I_X^H(\boldsymbol{r}) + 1 - \frac{1}{w_X^H(\boldsymbol{r})} = \frac{\delta \Delta S[\rho, \rho^0]}{\delta \rho(\boldsymbol{r})} = I(\boldsymbol{r}) + 1 - \frac{1}{w(\boldsymbol{r})} \equiv G(\boldsymbol{r}). \tag{6.1.15}$$

In the many-electron stockholder partition of Section 5.5 one encounters similar, *many*-electron equalization rules. Indeed, the equalization of the AIM k-cluster enhancement factors at the global k-electron level [Eq. (5.5.23)] implies the cluster-independent character of all their functions, e.g.. the cluster surprisals [Eq. (5.5.24)]. For example, in the most important *two*-electron case (Section 5.6),

$$\begin{aligned} w_{\alpha\beta}^S(\boldsymbol{r}, \boldsymbol{r}') &\equiv \rho_{\alpha\beta}^S(\boldsymbol{r}, \boldsymbol{r}')/\rho_{\alpha\beta}^0(\boldsymbol{r}, \boldsymbol{r}') = p_{\alpha\beta}^S(\boldsymbol{r}, \boldsymbol{r}')/p_{\alpha\beta}^0(\boldsymbol{r}, \boldsymbol{r}') \\ &= \rho_2(\boldsymbol{r}, \boldsymbol{r}')/\rho_2^0(\boldsymbol{r}, \boldsymbol{r}') = p_2(\boldsymbol{r}, \boldsymbol{r}')/p_2^0(\boldsymbol{r}, \boldsymbol{r}') \equiv w_2(\boldsymbol{r}, \boldsymbol{r}'), \end{aligned} \tag{6.1.16}$$

and hence:

$$\begin{aligned} I_{\alpha\beta}[w_{\alpha\beta}^S(\boldsymbol{r}, \boldsymbol{r}')] &\equiv I_{\alpha\beta}^S(\boldsymbol{r}, \boldsymbol{r}') = \log w_{\alpha\beta}^S(\boldsymbol{r}, \boldsymbol{r}') \\ &= \log w_2(\boldsymbol{r}, \boldsymbol{r}') = I_2[w_2(\boldsymbol{r}, \boldsymbol{r}')] \equiv I_2(\boldsymbol{r}, \boldsymbol{r}'), \qquad \alpha, \beta = A, B, \ldots \end{aligned} \tag{6.1.17}$$

6.2. ADDITIVITY OF INFORMATION DISTANCES

The overall information distance between the molecular and promolecular densities, $\Delta S[\rho\,|\,\rho^0]$, can be regarded as the *total* entropy deficiency functional for the trial subsystem densities satisfying the exhaustive division constraint (see Example 3.12),

$$\begin{aligned} \Delta S[\rho\,|\,\rho^0] &= \int \rho(\boldsymbol{r}) \log[\rho(\boldsymbol{r})/\rho^0(\boldsymbol{r})]\, d\boldsymbol{r} = \textstyle\sum_X \int \rho_X(\boldsymbol{r})]\, I(\boldsymbol{r})\, d\boldsymbol{r} \\ &= \int [\textstyle\sum_X \rho_X(\boldsymbol{r})] \log\{[\sum_Y \rho_Y(\boldsymbol{r})]/[\sum_Z \rho_Z^0(\boldsymbol{r})]\}\, d\boldsymbol{r} \equiv \Delta S^{total}[\boldsymbol{\rho}\,|\,\boldsymbol{\rho}^0]. \end{aligned} \tag{6.2.1}$$

It is preserved in all vertical divisions of the molecular electron density, which conserve $\rho(\boldsymbol{r}) = \sum_X \rho_X(\boldsymbol{r})$. The sum of the *intra*-subsystem information distances, $\Delta S[\boldsymbol{\rho}\,|\,\boldsymbol{\rho}^0]$, which we have used in Section 5.3 to derive the Hirshfeld partitioning in IT, determines the *additive* (*add*) part of this *total* cross-entropy functional in the subsystem resolution:

$$\Delta S[\boldsymbol{\rho}\,|\,\boldsymbol{\rho}^0] = \textstyle\sum_X \Delta S_X[\rho_X\,|\,\rho_X^0] \equiv \Delta S^{add}[\boldsymbol{\rho}\,|\,\boldsymbol{\rho}^0] = \sum_X \int \rho_X(\boldsymbol{r})]\, I_X(\boldsymbol{r})\, d\boldsymbol{r}. \tag{6.2.2}$$

Therefore, the difference

$$\Delta S^{nadd}[\boldsymbol{\rho}\,|\,\boldsymbol{\rho}^0] \equiv \Delta S^{total}[\boldsymbol{\rho}\,|\,\boldsymbol{\rho}^0] - \Delta S^{add}[\boldsymbol{\rho}\,|\,\boldsymbol{\rho}^0] = \sum_X \int \rho_X(\boldsymbol{r})\,[I(\boldsymbol{r}) - I_X(\boldsymbol{r})]\,d\boldsymbol{r}$$
$$= -\rho(\boldsymbol{r}) \sum_X \int d_X(\boldsymbol{r}) \log[d_X(\boldsymbol{r})/d_X^0(\boldsymbol{r})]\,d\boldsymbol{r} \equiv \Delta S^{nadd}[\boldsymbol{d}\,|\,\boldsymbol{d}^0] \equiv \int \Delta s^{nadd}[\boldsymbol{d}(\boldsymbol{r})\,|\,\boldsymbol{d}^0(\boldsymbol{r})]\,d\boldsymbol{r}$$
$$= -\rho(\boldsymbol{r}) \sum_X \Delta S_X[d_X\,|\,d_X^0] \le 0, \qquad (6.2.3)$$

measures the *non-additive* (*nadd*), "mixing" part of the molecular entropy deficiency in atomic resolution. It is non-positive, since both $\rho(\boldsymbol{r})$ and $\{\Delta S_X[d_X\,|\,d_X^0]\}$ are non-negative for the normalized conditional probabilities $\boldsymbol{d}(\boldsymbol{r}) = \{d_X(\boldsymbol{r}) = \pi(X\,|\,\boldsymbol{r})\}$ and $\boldsymbol{d}^0(\boldsymbol{r}) = \{d_X^0(\boldsymbol{r}) = \pi^0(X\,|\,\boldsymbol{r})\}$, $\sum_X d_X(\boldsymbol{r}) = \sum_X d_X^0(\boldsymbol{r}) = 1$, reaching the maximum value for $\boldsymbol{d}(\boldsymbol{r}) \equiv \boldsymbol{\rho}(\boldsymbol{r})/\rho(\boldsymbol{r}) = \boldsymbol{d}^0(\boldsymbol{r}) \equiv \boldsymbol{\rho}^0(\boldsymbol{r})/\rho^0(\boldsymbol{r}) = \boldsymbol{d}^H(\boldsymbol{r}) \equiv \boldsymbol{\rho}^H(\boldsymbol{r})/\rho(\boldsymbol{r})$, when the subsystem densities become the stockholder pieces of the molecular density. We shall call this missing information measure the *partition non-additivity* of the Kullback-Leibler functional. It reflects the *inter*-AIM, delocalization part of the total enetropy deficiency of Eq. (6.2.1). It follows from the foregoing equation that the non-additive entropy deficiency for a given molecular density $\rho(\boldsymbol{r})$ and the promolecular reference distributions $\boldsymbol{d}^0(\boldsymbol{r})$ is uniquely defined by ther AIM conditional probabilities $\boldsymbol{d}(\boldsymbol{r})$, which thus constitute the partition independent (local) degrees-of-freedom.

Thus, as we have already argued above and demonstrated in Example 3.12, this non-additive contribution exactly vanishes only for the unbiased Hirshfeld division (Nalewajski and Parr, 2001):

$$\Delta S^{nadd}[\boldsymbol{\rho}^H\,|\,\boldsymbol{\rho}^0] = \max_{\rho} \Delta S^{nadd}[\boldsymbol{\rho}\,|\,\boldsymbol{\rho}^0] = 0 \qquad \text{or}$$
$$\Delta s^{nadd}[\boldsymbol{d}^H(\boldsymbol{r})\,|\,\boldsymbol{d}^0(\boldsymbol{r})] = \max_{d} \Delta s^{nadd}[\boldsymbol{d}(\boldsymbol{r})\,|\,\boldsymbol{d}^0(\boldsymbol{r})] = 0, \qquad (6.2.4)$$

since then [see Eq. (5.2.1)]

$$\Delta S^{add}[\boldsymbol{\rho}^H|\boldsymbol{\rho}^0] = \sum_X \int \rho_X^H(\boldsymbol{r}) \log \frac{\rho_X^H(\boldsymbol{r})}{\rho_X^0(\boldsymbol{r})} d\boldsymbol{r} = \int \rho(\boldsymbol{r}) \log w_X^H(\boldsymbol{r}) [\sum_X d_X^H(\boldsymbol{r})] d\boldsymbol{r}$$
$$= \int \rho(\boldsymbol{r}) \log w(\boldsymbol{r})\, d\boldsymbol{r} = \Delta S^{total}[\boldsymbol{\rho}^H|\boldsymbol{\rho}^0] = \Delta S[\rho\,|\,\rho^0].$$

As explicitly indicated in the preceding equation this unique feature of the stockholder subsystems, the vanishing partition non-additivity of the entropy deficiency, is just another consequence of the local surprisal equalization and the exhaustive character of the Hirshfeld partition. Therefore, the stockholder fragments of a molecule mark an exact removal of the non-additive (mixing) component of the total molecular missing information in the subsystem resolution relative to the promolecular reference. It should be observed, however, that this zero level of the partition non-additivity part of the cross-enetropy represents its maximum value. The stockholder AIM can be thus viewed as those which *maximize* the delocalization entropy-deficiency in the molecule. Therefore, one could alternatively justify the Hirshfeld partition in IT using the local variational principle for the maximum of the non-additive entropy-deficiency density [Eq. (6.2.4)] subject to the relevant local normalization constraint of the conditional probabilities:

$$\delta\{\Delta s^{nadd}[\boldsymbol{d}(\boldsymbol{r})\,|\,\boldsymbol{d}^0(\boldsymbol{r})] - \lambda \textstyle\sum_X d_X(\boldsymbol{r})]\} = 0 \quad \text{or} \quad d_X^{opt}(\boldsymbol{r}) = C\, d_X^0(\boldsymbol{r}), \;\; \log C = \lambda, \tag{6.2.5}$$

where from the value of the constrait $C = 1$: $\boldsymbol{d}^{opt}(\boldsymbol{r}) = \boldsymbol{d}^H(\boldsymbol{r}) = \boldsymbol{d}^0(\boldsymbol{r})$.

The illustrative AIM information distances $\Delta S_X[p_X^H\,|\,p_X^0]$, between the shape (probability) factors of the bonded (Hirshfeld) and free atom X, are listed in Table 4.1. In general, all these missing information values are quite low, thus testifying that on average the bonded Hirshfeld atoms strongly resemble their free analogs. This is particularly true in heavier atoms (S, Cl), in which the most susceptible valence electron density constitutes a small part of the overall distribution. The largest information distances are observed for light atoms (H, Li), especially when they experience large CT, e.g., in lithium halides.

We have shown in the preceding section (see also the atomic entropy deficiencies listed in Table 4.1) that the Hirshfeld AIM densities are only slightly modified in comparison to the respective free-atoms:

$$|\Delta\rho_X^H(\boldsymbol{r})| = |\rho_X^H(\boldsymbol{r}) - \rho_X^0(\boldsymbol{r})| << \rho_X^H(\boldsymbol{r}) \approx \rho_X^0(\boldsymbol{r}) \qquad \text{or}$$
$$|\Delta\rho_X^H(\boldsymbol{r})|/\rho_X^H(\boldsymbol{r}) \cong |\Delta\rho_X^H(\boldsymbol{r})|/\rho_X^0(\boldsymbol{r}) \;\; (small), \tag{6.2.6}$$

since formation of the chemical bonds affects mainly the atomic valence-shells. Therefore, the atomic surprisals can be expanded in terms of the small relative density displacements of the preceding equation. A similar dominance of the molecular electron density $\rho(\boldsymbol{r})$ by the promolecular density $\rho^0(\boldsymbol{r})$, $\rho(\boldsymbol{r}) \approx \rho^0(\boldsymbol{r})$, has been exploited in Section 4.2. By expanding the logarithms of atomic surprisals $\{I_X[w_X(\boldsymbol{r})] = I_X(\boldsymbol{r})\}$ around $w_X(\boldsymbol{r}) = 1$, to the first-order in the relative displacement of the subsystem electron density, one obtains the following approximate relation between the Hirshfeld AIM surprisal and the corresponding density displacement:

$$I_X[w_X^H(\boldsymbol{r})] = I_X^H(\boldsymbol{r}) = \ln[\rho_X^H(\boldsymbol{r})/\rho_X^0(\boldsymbol{r})] \cong \Delta\rho_X^H(\boldsymbol{r})/\rho_X^0(\boldsymbol{r}), \tag{6.2.7}$$

This expansion also relates the integrands of the alternative information distance functionals of bonded atoms to the corresponding displacements in the AIM electron density:

$$\Delta s_X^H(\boldsymbol{r}) = \rho_X^H(\boldsymbol{r})\, I_X^H(\boldsymbol{r}) \cong \Delta\rho_X^H(\boldsymbol{r})\, w_X^H(\boldsymbol{r}) \approx \Delta\rho_X^H(\boldsymbol{r}), \tag{6.2.8}$$
$$\Delta D_X^H(\boldsymbol{r}) = \Delta\rho_X^H(\boldsymbol{r})\, I_X^H(\boldsymbol{r}) \cong [\Delta\rho_X^H(\boldsymbol{r})]^2/\rho_X^0(\boldsymbol{r}) \geq 0. \tag{6.2.9}$$

In this way one attributes the complementary information-theoretic interpretation to the AIM density displacements.

6.3. ENTROPY DISPLACEMENTS OF BONDED ATOMS

The global entropy displacements of the Hirshfeld AIM (in nats),

$$\mathcal{H}_X^H = S[\rho_X^H] - S[\rho_X^0] = -\int \rho_X^H(\boldsymbol{r}) \ln \rho_X^H(\boldsymbol{r})\, d\boldsymbol{r} + \int \rho_X^0(\boldsymbol{r}) \ln \rho_X^0(\boldsymbol{r})\, d\boldsymbol{r}$$
$$\equiv \int h_X^H(\boldsymbol{r})\, d\boldsymbol{r}, \qquad X = A, B, \ldots, \qquad (6.3.1)$$

for selected linear molecules of Chapter 4 are listed in Table 6.1 together with the corresponding stockholder- and free-atom values. The preceding equation also defines the corresponding entropy-displacement density $h_X^H(\boldsymbol{r})$. A reference to H_2 and N_2 entries of Tables 4.2 and 6.1 shows that the atomic entropy displacements for bonded hydrogen atoms are approximately additive: $2\mathcal{H}_X^H \cong \mathcal{H}[\rho(X_2)]$. A similar near additivity is observed for most of the remaining molecules, with the largest deviation from such an uncoupled (independent) behavior of changes in the AIM entropy being observed for the most ionic Li–F bond.

In general the bonded atom exhibits a lower degree of uncertainty compared to the free-atom value, a clear sign of the dominating effect of a relatively more compact distributions of bonded atoms. In strong electron acceptors, e.g., F in LiF and N in HCN, one detects positive displacements due to the dominating CT contribution, which should result in a softer atomic distribution of electrons, thus exhibiting more "disorder" (uncertainty).

Table 6.1. Displacements of the Hirshfeld AIM entropies (in bits) for representative linear molecules of Fig. 4.1. The molecular and promolecular entropy data are also reported. The corresponding molecular entropy shiftd are given in Table 4.2.

Molecule	AIM, X	$\mathcal{H}_X^H$	$S[\rho_X^H]$	$S[\rho_X^0]$
H_2	H	– 0.41	3.77	4.18
N_2	N	– 0.34	5.86	6.20
HF	H	– 1.09	3.09	4.18
	F	0.03	1.22	1.19
LiF	Li	– 4.02	3.87	7.89
	F	0.97	2.14	1.17
HCN	H	– 0.87	3.31	4.18
	C	– 0.73	7.29	8.03
	N	0.15	6.35	6.20
CNH	C	0.01	8.04	8.03
	N	– 0.44	5.76	6.20
	H	– 0.98	3.20	4.18

The atomic entropy displacements for HF and LiF indicate that the *donor* atom exhibits the dominating (negative) displacement, while the *acceptor* AIM only slightly increases its entropy (see also the net AIM charges in Table 4.1). The triatomic data in Table 6.1 provide an additional confirmation of this rule, with the exception of N[HNC]. A reference to the atomic charges reported in Table 4.1 again shows a strong sensitivity of the atomic entropy displacements to the magnitude of the atomic net charge transfer. As expected, a degree of this sensitivity to a change in the atomic overall electron population decreases with the overall number of electrons on atom in question. Indeed a given displacement in the AIM charge is seen to produce a relatively larger reconstruction of the free-atom electron distribution in H or Li, than in N or F.

Again, one can approximately relate the integrand $h_X^H(\boldsymbol{r})$ in Eq. (6.3.1) to the Hirshfeld AIM density difference $\Delta\rho_X^H(\boldsymbol{r}) = \rho_X^H(\boldsymbol{r}) - \rho_X^0(\boldsymbol{r})$ [compare Eq. (4.3.1)]:

$$h_X^H(\boldsymbol{r}) \cong -\Delta\rho_X^H(\boldsymbol{r})\,[1 + \ln \rho_X^0(\boldsymbol{r}) + \Delta\rho_X^H(\boldsymbol{r})/\rho_X^0(\boldsymbol{r})]. \qquad (6.3.2)$$

It should be emphasized that the sum of the AIM entropy displacements,

$$\Sigma_X\, \mathcal{H}_X^H \equiv \mathcal{H}^{add}[\boldsymbol{\rho}^H] = -\Sigma_X \int \Delta\rho_X^H(\boldsymbol{r}) \ln \rho_X^0(\boldsymbol{r})\, d\boldsymbol{r} \;-\Delta S^{total}[\boldsymbol{\rho}^H\,|\,\boldsymbol{\rho}^0], \qquad (6.3.3)$$

where $\Delta S^{total}[\boldsymbol{\rho}^H\,|\,\boldsymbol{\rho}^0] = \Delta S[\rho\,|\,\rho^0] = \Delta S[\boldsymbol{\rho}^H\,|\,\boldsymbol{\rho}^0] = \Delta S^{add}[\boldsymbol{\rho}^H\,|\,\boldsymbol{\rho}^0]$ [Eq. (6.2.5)], defining the additive part of the total entropy displacement in the Hirshfeld AIM resolution (see Example 3.11), $\mathcal{H}[\rho] = \mathcal{H}[\Sigma_X\, \rho_X^H] \equiv \mathcal{H}^{total}[\boldsymbol{\rho}^H]$ [see Eq. (4.3.2)], differs from the overall entropy displacement $\mathcal{H}[\rho]$ due to the non-additive entropy displacement:

$$\begin{aligned}\mathcal{H}^{nadd}[\boldsymbol{\rho}^H] &= \mathcal{H}^{total}[\boldsymbol{\rho}^H] - \mathcal{H}^{add}[\boldsymbol{\rho}^H]\\ &= \Sigma_X \int \Delta\rho_X^H(\boldsymbol{r}) \ln\rho_X^0(\boldsymbol{r})\, d\boldsymbol{r} \;- \int \Delta\rho(\boldsymbol{r}) \ln\rho^0(\boldsymbol{r})\, d\boldsymbol{r} \;= \Sigma_X \int \Delta\rho_X^H(\boldsymbol{r}) \ln d_X^0(\boldsymbol{r})\, d\boldsymbol{r}\\ &= \Sigma_X \int \rho_X^H(\boldsymbol{r}) \ln[\rho_X^H(\boldsymbol{r})/\rho(\boldsymbol{r})]\, d\boldsymbol{r} \;- \Sigma_X \int \rho_X^0(\boldsymbol{r}) \ln [\rho_X^0(\boldsymbol{r})/\rho^0(\boldsymbol{r})]\, d\boldsymbol{r}\\ &\equiv \Delta S[\boldsymbol{\rho}^H\,|\,\rho] - \Delta S[\boldsymbol{\rho}^0\,|\,\rho^0] = \Sigma_X \int \Delta\rho_X^H(\boldsymbol{r}) \ln d_X^H(\boldsymbol{r})\, d\boldsymbol{r},\end{aligned} \qquad (6.3.4)$$

where we have used the stockholder principle of Eq. (5.2.5). In the second line of the preceding equation the non-additive component has been expressed in terms of the atomic share factors (conditional probabilities) $\{d_X^0(\boldsymbol{r}) = d_X^H(\boldsymbol{r})\}$, while in the last line we have expressed this entropy displacement non-additivity as the difference between the two entropy deficiencies, each measuring the information distance between the constituent density fragments and the overall distribution.

In Figs. 6.1 and 6.2 we have compared the contour maps of the Hirshfeld AIM density-difference function $\Delta\rho_X^H(\boldsymbol{r})$ with the corresponding entropy displacement density $h_X^H(\boldsymbol{r})$, for the constituent AIM of the representative diatomic (Fig. 6.1) and triatomic (Fig. 6.2) linear molecules of Fig. 4.1. Let us first examine the H_2 diagrams shown in part *a* of Fig. 6.1. The density difference plot (left panel) for the "stockholder" hydrogen in H_2, denoted as H[H_2], exhibits changes already observed in the density profile of Fig. 5.1. The observed buildup of the electron density around the atomic nucleus and in the bond region between the nuclei, at the expense of the outer, mainly nonbonding region of the atomic density distribution, is due to the contraction of the hydrogen electron density near the nucleus and its polarization towards the other atom. This cylindrical, directional polarization of the atomic electron density is clearly seen in the left, electron density panel. A similar pattern, with a somewhat more emphasized bond polarization, is seen in the entropy difference plot of the right contour map. The bonding part of the AIM entropy displacement is positive, thus marking an increase in the local uncertainty due to the electron delocalization towards the bonding partner.

The maximum of this extra increase in the local electron uncertainty for $H[H_2]$ is found along the bond axis, in the direction of the other hydrogen. This feature of the atomic entropy displacement density indicates that it can serve as a sensitive diagnostic tool for monitoring the bonding condition of AIM. A reference to part *b* of Fig. 6.1, which is devoted to N in N_2, again shows a general similarity between the two compared contour diagrams. The displacements observed in the valence shell of both panels accord with the known density-deformation changes in the triply-bonded nitrogen: the $(2s,2p_\sigma)$-hybridization along the molecular axis, a transfer of the $2p_\pi$ electrons to the bond-charge region between the nuclei, with the buildup (lowering) of the electron density of the left panel giving rise to the associated increase (decrease) in the local electron uncertainty of the right panel. The entropy-difference map also reveals a rather complicated atomic 1*s*-core polarization. A reference to both these panels confirms the molecular character of the Hirshfeld atoms, with the atom displacement "tails" extending all over the molecule. This molecular origin of the stockholder AIM is due to the fact that each bonded atom effectively contributes to the valence-state promotion of all remaining atoms and vice versa, the state of a given bonded atom is determined by the atom whole molecular environment.

Next let us examine the contour maps for the constituent AIM of the two heteronuclear diatomics: HF (part *c*) and LiF (part *d*). In both density and entropy panels for H[HF], again strongly resembling one another, the hydrogen is seen to be strongly polarized towards the fluorine atom, with the transferred electron density being channeled to both the σ bond region and to the $2p_\pi$ lone-pair regions around fluorine. The valence-shell part of the F[HF] panels reveals a similar polarization of the fluorine towards the hydrogen, with the accompanying increase in the lone pair ($2p_\pi$) density, in the direction perpendicular to the bond axis. A complicated displacement pattern in the local information content of the promoted inner electrons is seen in the right F[HF] panel of the figure. The overall similarity of the H[HF] and F[HF] distribution patterns in the region between the two nuclei confirms a relatively high covalent character of the chemical bond between these two atoms. The localization of the fluorine bond-charge close to the proton position provides an additional support to this observation. A different charge-transfer pattern is found for the strongly ionic bond in LiF. The difference diagrams of the AIM density and entropy in Fig. 6.1*d* consistently show that the electrons removed from the peripheral part of the Li[LiF] electron distribution are transferred to F as a whole, giving rise to a much lower degree of localization of the CT density, particularly at lower contour values. The higher values of the positive contours in the F[LiF] plots show some concentration of the electronic charge between the nuclei, close to the fluorine nucleus. This reflects a partially covalent bond character, with much lower covalent (electron-sharing) contribution though, compared to that observed in HF. This observation accords with the relative hardness of the two atoms in these molecules: the two hard atoms in HF give rise to strongly covalent bond, while the soft (Li) and hard (F) atoms in LiF generate a relatively more ionic bond. The shape of the F[LiF] contours also reveals its weak ($2s$,$2p_\sigma$)-hybridization.

Next, let us examine the AIM density and entropy displacement plots of Fig. 6.2, for the two linear triatomics of Fig. 4.1. All these plots exhibit a truly molecular character of the bonded atoms, with the low magnitude displacements reflecting the shape of the whole molecule. The redistribution patterns observed on the corresponding left and right contour diagrams are very similar indeed, with the entropy displacement diagram additionally separating the outer (valence) and inner (core) effects. The hydrogen maps in both HCN and HNC resemble the corresponding diagrams of Figs. 6.1*a* and 6.1*c*. The contour diagrams for C and N in both isomers are typical for the valence-state displacements on atoms participating in both σ and π bonds. One detects in these plots the ($2s$,$2p_\sigma$)-hybridization on the triply bonded terminal heavy atoms, a relatively extended character of the resultant bond charge, due to the contributions from both the σ and π bonds, and the associated transfer of electrons from the outer regions of free atoms to the bond and lone-pair areas of the bonded atoms.

As a final illustration of this section we examine the bridgehead carbon atoms of the four propellanes, which have been investigated in Section 4.4 (see Fig. 4.5). The contour maps of Fig. 6.3 report the density difference function, $\Delta\rho_X^H$, the Kullback-Leibler missing information density, Δs_X^H, and the entropy displacement density, h_X^H. They fully support the conclusions already drawn from the molecular plots of Figs. 4.6 and 4.7. In Fig. 6.3 the three plots for the bridgehead carbon atoms are seen to be qualitatively similar, thus further validating the usefulness of all these quantities as alternative probes into changes the free atoms undergo in the molecule. These diagnostic quantities are seen to be to a large extent equivalent, thus testifying to the complementary character of the electronic density and entropy/information "distances" of AIM relative to their free analogs. For the [1.1.1] and [2.1.1] propellanes, which are lacking the "through-space" component of the central bond between the bridgehead carbons, the density buildups observed in the three panels indeed reflect the "through-bridge" chemical bonds only, with a distinct lowering of the AIM densities in the direction of the other bridgehead atom. For the [2.2.1] and [2.2.2] propellanes, in which the presence of the "through-space" component of the central bond has already been inferred from both the molecular density difference and the direct bond-order measures, the accumulation of the AIM electron density and the corresponding information distance/entropy densities is now observed in the central bond region of all the AIM density readjustment plots.

This analysis confirms that the "stockholder" atoms embody all major effects of the chemical bonds, which are contained in the molecular charge and information/entropy distributions. They are thus perceived as attractive concepts for chemical interpretation of the valence-state of AIM and the electronic/information origins of the chemical bonds.

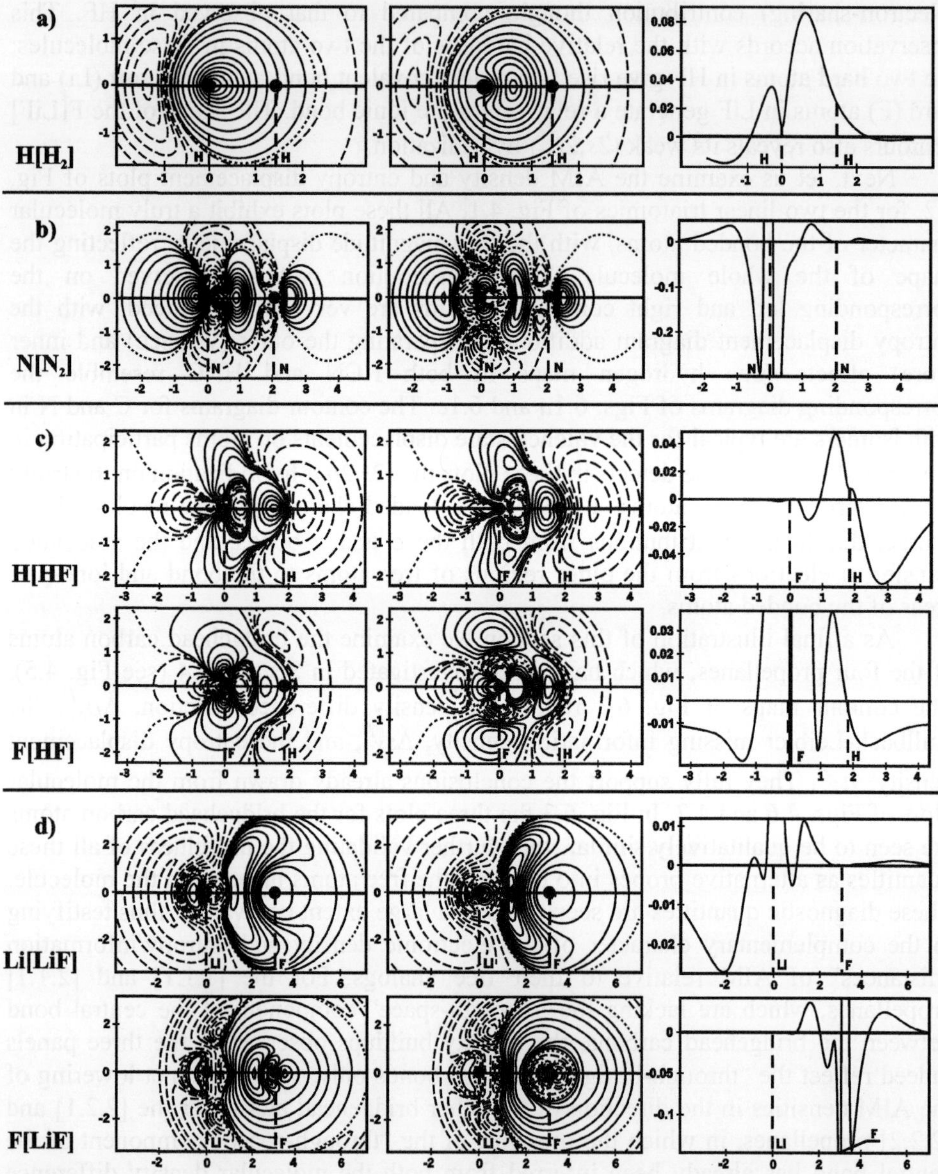

Figure 6.1. Representative contour maps of the atomic electron density-difference function $\Delta\rho_X^{\ H}(\boldsymbol{r})$ (first column) and the entropy-displacement density $h_X^{\ H}(\boldsymbol{r})$ (second column), for the constituent atoms of the diatomic molecules of Fig. 4.1: H_2 (*a*), N_2 (*b*), HF (*c*), and LiF (*d*). The contour values are not equidistant having been selected only for the purpose of revealing the topographic features of the compared quantities. In the third column the bond axis profiles of $h_X^{\ H}(\boldsymbol{r})$ [a.u.] are reported (Nalewajski and Broniatowska, 2003a).

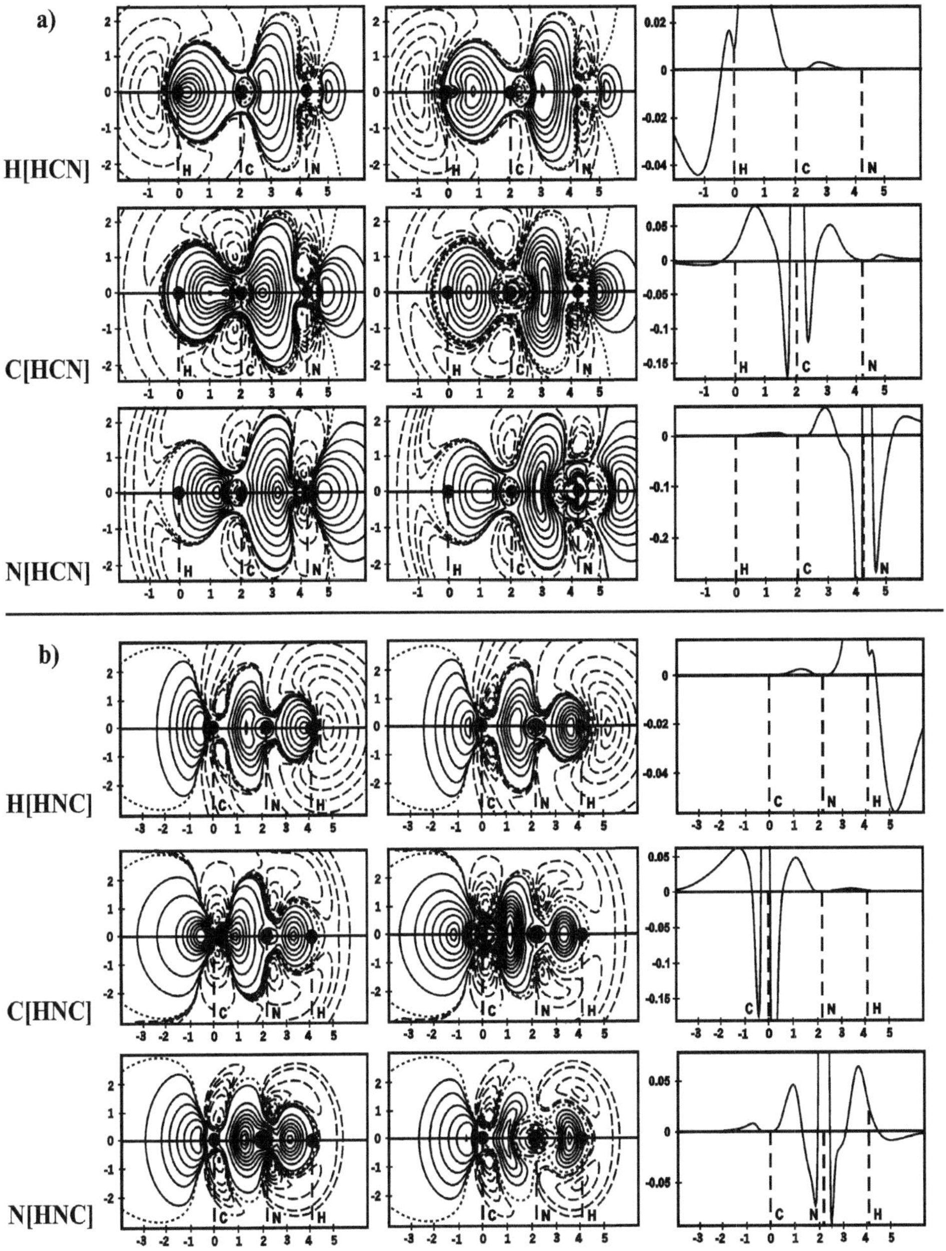

Figure 6.2. Same as in Fig. 6.1, for HCN (part a) and HNC (part b) (Nalewajski and Broniatowska, 2003a).

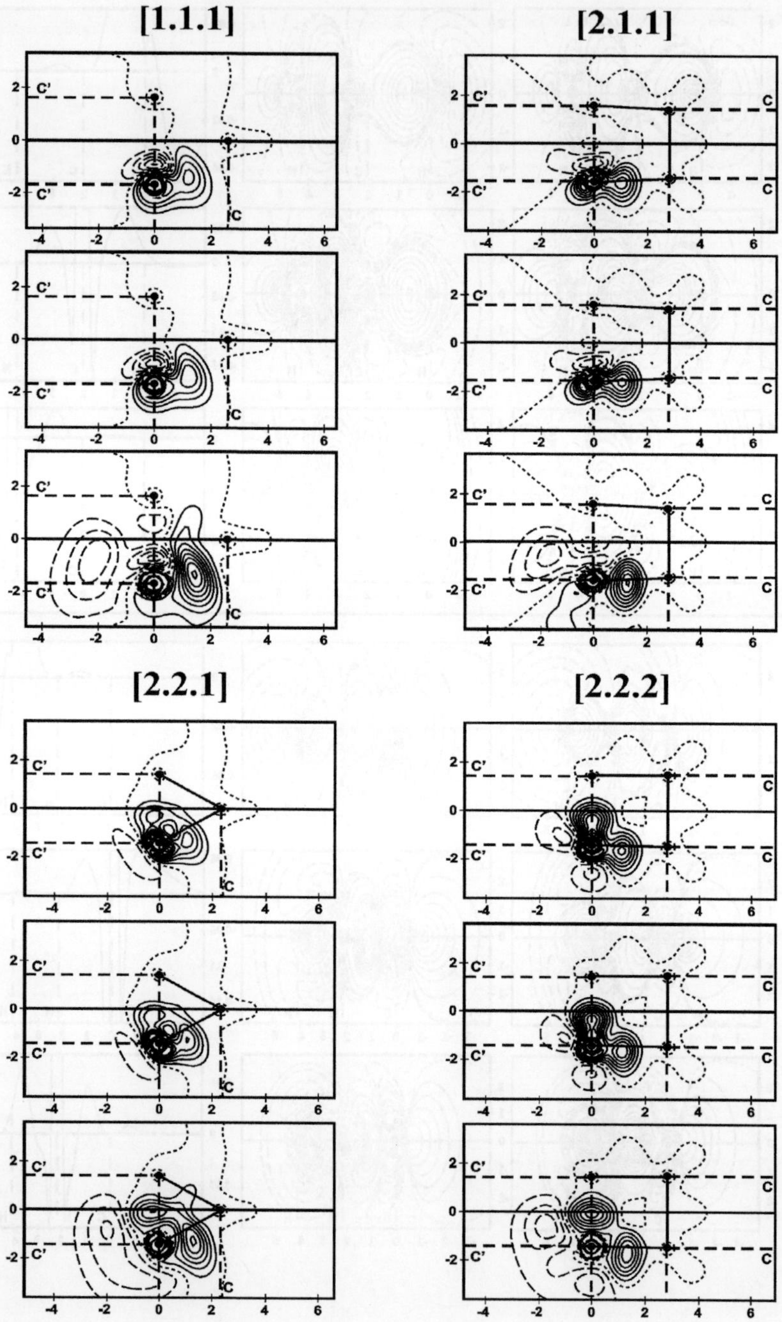

Figure 6.3. Contour diagrams of displacements in the electron density (upper panel), information distance density (medium panel) and Shannon entropy (lowest panel), for the stockholder bridgehead carbon atoms in the four propellanes of Fig. 4.5.

As we have demonstrated in the preceding chapter, the *localized* (single-cusped) densities of the stockholder subsystems result from the variational principles involving the free-atom/promolecule referenced cross-entropy in the AIM resolution, subject to the local constraints of the exhaustive partitioning of the molecular (multi-cusped) density. The auxiliary functional to be optimized in these variational searches contains the relative-entropy term, which plays the role of the information "penalty" for the AIM densities deviating from the corresponding free-atom electron distributions. These principles minimize the missing information between the two sets of atomic electron densities, thus maximizing the entropic similarity between them and enforcing the effective localization of the AIM electron distributions.

It should be observed that the additive entropy displacement in atomic resolution,

$$\mathcal{H}^{add}[\boldsymbol{\rho}] = \sum_X \mathcal{H}_X[\rho_X] = S^{add}[\boldsymbol{\rho}] - S^{add}[\boldsymbol{\rho}^0], \tag{6.3.5}$$

does not introduce any effective penalty for a lack of resemblance of the trial atomic distributions to the reference, free-atom densities. Indeed, when used in the variational principle for an optimum exhaustive division of the molecular electron density subject to the subsidiary condition of the exhaustive division,

$$\delta\{\mathcal{H}^{add}[\boldsymbol{\rho}] - \int \xi(\boldsymbol{r}) \sum_X \rho_X(\boldsymbol{r})\, d\boldsymbol{r}\} = \delta\{S^{add}[\boldsymbol{\rho}] - \sum_X \int \xi(\boldsymbol{r}) \sum_X \rho_X(\boldsymbol{r})\, d\boldsymbol{r}\} = 0, \tag{6.3.6}$$

where $\xi(\boldsymbol{r})$ is the Lagrange multiplier to be determined from the constraint $\sum_X \rho_X(\boldsymbol{r}) = \rho(\boldsymbol{r})$, it gives rise to the perfectly delocalized (*del*) subsystem densities, given by the equal division of the molecular density: $\rho_X^{del}(\boldsymbol{r}) = \rho(\boldsymbol{r})/m$, where m is the number of constituent atoms.

We therefore conclude that only the *relative*-entropy (entropy-deficiency, missing information) terms, involving the localized free-atom electron distributions of constituent atoms, can generate via the relevant extreme-information principle the localized electron distributions of AIM.

The localized character of the Hirshfeld solutions, giving rise to the constrained minimum of the entropy deficiency

$$\begin{aligned}\Delta S[\boldsymbol{\rho} \,|\, \boldsymbol{\rho}^0] &= \sum_X \int \rho_X(\boldsymbol{r}) \ln\rho_X(\boldsymbol{r})\, d\boldsymbol{r} - \sum_X \int \rho_X(\boldsymbol{r}) \ln\rho_X^0(\boldsymbol{r})\, d\boldsymbol{r} \\ &\equiv - S^{add}[\boldsymbol{\rho}] + S^{penalty}[\boldsymbol{\rho} \,|\, \boldsymbol{\rho}^0],\end{aligned} \tag{6.3.7}$$

is thus due to the combined effect of the additive information entropy and the penalty contribution $S^{penalty}[\boldsymbol{\rho} \,|\, \boldsymbol{\rho}^0]$ of $\Delta S[\boldsymbol{\rho} \,|\, \boldsymbol{\rho}^0]$, which couples the trial and reference atomic densities.

A similar localization influence is exerted by the *two*-electron entropy deficiency in the *two*-electron EED principle of Eq. (5.5.17), for $k = 2$, which gives rise to the stockholder partitioning of the molecular pair-density into the AIM 2-cluster components (see Sections 5.6 and 5.7).

6.4. VERTICAL AND HORIZONTAL DENSITY DISPLACEMENTS

In the preceding two chapters we have derived the stockholder AIM using the minimum entropy-deficiency principle involving the free-atom/promolecular reference and examined some remarkable properties of these entropy-equilibrium pieces of the molecular electron distributions. In this section we continue this exploration by examining their response properties in both the EF (softness) and EP (hardness) representations. In what follows we shall use the subscript X to label a single bonded atom, and the index α to identify a larger molecular fragment or reactant, i.e., a specific collection of AIM.

The traditional way of approaching the subsystem resolution (see, e.g., Nalewajski, 1993, 2002d. 2003a; Nalewajski *et al.*, 1996; Nalewajski and Korchowiec, 1977) is to regard the molecular fragment densities as the *mutually* closed components $\{X, Y, ...\}$, which can be either open or closed *externally*. In the first scenario each fragment α of the externally *open* (o) molecular system M_o is in contact with its own electron reservoir $\mathfrak{R}_\alpha$ in the combined super-system $\mathfrak{M}[M_o] \equiv (\mathfrak{R}_\alpha \,\vdots\, X \mid \mathfrak{R}_\beta \,\vdots\, Y \mid ...)$, with each reservoir generally exhibiting different levels of the chemical potential. In the second scenario the externally closed molecular fragments of the externally *closed* (c) molecule $M_c = (X \mid Y \mid ...)$, are not coupled to electron reservoirs thus conserving both the subsystem and overall numbers of electrons. Above, the two subsystems separated by the broken vertical line are free to exchange electrons; when they are separated by the vertical solid line these electron flows are not allowed. This approach gives rise to the *constrained* (internal) *equilibria* of molecular subsystems, which are summarized in Appendix D. In this approach each subsystem density-variable represents the truly *independent* local state-parameter of the composite molecular systems M_c or $\mathfrak{M}[M_c]$.

In such a *partial* treatment the Hirshfeld pieces $\boldsymbol{\rho}^H(\rho)$, originating from the specified (*fixed*) molecular electron density $\rho(\boldsymbol{r}) = \sum_X \rho_X^H(\boldsymbol{r})$, are subsequently treated as *independent* components, so that the density of one-subsystem can be changed for the "frozen" densities of the remaining subsystems giving rise to the *partial* functional derivatives

$$\partial\rho(\boldsymbol{r})/\partial\rho_X^H(\boldsymbol{r}') = \delta(\boldsymbol{r}' - \boldsymbol{r}) \qquad \partial\rho_Y^H(\boldsymbol{r})/\partial\rho_X^H(\boldsymbol{r}') = \delta_{X,Y}\,\delta(\boldsymbol{r}' - \boldsymbol{r}). \tag{6.4.1}$$

Clearly, by modifying a density of a single component X, $\delta\rho_X$, for the fixed densities of the remaining parts $Y \neq X$ of the molecule shifts the overall molecular density $\delta\rho = (\delta\rho_X)_{Y\neq X}$. It should be stressed, however, that this partial displacement violates the stockholder proportions with respect to the displaced overall density $\rho' = \rho + \delta\rho$.

Consider next the horizontal displacements of the exact Hirshfeld densities of subsystem, $\delta\boldsymbol{\rho}^H(\delta\rho)$, which accompany a given displacement $\delta\rho$ of the molecular electron density, $\delta\rho(\boldsymbol{r}) = \sum_X \delta\rho_X^H(\boldsymbol{r}) \neq 0$. In this case the subsystem density shifts are not independent, since each subsystem represents the respective free-atom share in the molecular density, which is fixed by the promolecular reference, Indeed, Eqs.

(5.2.1) and (5.2.4) identify the stockholder AIM "share" factors (conditional probabilities) $\boldsymbol{d}^H(\boldsymbol{r}) = \{d_X^H(\boldsymbol{r}) = \pi^H(X|\boldsymbol{r})\}$ as the local derivatives of the Hirshfeld AIM densities $\boldsymbol{\rho}^H(\boldsymbol{r}) = \boldsymbol{\rho}^H(\rho(\boldsymbol{r})) = \boldsymbol{\rho}^H[N, v; \boldsymbol{r}]$ with respect to the molecular ground-state density $\rho(\boldsymbol{r})$, called the "horizontal" derivatives:

$$\delta\rho_X^H(\boldsymbol{r}')/\delta\rho(\boldsymbol{r}) = \delta(\boldsymbol{r}' - \boldsymbol{r})\,[d\rho_X^H(\boldsymbol{r})/d\rho(\boldsymbol{r})] = \delta(\boldsymbol{r}' - \boldsymbol{r})\,d_X^H(\boldsymbol{r}) \qquad \text{or}$$
$$\delta\rho(\boldsymbol{r}')/\delta\rho_X^H(\boldsymbol{r}) = \delta(\boldsymbol{r}' - \boldsymbol{r})\,[d\rho(\boldsymbol{r})/d\rho_X^H(\boldsymbol{r})] = \delta(\boldsymbol{r}' - \boldsymbol{r})\,d_X^H(\boldsymbol{r})^{-1}, \quad X = A, B. \qquad (6.4.2)$$

since the AIM shares are solely promolecule-dependent.

This relation between the stockholder AIM densities and the molecular density further implies the related local horizontal dependence between the stockholder subsystem densities:

$$\delta\rho_Y^H(\boldsymbol{r}')/\delta\rho_X^H(\boldsymbol{r}) = \delta(\boldsymbol{r}' - \boldsymbol{r})[d\rho_Y^H(\boldsymbol{r})/d\rho_X^H(\boldsymbol{r})]$$
$$= \delta(\boldsymbol{r}' - \boldsymbol{r})\,[d\rho_Y^H(\boldsymbol{r})/d\rho(\boldsymbol{r})]/[d\rho_X^H(\boldsymbol{r})/d\rho(\boldsymbol{r})] = \delta(\boldsymbol{r}' - \boldsymbol{r})\,d_Y^H(\boldsymbol{r})/d_X^H(\boldsymbol{r}). \qquad (6.4.3)$$

Thus, the horizontal displacement of the Hirshfeld density of one component $\delta\rho_X^H(\delta\rho)$ is uniquely determined by the corresponding local displacement of any other component:

$$d\rho_Y^H(\boldsymbol{r}) = [d_Y^H(\boldsymbol{r})/d_X^H(\boldsymbol{r})]\,d\rho_X^H(\boldsymbol{r}) = d_Y^H(\boldsymbol{r})\,d\rho(\boldsymbol{r}). \qquad (6.4.4)$$

The preceding equation also implies that such a horizontal displacement of a single Hirshfeld subsystem density uniquely identifies the associated shift in the molecular electron density itself:

$$d\rho(\boldsymbol{r}) = d\rho_X^H(\boldsymbol{r})/d_X^H(\boldsymbol{r}). \qquad (6.4.5)$$

Finally, it should be observed that for the Hirshfeld partitioning the normalization of the subsystem local conditional probabilities $\sum_X d_X^H(\boldsymbol{r}) = 1$, automatically guarantees the overall closure relation for the *vertical* processes:

$$\sum_X \int d\rho_X^H(\boldsymbol{r}')\,[\partial\rho(\boldsymbol{r})/\partial\rho_X^H(\boldsymbol{r}')]\,d\boldsymbol{r}' = \sum_Y \sum_X d\rho_X^H(\boldsymbol{r})\,[\partial\rho_Y^H(\boldsymbol{r})/\partial\rho_X^H(\boldsymbol{r})]$$
$$= \sum_X d\rho_X^H(\boldsymbol{r}) = [\sum_X d_X^H(\boldsymbol{r})]\,d\rho(\boldsymbol{r}) = d\rho(\boldsymbol{r}) = 0. \qquad (6.4.6)$$

6.5. CHEMICAL POTENTIAL EQUALIZATION AND EFFECTIVE v-REPRESENTABILITY

For simplicity, let us again consider a diatomic molecule $M = AB$ and its ground-state density $\rho = \rho[N, v]$ representing the equilibrium distribution of electrons for a given overall number of electrons $N = N_A + N_B$ and the fixed external potential $v(\boldsymbol{R}_A, \boldsymbol{R}_B) = v_A(Z_A, \boldsymbol{R}_A) + v_B(Z_B, \boldsymbol{R}_B)$, due to the nuclei of both atoms in their fixed positions $\boldsymbol{R}_A$ and $\boldsymbol{R}_B$, respectively, with Z_X denoting the nuclear charge (a.u.) of atom X. The density corresponds to the (non-degenerate) ground-state $\Psi[\rho] = \Psi[N, v]$ and represents the optimum solution of the Hohenberg and Kohn (1964) density variational principle (1.5.20) for the *open* M in contact with the external electron reservoir $\mathcal{R}$ in the combined system $\mathcal{M} = (\mathcal{R}|M)$ (see, e.g., Parr and Yang, 1989; Nalewajski and Korchowiec, 1997),

$$\delta\{E_v[\rho] - \mu[\rho]\, N[\rho]\} = 0, \tag{6.5.1}$$

where the density functional for the system electronic energy [Eqs. (1.5.14), (1.5.15)], $E_v[\rho] = \int \rho(\boldsymbol{r})\, v(\boldsymbol{r})\, d\boldsymbol{r} + F[\rho]$, depends parametrically on the system overall number of electrons N, $E_v[\rho[N, v]] \equiv E[N, v]$; the broken vertical line in $\mathcal{M}$ symbolizes a freedom of the electron exchange between the two open subsystems. The universal functional $F[\rho]$, the expectation value of the v–independent operator $\hat{F}(N)$ [Eq. (1.5.4)] combining the electronic kinetic and repulsion terms, is defined by Levy's (1979) constrained search construction of Eq. (1.6.4), which generates the sum of the ground-state electronic kinetic ($T_e[\rho]$) and repulsion ($V_{ee}[\rho]$) energies.

In Eq. (6.5.1) the system global chemical potential $\mu[\rho]$ of Eq. (1.5.21) (see also, Parr *et al.*, 1978; Parr and Yang, 1989; Nalewajski and Korchowiec, 1997), the Lagrange multiplier associated with the constraint of the prescribed average number of electrons in the molecular system under consideration, $N[\rho] = \int \rho(\boldsymbol{r})\, d\boldsymbol{r} = N$, can be thermodynamically interpreted as the system "*intensive*" parameter fixed by the external electron reservoir (see Appendix D), $\mu[\rho] = \mu_{\mathcal{R}}$, determining the equilibrium state in the combined system $\mathcal{M}$. As we have already argued in Section 1.5, the Euler equation (1.5.22) resulting from the variational principle of Eq. (6.5.1) gives rise to the equalization of the *local* chemical potential $\mu(\boldsymbol{r})$ [Eq.(1.5.21)] at the *global* chemical potential level $\mu[\rho]$. This local, energy-conjugate "intensity" of the "extensive" state-parameter $\rho(\boldsymbol{r})$, is given by the functional derivative of the energy density-functional for the fixed molecular external potential:

$$\mu(\boldsymbol{r}) \equiv \mu[N, v; \boldsymbol{r}] = \{\delta E_v[\rho]/\delta\rho(\boldsymbol{r})\}_v = v(\boldsymbol{r}) + \delta F[\rho]/\delta\rho(\boldsymbol{r}) = \mu[\rho]. \tag{6.5.2}$$

Alternatively, the preceding equation can be viewed as the equalization principle for the local electronegativity $\chi(\boldsymbol{r}) = -\mu(\boldsymbol{r})$ [Eq. (1.5.23)] (Parr *et al.*, 1978) at the global electronegativity level $\chi[N, v] = \chi[\rho]$. This provides the DFT justification of the familiar *Electronegativity Equalization Principle* (EEP) of Sanderson (1951, 1976).

Let us examine from this perspective the subsystem partitioning of the molecular electronic energy (Nalewajski and Parr, 2001). We assume the exhaustive division of the molecular ground-state density: $\rho = \rho_A + \rho_B \equiv \boldsymbol{\rho}\boldsymbol{1}^{\mathrm{T}}$, into the two AIM components $\boldsymbol{\rho} = (\rho_A, \rho_B)$, e.g., using the Hirshfeld division scheme: $\boldsymbol{\rho}^H = (\rho_A{}^H, \rho_B{}^H)$. The ground-state energy of the system as a whole can be now interpreted as an equivalent functional of the subsystem densities of Eq. (1.6.11) (see Appendix D):

$$E[N, v] = E_v[\rho] \equiv \widetilde{E}_v[\boldsymbol{\rho}] = \int [\boldsymbol{\rho}(\boldsymbol{r})\, [v(\boldsymbol{r})\boldsymbol{1}]^{\mathrm{T}}\, d\boldsymbol{r} + \widetilde{F}[\boldsymbol{\rho}]$$

$$\equiv E_v[\boldsymbol{\rho}] \equiv \int [\boldsymbol{\rho}(\boldsymbol{r})\, \boldsymbol{v}(\boldsymbol{r})^{\mathrm{T}}\, d\boldsymbol{r} + \widetilde{F}[\boldsymbol{\rho}] \equiv E[\boldsymbol{N}, \boldsymbol{v}], \qquad (6.5.3)$$

where the unit row vector $\boldsymbol{1} = (1, 1)$, the molecular external potential vector $\boldsymbol{v}(\boldsymbol{r}) = v(\boldsymbol{r})\boldsymbol{1} = \{v_X^M(\boldsymbol{r})\}$, with $v_X^M(\boldsymbol{r}) = v(\boldsymbol{r})$ for each constituent atom, the AIM electron population vector $\boldsymbol{N} = (N_A, N_B)$, with $N_X = \int \rho_X(\boldsymbol{r})\, d\boldsymbol{r}$, and $\widetilde{F}[\boldsymbol{\rho}] = F[\rho_A + \rho_B]$.

Let us examine specific energy contributions in such a molecular subsystem resolution. Among others one distinguishes the AIM energy terms depending on the density of a single bonded atom [Eq. (1.6.13)],

$$E_v[\rho_X] = \int \rho_X(\boldsymbol{r})\, v(\boldsymbol{r})\, d\boldsymbol{r} + F[\rho_X]. \qquad (6.5.4)$$

They give rise to the subsystem *additive* part $\widetilde{E}_v^{add}[\boldsymbol{\rho}]$ of the molecular electronic energy in the subsystem resolution:

$$\widetilde{E}_v^{add}[\boldsymbol{\rho}] = E_v[\rho_A] + E_v[\rho_B] = \int \rho(\boldsymbol{r})\, v(\boldsymbol{r})\, d\boldsymbol{r} + \widetilde{F}^{add}[\boldsymbol{\rho}], \qquad (6.5.5)$$

where $\widetilde{F}^{add}[\boldsymbol{\rho}] = F[\rho_A] + F[\rho_B]$ stands for the additive part of the universal functional in the AIM resolution [Eq. (1.6.8)].

The complementary subsystem *non-additive* (n) part $\widetilde{F}^{nadd}[\boldsymbol{\rho}]$ of $\widetilde{F}[\boldsymbol{\rho}]$, $\widetilde{F}^{nadd}[\boldsymbol{\rho}] = \widetilde{F}[\boldsymbol{\rho}] - \widetilde{F}^{add}[\boldsymbol{\rho}]$, vanishes at infinite separation between the two atoms. It measures the electronic interaction energy, which determines the embedding correction to the subsystem effective external potential of Eq. (1.6.7). The embedded AIM energies of Eq. (1.6.13) ($\boldsymbol{E}_{vA}[\boldsymbol{\rho}], \boldsymbol{E}_{vB}[\boldsymbol{\rho}]$), of the bonded atoms in the presence of the other atom,

$$\boldsymbol{E}_{vA}[\boldsymbol{\rho}] = E_v[\rho_A] + \widetilde{F}^{nadd}[\boldsymbol{\rho}] \quad \text{and} \quad \boldsymbol{E}_{vB}[\boldsymbol{\rho}] = E_v[\rho_B] + \widetilde{F}^{nadd}[\boldsymbol{\rho}], \qquad (6.5.6)$$

include the embedding (non-additive) energy of Eq. (1.6.8). It should be observed that these energies do not sum up to the molecular energy $\widetilde{E}_v[\boldsymbol{\rho}]$ due to a double counting of the interaction energy.

These subsystem energies define the chemical potentials $\boldsymbol{\mu}[\boldsymbol{\rho}] = (\mu_A[\boldsymbol{\rho}], \mu_B[\boldsymbol{\rho}])$ of the bonded atomic fragments in M [Eq. (1.6.14)] (see also Appendix D),

$$\mu_X[\boldsymbol{\rho}] = \frac{\partial E[\boldsymbol{N}, \boldsymbol{v}]}{\partial N_X} = \partial \boldsymbol{E}_{vX}[\boldsymbol{\rho}]/\partial\rho_X(\boldsymbol{r}) = \partial \widetilde{E}_v[\boldsymbol{\rho}]/\partial\rho_X(\boldsymbol{r}) = v(\boldsymbol{r}) + \partial\widetilde{F}[\boldsymbol{\rho}]/\partial\rho_X(\boldsymbol{r})$$

$$= \{v(\boldsymbol{r}) + \partial\widetilde{F}^{nadd}[\boldsymbol{\rho}]/\partial\rho_X(\boldsymbol{r})\} + \partial\widetilde{F}^{add}[\boldsymbol{\rho}]/\partial\rho_X(\boldsymbol{r}) \equiv [v(\boldsymbol{r}) + v_X^e(\boldsymbol{r})] + \delta F[\rho_X]/\delta\rho_X(\boldsymbol{r})$$

$$\equiv v_X^{eff}(\boldsymbol{r}) + \delta F[\rho_X]/\delta\rho_X(\boldsymbol{r}), \qquad X = A, B, \qquad (6.5.7a)$$

where $v_X^{eff}(\boldsymbol{r})$ stands for the *effective* external potential of atom X in M, including the *embedding* (e) potential $v_X^e(\boldsymbol{r}) \equiv \partial \widetilde{F}^n[\boldsymbol{\rho}]/\partial\rho_X(\boldsymbol{r})$ due to the presence of the other atom $Y \neq X$. An algorithm has been proposed for determining the subsystem effective external potential from the embedding density of the complementary molecular environment (Ayers, 2000), in the spirit of the Zhao-Morrison-Parr (ZMP) procedure (Zhao and Parr, 1992, 1993; Zhao *et al.*, 1994).

The effective external potential $v_X^{eff}(\boldsymbol{r})$ for the *embedded* (bonded) atom X in $M = AB$ differs from the external potential due to the atom own nucleus by the external potential due to the nucleus of the other atom and the embedding potential due to the presence of the electrons in the complementary subsystem: $v_X^e(\boldsymbol{r})$. Each embeded, well behaved (v-representable) atomic density can thus be viewed in DFT as representing the ground-state density corresponding to the *separate* system defined by its effective external potential. In other words, such subsystem densities are in principle the v_X^{eff}-representable, Should the functional $\widetilde{F}^{nadd}[\boldsymbol{\rho}]$ be known,. This observation introduces an important element of causality into the subsystem description. Namely, each change in the electron density of a bonded atom or molecular fragment can now be interpreted as the *equilibrium* (ground-state) response to the concomitant displacement (perturbation) in the subsystem effective external potential. Moreover, the non-equilibrium set of the v-representable subsystem densities can be attributed in this way an effective equilibrium (ground-state) interpretation:

$$\rho_X = \rho_X[N_X[\rho_X], v_X^{eff}[\boldsymbol{\rho}]] \qquad \text{or} \qquad \boldsymbol{\rho} = \boldsymbol{\rho}[\boldsymbol{N}, \boldsymbol{v}^{eff}],$$

$$E_v[\rho] = E[N, v] = E[\boldsymbol{N}, \boldsymbol{v}] = E_{\boldsymbol{v}}[\boldsymbol{\rho}] = \widetilde{E}_v[\boldsymbol{\rho}] = \mathcal{E}[\boldsymbol{N}, \boldsymbol{v}^{eff}], \qquad \boldsymbol{v}(\boldsymbol{r}) = \{v_X^M(\boldsymbol{r}) = v(\boldsymbol{r})\},$$

$$\boldsymbol{E}_{vX}[\boldsymbol{\rho}] = E_X[N_X[\rho_X], v_X^{eff}[\boldsymbol{\rho}]], \quad \mu_X[\boldsymbol{\rho}] = \mu[N_X[\rho_X], v_X^{eff}[\boldsymbol{\rho}]], \text{ etc., } X = A, B. \qquad (6.5.8)$$

This is vital for a thermodynamic-like description of various reconstructions of electron distributions of molecular subsystems (Nalewajski and Parr, 2001; Nalewajski, 2002c, 2003a,b, 2004a). More specifically, the displacements of the effective external potentials $\{v_X^{eff}[\boldsymbol{\rho}]\}$ from those characterizing the *equilibrium* density pieces, $\{v_X^H(\boldsymbol{r}) = v_X^{eff}[\boldsymbol{\rho}^H]\}$, which mark the *equilibrium* Hirshfeld division $\boldsymbol{\rho}^H[\rho]$, can now be viewed as thermodynamic "constraints" associated with the particular non-equilibrium (trial) division of ρ into $\boldsymbol{\rho}$. The basic problem of "thermodynamics" of such a *vertical* (ρ-preserving) partition is then a determination of the *equilibrium* subsystem distributions that are reached after a removal of such internal (effective external potential) constraints in the externally closed composite molecular system in atomic resolution, characterized by the fixed molecular electronic density ρ, when all embedded external potentials become exactly the *Hirshfeld effective potentials* of AIM:

$$\boldsymbol{v}^H(\boldsymbol{r}) = \{v_X^H(\boldsymbol{r}) \equiv v_X^{eff}[\boldsymbol{\rho}^H[\rho]; \boldsymbol{r}] = v(\boldsymbol{r}) + \partial \widetilde{F}^{add}[\boldsymbol{\rho}^H]/\partial\rho_X^H(\boldsymbol{r})\}. \qquad (6.5.9)$$

Speaking now in terms of quite general associations, the molecular external potential $v = v_A + v_B$, effectively determines the molecular "volume", in which all $N = N_A + N_B$ electrons are "confined" (Nalewajski, 2003e). Similarly, the atomic effective external potentials can be considered as determining the "thermodynamic" volumes for these subsystems. It should be realized that the atomic density $\rho_X(\boldsymbol{r})$, represents the embedded energy conjugate of the atom effective external potential $v_X^{eff}(\boldsymbol{r})$, by the subsystem Hellmann-Feynman theorem [Eq. (D.2.9), compare Eq. (1.5.12)],

$$\partial E_X[N_X, v_X^{eff}]/\partial v_X^{eff}(\boldsymbol{r}) = \partial E[N, \boldsymbol{v}]/\partial v_X(\boldsymbol{r}) = \rho_X(\boldsymbol{r}), \qquad X = A, B. \tag{6.5.10}$$

The chemical potentials of Eq. (6.5.7a) represent the energy conjugates of the AIM average electron populations $\boldsymbol{N} = \{N_X[\rho_X] = \int \rho_X(\boldsymbol{r})\, d\boldsymbol{r}\}$. One could thus regard the AIM densities as reminiscent of the thermodynamic partial "pressures" of atomic subsystems.

The differential of the system energy in the AIM/subsystem resolution, $\mathcal{E}[\boldsymbol{N}, \boldsymbol{v}^{eff}]$, reads (Nalewajski, 1993, 1995a,b, 1997a,b, 2002d, 2003a; Nalewajski *et al.*, 1996; Nalewajski and Korchowiec, 1997):

$$\begin{aligned} d\mathcal{E}[\boldsymbol{N}, \boldsymbol{v}^{eff}] &= \{\partial\mathcal{E}[\boldsymbol{N}, \boldsymbol{v}^{eff}]/\partial\boldsymbol{N}\}\, d\boldsymbol{N}^{\mathrm{T}} + \int \{\partial\mathcal{E}[\boldsymbol{N}, \boldsymbol{v}^{eff}]/\partial\boldsymbol{v}^{eff}(\boldsymbol{r})\}\, d\boldsymbol{v}^{eff}(\boldsymbol{r})^{\mathrm{T}}\, d\boldsymbol{r} \\ &= \boldsymbol{\mu}\, d\boldsymbol{N}^{\mathrm{T}} + \int \boldsymbol{\rho}(\boldsymbol{r})\, d\boldsymbol{v}^{eff}(\boldsymbol{r})^{\mathrm{T}}\, d\boldsymbol{r} \\ &= \{\partial E[\boldsymbol{N}, \boldsymbol{v}]/\partial\boldsymbol{N}\}\, d\boldsymbol{N}^{\mathrm{T}} + \int\{\partial E[\boldsymbol{N}, \boldsymbol{v}]/\partial\boldsymbol{v}(\boldsymbol{r})\}\, d\boldsymbol{v}(\boldsymbol{r})^{\mathrm{T}}\, d\boldsymbol{r} \\ &= \boldsymbol{\mu}\, d\boldsymbol{N}^{\mathrm{T}} + \int \boldsymbol{\rho}(\boldsymbol{r})\, d\boldsymbol{v}(\boldsymbol{r})^{\mathrm{T}}\, d\boldsymbol{r} = dE[\boldsymbol{N}, \boldsymbol{v}] \end{aligned} \tag{6.5.11}$$

The AIM chemical potentials $\boldsymbol{\mu} = \{\mu_X[\boldsymbol{\rho}]\}$ are the Lagrange multipliers associated with the constraints $\{N_X[\rho_X] = N_X^0\}$ of the prescribed electron populations on atomic fragments in the molecule [see Eq. (D.1.8)]:

$$\delta\{E_v[\boldsymbol{\rho}] - \Sigma_X\, \mu_X[\boldsymbol{\rho}]\, N[\rho_X]\} = 0. \tag{6.5.12}$$

They are *inter*-subsystem equalized when both AIM fragments reach the *global* (*g*) equilibrium, by being coupled to a common *external* electron reservoir $\mathfrak{R}$ in the composite molecular system $\mathcal{M}[M_g] \equiv (\mathfrak{R}\,\vert\, A\,\vert\, B)$, in which the two AIM are mutually open: $M_g = (A\vert B)$.

Similarly the system electron density $\rho(\boldsymbol{r})$ represents the energy conjugate of $v(\boldsymbol{r})$, by the *molecular* Hellmann-Feynman theorem (1.5.12). It thus represents an analogue of the local intensity ("pressure") of the whole molecule, which is associated with the "volume" measure provided by the molecular external potential. The corresponding differential of $E[N, v]$ [see Eq. (2.3.11)] reads:

$$\begin{aligned} dE[N, v] &= \{\partial E[N, v]/\partial N\}\, dN + \int\{\partial E[N, v]/\partial v(\boldsymbol{r})\}\, dv(\boldsymbol{r})\, d\boldsymbol{r} \\ &= \mu\, dN + \int\rho(\boldsymbol{r})\, dv(\boldsymbol{r})\, d\boldsymbol{r}. \end{aligned} \tag{6.5.13}$$

To summarize, a given partitioning $\rho = \rho_A + \rho_B$ of the molecular electron density is uniquely identified by the corresponding effective external potentials of AIM: $\boldsymbol{\rho}[\rho] \equiv \{\rho_A[\rho], \rho_B[\rho]\} \leftrightarrow \{v_A^{eff}[\rho], v_B^{eff}[\rho]\} \equiv \boldsymbol{v}^{eff}[\rho]$. Moreover, as we have already demonstrated in Chapter 5, the equilibrium (Hirshfeld) electron densities of subsystems are identified by the constrained entropy-deficiency variational principles, e.g., $\Delta S[\boldsymbol{\rho}|\boldsymbol{\rho}^0]$. Therefore, in the molecular ground-state the following mapping relations exist between the Hirshfeld AIM densities, the associated effective potentials of bonded atoms, the global entropy deficiency, and the embedded subsystem energies:

$$\boldsymbol{\rho}^H[\rho] \leftrightarrow \boldsymbol{v}^{eff}[\boldsymbol{\rho}^H[\rho]] = \boldsymbol{v}^H[\rho] \leftrightarrow \Delta S[\boldsymbol{\rho}^H[\rho]|\boldsymbol{\rho}^0] \leftrightarrow \{\boldsymbol{E}_{vX}[\boldsymbol{\rho}^H[\rho]]\} \leftrightarrow E_v[\boldsymbol{\rho}^H] = \mathcal{E}[\boldsymbol{N}^H, \boldsymbol{v}^H]. \quad (6.5.14)$$

It should be realized, however, that irrespective of the adopted density partition scheme the chemical potentials of the mutually *open* AIM pieces of the molecular ground-state density are always equalized at the global value of the chemical potential, for the system as a whole. Indeed, for any exhaustive partitioning of the molecular ground-state density one obtains, via a straightforward chain-rule transformation [see Eq.(6.4.1)],

$$\mu_X[\boldsymbol{\rho}] = \{\partial E_v[\boldsymbol{\rho}]/\partial \rho_X(\boldsymbol{r})\}_v = \{\delta E_v[\rho]/\delta\rho(\boldsymbol{r})\}\,[\partial\rho(\boldsymbol{r})/\partial\rho_X(\boldsymbol{r})] = \delta E_v[\rho]/\delta\rho(\boldsymbol{r}) = \mu. \quad (6.5.15)$$

This chemical potential equalization should indeed be expected, by an analogy to the ordinary thermodynamics, for any set of the mutually open subsystems, i.e., for arbitrary shapes of the atomic pieces of the molecular electron density. Therefore, the fragment chemical potential, representing in the energy representation the subsystem "intensity" associated with the subsystem electron density, does not discriminate between alternative divisions of the molecular electron density. In other words, such quantities are insensitive to all admissible "vertical" displacements in the electronic structure, which preserve the specified ground-state density of the molecule as a whole.

Recognizing this chemical potential equalization in the equilibrium (ground) state and introducing the subsystem relative external potentials,

$$u_X^{eff}(\boldsymbol{r}) = v_X^{eff}(\boldsymbol{r}) - \mu_\alpha[\boldsymbol{\rho}] = v(\boldsymbol{r}) + v_X^{e}(\boldsymbol{r}) - \mu, \quad (6.5.7b)$$

allows one to write the subsystem Euler equation (6.5.7a) in two alternative forms:

$$u_X^{eff}(\boldsymbol{r}) = -\frac{\delta F[\rho_X]}{\delta\rho_X(\boldsymbol{r})} = -\frac{\partial \tilde{F}^{add}[\boldsymbol{\rho}]}{\partial\rho_X(\boldsymbol{r})},\quad u(\boldsymbol{r}) = -\frac{\partial \tilde{F}[\boldsymbol{\rho}]}{\partial\rho_X(\boldsymbol{r})} = -\frac{\partial\{\tilde{F}^{add}[\boldsymbol{\rho}] + \tilde{F}^{nadd}[\boldsymbol{\rho}]\}}{\partial\rho_X(\boldsymbol{r})}. \quad (6.5.7c)$$

The former identifies the negative effective external potential of atom X, relative to the system global chemical potential, as the $\tilde{F}^{add}[\boldsymbol{\rho}]$-conjugate of the subsystem electron density. The latter form implies that the full $\tilde{F}[\boldsymbol{\rho}]$-conjugate of the molecular fragment density defines the molecular relative external potential.

In order to determine the *equilibrium* partition one thus needs the *entropic* principle of IT, to uniquely characterize the states of such embedded open AIM. As we have already shown in Chapter 5, the alternative extremum principles of the entropy-deficiency (cross-entropy), which give rise to the local inter-subsystem equalization rules for the information densities, are indeed required to identify the Hirshfeld AIM as the optimum (equilibrium) atomic fragments, which exhibit the least information distance from the corresponding free atoms.

It follows from Eq. (5.2.1) that any ("horizontal") displacement of the molecular ground-state density $d\rho$ is uniquely partitioned into the corresponding displacements of the "stockholder" AIM, $d\boldsymbol{\rho}^H[d\rho] = \{d\rho_X^H(\boldsymbol{r}) = d_X^H(\boldsymbol{r})d\rho(\boldsymbol{r})\}$ (see Section 6.4). This allows one to interpret the local "share" factors (conditional probabilities) $\{d_\alpha^H(\boldsymbol{r}) = \pi^H(\alpha|\boldsymbol{r})\}$ and their inverses as corresponding local derivatives [Eq. (6.4.2)]:

$$d\rho_X^H(\boldsymbol{r})/d\rho(\boldsymbol{r}) = d_X^H(\boldsymbol{r}) \quad \text{and} \quad d\rho(\boldsymbol{r})/d\rho_X^H(\boldsymbol{r}) = 1/d_X^H(\boldsymbol{r}), \quad X = A, B. \tag{6.5.16}$$

Let us examine the overall electronic energy in the stockholder AIM resolution [see Eqs. (6.5.5) and (6.5.8)]:

$$\tilde{E}_v[\boldsymbol{\rho}^H] \equiv E_v[\rho] = E_v[\rho_A{}^H + \rho_B{}^H] = \tilde{E}_v^{add}[\boldsymbol{\rho}^H] + \tilde{F}^{nadd}[\boldsymbol{\rho}^H]. \tag{6.5.17}$$

A reference to Eq. (6.5.15), shows that for the ground-state electron density one indeed obtains the equalized local chemical potentials of the Hirshfeld bonded atoms, defined by the partial derivatives:

$$\partial \tilde{E}_v[\boldsymbol{\rho}^H]/\partial\rho_X^H(\boldsymbol{r}) = \partial\mathcal{E}[\boldsymbol{N}^H, \boldsymbol{v}^H]/\partial N_X^H = \mu_X^H = \{\delta E_v[\rho]/\delta\rho(\boldsymbol{r})\}_v = \mu. \tag{6.5.18}$$

However, since the molecular density can be expressed in terms of any *single* Hirshfeld component, $\rho = \rho(\rho_X^H)$, the overall electronic energy also represents the unique functional of the chosen subsystem density:

$$\bar{E}_v[\rho_X^H] = \int \rho_X^H(\boldsymbol{r})\, V_X^H(\boldsymbol{r})\, d\boldsymbol{r} + \bar{F}[\rho_X^H] \equiv \bar{E}[\rho_X^H, V_X^H] \equiv E_v[\rho] = \int \rho(\boldsymbol{r})\, v(\boldsymbol{r})\, d\boldsymbol{r} + F[\rho], \tag{6.5.19}$$

where $\bar{F}[\rho_X^H] = F[\rho]$ and the *Hirshfeld external potential*

$$V_X^H(\boldsymbol{r}) = v(\boldsymbol{r})/d_X^H(\boldsymbol{r}). \tag{6.5.20}$$

Equation (6.5.19) expresses the *full* dependence of the molecular energy on the selected Hirshfeld AIM density, including a dependence due to $\rho_{Y(\neq X)}{}^H = \rho_Y{}^H(\rho_X{}^H)$.

Therefore, the functional derivative of $E_v[\rho_X^H]$ does not represent the *partial* derivative of the molecular energy with respect to ρ_X^H, for the fixed electron density of the other AIM, and as such cannot be equalized. Indeed, using the relevant local chain rule transformation gives for the total energy conjugate of the Hirshfeld subsystem density, called the *Hirshfeld potential*:

$$\delta \overline{E}_v[\rho_X^H]/\delta\rho_X^H(\mathbf{r}) = \{\delta E_v[\rho]/\delta\rho(\mathbf{r})\}[d\rho(\mathbf{r})/d\rho_X^H(\mathbf{r})] = \mu/d_X^H(\mathbf{r}) \equiv m_X^H(\mathbf{r}) \neq m_X^H(\mathbf{r}') \ldots \quad (6.5.21)$$

In the externally closed diatomic consisting of two *mutually* open (bonded) atoms A and B, $M_g = (A|B)$, for which $N = N_A + N_B =$ const., the fixed ground-state density condition, $d\rho = 0$, implies a local relation between the "vertical" inter-subsystem density displacements: $[d\rho_A(\mathbf{r})]_\rho = -\ [d\rho_B(\mathbf{r})]_\rho$. Therefore, the *in situ* CT derivative of $\overline{E}_v[\rho_X^H]$ for constant ρ, reads:

$$\{\delta\overline{E}_v[\rho_X^H]/\delta\rho_X(\mathbf{r})\}_\rho = \Sigma_Y\{\partial\widetilde{E}_v[\boldsymbol{\rho}]/\partial\rho_Y(\mathbf{r})\}\,[\partial\rho_Y(\mathbf{r})/\partial\rho_X(\mathbf{r})]_\rho = \mu_X - \mu_{Y\neq X} = 0. \quad (6.5.22)$$

It should be stressed that this *internal* equilibrium condition is satisfied by any exhaustive partitioning of a given ground-state molecular density into the mutually open fragments. It confirms that the chemical potential quantities of subsystems cannot be used to identify the equilibrium partitioning of ρ into densities of molecular subsystems.

This failure of the energetic equilibrium criterion to identify the equilibrium partition among all admissible, exhaustive divisions of a given ground-state molecular density is because the overall electronic energy in the subsystem resolution is the same for all partitioning schemes:

$$E_v[\boldsymbol{\rho}^H] = E_v[\boldsymbol{\rho}] = E_v[\rho], \qquad \Sigma_X\,\rho_X^H(\mathbf{r}) = \Sigma_X\,\rho_X[\rho;\mathbf{r}] = \rho(\mathbf{r}). \quad (6.5.23)$$

This observation further confirms a need for the entropy/information descriptors of molecular subsystems. As we have already argued in Chapter 5, the equilibrium electron distributions of molecular fragments are uniquely defined in terms of a variational principle only in the complementary, *entropy-deficiency* representation of IT, which involves the promolecular reference. We have shown in the preceding chapter that for many-electron free-atom reference the atomic densities resulting from the stockholder division of *one*- and *two*-electron densities are for all practical purposes identical, thus testifying to the quasi-invariant character of the AIM concept in IT.

In Chapter 2 we have examined the charge sensitivities characterizing the equilibrium distribution of electrons in the molecule as a whole. A similar development for the mutually-closed molecular fragments, externally closed or open, is presented in Appendix D. In the next two sections we shall examine specific examples of CS of the one-electron "stockholder" (1-*S*) AIM, resulting from the

Hirshfeld division scheme, which represent the mutually-open subsystems. Following the molecular development of Chapter 2 we shall investigate such second-derivative properties in both the chemical softness (EF) and hardness (EP) representations, in which the local nuclear and electronic parameters, respectively, provide the system state-variables. These perspectives are in spirit of the familiar BO approximation and Hellmann-Feynman theorem of quantum chemistry, respectively.

6.6. CHARGE SENSITIVITIES

In this section we shall examine in some detail the second-derivative, LR properties (see: Nalewajski, 1993, 1995a,b, 1997a,b, 1999, 2002d,e, 2003a; Nalewajski *et al.*, 1996; Nalewajski and Korchowiec, 1997; Appendix D), which define the density- and potential-responses of the stockholder AIM. Consider the *Electron-Following* (EF) Perspective, of the *Chemical Softness* representation (see Section 2.3), in which the external potential due to the nuclei or the *effective* external potentials of molecular fragments define the controlled *local* state-variables of the molecular system in the fragment resolution. For simplicity, we again assume a diatomic molecule consisting of two Hirshfeld AIM: either mutually closed, in $M_c = (A^H | B^H)$, or open, in $M_g = (A^H \mid B^H)$.

As we have already stressed in Section 6.4, there are two strategies of approaching the second-order derivative properties of the *originally* stockholder molecular fragments, which we shall call the *partial* and *total* treatments. The *partial* derivatives of Eq. (6.4.1), resulting from the horizontal-displacement relation $d\rho(\boldsymbol{r}) = \sum_X d\rho_X(\boldsymbol{r})$,

$$\left.\frac{\partial\rho(\boldsymbol{r})}{\partial\rho_X(\boldsymbol{r})}\right|_{\rho=\rho^H} = 1, \qquad X = A, B \tag{6.6.1}$$

treat each subsystem component as *independent* electronic variable. Therefore, in this approach one examines independent variations in the electronic densities of molecular subsystems, which have initially originated from the Hirshfeld partitioning of the system overall electron density. They are then considered as the *mutually* closed atomic components, which may be *externally* open relative to the corresponding (separate) electron reservoirs. Such an approach to charge sensitivities of molecular fragments is described in Appendix D. It should be realized, however, that a given displacement of the electron density of one Hirshfeld AIM, for the fixed density of the other bonded atom, violates the local stockholder proportions between subsystem densities, so that the independently displaced molecular fragments no longer represent the Hirshfeld pieces of the displaced molecular density.

Indeed, only the *total* derivatives of Eqs. (6.4.2) and (6.5.16) imply the strict preservation of the stockholder proportions of the displaced molecular densities,

$$\frac{d\rho(\boldsymbol{r})}{d\rho_X^H(\boldsymbol{r})} = \frac{1}{d_X^H(\boldsymbol{r})}, \quad X = A, B, \tag{6.6.2}$$

resulting from the Hirshfeld partitioning of the displaced molecular density, $\sum_X d\rho_X{}^H[\rho; \boldsymbol{r}] = d\rho(\boldsymbol{r})$, with a single density of the Hirshfeld AIM uniquely specifying the molecular density. Therefore, the molecular Euler equation for the ground-state density [Eq. (1.5.22)],

$$u(\boldsymbol{r}) \equiv v(\boldsymbol{r}) - \mu = -\delta F[\rho]/\delta\rho(\boldsymbol{r}), \tag{6.6.3}$$

can be transcribed as the corresponding equation for its single Hirshfeld component. As explicitly stated in Eqs. (6.5.7c) and (D.1.5), the $F[\rho[\rho_X{}^H]] \equiv \bar{F}[\rho_X^H]$-conjugates of the stockholder AIM densities $\{\rho_X{}^H\}$ derived from the ground-state density ρ, for which $\boldsymbol{\mu} = \mu\boldsymbol{1}$, read [see Eqs. (6.5.19) and (6.5.20)]:

$$\begin{aligned} -\boldsymbol{U}^H(\boldsymbol{r}) &= -\{U_X{}^H(\boldsymbol{r})\} \equiv \boldsymbol{m}^H(\boldsymbol{r}) - \boldsymbol{V}^H(\boldsymbol{r}) \equiv \{m_X{}^H(\boldsymbol{r}) - V_X{}^H(\boldsymbol{r})\} \\ &= \{\frac{\delta\bar{F}[\rho_X^H]}{\delta\rho_X^H(\boldsymbol{r})} = \frac{\delta F[\rho]}{\delta\rho(\boldsymbol{r})}\frac{d\rho(\boldsymbol{r})}{d\rho_X^H(\boldsymbol{r})} = \frac{u(\boldsymbol{r})}{d_X^H(\boldsymbol{r})} = \mu/d_X{}^H(\boldsymbol{r}) - v(\boldsymbol{r})/d_X{}^H(\boldsymbol{r})\}. \end{aligned} \tag{6.6.4}$$

It follows from this Euler equation for the Hirshfeld AIM density that $\rho_X{}^H = \rho_X{}^H[U_X{}^H]$ and $U_X{}^H = U_X{}^H[\rho_X{}^H]$, so that $\bar{F}[\rho_X^H] = \tilde{F}[U_X^H]$.

We again stress that the *vertical* Hirshfeld potential $m_X{}^H(\boldsymbol{r})$ in the preceding equation is not equalized throughout the space [see also Eq. (6.5.19)]. This is contrary to the corresponding *horizontal* partial energy conjugates, representing the true AIM chemical potentials, which are equalized at the global chemical potential level [Eq. (6.5.15)]. The latter thus signifies the chemical potential of the external electron reservoir, common to all molecular subsystems.

Equation (6.6.4) is equivalent to the molecular Euler equation (6.6.3), which is recovered by multiplying both sides of Eq. (6.6.4) by the atomic share factor $d_X{}^H(\boldsymbol{r})$. This should be expected, since for the fixed promolecular reference in the Hirshfeld scheme the subsystem density uniquely specifies the molecular distribution of electrons and, conversely, the overall density uniquely determines the densities of subsystems.

In terms of the subsystem quantities defined in the last equation the ground-state electronic energy $E_v[\rho]$ of the whole molecular system can be expressed as the related functional of any single Hirshfeld AIM component [Eq. (6.5.19)]:

$$\begin{aligned} E_v[\rho] &= E[N, v] = \int \rho(\boldsymbol{r})\, v(\boldsymbol{r})\, d\boldsymbol{r} + F[\rho] \\ &= \int \rho_X{}^H(\boldsymbol{r})\, V_X{}^H(\boldsymbol{r})\, d\boldsymbol{r} + \bar{F}[\rho_X^H] \equiv \bar{E}[\rho_X^H, V_X^H], \end{aligned} \tag{6.6.5}$$

giving rise to the associated first-differential expression:

$$d\bar{E}[\rho_X^H, V_X^H] = \int \left(\frac{\partial \bar{E}[\rho_X^H, V_X^H]}{\partial \rho_X^H(\boldsymbol{r})} \right)_{V_X^H} d\rho_X^H(\boldsymbol{r}) d\boldsymbol{r} + \int \left(\frac{\partial \bar{E}[\rho_X^H, V_X^H]}{\partial V_X^H(\boldsymbol{r})} \right)_{\rho_X^H} dV_X^H(\boldsymbol{r}) d\boldsymbol{r}$$

$$= \int m_X^H(\boldsymbol{r}) d\rho_X^H(\boldsymbol{r}) d\boldsymbol{r} + \int \rho_X^H(\boldsymbol{r}) dV_X^H(\boldsymbol{r}) d\boldsymbol{r} = \mu dN + \int \rho(\boldsymbol{r}) dv(\boldsymbol{r}) d\boldsymbol{r} = dE[N, v]. \quad (6.6.6)$$

In the foregoing equation we have identified the local intensity $m_X^H(\boldsymbol{r})$ as the $\bar{E}[\rho_X^H, V_X^H]$-conjugate of $\rho_X^H(\boldsymbol{r})$ with the latter representing the $\bar{E}[\rho_X^H, V_X^H]$-conjugate of $V_X^H(\boldsymbol{r})$ (the Hirshfeld-AIM Hellmann-Feynman theorem).

One similarly re-interprets the other Legendre-transforms of the system energy as the associated functionals of any single Hirshfeld component of the molecular electron density [see Eqs. (2.3.15), (2.4.1) and (2.4.2)]:

$$\begin{aligned} \Omega_u[\rho] &= \Omega[\mu, v] = \int \rho(\boldsymbol{r})\, u(\boldsymbol{r})\, d\boldsymbol{r} + F[\rho] = \Omega[u] \\ &= \int \rho_X^H(\boldsymbol{r})\, U_X^H(\boldsymbol{r})\, d\boldsymbol{r} + \bar{F}[\rho_X^H] \\ &\equiv \bar{\Omega}[\rho_X^H, U_X^H] \equiv \bar{\Omega}_{U_X^H}[\rho_X^H] \equiv \bar{\Omega}[m_X^H, V_X^H] \equiv \tilde{\Omega}[U_X^H], \end{aligned} \quad (6.6.7)$$

$$F[\rho] = \mathsf{F}[N, \rho] = \bar{F}[\rho_X^H] \equiv \tilde{F}[U_X^H], \quad (6.6.8)$$

$$\begin{aligned} R_\mu[\rho] &= \mathsf{R}[\mu, \rho] = F[\rho] - \mu N[\rho] \\ &= \bar{F}[\rho_X^H] - \int m_X^H(\boldsymbol{r})\, \rho_X^H(\boldsymbol{r})\, d\boldsymbol{r} \equiv \bar{R}[m_X^H, \rho_X^H], \end{aligned} \quad (6.6.9)$$

The associated differential expressions read:

$$\begin{aligned} d\bar{\Omega}[m_X^H, V_X^H] &= \int \left(\frac{\partial \bar{\Omega}[m_X^H, V_X^H]}{\partial m_X^H(\boldsymbol{r})} \right)_{V_X^H} dm_X^H(\boldsymbol{r}) d\boldsymbol{r} + \int \left(\frac{\partial \bar{\Omega}[m_X^H, V_X^H]}{\partial V_X^H(\boldsymbol{r})} \right)_{m_X^H} dV_X^H(\boldsymbol{r})\, d\boldsymbol{r} \\ &= -\int \rho_X^H(\boldsymbol{r}) dm_X^H(\boldsymbol{r}) d\boldsymbol{r} + \int \rho_X^H(\boldsymbol{r}) dV_X^H(\boldsymbol{r}) d\boldsymbol{r} = \int \rho_X^H(\boldsymbol{r}) dU_X^H(\boldsymbol{r}) d\boldsymbol{r} \\ &= -N d\mu + \int \rho(\boldsymbol{r}) dv(\boldsymbol{r}) d\boldsymbol{r} = d\Omega[\mu, v] = \int \rho(\boldsymbol{r}) du(\boldsymbol{r}) d\boldsymbol{r} = d\Omega[u], \end{aligned} \quad (6.6.10)$$

$$d\bar{F}[\rho_X^H] = \int \frac{\delta \bar{F}[\rho_X^H]}{\delta \rho_X^H(\boldsymbol{r})} d\rho_X^H(\boldsymbol{r}) d\boldsymbol{r} = -\int U_X^H(\boldsymbol{r}) d\rho_X^H(\boldsymbol{r}) d\boldsymbol{r} = -\int u(\boldsymbol{r}) d\rho(\boldsymbol{r}) d\boldsymbol{r} = dF[\rho], \quad (6.6.11)$$

$$\begin{aligned} d\bar{R}[m_X^H, \rho_X^H] &= \int \left(\frac{\partial \bar{R}[m_X^H, \rho_X^H]}{\partial m_X^H(\boldsymbol{r})} \right)_{\rho_X^H} dm_X^H(\boldsymbol{r}) d\boldsymbol{r} + \int \left(\frac{\partial \bar{R}[m_X^H, \rho_X^H]}{\partial \rho_X^H(\boldsymbol{r})} \right)_{m_X^H} d\rho_X^H(\boldsymbol{r}) d\boldsymbol{r} \\ &= -\int \rho_X^H(\boldsymbol{r}) dm_X^H(\boldsymbol{r}) d\boldsymbol{r} - \int V_X^H(\boldsymbol{r}) d\rho_X^H(\boldsymbol{r}) d\boldsymbol{r} = -N d\mu - \int v(\boldsymbol{r}) d\rho(\boldsymbol{r}) d\boldsymbol{r} = d\mathsf{R}[\mu, \rho] \end{aligned} \quad (6.6.12)$$

The second functional derivatives of $\overline{E}[\rho_X^H, V_X^H]$ and $\overline{\Omega}[m_X^H, U_X^H] \equiv \overline{\Omega}_{U_X^H}[\rho_X^H]$ with respect to the stockholder subsystem density, calculated for the fixed external potential, define the AIM *diagonal* (one-centre) hardness kernel:

$$\left(\frac{\delta^2 \overline{E}[\rho_X^H, V_X^H]}{\delta\rho_X^H(\boldsymbol{r})\delta\rho_X^H(\boldsymbol{r}')}\right)_{V_X^H} = \frac{\delta^2 \Omega_{U_X^H}[\rho_X^H]}{\delta\rho_X^H(\boldsymbol{r})\delta\rho_X^H(\boldsymbol{r}')} = \frac{\delta^2 \overline{F}[\rho_X^H]}{\delta\rho_X^H(\boldsymbol{r})\delta\rho_X^H(\boldsymbol{r}')}$$
$$\equiv \eta_{X,X}^H(\boldsymbol{r},\boldsymbol{r}') = -\frac{\delta U_X^H(\boldsymbol{r}')}{\delta\rho_X^H(\boldsymbol{r})} = \frac{\eta(\boldsymbol{r},\boldsymbol{r}')}{d_X^H(\boldsymbol{r})d_X^H(\boldsymbol{r}')}. \qquad (6.6.13a)$$

Since the specification of any Hirshfeld component is equivalent to the specification of the molecular ground-state density, $\overline{\Omega}_{U_X^H}[\rho_X^H] = \overline{\Omega}_{U_Y^H}[\rho_Y^H]$, one could similarly define the AIM *off-diagonal* (two-centre) hardness kernels. Together, they determine the corresponding matrix of the hardness kernels in the Hirshfeld AIM resolution:

$$\frac{\delta^2 \widetilde{F}[\boldsymbol{\rho}^H]}{\delta\boldsymbol{\rho}^H(\boldsymbol{r})\delta\boldsymbol{\rho}^H(\boldsymbol{r}')} = -\frac{\delta \boldsymbol{U}^H(\boldsymbol{r}')}{\delta\boldsymbol{\rho}^H(\boldsymbol{r})} \equiv \boldsymbol{\eta}^H(\boldsymbol{r},\boldsymbol{r}') = \left\{\eta_{X,Y}^H(\boldsymbol{r},\boldsymbol{r}') = -\frac{\delta U_Y^H(\boldsymbol{r}')}{\delta\rho_X^H(\boldsymbol{r})} = \frac{\eta(\boldsymbol{r},\boldsymbol{r}')}{d_X^H(\boldsymbol{r})d_Y^H(\boldsymbol{r}')}\right\}, \qquad (6.6.13b)$$

with each element representing the corresponding fraction of the molecular kernel determined by the product of atomic share factors. Its "inverse" determines the corresponding matrix of the softness kernels in the stockholder-AIM resolution:

$$\boldsymbol{\sigma}^H(\boldsymbol{r},\boldsymbol{r}') = -\frac{\delta\boldsymbol{\rho}^H(\boldsymbol{r}')}{\delta\boldsymbol{U}^H(\boldsymbol{r})} \equiv \boldsymbol{\eta}^H(\boldsymbol{r},\boldsymbol{r}')^{-1} = \left\{\sigma_{X,Y}^H(\boldsymbol{r},\boldsymbol{r}') = -\frac{\delta\rho_Y^H(\boldsymbol{r}')}{\delta U_X^H(\boldsymbol{r})} = d_X^H(\boldsymbol{r})\sigma(\boldsymbol{r},\boldsymbol{r}')d_Y^H(\boldsymbol{r}')\right\}. \qquad (6.6.14)$$

They satisfy the modified reciprocity relation [compare Eq. (D.3.6)],

$$\int \eta_{X,Z}^H(\boldsymbol{r},\boldsymbol{r}'')\sigma_{Z,Y}^H(\boldsymbol{r}'',\boldsymbol{r}')\,d\boldsymbol{r}'' = \delta(\boldsymbol{r}'-\boldsymbol{r})\frac{d_Y^H(\boldsymbol{r}')}{d_X^H(\boldsymbol{r})} = \frac{\delta\rho_Y^H(\boldsymbol{r}')}{\delta\rho_X^H(\boldsymbol{r})}, \qquad X, Y, Z = A, B, \quad (6.6.15)$$

where we have used Eq. (2.3.30). It reflects the "vertical" dependence between the Hirshfeld components of Eq. (6.4.3).

Similar interpretations of other linear response properties of the Hirshfeld AIM as stockholder fractions of the corresponding molecular quantities follow from the relevant chain-rule manipulation of the defining derivatives. We recall [see Section 2.3.1)] that the two-point molecular kernel defined by the second (partial) functional derivative of the ground-state energy $E_v[\rho] = E[N, v]$ with respect to the molecular external potential [Eq. (2.3.10)] represents the molecular *Linear-Response* (LR) function (see also Appendix B):

$$\beta(\boldsymbol{r}, \boldsymbol{r}') \equiv \{\partial^2 E[N, v]/\partial v(\boldsymbol{r})\partial v(\boldsymbol{r}')\}_N = [\partial\rho(\boldsymbol{r}')/\partial v(\boldsymbol{r})]_N. \tag{6.6.16a}$$

In the externally *closed* molecular systems, for the fixed overall number of electrons N, it transforms the external potential perturbation Δv into the associated linear response of the system ground-state density, $[\Delta\rho]_N$:

$$[\Delta\rho(\boldsymbol{r})]_N = \int \Delta v(\boldsymbol{r}')\beta(\boldsymbol{r}', \boldsymbol{r})\, d\boldsymbol{r}'. \tag{6.6.16b}$$

It follows from Eq. (6.6.6) that the row vector of the *singly* AIM-resolved LR kernels defined by the mixed second partials of the electronic energy $E[N,v] = \mathcal{E}[\boldsymbol{N}^H, \boldsymbol{V}^H(v)]$ with respect to the molecular and the Hirshfeld external potential [Eqs. (6.5.20) and (6.6.4)], respectively, reads:

$$\boldsymbol{\beta}^H(\boldsymbol{r},\boldsymbol{r}') = \left(\frac{\partial}{\partial v(\boldsymbol{r})}\left(\frac{\partial \mathrm{E}[\boldsymbol{N}^H,\boldsymbol{V}^H]}{\partial \boldsymbol{V}^H(\boldsymbol{r}')}\right)_{\boldsymbol{N}^H}\right)_N = \left(\frac{\partial \boldsymbol{\rho}^H(\boldsymbol{r}')}{\partial v(\boldsymbol{r})}\right)_N = \left(\frac{\partial \rho(\boldsymbol{r})}{\partial \boldsymbol{V}^H(\boldsymbol{r}')}\right)_{\boldsymbol{N}^H}$$

$$= \boldsymbol{d}^H(\boldsymbol{r}')\left(\frac{\partial \rho(\boldsymbol{r}')}{\partial v(\boldsymbol{r})}\right)_N = \boldsymbol{d}^H(\boldsymbol{r}')\beta(\boldsymbol{r},\boldsymbol{r}') = \{\beta_X^H(\boldsymbol{r},\boldsymbol{r}') = \beta(\boldsymbol{r},\boldsymbol{r}')d_X^H(\boldsymbol{r}')\}. \tag{6.6.17}$$

Here we have used the local chain-rule transformation

$$\beta_X^H(\boldsymbol{r},\boldsymbol{r}') = \left(\frac{\partial \rho_X^H(\boldsymbol{r}')}{\partial v(\boldsymbol{r})}\right)_N = \left(\frac{\partial \rho(\boldsymbol{r}')}{\partial v(\boldsymbol{r})}\right)_N \frac{d\rho_X^H(\boldsymbol{r}')}{d\rho(\boldsymbol{r}')} = \beta(\boldsymbol{r},\boldsymbol{r}')\, d_X^H(\boldsymbol{r}') = \left(\frac{\partial \rho(\boldsymbol{r})}{\partial V_X^H(\boldsymbol{r}')}\right)_{\rho^H}$$

$$\Sigma_X\, \beta_X^H(\boldsymbol{r}, \boldsymbol{r}') = [\partial\rho(\boldsymbol{r}')/\partial v(\boldsymbol{r})]_N = \beta(\boldsymbol{r}, \boldsymbol{r}'). \tag{6.6.18}$$

These kernels transform the displacement in the molecular or AIM external potential state-parameters into the conjugated responses of the electron densities:

$$[\Delta\rho(\boldsymbol{r})]_N = \int \Delta v(\boldsymbol{r}')\boldsymbol{\beta}^H(\boldsymbol{r}',\boldsymbol{r})\, d\boldsymbol{r}' \quad \text{or} \quad [\Delta\rho(\boldsymbol{r})]_{\boldsymbol{N}^H} = \int \boldsymbol{\beta}^H(\boldsymbol{r},\boldsymbol{r}')\, \Delta\boldsymbol{V}^H(\boldsymbol{r}')^{\mathrm{T}}\, d\boldsymbol{r}'. \tag{6.6.19}$$

One could also define the square matrix of the *doubly* AIM-resolved kernels, of the second partial derivatives with respect to the AIM external potentials:

$$\boldsymbol{\beta}^H(\boldsymbol{r}, \boldsymbol{r}') = \left(\frac{\partial^2 \mathrm{E}[\boldsymbol{N}^H,\boldsymbol{V}^H]}{\partial \boldsymbol{V}^H(\boldsymbol{r})\, \partial \boldsymbol{V}^H(\boldsymbol{r}')}\right)_{\boldsymbol{N}^H} = \left(\frac{\partial \boldsymbol{\rho}^H(\boldsymbol{r}')}{\partial \boldsymbol{V}^H(\boldsymbol{r})}\right)_{\boldsymbol{N}^H} = \left\{\beta_{X,Y}^H(\boldsymbol{r},\boldsymbol{r}') = \left(\frac{\partial \rho_Y^H(\boldsymbol{r}')}{\partial V_X^H(\boldsymbol{r})}\right)_{\boldsymbol{N}^H}\right\}. \tag{6.6.20}$$

This matrix kernel transforms the displacements of the atomic effective external potentials $\Delta\boldsymbol{V}^H$ into the linear responses of the electron densities of the externally closed AIM:

$$[\Delta\boldsymbol{\rho}^H(\boldsymbol{r})]_{\boldsymbol{N}^H} = \int \Delta\boldsymbol{V}^H(\boldsymbol{r}')\boldsymbol{\beta}^H(\boldsymbol{r}',\boldsymbol{r})d\boldsymbol{r}'. \tag{6.6.21}$$

The double chain-rule transformation of the $\beta_{X,Y}{}^H(\boldsymbol{r}, \boldsymbol{r}')$ derivative gives:

$$\beta^H_{X,Y}(\boldsymbol{r},\boldsymbol{r}') = \left(\frac{\partial \rho^H_Y(\boldsymbol{r}')}{\partial v^H_X(\boldsymbol{r})}\right)_{N^H} = \frac{dv(\boldsymbol{r})}{dv^H_X(\boldsymbol{r})}\left(\frac{\partial \rho(\boldsymbol{r}')}{\partial v(\boldsymbol{r})}\right)_N \frac{d\rho^H_Y(\boldsymbol{r}')}{d\rho(\boldsymbol{r}')} = d^H_X(\boldsymbol{r})\beta(\boldsymbol{r},\boldsymbol{r}')d^H_Y(\boldsymbol{r}')$$

$$= d_X{}^H(\boldsymbol{r})\, \beta_Y{}^H(\boldsymbol{r}, \boldsymbol{r}') = \beta_X{}^H(\boldsymbol{r}, \boldsymbol{r}')\, d_Y{}^H(\boldsymbol{r}'), \qquad (6.6.22)$$

where: $\sum_Y \beta_{X,Y}{}^H(\boldsymbol{r}, \boldsymbol{r}') = \beta_X{}^H(\boldsymbol{r}, \boldsymbol{r}')$ and $\sum_X \beta_{X,Y}{}^H(\boldsymbol{r}, \boldsymbol{r}') = \beta_Y{}^H(\boldsymbol{r}, \boldsymbol{r}')$.

Consider next the FF indices of the Hirshfeld atoms, the bonded-atom analogs of the molecular FF [Eq. (2.3.1)],

$$f(\boldsymbol{r}) = \partial^2 E[N, v]/\partial N\, \partial v(\boldsymbol{r}) = [\partial \rho(\boldsymbol{r})/\partial N]_v = [\partial \mu/\partial v(\boldsymbol{r})]_N. \qquad (6.6.23)$$

The row vector of the *singly* AIM-resolved FF quantities reads:

$$\boldsymbol{f}^H(\boldsymbol{r}) = \partial^2 \mathcal{E}[N, \boldsymbol{V}^H]/\partial N\, \partial \boldsymbol{V}^H(\boldsymbol{r}) = \{ f_X{}^H(\boldsymbol{r}) = [\partial \rho_X{}^H(\boldsymbol{r})/\partial N]_v = [\partial \mu/\partial V_X{}^H(\boldsymbol{r})]_N\}. \qquad (6.6.24)$$

Expressing the AIM quantities in terms of the associated molecular properties then gives:

$$f_X{}^H(\boldsymbol{r}) = [\partial \mu/\partial v_X{}^H(\boldsymbol{r})]_N = f(\boldsymbol{r})\, d_X{}^H(\boldsymbol{r}). \qquad (6.6.25)$$

One could also introduce the square matrix of the Hirshfeld *doubly* AIM–resolved FF contributions, $\mathbf{f}^H(\boldsymbol{r}) = \{f_{X,Y}{}^H(\boldsymbol{r})\}$:

$$\mathbf{f}^H(\boldsymbol{r}) = \frac{\partial^2 \bar{E}[\rho^H_X, V^H_X]}{\partial \rho^H_Y(\boldsymbol{r})\, \partial V^H_X(\boldsymbol{r}')} = \frac{1}{d^H_Y(\boldsymbol{r})}\left(\frac{\partial \mu}{\partial v(\boldsymbol{r}')}\right)_N d^H_X(\boldsymbol{r}') = f(\boldsymbol{r}')\, d^H_X(\boldsymbol{r}')/d^H_Y(\boldsymbol{r}), \qquad (6.6.26)$$

where we have recognized the local chemical potential equalization at the ground-state distribution of electrons: $\mu(\boldsymbol{r}) = \mu$.

The Hirshfeld AIM contributions of Eq. (6.6.25) define additive atomic contributions to the molecular FF: $\sum_X f_X{}^H(\boldsymbol{r}) = f(\boldsymbol{r})$. By integrating these atomic FF one finds the associated condensed FF index of *X*th Hirshfeld AIM:

$$F_X{}^H \equiv [\partial N_X{}^H/\partial N]_v = \int f_X{}^H(\boldsymbol{r})\, d\boldsymbol{r} = \int f(\boldsymbol{r})\, d_X(\boldsymbol{r})\, d\boldsymbol{r}, \qquad (6.6.27)$$

measuring the linear response in the subsystem electron population $N_X{}^H$ per unit shift in the global number of electrons *N*.

As we have already indicated in Eq. (6.6.6), the first partials of $\bar{E}[\rho^H_X, v^H_X]$ reproduce the first differential of the molecular electronic energy [see Eq. (2.3.12)]. One can similarly show that the charge sensitivities of the Hirshfeld AIM, which we have introduced in this section, also recover the second differential of the molecular electronic energy. In the following demonstration we shall use the relevant definitions and relations, which have been summarized in Chapter 2. Moreover, we shall recognize that for the "frozen" molecular external potential v of the BO approximation, the shift of the Hirshfeld AIM density is determined by its FF:

$$[d\rho_X^H(\boldsymbol{r})]_v = dN\,[\partial\rho_X^H(\boldsymbol{r})/N]_v = dN\,f_X^H(\boldsymbol{r}) = dN\,d_X^H(\boldsymbol{r})\,f(\boldsymbol{r}). \tag{6.6.28}$$

The second differential of $\overline{E}[\rho_X^H, V_X^H]$ then gives:

$$\begin{aligned}
d^2\overline{E}[\rho_X^H, V_X^H] = \frac{1}{2}\Bigg[& \iint [d\rho_X^H(\boldsymbol{r}')]_v \left(\frac{\partial m_X^H(\boldsymbol{r})}{\partial \rho_X^H(\boldsymbol{r}')}\right)_v [d\rho_X^H(\boldsymbol{r})]_v \, d\boldsymbol{r}\, d\boldsymbol{r}' \\
& + \iint dV_X^H(\boldsymbol{r}') \left(\frac{\partial m_X^H(\boldsymbol{r})}{\partial V_X^H(\boldsymbol{r}')}\right)_N [d\rho_X^H(\boldsymbol{r})]_v \, d\boldsymbol{r}\, d\boldsymbol{r}' \\
& + \iint dV_X^H(\boldsymbol{r}') \left(\frac{\partial \rho_X^H(\boldsymbol{r})}{\partial V_X^H(\boldsymbol{r}')}\right)_N dV_X^H(\boldsymbol{r}) \, d\boldsymbol{r}\, d\boldsymbol{r}' \\
& + \iint [d\rho_X^H(\boldsymbol{r}')]_v \left(\frac{\partial \rho_X^H(\boldsymbol{r})}{\partial \rho_X^H(\boldsymbol{r}')}\right)_v dV_X^H(\boldsymbol{r}) \, d\boldsymbol{r}\, d\boldsymbol{r}' \Bigg] \\
= \frac{1}{2}\Big[& (dN)^2 \iint f_X^H(\boldsymbol{r}')\eta_{X,X}^H(\boldsymbol{r}',\boldsymbol{r}) f_X^H(\boldsymbol{r}) \, d\boldsymbol{r}\, d\boldsymbol{r}' \\
& + dN \iint \frac{d\rho_X^H(\boldsymbol{r})}{d_X^H(\boldsymbol{r})} f_X^H(\boldsymbol{r}') dV_X^H(\boldsymbol{r}') f_X^H(\boldsymbol{r}) d\boldsymbol{r}\, d\boldsymbol{r}' \\
& + \iint dV_X^H(\boldsymbol{r}')\beta_{X,X}^H(\boldsymbol{r}',\boldsymbol{r}) dV_X^H(\boldsymbol{r}) \, d\boldsymbol{r}\, d\boldsymbol{r}' \\
& + dN \iint f_X^H(\boldsymbol{r}') \frac{d_X^H(\boldsymbol{r})}{d_X^H(\boldsymbol{r}')} \delta(\boldsymbol{r}-\boldsymbol{r}') \, dV_X^H(\boldsymbol{r}) \, d\boldsymbol{r}\, d\boldsymbol{r}' \Big] \\
= \frac{1}{2}\Big[& (dN)^2\eta + 2dN \int f(\boldsymbol{r}) dv(\boldsymbol{r}) \, d\boldsymbol{r} + \iint dv(\boldsymbol{r}')\beta(\boldsymbol{r}',\boldsymbol{r}) dv(\boldsymbol{r}) \, d\boldsymbol{r}\, d\boldsymbol{r}'' \Big] \\
= d^2 & E[N,v].
\end{aligned} \tag{6.6.29}$$

6.7. CONCLUSION

We have explored in this chapter several additional properties of the Hirshfeld (1-*S*) AIM/subsystems, which make them attractive concepts for chemical applications. These entropy-equilibrium pieces of the molecular electron density were shown to give rise to the local inter-fragment equalization of the information-distance densities relative to the atomic-promolecule reference. The role of the entropy-penalty term in the variational principle of the entropy-deficiency in the atomic resolution, which is responsible for the Hirshfeld density localization around the atomic nucleus, has been examined and the additivity of the information distances of Hirshfeld AIM has been established. In addition to the variational principle of the *minimum* (additive) entropy-deficiency of atomic components of the molecular electron density (see Section 5.3) the complementary variational rule has been formulated of the *maximum* non-additive missing-information in atomic densities relative to the free-atom distributions. The latter has been shown to also give rise to the stockholder partition of the molecular electron distribution into atomic fragments.

Displacements of the electron density of bonded atoms from the corresponding free-atom references and the associated missing-information densities and changes in the Shannon entropy density, have all been used to diagnose the atomic promotion to their valence-states in molecules. All these probes were shown to give a consistent diagnosis of the main changes the atoms undergo, when they form the chemical bonds. These novel entropic concepts provide the atomic density-difference function the complementary information interpretation. These illustrative applications demonstrate the potential of IT in extracting the chemical interpretation from the known molecular electron distributions and reflect the information origins of the chemical bonds.

It has been argued that the Hirshfeld atoms equalize their chemical potentials and are in principle effective-external-potential–representable. This adds the physical meaning to these density-constructs and introduces an element of causality in treating their displacements in a changed molecular environment.

The partial and total approaches to displacements of the electron densities of the Hirshfeld-AIM have been distinguished. The Euler equation in the total perspective, in which the shifted AIM densities originate from the ground-state density of the molecule, has been examined. The charge sensitivities of the stockholder (1-*S*) atoms have been expressed in terms of the share-factors of atomic fragments as fractions of the corresponding molecular CS. An overview of such *chemical softness* (density response) and *hardness* (potential response) properties of the Hirshfeld atoms has been given.

An important part of the chemical understanding of molecules comes from various descriptors of a network of *chemical bonds* connecting AIM. In the next two chapters we shall demonstrate how the theory of communication systems (Shannon, 1948; Shannon and Weaver, 1949; Abramson, 1963; see also Chapter 3) can be used to index the chemical bond multiplicity and its covalent/ionic composition.

7. COMMUNICATION THEORY OF THE CHEMICAL BOND

The elements of the *Communication Theory* of the chemical bond are presented. Molecular systems are interpreted as information channels in atomic resolution, in which the *"signals"* of electron allocations to constituent atoms are propagated from the molecular *input* ("source"), determined by either the free-atoms of the promolecule or the bonded-atoms in the molecule, to the molecular *output* ("receiver") consisting of AIM. The transmission network of the system chemical bonds is determined by the molecular two-electron (conditional) probabilities in atomic resolution. The electron delocalization throughout the system is responsible for the effective "*noise*" affecting the transmission of such atom-assignment signals, which inhibits the amount of information passing through the communication channel. These complementary facets of the molecular information systems are characterized by the *conditional entropy* and *mutual information* descriptors, respectively. It is argued that they accordingly provide the adequate (ground-state) IT- indices of the system overall *covalent* and *ionic* bond components. Their variational principles are explored and an illustrative application of this novel IT-approach to the Heitler-London theory of the chemical bond in H_2 is reported. The communication systems corresponding to the VB-structures and their VB-*covalent* and VB-*ionic* combinations are examined for their information-ionicity and entropy-covalency content. The purely IT-covalent and purely IT-ionic combinations of elementary VB-structures are identified thus providing a *dichotomous* separation of these two bond components and a new perspective on their competition in the chemical bond. A distinction between the *electron-sharing* (covalent) and *pair-sharing* (coordination) bond origins is explored and the illustrative application to the two-orbital model of the ground-state of two-electrons in a diatomic system is given. The overall bond-order conservation due to a competition between the bond IT-*ionic* and IT-*covalent* components is examined and the entropy/information descriptors of the model singlet spin-channel are discussed. Finally, the ground-state IT-indices of π bonds in ethylene, allyl, butadiene and benzene are determined for the information channels generated by the Hückel theory.

7.1. INTRODUCTION

In preceding chapters we have used the Information Theory in an exploration of the effective states of bonded-atoms in the molecule relative to those characterizing the corresponding free-atoms of the Atomic Promolecule (AP) reference. We have also examined the molecular displacements of the entropy/information quantities due to formation of the chemical bonds. A network of the chemical bonds provides channels for the electron delocalization throughout the whole molecular system. It generates the associated flow of the entropy/information contained in the electron probability distributions. As we have demonstrated, these changes in the information content of the electron distribution can indeed be used as reliable diagnostic tools for probing the chemical bonds in a molecule. The contour diagrams of these IT-quantities complement the familiar density-difference maps of quantum chemistry.

The reconstruction of the initial densities of free atoms is effected through the polarization (P) of the constituent atoms and the charge-transfer (CT) between them. In accordance with this prevailing perspective on the bond-formation process, it is said that the free atoms of AP are "promoted" to their effective "valence-states" in the molecule. These flows of electrons, both inside and between constituent atoms, modify the electron densities and the associated probability distributions relative to the corresponding free-atom quantities. The resulting subtle displacements in the molecular electronic structure, mainly in the valence shell, determine all properties of the system chemical bonds. The DFT indeed assures us that the molecular (ground-state) electron density determines all molecular properties, including those of the chemical bonds, which represent the AIM "connectivities" in the system under consideration.

The AP-reference, of the non-bonded atoms for the molecular configuration of the system nuclei, constitutes a natural "initial" stage of the bond-formation process. It should be used to determine the aspects of the chemical bond, which depend upon its "history". Since the chemical bonds are defined in the framework of an underlying atomic "discretization", which gives rise to the associated "condensed" (*coarse*-grained) representation of the continuous (*fine*-grained) electron distributions, some type of AIM resolution of the molecular densities is called for in the theory of bond "orders". Indeed, the very notion of the chemical bond *multiplicity* can be formulated only when a definite reference to AIM is established within the adopted atomic resolution of the molecular electronic structure.

As we have already observed before, the molecular relaxation of the electronic structure of the free-atoms also implies the associated redistribution of the entropy/information densities throughout the system in question. In such an information-theoretic outlook (Nalewajski, 2000c, 2004b-e, 2005a-c; Nalewajski and Jug, 2002) one perceives a molecule as a kind of the information "transmission" channel, in which the bonded atoms "communicate" between themselves "messages" about the instantaneous allocations of electrons to the constituent atoms. In such a molecular communication system the chemical bonds provide a "network" for the propagation of these AIM assignment "signals".

The molecular system is thus viewed in this IT approach as the "communication" channel (see Section 3.7), in which the *"signals"* conveying a message about the electron distribution among the atomic subsystems are transmitted from the molecular *"source"* (input, **A**), represented either by AIM or free atoms of AP, to the molecular *"receiver"* (output, **B**), consisting of the bonded atoms in the molecule. Elements of this communication theory of the chemical bond have recently been developed and the information quantities of the communication theory have been successfully used as new tools for probing the overall chemical bond multiplicities in molecules or their fragments, and their ionic and covalent components (Nalewajski, 2000c, 2004b-e, 2005a-c; Nalewajski and Jug, 2002).

When examining a connection between the electronic structure of molecular systems and the information theory of communication systems the following questions naturally arise:

In what sense can we treat molecules as "communication" channels?

Which resolution of the electron probabilities is required to grasp specific bond characteristics?

What kind of *signals* ("messages") are to be considered to probe the chemical bond?

What constitutes the *input* (source) and the *output* (receiver) of these signals in molecular channels?

What "amount" of information flows through the molecular channel and how various factors of the electronic structure influence its information "capacity"?

What role is played by the electron spatial (orbital) and spin probability distributions?

How do the entropy/information quantities reflect upon the familiar aspects of the chemical bond, e.g., its overall "multiplicity", the ionic/covalent composition and the coordination (donor-acceptor) or electron-sharing (covalent) origins?

How to extract from the molecular distributions of electrons the adequate indices of the *localized* bonds, between a given pair of bonded atoms, and molecular fragments?

How to separate the *internal* (inside the molecular fragment) and *external* (between the molecular fragment and its environment) bond characteristics of molecular subsystems?

How does the channel *reduction*, by changing the resolution level of the input and/or output probabilities, affects the IT bond descriptors?

It is the main goal of this chapter to examine some of these intriguing questions and to establish basic elements of the *global* Communication Theory of the chemical bond, generating the entropy/information bond indices for the molecule as a whole. The alternative approaches to molecular subsystems will be the subject of the next chapter. The entropy/information bond orders will be compared with both the chemical intuitive expectations and other successful measures of bond multiplicity formulated in the MO theory (e.g.: Wiberg, 1968; Gopinathan and Jug, 1983; Mayer, 1983; Jug and Gopinathan, 1990; Nalewajski *et al.*, 1993; Nalewajski and Mrozek,

1994, 1996; Nalewajski *et al.*, 1994, 1996, 1997; Mrozek *et al.*, 1998; Nalewajski, 2004b).

7.2. MOLECULES AS "COMMUNICATION" SYSTEMS

7.2.1. Atomic Discretization of Electron Distributions

The distribution of electrons in a molecule $M = (\ldots|X|Y|\ldots)$ can be described using both the *local* (fine-grained) or *discrete* (coarse-grained) resolutions. An example of a continuous approach is the molecular *one*-electron probability distribution $p(\boldsymbol{r}) = \rho(\boldsymbol{r})/N$. It represents the shape-factor of the molecular electron density $\rho(\boldsymbol{r})$, which integrates to the overall number of electrons $N = \int \rho(\boldsymbol{r})\, d\boldsymbol{r}$. The atomic resolution in the continuous treatment is obtained by an appropriate division of the molecular distribution into the corresponding AIM pieces, e.g., using alternative schemes of the familiar population analysis, the topological division of the electron density (Bader, 1994), or the "stockholder" (H) partition (Hirshfeld, 1977), into contributions from the bonded atomic fragments. For example, in the Hirshfeld partition:

$$p(\boldsymbol{r}) = \Sigma_X\, p_X^H(\boldsymbol{r}) \qquad \text{or} \qquad \rho(\boldsymbol{r}) = \Sigma_X\, \rho_X^H(\boldsymbol{r}) = N\, \Sigma_X\, p_X^H(\boldsymbol{r}). \tag{7.2.1}$$

Here, $p_X^H(\boldsymbol{r}) = p(\boldsymbol{r})[p_X^0(\boldsymbol{r})/p^0(\boldsymbol{r})] \equiv p(\boldsymbol{r})\pi^0(X|\boldsymbol{r}) = p_X^0(\boldsymbol{r})[p(\boldsymbol{r})/p^0(\boldsymbol{r})] \equiv p_X^0(\boldsymbol{r})w(\boldsymbol{r})$, where $\pi^0(X|\boldsymbol{r}) \equiv d_X^0(\boldsymbol{r}) = p_X^0(\boldsymbol{r})/p^0(\boldsymbol{r}) = p_X^H(\boldsymbol{r})/p(\boldsymbol{r}) = \pi^H(X|\boldsymbol{r}) \equiv d_X^H(\boldsymbol{r})$ denotes the local conditional probability (share) of the stockholder atom. It corresponds to the promolecular (or molecular) event that an electron of the molecule found at $\boldsymbol{r}$ originates from atom X. The ratio $w(\boldsymbol{r})$ stands for the unbiased (subsystem independent) local enhancement factor [see Eq. (5.2.1)].

The discrete AIM representation of the molecular one-electron probability density obtained from the Hirshfeld partition of the molecular distribution is then given by the condensed one-electron probabilities of the "stockholder" atoms,

$$\boldsymbol{P}^H = \{P_X^H = \int p_X^H(\boldsymbol{r})\, d\boldsymbol{r} = N_X^H/N\}, \qquad \Sigma_X P_X^H = 1, \tag{7.2.2}$$

where $N_X^H = \int \rho_X^H(\boldsymbol{r})\, d\boldsymbol{r}$ is the average number of electrons on X^H, satisfying the closure relation: $\Sigma_X N_X^H = N$.

As we have already demonstrated in the preceding chapters, this exhaustive, local partition has a solid IT basis by corresponding to the minimum of the entropy deficiency between the AIM pieces of the molecular electron density and the corresponding electron densities of free atoms of the isoelectronic promolecule. The electron density $\rho^0(\boldsymbol{r})$ of AP $M^0 = (\ldots|X^0|Y^0|\ldots)$, consisting of the free-atom densities $\boldsymbol{\rho}^0(\boldsymbol{r}) = \{\rho_X^0(\boldsymbol{r})\}$ shifted to the actual positions of atoms in M, $\rho^0(\boldsymbol{r}) = \Sigma_X\, \rho_X^0(\boldsymbol{r}) = N^0 p^0(\boldsymbol{r})$, where $N^0 = \Sigma_X N_X^0 = N$ and the row vector $\boldsymbol{N}^0 = \{N_X^0\}$ groups the numbers of electrons of the constituent free atoms, is also used as the non-bonded reference in the density-difference diagrams.

The associated AP-*probability* data include the atomic *one*-electron probability densities $\boldsymbol{p}^0(\boldsymbol{r}) = \{p_X{}^0(\boldsymbol{r}) = \rho_X{}^0(\boldsymbol{r})/N\}$, giving rise to the promolecule probability density $p^0(\boldsymbol{r}) = \sum_X p_X{}^0(\boldsymbol{r}) = \sum_X \rho_X{}^0(\boldsymbol{r})/N = \rho^0(\boldsymbol{r})/N$. In this atomic resolution the square matrix of the *two*-electron *joint*-distributions in AP [Eq. (5.6.3)], $\mathbf{p}_2{}^0(\boldsymbol{r}, \boldsymbol{r}') = \{p_{XY}{}^0(\boldsymbol{r}, \boldsymbol{r}') = \rho_{XY}{}^0(\boldsymbol{r}, \boldsymbol{r}')/[N(N-1)]\}$, similarly determines the overall promolecular distribution:

$$p_2{}^0(\boldsymbol{r}, \boldsymbol{r}') = \sum_X\sum_Y p_{XY}{}^0(\boldsymbol{r}, \boldsymbol{r}') = \sum_X\sum_Y \rho_{XY}{}^0(\boldsymbol{r})/[N(N-1)] = \rho_2{}^0(\boldsymbol{r}, \boldsymbol{r}')/[N(N-1)], \quad (7.2.3a)$$

where [see Eq. (5.6.3)]:

$$p_{XY}{}^0(\boldsymbol{r}, \boldsymbol{r}') = P_{X,X}{}^0\, \pi_{2,X}{}^0(\boldsymbol{r}, \boldsymbol{r}')\delta_{X,Y} + P_{X,Y}{}^0\pi_X{}^0(\boldsymbol{r})\pi_Y{}^0(\boldsymbol{r}')(1 - \delta_{X,Y}), \quad (7.2.3b)$$

since the electron distributions on different free-atoms are mutually independent. The condensed two-electron probabilities $\mathbf{P}^0 = \{P_{X,Y}{}^0\}$ satisfy the usual normalization condition $\sum_X \sum_Y P_{X,Y}{}^0 = \sum_X P_X{}^0 = \sum_Y P_Y{}^0$ while the probability densities of the separate free atom X^0 are unity-normalized: $\iint \pi_{2,X}{}^0(\boldsymbol{r}, \boldsymbol{r}')\, d\boldsymbol{r}\, d\boldsymbol{r}' = \int \pi_X{}^0(\boldsymbol{r})\, d\boldsymbol{r} = 1$. This gives rise to the AP overall normalization relation in atomic resolution:

$$\iint p_2{}^0(\boldsymbol{r}, \boldsymbol{r}')\, d\boldsymbol{r}\, d\boldsymbol{r}' = \sum_X\sum_Y \iint p_{XY}{}^0(\boldsymbol{r}, \boldsymbol{r}')\, d\boldsymbol{r}\, d\boldsymbol{r}' = \sum_X\sum_Y P_{X,Y}{}^0 = 1. \quad (7.2.3c)$$

The probability distributions $\boldsymbol{p}^0(\boldsymbol{r})$ of the free-atoms $\{X^0\}$ in AP are accordingly normalized to the condensed probabilities of free-atoms in M^0 (a row vector):

$$\boldsymbol{P}^0 = \{P_X{}^0 = \int p_X{}^0(\boldsymbol{r})\, d\boldsymbol{r} = \sum_Y \iint p_{XY}{}^0(\boldsymbol{r}, \boldsymbol{r}')\, d\boldsymbol{r}\, d\boldsymbol{r}' = \sum_Y P_{X,Y}{}^0 = N_X{}^0/N\}, \;\; \sum_X P_X{}^0 = 1. \quad (7.2.4)$$

The probabilities $\mathbf{P}^0$ and $\boldsymbol{P}^0$ provide the discrete (AIM) representation of the two- and one-electron distributions in the AP-reference.

The continuous description of the two-electron localization events in molecules calls for the AIM division of the system two-electron probability distribution $p_2(\boldsymbol{r}, \boldsymbol{r}')$ or the associated *two*-electron (pair-) density $\rho_2(\boldsymbol{r}, \boldsymbol{r}') = N(N-1)\, p_2(\boldsymbol{r}, \boldsymbol{r}')$. Their S-partition in atomic resolution (Nalewajski, 2002a; Nalewajski and Broniatowska, 2005b) has been described in Sections 5.6 and 5.7:

$$p_2(\boldsymbol{r}, \boldsymbol{r}') = \sum_X \sum_Y p_{X,Y}{}^S(\boldsymbol{r}, \boldsymbol{r}') \quad \text{or} \quad \rho_2(\boldsymbol{r}, \boldsymbol{r}') = \sum_X \sum_Y \rho_{X,Y}{}^S(\boldsymbol{r}, \boldsymbol{r}') = N(N-1)\, p_2(\boldsymbol{r}, \boldsymbol{r}'). \quad (7.2.5)$$

It defines the square matrices of the condensed two-electron probabilities:

$$\mathbf{P}^S = \{P_{X,Y}{}^S = \iint p_{X,Y}{}^S(\boldsymbol{r}, \boldsymbol{r}')\, d\boldsymbol{r}\, d\boldsymbol{r}'\}, \quad \sum_X\sum_Y P_{X,Y}{}^S = 1, \quad (7.2.6a)$$

and the associated matrix of the electronic pair-populations in the AIM resolution,

$$\mathbf{\Gamma}^S = \{\Gamma_{X,Y}{}^S = \iint \rho_{X,Y}{}^S(\boldsymbol{r}, \boldsymbol{r}')\, d\boldsymbol{r}\, d\boldsymbol{r}'\}, \qquad \sum_X\sum_Y \Gamma_{X,Y}{}^S = N(N-1). \quad (7.2.6b)$$

The promolecular two-electron distributions in $\mathbf{p}_2^0(\boldsymbol{r}, \boldsymbol{r}')$ [Eqs. (7.2.3a-c)] similarly give rise to the condensed probabilities and electron-pair populations in the AP-reference. Integration of the probability kernels gives:

$$\mathbf{P}^0 = \{P_{X,Y}^0 = \iint p_{XY}^0(\boldsymbol{r}, \boldsymbol{r}')\, d\boldsymbol{r}\, d\boldsymbol{r}'\}, \quad \Sigma_X \Sigma_Y P_{X,Y}^0 = \Sigma_X P_X^0 = \Sigma_Y P_Y^0 = 1, \qquad (7.2.7a)$$

where:

$$P_{X,X}^0 = N_X^0(N_X^0 - 1)/[N(N-1)] \quad \text{and} \quad P_{X,Y}^0 = N_X^0 N_Y^0/[N(N-1)],\ X \neq Y. \quad (7.2.7b)$$

The integration of the associated *two*-electron densities in AP, $\boldsymbol{\rho}_2^0(\boldsymbol{r}, \boldsymbol{r}') = \{\rho_{X,Y}^S(\boldsymbol{r}, \boldsymbol{r}') = N(N-1)\, p_{X,Y}^S(\boldsymbol{r}, \boldsymbol{r}')\}$, generates the cotrresponding pair-population matrix:

$$\boldsymbol{\Gamma}^0 = \{\Gamma_{X,Y}^0 = \iint \rho_{X,Y}^0(\boldsymbol{r}, \boldsymbol{r}')\, d\boldsymbol{r}\, d\boldsymbol{r}'\}, \quad \Sigma_X \Sigma_Y \Gamma_{X,Y}^0 = N(N-1), \qquad (7.2.8)$$

where $\Gamma_{X,X}^0 = N_X^0(N_X^0 - 1)$ and $\Gamma_{X,Y}^0 = N_X^0 N_Y^0$, for $X \neq Y$.

It should be emphasized that atoms in the molecule or promolecule are fragments of a larger system. This is reflected by the normalization of subsystem probability distributions to the corresponding condensed probabilities of atomic fragments in the whole molecular system. Indeed, the set of the mutually exclusive one-electron events in the atomic resolution is determined by all possible values of the atomic labels $X \in \boldsymbol{X} = (A, B, \ldots)$ and all positions $\boldsymbol{r}$ in space. The collections of the atomic (subsystem) probability densities $\{p_X(\boldsymbol{r})\}$ and $\{p_X^0(\boldsymbol{r})\}$ then determine the overall (system) distributions in atomic resolution: molecular, $\boldsymbol{p}(\boldsymbol{X}, \boldsymbol{r}) = \{p_X(\boldsymbol{r})\}$, and promolecular, $\boldsymbol{p}^0(\boldsymbol{X}, \boldsymbol{r}) = \{p_X^0(\boldsymbol{r})\}$, which depend on both the discrete and continuous event-arguments. Only the complete collections of these distributions should be unity-normalized, since their normalization cvondition involves summation and integration over the *complete* set of event-arguments, combining the discrete and continuous degrees-of-freedom of the *one*-electron probabilities in atomic resolution:

$$\Sigma_X \int p(X, \boldsymbol{r})\, d\boldsymbol{r} = \Sigma_X \int p^0(X, \boldsymbol{r})\, d\boldsymbol{r} = 1. \qquad (7.2.9)$$

These atomic distributions satisfy the following partial normalizations:

$$\Sigma_X p(X, \boldsymbol{r}) = p(\boldsymbol{r}),\ \int p(X, \boldsymbol{r})\, d\boldsymbol{r} = P_X,\ \Sigma_X p^0(X, \boldsymbol{r}) = p^0(\boldsymbol{r}),\ \int p^0(X, \boldsymbol{r})\, d\boldsymbol{r} = P_X^0. \quad (7.2.10)$$

Similarly, the complete set of the *two*-electron events in atomic discretization involves all possible atomic allocations, $(\boldsymbol{X}, \boldsymbol{Y}) = \{X, Y\}$, and all positions $(\boldsymbol{r}, \boldsymbol{r}')$ of the two electrons. Therefore, only the system two-electron probability distributions in atomic resolutions, generated by the complete collections of the corresponding subsystem distributions, molecular $\{p_{XY}(\boldsymbol{r}, \boldsymbol{r}') \equiv p_2(X, \boldsymbol{r}; Y, \boldsymbol{r}')\} = \mathbf{p}_2(\boldsymbol{r}, \boldsymbol{r}')$ and promolecular $\{p_{XY}^0(\boldsymbol{r}, \boldsymbol{r}') \equiv p_2^0(X, \boldsymbol{r}; Y, \boldsymbol{r}')\} = \mathbf{p}_2^0(\boldsymbol{r}, \boldsymbol{r}')$, should be unity-normalized:

$$\Sigma_X \Sigma_Y \iint p_2(X, \boldsymbol{r}; Y, \boldsymbol{r}')\, d\boldsymbol{r}\, d\boldsymbol{r}' = \Sigma_X \Sigma_Y \iint p_2^0(X, \boldsymbol{r}; Y, \boldsymbol{r}')\, d\boldsymbol{r}\, d\boldsymbol{r}' = 1. \qquad (7.2.11)$$

The relevant partial normalizations then read [see Eqs. (7.2.3c) and (7.2.7a)]:

$$\sum_X \sum_Y p_2(X, \boldsymbol{r}; Y, \boldsymbol{r}') = p_2(\boldsymbol{r}, \boldsymbol{r}'), \qquad \iint p_2(X, \boldsymbol{r}; Y, \boldsymbol{r}')\, d\boldsymbol{r}\, d\boldsymbol{r}' = P_{X,Y};$$
$$\sum_X \sum_Y p_2^0(X, \boldsymbol{r}; Y, \boldsymbol{r}') = p_2^0(\boldsymbol{r}, \boldsymbol{r}'), \quad \iint p_2^0(X, \boldsymbol{r}; Y, \boldsymbol{r}')\, d\boldsymbol{r}\, d\boldsymbol{r}' = P_{X,Y}^0. \qquad (7.2.12)$$

7.2.2. Molecular Information Channels

In what follows we shall assume that the specific AIM discretization of the molecular electron densities and the associated probability distributions has already been carried out. The complete set of the *one*-electron (spatially condensed), mutually exclusive events in such a description of the electron distributions in the *molecular input* **A**, $\boldsymbol{a} = \{a_i\}$ (or *output* **B**, $\boldsymbol{b} = \{b_j\}$) (see Scheme 7.1 and Section 3.7), contains all events of locating an electron on atoms $i \in \boldsymbol{X}$ (or $j \in \boldsymbol{Y}$) with probabilities $\boldsymbol{P}(\mathbf{A}) \equiv \{P(a_i) = P_i\} = \boldsymbol{P}(\mathbf{B}) \equiv \{P(b_j) = P_j\} = \boldsymbol{P}$. The spatially-condensed *two*-electron events, of simultaneously finding a pair of electrons on specific AIM $i \in \boldsymbol{X}$ and $j \in \boldsymbol{Y}$, occur with the joint probabilities $\mathbf{P}(\mathbf{AB}) = \{P(a_i b_j) = P_{i,j}\} = \mathbf{P}$. Clearly, the set of *one*-electron events, identified by labels of the system constituent atoms, is the same in both the molecular (or promolecular) "input" and the molecular "output". The electron interaction influences the probability distribution of the *two*-electron events, so that in general $\mathbf{P} \neq \mathbf{P}^{ind} \equiv \boldsymbol{P} \otimes \boldsymbol{P} = \{P_{i,j}^{ind} = P_i P_j\}$, where $\mathbf{P}^{ind}$ stands for the *two*-electron probability distribution of independent AIM. The electron delocalization throughout the molecular system, via a network of chemical bonds, introduces an effect of the molecular communication "noise", a characteristic feature of ordinary communication systems.

One can then regard a given molecular system as the AIM-resolved communication system shown in Scheme 7.1, in which the signals of electron allocations to constituent atoms are being transmitted from the molecular (or promolecular) input (source) **A** (or $\mathbf{A}^0$) to the *molecular* output (receiver) **B** (Nalewajski, 2000c, 2004c-e, 2005a-c). We call such a unit AIM-allocation signal the *message*. When a signal conveying a message is received in the output, it is known that one of possible input signals has been sent. Te molecular communication channel is characterized by disturbances of a random character, which perturb the transmitted signal. This noise in ascribing electrons to atoms originates from the quantum-mechanical uncertainty in the distribution of electrons in the molecule: it is not known with certainty, on which atom an electron will be found in the molecular output, when its atomic origin in the input is known. This is a result of forming the chemical bonds, which provide channels for the electron delocalization throughout the whole system. In the molecule $M = (A\!\mid\! B\!\mid\! C\!\mid\! \ldots)$ the constituent AIM are mutually bonded (open), being free to delocalize their electrons throughout the molecular environment, as symbolized by the vertical *broken* lines between the AIM labels.

In the AIM resolution the elements of the molecular conditional probability matrix of the molecular outputs given the molecular inputs,

Input (Molecular): **A** $\qquad$ **P(B|A)** $\qquad$ *Output* (Molecular): **B**

$$P_i \longrightarrow a_i \longrightarrow P(b_j|a_i) \equiv P(j|i) \longrightarrow b_j \longrightarrow P_j$$

I) $\boldsymbol{P}(\mathbf{A}) \longrightarrow \mathbf{P}(\mathbf{B}|\mathbf{A}) \longrightarrow \boldsymbol{P}(\mathbf{B})$

Input (Promolecular): $\mathbf{A}^0$ $\qquad$ **P(B|A)** $\qquad$ *Output* (≈Molecular): $\tilde{\mathbf{B}}$

$$P_i^0 \longrightarrow a_i \longrightarrow P(j|i) \longrightarrow b_j \longrightarrow \tilde{P}_j = \sum_i P_i^0 P(j|i) \approx P_j$$

II) $\boldsymbol{P}(\mathbf{A}^0) \longrightarrow \mathbf{P}(\mathbf{A}|\mathbf{B}) \longrightarrow \boldsymbol{P}(\tilde{\mathbf{B}}) \approx \boldsymbol{P}$

Input (Promolecular): $\mathbf{A}^0$ $\qquad$ $\mathbf{P}^0(\mathbf{B}^0|\mathbf{A}^0)$ $\qquad$ *Output* (Promolecular): $\mathbf{B}^0$

$$P_i^0 \longrightarrow a_i \longrightarrow 1 \longrightarrow b_i \longrightarrow P_i^0$$

$$P_j^0 \longrightarrow a_j \longrightarrow 1 \longrightarrow b_j \longrightarrow P_j^0$$

III) $\boldsymbol{P}(\mathbf{A}^0) \longrightarrow \mathbf{I} \longrightarrow \boldsymbol{P}(\mathbf{B}^0)$

Input (SAL): $\mathbf{A}^s$ $\qquad$ $\mathbf{P}^s(\mathbf{B}^s|\mathbf{A}^s)$ $\qquad$ *Output* (SAL): $\mathbf{B}^s$

$$1 \longrightarrow a_i \longrightarrow 1 \longrightarrow b_i \longrightarrow 1$$

$$1 \longrightarrow a_j \longrightarrow 1 \longrightarrow b_j \longrightarrow 1$$

IV) $\boldsymbol{1}(\mathbf{A}^s) \longrightarrow \mathbf{I} \longrightarrow \boldsymbol{1}(\mathbf{B}^s)$

Scheme 7.1. Probability distributions of the molecular $M = (\ldots\,\vert\, i \,\vert\, j \,\vert\ldots)$ and promolecular $M^0 = (\ldots|i^0|j^0|\ldots)$ communication systems in atomic resolution. In the *stationary* molecular channel (Part I) the AIM *input* events $\boldsymbol{a} = \{a_i\}$ occur with the condensed one-electron probabilities $\boldsymbol{P}(\mathbf{A}) \equiv \boldsymbol{P}(\boldsymbol{a}) = \{P(a_i) = P_i\} = \boldsymbol{P}$ and the molecular *output* events $\boldsymbol{b} = \{b_j\}$ occur with probabilities $\boldsymbol{P}(\mathbf{B}) \equiv \boldsymbol{P}(\boldsymbol{b}) = \{P(b_j) = P_j\} = \boldsymbol{P}$. These two probability distributions are transformed one into the other using the conditional probabilities $\mathbf{P}(\mathbf{B}|\mathbf{A}) = \{P(b_j|a_i) \equiv P(j|i)\}$: $\boldsymbol{P}(\mathbf{B}) = \boldsymbol{P}(\mathbf{A})\,\mathbf{P}(\mathbf{B}|\mathbf{A})$. In the *non-stationary* molecular channel of Part II the promolecular input probabilities $\boldsymbol{P}(\mathbf{A}^0) \equiv \boldsymbol{P}^0(\boldsymbol{a}) = \{P^0(a_i) = P_i^0\} = \boldsymbol{P}^0$ are transformed into output probabilities $\boldsymbol{P}(\tilde{\mathbf{B}}) = \boldsymbol{P}(\mathbf{A}^0)\,\mathbf{P}(\mathbf{B}|\mathbf{A}) \approx \boldsymbol{P}$. The molecular channels of Parts I and II are collections of the mutually *open* AIM, $M = (A\vert B\vert C\vert\ldots)$, in which electrons are free to delocalize into (communicate with) the remaining AIM via the system chemical bonds. Parts III and IV, corresponding to the *non*-bonded (free) atoms in the channel input and output, represent collections of the *disconnected* atomic channels (deterministic, noiseless). They describe the mutually *closed* free atoms of the promolecule $M^0 = (A^0\,|\,B^0\,|\,C^0\,|\ldots)$ (Part III) or a set of the isolated atoms in the *Separated Atoms Limit* (SAL) (Part IV), $M^s = A^0 + B^0 + C^0 + \ldots$, which do not allow for the electron delocalization into other atoms, due to the assumed (promolecule) or real (SAL) absence of chemical bonds linking the atomic subsystems. The conditional probability matrix in the disconnected channels III and IV is thus given by the identity matrix: $\mathbf{P}^0(\mathbf{B}^0|\mathbf{A}^0) = \mathbf{P}^s(\mathbf{B}^s|\mathbf{A}^s) = \{\delta_{i,j}\} \equiv \mathbf{I}$.

$$\mathbf{P}(\mathbf{B}\,|\,\mathbf{A}) = \{P(j\,|\,i) = P(a_i b_j)/P(a_i) \equiv P_{i,j}/P_i\}, \qquad \sum_j P(j\,|\,i) = 1, \tag{7.2.13}$$

determine the communication networks of Schemes 7.1.I and 7.1.II. Here, $P(j\,|\,i)$ denotes the condensed conditional probability of finding in a molecule an electron on atom j, when we know for sure that another electron has already been located on ith atom. When atomic fragments do not form the chemical bonds with the remaining atoms of their molecular environment, e.g., in the promolecular reference state (Scheme 7.1.III) or in the *Separated* (dissociated) *Atoms Limit* (SAL) (Scheme 7.1.IV), the atom allocation signal reaching the free (isolated) atom X^0 of the promolecular (or SAL) input $\mathbf{A}^0$ (or $\mathbf{A}^s$) is transmitted with certainty to the same free atom X^0 in the promolecular (or SAL) output $\mathbf{B}^0$ (or $\mathbf{B}^s$). The conditional probability matrix of such a collection of the non-bonded (separate) atomic channels is thus defined by the identity matrix:

$$\mathbf{P}^0(\mathbf{A}^0\,|\,\mathbf{B}^0) = \mathbf{P}^0(\mathbf{B}^0\,|\,\mathbf{A}^0) = \mathbf{P}^s(\mathbf{A}^s\,|\,\mathbf{B}^s) = \mathbf{P}^s(\mathbf{B}^s\,|\,\mathbf{A}^s) = \{\delta_{i,j}\} = \mathbf{I}. \tag{7.2.14}$$

The promolecular network in atomic resolution is thus represented by a collection of the separate (disconnected) channels for each free atom, with the input and output probabilities equal to $\boldsymbol{P}(\mathbf{A}^0) = \boldsymbol{P}^0$. The corresponding SAL network is determined by the same identity matrix, as in the promolecular case, with the unit AIM probabilities $\boldsymbol{P}^s = \{P_i^s = 1\} = \boldsymbol{1}$ in the input and output of the channel. This is because each infinitely separated atom constitutes a different system itself, rather than being a part of a collection of disconnected atoms, which form the AP system.

Therefore, each atomic sub-channel of the promolecular channel of Scheme 7.1.III represents the noiseless, deterministic channel, exhibiting a zero loss of the information content, when the signal is transmitted from the input $\boldsymbol{X}^0$ to the output $\boldsymbol{X}^0$. This *identity* channel represents a collection of the mutually non-bonded (closed) atomic *parts* of the promolecular communication system $M^0 = (A^0|B^0|C^0|\ldots)$, where the solid vertical lines separating the constituent free atoms signify that the inter-atomic delocalization of electrons, i.e., forming the chemical bonds between atoms, is not allowed.

This delocalization is also non-existent in the SAL $M^s = A^0 + B^0 + C^0 + \ldots$, with each isolated atom defining a *different* communication channel. In other words each (separated) free-atom in the SAL itself constitutes a separate communication system, while the same free atom in the promolecule constitutes a disconnected part of a larger communication system. This basic difference is reflected by the input and output atomic probabilities in Schemes 7.1.III and 7.1.IV.

As we shall argue later in this chapter, the inputs of the molecular channel determine the *origins* of the chemical bond, while the outputs similarly reflect the bonds being "counted". Using the molecular AIM input and output generates the *stationary* molecular channel, the entropic descriptors of which reflect the molecular "equilibrium". In such channels one makes no reference to the "initial" stage of the bond-formation process, viz., the promolecular prototype, thus focusing on the purely molecular aspects of chemical bonds, i.e., their overall covalency.

In order to characterize the "distance" aspect of the chemical bonds in the molecule, measuring the information "displacement" between the final (molecular) and initial (promolecular) electronic structures, one requires the *non*-bonded input $\boldsymbol{P}^0$ of Scheme 7.1.III and the molecular conditional probabilities of Eq. (7.2.13). This channel corresponds to the modified output probability distribution,

$$\boldsymbol{P}(\tilde{\mathbf{B}}) = \boldsymbol{P}(\mathbf{A}^0)\,\mathbf{P}(\mathbf{B}\,|\,\mathbf{A}) \equiv \tilde{\boldsymbol{P}} \cong \boldsymbol{P}, \tag{7.2.15}$$

which should strongly resemble the molecular condensed probabilities since $\boldsymbol{P}(\mathbf{A}^0) \cong \boldsymbol{P}(\mathbf{A}) = \boldsymbol{P}(\mathbf{B}) = \boldsymbol{P}$. In the independent AIM approximation, $\mathbf{P} = \{P_{i,j}\} \cong \mathbf{P}^{ind} = \{P_i Pj\}$, which can be considered as the zeroth-order approximation of the correlated electron pair-distribution in a molecule, one indeed obtains:

$$\tilde{P}_j = \sum_i P_i^0 P(j|i) = \sum_i P_i^0 P_{i,j} / P_i \cong (\sum_i P_i^0)\, P_j = P_j\,. \tag{7.2.16}$$

It should be emphasized, that by modifying the input signal in Schemes 7.2.I and 7.2.II the molecular network, defined by the conditional probabilities $\tilde{\mathbf{P}}(\tilde{\mathbf{B}}|\mathbf{A}^0) = \{\tilde{P}(j|i)\}$ remains unchanged:

$$P(j|i) = P_{i,j}/P_i = \tilde{P}_{i,j} / P_i^0 \equiv \tilde{P}(j|i)\,, \tag{7.2.17}$$

where the modified *two*-electron probabilities in AIM resolution, $\tilde{\mathbf{P}} \equiv \{\tilde{P}_{i,j}\} \cong \mathbf{P}$,

$$\tilde{P}_{i,j} = P_{i,j}(P_i^0 / P_i) \cong P_{i,j}\,, \tag{7.2.18}$$

assures the required partial normalizations to the corresponding output and input probabilities:

$$\sum_i \tilde{P}_{i,j} = \tilde{P}_j \cong P_j, \quad \sum_j \tilde{P}_{i,j} = P_i^0\,. \tag{7.2.19}$$

As indicated in Eq. (7.2.18) the whole *i*th row of $\mathbf{P}$, for $j = A, B, \ldots$, is scaled using the same scale factor $\kappa_i = P_i^0/P_i \cong 1$ to give the *i*th row of $\tilde{\mathbf{P}}$.

These atomic events and the associated molecular probabilities define the corresponding input and output *one*-electron probability *schemes* grouping events and their probabilities:

$$\mathbf{A} = \{\boldsymbol{a}, \boldsymbol{P}(\boldsymbol{a})\},\ \mathbf{B} = \{\boldsymbol{b}, \boldsymbol{P}(\boldsymbol{b})\},\ \tilde{\mathbf{B}} = \{\boldsymbol{b}, \tilde{\boldsymbol{P}}(\boldsymbol{b})\} \cong \mathbf{B}, \mathbf{A}^0 = \{\boldsymbol{a}, \boldsymbol{P}^0(\boldsymbol{a})\},\ \mathbf{B}^0 = \{\boldsymbol{b}, \boldsymbol{P}^0(\boldsymbol{b})\}, \tag{7.2.20}$$

In what follows we shall provide examples of the molecular information channels for simple orbital models of the chemical bond in two-electron diatomic system, which result from its representative wave-functions. For example, in the Heitler-London

(1927) theory of the hydrogen molecule, the VB structures and their combinations can be interpreted as describing specific information channels.

7.2.3. Example: Communication Channels of the VB-Structures in H_2

Consider now the classical VB description of the covalent bond in H_2, formed by the two (ortho-normal) 1*s* orbitals, $\boldsymbol{\chi}(\boldsymbol{r}) = [a(\boldsymbol{r}), b(\boldsymbol{r})]$, contributed by the two hydrogen atoms, $A = H^{(1)}$ and $B = H^{(2)}$, respectively. These two AO define the minimum basis set for this system. The products of these two *one*-electron basis functions describing the two electrons in H_2, $\mathbf{S}(1, 2) \equiv \{S_{m,n}(1, 2) \equiv \chi_m(1)\chi_n(2);\ m, n \in (a, b)\}$, then define the 2×2 matrix of the elementary VB-structures, which form the *two*-electron basis set for approximating the molecular wave-function $\Psi(1, 2)$. The four elementary VB-structures of $\mathbf{S}(1, 2)$ define the complete set of the elementary two-electron "events" in the minimum basis set description of the hydrogen molecule since in each elementary VB-structure the two electrons are uniquely ascribed to specific atoms.

In this familiar VB-description one identifies the *diagonal* (*one*-centre) products $\{S_{m,m}(1, 2)\}$, when the two electrons are located on the same atom m, as the *VB-ionic* structures. They indeed correspond to the ion-pair electron configurations of the molecule, marking the CT of a single electron between the two atoms. The remaining, *off-diagonal* (*two*-centre) products $\{S_{m\neq n}(1, 2)\}$, when two electrons are equally "shared" between the bond partners, for the vanishing internal CT between the two hydrogens, are conventionally classified as *VB-covalent*. This approach emphasizes the crucial role played by the two-electron VB basis functions derived from AO, in distinguishing the bond components. In what follows we shall compare this description with the communication theory perspective (Nalewajski, 2005f).

The *singlet* (*S*) ground-state of H_2 is described by the two spin-paitred electrons occupying the bonding MO,

$$\varphi_b(\boldsymbol{r}) = [a(\boldsymbol{r}) + b(\boldsymbol{r})]/\sqrt{2}\,. \qquad (7.2.21)$$

This electron configuration defines the symmetric spatial part

$$\begin{aligned}\Phi_S(\boldsymbol{r}_1, \boldsymbol{r}_2) \equiv \Phi_S(1, 2) &= \varphi_b(\boldsymbol{r}_1)\varphi_b(\boldsymbol{r}_2) \equiv \varphi_b(1)\varphi_b(2)\\ &= \tfrac{1}{2}\{[a(1)\,a(2) + b(1)\,b(2)] + [a(1)\,b(2) + b(1)\,a(2)]\}\\ &\equiv \tfrac{1}{2}\{[S_{a,a}(1, 2) + S_{b,b}(1, 2)] + [S_{a,b}(1, 2) + S_{b,a}(1, 2)]\}\\ &\equiv 1/\sqrt{2}\,[\Phi_{ion}(1, 2) + \Phi_{cov}(1, 2)], \qquad (7.2.22a)\end{aligned}$$

of the anti-symmetric wave-function

$$\Psi_S(1, 2) = \Phi_S(1, 2)\,\Xi_S(\sigma_1, \sigma_2) \equiv \Phi_S(1, 2)\,\Xi_S(1, 2). \qquad (7.2.23)$$

Here the anti-symmetric spin-function for the singlet state of two-electrons

$$\Xi_S(1,2) = [\alpha(1)\beta(2) - \beta(1)\alpha(2)]/\sqrt{2} \qquad (7.2.22b)$$

and the spin functions $\alpha(i)$ and $\beta(i)$ represent the spin-up and spin-down states of ith electron, respectively. Therefore, the ground-state of H_2 represents the equi-weight combination of the VB-covalent $[\Phi_{cov}(1, 2)]$ and VB-ionic $[\Phi_{ion}(1, 2)]$ two-electron basis functions, which themselves represent the equi-probability mixtures of the two off-diagonal and diagonal structures, respectively.

The purely *covalent* bond in H_2 is thus marked by equal contributions of the VB-ionic and VB-covalent structures. Hence, the *chemical* bond-ionicity in the VB-description is manifested by a deviation from this "perfect" balance exhibited by the H_2 prototype of the covalent bond. It also follows from the preceding equation that all four elementary two-electron "events" in AO resolution are equally probable. Indeed, using the familiar *superposition principle* of quantum mechanics generates the following *two*-electron probabilities: $\mathbf{P} = \{P_{m,n} = ¼\}$, which in turn give rise to equal *one*-electron probabilities on atoms $\boldsymbol{P} = (½, ½) = \boldsymbol{P}^0$. Hence, the associated matrix of the molecular ground-state *conditional* probabilities also involves equal elements: $\mathbf{P}(\mathbf{B}|\mathbf{A}) = \{P(n|m) = P_{m,n}/P_m = ½\}$. These probabilities define the *stationary* molecular communication channel shown in Scheme 7.2.

Input **A**: $\boldsymbol{P}^0 = \boldsymbol{P}$ $\quad$ $\mathbf{P}_S(\mathbf{B}|\mathbf{A})$ $\quad$ Output **B**: $\boldsymbol{P}$ $\quad$ $\mathbf{P}_S = \begin{bmatrix} 0.25 & 0.25 \\ 0.25 & 0.25 \end{bmatrix}$

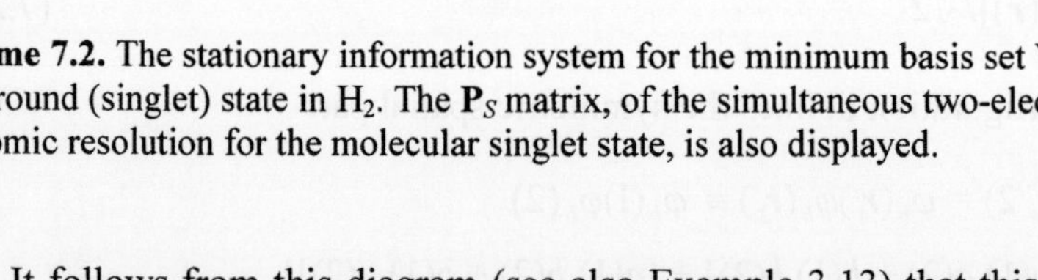

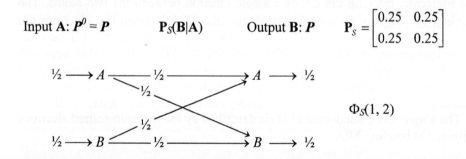

Scheme 7.2. The stationary information system for the minimum basis set VB-description of the ground (singlet) state in H_2. The $\mathbf{P}_S$ matrix, of the simultaneous two-electron probabilities in atomic resolution for the molecular singlet state, is also displayed.

It follows from this diagram (see also Example 3.13) that this system exhibits the zero amount of the information flowing through the communication network, as reflected by the vanishing mutual-information descriptor $I(\mathbf{A}:\mathbf{B}) = 0$. Indeed, in this 2-AO VB-model the input and output one-electron probabilities are independent. This is reflected by the simultaneous *two*-electron probabilities, given by the Cartesian product of *one*-electron probabilities. This channel corresponds to the maximum amount of the average communication noise in the molecular communication system, measured by the conditional entropy of the molecular output given the molecular input: $S(\mathbf{B}|\mathbf{A}) = 1$ bit.

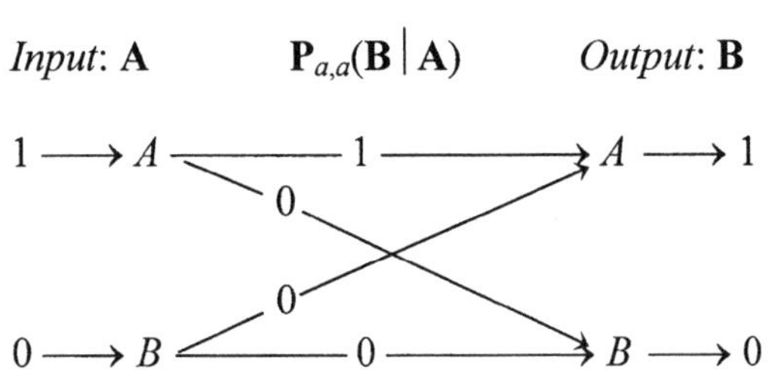

$$\mathbf{P}_{a,a} = \begin{bmatrix} 1 & 0 \\ 0 & 0 \end{bmatrix}$$

$$S_{a,a}(1, 2) = a(1)\, a(2)$$

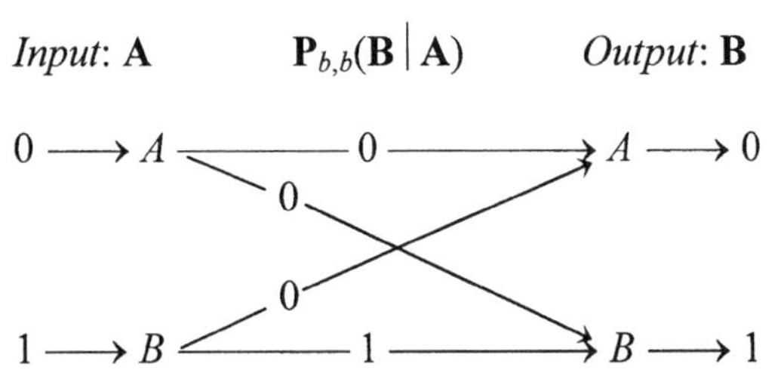

$$\mathbf{P}_{b,b} = \begin{bmatrix} 0 & 0 \\ 0 & 1 \end{bmatrix}$$

$$S_{b,b}(1, 2) = b(1)\, b(2)$$

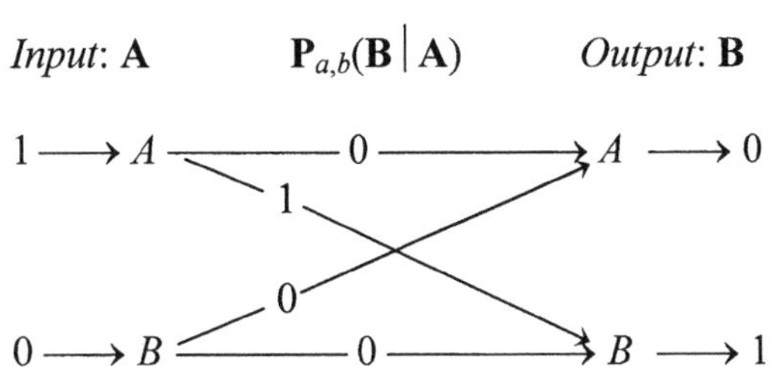

$$\mathbf{P}_{a,b} = \begin{bmatrix} 0 & 1 \\ 0 & 0 \end{bmatrix}$$

$$S_{a,b}(1, 2) = a(1)\, b(2)$$

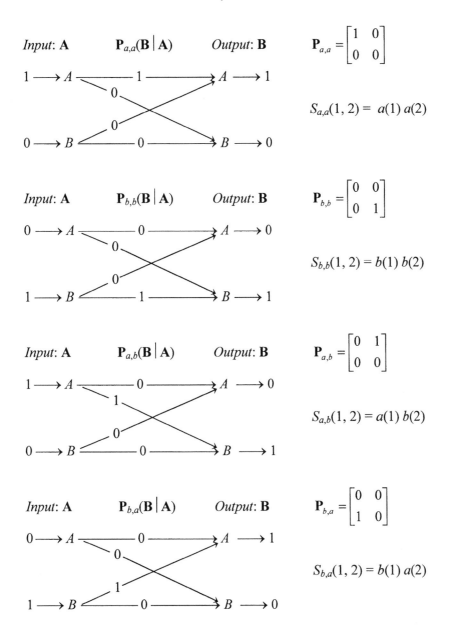

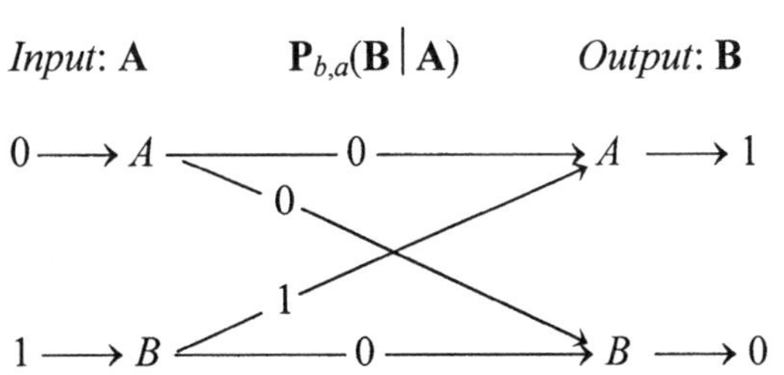

$$\mathbf{P}_{b,a} = \begin{bmatrix} 0 & 0 \\ 1 & 0 \end{bmatrix}$$

$$S_{b,a}(1, 2) = b(1)\, a(2)$$

Scheme 7.3. The noiseless (deterministic) information channels corresponding to four elementary VB-structures of the minimum basis set description of the chemical bond in H_2. For all these communication networks $S(\mathbf{B}|\mathbf{A}) = I(\mathbf{A}:\mathbf{B}) = 0$. Therefore, each separate VB-structure represents the perfectly *IT–non-bonded* (*IT-n*) information system, in which there is no communication noise and a loss of information in the output compared to the information content in the channel input. The corresponding two-electron joint-probability matrices $\{\mathbf{P}_{m,n}\}$ are also displayed.

These properties of this prototype purely-covalent H_2 system suggest that one should associate its single covalent bond with the average noise in the molecular channel. This is indeed consistent with the electron-sharing (delocalization) intuition, which is conventionally associated with bond-covalency. This demonstration also implies that the zero ionicity of this single bond should be linked to the vanishing mutual-information descriptor of the molecular communication system, which measures in IT a degree of the dependence between the channel input and output probability distributions (see Fig. 3.1).

Clearly, this somewhat surprising association between the bond information-ionicity and a degree of the independence of the one-electron distributions in the molecule requires some additional insight and comment. Let us first examine the communication systems representing the four separate VB-structures of the minimum basis set description of the chemical bond in H_2, which are reported in Scheme 7.3. One observes that each structure $S_{m,n}(1, 2)$ uniquely ascribes the two electrons to the specified pair (m, n) of AO (atoms), as indeed reflected by the unit conditional probability linking this AO-pair and the vanishing probabilities of the remaining communication connections in the channel diagram. Therefore, each separate structure implies the noiseless, deterministic information system (see Section 3.7), for which $S(\mathbf{B}|\mathbf{A}) = I(\mathbf{A}:\mathbf{B}) = 0$, since its input and output probabilities consist are permutations of (1, 0). In other words, each separate VB-structure represents the perfectly *IT–non-bonding* (*IT-n*) communication channel in the information-theoretic sense of the word.

The *purely covalent* channel of Scheme 7.2 is thus obtained as a result of equal "mixing" of these separately non-bonded elementary VB-channels into the "equi-ensemble" implied by the ground-state wave-function of Eq. (7.2.22a), with equal "weights" of each member of the ensemble of VB-structures: $\mathbf{P}_S = \{P_{m,n} = ¼\}$. Therefore, the 1 bit of the entropy covalency of the H–H bond in this minimum-basis-set perspective originates from equal weights of these four elementary, separately nonbonding VB-channels.

Consider next the partial information channels corresponding to the $\Phi_{ion}(1, 2)$ and $\Phi_{cov}(1, 2)$ components of the singlet wave-function $\Phi_S(1, 2)$ [Eq. (7.2.22a)]. They are reported in Scheme 7.4. A reference to these diagrams shows that they both exhibit the vanishing conditional-entropy index with 1 bit of the mutual-information descriptor, marking the purely IT-ionic partial channels. Indeed, the noiseless/ deterministic character of both these communication systems implies that each of them gives rise to the one-to-one correspondence between the input and output probability distributions, so that there is no loss of information in the channel output compared to its input. In both the VB-covalent and VB-ionic channels of Scheme 7.4 there is a maximum (complete) dependence between these two molecular probability schemes, as indeed reflected by the 1 bit of the mutual-information index. Therefore, within the IT approach both these two partial molecular configurations must be classified as being purely *IT-ionic* in character. The covalent bond in H_2 thus arises as a result of the equal-weight mixing of both these separately purely IT-ionic wave-function components of $\Phi_S(1, 2)$.

$$\Phi_{ion}(1, 2) = [S_{a,a}(1, 2) + S_{b,b}(1, 2)]/\sqrt{2} : \mathbf{P}_{ion} = \begin{bmatrix} 0.5 & 0 \\ 0 & 0.5 \end{bmatrix}$$

Input: **A** $P_{ion}(\mathbf{B}|\mathbf{A})$ *Output*: **B**

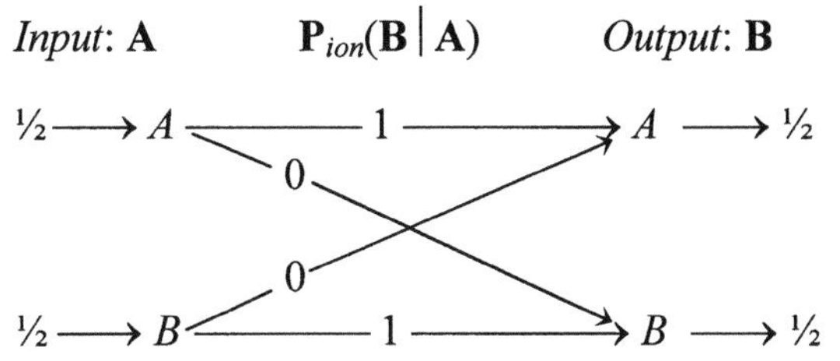

$$\Phi_{cov}(1, 2) = [S_{a,b}(1, 2) + S_{b,a}(1, 2)]/\sqrt{2} : \mathbf{P}_{cov} = \begin{bmatrix} 0 & 0.5 \\ 0.5 & 0 \end{bmatrix}$$

Input: **A** $P_{cov}(\mathbf{B}|\mathbf{A})$ *Output*: **B**

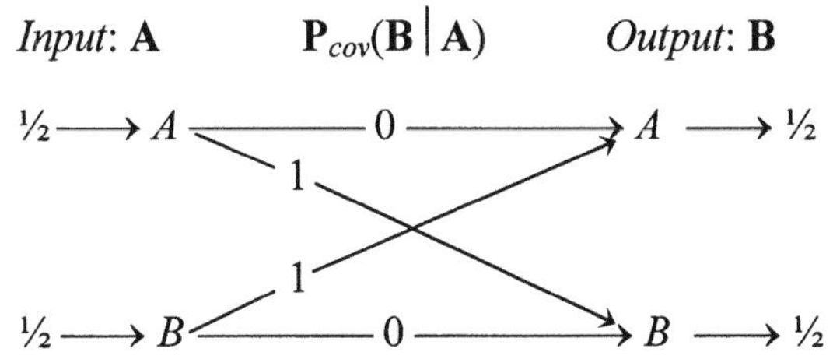

Scheme 7.4. The partial information channels corresponding to the $\Phi_{ion}(1, 2)$ and $\Phi_{cov}(1, 2)$ components of $\Phi_S(1, 2)$ [Eq. (7.2.22a)]. For both these communication systems $S(\mathbf{B}|\mathbf{A}) = 0$ and $I(\mathbf{A}:\mathbf{B}) = 1$ bit. Therefore, in the information-theoretic approach both the *VB-ionic* and *VB-covalent* components describe the "single" *IT-ionic* bond, with exactly vanishing entropy covalency. The relevant two-electron probability matrices in atomic resolution, $\mathbf{P}_{ion}$ and $\mathbf{P}_{cov}$, are also displayed.

The intriguing question then arises: what combinations of the elementary VB-structures do constitute the purely *IT-covalent* partial channels? To answer this question one could examine the two remaining ways of combining the elementary VB-structures of the singlet wave-function of Eq. (7.2.22a):

$$\Phi_S(1, 2) = 1/\sqrt{2}\ \{1/\sqrt{2}\ [S_{a,a}(1, 2) + S_{a,b}(1, 2)] + 1/\sqrt{2}\ [S_{b,b}(1, 2) + S_{b,a}(1, 2)]\}$$

$$\equiv 1/\sqrt{2}\ [\Phi_A^{AB}(1, 2) + \Phi_B^{AB}(1, 2)]$$

$$= 1/\sqrt{2}\ \{1/\sqrt{2}\ [S_{a,a}(1, 2) + S_{b,a}(1, 2)] + 1/\sqrt{2}\ [S_{b,b}(1, 2) + S_{a,b}(1, 2)]\}$$

$$\equiv 1/\sqrt{2}\ [\Phi_{AB}^A(1, 2) + \Phi_{AB}^B(1, 2)]. \tag{7.2.24}$$

The communication systems associated with these two additional sets of partial VB-components of the molecular wave-function $\Phi_S(1, 2)$ and the relevant two-electron probability matrices are displayed in Schemes 7.5 and 7.6, respectively.

$$\Phi_A^{AB}(1, 2) = [S_{a,a}(1, 2) + S_{a,b}(1, 2)]/\sqrt{2} : \mathbf{P}_A^{AB} = \begin{bmatrix} 0.5 & 0.5 \\ 0 & 0 \end{bmatrix}$$

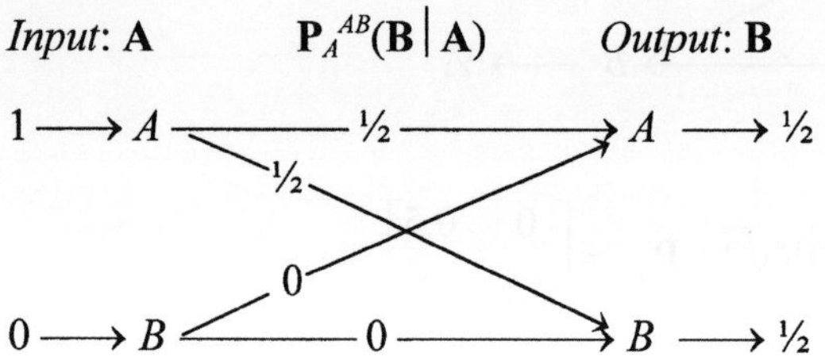

$$\Phi_B^{AB}(1, 2) = [S_{b,b}(1, 2) + S_{b,a}(1, 2)]/\sqrt{2} : \mathbf{P}_B^{AB} = \begin{bmatrix} 0 & 0 \\ 0.5 & 0.5 \end{bmatrix}$$

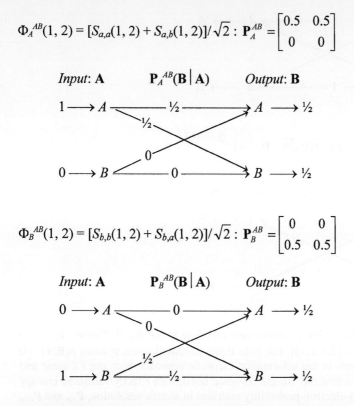

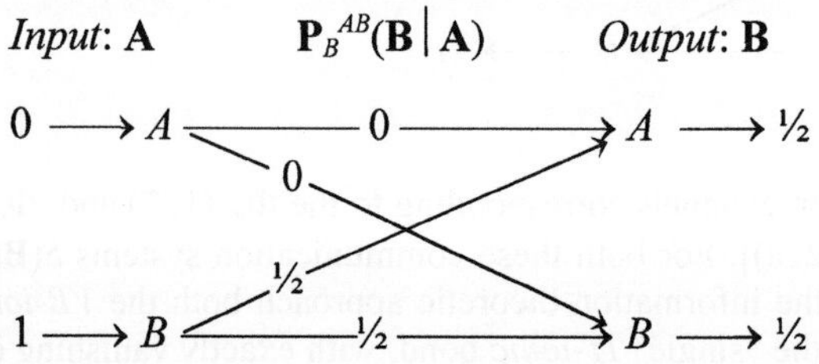

Scheme 7.5. The partial information channels corresponding to the $\Phi_X^{AB}(1, 2)$ components of the ground-state wave-function $\Phi_S(1, 2)$, $X = A, B$ [Eq. (7.2.24)]. For both these communication systems $S(\mathbf{B}|\mathbf{A}) = 1$ bit and $I(\mathbf{A}:\mathbf{B}) = 0$. Therefore, in the IT-approach each of these components gives rise to the "single", purely-covalent communication channel, with the exactly vanishing information ionicity. They can be thus classified as *IT-covalent*. The relevant two-electron probability matrices in atomic resolution $\{\mathbf{P}_X^{AB}\}$ are also reported.

It follows from Scheme 7.5 that the partial channels corresponding to the $\Phi_A^{AB}(1, 2)$ and $\Phi_B^{AB}(1, 2)$ wave-function components represent the maximum, single-bond entropy-covalency and vanishing information-ionicity. Therefore, they constitute the separately *IT-covalent* functions of the minimum basis set description of the chemical bond in H_2.

$$\Phi_{AB}^{A}(1,2) = [S_{a,a}(1,2) + S_{b,a}(1,2)]/\sqrt{2}: \ \mathbf{P}_{AB}^{A} = \begin{bmatrix} 0.5 & 0 \\ 0.5 & 0 \end{bmatrix}$$

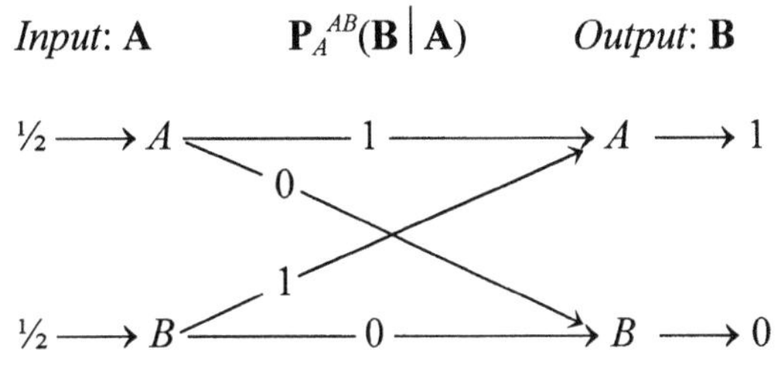

$$\Phi_{AB}^{B}(1,2) = [S_{b,b}(1,2) + S_{a,b}(1,2)]/\sqrt{2}: \ \mathbf{P}_{AB}^{B} = \begin{bmatrix} 0 & 0.5 \\ 0 & 0.5 \end{bmatrix}$$

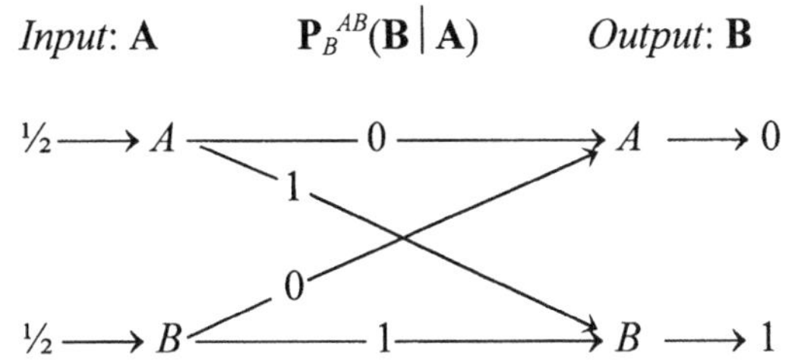

Scheme 7.6. The partial information channels corresponding to the $\Phi_{AB}^{X}(1, 2)$ components of $\Phi_S(1, 2)$, $X = A, B$ [Eq. (7.2.24)]. For both these communication systems $S(\mathbf{B}|\mathbf{A}) = I(\mathbf{A}:\mathbf{B}) = 0$. Therefore, in the information-theoretic interpretation each of these components generates the *IT-n* channel. The corresponding two-electron probability matrices in atomic resolution $\{\mathbf{P}_{AB}^{X}\}$ are also reported.

Finally, a reference to Scheme 7.6 shows that the partial communication channels associated with the $\Phi_{AB}^{A}(1, 2)$ and $\Phi_{AB}^{B}(1, 2)$ combinations of elementary VB-structures are also *IT-n* in character. Indeed, there is no uncertainty left about the input probabilities in these communication systems once the output probabilities have been determined, and vice versa, the given input probabilities uniquely determine the channel output probabilities.

This IT analysis of the VB-description of the chemical bond in H_2 allows one to draw some general conclusions about the essence of the bond information-ionicity and its entropy-covalency. Let us first examine the information-theoretic conditions for the complete absence of the chemical bond in the communication systems describing this model homonuclear diatomic. The truly *IT-n* channels of Schemes 7.3 and 7.6 are noiseless and deterministic in character and they all exhibit the vanishing output entropy. Notice that the input probability equalization (Scheme 7.6) or its atomic localization (Scheme 7.3) both give rise to the non-bonded information channel. This indicates that a degree of the probability delocalization in the output of the molecular channel determines the chemical bonds being counted (see also Nalewajski, 2005a). Obviously, the non-zero probabilities limited to a single atom then directly imply the absence of chemical bonds between the twom AIM. Indeed, one can easily verify that this prediction remains valid for any input probability distribution $\boldsymbol{P}(\mathbf{A}) = (x, 1-x)$, $0 \leq x \leq 1$, in the *IT-n* channels of Schemes 7.3 and 7.6.

Next, let us explore the IT-ionic and IT-covalent molecular channels of Schemes 7.4 and 7.5, respectively. It is somewhat surprising to discover that both the VB-ionic and VB-covalent structures of the Heitler-London description have the same information-theoretic classification as being purely-ionic. Indeed, the wave-functions $\Phi_{ion}(1, 2)$ and $\Phi_{cov}(1, 2)$ are not dichotomous in separating the mutually exclusive ionic/covalent effects of the chemical bond, since only their equal mixture in the ground-state wave-function for H_2 implies the purely-covalent bond. The present analysis establishes the *dichotomous* pure IT-covalent combinations of the elementary VB-structures, $\Phi_A^{AB}(1, 2)$ and $\Phi_B^{AB}(1, 2)$, which are free from a partial IT-ionic "contamination" from the VB-components $\Phi_{ion}(1, 2)$ and $\Phi_{cov}(1, 2)$.

The information-ionicity measured by the mutual information quantity reflects a degree of a dependence between the input and output probabilities of the molecular communication system. It reaches the maximum value of 1 bit in the deterministic, noiseless channels of Scheme 7.4, in which the input and output probabilities, extending on both AIM, are in one-to-one correspondence. For the given atomic input a strong ionic character in the channel conditional probabilities is manifested by the strongly localized output probabilities, say on a single atom, since this implies a strong correlation between the molecular channel inputs and outputs. Accordingly, a high degree of equalization of the conditional probabilities in the output for the specified input, i.e., their delocalization among many AIM outputs, is a clear symptom of a strong entropy-covalency (communication noise) in the chemical bond. Indeed, both the IT-ionic channels of Scheme 7.4 represent the perfect correlation (localization) in distributing the given input probability among the output atoms, while in the IT-covalent channels shown in Schemes 7.2 and 7.5 the active input-atom probability is equally divided between the two atomic outputs.

It should be observed that the same $S(\mathbf{B}|\mathbf{A}) = 1$ bit measure of the bond covalency follows from both the overall molecular channel of Scheme 7.2 and separately from the two partial channels of Scheme 7.5. They all exhibit the exact equalization of the output probabilities, an indication that the H-H bond is being "counted". However, in the partial communication systems only a single input atom exhibits a non-zero (unit) probability, thus being *active* in the bond-formation process. The covalency indices of the two *partial* channels of Scheme 7.5 can be thus interpreted as reflecting the active-atom *valence* in the molecule, i.e., their propensities to form the covalent bonds with their molecular partner, counting the *active* valence-electrons involved in forming the chemical bond.

The equalized input probabilities in the H_2 information channel, manifesting the two active atoms, correspond to the *electron-sharing* bond origin, when both atoms contribute a single electron to form the *covalent*-bond in the hydrogen molecule. Similarly, the localized (unit probability) input on a single active atom signifies the *pair-sharing* input, when the electron lone-pair of the active atom is used to form the *coordination*-bond. It can be straightforwardly verified that using the coordination input distributions $\boldsymbol{P}(\mathbf{A}) = (1, 0)$ or $\boldsymbol{P}(\mathbf{A}) = (0, 1)$ in the molecular channel of Scheme 7.2 gives identical predictions of a single, purely covalent coordination bonds $A \rightarrow B$ or $B \rightarrow A$, respectively.

To summarize these VB information-systems for the hydrogen molecule, we conclude that the channel input reflects the atomic origins and a "history" of forming the chemical bonds between the atomic units of the molecular communication system. It provides the probability reference, against which the molecular distribution is evaluated. In what follows we shall further explore the specificity of the entropy/information description of the chemical bonds in model orbital and spin channels of the polarized diatomic (two-electron) systems.

7.3. GROUND-STATE INDICES OF CHEMICAL BONDS

As we have already observed above, different input signals have to be used to probe different aspects of the chemical bond. On one hand, the chemical bond *covalency*, which reflects the *molecular* degree of the electron sharing between the bonded atoms without any reference to the bond atomic origins, requires the molecular input signal $\boldsymbol{P}(\mathbf{A}) = \boldsymbol{P}$. On the other hand, both the bond-*ionicity* and the *covalent* component of the *coordination* (donor-acceptor) bond, depend on the initial, promolecular distribution of electrons $\boldsymbol{P}(\mathbf{A}^0) = \boldsymbol{P}^0$, since they reflect the *displacement* aspect of the electronic structure in the molecule, relative to the free-atom reference. Therefore, a use of the molecular input should facilitate an "extraction" of the purely molecular, covalent bond descriptors, while the promolecular input signal is required for a determination of the entropy/information quantities characterizing the information-"distance" descriptors of the *relative* (ionic and coordination-covalent) bond components, which invoke the initial distributions of electrons in free atoms.

In atomic resolution the average amount of uncertainty about the occurrence of the molecular (output) events, given that the molecular input events are known to have occurred, measures a degree of the "communication" noise due to the electron delocalization via a network of chemical bonds. The molecular quantity, which reflects this uncertainty, is the *conditional entropy* $S(\mathbf{B}|\mathbf{A}) = S(\boldsymbol{P}|\boldsymbol{P})$, providing the information-theoretic measure of the overall bond *covalency* in the system under consideration. Indeed, the latter is intuitively associated by chemists with a degree of the *electron sharing* between AIM. The electron delocalization among all constituent AIM implies a lowering of $S(\mathbf{B}|\mathbf{A})$ relative to $S(\mathbf{B})$, since the knowledge of the occurrence of the input events beforehand cannot raise the uncertainty about the occurrence of the output events.

The information measure of the bond-*ionicity*, representing the *displacement* aspect of the chemical bond, relative to the promolecular input probabilities $\boldsymbol{P}(\mathbf{A}^0) = \boldsymbol{P}^0$, complementary to bond-*covalency*, should be related to the *mutual-information* between the molecular and promolecular condensed probability distributions of electrons in atomic resolution: $I(\mathbf{A}^0{:}\mathbf{B}) = I(\boldsymbol{P}^0{:}\boldsymbol{P})$.

It follows from Fig. 3.1 that an increase in the conditional entropy $S(\mathbf{B}|\mathbf{A})$ marks an accompanying decrease in the complementary *molecular* mutual information quantity $I(\mathbf{A}{:}\mathbf{B})$, which measures the "overlap" between the entropies of the dependent input and output one-electron distributions in the molecule. Therefore, for the known molecular output probabilities $\boldsymbol{P}(\mathbf{B})$, an increase in $S(\mathbf{B}|\mathbf{A})$ implies an associated decrease in the molecular mutual information $I(\mathbf{A}{:}\mathbf{B}) = S(\mathbf{B}) - S(\mathbf{B}|\mathbf{A}) \geq 0$, measuring the amount of information flowing through the stationary molecular communication channel in atomic resolution.

One should indeed expect a *competition* between these two complementary aspects of the chemical bond, since a strongly covalent character of the chemical bond, i.e., a high degree of a common possession of electrons by the bonded atoms, implies a small amount of charge-transfer between the bond partners, and hence a weakly ionic bond. This competition effect is also reflected by the two-electron (difference) indices of the chemical bond-multiplicity (Nalewajski *et al.*, 1993; Nalewajski and Mrozek, 1994, 1996; Nalewajski *et al*., 1994, 1996, 1997; Mrozek *et al.*, 1998; Nalewajski, 2004b).

As shown in Scheme 7.1, the molecular conditional probabilities $\mathbf{P}(\mathbf{B}|\mathbf{A})$ transform the AIM input probabilities $\boldsymbol{P}(\mathbf{A}) = \{\sum_j P_{i,j} \equiv P_i\} = \boldsymbol{P}$ into the molecular AIM output distribution $\boldsymbol{P}(\mathbf{B}) = \{\sum_i P_{i,j} \equiv P_j\} = \boldsymbol{P}$: $\boldsymbol{P}(\mathbf{B}) = \boldsymbol{P}(\mathbf{A})\,\mathbf{P}(\mathbf{B}|\mathbf{A})$. The other conditional probability matrix $\mathbf{P}(\mathbf{A}|\mathbf{B}) = \{P(i|j) = P_{i,j}/P_j\}$ similarly transforms the molecular output probabilities into the molecular input probabilities: $\boldsymbol{P}(\mathbf{A}) = \boldsymbol{P}(\mathbf{B})\,\mathbf{P}(\mathbf{A}|\mathbf{B})$. Since the sets of the constituent AIM in the input and output of the molecular communication channel are identical, the "stationary" molecular channel exhibits the same input and output probability schemes. Such one-electron probabilities also characterize the *separable* molecular channel, defined the independent-atom *two*-electron probability distribution $\mathbf{P}^{ind} = \boldsymbol{P}\otimes\boldsymbol{P} = \{P_{i,j}{}^{ind} = P_iP_j\}$ and gives rise to *equal* input and output one-electron Shannon entropies: $S(\mathbf{A}) = S(\mathbf{B}) = -\sum_j P_j \log P_j = S(\boldsymbol{P})$.

It should be realized, however, that the input probabilities, which specify how the communication channel is used, can be freely manipulated. By changing the input signal of the molecular communication channel one can probe different aspects of the chemical bond. For example, when diagnosing the entropy/information displacements accompanying a formation of the chemical bond the input probabilities $\boldsymbol{P}(\mathbf{A}) = \boldsymbol{P}^0$ of the atomic promolecule M^0, consisting of the free (non-bonded) constituent atoms shifted to their positions in molecule M, are traditionally used as the reference for separating the effects due to bond formation.

The Shannon entropy of the molecular joint (*input*, *output*)–events in atomic resolution is determined by the two-electron probabilities $\mathbf{P}$, $S(\mathbf{AB}) = -\sum_i \sum_j P_{i,j} \log P_{i,j}$

$= S(\mathbf{P})$. Hence, the conditional entropy (see Section 3.7) of the molecular output, given the molecular input in Scheme 7.1/I reads:

$$S(\mathbf{B}\,|\,\mathbf{A}) = -\Sigma_i \Sigma_j P_{i,j} \log(P_{i,j}/P_i) = -\Sigma_i P_i \{\Sigma_j P(j|i) \log P(j|i)\} = S(\mathbf{P}) - S(\boldsymbol{P}). \qquad (7.3.1)$$

Clearly, the same result is obtained for the conditional entropy of the molecular input, given the molecular output:

$$S(\mathbf{A}\,|\,\mathbf{B}) = -\Sigma_i \Sigma_j P_{i,j} \log(P_{i,j}/P_j) = -\Sigma_j P_j \{\Sigma_i P(i|j) \log P(i|j)\} = S(\mathbf{B}\,|\,\mathbf{A}). \qquad (7.3.2)$$

It should be observed that the conditional entropy for the mutually non-bonded, free atoms of the promolecule and the separated atoms in the SAL, for which the two-electron probability matrices $\mathbf{P}^0 = \{P_{i,j}^{\ 0} = P_i^{\ 0}\delta_{i,j}\}$ and $\mathbf{P}^s = \{P_{i,j}^{\ s} = \delta_{i,j}\}$, respectively, must identically vanish, since $\mathbf{P}(\mathbf{B}^0\,|\,\mathbf{A}^0) = \mathbf{P}(\mathbf{B}^s\,|\,\mathbf{A}^s) = \mathbf{I}$:

$$S(\mathbf{B}^0\,|\,\mathbf{A}^0) = -\Sigma_i P_i^{\ 0} \log 1 = S(\mathbf{B}^s\,|\,\mathbf{A}^s) = -\Sigma_i 1 \log 1 = 0. \qquad (7.3.3)$$

This correctly identifies the zero entropy-covalency level for these two communication channels shown in Schemes 7.1/III and 7.1/IV, respectively.

The *mutual information* quantity (see Section 3.7), measuring the amount of information flowing through the molecular channel of Scheme 7.1/II, which exhibits the promolecular AIM probabilities $\boldsymbol{P}(\mathbf{A}^0) = \boldsymbol{P}^0$ in the input and the molecular probabilities $\boldsymbol{P}(\mathbf{B}) = \boldsymbol{P}$ in the output of the communication network, can be similarly expressed in terms of the relevant Shannon entropies, conditional entropies, and the entropy deficiency of Kullback and Leibler between the molecular (correlated) and the independent two-electron probabilities. For example, for the stationary molecular channel, when $\boldsymbol{P}^0 = \boldsymbol{P}$, one obtains:

$$\begin{aligned} I(\mathbf{A}{:}\mathbf{B}) &= \Sigma_i \Sigma_j P_{i,j} \log\,[P_{i,j}/(P_i P_j)] = \Delta S(\mathbf{P}\,|\,\mathbf{P}^{ind}) \\ &= S(\mathbf{B}) - S(\mathbf{B}\,|\,\mathbf{A}) = S(\mathbf{A}) - S(\mathbf{A}\,|\,\mathbf{B}) = S(\mathbf{A}) + S(\mathbf{B}) - S(\mathbf{AB}). \end{aligned} \qquad (7.3.4a)$$

In a general case of $\boldsymbol{P}^0 \neq \boldsymbol{P}$ one finds:

$$\begin{aligned} I(\mathbf{A}^0{:}\mathbf{B}) &= \Sigma_i \Sigma_j P_{i,j} \log\,[P_{i,j}/(P_i^{\ 0} P_j)] = \Sigma_i \Sigma_j P_{i,j} \log\,\{[P_{i,j}/(P_i P_j)](P_i/P_i^{\ 0})\} \\ &= \Delta S(\mathbf{P}\,|\,\mathbf{P}^{ind}) + \Delta S(\boldsymbol{P}\,|\,\boldsymbol{P}^0). \end{aligned} \qquad (7.3.4b)$$

Therefore, the mutual information descriptor of the channel shown in Scheme 7.1/II measures the sum of the two-electron information distance $\Delta S(\mathbf{P}\,|\,\mathbf{P}^{ind})$, between the molecular correlated and independent (non-interacting) electrons, and the entropy deficiency $\Delta S(\boldsymbol{P}|\boldsymbol{P}^0)$ between the one-electron distributions in the molecule and promolecule. This amount of information flowing through a communication system depends on channel's conditional probability network (property of the channel itself) and probabilities of the input signals, which describe the way the channel is exploited and provide the reference for the output signals. By changing the input probabilities one can tune the flow of information from the minimum level up to the *channel capacity* (see Section 3.7).

A detection of signals in the channel output in general gives a non-vanishing information about the channel input and *vice versa*. The non-negative character of the *molecular* mutual information quantities of Eqs. (7.3.4a) and (7.3.4b) directly follows from the non-negativity of the two cross-entropies in these expressions.

The deterministic and noiseless promolecular channel of Scheme 7.1/III, corresponding to $\boldsymbol{P}(\mathbf{A}^0) = \boldsymbol{P}(\mathbf{B}^0) = \boldsymbol{P}^0$ and $\mathbf{P}(\mathbf{A}^0\mathbf{B}^0) = \{P_i^0\delta_{i,j}\}$, gives rise to the mutual information determined by the equal entropies of the system input and output: $I(\mathbf{A}^0{:}\mathbf{B}^0) = S(\mathbf{B}^0) = S(\mathbf{A}^0) = S(\boldsymbol{P}^0)$ (see also Section 3.7). For the other disconnected channel of Scheme 7.1/IV, which represents the free atoms in the SAL, $\boldsymbol{P}(\mathbf{A}^s) = \boldsymbol{P}(\mathbf{B}^s) = \mathit{1}$ and $\mathbf{P}(\mathbf{A}^s\mathbf{B}^s) = \mathbf{I}$. Hence, in this dissociation limit one predicts the vanishing mutual information: $I(\mathbf{A}^s{:}\mathbf{B}^s) = 0$. These two results underline the basic difference between the promolecular- and SAL-references and their communication channels, both describing the non-bonded (non-communicating, disconnected, free) atoms.

The competition between the two complementary bond components of the stationary molecular channel is also seen in the deterministic and noiseless promolecular channel of Scheme 7.1/III, representing the limiting case of the vanishing covalency and the maximum ionicity $I(\mathbf{A}^0{:}\mathbf{B}^0) = S(\mathbf{B}^0) = S(\mathbf{A}^0) = S(\boldsymbol{P}^0)$.

Another displacement descriptor of the chemical bonding in the donor-acceptor interactions is represented by the *coordination*-covalency, which results from sharing by the two bond-partners the electron pair donated by the *basic* atom. This CT-covalency should be adequately measured by the conditional entropy $S(\mathbf{B}|\mathbf{A}^0) = S(\boldsymbol{P}|\boldsymbol{P}^0)$, of the *molecular* output probabilities given the *promolecular* input probabilities (Nalewajski, 2004c-e):

$$S(\mathbf{B}|\mathbf{A}^0) = -\Sigma_i \Sigma_j P_{i,j} \log(P_{i,j}/P_i^0) = -\Sigma_i \Sigma_j P_{i,j} \log(P_{i,j}/P_i) - \Sigma_i P_i \log(P_i/P_i^0)$$
$$= S(\mathbf{B}|\mathbf{A}) - \Delta S(\boldsymbol{P}|\boldsymbol{P}^0). \qquad (7.3.5)$$

A comparison between Eqs. (7.3.1) and (7.3.5) reveals that this modified entropy-covalency of the coordination bond involves a correction to the ordinary molecular covalent component $S(\mathbf{B}|\mathbf{A})$ given by the entropy deficiency term $\Delta S(\boldsymbol{P}|\boldsymbol{P}^0)$, representing the information distance between the molecular and promolecular one-electron probabilities in atomic resolution.

To summarize, in the AIM resolved molecular communication system the conditional entropy and mutual information quantities have been identified as promissing information-theoretic descriptors of the complementary covalent and ionic components of the chemical bond, with the promolecular probabilities providing the reference for the entropy/information descriptors of the bond displacement aspect, reflecting the bond atomic origins. We shall thus adopt the two-input approach to the molecular ground-state information channel, which uses the molecular input, to determine the bond entropy-covalency, and the promolecular input, to evaluate the bond information-ionicity and coordination-covalency.

It should be recalled that thwe formation of chemical bonds affects mainly the valence-shell electrons of constituent atoms, with the inner, "core" electron distributions of the free atoms (or ions) remaining practically unchanged. Therefore, in qualitative and model bond-order considerations one usually explicitly considers the valence electrons only.

7.4. VARIATIONAL PRINCIPLES

As we have already remarked before, the input-signal probabilities $\boldsymbol{P}(\mathbf{A}) \equiv \boldsymbol{Q}$ can be appropriately modified, to probe various aspects of the molecular communication channel. Therefore, it is of interest to determine the normalized (trial) input probabilities, for which the amount of information flowing through the communication system exhibiting the output probabilities,

$$\boldsymbol{P}(\mathbf{B}) = \boldsymbol{P}(\boldsymbol{Q}) = \{P_j(\boldsymbol{Q}) = \sum_i Q_i P(j|i)\}, \tag{7.4.1}$$

measured by the channel IT-ionicity function,

$$I(\boldsymbol{Q}: \boldsymbol{P}(\boldsymbol{Q})) = \sum_i \sum_j Q_i P(j|i) \log[P(j|i)/P_j(\boldsymbol{Q})], \tag{7.4.2}$$

reaches the extremum value. We have assumed in the above expression that the molecular conditional probability matrix $\mathbf{P}(\mathbf{B}|\mathbf{A})$, which defines the molecular channel in atomic resolution and determines the transformation $\boldsymbol{P}(\mathbf{B}) = \boldsymbol{P}(\mathbf{A})\, \mathbf{P}(\mathbf{B}|\mathbf{A})$ (see Scheme 7.1/I), remains fixed in this constrained search.

It can be demonstrated (Nalewajski, 2004c), by solving the appropriate variational principle in the information-entropy representation, that the minimum of this mutual information functional is reached for the *molecular* input probability scheme, which marks the *stationary* channel,

$$\boldsymbol{Q}^{opt} = \boldsymbol{P} = \{P_j = \sum_i P_i P(j|i) = \sum_i P_{i,j}\}, \tag{7.4.3}$$

$$\min_{\boldsymbol{Q}} I(\boldsymbol{Q} : \boldsymbol{P}(\boldsymbol{Q})) = I(\boldsymbol{P} : \boldsymbol{P}). \tag{7.4.4}$$

The stationary molecular channel also minimizes to zero the entropy deficiency $\Delta S(\boldsymbol{P}|\boldsymbol{Q})$ of Eq. (7.3.4b):

$$\Delta S(\boldsymbol{P}(\boldsymbol{Q}^{opt})|\boldsymbol{Q}^{opt}) = \sum_j P_j(\boldsymbol{Q}^{opt}) \log[P_j(\boldsymbol{Q}^{opt})/Q_j^{opt}] = \sum_j P_j \log(P_j/P_j) = 0. \tag{7.4.5}$$

A reference to this equation then gives a direct confirmation of the above principle of the *minimum information-ionicity* in the stationary molecular communication system.

Indeed, since both $\Delta S(\mathbf{P}\,|\,\mathbf{P}^{ind}) \geq 0$ and $\Delta S(\boldsymbol{P}(\boldsymbol{Q})\,|\,\boldsymbol{Q}) \geq 0$ reach the minimum values for the stationary solution $\boldsymbol{Q} = \boldsymbol{P}$, the latter also minimizes their sum $I(\boldsymbol{Q}{:}\boldsymbol{P}(\boldsymbol{Q}))$.

This result indicates that the stationary molecular communication system, with the molecular probabilities in the input and output of the information channel, corresponds to the *minimum* of the overall information ionicity in the electron distribution. As we have already observed [see Eq. (7.3.4a)], this minimum value of the mutual information measures the information distance (missing information) between the two-electron distribution **P** in the molecular (*interacting*) system, with respect to $\mathbf{P}^s$ of the hypothetical Kohn-Sham (1965) (*non-interacting*) system (see Appendix C), thus also providing the information index of the electron correlation in the AIM-resolved molecular system.

A reference to Fig. 3.1 implies that the minimum of the mutual information (common area of two circles) corresponds to the maximum conditional entropy of the output probabilities with respect to the probability distribution of the channel input, represented by complementary part of the molecular-output entropy circle in this diagram. This complementarity implies that the stationary communication system for $\boldsymbol{Q} = \boldsymbol{P}$ corresponds to the maximum value of the overall entropy-covalency in the distribution:

$$\max_{\boldsymbol{Q}} S(\boldsymbol{P}(\boldsymbol{Q})\,|\,\boldsymbol{Q}) = S(\boldsymbol{P}\,|\,\boldsymbol{P}). \tag{7.4.6}$$

Therefore, the promolecule→molecule transition of the system electronic structure represents a transformation from the *maximum ionicity* (*minimum covalency*) promolecular channel of Scheme 7.1/III to the *maximum covalency* (*minimum ionicity*) molecular (stationary) channel of Scheme 7.1/I. Similarly, the SAL→molecule transition similarly signifies a displacement from the *zero*-bond communication system of Scheme 7.1/IV to the molecular, maximum-covalency composition of the system chemical bonds characterizing the stationary channel of Scheme 7.1/I.

7.5. TWO-ORBITAL MODEL OF THE CHEMICAL BOND

In the *Molecular Orbital* (MO) approximation the (normalized) *two*-electron wave-function $\Psi(\mathbf{q}) \equiv \Psi(\boldsymbol{q}_1, \boldsymbol{q}_2) \equiv \Psi(1, 2)$, where the electron arguments $\boldsymbol{q} = (\boldsymbol{r}, \sigma)$ group the electron position ($\boldsymbol{r}$) and spin ($\sigma = \pm\, ½$) coordinates, is given by the product of the *two*-electron *spatial* and *spin* wave-functions, $\Phi(\boldsymbol{r}_1, \boldsymbol{r}_2) \equiv \Phi(1, 2)$ and $\Xi(\sigma_1, \sigma_2) \equiv \Xi(1, 2)$, respectively [see, e.g., Eq. (7.2.23)]: $\Psi(1, 2) = \Phi(1, 2)\, \Xi(1, 2)$. It thus gives rise to the *independent* spatial and spin *two*-electron probability densities: $p_2(\boldsymbol{r}_1, \boldsymbol{r}_2) = |\Phi(1, 2)|^2 \equiv p_2(1, 2)$ and $\pi_2(\sigma_1, \sigma_2) = |\Xi(1, 2)|^2 \equiv \pi_2(1, 2)$, respectively. Their product defines the overall *spatial-spin* probability distribution of two electrons: $\Pi_2(1, 2) = |\Psi(1, 2)|^2 = p_2(1, 2)\pi_2(1, 2)$. They satisfy the relevant normalization requirements:

$$\iint p_2(\boldsymbol{r}_1, \boldsymbol{r}_2)\, d\boldsymbol{r}_1\, d\boldsymbol{r}_2 = \int p(\boldsymbol{r}_1)\, d\boldsymbol{r}_1 = 1 \text{ and } \sum_{\sigma_1}\sum_{\sigma_2} \pi_2(\sigma_1, \sigma_2) = \sum_{\sigma_1} \pi(\sigma_1) = 1, \qquad (7.5.1)$$

with $\pi(\sigma)$ denoting the *one*-electron spin probability distribution. In atomic resolution this factorization of the overall *two*-electron probability matrix reads:

$$\mathbf{\Pi}_2(\boldsymbol{X}, \boldsymbol{X}'; \boldsymbol{\sigma}, \boldsymbol{\sigma}') \equiv \{P_{X,X'}\, \pi_2(\sigma, \sigma')\} = \mathbf{P}(\boldsymbol{X}, \boldsymbol{X}') \otimes \boldsymbol{\pi}_2(\boldsymbol{\sigma}, \boldsymbol{\sigma}') \equiv \mathbf{\Pi}_2, \qquad (7.5.2)$$

where $\mathbf{P}(\boldsymbol{X}, \boldsymbol{X}') = \{P_{X,X'}\} \equiv \mathbf{P}$ combines the condensed spatial probabilities of the joint *two*-electron events in the AIM resolution, with $P_{X,X'}$ standing for the probability of simultaneously finding two electrons on atoms X and X', $\boldsymbol{\pi}_2(\boldsymbol{\sigma}, \boldsymbol{\sigma}') = \{\pi_2(\sigma, \sigma')\} \equiv \boldsymbol{\pi}_2$ groups the corresponding probabilities $\pi_2(\sigma, \sigma')$ of the simultaneous spin orientations of two-electrons, $\sigma_1 = \sigma$ and $\sigma_2 = \sigma'$. The relevant normalization condition of these discrete spatial distribution reads: $\sum_X \sum_Y P_{X,Y} = \sum_X P_X = 1$. The *one*-electron spatial-spin probability vector is defined by the partial summation of the matrix elements of the joint *two*-electron probability distribution:

$$\boldsymbol{\Pi}_1(\boldsymbol{X}; \boldsymbol{\sigma}) \equiv \{\textstyle\sum_{X'}\sum_{\sigma'} P_{X,Y}\, \pi_2(\sigma, \sigma') = P_X\, \pi(\sigma)\} \equiv \boldsymbol{\Pi}_1. \qquad (7.5.3)$$

The MO's are combinations of the (real) normalized and orthogonal *Atomic Orbitals* (AO's) centered on the nuclei. In this section we consider the simplest model of the *A–B* bond, in which the MO are obtained from combining the two orthonormal AO, $\chi_A(\boldsymbol{r}) \equiv a(\boldsymbol{r})$ and $\chi_B(\boldsymbol{r}) \equiv b(\boldsymbol{r})$, originating from atoms A and B, respectively (see also Section 7.2.3). The complete set of mutually exclusive *one*-electron events in the AIM/AO resolution thus includes 4 *orbital-spin* events, of detecting an electron on one of the two AO's (or atoms) in one of its two alternative spin orientations. Accordingly, there are 16 distinct *two*-electron orbital-spin events, of simultaneously detecting two electrons on the specified pairs of atomic orbitals, with each electron in one of its two allowed spin orientations. It should be observed, however, that the Pauli exclusion principle, demanding the antisymmetry of the molecular state $\Psi(1, 2)$ with respect to exchanging the electronic labels, $1 \leftrightarrow 2$, i.e., their spatial-spin coordinates, puts some restrictions on which of these elementary spatial two-electron events can be actually associated with the elementary two-electron spin events into an allowed spatial-spin event of two-electrons, represented by the corresponding antisymmetric wave function of two fermions. Therefore, one actually generates these probability distributions from an admissible wave-function of the two-electron system (see Section 7.2.3 and Appendix E).

In what follows we shall examine the ground-state probabilities of the AIM/AO, spin, and the joint AIM/AO-spin events of this 2-AO model. We shall generate the relevant entropy/information descriptors of the communication channels of a two-orbital diatomic molecule *AB* consisting of two electrons in the bonding, *singlet* state [Eqs. (7.2.2a,b) and (7.2.23)], when two spin-paired electrons occupy the *polarized* bonding MO. The excited configurations of the model, including the single- and double-excitations from the bonding to anti-bonding MO, are discussed in Appendix E.

A general shape of the bonding MO is admitted, covering the whole range of polarizations, from the symmetrical, perfectly delocalized shapes of $\varphi_b(\boldsymbol{r})$ [Eq.(7.2.21) and the corresponding anti-bonding MO,

$$\varphi_a(\boldsymbol{r}) = [a(\boldsymbol{r}) - b(\boldsymbol{r})]/\sqrt{2}\,, \qquad (7.5.4)$$

to the non-bonding shapes of the separate AO. Therefore, in what follows we assume the following general forms of these two MO,

$$\varphi_b(\boldsymbol{r}) = Ca(\boldsymbol{r}) + Db(\boldsymbol{r})\,, \quad \varphi_a(\boldsymbol{r}) = -Da(\boldsymbol{r}) + Cb(\boldsymbol{r})\,,\ C^2 + D^2 \equiv P + Q = 1, \qquad (7.5.5)$$

which automatically guarantee their normalization and mutual orthogonality, $\langle \varphi_i \mid \varphi_j \rangle = \delta_{i,j}$, $(i, j) \in \{a, b\}$. Here the orbital probabilities (P, $Q = 1 - P$) stand for the probabilities of finding an electron occupying the *bonding* MO on atoms A and B, respectively; alternatively, they represent the probabilities of finding an electron occupying the *anti-bonding* MO on atoms B and A, respectively. Thus, a single LCAO coefficient C or D, or a single orbital probability P or Q, uniquely specifies the shapes of both MO. To get a better physical insight into the role of CT in the AB communication channels we shall fix this independent variable to reproduce the specified electron redistribution between the two bonded atoms, which is directly reflected by the probability parameter P.

The main purpose of this analysis is to obtain the communication information/entropy "fingerprints" ("signatures") in the prototype bonding (singlet) ground-state of two electrons of the AB system: in the AO, spin, and AO-spin resolutions, respectively. As we have already demonstrated in Section 7.2.3, the relevant communication networks between the system two atoms directly follow from expressing the *two*-electron spatial wave-functions in terms of AO, $\{a(i), b(i)\}$ and the two-electron spin-wave functions in terms of the *one*-electron spin-states $\{\alpha(i), \beta(i)\}$.

7.5.1. Orbital Channel

Let us first examine the entropies in the bonding singlet-state of the model, when two electrons with opposite spins occupy the bonding MO of Eq. (7.5.5). This electron configuration gives rise to the singlet wave-function consisting of the symmetric orbital part $\Phi_S(1, 2)$ being combined with the *anti*-symmetric (singlet) spin part $\Xi_S(1, 2)$ of Eq. (7.2.22b). This state generates the independent (uncorrelated) spatial distributions of two electrons and their *correlated* spin distribution.

The four elements of the familiar *Charge-and-Bond-Order* (CBO) matrix in the AO representation, $\gamma^S = \{\gamma_{m,n}{}^S\}$, can then be expressed as functions of only one independent parameter controlling the model electronic structure, e.g., the electron population of orbital a, $q_a{}^S = \gamma_{a,a}{}^S \equiv q$: $q_b{}^S(q) \equiv \gamma_{b,b}{}^S(q) = 2 - q$ and $\gamma_{a,b}{}^S(q) = \gamma_{b,a}{}^S(q)$

$= [(2-q)q]^{1/2}$. These matrix elements determine the vector $\boldsymbol{P}^S = (P_A{}^S, P_B{}^S) \equiv (P, Q)$ of the molecular, AO-resolved one-electron probabilities:

$$P_A{}^S = C^2 = q/2 \equiv P, \quad P_B{}^S = D^2 = (2-q)/2 \equiv Q, \qquad P + Q = 1. \tag{7.5.6}$$

For $q = (0, 2)$, i.e., $P = (0, 1)$, both electrons occupy a single AO (lone-pair configuration) giving rise to the purely *ionic* VB-structures of Section 7.2.3, when $\gamma_{a,b}{}^S = 0$, while the maximum delocalization of two electrons, i.e., the maximum bond covalency, is found for the symmetric case of $q = 1$ ($P = ½$) representing the bonding MO of Eq. (7.2.21), for the perfectly balanced mixture of both the covalent and ionic VB-structures of Eq. (7.2.23), when $\gamma_{a,b}{}^S = 2(PQ)^{1/2} = 1$.

Expressing the spatial wave-function $\Phi_S(1, 2)$ in terms of AO gives:

$$\Phi_S(1,2) = C^2\, a(1)a(2) + D^2\, b(1)b(2) + CD\,[a(1)b(2) + b(1)a(2)]$$
$$\equiv \Theta_{ion}(1,2) + \Theta_{cov}(1,2), \tag{7.5.7a}$$

where the *polarized* VB-ionic component now represents the weighted average of the elementary VB-ionic valence structures of Section 7.2.3,

$$\Theta_{ion}(1,2) \equiv C^2 S_{a,a}(1,2) + D^2 S_{b,b}(1,2) \equiv P[A^{-1}B^{+1}] + Q[A^{+1}B^{-1}], \tag{7.5.7b}$$

and the *polarized* VB-covalent component combines the elementary VB-covalent valence structures:

$$\Theta_{cov}(1,2) \equiv CD[S_{a,b}(1,2) + S_{b,a}(1,2)] \equiv (PQ)^{1/2}[A\text{—}B]. \tag{7.5.7c}$$

As we have already observed in Section 7.2.3, in the Heitler-London theory the elementary ionic components of the system wave-function are associated with the lone-pair configurations, when both electrons occupy a single atomic orbital, i.e., with an effective *ion-pair* resulting from the electron transfer relative to the reference state of two free atoms contributing a single electron each. The elementary covalent components of the system wave-function are similarly linked to the *shared-electron* configurations, of the two electrons effectively occupying both AO, i.e., to the simultaneous events of two electrons being found on different AO. We also recall that the purely-covalent situation in H_2, when $P = Q = ½$ and $\gamma_{a,b}{}^S = 1$, is marked by the equal participation of the VB-covalent and VB-ionic components.

It directly follows from Eq. (7.5.7a), by the familiar superposition principle of quantum mechanics, that the joint AO/AIM probabilities in the singlet state, $\mathbf{P}^S \equiv \{P_{X,X'}{}^S\}$, of simultaneously observing in the bonding state $\Psi_S(1, 2)$ a pair of electrons on atomic orbitals $m(\boldsymbol{r})$ on X and $n(\boldsymbol{r})$ on X', $m, n \in \{a, b\}$, are given by the products of the corresponding one-electron probabilities, $\mathbf{P}^S = \{P_{X,X'}{}^S = P_X{}^S P_{X'}{}^S\} = \boldsymbol{P}^S \otimes \boldsymbol{P}^S$:

$$P_{A,A}{}^S = (P_A{}^S)^2 = q^2/4 = P^2, \qquad P_{A,B}{}^S = P_{B,A}{}^S = P_A{}^S P_B{}^S = q(2-q)/4 = PQ,$$
$$P_{B,B}{}^S = (P_B{}^S)^2 = (2-q)^2/4 = Q^2. \tag{7.5.8}$$

Input: $\mathbf{A}^0$ (promolecular) $\quad \mathbf{P}_S(\mathbf{B}\,|\,\mathbf{A}^0)$ $\quad$ *Output*: $\mathbf{B}$ (molecular)

I)

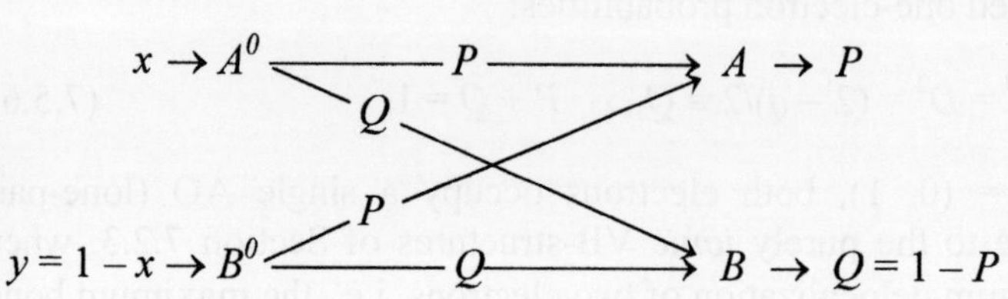
$x \rightarrow A^0$
P
$A \rightarrow P$
Q
P
$y = 1 - x \rightarrow B^0$
Q
$B \rightarrow Q = 1 - P$

Input: $\mathbf{A}$ (molecular) $\quad \mathbf{P}_S(\mathbf{B}\,|\,\mathbf{A})$ $\quad$ *Output*: $\mathbf{B}$ (molecular)

II)

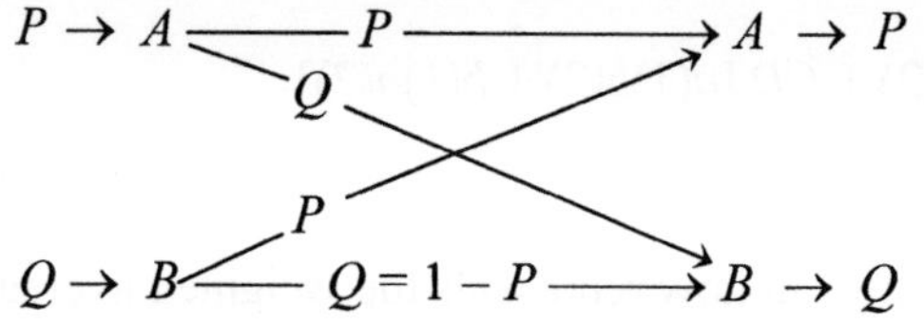
$P \rightarrow A$
P
$A \rightarrow P$
Q
P
$Q \rightarrow B$
$Q = 1 - P$
$B \rightarrow Q$

Input: $\mathbf{A}^0$ (promolecular) $\quad \mathbf{P}^0(\mathbf{B}^0\,|\,\mathbf{A}^0) = \mathbf{I}$ $\quad$ *Output*: $\mathbf{B}^0$ (promolecular)

III)

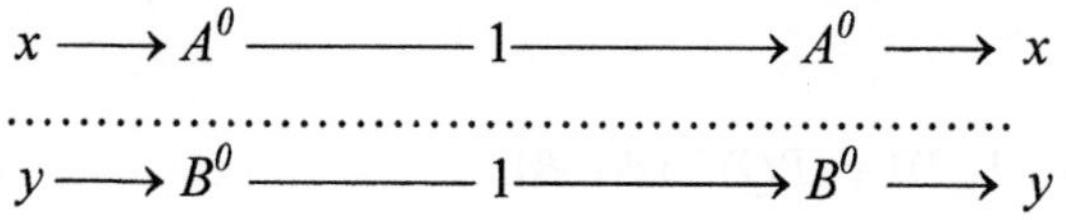
$x \longrightarrow A^0$
1
$A^0 \longrightarrow x$
$y \longrightarrow B^0$
1
$B^0 \longrightarrow y$

Input: $\mathbf{A}^s$ (SAL) $\quad \mathbf{P}^s(\mathbf{B}^s\,|\,\mathbf{A}^s) = \mathbf{I}$ $\quad$ *Output*: $\mathbf{B}^s$ (SAL)

IV)

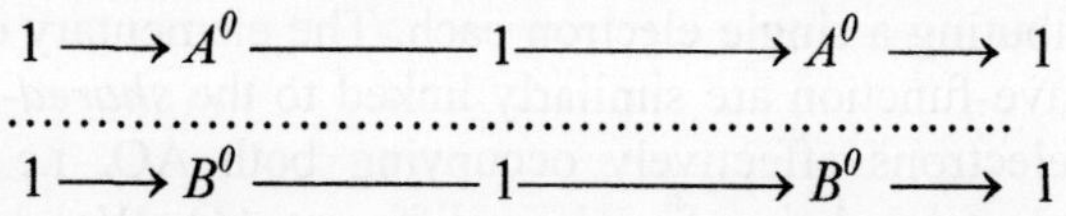
$1 \longrightarrow A^0$
1
$A^0 \longrightarrow 1$
$1 \longrightarrow B^0$
1
$B^0 \longrightarrow 1$

Scheme 7.7. The binary channels for the 2-AO model of the singlet state of two electrons occupying the bonding MO. The *non-stationary* molecular channel of Part I exhibits the promolecular input and molecular output probabilities, $\boldsymbol{P}^0(\mathbf{A}^0) = (x, y)$ and $\boldsymbol{P}(\mathbf{B}) = (P, Q) = \boldsymbol{P}^S(P)$, respectively, while the *stationary* molecular system of Part II exhibits the molecular input and output probabilities. In both cases the same molecular two-electron conditional probability matrix $\mathbf{P}(\mathbf{B}\,|\,\mathbf{A}^0) = \mathbf{P}(\mathbf{B}\,|\,\mathbf{A})$ has been used. The promolecular (Part III) and SAL (Part IV) channels consist of the separate (disconnected) channels for each free atom, with the combined conditional probability matrix defined by the identity matrix $\mathbf{P}^0(\mathbf{B}^0\,|\,\mathbf{A}^0) = \mathbf{P}^s(\mathbf{B}^s\,|\,\mathbf{A}^s) = \mathbf{I}$.

Together with the molecular one-electron probabilities $\boldsymbol{P}^S$ they determine the molecular conditional probability matrix,

$$\mathbf{P}_S(\mathbf{B}\,|\,\mathbf{A}) = \{P_S(X'\,|X) = P_X{}^S P_{X'}{}^S / P_X{}^S = P_{X'}{}^S\}, \tag{7.5.9}$$

grouping probabilities $P_S(X'\,|X)$ of observing in the ground-state an electron on atom X', when the other electron has already been found on atom X. This matrix determines the *non*-symmetric *binary channel* of Schemes 7.7/(I, II), for a general (promolecular) $\boldsymbol{P}^0 = (P_A{}^0, P_B{}^0) = (x, 1-x)$ (Part I) and the molecular $\boldsymbol{P}^S = (P, Q)$ (Part II) input probabilities. In Parts III and IV of this scheme the corresponding promolecular and SAL communication channels are displayed.

Bond Covalency

The one-electron entropies (in bits) of the condensed input and output probabilities in atomic resolution are determined by the *binary entropy function* of Section 3.7,

$$S(\mathbf{A}^0) = S(\boldsymbol{P}^0) = -x\log_2 x - (1-x)\log_2(1-x) = H(x),$$
$$S_S(\mathbf{A}) = S_S(\mathbf{B}) = S(\boldsymbol{P}) = -P\log_2 P - Q\log_2 Q = H(P), \tag{7.5.10}$$

while the molecular joint *two*-electron probabilities $\mathbf{P}^S(\mathbf{AB}) = \mathbf{P}^S(P)$ give rise to the two-electron and conditional entropies:

$$S_S(\mathbf{AB}) = S(\mathbf{P}_S(P)) = -\Sigma_{X\in\mathbf{A}}\Sigma_{Y\in\mathbf{B}}\, P_{X,Y}{}^S \log_2 P_{X,Y}{}^S = 2H(P),$$
$$S_S(\mathbf{B}\,|\,\mathbf{A}) = S_S(\mathbf{AB}) - S_S(\mathbf{B}) = H(P) \equiv N_S{}^{cov}(P). \tag{7.5.11}$$

As we have already observed above, the molecular input of Scheme 7.7/II is required to extract the very facets of the chemical bond, which represent the purely *molecular* phenomena, e.g., the system entropy-covalency (communication noise). Therefore, we apply this molecular communication system, which represents the *stationary* probability network in the molecule, with identical ground-state probabilities in the input and output of the information channel, to determine the entropic measure of the bond *covalency* provided by the conditional entropy of Eq. (7.5.11). The chemical interpretation of this bond component is indeed linked to a common possession (sharing) by the two bonded atoms of two electrons originating from different atoms of the promolecule, i.e., – to the quantum mechanical electron delocalization (mixing) effect. The maximum covalency is expected for the symmetrical bonding MO [Eq. (7.2.21)], when $(CD)^2 = PQ =$ maximum, i.e., for $P = Q = ½$, when the diatomic bond-order $\gamma_{a,b}{}^S = 1$.

Most of the previous treatments of the bond multiplicity problem have indeed identified this model situation as a single, purely covalent (electron sharing) bond (Wiberg, 1968; Gopinathan and Jug, 1983; Mayer, 1983; Jug and Gopinathan, 1990; Nalewajski *et al.*, 1993; Nalewajski and Mrozek, 1994, 1996; Nalewajski *et al.*,

1994, 1996, 1997; Mrozek *et al.*, 1998; Nalewajski, 2004b). Indeed, the same qualitative behavior is observed by alternative bond-indices in the ground-state of the 2-AO model. This is the case for the Wiberg (1968) measure of the covalent component in the model,

$$\mathfrak{I}_{a,b}{}^{W}(P) = (\gamma_{a,b}{}^{S})^{2} = 4[P(1-P)], \tag{7.5.12}$$

which is identical with the sum of the covalent and one-centre–ionic bond indices of the difference approach (Nalewajski *et al.*, 1993; Nalewajski and Mrozek, 1994, 1996; Nalewajski *et al.*, 1994, 1996, 1997; Mrozek *et al.*, 1998; Nalewajski, 2004b):

$$\mathcal{B}(P) = \mathfrak{I}^{cov}(P) + \mathfrak{I}_{oc}{}^{ion}(P) = \mathfrak{I}(P) - \mathfrak{I}_{tc}{}^{ion}(P) = \mathfrak{I}_{a,b}{}^{W}(P). \tag{7.5.13}$$

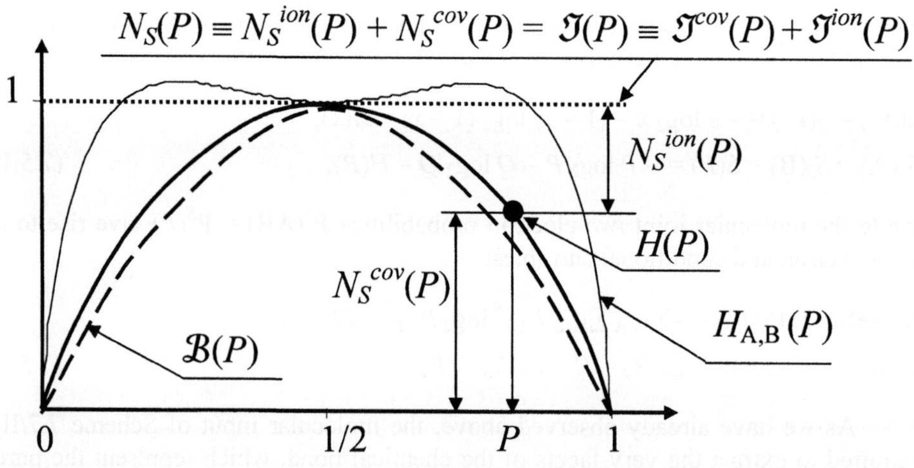

Figure 7.1. The conservation of the overall entropy/information bond-order $N_S(P) = N_S{}^{cov}(P) + N_S{}^{ion}(P) = 1$ bit and variations in the covalent/ionic composition of the orbital (spatial) entropic bond indices in the singlet (bonding) state of the 2-AO model, relative to the *electron-sharing* promolecular reference ($x = ½$), when two free atoms contribute a single valence electron to form the chemical bond. A similar conservation of the overall single bond is predicted by the quadratic Difference Approach (Nalewajski *et al.*, 1993; Nalewajski and Mrozek, 1994; Nalewajski, 2004b; see also Section 8.7), $\mathfrak{I}(P) = \mathfrak{I}^{cov}(P) + \mathfrak{I}^{ion}(P) = \mathfrak{I}^{cov}(P) + [\mathfrak{I}_{tc}{}^{ion}(P) + \mathfrak{I}_{oc}{}^{ion}(P)] = 1$, with the (broken line) bond-order quantity of Eq. (7.5.13), $\mathcal{B}(P) = \mathfrak{I}^{cov}(P) + \mathfrak{I}_{oc}{}^{ion}(P) = \mathfrak{I}(P) - \mathfrak{I}_{tc}{}^{ion}(P) = \mathfrak{I}_{a,b}{}^{W}(P) = 4PQ$, defining its division into the two-center quadratic ionicity $\mathfrak{I}_{tc}{}^{ion}(P)$ (upper area, above the parabola) and the sum of bond covalency $\mathfrak{I}^{cov}(P)$, and the one-center ionicity $\mathfrak{I}_{oc}{}^{ion}(P)$ (lower area, inside the parabola). This function also represents the Wiberg covalency index of Eq. (7.5.12). For comparison the bond entropy function $H_{A,B}(P) = -2\Gamma_{A,B}(P)\log_2\Gamma_{A,B}(P) = -4PQ\log_2(2PQ)$ (see Section 8.8) is also shown in the diagram.

As shown in Fig. 7.1, the complementary contribution $\mathfrak{I}_{tc}^{ion}(P)$ in Eq. (7.5.13), representing the two-centre ionicity of the quadratic difference approach, generates together with $\mathcal{B}(P)$ the conserved overall bond-multiplicity $\mathfrak{I}(P) = 1$.

The entropy quantity, which vanishes for the lone-pair configurations $P = (0, 1)$, and reaches the maximum value of 1 bit for $P = Q = ½$, thus exhibiting a qualitatively correct behavior expected of an adequate measure of the orbital (spatial) covalency N_S^{cov}, is thus identified by the average conditional entropy:

$$N_S^{cov}(P) \equiv S_S(\mathbf{B}\,|\,\mathbf{A}) = S(\mathbf{P}_S(P)) - S(\boldsymbol{P}_S(P)) = 2H(P) - H(P) = H(P). \qquad (7.5.14)$$

This conditional entropy function is also shown in Fig. 7.1. It measures the extra uncertainty in ascribing electrons to bonded atoms in *AB*, relative to that already present in the molecular input probabilities. Therefore, this quantity represents the average measure of the communication *noise* in the singlet molecular channel, due to the electron delocalization throughout the bonding MO.

Bond Ionicity

Let us examine next the information measure of bond ionicity in the molecular bonding (singlet) state of the 2-AO model. The average mutual information of Eq. (7.3.4b) between the input and output probabilities in Scheme 7.7/I, measures the amount of information flowing through this non-stationary channel and provides the information index of the complementary bond component - its orbital (spatial) IT-ionicity.

As in real communication devices, one freezes the molecular two-electron joint and conditional probabilities, when manipulating the channel input signal:

$$\mathbf{P}^S(\mathbf{A}^0\mathbf{B}) = \{ P^S_{m^0,n} \} \equiv \mathbf{P}^S(\mathbf{AB}) = \mathbf{P}^S = \{P_{m,n}{}^S\}. \qquad (7.5.15)$$

$$\mathbf{P}_S(\mathbf{A}^0\,|\,\mathbf{B}) = \{ P^S_{m^0,n} / P_n{}^S = P(m^0\,|\,n)\} \equiv \mathbf{P}_S(\mathbf{A}\,|\,\mathbf{B}) = \{P_{m,n}{}^S / P_n{}^S = P_S(m\,|\,n)\}, \qquad (7.5.16)$$

$$\begin{aligned}\mathbf{P}_S(\mathbf{B}\,|\,\mathbf{A}^0) = \{ P^S_{m^0,n} / P_{m^0} &= P(n\,|\,m^0) = (P_{m,n}{}^S / P_m{}^S)\,(P_m{}^S / P_{m^0}) \\ &= P_S(n\,|\,m)\,(P_m{}^S / P_{m^0})\}. \end{aligned} \qquad (7.5.17)$$

Therefore, the molecular communication connections are fixed, when the input probabilities of the stationary channel $\boldsymbol{P}^S(\mathbf{A}) = \boldsymbol{P}^S(P) = \boldsymbol{P}^S(\mathbf{B})$ of Scheme 7.7/II are displaced to their promolecular values $\boldsymbol{P}^S(\mathbf{A}^0) = \boldsymbol{P}(x)$ in Scheme 7.7/I. Using relations summarized in Fig. 3.1 then gives:

$$\begin{aligned}N_S^{ion}(x, P) &\equiv I(\mathbf{A}^0{:}\mathbf{B}) = S(\mathbf{A}^0) - S(\mathbf{A}^0\,|\,\mathbf{B}) = S(\mathbf{A}^0) + S(\mathbf{B}) - S(\mathbf{A}^0\mathbf{B}) \\ &= S(\mathbf{A}^0) + S(\mathbf{B}) - S(\mathbf{AB}) = H(x) + H(P) - 2H(P) = H(x) - H(P). \end{aligned} \qquad (7.5.18)$$

It should be recalled that the *channel capacity* (C) of Section 3.7 is determined by the input probabilities $\boldsymbol{P}^0(x_C) \equiv \boldsymbol{P}_C{}^0$, for which $I_S(\mathbf{A}^0\mathbf{:B})$ reaches the highest value:

$$C_S(P) = \sup_x N_S{}^{ion}(x, P) \equiv N_S{}^{ion}(x_C, P) \equiv N_S{}^{ion}(P). \tag{7.5.19}$$

It defines the input-independent measure of the IT-ionicity of the chemical bond in the model. The channel capacity of 2-AO model (in bits), for $x_C = ½$, i.e., for equal distribution of input signals, i.e., for $H(x_C) = 1$ bit, thus becomes

$$N_S{}^{ion}(½, P) = 1 - H(P) \equiv N_S{}^{ion}(P). \tag{7.5.20}$$

It should be recalled that the same formula applies to the SBC channel of Scheme 3.2 [see Eq. (3.7.6)].

Therefore, the sum of two IT-components of the chemical bond in the 2-AO model is conserved at the input-entropy level:

$$N_S(x, P) \equiv N_S{}^{cov}(P) + N_S{}^{ion}(x, P) = H(x) \qquad \text{or}$$
$$N_S(P) \equiv N_S{}^{cov}(P) + N_S{}^{ion}(P) = 1 \text{ bit}. \tag{7.5.21}$$

The second, input independent quantity, corresponds to the *electron-sharing* promolecular reference of the 2-AO model, when both free atoms contribute a single electron each to form the chemical bond.

Therefore, the mutual-information measure of bond ionicity [Eq. (7.5.20)] identifies the electron-sharing promolecular input, for $x = ½$, as the one for which the associated amount of information flowing through the 2-AO communication system reaches the capacity level. These input probabilities represent the most favorable "valence" (promoted) state of two atoms, to eventually form the covalent bond. For this promolecular input the information capacity index vanishes identically for $P = Q = ½$, i.e., for $\gamma_{A,B} = 1$ (see Scheme 7.2), when the probability of a *"survival"* of an electron in a given AO and that of its being *"scattered"* to the other AO are equal. This implies no flow of information in the symmetrical MO configuration, when $P = Q = ½$, e.g., in the ground-state of H_2 or the π-electron channel of ethylene. Indeed, as we have already stressed before, such probabilities mark the purely covalent, single-bond in the orbital resolution, with the vanishing multiplicity of the *ionic* bond component (see also Section 7.2.3 and Fig. 7.1).

Next, let us examine changes in the singlet 2-AO channel, when the bonding MO becomes polarized (non-symmetrical), for $P \neq Q$. Consider first the *lone-pair* limit, for $P = 0$ or $P = 1$, when both electrons in the molecule are localized on a single AO on atoms B and A, respectively, and $\gamma_{A,B} = 0$ (no electron "sharing", zero orbital covalency). An example of such an extremally polarized molecular channel for $P = 0$ is shown in Scheme 7.8a. It exhibits the same conditional probability network as that generated by the $\Phi_{AB}{}^B(1, 2)$ wave-function in Scheme 7.6.

In part *b* of this scheme both electrons originate from the electron-pair located on atom B. This is consistent with the molecular electronic structure determining this channel. Therefore, Scheme 7.8*b* represents the stationary lone-pair information system in the 2-AO model.

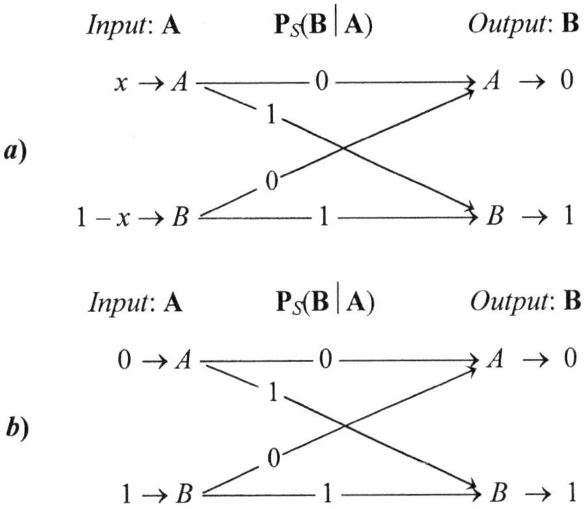

Scheme 7.8. The communication channel for the limiting polarization of the bonding MO corresponding to the lone-pair on atom B ($P = 0$ and $Q = 1$) in the 2-AO model of the chemical bond, for a general promolecular input (Part *a*) and for the lone-pair input on atom B ($x = 0$) (stationary channel).

Since for $P = 0$ the conditional probabilities in this lone-pair molecular channel are either 1 or 0, the average conditional entropy vanishes identically, $H(0) = 0$, so that the chemical bond is of the purely-ionic character:

$$N_S^{ion}(0, 0) = 0 \qquad \text{and} \qquad N_S^{ion}(½, 0) = H(½) = 1 \text{ bit}.$$

Thus the $P = 0$ electronic structure in the molecule, identical with that of the $x = 0$ donor-acceptor promolecule, when the lone-pair is located on the donor (B) atom, is correctly identified as giving rise to the non-bonding information channel. The second part of the preceding equation also correctly identifies the lone-pair ($P = 0$) electron configuration of the molecule as representing the $A \rightarrow B$ electron transfer (ion-pair) relative to the symmetrical, electron-sharing promolecule of the free atoms for $x = ½$, when they both contribute a single electron to form the chemical bond, thus indeed exhibiting a single ionic bond.

One could also combine the orbital (spatial) bond components of Eqs. (7.5.11) and (7.5.21) into the corresponding *vector* entropic "signature" (in bits):

$$\mathbb{N}_S^{orb}(P; x) = [N_S^{cov}(P),\ N_S^{ion}(x, P)] = [H(P), H(x) - H(P)]. \qquad (7.5.22)$$

For the electron sharing reference $x = ½$ the purely covalent bond, for $P = Q = ½$, is thus identified by the 1 bit of the orbital-covalency and zero orbital-ionicity, $\aleph_S^{orb}(½; ½) = (1, 0)$, while the molecular lone-pair configuration for $P = 0$ gives for this reference the vanishing entropy-covalency and 1 bit of information ionicity: $\aleph_S^{orb}(0; ½) = (0, 1)$. Finally for the identical promolecular and molecular lone-pair configurations one correctly predicts $\aleph_S^{orb}(0; 0) = (0, 0)$.

As we have already observed in Eq. (7.5.21), the orbital multiplicities N_S^{ion} and N_S^{cov} of the singlet state imply that the total orbital bond entropy $H(½) = 1$ bit for the electron sharing input $x = ½$ is conserved (see Fig. 7.1), when the orbital probability P changes in the full range $0 \leq P \leq 1$:

$$N_S(P) = N_S^{cov}(P) + N_S^{ion}(P) = 1. \tag{7.5.23}$$

However, its covalent/ionic composition changes rather dramatically in such a hypothetical transformation of the occupied MO. This overall bond *conservation principle* reflects the *competition effect* between these complementary bond components, which is also manifested by other measures of the bond multiplicity formulated in the MO theories (see also Fig. 7.1) (Nalewajski *et al.*, 1993; Nalewajski and Mrozek, 1994, 1996; Nalewajski *et al.*, 1994, 1996, 1997; Mrozek *et al.*, 1998; Nalewajski, 2004b). It accords with an intuitive expectation that the valence electrons strongly involved in one bond component become less available to participate in the other component.

Coordination Bond

The previously considered electron-sharing input $\boldsymbol{P}^0(x = ½) = (½, ½)$, for which the mutual information $N_S^{ion}(x,P)$ of Scheme 7.7/I reaches the capacity level $C_S(P) = 1 - H(P)$, indicates that both atoms contribute equally, a single electron each, to form the chemical bond. To account for the pair-sharing, *coordination* (c) chemical bond between the donor (*B*, *base*) and acceptor (*A*, *acid*) atoms, when the electrons of the bonding pair originate from B^0, one selects the coordination input probabilities $\mathbf{A}_c^0$, for, $\boldsymbol{P}_c^0 = \boldsymbol{P}^0(x = 0) = (0, 1)$, for which $S(\mathbf{A}_c^0) = S(\boldsymbol{P}_c^0) = H(0) = 0$.

The two-channel approach of the preceding section, with the molecular stationary network of Scheme 7.7.II determining the covalent component $N_c^{cov}(P)$ and the network of Scheme 7.7/I being used to extract the bond ionicity $N_S^{ion}(x,P)$, gives the same estimate of the reference-independent bond covalency, $N_c^{cov}(P) = H(P)$, and the *negative* IT-ionicity index of the coordination bond:

$$N_c^{ion}(P) = I_S(\mathbf{A}_c^0:\mathbf{B}) \equiv N_S^{ion}(x = 0, P) = -H(P). \tag{7.5.24}$$

This negative sign distinguishes the information-ionicity index of the coordination (*pair-sharing*, donor-acceptor) origin from the positive sign of the ionicity index of the chemical bond resulting from the covalent (*electron-sharing*) origin. The quantity $N_c^{ion}(P) = I_S(\mathbf{A}_c^0{:}\mathbf{B})$ measuring the amount of information flowing between the coordination input and the molecular output provides the information about the occurrence of events in one set of the electron detection events in atomic resolution, provided by the occurrence of the events in the other set. Therefore, a negative value of this quantity implies that the *occurrence* of events in one set makes the *non-occurrence* of events in the other set more likely. Its negative value is indeed a characteristic feature of the *coordination* bond, since finding an electron on the acceptor makes it more likely that this electron will be absent on the donor.

Since the information measure of the ionic bond-multiplicity, $N_S^{ion}(x, P) = H(x) - H(P)$, is reference-dependent, the same electron configuration has different information signature, when the ionic index is estimated with respect to the coordination promolecule, for $x = 0$, with both electrons of the model originating from the *donor* (basic) atom B^0, when $H(x = 0) = 0$, compared to that for the electron-sharing reference $x = ½$: $\mathbb{N}_S^{orb}(P; x = 0) = [H(P), -H(P)]$. The single *coordination* bond, when the lone electron pair of the donor atom B^0 is fully shared in the molecule with the acceptor atom A for $P = Q = ½$, is thus identified by the information signature $\mathbb{N}_S^{orb}(½; 0) = (1, -1)$. Accordingly, the entropy/information signatures of the *lone*-pair configuration $P = 0$, relative to the pair-sharing, coordination input, then reads $\mathbb{N}_S^{orb}(0; x = 0) = (0, 0)$, thus correctly identifying the non-bonding character of such a promolecular distribution of electrons.

The maximum-covalency electron configuration in the molecule, $P = Q = ½$, which was previously identified as representing a single, purely covalent bond relative to the *electron-sharing* reference, is diagnosed as representing 1 bit of the orbital-covalency and 1 bit of the *coordination* orbital-ionicity (negative). This description agrees with the AIM equal electron populations in this state, $q_A = q_B = 1$, which imply an electron transfer $B \rightarrow A$ relative to the coordination promolecular reference, in which the two electrons have been located on B^0.

Therefore, in the coordination-bond case an increase in the orbital entropy of the covalent component implies an associated increase in the magnitude of the information index of the coordination ionic component. In the coordination-bond case the two components are indeed *linearly dependent*, since any degree of the electron delocalization (bond covalency) $B \rightarrow A$ generates the associated CT (bond ionicity). In other words, the larger sharing of the donor electron pair with the acceptor bond-partner, the more orbitally covalent is the coordination bond and the larger electron transfer from the donor to acceptor, i.e., the greater coordination ionicity.

7.5.2. Spin Channel

The entropy/information quantities of molecular systems depend on the whole set of the electron-event arguments of the input and output probability schemes in atomic resolution. In the preceding two sections we have explored the entropy/information quantities of the ground-state probabilities of finding electrons on specified atoms, for the spatial (orbital) events. In this section we shall briefly examine the entropy/information descriptors of the electron spin-events. As we have already observed, the events of finding electrons on specified AIM and those of detecting their specific spin orientations are independent in the present *two*-electron model. Therefore, in this case the joint *orbital-spin* probability distributions are the Kronecker products of the separate *orbital* and *spin* probabilities in atomic resolution, thus giving rise to the *additivity* of their associated entropy/information indices.

Consider the illustrative example the singlet-state of two electrons. The antisymmetric spin-function $\Xi_S(\sigma, \sigma')$ of Eq. (7.2.22b) gives rise to the correlated distribution of spins:

$$\boldsymbol{\pi}_2^S = \{\pi_2^S(\sigma, \sigma') = |\Xi_S(\sigma, \sigma')|^2 = ½(1 - \delta_{\sigma,\sigma'}),\ \sigma, \sigma' \in (-½, ½)\}$$
$$= \begin{bmatrix} 0 & 1/2 \\ 1/2 & 0 \end{bmatrix}. \qquad (7.5.25)$$

This matrix generates the one-electron spin probability vector $\boldsymbol{\pi}_1^S = \{\pi_1^S(\sigma) = \sum_{\sigma'} \pi_2^S(\sigma, \sigma') = ½\}$. The resulting singlet *one*- and *two*-electron spin entropies (in bits) are: $S(\boldsymbol{\pi}_1^S) = 1$ and $S(\boldsymbol{\pi}_2^S) = 1$. The two-electron spin-probabilities $\boldsymbol{\pi}_2^S$ give rise to the associated conditional probability matrix $\boldsymbol{\pi}_S(\boldsymbol{\Sigma}_B|\boldsymbol{\Sigma}_A) = \{\pi_S(\sigma'|\sigma) = 1 - \delta_{\sigma,\sigma'}\}$, which determines the symmetric, deterministic spin-channel of Scheme 7.9, reminiscent of the network for the VB-covalent wave-function of Scheme 7.4.

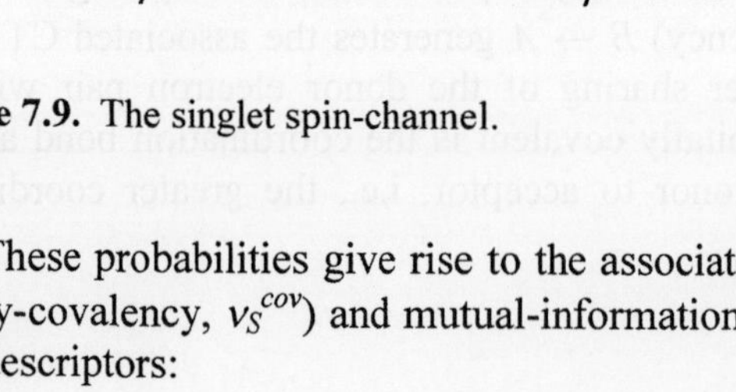

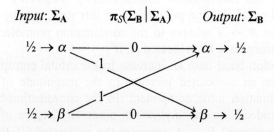

Scheme 7.9. The singlet spin-channel.

These probabilities give rise to the associated spin conditional-entropy (singlet entropy-covalency, v_S^{cov}) and mutual-information (singlet information-ionicity, v_S^{ion}) bond descriptors:

$$S_S(\boldsymbol{\Sigma}_B|\boldsymbol{\Sigma}_A) = -\sum_\sigma \sum_{\sigma'} \pi_2^S(\sigma, \sigma') \log_2[\pi_2^S(\sigma, \sigma')/\pi_1^S(\sigma)]$$
$$= -\sum_\sigma \pi_1^S(\sigma) [\sum_{\sigma'} \pi_S(\sigma'|\sigma) \log_2 \pi_S(\sigma'|\sigma)] = S(\boldsymbol{\pi}_2^S) - S(\boldsymbol{\pi}_1^S) \equiv v_S^{cov} = 0, \qquad (7.5.26)$$

$$I_S(\boldsymbol{\Sigma}_\mathbf{A}: \boldsymbol{\Sigma}_\mathbf{B}) = \sum_\sigma \sum_{\sigma'} \pi_2^S(\sigma, \sigma') \log_2\{\pi_2^S(\sigma, \sigma')/[\pi_1^S(\sigma)\, \pi_1^S(\sigma')]\}$$
$$= -\sum_\sigma \pi_1^S(\sigma)\, \{\sum_{\sigma'} \pi_S(\sigma' | \sigma) \log_2[\pi_S(\sigma' | \sigma)/\pi_1^S(\sigma')]\}$$
$$= S(\boldsymbol{\pi}_1^S) - S_S(\boldsymbol{\Sigma}_\mathbf{B}|\boldsymbol{\Sigma}_\mathbf{A}) \equiv \nu_S^{ion} = 1 \text{ bit.} \qquad (7.5.27)$$

The vanishing spin-covalency of the singlet state reflects the fact that the spin channel of Scheme 7.9 does not introduce any extra uncertainty relative to that already present in the input spin-probabilities, due to its deterministic, noiseless character. Similarly, the 1 bit of the spin-ionicity measures the statistical correlation embedded in the antisymmetric spin-function of two electrons.

The spin channel deals with probabilities of different spin orientations of two electrons without any reference to atoms. Therefore, they provide the overall (molecular) entropic descriptors of the information content in the spin probability distribution of the system electrons, complementary to that contained in the atomically resolved spatial distribution. Although interesting for the complete description of the entropy flow in the molecular communication systems (Nalewajski, 2004c) such spin-channels do not constitute an adequate basis for the entropy/information indexing of the chemical bonds, between the constituent AIM. Therefore, when indexing the chemical bonds (AIM connectivities) we shall focus our considerations only on the *spatial* communication channels.

7.6. MULTIPLICITIES OF π BONDS

Consider now, as an illustrative application to the ground-state in polyatomics, the entropic multiplicities of the conjugated π bonds in allyl, butadiene and benzene, generated by the AIM-resolved probability distributions of π electrons from the Hückel theory. The relevant ground-state communication channels are shown in Schemes 7.10–7.12. The corresponding information network for ethylene is identical with that of the 2-AO model of Section 7.5 (see Scheme 7.7) for $P = Q = x = ½$, predicting a single, purely covalent π bond with the IT descriptors (in bits): $N^{ion,\pi} = I(\mathbf{A}^0{:}\mathbf{B}) = 0$, $N^{cov,\pi} = S(\mathbf{B}|\mathbf{A}) = N(\mathbf{A};\mathbf{B}) = N^\pi = N^{cov,\pi} + N^{ion,\pi} = 1$, in perfect agreement with the chemical intuition. A similar conclusion follows from Scheme 7.10, where the information channel and IT bond indices for allyl are reported.

In all these *alternant* hydrocarbons the probability of finding an electron on *i*th carbon in a molecule is equal to that in the atomic promolecule: $\boldsymbol{P}^\pi = \{P_i^\pi = P_i^{0,\pi} = 1/N\} = \boldsymbol{P}^{0,\pi}$, $i = 1, 2, ..., N$, where N stands for the number of π electrons, equal to the number of carbon atoms. Therefore, the orbital input and output entropies (in bits),

$$S(\mathbf{A}) = S(\mathbf{B}) \equiv S(\boldsymbol{P}^\pi) = S(\mathbf{A}^0) = S(\mathbf{B}^0) = S(\mathbf{A}^s) = S(\mathbf{B}^s) \equiv S(\boldsymbol{P}^{0,\pi}) = \log_2 N,$$

for these representative π-electron "communication" systems are: 1 in ethylene, 1.585 in allyl, 2 in butadiene, and 2.585 in benzene. Again, in the SAL reference system $M^s = i^0 + j^0 + k^0 + ...$, consisting of the infinitely-separated (dissociated) free-carbon atoms, all entropy/information bond indices vanish. In the AP reference, $M^0 =$

$(i^0|j^0|k^0|\ldots)$, consisting of the free-carbons placed at molecular positions, the overall entropy-information bond-order of $N^\pi = \log_2 N$ bits is exclusively of the information-ionic character measured by the channel mutual-information quantity $N^\pi = N^{ion,\pi}$, with the identically vanishing conditional-entropy measure of the overall bond covalency $N^{cov,\pi} = 0$.

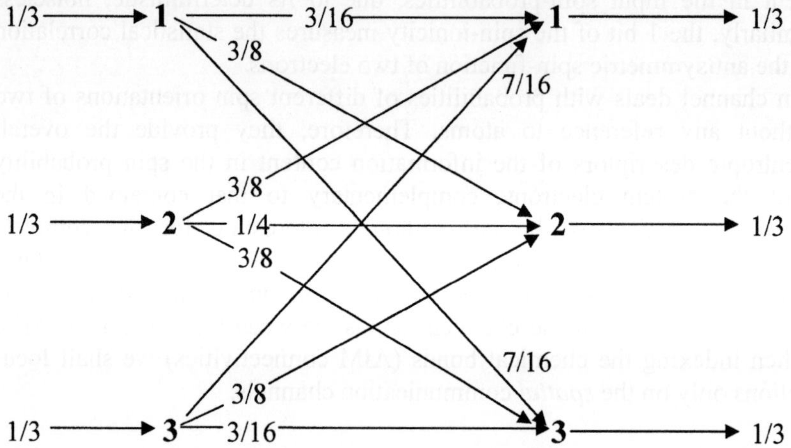

$$S(\mathbf{B}\,|\,\mathbf{A}) = 1.52, \quad I(\mathbf{A}^0\!:\!\mathbf{B}) = 0.06, \quad N(\mathbf{A};\mathbf{B}) = S(\mathbf{B}\,|\,\mathbf{A}) + I(\mathbf{A}^0\!:\!\mathbf{B}) = 1.58$$

Scheme 7.10. The information channel for π electrons in allyl (Hückel theory). The entropy/information bond indices (in bits): $S(\mathbf{B}|\mathbf{A}) = N^{cov,\pi} = 1.524$, $I(\mathbf{A}^0\!:\!\mathbf{B}) = N^{ion,\pi} = 0.061$, $N^\pi = N^{cov,\pi} + N^{ion,\pi} = 1.585$ reflect the chemical intuition of about 3/2 bond multiplicity in the allyl π-bond system.

Since in these model Hückel systems the elements of each row of the conditional probability matrix in atomic resolution, $\mathbf{P}^\pi(\mathbf{B}|\mathbf{A})$, are permutations of the elements in the first row, all these π electron channels are *uniform* (see Section 3.7). For such systems [see Eq. (3.7.8)] the mutual information between the input and output probability distributions is given by the difference of the output entropy $S(\mathbf{B})$ and the Shannon entropy $S(\mathbf{B}|i) = S[\boldsymbol{P}^\pi(\mathbf{B}|i)] = S(\mathbf{B}|\mathbf{A})$ of any (say ith) row $\boldsymbol{P}^\pi(\mathbf{B}|i)$ of the conditional probability matrix $\mathbf{P}^\pi(\mathbf{B}|\mathbf{A})$, thus predicting the channel capacity (in bits) [Eq.(3.7.9)]: $C^\pi = \log_2 N - S(\mathbf{B}|i)$.

In the Hückel approximation the joint two-electron π electron probabilities $\mathbf{P}^\pi(\mathbf{AB}) = \{P_{i,j}^{\ \pi}\}$ in ethylene are the same as in the 2-AO model of Section 7.5 for $P = Q = ½$. For allyl radical one obtains: $P_{1,1} = P_{3,3} = 1/16$, $P_{2,2} = 1/12$, $P_{1,2} = P_{2,3} = 1/8$, and $P_{1,3} = 7/48$. In butadiene one similarly finds: $P_{i,i}^{\ \pi} = 1/24$, $i = 1, \ldots, 4$; $P_{1,2}^{\ \pi} = P_{3,4}^{\ \pi} = 1/20$; $P_{2,3}^{\ \pi} = P_{1,4}^{\ \pi} = 3/40$; and $P_{1,3}^{\ \pi} = P_{2,4}^{\ \pi} = 1/12$, for the consecutive numbering of four carbons, starting from the terminal atom. The corresponding data for benzene read: $P_{i,i}^{\pi} = 1/60$, $i = 1, ..., 6$; $P_{i,i+1}^{\pi} = 7/270$ (12 contributions); $P_{i,i+2}^{\pi} = 1/30$ (12 contributions); and $P_{i,i+3}^{\pi} = 17/540$ (6 contributions). These two electron probabilities give rise to the conditional probabilities $\mathbf{P}^{\pi}(\mathbf{B}|\mathbf{A}) = \{P^{\pi}(j|i) = P_{i,j}^{\pi}/P_i^{\pi}\}$, of finding one π electron on carbon j, when another π electron is known to have been found on atom i, which determine the communication channel for the system π electrons. They are listed in the π communication systems for these molecules.

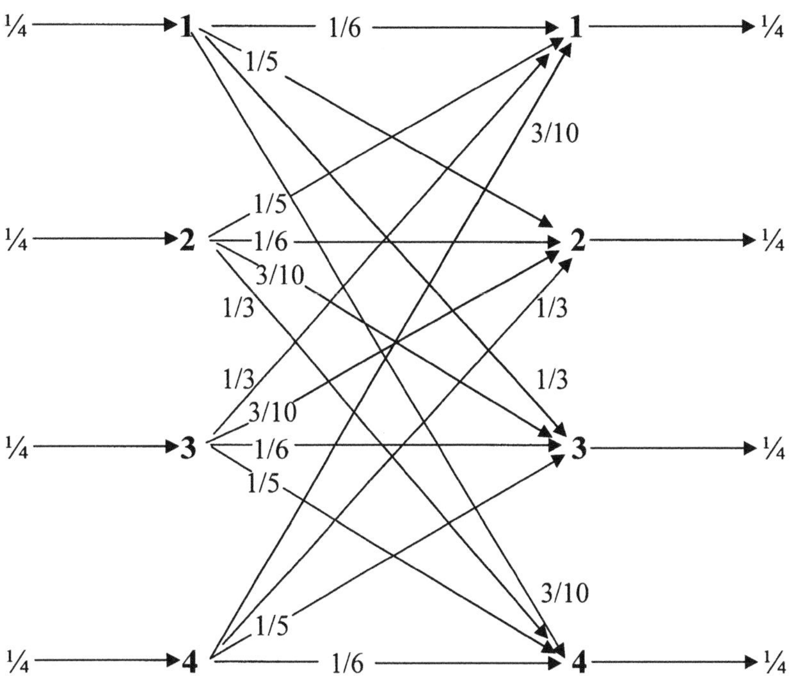

Scheme 7.11. As in Scheme 7.10 for butadiene (Hückel theory). The IT-bond indices (in bits) read: $S(\mathbf{B}|\mathbf{A}) = N^{cov,\pi} = 1.944$, $I(\mathbf{A{:}B}) = N^{ion,\pi} = 0.056$, $N^{\pi} = N^{cov,\pi} + N^{ion,\pi} = 2$.

The π electron probabilities give rise to the corresponding average two-electron entropies (in bits): $S(\mathbf{P}^{\pi}) = 3.231$ (allyl), $S(\mathbf{P}^{\pi}) = 3.944$ (butadiene) and $S(\mathbf{P}^{\pi}) = 5.136$ (benzene). Hence the corresponding average conditional entropies, $S(\mathbf{B}|\mathbf{A}) = S(\mathbf{AB}) - S(\mathbf{B}) = N^{cov,\pi}$, measuring the overall entropy-covalency of π electrons in these molecules are: $N^{cov,\pi} = 1$ (ethylene), $N^{cov,\pi} = 1.524$ (allyl), $N^{cov,\pi} = 1.944$ (butadiene) and $N^{cov,\pi} = 2.551$ (benzene). They give rise to the associated amounts of the information flowing through these two communication systems, $I(\mathbf{A{:}B}) = S(\mathbf{B}) - S(\mathbf{B}|\mathbf{A}) = N^{ion,\pi}$, which reflects the overall information-ionicity of π electrons in the molecular ground-state: $N^{ion,\pi} = 0$ (ethylene), $N^{ion,\pi} = 0.061$ (allyl), $N^{ion,\pi} = 0.056$ (butadiene), and $N^{ion,\pi} = 0.034$ (benzene).

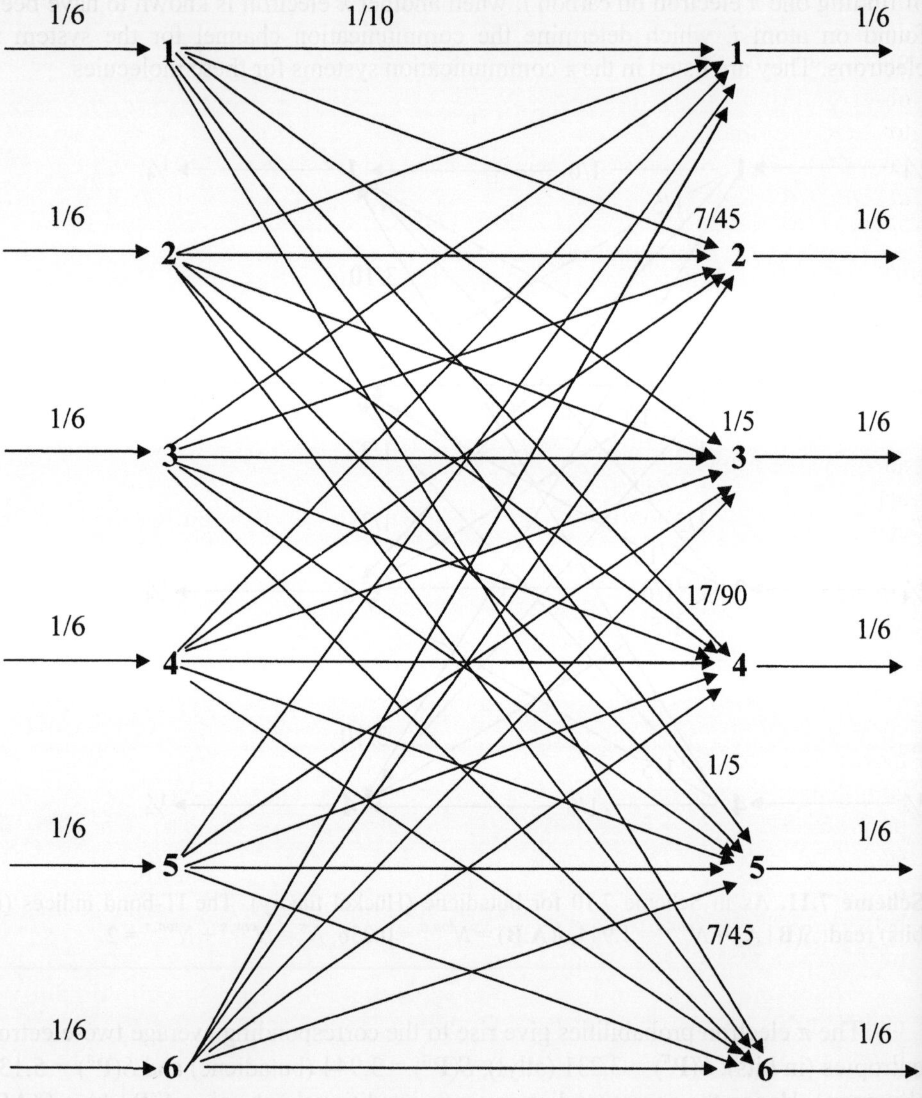

Scheme 7.12. As in Scheme 7.10 for benzene. Only the first row of the conditional probability matrix **P**(**B**|**A**) is listed in the diagram; the remaining rows are permutations of the first row (the uniform channel). Bond indices in bits: $S(\mathbf{B} \mid \mathbf{A}) = N^{cov,\pi} = 2.551$, $I(\mathbf{A{:}B}) = N^{ion,\pi} = 0.034$, $N^{\pi} = N^{cov,\pi} + N^{ion,\pi} = 2.585$.

The numerical values of the corresponding total orbital indices of the system π bonds, $N^{\pi} = N^{cov,\pi} + N^{ion,\pi}$, are equal to the corresponding one-electron entropies of the molecular channel: 1 (ethylene), 1.58 (allyl), 2 (butadiene), and 2.58 (benzene). Their numerical values qualitatively agree with the intuitive chemical multiplicities of π bonds for these systems: a single bond in ethylene, around 3/2 bond in allyl, the double conjugated bond in butadiene and less than 3 bonds in benzene (Shaik, 1989; Shaik and Hiberty, 1991, 1995, 2004; Jug and Köster, 1990). It should be also observed that the predicted π-bond ionicities of these two uniform channels are equal to their information *capacities*. These entropy/information bond indices compare favorably with the alternative bond-order measures from the two-electron *difference approach* (Nalewajski *et al.*, 1993; Nalewajski and Mrozek, 1994, 1996; Nalewajski *et al.*, 1994, 1996, 1997; Mrozek *et al.* 1998; Nalewajski, 2004b), based upon the displacements in the system two-electron density matrix relative to the SAL values.

7.7. CONCLUSION

We have demonstrated in this chapter that the key issue in chemistry, viz., an adequate description of the chemical bonds in molecular systems, can be successfully approached using concepts and techniques of the information theory of communication systems. This IT approach provides the adequate information/entropy descriptors of the prototype molecular systems, which in general agree with chemical intuitive expectations. This *global* analysis of the bond descriptors in the molecular system as a whole confirms the conjecture (Nalewajski, 2000c, 2004b-e, 2005a-c; Nalewajski and Jug, 2002) that in atomic resolution the molecular indeterminacy of the electron probability output relative to input, measured by the channel *conditional entropy*, i.e., the average communication noise in the stationary molecular channel, provides a realistic index of the *covalent* bond component. The complementary quantity, the *mutual information* between the molecular output and promolecular input, i.e., the amount of information flowing through the non-stationary molecular channel, has been found to generate an adequate representation of the *ionic* bond component. It reflects a degree of dependence (correlation) between the promolecular input and molecular output probabilities of the molecular information channel.

In the 2-AO model of the chemical bond these entropy/information bond-multiplicities give rise to the conservation of the overall IT–bond-order in the bonding state of the spin-paired electrons, thus correctly representing the intuitively expected competition between the covalent and ionic components of the chemical bond. This two-electron entropic development naturally connects to the Valence-Bond perspective of Heitler and London, thus representing a direct continuation of these classical ideas of quantum chemistry in IT. The comparison with VB theory has also revealed important differences between these two approaches in defining the bond ionicity and covalency, with the IT treatment being shown to provide the *dichotomous* framework for indexing these complementary bond components.

A successful entropy/information indexing of the chemical bond should bring a better understanding of the molecular electron distributions and the information origins of the chemical bond. The intriguing question of the electronic structure theory has allways been the role played by the *one*- and *two*-electron effects in the bond covalency and ionicity. The conventional approach to this problem adopts the *one*-electron perspective, based upon the polarization of the localized MO responsible for the chemical bond in question. The *two*-electron *VB*-perspective distinguishes the elementary two-electron wave-function components corresponding to the electron-sharing (covalent) and lone-pair (ionic) situations, but gives rise to the *non-dichotomous* description of the purely covalent bond in H_2 marked by the equal contributions of the *VB*-covalent and *VB*-ionic structures to the ground-state wave-function in H_2. The present communication theory approach to the chemical bond in 2-AO model was shown to give rise to a novel two-electron description, which generates the *dichotomous* bond indices, which explicitly reflect the covalent-ionic competition and give rise to the conservation of the IT overall bond-order. It also brings about a new perspective on bond-ionicity as reflecting the electron-correlation in molecules. In the spirit of the KS theory of electronic structure the mutual-information descriptor of bond ionicity has indeed been interpreted as the information-distance between the correlated (molecular) and independent *two*-electron distributions, which describe the real (interacting) and hypothetical (non-interacting) systems, respectively.

It has been demonstrated that the global bonding patterns emerging from the communication-theory for all illustrative molecular systems are generally in accord with the chemical intuition. For example, it has been shown that in the benzene carbon ring the total bond index is lower than 3 bits value expected for the *triple* conjugated π-bond, which characterizes (see the next chapter) the mutually closed diatomic fragments of cyclohexatriene. In other words, the aromaticity of π electrons manifests itself in *lowering* the molecular entropy/information index relative to the reference of 3 bits level predicted for the three separate π bonds. Therefore, the natural tendency of the delocalized π electrons in benzene is to destabilize the regular hexagonal structure towards the distorted, alternated system of hexatriene. This is in accord with the modern outlook on the influence of σ and π electrons on aromaticity (Shaik, 1989; Shaik and Hiberty, 1991, 1995, 2004; Jug and Köster, 1990). These modern MO and VB approaches indeed predict the π bonds to favor the distorted (cyclohexatriene-like) structure, while the dominating σ-bond stabilization is responsible for the regular hexagon structure of the benzene ring.

One concludes from the present IT analysis that the molecular systems can indeed be regarded as information networks exhibiting the quantum-mechanical noise in propagating the AIM-assignment signals of the valence electrons, due to the electron delocalization via the network of chemical bonds. We have identified the right information quantities which can be used to characterize chemical bonds and their ionic and covalent components in such communication channels. It has been demonstrated that the minimum value of the mutual information (bond ionicity), i.e., the maximum value of the conditional entropy (bond covalency), is obtained for the equal (molecular) input and output probabilities of the molecular information channel, which identify the *stationary* information channel.

Additional applications to the excited electron configurations in the 2-AO model are reported in Appendix E, where the problem of AO-*phases* in MO is briefly addressed. In fact the entropy/information indices of the chemical bond, which have been established in this chapter to treat the ground-state (equilibrium) electron distributions in molecular systems, have to be modified to adequately diagnose the excited configurations (Nalewajski, 2005g). More specifically, in Appendix E we examine the excited *non*-bonding and *anti*-bonding configurations in the 2-AO, when the two electrons occupy partially or completely the *anti*-bonding MO of Eq. (7.5.4). The new problem one then encounters is that of *phases* of the LCAO MO coefficients, which are lost, when the *one*- or *two*-electron probabilities are generated using the familiar *superposition principle* of quantum mechanics. This analysis demonstrates the importance of recognizing the relative phases of AO in the excited MO, in a proper diagnosis of all intricacies of chemical bonds in the model excited states. This information is lost in the phase-independent communication treatment of the model. However, by generating the molecular channels from the known wave-function in AO representation, this information can be directly "fed" into the information system by an appropriate definition of inputs defining the *directed* information flows in the molecule. These *partial* information flows through the molecular channel give rise to the *generalized* entropy/information descriptors of the chemical bond covalency and ionicity, which generally agree with both the elementary MO description and chemical intuition associated with the non-bonding and anti-bonding electron configurations of this prototype 2-AO framework for the chemical bond interactions.

However, the global bond descriptors introduced in this chapter, representing the overall entropy/information bond "orders" in the system as a whole, are insufficient for many chemical considerations, which also call for the corresponding indices of chemical bonds in molecular fragments. Both internal (inside the fragment in question) and external (between the fragment and the rest of the molecule) bond indices are needed to describe the chemical condition of a molecular fragment. In the next chapter we shall present three altrernative communication-theory approaches to this problem. We shall also examine general combinational (grouping) rules within these alternative subsystem developments, for combining the intra- and inter-group entropy/information bond descriptors into those characterizing the molecule as a whole (see Section 3.6).

Finally, it should be emphasized that both the chemical bonds and bonded atoms are not quantum-mechanical observables. Therefore, their descriptors are not unique. Indeed, these crucial chemical concepts of the molecular electronic structure can be best classified as Kantian *"noumenons"* of chemistry (Parr *et al.*, 2005). The information-theoretic approach to this classical problem of chemistry only guarantees that the extraction of these chemical concepts from the known molecular distributions of electrons is done in the most *objective* (least biased) manner possible.

8. ENTROPY/INFORMATION INDICES OF MOLECULAR FRAGMENTS

Three alternative strategies for generating the IT bond indices of molecular fragments are proposed within the communication theory approach. They respectively involve the *disconnected* (renormalized) communication channels of the mutually non-bonded parts of the molecule, and the *connected* (mutually bonded) *partial* and *reduced* information channels of molecular fragments, constructed from the molecular channel in atomic resolution. The entropy/information indices of these *sub-channels* are used as IT measures of the fragment bond-orders, both *internal* (intra-subsystem) and *external* (between the subsystem and its complementary molecular reminder) and their covalent/ionic composition. The important issue of combining the subsystem data into the corresponding global quantities describing the system as a whole is addressed. Both approximate and exact combination (grouping) rules for the entropy/information bond indices are derived and tested in model systems. The renormalized channels of the mutually non-bonded (externally disconnected, "closed") molecular subsystems are used to generate the fragment internal IT bond indices. They are subsequently combined into the corresponding global indices using the external bond contributions estimated in the *Independent Fragment Approximation*. The partial *row* and *column* communication channels resulting from the *additive* decomposition of the global bond descriptors are introduced for the *bonded* (externally connected, "open") groups of AIM. The channel reduction in the subsystem resolution is used to extract the internal and external bond indices, and to combine them into the corresponding global bond descriptors. The atomic resolution of the molecular bond indices is examined and the *bond entropy* concept is introduced, which provides an adequate representation of the bond-orders generated by the MO theory. Illustrative applications to simple orbital models, including 2-AO model and the π-bond systems in butadiene and benzene (from the Hückel approximation), are used to illustrate the concepts proposed and to compare the IT bond indices with the parallel predictions from the MO theory.

8.1. INTRODUCTION

Chemistry is the science about molecules and their fragments, e.g., pairs of bonded atoms or larger collections of AIM, functional groups, reactants in bimolecular reactive system, σ and π subsystems in benzene, adsorbate and substrate in the chemisorption system of heterogeneous catalysis, etc. Therefore, besides *global* bond indices, characterizing the molecular system as a whole, of interest in chemistry also are the entropy/information descriptors of various molecular subsystems. In this chapter we shall approach this classical problem using three alternative strategies within the communication theory approach which has been outlined in the preceding chapter. The *additive* decomposition of the global bond descriptors into the corresponding molecular fragment indices will be investigated. The *bond entropy* concept will be introduced and shown to provide the information theoretic bond indices, which closely parallel the bond-order measures from the molecular orbital theory.

The key issue of the subsystem development is an extraction of the information channels of molecular fragments, called *sub-channels*, from the known *molecular* communication system. We shall explore in this chapter three alternative strategies for constructing such partial information networks describing specific parts of a molecule (Nalewajski, 2005a-c). First, the *internal* entropy/information data of the *separate* (mutually-"closed", externally-disconnected, non-bonded) parts of the molecule will be examined. The *external* communication coupling in the molecule of such separated molecular subsystems will then be estimated using the *Independent Fragment Approximation* (IFA) (Nalewajski, 2005c). The information bond indices of such separate, independent molecular subsystems will be combined into the corresponding global bond descriptors using the approximate combination rule related to the *Grouping Axiom* of IT (Shannon, 1948), which we have summarized in Section 3.6. This simple scheme will be shown to reproduce to a remarkably high accuracy the known global indices of model systems.

A more realistic treatment of molecular subsystems in the communication theory has to take into account their open (bonded) character relative to the rest of the molecule. This is manifested by the communication connections between the constituent AIM of the fragment in question and all remaining atoms determining the subsystem molecular environment. The two types of the *partial* channels of molecular fragments, which recognize this information coupling in molecule, the so called *column* and *row* channels suggested by the *additive* decomposition of the global entropy/information descriptors, will be proposed and illustrated for model systems (Nalewajski, 2005b). Their entropy/information descriptors will then be used as probes of the bonded subsystems in model systems.

Finally, the channel *reductions*, carried out by combining several atomic inputs and/or outputs into a single unit, will be used to diagnose the internal and external chemical bonds of molecular fragments (Nalewajski, 2005a). The so called *complementary* reductions will be shown to allow a formulation of the exact combination rules for the molecular entropy/information bond indices.

8.2. RENORMALIZED CHANNELS OF SEPARATE DIATOMICS-IN-MOLECULES

By hypothetically closing a given subset of constituent atoms of the molecular system M, relative to the rest of the molecule, one effectively disregards the chemical bonds between the fragment of interest and its molecular environment. In other words, in such a *Separate Fragment Approximation* (SFA) one counts only the *internal* (intra-fragment) bonds and ignores the chemical interactions between the molecular subsystem and its complementary molecular environment (Nalewajski, 2005c).

Consider, for example, the *renormalized* (r) diatomic sub-channel shown in Scheme 8.1, for the separate bond i—j in the hypothetical molecular system $M^*(i, j) = (\ldots \mid l \mid i \,\vdots\, j \mid k \mid \ldots)$, in which only atoms i and j are mutually open, with all remaining constituent atoms being mutually closed (disconnected).

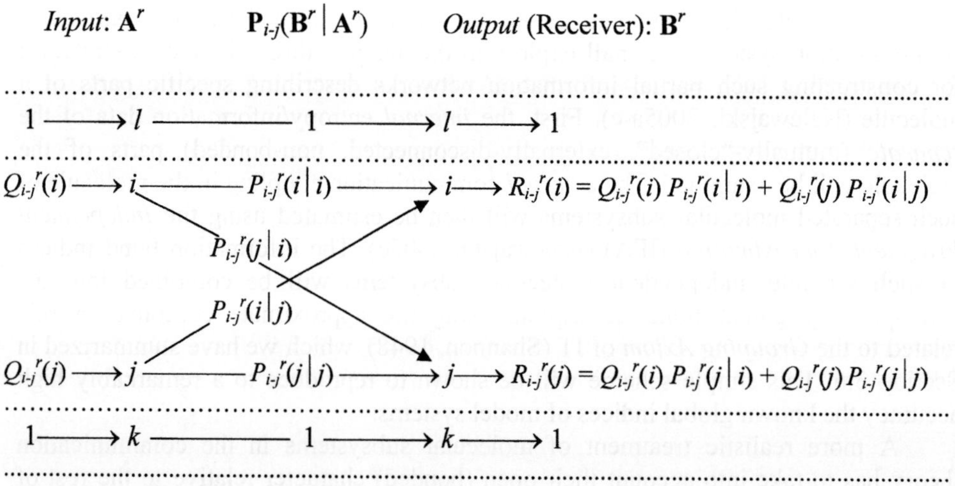

Scheme 8.1. Schematic diagram of the communication sub-channel for the separated diatomic (i, j) representing the internal (localized) bond between atoms i and j in $M^*(i, j) = (\ldots \mid l \mid i \,\vdots\, j \mid k \mid \ldots)$.

Its (renormalized) probabilities have to satisfy the following closure relations:

$$Q_{i\text{-}j}{}^{r}(i) + Q_{i\text{-}j}{}^{r}(j) = R_{i\text{-}j}{}^{r}(i) + R_{i\text{-}j}{}^{r}(j) = \sum_{m=(i,j)} P_{i\text{-}j}{}^{r}(m \mid n) = 1, \quad n \in (i, j). \qquad (8.2.1)$$

They are obtained from the (i, j)-block, $\mathbf{P}_{i\text{-}j}$, of the molecular condensed two-electron probabilities $\mathbf{P}$, which determines the following partial sums for the diatomic under consideration:

$$M^*(A, B) = (A \mid B|C|D)$$

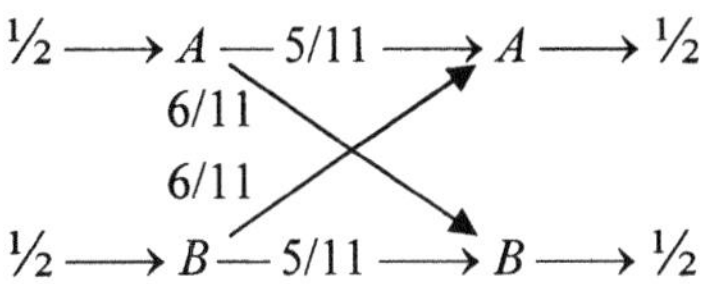

$S_{A\text{-}B}(\mathbf{B}^r \mid \mathbf{A}^r) = 0.994,\ I_{A\text{-}B}(\mathbf{A}^r{:}\mathbf{B}^r) = 0.006,$

$N_{A\text{-}B}{}^r = 1;$

a)

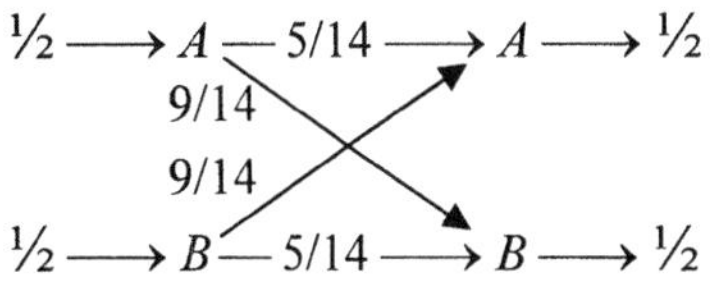

$S_{A\text{-}B}(\mathbf{B}^r \mid \mathbf{A}^r) = 0.881,\ I_{A\text{-}B}(\mathbf{A}^r{:}\mathbf{B}^r) = 0.119,$

$N_{A\text{-}B}{}^r = 1;$

b)

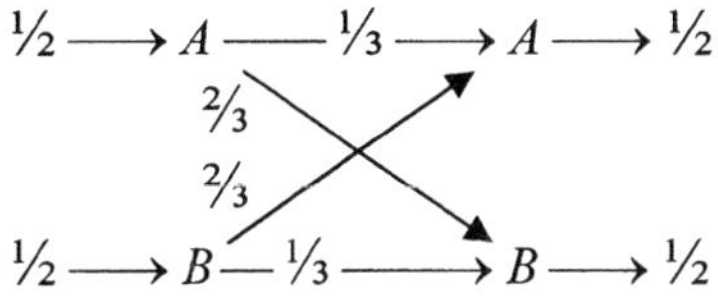

$S_{A\text{-}B}(\mathbf{B}^r \mid \mathbf{A}^r) = 0.918,\ I_{A\text{-}B}(\mathbf{A}^r{:}\mathbf{B}^r) = 0.082,$

$N_{A\text{-}B}{}^r = 1.$

c)

Scheme 8.2. The renormalized communication systems for the separate (A-B)–bond in the biradical butadiene $M^*(A, B) = (A \mid B|C|D)$ in the Hückel approximation: $(A\text{-}B) = \{(1\text{-}2)$ or $(3\text{-}4)\}$ (Panel *a*), $\{(2\text{-}3)$ or $(1\text{-}4)\}$ (Panel *b*), and $\{(1\text{-}3)$ or $(2\text{-}4)\}$ (Panel *c*). The closed carbon atoms C and D, which are hypothetically forbidden to form the π-bonds, are not shown in these diagrams. The reported entropy/information indices (in bits) represent the "reduced" quantities, per overall unit input-probability of the diatomic fragment D. The corresponding "absolute" data, for D in butadiene, are obtained by multiplying these numerical results by the probability $P_D = ½$ of D in the system as a whole.

$$\sum_{k=(i,j)} \sum_{l=(i,j)} P_{k,l} = \wp_{i\text{-}j},\quad \sum_{k=(i,j)} P_{k,l} = \wp_{i\text{-}j}(l),\quad \sum_{l=(i,j)} P_{k,l} = \wp_{i\text{-}j}(k),\quad k, l \in (i, j). \quad (8.2.2a)$$

The renormalized diatomic-channel probabilities then read:

$$\{Q_{i\text{-}j}{}^r(n) = \wp_{i\text{-}j}(n)/\wp_{i\text{-}j}\} = \boldsymbol{P}_{i\text{-}j}(\mathbf{A}^r) = \boldsymbol{P}_{i\text{-}j}(\mathbf{B}^r),\quad \{P_{i\text{-}j}{}^r(m, n) = P_{m,n}/\wp_{i\text{-}j}\} = \mathbf{P}_{i\text{-}j}(\mathbf{A}^r\mathbf{B}^r),$$
$$\{P_{i\text{-}j}{}^r(m \mid n) = P_{i\text{-}j}{}^r(m, n)/Q_{i\text{-}j}{}^r(n)\} = P_{i\text{-}j}(\mathbf{B}^r \mid \mathbf{A}^r);\quad (n, m) \in (i, j). \quad (8.2.2b)$$

As an illustration, let us examine such separated diatomic π-bond channels in the hypothetical *biradical* butadiene, $M^*(A, B) = (A \mid B|C|D)$, with the $2p_\pi$-electrons of the two (closed) carbon atoms C and D being excluded from the π-bond system. We are interested in the entropy/information descriptors of the covalent/ionic composition of three sets of the equivalent π-bonds in the Hückel approximation:

{(1-2), (3-4)}, {(2-3), (1-4)}, and {(1-3), (2-4)}. The corresponding renormalized communication channels are shown in Scheme 8.2, where the relevant entropy/information bond indices (in bits) are also reported. The conserved value of the overall information index of 1 bit confirms that these channels provide the "normalized" bond components, calculated per unit input-probability in the diatomic fragment under consideration (see Scheme 8.2). In what follows we such quantities the "reduced" bond-indices. The actual ("absolute") entropy/information descriptors of the chemical bond of the diatomic fragment D in the molecule are obtained by multiplying the entries of this scheme by the fragment probability P_D = ½ in the butadiene π-system. It follows from the IT-data reported in Scheme 8.2 that the terminal (1-2) and (3-4) π-bonds in the carbon chain exhibit practically purely IT-covalent character (Panel a), while the (2-3) and (1-4) π-bonds, between the middle and terminal carbons, respectively, have distinctly higher IT-ionic content (about 12 %) (Panel b). A comparable level of ionicity (around 8%) is predicted for the (1-3) and (2-4) π-bonds (Panel c).

$$M^*(A, B) = (A{\mid}B|C|D|E|F)$$

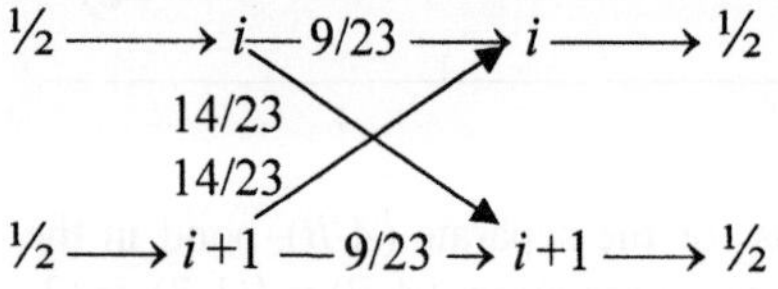

$S_{A\text{-}B}(\mathbf{B}^r | \mathbf{A}^r) = 0.966$, $I_{A\text{-}B}(\mathbf{A}^r{:}\mathbf{B}^r) = 0.034$, $N_{A\text{-}B}{}^r = 1$;

a) ortho: $M^*(i{\mid}i+1) = (i{\mid}i+1|k|l|...)$

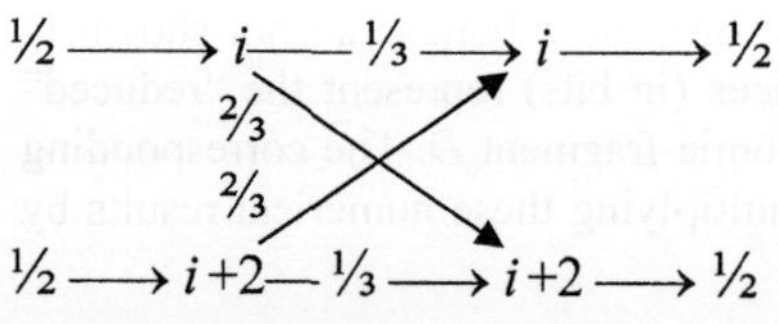

$S_{A\text{-}B}(\mathbf{B}^r | \mathbf{A}^r) = 0.918$, $I_{A\text{-}B}(\mathbf{A}^r{:}\mathbf{B}^r) = 0.082$, $N_{A\text{-}B}{}^r = 1$;

b) meta: $M^*(i, i+2) = (i{\mid}i+2|k|l|...)$

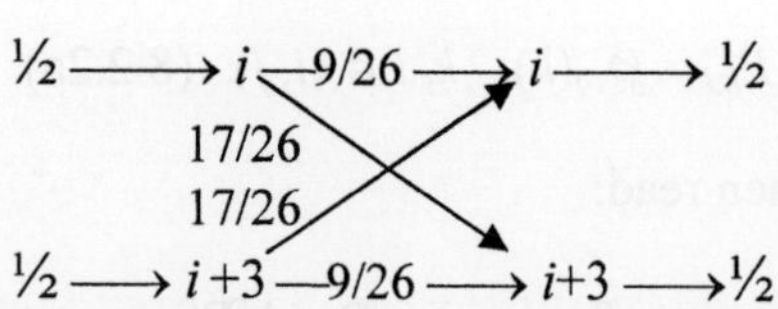

$S_{A\text{-}B}(\mathbf{B}^r | \mathbf{A}^r) = 0.931$, $I_{A\text{-}B}(\mathbf{A}^r{:}\mathbf{B}^r) = 0.069$, $N_{A\text{-}B}{}^r = 1$;

c) para: $M^*(i{\mid}i+3) = (i{\mid}i+3|k|l|...)$

Scheme 8.3. Renormalized sub-channels for the separated π bonds i-j in benzene, for $M^*(i, j) = (i{\mid}j|k|l|...)$ and j = {i+1, *ortho* (Panel a), i+2, *meta* (Panel b), and i+3, *para* (Panel c)}, in the Hückel approximation. The closed carbon atoms $k \neq (i, j)$, which are hypothetically excluded from forming π-bonds with the system remainder, are removed from the diagrams. The consecutive numbering of carbons in the ring is adopted. The "absolute" IT-indices are obtained by multiplying the reported "reduced" data by P_D = 1/3.

Of interest also are the corresponding channels for alternative pairs of carbon atoms in the benzene ring, shown in Scheme 8.3. It should be recalled that these channels are defined by a single pair of bonded (open) ring atoms, thus corresponding to a hypothetical state of the localized (single) π electron on each of the four non-bonded (closed) carbon atoms of the benzene ring. A reference to Scheme 8.3 indicates that all these separated π bonds in the benzene ring are strongly covalent, exhibiting the residual information-ionicity below 10%. The highest entropy-covalency (lowest information-ionicity) is found for the mutual *ortho*-position of the two carbon atoms, while the lowest covalency (highest ionicity) is observed for the *meta*-pairs of the ring carbons. The *para* C–C bond exhibits intermediate levels of the two bond components, being somewhat closer to the *meta* than *ortho* IT indices.

8.3. COMMUNICATION CHANNELS OF THE MUTUALLY SEPARATE GROUPS OF AIM

One can also consider several externally (inter-group) closed groups $\{G_i\}$ of the internally (intra-group) open AIM, when the chemical bonds are allowed exclusively within each group of the system constituent atoms, e.g., in $M^*(G_1|G_2|G_3) = (i\,\vdots\,j\,\vdots\ldots|k\,\vdots\,l\,\vdots\ldots|m\,\vdots\,n\,\vdots\ldots) \equiv M^*(i, j, \ldots|k, l, \ldots|m, n, \ldots)$. Examples of such generalized divisions of the π-bond systems in butadiene and benzene are shown in Schemes 8.4 and 8.5, respectively, where the relevant sub-channels in the Hückel approximation and the corresponding "reduced" entropy/information indices are also reported. It should be observed that the overall entropic indices of such hypothetical collections of the mutually-separated (non-bonded, disconnected, closed) groups of AIM are the sums of the internal subsystem contributions.

It follows from Scheme 8.4 that the entropy/information indices of the overall intra-group bond multiplicity from IT and its ionic/covalent composition generally agree with intuitive chemical estimates for these selected valence structures in butadiene. For example, a diminished entropic bond-order of $M^*(1, 2, 3\,|\,4) = M^*(G_1\,|\,4)$ (Panel A*b*) agrees with about 3/2-bond expected for the allyl fragment G_1 (see Scheme 7.10) with the remaining structures (Panels A*c-e*), involving divisions into separate diatomic subsystems, all giving rise to the conserved, double total entropic π-bond index.

The differences in "strengths" of the corresponding separated bonds are well reflected by the bond covalent/ionic composition. The lowest covalent component $N^{cov} = 1.76$ bits is reported in Panel A*d*, for two π-bonds between the terminal and middle carbon atoms, respectively, while the two pairs of neighboring atoms in Panel A*c* generate almost purely covalent bonds: $N^{cov} = 1.99$. The second-neighbor pairs of AIM in Panel A*e* give rise to the intermediate bond composition between these two extreme "valence" structures.

a) $M(1, 2, 3, 4)$: $N^{cov} = 1.94,\ N^{ion} = 0.06,\ N = 2;$ **A)**

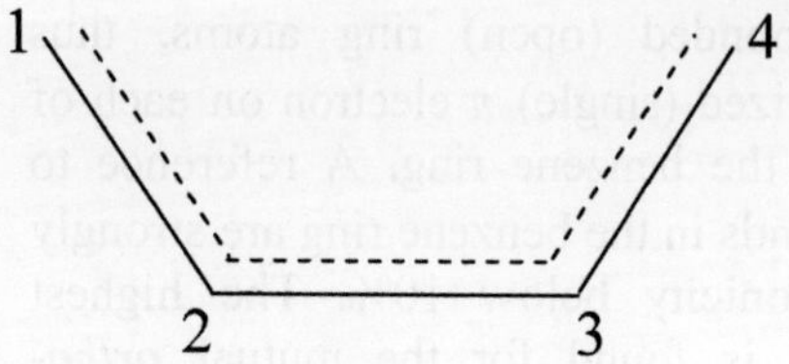

b) $M^*(1, 2, 3 \,|\, 4) \equiv M^*(G_1 \,|\, 4)$: $N^{cov} = 1.53,\ N^{ion} = 0.05,\ N = 1.58;$

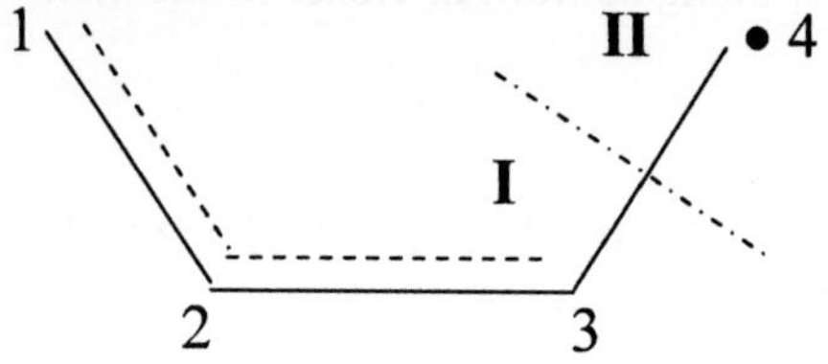

c) $M^*(1, 2 \,|\, 3, 4) \equiv M^*(G_2 \,|\, G_3)$: $N^{cov} = 1.99,\ N^{ion} = 0.01,\ N = 2;$

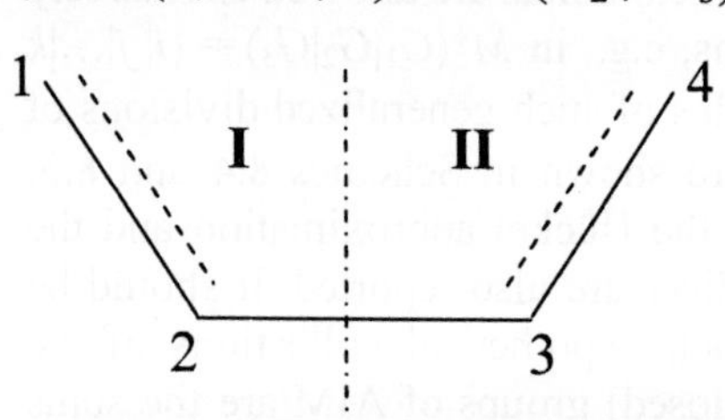

d) $M^*(1, 4 \,|\, 2, 3) \equiv M^*(G_4 \,|\, G_5)$: $N^{cov} = 1.76,\ N^{ion} = 0.24,\ N = 2;$

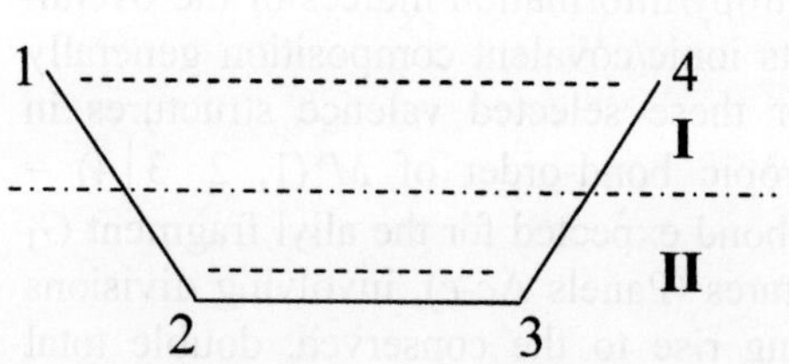

e) $M^*(1, 3 \,|\, 2, 4) \equiv M^*(G_6 \,|\, G_7)$: $N^{cov} = 1.84,\ N^{ion} = 0.16,\ N = 2;$

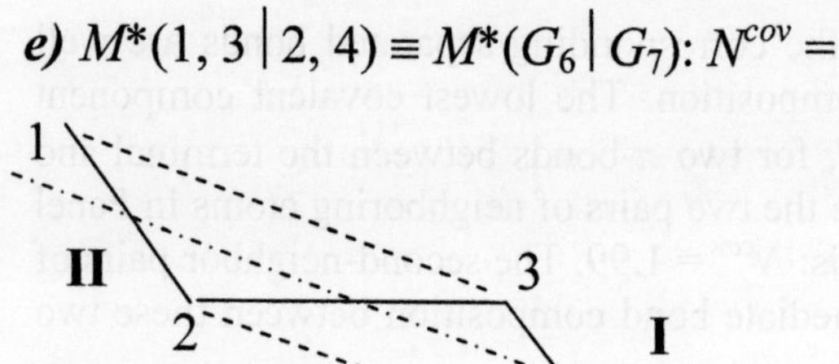

$M^*(1, 2, 3 \mid 4) \equiv M^*(G_1 \mid 4)$: **B)**

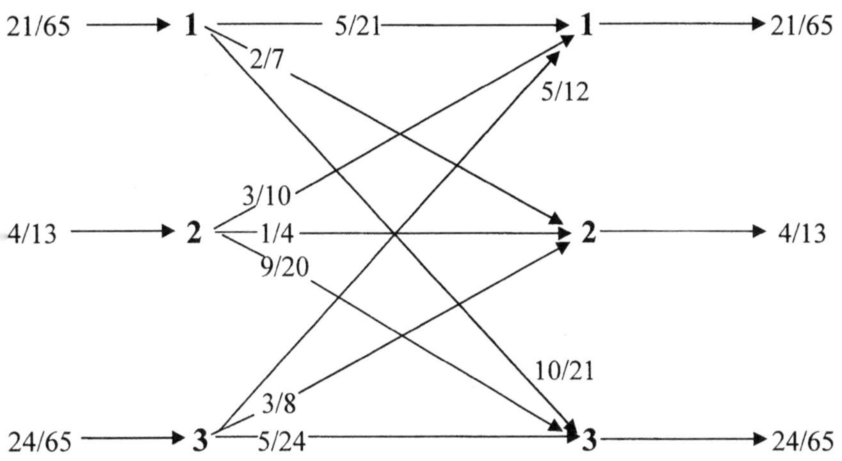

$$S(\mathbf{B}^r \mid \mathbf{A}^r) = 1.53, \quad I(\mathbf{A}^r{:}\mathbf{B}^r) = 0.05, \quad N^r = S(\mathbf{B}^r \mid \mathbf{A}^r) + I(\mathbf{A}^r{:}\mathbf{B}^r) = 1.58$$

Scheme 8.4. Separated groups of π-bonds in butadiene (Part A) and the predicted entropy/information indices (in bits). In Part B the renormalized channel for the separate triatomic group $G_1 = (1, 2, 3)$ in butadiene is shown. It is seen to strongly resemble the allyl communication system of Scheme 7.10, thus exhibiting similar IT bond indices. The corresponding channels for the diatomic fragments are reported in Scheme 8.2.

Similar conclusions follow for the valence structures in benzene (Scheme 8.5). Limiting the electron delocalization in the benzene ring to the consecutive four and two carbon atoms, respectively, in $M^*(1, 2, 3, 4 \mid 5, 6) = M^*(G_1 \mid G_2)$ (Panel A*b*) gives the overall 3-bits ("triple") π-bond in such a two-group system, in full agreement with an intuitive chemical expectation for these separate butadiene- and ethylene-like fragments.

The bond composition in group G_1 of four neighboring carbons in benzene ring is indeed almost identical with that predicted for butadiene, with the system of two conjugated bonds being almost purely covalent; the single bond in G_2 is also seen to be equally covalent in character, again approximating rather accurately the purely covalent π-bond in ethylene.

A comparison of these predictions for the separated, complementary fragments of benzene ring with the corresponding entropy/information indices for the system as a whole (listed in Panel A*a* of the diagram) indicates that the overall index, of almost exclusively IT covalent origin, becomes lowered, when the hypothetical barriers for the electron delocalization in the carbon-ring are lifted.

a) $M(1, 2, 3, 4, 5, 6)$: $N^{cov} = 2.55$, $N^{ion} = 0.03$, $N = 2.58$ **A)**

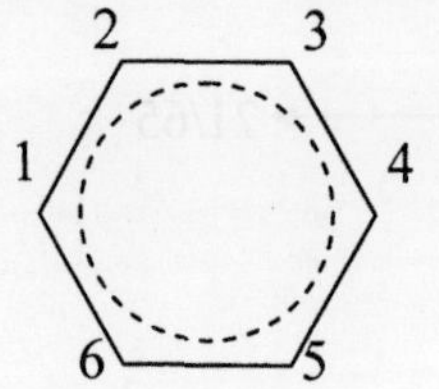

b) $M^*(1, 2, 3, 4|5, 6) \equiv M^*(G_1|G_2)$: $N^{cov} = 2.92$, $N^{ion} = 0.08$, $N = 3.00$

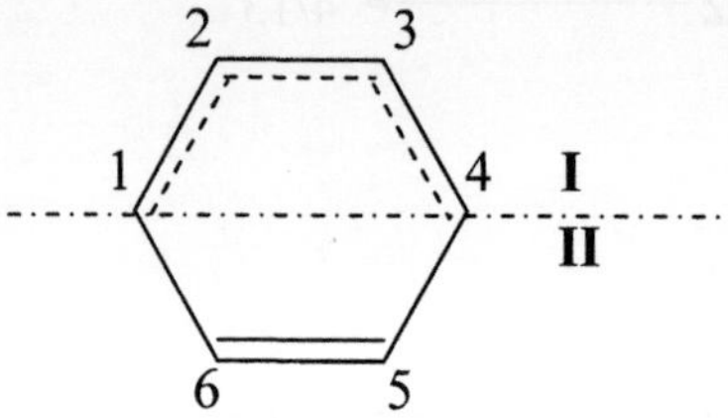

$N_I^{cov} = 1.96$, $N_I^{ion} = 0.04$, $N_I = 2.00$

$N_{II}^{cov} = 0.97$, $N_{II}^{ion} = 0.03$, $N_{II} = 1.00$

c) $M^*(1, 2|3, 4|5, 6) \equiv M^*(G_3|G_4|G_2)$: $N^{cov} = 2.90$, $N^{ion} = 0.10$, $N = 3.00$

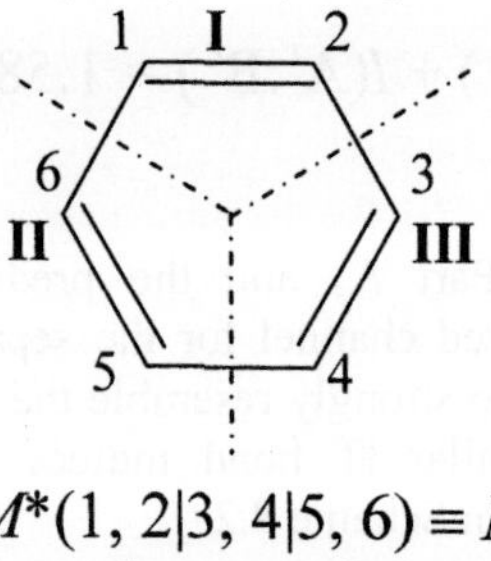

$N_I^{cov} = 0.97$, $N_I^{ion} = 0.03$, $N_I = 1.00$

d) $M^*(1, 2|3, 4|5, 6) \equiv M^*(G_3|G_5|G_6)$: $N^{cov} = 2.86$, $N^{ion} = 0.14$, $N = 3.00$

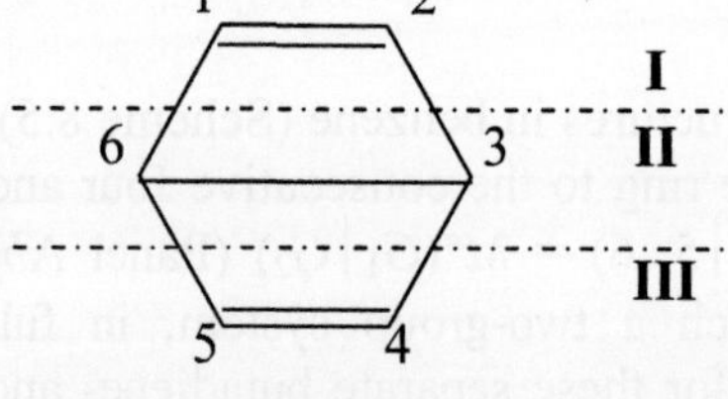

$N_I^{cov} = 0.97$, $N_I^{ion} = 0.03$, $N_I = 1.00$

$N_{II}^{cov} = 0.93$, $N_{II}^{ion} = 0.07$, $N_{II} = 1.00$

e) $M^*(1, 3\,|\,2\,|\,4, 6\,|\,5) \equiv M^*(G_7\,|\,2\,|\,G_8\,|\,5)$:

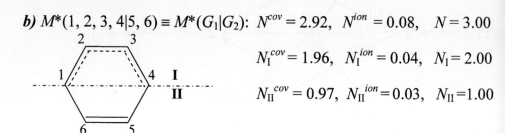

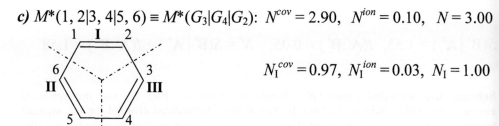

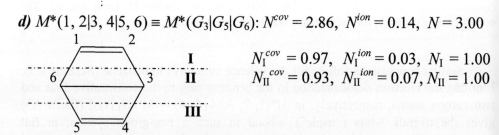

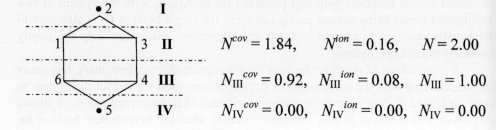

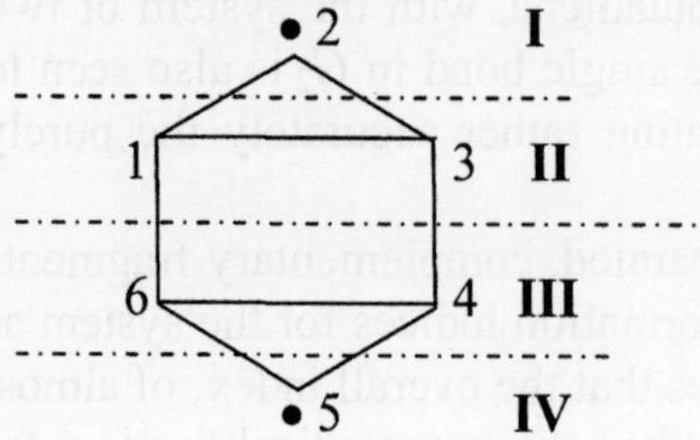

$N^{cov} = 1.84$, $N^{ion} = 0.16$, $N = 2.00$

$N_{III}^{cov} = 0.92$, $N_{III}^{ion} = 0.08$, $N_{III} = 1.00$

$N_{IV}^{cov} = 0.00$, $N_{IV}^{ion} = 0.00$, $N_{IV} = 0.00$

$M^*(1, 2, 3, 4 \mid 5, 6) \equiv M^*(G_1 \mid G_2)$: **B)**

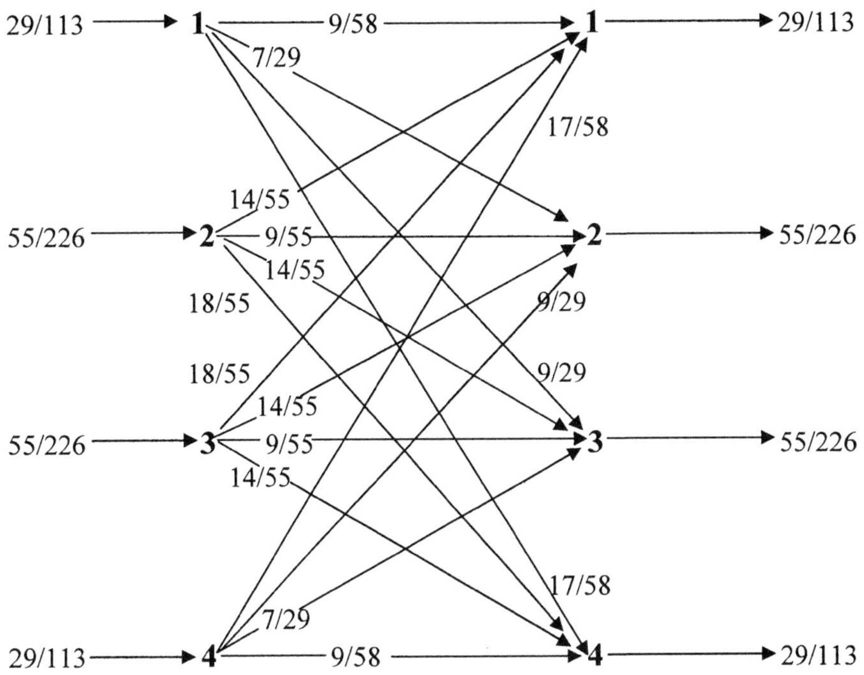

$S(\mathbf{B}^r \mid \mathbf{A}^r) = 1.96, \quad I(\mathbf{A}^r{:}\mathbf{B}^r) = 0.04, \quad N^r = S(\mathbf{B}^r \mid \mathbf{A}^r) + I(\mathbf{A}^r{:}\mathbf{B}^r) = 2.00$

Scheme 8.5. Valence structures of the separate π-bonds in benzene (Part A) and the renormalized communication channel (Part B) for the four-atom fragment G_1 of Scheme A*b* (Part B). The relevant diatomic channels of the separate bonds are summarized in Scheme 8.3. In Part A the predicted overall (Panel *a*) and group entropy/information indices (in bits) of the covalent and ionic bond components and the resulting total internal bond-orders are also listed.

This is in accord with the modern outlook on the role of σ and π bonds in aromatic systems (Shaik, 1989; Jug and Köster, 1990; Shaik and Hiberty, 1991, 1995, 2004). Indeed, the regular hexagon structure of the benzene ring is due to dominating stabilizing influence of the σ bonds, which counter the bond alternation, while the π system favors the localized bonds, which reach the maximum strength in the bond-alternated hexatriene. The entropy/information indices predicted for the remaining partitions of Panels A(*c*-*e*), into the separated diatomic and atomic subsystems, are also in accord with the chemical intuition reflected by the structural formulas of these "valence" structures. The diatomic interactions across the benzene ring are diagnosed as generating a slightly higher information ionicity [Panels A(*c*, *d*)] in comparison to those between nearest neighbors [Panels A(*b*, *c*)].

8.4. COMBINING SUBSYSTEM INDICES INTO GLOBAL INFORMATION DESCRIPTORS

The renormalized input/output probabilities of the separate molecular subsystems are "conditional" in character, because they represent the probabilities of the intra-group outcomes, conditional upon the fact that the outcome has already been located in a given molecular fragment. The sum of Eq. (8.2.2a),

$$\wp_{i\text{-}j} \equiv \wp_K = \wp_K(i) + \wp_K(j), \qquad (8.4.1)$$

represents the (diagonal) condensed probability $P_{K,K} = \wp_K$ that a pair of electrons has been simultaneously located on the diatomic fragment $K = (i, j)$ in the molecule. The partial sum $P_{K,n} = P_{n,K} = \wp_K(n)$ similarly represents the effective one-electron probability that an electron of the electronic pair of the diatomic fragment K in the molecule has been located on atom $n \in K$. Therefore, the group input-probability of Scheme 8.1 [Eq. (8.2.2b)] denotes the *conditional* effective one-electron probability that an electron will be located on atom n in K, provided that it has originated from an electronic pair attributed in the molecule to this molecular subsystem:

$$Q_K^r(m) = \wp_K(m)/\wp_K = P_{m,K}/P_{K,K} = P(m, K \mid K) \equiv P(m \mid K),\ m \in K;$$
$$\textstyle\sum_{m\in K} P(m \mid K) = P(K \mid K) = 1, \qquad (8.4.2)$$

The conditional character of this subsystem-normalized distribution is already reflected by the normalization condition in the preceding equation, which involves only the summation over the subsystem constituent atomic labels (variables), for the fixed fragment identity (a parameter). A similar conditional interpretation can be attributed to the output-probabilities of atoms $n \in K$ in Scheme 8.1:

$$R_K^r(n) = Q_K^r(i)\, P_K^r(n \mid i) + Q_K^r(j)\, P_K^r(n \mid j) = P_K^r(n, i) + P_K^r(n, j) = P(n \mid K). \qquad (8.4.3)$$

Also the subsystem renormalized two-electron probabilities of Eq. (8.2.2b) have a conditional meaning. More specifically, for $\{m, n\} \in K$,

$$P_K^r(m, n) = P_{m,n}/\wp_K \equiv P(m, n \mid K), \qquad \textstyle\sum_{m\in K}\sum_{n\in K} P(m, n \mid K) = 1. \qquad (8.4.4)$$

This conditional interpretation of the renormalized probabilities of the separate diatomic subsystems naturally connects this separated-subsystem development to the scenario of the Grouping Axioms in IT, which we have summarized in Section 3.6. Let us recall that in a discrete case of the probability vector $\boldsymbol{p} = \{p_i\} \equiv \{\boldsymbol{p}_K\}$, exhaustively divided into probabilities $\{\boldsymbol{p}_K = \{p_{i\in K}\}\}$ of the exclusive groups $\boldsymbol{G} = \{G_K \equiv K\}$ of outcomes, the overall normalization condition reads:

$$\textstyle\sum_i p_i = \sum_K (\sum_{i\in K} p_i) \equiv \sum_K P_K = 1. \qquad (8.4.5)$$

Here, the row vector $\boldsymbol{P}^G = \{P_K\}$ contains the condensed probabilities of the molecular outcome in group K, which further define the conditional intra-group probabilities:

$$\boldsymbol{\pi}_K = \{\pi(i \in K \mid K) = p_{i\in K}/P_K\}, \qquad \sum_{i\in K} \pi(i \mid K) = 1. \tag{8.4.6}$$

The Grouping Axiom for the Shannon entropy [Eq. (3.6.1)] states that the inter- and intra-group entropies should be combined into the entropy of the whole probability distribution in the following way:

$$S(\boldsymbol{p}) = S(\boldsymbol{P}^G) + \sum_K P_K S(\boldsymbol{\pi}_K). \tag{8.4.7}$$

Here the entropy of the overall distribution $\boldsymbol{p}$ is expressed as the sum of the *external* (group) entropy $S(\boldsymbol{P}^G)$ of the condensed probabilities of the groups of outcomes and the mean of the *internal* entropies $\{S(\boldsymbol{\pi}_K)\}$ of the conditional (intra-group) probabilities $\{\boldsymbol{\pi}_K\}$ defined by the $\boldsymbol{P}^G$ "weights".

We now seek similar rules for combining the separate-fragment data into the global entropy/information indices of all chemical bonds in the molecular system as a whole, when all constituent atoms are considered to be mutually-open. It should be realized that the *Separate Fragment Approximation* (SFA), to which Scheme 8.6*a* applies, completely ignores the inter-subsystem communications responsible for the external entropy term in the foregoing equation. In order to generate these missing conditional-entropy contributions, we shall adopt the IFA, in which the inter-subsystem two-electron probabilities in the condensed subsystem resolution are products of the condensed one-electron probabilities of the independent fragments (see Scheme 8.6*b*).

When combining, i.e., mutually "opening", the molecular subsystems in the *inter*-subsystem stage of Scheme 8.6*b*, this approximate approach uses the molecular renormalized (conditional) intra-subsystem two-electron probabilities at the separate subsystem stage of Scheme 8.6*a*, and it views the inter-subsystem events, as involving only the statistically *independent* molecular fragments, treated as whole (condensed, reduced) units, which are fully characterized by the group one-electron probabilities in the molecule, $\boldsymbol{P}^G = \{P_K\}$. Indeed, for independent subsystems the corresponding (condensed) two-electron probabilities are given by the products of the condenased one-electron probabilities, $\mathbf{P}^{IFA}(\boldsymbol{G}, \boldsymbol{G}') = \{P^{IFA}(K, L) = P_K P_L\}$, and hence the corresponding conditional probabilities: $\mathbf{P}^{IFA}(\boldsymbol{G} \mid \boldsymbol{G}') = \{P^{IFA}(L \mid K) = P^{IFA}(K, L)/P_K = P_L\}$.

The "cascade" of the two molecular channels, shown in Scheme 8.6*c*, then determines the desired combination rule for generating the molecular bond indices from the condensed group probabilities and the intra-group entropies of the separated subsystems. The Schemes 8.6(*a*, *b*) reflect the two stages of acquiring the information about the molecular system as a whole. The AIM resolved SFA channel of the first panel gathers the internal information, about the separate subsystems. It accounts for the internal communications in each detached fragment. The subsystem resolved IFA channel of the second panel takes into account, be it in an approximate manner, the effects due to communications between the whole ("reduced") molecular fragments and generates an approximate group entropy, i.e., the external information, due to the inter-fragment communications.

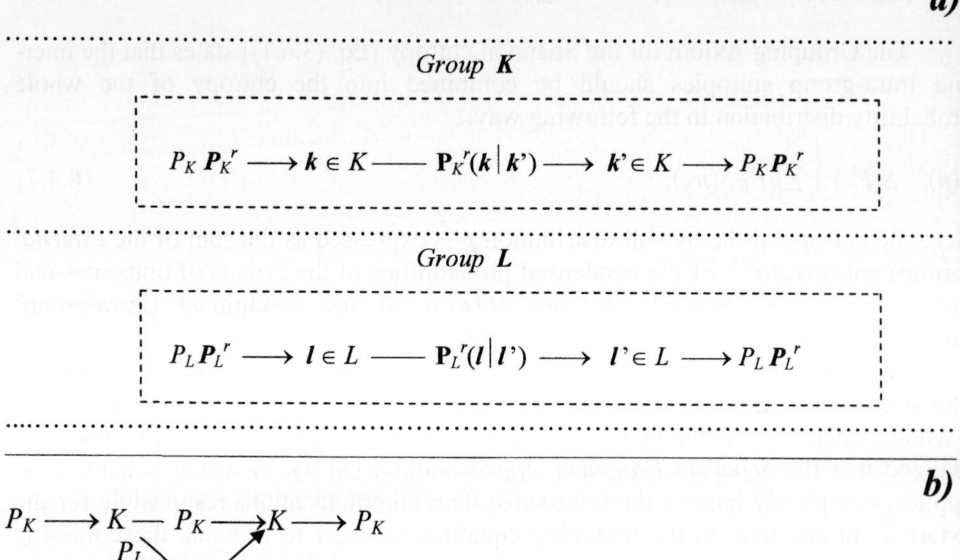

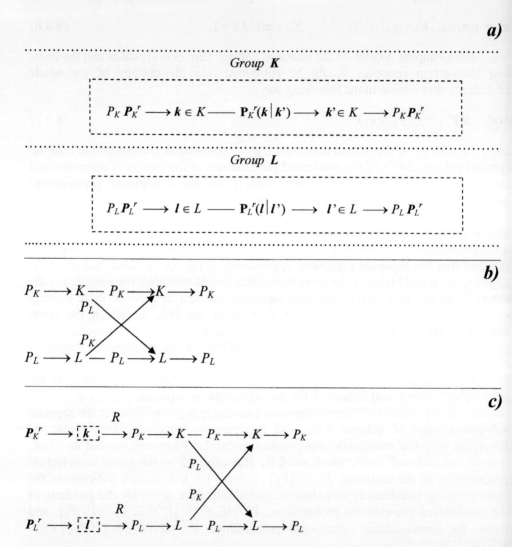

Scheme 8.6. Combining the AIM-resolved channels of the separate fragments in $\boldsymbol{M}^* = (\boldsymbol{k}|\boldsymbol{l}) \equiv (\boldsymbol{K}|\boldsymbol{L})$ (in boxes of Panel *a*) with the subsystem-resolved channel of independent molecular fragments in $M = (K, L)$ into an effective molecular information "cascade" (Panel *c*). The effective channel in Panel *c* allows one to use the internal subsystem entropy/information indices to generate the approximate global bond indices, of mutually open atoms in $\boldsymbol{M} = (\boldsymbol{k}, \boldsymbol{l})$, by supplementing the internal entropies of molecular fragments with the approximate external-entropy contribution from the IFA. The symbol *R* in the last panel denotes the channel "reduction", in which the separate AIM outputs of a given molecular fragment are combined into a single (*reduced*) input, of the subsystem as a whole.

Let us first examine the combination rule for the conditional-entropy measure of the IT bond covalency. Here the constituent AIM $\boldsymbol{k} = (k, k', \ldots)$ in fragment K of the molecular system M are alternatively denoted by the bold fragment label $\boldsymbol{K}$. In the same spirit the complete set of bonded atoms in M, $(i, j, \ldots) = \boldsymbol{i}$ is also denoted by $\boldsymbol{M}$. In accordance with the SFA we exhaustively divide the complete set of the constituent atoms $\boldsymbol{M} = (i, i', \ldots) = \boldsymbol{i}$ among the exclusive groups $\{\ldots, \boldsymbol{k} = (k, k', \ldots) \equiv \boldsymbol{K}, \boldsymbol{l} = (l, l', \ldots) \equiv \boldsymbol{L}, \ldots\}$ in $\boldsymbol{M}^* = (\ldots|\boldsymbol{k}|\boldsymbol{l}|\ldots) \equiv (\ldots|\boldsymbol{K}|\boldsymbol{L}|\ldots)$, representing the separate molecular fragments $\boldsymbol{G} = \{.., K, L, \ldots\}$ of $M^* = (\ldots|K|L|\ldots)$ (see Scheme 8.6*a*). The associated division of the molecular one- and two-electron probabilities into the corresponding blocks of the fragment-resolution gives:

$$\boldsymbol{P} = [\ldots, \boldsymbol{P}_K = \{P_{k\in K}\} \equiv \boldsymbol{P}(\boldsymbol{k}), \boldsymbol{P}_L = \{P_{l\in L}\} \equiv \boldsymbol{P}(\boldsymbol{l}), \ldots],$$
$$\mathbf{P} = \{\mathbf{P}_{K,L} = \{P_{k\in K,\, l\in L}\} \equiv \mathbf{P}(\boldsymbol{k}, \boldsymbol{l})\}. \tag{8.4.8}$$

This partition also defines the corresponding *condensed* probabilities of molecular fragments: $\boldsymbol{P}^G = \{P_K = \sum_{k\in K} P_k\} \equiv \boldsymbol{P}(\boldsymbol{G})$ and $\mathbf{P}^G = \{P_{K,L} = \sum_{k\in K}\sum_{l\in L} P_{k,l}\} = \mathbf{P}(\boldsymbol{G}, \boldsymbol{G}')$.

The global conditional-entropy index, for the system as a whole, can then be transformed into the equivalent subsystem-resolved expression:

$$S(\mathbf{B}|\mathbf{A}) \equiv S(\boldsymbol{j}|\boldsymbol{i}) = -\sum_i\sum_j P_{i,j}\log(P_{i,j}/P_i)$$
$$= -\sum_K\sum_L P_{K,L}\sum_{k\in K}\sum_{l\in L}(P_{k,l}/P_{K,L})\log(P_{k,l}/P_k) \tag{8.4.9}$$

The SFA channel for $\boldsymbol{M}^* = (\ldots|\boldsymbol{k}|\boldsymbol{l}|\ldots)$ (Scheme 8.6*a*), representing the first stage of the information cascade of Scheme 8.6*c*, consists of the AIM-resolved separate (mutually closed, disconnected) group-channels, thus neglecting all the inter-group two-electron connections: $\{P_{K,L}{}^{SFA} = P_{K,K}\delta_{K,L}\}$ and $\{P_{k,l}{}^{SFA} \equiv P_{k,k'}\delta_{l,k'}\}$. This reduces the foregoing expression to the following SFA quantity:

$$S^{SFA}(\mathbf{B}|\mathbf{A}) = -\sum_K P_{K,K}\sum_{k\in K}\sum_{k'\in K}(P_{k,k'}/P_{K,K})\log(P_{k,k'}/P_k)$$
$$\equiv -\sum_K P_{K,K}\sum_{k\in K}\sum_{k'\in K} P(k, k'|K)\log[P(k, k'|K)/P(k|K)],$$
$$= \sum_K P_K\, S_K(\boldsymbol{k}|\boldsymbol{k}'). \tag{8.4.10}$$

Here the intra-fragment conditional probability of finding an electron of group K on its constituent AIM k $P(k|K) = P_k/P_{K,K}$, while $P(k, k'|K) = P_{k,k'}/P_{K,K}$ stands for the conditional two-electron probability of freagment K.

Therefore, the conditional entropy $S^{SFA}(\mathbf{B}|\mathbf{A})$, measuring the overall IT-covalency in a collection of the separate fragments of Scheme 8.6*a*, is the mean of the intra-subsystem covalencies, which are weighted in accordance with the molecular group-probabilities $\boldsymbol{P}^G$. In other words, this AIM-resolved SFA channel represents an ensemble of the separate fragment-channels, with the subsystem condensed probabilities determining the "weighting" factors in determining the ensemble-average measure of the conditional-entropy in M^*.

It should be observed that Eq. (8.4.10) represents the conditional entropy analog of the second, *internal* term in the Shannon grouping theorem of Eq. (8.4.7). Its first, *external* term is generated by the IFA network, shown in Scheme 8.6*b*, the second part of the information cascade of Scheme 8.6*c*,

$$S^{IFA}(\boldsymbol{G}\,|\,\boldsymbol{G'}) = -\sum_K \sum_L P^{IFA}(K, L)\log[P^{IFA}(K, L)/P_K]$$
$$= -\sum_L P_L(\sum_K P_K)\log P_L = -\sum_L P_L \log P_L = S(\boldsymbol{G}). \qquad (8.4.11)$$

To summarize, the SFA/IFA combination rule for the global conditional-entropy (IT-covalency) index in terms of the relevant quantities characterizing molecular fragments reads [compare Eq. (8.4.7)]:

$$S(\mathbf{B}\,|\,\mathbf{A}) = S(\boldsymbol{j}\,|\,\boldsymbol{i}) \cong S^{IFA}(\boldsymbol{G}\,|\,\boldsymbol{G'}) + S^{SFA}(\mathbf{B}\,|\,\mathbf{A}) = S(\boldsymbol{P}^G) + \sum_K P_K S_K(\boldsymbol{k}\,|\,\boldsymbol{k'}). \qquad (8.4.12)$$

The corresponding combination rule for the mutual-information measure the overall bond ionicity in the whole molecule can be derived in a similar way. We start from the exact definition in terms of the molecular probabilities in atomic resolution:

$$I(\mathbf{A}{:}\mathbf{B}) = I(\boldsymbol{i}{:}\boldsymbol{j}) = -\sum_i \sum_j P_{i,j} \log[P_{i,j}/(P_i P_j)]$$
$$= -\sum_K \sum_L P_{K,L} \sum_{k\in K} \sum_{l\in L} (P_{k,l}/P_{K,L}) \log\{(P_{k,l}/P_{K,L})/(P_k P_l/P_{K,L})]\}$$
$$\equiv -\sum_K \sum_L P_{K,L} \sum_{k\in K} \sum_{l\in L} P(k, l|K, L) \log[P(k, l|K, L)/P^{ind}(k, l|K, L)], \qquad (8.4.13)$$

where $P^{ind}(k, l|K, L)$ stands for the molecular joint probabilities of *independent* AIM, conditional upon the specified molecular fragment origins of the two electrons. Invoking next the SFA of Scheme 8.6*a* gives:

$$I(\mathbf{B}{:}\mathbf{A}) \cong I^{SFA}(\mathbf{B}{:}\mathbf{A}) = -\sum_K P_{K,K} \sum_{k\in K} \sum_{k'\in K} P(k, k'|K) \log[P(k, k'|K)/P^{ind}(k, k'|K)]$$
$$= \sum_K P_K I_K(\boldsymbol{k}{:}\boldsymbol{k'}), \qquad (8.4.14)$$

where $I_K(\boldsymbol{k}{:}\boldsymbol{k'})$ denotes the amount of information flowing through the communication channel of the separate fragment K.

It should be observed that the *inter*-subsystem contribution to the global mutual-information corresponding to Scheme 8.6*b* vanishes identically in the IFA, since the condensed probabilities of molecular fragments in the input and output of the IFA-channel are independent. Therefore, the preceding equation represents the final SFA/IFA combination rule for the bond information ionicity. The global index is thus approximated by the mean value of the *intra*-subsystem IT-ionicities, with the weights provided by the condensed one-electron probabilities of the molecular fragments of the partition in question.

In Table 8.1 we have tested the performance of the combination rules of Eqs. (8.4.12) and (8.4.14) for the π-bonds in butadiene and benzene in the Hückel approximation. The relevant partitions of the bonded carbon atoms into subsystems, the associated communication channels of the separated subsystems and their entropy/information indices have already been summarized in Schemes 8.2-8.5. In these alternant hydrocarbons the one-electron probabilities of all carbon atoms are identical, so that the group probabilities are proportional to the subsystem number of atoms participating in the π-bond system in the molecule. Thus, for the fragment K of the π-system consisting of N_K carbon atoms, $P_K = N_K/N$, where N denotes the overall number of atoms in the molecular π-bond system. The reference global entropy/information data reported in Table 8.1, for the molecule as a whole, correspond to Panels *a* in Schemes 8.4A and 8.5A, which involve no internal partitions of the molecular π-bond system.

Table 8.1. The performance of the SFA/IFA approximate combination rules for predicting the global entropy/information bond indices (in bits) from those characterizing the molecular fragments, for butadiene and benzene in the Hückel theory approximation. The symbols of partitions refer to those shown in Schemes 8.4A and 8.5A.

Bond Index	Butadiene (Scheme 8.4A)					Benzene (Scheme 8.5A)				
	Partition: *a*	*b*	*c*	*d*	*e*	*a*	*b*	*c*	*d*	*e*
N^{cov}	1.944	1.958	1.994	1.881	1.918	2.551	2.545	2.551	2.539	2.530
N^{ion}	0.056	0.042	0.006	0.119	0.082	0.034	0.040	0.034	0.046	0.055
N	2.000	2.000	2.000	2.000	2.000	2.585	2.585	2.585	2.585	2.585

An inspection of Table 8.1 shows that the combination rules for obtaining the global information indices of chemical bonds in the molecule, from the separate subsystem entropy/information descriptors, work quite well. They exactly predict the global measure of all chemical bonds in the system and generate semi-quantitatively its covalent/ionic components. It should be recalled that these simple rules have been derived using the SFA/IFA approximation, in which the molecular fragment data, acquired by treating each subsystem separately, are combined as the overall, independent units of the approximate molecular communication system. A generally good performance of these grouping rules indicates that these fragments of π-electron systems are indeed to a good approximation independent of each other. This may indicate that their entropy/information indices may be to a large extent "transferable" (in the combination-rule sense of the word) from one molecular environment to another.

8.5. ADDITIVE DECOMPOSITION OF MOLECULAR BOND INDICES

The separate (disconnected) channels of molecular fragments in the SFA correspond to rather artificial, hypothetical state, in which the constituent molecular subsystems are forbidden to form chemical bonds between themselves. The inter-fragment communication coupling has then been added, at the IFA stage of Scheme 8.6*b*, in an approximate way by treating molecular fragments as independent units in a more "coarse" communication system, in which subsystems are regarded as whole (condensed, "reduced") blocks. In this section we shall examine the *partial* channels of molecular fragments (Nalewajski, 2005b), which take into account their probability connections in a molecule, both between the subsystem constituent AIM and with the remaining atoms defining the fragment molecular environment. These channels thus correspond to the communication-*embedded* parts of the molecule. The partial-channel approach thus constitutes an exact treatment in atomic resolution of both the *intra*- and *inter*-group probability couplings in the molecule. Together the partial-channel description is equivalent to that provided by the full AIM resolved molecular channel itself. Indeed, the partial communication channels of molecular fragments define by construction the *additive* contributions to the global entropy/information bond indices defined by the full AIM-resolved molecular channel.

As before, the bold molecule/fragment symbols will be used to indicate the atomic resolution of the system/subsystem in question. It follows from the expressions for the global bond indices in atomic resolution that they can be naturally decomposed in terms of the *additive* contributions from the exclusive groups of AIM in M, $\boldsymbol{G} = (K, L, \ldots)$, $\boldsymbol{M} = (i, i', \ldots) \equiv (\boldsymbol{i}) = (\ldots; k, k', \ldots; l, l', \ldots; \ldots) = (\ldots, \boldsymbol{k}, \boldsymbol{l}, \ldots) \equiv (\ldots, \boldsymbol{K}, \boldsymbol{L}, \ldots)$, defining a given partition of the molecule M into constituent fragments (groups of AIM), $\boldsymbol{K} \equiv \boldsymbol{k} = \{k \in K\}$, $\boldsymbol{L} \equiv \boldsymbol{l} = \{l \in L\}$, etc.:

$$\begin{aligned} S(\mathbf{B}\,|\,\mathbf{A}) &= -\textstyle\sum_{a\in\mathbf{A}}\sum_{b\in\mathbf{B}} P_{a,b}\log(P_{a,b}/P_a) \equiv \sum_{a\in\mathbf{A}} S(\boldsymbol{M}\,|\,a) \equiv \sum_{b\in\mathbf{B}} S(b\,|\,\boldsymbol{M}) \\ &= -\textstyle\sum_{K\in\mathbf{A}}\sum_{L\in\mathbf{B}}\sum_{k\in K}\sum_{l\in L} P_{k,l}\log(P_{k,l}/P_k) \\ &\equiv \textstyle\sum_K\sum_L S(\boldsymbol{L}\,|\,\boldsymbol{K}) \equiv \sum_K S(\boldsymbol{M}\,|\,\boldsymbol{K}) \equiv \sum_L S(\boldsymbol{L}\,|\,\boldsymbol{M}), \end{aligned} \tag{8.5.1}$$

$$\begin{aligned} I(\mathbf{A}^0{:}\mathbf{B}) &= \textstyle\sum_{a\in\mathbf{A}}\sum_{b\in\mathbf{B}} P_{a,b}\log[P_{a,b}/(P_a^{\,0} P_b)] = \sum_{a\in\mathbf{A}}\sum_{b\in\mathbf{B}} P_{a,b}\log[P(a|b)/P_a^{\,0}] \\ &\equiv \textstyle\sum_{a\in\mathbf{A}} I(a^0{:}\boldsymbol{M}) \equiv \sum_{b\in\mathbf{B}} I(\boldsymbol{M}^0{:}b) \\ &= \textstyle\sum_{K\in\mathbf{A}}\sum_{L\in\mathbf{B}}\sum_{k\in K}\sum_{l\in L} P_{k,l}\log[P_{k,l}/(P_k^{\,0} P_l)] \\ &\equiv \textstyle\sum_K\sum_L I(\boldsymbol{K}^0{:}\boldsymbol{L}) \equiv \sum_K I(\boldsymbol{K}^0{:}\boldsymbol{M}) \equiv \sum_L I(\boldsymbol{M}^0{:}\boldsymbol{L}). \end{aligned} \tag{8.5.2}$$

$$\begin{aligned} N(\mathbf{A};\,\mathbf{B}) &= \textstyle\sum_{a\in\mathbf{A}}\sum_{b\in\mathbf{B}} P_{a,b}\log[P_a/(P_a^{\,0} P_b)] \equiv \sum_{a\in\mathbf{A}} N(a;\,\boldsymbol{M}) \equiv \sum_{b\in\mathbf{B}} N(\boldsymbol{M};\,b) \\ &= \textstyle\sum_{K\in\mathbf{A}}\sum_{L\in\mathbf{B}}\sum_{k\in\boldsymbol{K}}\sum_{l\in\boldsymbol{L}} P_{k,l}\log[P_k/(P_k^{\,0} P_l)] \\ &\equiv \textstyle\sum_K\sum_L N(\boldsymbol{K};\,\boldsymbol{L}) \equiv \sum_K N(\boldsymbol{K};\,\boldsymbol{M}) \equiv \sum_L N(\boldsymbol{M};\,\boldsymbol{L}) \end{aligned} \tag{8.5.3}$$

a) AIM *row* channel:	$1 \longrightarrow a \text{---} \boldsymbol{P}(\boldsymbol{j}\vert a) \longrightarrow \boldsymbol{j} \longrightarrow \boldsymbol{P}(\boldsymbol{j}\vert a)$ $\Rightarrow \{S^r(\boldsymbol{M}\vert a),\ I^r(a^0{:}\boldsymbol{M}),\ N^r(a;\boldsymbol{M})\}$
b) AIM *column* channel:	$\boldsymbol{P}(\boldsymbol{i}) \longrightarrow \boldsymbol{i} \text{---} \boldsymbol{P}(b\vert\boldsymbol{i}) \longrightarrow b \longrightarrow P_b$ $\Rightarrow \{S(b\vert\boldsymbol{M}),\ I(\boldsymbol{M}^0{:}b),\ N(\boldsymbol{M};b)\}$
c) fragment *row* channel:	$\boldsymbol{P}(\boldsymbol{k}\vert K) \longrightarrow \boldsymbol{k} \text{---} \mathbf{P}(\boldsymbol{j}\vert\boldsymbol{k}) \longrightarrow \boldsymbol{j} \longrightarrow \boldsymbol{P}(\boldsymbol{k}\vert K)\,\mathbf{P}(\boldsymbol{j}\vert\boldsymbol{k})$ $\Rightarrow \{S^r(\boldsymbol{M}\vert\boldsymbol{K}),\ I^r(\boldsymbol{K}^0{:}\boldsymbol{M}),\ N^r(\boldsymbol{K};\boldsymbol{M})\}$
d) fragment *column* channel:	$\boldsymbol{P}(\boldsymbol{i}) \longrightarrow \boldsymbol{i} \text{---} \mathbf{P}(\boldsymbol{l}\vert\boldsymbol{i}) \longrightarrow \boldsymbol{l} \longrightarrow \boldsymbol{P}(\boldsymbol{l})$ $\Rightarrow \{S(\boldsymbol{L}\vert\boldsymbol{M}),\ I(\boldsymbol{M}^0{:}\boldsymbol{L}),\ N(\boldsymbol{M};\boldsymbol{L})\}$
e) fragment *diagonal* channel:	$\boldsymbol{P}(\boldsymbol{k}\vert K) \rightarrow \boldsymbol{k} \text{---} \mathbf{P}(\boldsymbol{k}\vert\boldsymbol{k}) \longrightarrow \boldsymbol{k}' \longrightarrow \boldsymbol{P}(\boldsymbol{k}\vert K)\,\mathbf{P}(\boldsymbol{k}'\vert\boldsymbol{k})$ $\Rightarrow \{S^r(\boldsymbol{K}\vert\boldsymbol{K}),\ I^r(\boldsymbol{K}^0{:}\boldsymbol{K}),\ N^r(\boldsymbol{K};\boldsymbol{K})\}$
f) fragment *off-diagonal* channel:	$\boldsymbol{P}(\boldsymbol{k}\vert K) \rightarrow \boldsymbol{k} \text{---} \mathbf{P}(\boldsymbol{l}\vert\boldsymbol{k}) \longrightarrow \boldsymbol{l} \longrightarrow \boldsymbol{P}(\boldsymbol{k}\vert K)\,\mathbf{P}(\boldsymbol{l}\vert\boldsymbol{k})$ $\Rightarrow \{S^r(\boldsymbol{L}\vert\boldsymbol{K}),\ I^r(\boldsymbol{K}^0{:}\boldsymbol{L}),\ N^r(\boldsymbol{K};\boldsymbol{L})\}$

Scheme 8.7. The partial communication systems of bonded atoms and molecular fragments.

The entropy/information contributions defined in the preceding three equations characterize various *partial* communication systems in the molecular fragment resolution, which can be derived from the molecular conditional probability matrix in atomic resolution, $\mathbf{P}(\mathbf{B}\,|\,\mathbf{A}) \equiv \mathbf{P}(\boldsymbol{j}\,|\,\boldsymbol{i})$. They are summarized in Scheme 8.7.

The conditional entropy

$$S(\boldsymbol{M}\,|\,a) = -\sum_{b\in\mathbf{B}} P_{a,b}\log(P_{a,b}/P_a) = -\sum_{b\in\mathbf{B}} P_{a,b}\log P(b|a)$$
$$= P_a\{-\sum_{b\in\mathbf{B}} P(b|a)\log P(b|a)\} \equiv P_a \sum_{b\in\mathbf{B}} S^r(b\,|\,a) \equiv P_a\, S^r(\boldsymbol{M}\,|\,a) \qquad (8.5.4)$$

characterizes the AIM *row*-channel of Panel *a* in Scheme 8.7, consisting of atom *a* in the input and all AIM $\boldsymbol{j} = \boldsymbol{M}$ of M in the output, with the *reduced* atomic quantities $S^r(\boldsymbol{M}\,|\,a)$ and $S^r(b\,|\,a)$ reflecting the conditional entropy indices per atom unit input probability. This quantity is defined by the *a*th row $\boldsymbol{P}(\mathbf{B}\,|\,a) = \boldsymbol{P}(\boldsymbol{j}\,|\,a)$ of the molecular conditional probability matrix $\mathbf{P}(\mathbf{B}|\mathbf{A})$, of the atomic outputs **B** given the atomic inputs **A**. In such a partial atomic *row*-channel the input signal thus enters the specified atom and is "scattered" to all possible atomic outputs in the molecule. The quantity of the preceding equation determines the atomic contribution to the global covalent component in the molecule.

The AIM mutual information contribution $I(a^0:\boldsymbol{M})$ can be similarly interpreted:

$$I(a^0:\boldsymbol{M}) = \sum_{b\in\mathbf{B}} P_{a,b} \log[P_{a,b}/(P_a^{\,0} P_b)] = P_a\{\sum_{b\in\mathbf{B}} P(b|a) \log[P(a|b)/P_a^{\,0}]\}$$
$$\equiv P_a \sum_{b\in\mathbf{B}} I^r(a^0:b) \equiv P_a I^r(a^0:\boldsymbol{M}), \qquad (8.5.5)$$

where the reduced mutual information indices $I^r(a^0:\boldsymbol{M})$ and $I^r(a^0:b)$ provide measures per unit input signal on atom a.

Therefore, the global indices $S(\mathbf{B}\,|\,\mathbf{A})$ and $I(\mathbf{A}^0:\mathbf{B})$ are the *input*-ensemble averages of the reduced quantities characterizing the atomic row channels,

$$S(\mathbf{B}\,|\,\mathbf{A}) = S(\boldsymbol{M}) = \sum_{a\in\mathbf{A}} P_a\, S^r(\boldsymbol{M}\,|\,a), \qquad I(\mathbf{A}^0:\mathbf{B}) \equiv I(\boldsymbol{M}) = \sum_{a\in\mathbf{A}} P_a I^r(a^0:\boldsymbol{M}), \qquad (8.5.6)$$

with the atomic input probabilities providing weights in the atomic row-channel.

The sum of these two entropic indices of the bonded atom a,

$$N(a;\boldsymbol{M}) = S(\boldsymbol{M}\,|\,a) + I(a^0:\boldsymbol{M}) \equiv P_a N^r(a;\boldsymbol{M}), \qquad (8.5.7)$$

where

$$N^r(a;\boldsymbol{M}) = \sum_{b\in\mathbf{B}} P(b|a)\, \{\log[P(a|b)/P(b|a)] - \log P_a^{\,0}\}$$
$$= \log(P_a/P_a^{\,0}) - \sum_{b\in\mathbf{B}} P(b|a) \log P_b, \qquad (8.5.8)$$

provides a reduced measure of an overall participation of the valence electrons originating from atom a in all chemical bonds in the molecule, thus reflecting the IT bond-propensity of a given atom in M. The conditional entropy contribution $S(\boldsymbol{M}\,|\,a)$ measures the IT-covalent affinity of atom a, while the mutual information quantity $I(a^0:\boldsymbol{M})$ provides the index of the complementary IT-ionic contribution of this bonded atom. Hence, the global entropy/information bond-order of the molecule is given by the sum of bond propensities of all constituent atoms:

$$N(\boldsymbol{M}) = S(\boldsymbol{M}) + I(\boldsymbol{M}) = \sum_{a\in\mathbf{A}} N(a;\boldsymbol{M}) = \sum_{a\in\mathbf{A}} P_a N^r(a;\boldsymbol{M}). \qquad (8.5.9)$$

As an illustration let us reexamine the 2-AO model (Section 7.5) of a single chemical bond resulting from the interaction of two (orthonormal) AO from atoms A and B, respectively. The relevant communication systems of the model, for the promolecular $\boldsymbol{P}^0 = (x, y = 1 - x)$ and molecular $\boldsymbol{P} = (P, Q = 1 - P)$ input probabilities, respectively, are shown in Schemes 7.7/(I, II). The global entropy/information indices for this channel are defined in terms of the binary-entropy function $H(z)$: $S(\mathbf{B}\,|\,\mathbf{A}) = H(P)$, $I(\mathbf{A}^0:\mathbf{B}) = H(x) - H(P)$, and $N(\mathbf{A};\mathbf{B}) = S(\mathbf{B}\,|\,\mathbf{A}) + I(\mathbf{A}^0:\mathbf{B}) = H(x)$. In Scheme 8.8 the row channels of the constituent AIM are displayed with the relevant expressions for the entropy/information indices of the model chemical bond.

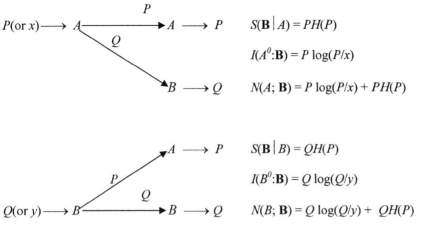

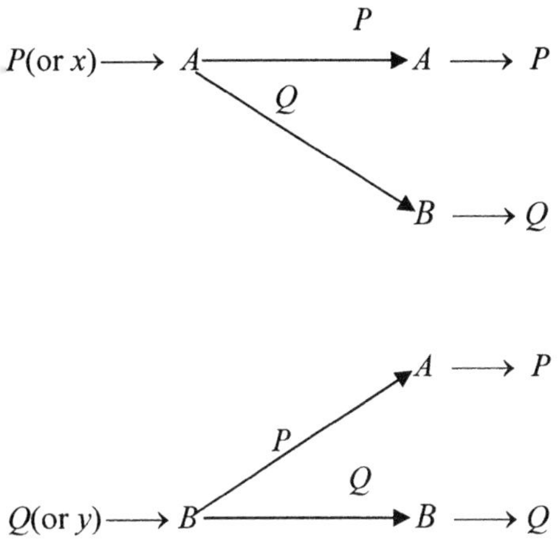

$S(\mathbf{B}\,|\,A) = PH(P)$

$I(A^0{:}\mathbf{B}) = P \log(P/x)$

$N(A;\,\mathbf{B}) = P \log(P/x) + PH(P)$

$S(\mathbf{B}\,|\,B) = QH(P)$

$I(B^0{:}\mathbf{B}) = Q \log(Q/y)$

$N(B;\,\mathbf{B}) = Q \log(Q/y) + QH(P)$

Scheme 8.8. The partial *row* channels in the 2-AO model of a single chemical bond *A*—*B* and their entropy/information indices.

Therefore, the electron-sharing AP reference $x = y = ½$, in which both free atoms contribute a single electron each to form the chemical bond, gives the conserved overall information-theoretic bond order of 1 bit, marking a total single bond irrespectively of the current value of the molecular probability parameter $0 \le P \le 1$ measuring the MO polarization (or CT) (see also Fig. 7.1). This overall bond-order preservation is a result of the competition between the bond entropy-covalency and its information-ionicity. These overall bond indices also predict the zero bond-order for the limiting configuration $P = x = 0$, when the identical electron configurations of the molecule and promolecule exhibit the lone electron pair on the donor atom *B* of the coordination (pair-sharing) promolecular reference. For a comparison between these IT bond-order contributions of constituent atoms and the corresponding predictions from the MO theory the reader is referred to Appendix F.

Next, let us briefly comment upon the other type of the entropy/information bond contributions in Eqs. (8.5.1)- (8.5.3), which characterize a single bonded atom of the molecular *output* (see Scheme 8.9): $\{S(b\,|\,\boldsymbol{M}),\ I(\boldsymbol{M}^0{:}b),\ N(\boldsymbol{M};\ b)\}$. They are generated by the AIM *column* channel shown in Panel *b* of Scheme 8.7. Here the input signal, entering through all atomic inputs of the molecular system, distributed in accordance with the molecular/promolecular AIM probabilities $\boldsymbol{P} = \boldsymbol{P}(\boldsymbol{i})$ is scattered into a single atom *b* of the molecular output, in accordance with the column $\boldsymbol{P}(b|\boldsymbol{i}) = \boldsymbol{P}(b|\mathbf{A})$ of the molecular conditional probability matrix $\mathbf{P(B|A)}$:

$$S(b\,|\,\boldsymbol{M}) = -\textstyle\sum_{a\in\mathbf{A}} P_{a,b} \log(P_{a,b}/P_a) = -\sum_{a\in\mathbf{A}} P_a\, P(b|a) \log P(b|a)$$
$$= \textstyle\sum_{a\in\mathbf{A}} P_a\, S^r(b\,|\,a), \qquad (8.5.10)$$

$$I(\boldsymbol{M}^0{:}b) = \Sigma_{a\in\mathbf{A}}\, P_{a,b}\log[(P_{a,b}/P_b)/P_a^{\,0}] = \Sigma_{a\in\mathbf{A}}\, P_a P(b|a)\log[P(a|b)/P_a^{\,0}]$$
$$= \Sigma_{a\in\mathbf{A}}\, P_a I^{r}(a^0{:}b). \qquad (8.5.11)$$

Therefore, this pair of the complementary atomic indices is defined by the *input*-ensemble average of the corresponding reduced quantities coupling a current atomic input with the channel fixed AIM output. By normalization of the atomic input probabilities, $\Sigma_{a\in\mathbf{A}}\, P_a = 1$, the overall quantities $S(b\,|\,\boldsymbol{M})$ and $I(\boldsymbol{M}^0{:}b)$ defined in two preceding equations are themselves the *reduced* entropy/information indices of the column-channel for atom b.

The IT components of the preceding two equations give rise to the total IT index of the atomic column channel:

$$N(\boldsymbol{M};\, b) = S(b\,|\,\boldsymbol{M}) + I(\boldsymbol{M}^0{:}b) = -\,\Sigma_{a\in\mathbf{A}}\, P_{a,b}\,\log(P_a/P_a^{\,0}) - P_b\log P_b. \qquad (8.5.12)$$

Let us recall that the output of the molecular information channel identifies the bonds being counted, while the inputs reflect the atomic bond origins. Therefore, in the particular case of the atomic column channel, the single atom in the output does not define any chemical bond. Instead, the IT indices of the atomic column channel reflect on the *valences* of the bonded atom b, promoted by its interaction with the remaining atoms in a molecule. The overall IT-index of the atomic column-channel reflects a degree of the atom involvement in the bond formation process, i.e., the number of its bond-active valence electrons in M. Accordingly, the conditional-entropy and mutual-information components of this overall atomic valence measure the atom involvement in all covalent and ionic bonds in the molecule, respectively.

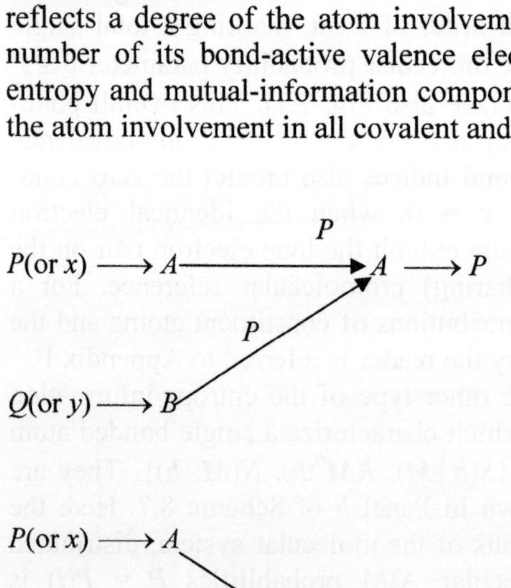

$S(A\,|\,\mathbf{A}) = PH(P)$

$I(\mathbf{A}^0{:}A) = P\,[P\log(P/x) + Q\log(Q/y)]$

$N(\mathbf{A};\, A) = P(-\,P\log x - Q\log y)$

$S(B\,|\,\mathbf{A}) = QH(P)$

$I(\mathbf{A}^0{:}B) = Q\,[P\log(P/x) + Q\log(Q/y)]$

$N(\mathbf{A};\, B) = Q(-\,P\log x - Q\log y)$

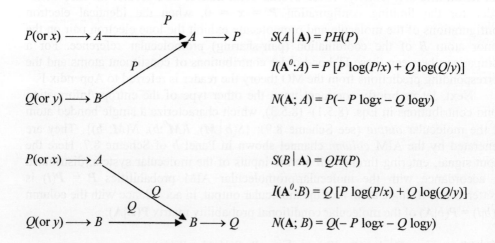

Scheme 8.9. The partial atomic *column*-channels in the 2-AO model of a single chemical bond A—B and their entropy/information indices.

The illustrative partial *column* channels in the 2AO model are displayed in Scheme 8.9 together with the corresponding entropy/information quantities of Eqs. (8.5.10)-(8.5.12). In Appendix F the plots of these atomic valence numbers from IT are reported together with their covalent/ionic components. They are compared with the relevant predictions from the MO theory. This comparison shows that both the IT and MO approaches give qualitatively similar partition of the atomic contributions to the conserved overall bond-order and valence number in the 2-AO model, thus providing an independent validation of these intuitive chemical concepts.

Reversing the roles of the outputs and inputs in the molecular channel, for the fixed (molecular) conditional-probability connections between atoms, indicates that the entropy/information indices of the specified atomic input or output should be similar, particularly for the identical molecular and promolecular atomic probability distributions $\boldsymbol{P} = \boldsymbol{P}^0$. This is the case in π-bond systems of butadiene and benzene in the Hückel theory approximation. Indeed, the entropic descriptors (in bits) of such atomic channels are identical for all carbon atoms in these alternant hydrocarbons:

butadiene: $S(\boldsymbol{M}|a) = S(b|\boldsymbol{M}) = ¼\, S(\mathbf{B}|\mathbf{A}) = 0.486$,

$I(a^0: \boldsymbol{M}) = I(\boldsymbol{M}^0: b) = ¼\, I(\mathbf{A}^0: \mathbf{B}) = 0.014$,

$N(a; \boldsymbol{M}) = N(\boldsymbol{M}; b) = ¼\, N(\mathbf{A}; \mathbf{B}) = 0.500$;

benzene: $S(\boldsymbol{M}|a) = S(b|\boldsymbol{M}) = S(\mathbf{B}|\mathbf{A})/6 = 0.425$,

$I(a^0: \boldsymbol{M}) = I(\boldsymbol{M}^0: b) = I(\mathbf{A}^0: \mathbf{B})/6 = 0.006$,

$N(a; \boldsymbol{M}) = N(\boldsymbol{M}; b) = N(\mathbf{A}; \mathbf{B})/6 = 0.431$.

It follows from these information indices that - as far as the entropy/information indices are concerned - the carbon atoms in these π-electrons systems are used somewhat more efficiently in butadiene, compared to benzene. This agrees with the observed preference of π-bonds to favor the bond alternation (e.g., Shaik 1989; Jug and Köster, 1990; Shaik and Hiberty, 1991, 1995, 2004).

For a larger *input*-fragment $\boldsymbol{K}$ consisting of several AIM $\boldsymbol{k}$ the entropy/ information indices $S(\boldsymbol{M}|\boldsymbol{K}) = P_K\, S^r(\boldsymbol{M}|\boldsymbol{K})$, $I(\boldsymbol{K}^0{:}\boldsymbol{M}) = P_K\, I^r(\boldsymbol{K}^0{:}\boldsymbol{M})$, and $N(\boldsymbol{K}{:}\boldsymbol{M}) = S(\boldsymbol{M}|\boldsymbol{K}) + I(\boldsymbol{K}^0{:}\boldsymbol{M}) = P_K\, N^r(\boldsymbol{K}{:}\boldsymbol{M})$, where $N^r(\boldsymbol{K}{:}\boldsymbol{M}) = S^r(\boldsymbol{M}|\boldsymbol{K}) + I^r(\boldsymbol{K}^0{:}\boldsymbol{M})$ and P_K stands for the condensed probability of the input fragment K in M, correspond to the subsystem *row*-channel of Panel *c* in Scheme 8.7, in which the atomic inputs are limited to all constituent atoms $\boldsymbol{k}$ of K, with the rows $\mathbf{P}(\,\boldsymbol{j}|\boldsymbol{k})$ of the overall conditional probability matrix $\mathbf{P}(\mathbf{B}|\mathbf{A})$ now determining the communication network linking atoms $\boldsymbol{k}$ in the input to all possible atomic outputs $\boldsymbol{j}$ in the molecular channel.

Another set of indices $\{S(\boldsymbol{L}|\boldsymbol{M}), I(\boldsymbol{M}^0{:}\boldsymbol{L}), N(\boldsymbol{M}{:}\boldsymbol{L}) = S(\boldsymbol{L}|\boldsymbol{M}) + I(\boldsymbol{M}^0{:}\boldsymbol{L})\}$, refers to the partial *column*-channel of the AIM-resolved *output* fragment $\boldsymbol{L}$ (Panel 8.7.*d*) involving all atomic inputs $\boldsymbol{i}$ in M and the AIM outputs $\boldsymbol{l}$ of fragment L in M, defined by the columns $\mathbf{P}(\boldsymbol{l}|\boldsymbol{i})$ of the molecular conditional-probability matrix $\mathbf{P}(\mathbf{B}|\mathbf{A}) = \mathbf{P}(\boldsymbol{j}|\boldsymbol{i})$. Finally, the condensed entropy/information matrix elements, $\{S(\boldsymbol{L}|\boldsymbol{K})$,

$I(\boldsymbol{K}^0:\boldsymbol{L})$, $N(\boldsymbol{K}:\boldsymbol{L}) = S(\boldsymbol{L}|\boldsymbol{K}) + I(\boldsymbol{K}^0:\boldsymbol{L})\}$, provide the IT bond indices of the of the partial channels shown in Panels *e* ($\boldsymbol{K} = \boldsymbol{L}$, *diagonal*–channel) and *f* ($\boldsymbol{K} \neq \boldsymbol{L}$, *off-diagonal*–channel) of Scheme 8.7, with the atomic inputs of atoms $\boldsymbol{k} \in K$, and the AIM outputs $\boldsymbol{l} \in L$. It should be observed that the singly-resolved partial channels of Panels *c* i *d* of this scheme are combinations of the doubly-resolved channels of Panels *e* and *f*. In the next section we shall examine in a more detail the illustrative examples of such partial channels of molecular fragments in the π-systems of butadiene and benzene.

8.6. ILLUSTRATIVE PARTIAL CHANNELS OF MOLECULAR FRAGMENTS

We have introduced in the preceding section two types of partial channels, which describe the embedded molecular subsystems: the so called "column" and "row" communication systems, defined by a selection of the corresponding columns and rows of the molecular conditional probability matrix in atomic resolution. To describe the localized-bond effects, we shall often consider the diatomic subsystems $\boldsymbol{k} = (k, k') \equiv \boldsymbol{K}$ or $\boldsymbol{l} = (l, l') \equiv \boldsymbol{L}$, in the input or output of the molecular communication system, respectively, but an extension to larger groups of AIM, e.g., the bond complementary environment in the molecule, is straightforward.

In general all atoms $\boldsymbol{i}$ of the promolecular/molecular input in the molecular information system contribute to the entropy/information indices involving a given pair of bonded atoms $\boldsymbol{l} \in L$ in the molecular output. Therefore, in a search for the entropy/information characteristics of the overall bond, external and internal, in which this atomic pair is involved in the molecule, one constructs the partial column channel consisting of all constituent atoms in the molecular/promolecular input $\mathbf{A}$, and the specified pair $\boldsymbol{l} = (l, l')$ of bonded atoms in the molecular output, $\mathbf{B}(l, l') \equiv \mathbf{B}_L$. This diatomic column channel in M (see Scheme 8.7*d*) is defined by two columns $\mathbf{P}(\mathbf{B}_L|\mathbf{A}) = \mathbf{P}(\boldsymbol{l}|\mathbf{A})$ of the overall molecular conditional probability matrix in atomic resolution, $\mathbf{P}(\mathbf{B}|\mathbf{A}) = \mathbf{P}(\boldsymbol{j}|\boldsymbol{i})$. For a larger than diatomic fragment of a molecule one constructs the partial channel involving all columns of $\mathbf{P}(\mathbf{B}|\mathbf{A})$ corresponding to the constituent AIM of this subsystem. As already argued in the preceding section, in the limiting atomic case, $L = b$, the entropic indices of the associated single-column conditional probability network $\mathbf{P}(\mathbf{B}_b|\mathbf{A}) \equiv \mathbf{P}(b|\mathbf{A})$ reflect on the *i*th AIM valence-number, measuring the atom participation (number of active valence electrons) in chemical bonds with all remaining atoms in the molecule.

The conditional-entropy of the partial column-channel, for the specified output fragment $\boldsymbol{L}$, given the full molecular input $\mathbf{A}$,

$$S(\boldsymbol{L}|\boldsymbol{M}) = S(\mathbf{B}_L|\mathbf{A}) = -\Sigma_{a\in\mathbf{A}}\Sigma_{l\in l} P_{a,l} \log(P_{a,l}/P_a)] = -\Sigma_{a\in\mathbf{A}} P_a \Sigma_{l\in l} P(l|a) \log P(l|a)]$$
$$= \Sigma_{a\in\mathbf{A}} P_a [\Sigma_{l\in l} S^{*}(l|a)] \equiv \Sigma_{a\in\mathbf{A}} P_a S^{*}(\boldsymbol{L}|a) = \Sigma_{l\in l} S(l|\boldsymbol{M}), \qquad (8.6.1)$$

and the associated mutual-information for the input (promolecular) probabilities $\boldsymbol{P}(\mathbf{A}^0) = \boldsymbol{P}^0$,

$$I(\boldsymbol{M}^0{:}\boldsymbol{L}) = I(\mathbf{A}^0{:}\mathbf{B}_L) = \sum_{a\in\mathbf{A}}\sum_{l\in\boldsymbol{l}} P_{a,l}\log[P_{a,l}/(P_a^{\,0}P_l)] = \sum_{a\in\mathbf{A}} P_a \sum_{l\in\boldsymbol{l}} P(l|a)\log[P(a|l)/P_a^{\,0}]$$
$$= \sum_{a\in\mathbf{A}} P_a^{\,0} \sum_{l\in\boldsymbol{l}} I^r(a^0{:}l) = \sum_{a\in\mathbf{A}} P_a^{\,0}\, I^r(a^0{:}\boldsymbol{L}) = \sum_{l\in\boldsymbol{l}} I(\boldsymbol{M}^0;\, l), \qquad (8.6.2)$$

give rise to the overall index:

$$N(\boldsymbol{M};\boldsymbol{L}) = S(\boldsymbol{L}\,|\,\boldsymbol{M}) + I(\boldsymbol{M}^0{:}\boldsymbol{L}) = N(\mathbf{A};\mathbf{B}_L) = \sum_{a\in\mathbf{A}}\sum_{l\in\boldsymbol{l}} P_{a,l}\log[P_a/(P_a^{\,0}P_l)]$$
$$= \sum_{a\in\mathbf{A}} P_a \sum_{l\in\boldsymbol{l}} P(l|a)\log\{P(a|l)/[P(l|a)P_a^{\,0}]\}$$
$$= \sum_{a\in\mathbf{A}} P_a \sum_{l\in\boldsymbol{l}} N^r(a;\, l) = \sum_{a\in\mathbf{A}} P_a N^r(a;\, \boldsymbol{L}) = \sum_{l\in\boldsymbol{l}} N(\boldsymbol{M};\, l). \qquad (8.6.3)$$

This set of IT-indices reflects the average covalency, ionicity, and the overall bond-order of fragment L originating from all atoms in the molecule. These quantities are seen to represent the mean-values of the reduced entropy/information indices $S^r(\boldsymbol{L}\,|\,a)$, $I^r(a^0{:}\boldsymbol{L})$, and $N^r(a;\,\boldsymbol{L})$, calculated per unit atomic input probability. In the "stationary" molecular channel, when $\boldsymbol{P} = \boldsymbol{P}^0$, the total entropy/information index of Eq. (8.6.3) for all atoms in the output subsystem L of M, becomes equal to the output-fragment entropy: $N(\boldsymbol{M};\,\boldsymbol{L}) = -\sum_{a\in\mathbf{A}}\sum_{l\in\boldsymbol{l}} P_{a,l}\log P_l = -\sum_{l\in\boldsymbol{l}} P_l \log P_l = S(\mathbf{B}_L)$.

One could also examine the IT-characteristics of the partial *row*-channel for the specified *input*-fragment $\boldsymbol{K} = \boldsymbol{k}$, which definines the partial-input $\mathbf{A}_K$, and the complete molecular output $\mathbf{B}$ in atomic resolution (Scheme 8.7*c*). The single atom input channel of Scheme 8.7*a*, for $K = a$, defined by the *i*th row $\boldsymbol{P}(\mathbf{B}\,|\,\mathbf{A}_i) \equiv \boldsymbol{P}(\mathbf{B}\,|\,i)$ of the molecular conditional probability matrix $\mathbf{P}(\mathbf{B}\,|\,\mathbf{A})$, represents a limiting case of such a general subsystem row channel. In a general case the relevant conditional probability matrix $\mathbf{P}(\mathbf{B}\,|\,\mathbf{A}_K)$ is defined by the rows $\mathbf{P}(\mathbf{B}\,|\,\boldsymbol{k})$ of $\mathbf{P}(\mathbf{B}\,|\,\mathbf{A})$. This fragment row-channel provides the complementary set of the entropy/information indices of all chemical bonds in the molecule, which originate from all constituent atoms of the input fragment:

$$S(\boldsymbol{M}\,|\,\boldsymbol{K}) = -\sum_{k\in\boldsymbol{k}}\sum_{b\in\mathbf{B}} P_{k,b}\log(P_{k,b}/P_k) = -P_K\sum_{k\in\boldsymbol{k}}(P_k/P_K)\sum_{b\in\mathbf{B}}(P_{k,b}/P_k)\log P(b|k)]\}$$
$$\equiv P_K\sum_{k\in\boldsymbol{k}} P(k|K)\,[-\sum_{b\in\mathbf{B}} P(b|k)\log P(b|k)]$$
$$= P_K\sum_{k\in\boldsymbol{k}} P(k|K)\, S^r(\boldsymbol{M}\,|\,k)] \equiv P_K\, S^r(\boldsymbol{M}\,|\,\boldsymbol{K}), \qquad (8.6.4)$$

$$I(\boldsymbol{K}^0{:}\boldsymbol{M}) = \sum_{k\in\boldsymbol{k}}\sum_{b\in\mathbf{B}} P_{k,b}\log[P_{k,b}/(P_k^{\,0}P_b)] = P_K\sum_{k\in\boldsymbol{k}} P(k|K)\{\sum_{b\in\mathbf{B}} P(b|k)\log[P(k|b)/P_k^{\,0}]\}$$
$$= P_K\,[\sum_{k\in\boldsymbol{k}} P(k|K)\, I^r(k^0{:}\boldsymbol{M})] \equiv P_K\, I^r(\boldsymbol{K}^0{:}\boldsymbol{M}), \qquad (8.6.5)$$

$$N(\boldsymbol{K};\,\boldsymbol{M}) = S(\boldsymbol{M}\,|\,\boldsymbol{K}) + I(\boldsymbol{K}^0{:}\boldsymbol{M}) = P_K\,[S^r(\boldsymbol{M}\,|\,\boldsymbol{K}) + I^r(\boldsymbol{K}^0{:}\boldsymbol{M})] \equiv P_K\, N^r(\boldsymbol{K};\,\boldsymbol{M})$$
$$= P_K\sum_{k\in\boldsymbol{k}} P(k|K)\sum_{b\in\mathbf{B}} P(b|k)\log\{P(k|b)/[P(b|k)P_k^{\,0}]\}. \qquad (8.6.6)$$

Here $P_K = \sum_{k\in K} P_k$ stands for the condensed probability of finding an electron of M on its fragment K, $P(k|K) = P_k / P_K$ is the conditional probability of finding an electron of K on its constituent atom k, satisfying the relevant normalization $\sum_{k\in \boldsymbol{k}} P(k|K) = 1$ [Eq. (8.4.6)], and the input-fragment reduced indices $\{S^r(\boldsymbol{M}\,|\,\boldsymbol{K}), I^r(\boldsymbol{K}^0{:}\boldsymbol{M}), N^r(\boldsymbol{K}{:}\boldsymbol{M})\}$ are defined per unit overall input probability of the row channel for fragment K.

It should be stressed that these entropy/information indices reflect both the *intra*-fragment chemical bonds $\boldsymbol{l}$–$\boldsymbol{l}'$ and the *inter*-fragment (*external*) chemical interactions $\boldsymbol{l}$-$\boldsymbol{k}$ between atoms $\{l\in\boldsymbol{l}\}$ and the remaining atoms $\{k\notin\boldsymbol{l}\}$. These IT-descriptors of molecular fragments can be straightforwardly partitioned into the contributions from the *internal* probability connections and the associated *external* term from the subsystem-environment communications, between the fragment under consideration and the rest of the molecule. The former describe the effect of the intra-fragment electron delocalization, while the latter characterize the IT bond-coupling between the given fragment and the remaining AIM. For a given *output* collection $\boldsymbol{l} = \boldsymbol{L}$ of the system constituent atoms, which definines the molecular fragment L, this additive separation of the internal and external IT bond contributions of the fragment *column*-channel gives:

$$S(\boldsymbol{L}\,|\,\boldsymbol{L}) \equiv S^{int}(\boldsymbol{L}) = \sum_{l\in\boldsymbol{l}} P_l \sum_{l'\in\boldsymbol{l}} S^r(l'\,|\,l),$$

$$I(\boldsymbol{L}^0{:}\boldsymbol{L}) \equiv I^{int}(\boldsymbol{L}) = \sum_{l\in\boldsymbol{l}} P_l \sum_{l'\in\boldsymbol{l}} I^r(l^0{:}l'),$$

$$N(\boldsymbol{L};\boldsymbol{L}) \equiv N^{int}(\boldsymbol{L}) = S^{int}(\boldsymbol{L}) + I^{int}(\boldsymbol{L}) = \sum_{l\in\boldsymbol{l}} P_l \sum_{l'\in\boldsymbol{l}} [S^r(l'\,|\,l) + I^r(l^0{:}l')]$$
$$= \sum_{l\in\boldsymbol{l}} P_l \sum_{l'\in\boldsymbol{l}} N^r(l;l');$$

$$S(\boldsymbol{L}\,|\,\notin\boldsymbol{L}) \equiv S^{ext}(\boldsymbol{L}) = \sum_{k\notin\boldsymbol{l}} P_k \sum_{l\in\boldsymbol{l}} S^r(l\,|\,k),$$

$$I(\notin\boldsymbol{L}^0{:}\boldsymbol{L}) \equiv I^{ext}(\boldsymbol{L}) = \sum_{k\notin\boldsymbol{l}} P_k \sum_{l\in\boldsymbol{l}} I^r(k^0{:}l),$$

$$N(\notin\boldsymbol{L};\boldsymbol{L}) \equiv N^{ext}(\boldsymbol{L}) = S^{ext}(\boldsymbol{L}) + I^{ext}(\boldsymbol{L}) = \sum_{k\notin\boldsymbol{l}} P_k \sum_{b\in\boldsymbol{l}} N^r(a;b). \qquad (8.6.7)$$

It partitions the IT bond indices of the embedded fragment L in M into the bond-order contributions originating from the fragment own constituent AIM and the rest of the molecule, respectively.

A similar division can be also carried out in the corresponding quantities for the *row*-channel of the molecular fragment K consisting of atoms $\boldsymbol{K}$ to separate the intra-subsystem contributions from the external (subsystem-environment) terms due to the subsystem complementary reminder:

$$S(\boldsymbol{K}\,|\,\boldsymbol{K}) \equiv S^{int}(\boldsymbol{K}), \quad I(\boldsymbol{K}^0{:}\boldsymbol{K}) \equiv I^{int}(\boldsymbol{K}), \quad N(\boldsymbol{K};\boldsymbol{K}) = S^{int}(\boldsymbol{K}) + I^{int}(\boldsymbol{K}) \equiv N^{int}(\boldsymbol{K}),$$

$$S(\notin\boldsymbol{K}\,|\,\boldsymbol{K}) \equiv S^{ext}(\boldsymbol{K}),\ I(\boldsymbol{K}^0{:}\notin\boldsymbol{K}) \equiv I^{ext}(\boldsymbol{K}),\ N(\boldsymbol{K};\notin\boldsymbol{K}) = S^{ext}(\boldsymbol{K}) + I^{ext}(\boldsymbol{K}) \equiv N^{ext}(\boldsymbol{K}). \quad (8.6.8)$$

It separates the overall bond-orders of M due to atoms $\boldsymbol{K}$ into the corresponding *internal* (intra-K) and the *external* contributions, between atoms $\boldsymbol{K}$ and the rest of the molecule.

Table 8.2. The entropy/information indices (in bits) of the diatomic (column) channels of the π-electron systems in butadiene and benzene from the Hückel theory.

Molecule:	Butadiene			Benzene		
$\boldsymbol{L}$:	(1, 2), (3,4)	(1, 3), (2, 4)	(2, 3), (1, 4)	(i, i+1)	(i, i+2)	(i, i+3)
$S^{int}(\boldsymbol{L})$	0.448	0.480	0.476	0.250	0.266	0.262
$S^{ext}(\boldsymbol{L})$	0.525	0.493	0.496	0.600	0.585	0.588
$S(\boldsymbol{L}\,|\,\boldsymbol{M})$	0.972	0.972	0.972	0.850	0.850	0.850
$I^{int}(\boldsymbol{L})$	− 0.081	0.020	−0.009	− 0.030	−0.007	−0.013
$I^{ext}(\boldsymbol{L})$	0.109	0.007	0.037	0.041	0.019	0.025
$I(\boldsymbol{M}^0{:}\boldsymbol{L})$	0.028	0.028	0.028	0.012	0.012	0.012
$N^{int}(\boldsymbol{L})$	0.367	0.500	0.467	0.221	0.259	0.249
$N^{ext}(\boldsymbol{L})$	0.634	0.500	0.533	0.641	0.603	0.613
$N(\boldsymbol{M};\,\boldsymbol{L})$	1.000	1.000	1.000	0.862	0.862	0.862

In Table 8.2 we have listed representative entropy/information descriptors of the diatomic *output* fragments of the π-electron systems in butadiene and benzene, derived from the corresponding *column* channels in the Hückel approximation. A reference to these illustrative predictions shows that the total conditional entropy $S(\boldsymbol{L}\,|\,\boldsymbol{M})$, mutual information $I(\boldsymbol{M}^0{:}\boldsymbol{L})$ and overall bond index $N(\boldsymbol{M};\,\boldsymbol{L})$ are conserved for all embedded pairs of carbon atoms in these two alternant π-systems. Thus, on average all diatomic parts of the carbon chain/ring participate equally in the forming the internal/external π-bonds in these model systems. The overall index is again slightly higher in butadiene than in benzene. A reference to this table shows that the overall IT bond-order in which a given diatomic fragment is engaged in the molecular π-system is higher in butadiene (1 bit per diatomic fragment) compared to benzene (0.862 bits per diatomic fragment), thus confirming a general tendency of π-bonds to localize (alternate). These bond-order estimates are almost purely entropy-covalent in character, with only a marginal information-ionicity component, which is seen to be relatively smaller in benzene.

One also observes that the internal/external composition of a given overall entropy/information index is only weakly dependent on the selected diatomic fragment. In butadiene about 1 bit of the total conditional entropy (bond-covalency) is seen to be almost equally divided between the internal and external components, while in benzene, where the complementary subsystem is twice as large as the diatomic fragment, the external (coupling) term dominates over the internal entropy index. In both cases, however, these entropy-covalency terms only slightly differentiate between alternative diatomics-in-molecules. A stronger distinction of these subsystem communication systems is detected in the internal and external components of a generally small mutual-information (bond-ionicity) index, measuring the amount of information flowing through the respective parts of the fragment channel. An inspection of these entries in the table reveals that the largest magnitude of the ionic terms is predicted for the nearest neighbor interactions in benzene and for the peripheral, nearest-neighbor diatomic subsystems in butadiene.

This approximate "equalization" of the molecular fragment IT descriptors reflects the uniform distribution of the input and output probabilities in the molecular channel, from which the fragment channels were generated. It also confirms that the $2p_\pi$ orbitals of these constituent two-carbon fragments altogether participate equally in all π bonds in these two alternant hydrocarbons. A degree of this involvement of AO in bonding MO is predicted to be higher in butadiene compared to benzene. These almost equalized, overall entropy/information quantities of diatomic fragments of the two conjugated π-bond systems are in contrast the corresponding MO descriptors. The latter strongly differentiate between alternative diatomic fragments in these two π-systems.

The overall index $N(\boldsymbol{M};\boldsymbol{L}) \equiv N_c^{\pi}(\boldsymbol{L})$ can be compared with the corresponding average π-bond index per atomic pair, $N_{av}^{\pi} \equiv 2N^{\pi}(M)/[N(N-1)]$, where N is the system global number of the valence π-electrons (carbon atoms) and $N^{\pi}(\boldsymbol{M}) = N(\boldsymbol{M};\boldsymbol{M})$ stands for the overall entropy/information index of the whole π-electron system: $N_{av}^{\pi} = ⅓$ in butadiene and $N_{av}^{\pi} = 0.17$ in benzene. The deviations $N_c^{\pi}(\boldsymbol{L}) - N_{av}^{\pi}$ then read: ⅔ in butadiene and 0.69 in benzene. They measure the effect of the bond-conjugation (non-additivity) effect in these two molecules, with a slightly higher value of this deviation being predicted for benzene.

In Table 8.3 we have displayed the *absolute* entropy/information descriptors of alternative diatomic *input*-fragments of butadiene and benzene. A comparison with the corresponding predictions from the associated column channels (Table 8.2) reveals that the overall conditional entropy (covalency) indices resulting from these two types of the partial channels of diatomic fragments are identical, with only minor changes being observed in the mutual-information (ionicity) components, which are also reflected in the total fragment indices. The entropy/information descriprors of these diatomic row-channels are again seen to be roughly equalized in these illustrative π-electron systems. This implies that the alternative pairs of carbon atoms participate equally in chemical bonds in the molecule. In other words they exhibit almost identical (effective) *diatomic-valences* in these two alternant hydrocarbons. The chemical implications of the table entries are consistent with the diatomic *bond-orders* of the diatomic column-channel data in Table 8.2.

Table 8.3. The absolute entropy/information indices (in bits) of the diatomic row-channels in the π-electron systems of butadiene and benzene from the Hückel theory.*

Molecule:	Butadiene			Benzene		
$\boldsymbol{K}$:	(1, 2), (3,4)	(1, 3), (2, 4)	(2, 3), (1, 4)	$(i, i+1)$	$(i, i+2)$	$(i, i+3)$
$S^{int}(\boldsymbol{K})$	0.448	0.480	0.476	0.250	0.266	0.262
$S^{ext}(\boldsymbol{K})$	0.525	0.493	0.496	0.600	0.585	0.588
$S(\boldsymbol{M}\vert\boldsymbol{K})$	0.972	0.972	0.972	0.850	0.850	0.850
$I^{int}(\boldsymbol{K})$	−0.001	0.020	0.014	0.003	0.008	0.007
$I^{ext}(\boldsymbol{K})$	0.001	0.007	0.012	0.010	0.001	0.003
$I(\boldsymbol{K}^0:\boldsymbol{M})$	0.002	0.028	0.026	0.013	0.009	0.009
$N^{int}(\boldsymbol{K})$	0.447	0.500	0.490	0.253	0.274	0.269
$N^{ext}(\boldsymbol{K})$	0.528	0.500	0.508	0.610	0.585	0.591
$N(\boldsymbol{K};\boldsymbol{M})$	0.974	1.000	0.998	0.863	0.859	0.859

* The corresponding "reduced" quantities can be obtained by dividing entries of this table by the diatomic fragment probability in the system as a whole: $P_K = ½$ (butadiene) and $P_K = ⅓$ (benzene).

As a final illustration let us examine the *diagonal* and *off-diagonal* entropy-information indices of Schemes 8.7(e, f), for both the atomic and diatomic fragments of these two π-systems. In butadiene we divide the four constituent carbon atoms into two complementary pairs of atoms. The resulting absolute entropy/information indices for alternative partitions are listed in Table 8.4. The corresponding reduced quantities, per unit input-probability of the fragment in question, are obtained by dividing the table entries by the atomic or diatomic input group probability in M, P_i = ¼ or P_K= ½, respectively.

It should be again recalled that the input-fragment specifies the orbitals contributing to all chemical bonds in which the output-fragment is involved. Therefore, the fragment-diagonal IT-indices describe the internal bonds in subsystem L, due to its own constituent atoms. Accordingly, the *off*-diagonal indices in the molecular fragment resolution reflect the external contributions to the bonds of the output fragment L in M, due to the constituent atoms of the input-fragment $K \neq L$. A reference to Table 8.4 shows that for the complementary diatomic fragments of butadiene the diagonal and off-diagonal entropy/information contributions to bond indices are comparable, of roughly a half of a bit each. They are being seen to be dominated by the IT-covalent contributions. The entropy/information characteristics of the intra-group communications reflected by the diagonal indices are again very similar for all choices of diatomic subsystem. A similar conclusion follows from examining the IT-descriptors of the off-diagonal channels.

Table 8.4. The absolute *diagonal* and *off-diagonal* entropy/information indices (in bits) describing chemical interaction between atoms *i* and *j*, and between the specified diatomic fragments *K* and *L* of the carbon chain in butadiene (the Hückel theory).

Index	Atomic (*input*, *output*) fragments: (i, j)			
	(i, i)	(1, 2), (3,4)	(1, 3), (2, 4)	(2, 3), (1, 4)
$S(j \mid i)$	0.108	0.116	0.132	0.145

Index	Diatomic (*input*, *output*) fragments $(\boldsymbol{K}, \boldsymbol{L})$: *diagonal* $(\boldsymbol{K} = \boldsymbol{L})$		
	(1, 2), (3, 4)	(1, 3), (2, 4)	(2, 3), (1, 4)
$S(\boldsymbol{L} \mid \boldsymbol{L})$	0.448	0.480	0.476
$I(\boldsymbol{L}^0{:}\boldsymbol{L})$	0.001	0.020	0.014
$N(\boldsymbol{L};\boldsymbol{L})$	0.449	0.500	0.490

Index	Diatomic (*input*, *output*) fragments $\boldsymbol{K}, \boldsymbol{L}$: *off-diagonal* $(\boldsymbol{K} \neq \boldsymbol{L})$		
	$[\boldsymbol{K} = (1, 2),\ \boldsymbol{L} = (3, 4)]$ $[\boldsymbol{L} = (1, 2),\ \boldsymbol{K} = (3, 4)]$	$[\boldsymbol{K} = (1, 3),\ \boldsymbol{L} = (2, 4)]$ $[\boldsymbol{L} = (1, 3),\ \boldsymbol{K} = (2, 4)]$	$[\boldsymbol{K} = (2, 3),\ \boldsymbol{L} = (1, 4)]$ $[\boldsymbol{L} = (2, 3),\ \boldsymbol{K} = (1, 4)]$
$S(\boldsymbol{L} \mid \boldsymbol{K})$	0.525	0.493	0.493
$I(\boldsymbol{K}^0{:}\boldsymbol{L})$	0.001	0.007	0.012
$N(\boldsymbol{K};\boldsymbol{L})$	0.525	0.500	0.505

In butadiene the strongest IT-covalent interactions between single atoms are predicted for the middle-bond, between atoms 2 and 3, and for the pair of terminal carbons 1 and 4. This is contrary to the familiar MO predictions. Consider, for example, the Wiberg (1968) index of the bond covalency. In the *spin-restricted* Hartree-Fock (RHF) theory it is determined by a the sumof squares of the off-diagonal CBO matrix elements between AO of atoms *A* and *B*, respectively,

$$\Im_{A,B}{}^{W} = \sum_{a\in A} \sum_{b\in B} \gamma_{a,b}{}^{2}. \tag{8.6.9a}$$

For the *closed-shell* electron configuration defined by the row-vector of the doubly occupied MO, $\boldsymbol{\varphi}^{occ} = \boldsymbol{\chi}\mathbf{C} = \{\varphi_s\}$, where $\mathbf{C}$ groups the LCAO MO coefficients of the AO basis set $\boldsymbol{\chi} = \{\chi_i\}$:

$$\boldsymbol{\gamma} = 2\mathbf{C}\mathbf{C}^{\dagger} = \{\gamma_{a,b} = 2\sum_s C_{a,s} C_{b,s}^{*}\}. \qquad (8.6.10)$$

Its spin-resolved form, valid for the spin components $\gamma_{a,b}^{\sigma}$, $\sigma = \alpha, \beta$, of the spinless CBO matrix element $\gamma_{a,b} = \sum_{\sigma=\alpha,\beta} \gamma_{a,b}^{\sigma}$ in the *spin-unrestricted* HF (UHF) theory, this index reads:

$$\mathfrak{I}_{A,B}^{W,UHF} = 2\sum_{a\in A}\sum_{b\in B}\sum_{\sigma=\alpha,\beta}(\gamma_{a,b}^{\sigma})^2,\ \boldsymbol{\gamma}^{\sigma} = \mathbf{C}^{\sigma}\mathbf{C}^{\sigma\dagger} = \{\gamma_{a,b}^{\sigma} = \sum_s C_{a,s}^{\sigma} C_{b,s}^{\sigma *}\}. \quad (8.6.9b)$$

Here $\mathbf{C}^{\sigma} = \{C_{a,s}^{\sigma}\}$ groups the LCAO MO coefficients of the occupied MO for the spin-up and spin-down electrons: $\boldsymbol{\varphi}^{\sigma,occ} = \boldsymbol{\chi}\mathbf{C}^{\sigma} = \{\varphi_s^{\sigma}\}$, $\sigma = \alpha, \beta$.

The Wiberg index predicts the following π-bond (covalent) multiplicities in butadiene: $\mathfrak{I}_{1,2}^{W} = \mathfrak{I}_{3,4}^{W} = 0.80$, $\mathfrak{I}_{1,3}^{W} = \mathfrak{I}_{2,4}^{W} = 0$ and $\mathfrak{I}_{2,3}^{W} = \mathfrak{I}_{1,4}^{W} = 0.27$, thus identifying the strongest covalent interaction in the two peripheral bonds, and giving rise to somewhat unphysical result of the vanishing bond-order in the fragments (1, 3) and (2, 4). The above information measures do not differentiate so strongly between different partitions. They instead diagnose a high degree of equalization of the fragment IT-indices in the molecule, thus reflecting the entropy-equilibrium character of these mutually open molecular subsystems.

Table 8.5. The absolute entropy/information indices (in bits) describing the chemical π-interactions between atoms and diatomic fragments of the carbon ring in benzene (Hückel theory).

Index	Atomic (*input*, *output*) fragments: i, j			
	$j = i$	$j = i + 1$	$j = i + 2$	$j = i + 3$
$S(j \mid i)$	0.055	0.070	0.072	0.076
Index	Diatomic (*input*, *output*) fragments: $\boldsymbol{K} = (i, i + 1)$, $\boldsymbol{L} = (j, j + 1)$			
	$j = i$	$j = i + 1$	$j = i + 2$	$j = i + 3$
$S(\boldsymbol{L} \mid \boldsymbol{K})$	0.250	0.272	0.300	0.306
$I(\boldsymbol{K}^0{:}\boldsymbol{L})$	0.003	−0.072	−0.101	−0.074
$N(\boldsymbol{K}; \boldsymbol{L})$	0.253	0.200	0.199	0.232

* The corresponding "reduced" quantities can be obtained by dividing the entries of this table by the fragment probability in the system as a whole: P_i = 1/6 (atomic) and PK = ⅓ (diatomic).

Turning now to a similar comparison for benzene, we first examine the IT indices for the atomic input and output fragments reported in Table 8.5. One detects a slowly increasing trend exhibited by the entropy-covalency between a pair of increasingly distant AIM in the carbon ring, with the strongest covalent coupling found for the two carbons in the mutual *para*-position. The Wiberg orbital covalency index differentiates strongly between the relative positions of two carbons: $\mathfrak{I}_{i,i+1}{}^{W}$ = 0.44, $\mathfrak{I}_{i,i+2}{}^{W}$ = 0, and $\mathfrak{I}_{i,i+3}{}^{W}$ = 0.11. This is contrary to the trend exhibited by the atomic off-diagonal entropy-covalency descriptors. Again, the zero value of the second Wiberg index seems unphysical. It should be realized, however, that these IT bond-orders do not represent the total bond-orders of molecular subsystems, but only specific subsystem contributions to the overall IT bond-multiplicities, which were reported in Table 8.2. The IT-indices of Table 8.5 which reflect the entropy/information coupling between the increasingly distant pairs of neighboring carbon atoms in the benzene ring, $\boldsymbol{K} = (i, i + 1)$ and $\boldsymbol{L} = (j, j + 1)$, for $j = i, i + 1, .., i + 3$ again exhibit almost equalized total indices $N(\boldsymbol{K}; \boldsymbol{L})$. The highest entropy-covalency $S(\boldsymbol{L}\,|\,\boldsymbol{K})$ is observed for the most distant carbon pairs, when $j = i + 3$, and the lowest value is predicted for identical AIM pairs ($j = i$). In the latter case the interaction is practically purely covalent, while other carbon pairs are seen to give rise to an appreciable (negative) information-ionicity.

8.7. ATOMIC RESOLUTION OF GLOBAL ENTROPY/INFORMATION BOND INDICES

As we have seen in Chapters 8.5 and 8.6, the global entropy-covalency $N^{cov}(M) = S(\mathbf{B}\,|\,\mathbf{A})$ and information-ionicity $N^{ion}(M) = I(\mathbf{A}^0{:}\mathbf{B})$, and hence also the total IT bond index $N(M) = N(\mathbf{A};\mathbf{B})$, can all be partitioned into the additive contributions from the constituent AIM, atomic pairs, etc.. In fact, this division can be considered as a special case of the *mono*-atomic subsystem partitioning of Eqs. (8.5.1)-(8.5.3) (see also Panels *e* and *f* of Scheme 8.7), when the molecular fragments in the input and output of the partial communication systems are limited to a single atom. In this section we examine specific contributions resulting from such an ultimate division of bond-indices of molecular communication channels in atomic resolution.

The global conditional entropy $S(\mathbf{B}\,|\,\mathbf{A}) = S(\boldsymbol{M}\,|\,\boldsymbol{M})$ measures the average indeterminancy (communication noise) of the electron distributions among constituent atoms $\boldsymbol{M}$ in the molecular system M, thus reflecting a loss of a "memory" in the AIM-resolved molecular *output* about the atomic origins of electrons in the molecular/promolecular *input*. It can be naturally divided into the *one*- and *two-centre* entropy-covalency terms, $N_{oc}{}^{cov}(M)$ and $N_{tc}{}^{cov}(M)$, respectively:

$$\begin{aligned} N^{cov}(M) &= -\textstyle\sum_{a\in\mathbf{A}}\sum_{b\in\mathbf{B}} P_{a,b}\log(P_{a,b}/P_a) = -\sum_{a\in\mathbf{A}}\sum_{b\in\mathbf{B}} P_{a,b}\log P(b\,|\,a) \\ &= -\textstyle\sum_X P_{X,X}\log P(X|X) - \sum\sum_{X<Y} P_{X,Y}\log[P(Y|X)\,P(X|Y)] \\ &\equiv \{\textstyle\sum_X N^{cov}(X)\} + \{\sum\sum_{X<Y} N^{cov}(X,Y)\} \equiv N_{oc}{}^{cov}(M) + N_{tc}{}^{cov}(M). \end{aligned} \qquad (8.7.1)$$

The one-centre contribution $N^{cov}(X)$ measures the *intra*-atom conditional entropy (covalency) due to the atom "promotion" to its "valence-state" in the molecule, while the two-centre quantity $N^{cov}(X, Y)$ represents the *inter*-atom IT-covalency index for the specified pair of AIM. The additive partitioning of the global mutual-information (ionicity) index, measuring the amount of information flowing through the molecular communication system, when the promolecular electron probabilities of free-atoms define the channel input, similarly gives:

$$\begin{aligned} N^{ion}(M) &= \sum_{a\in\mathbf{A}}\sum_{b\in\mathbf{B}} P_{a,b}\log[P_{a,b}/(P_a^{\,0}P_b)] = \sum_{a\in\mathbf{A}}\sum_{b\in\mathbf{B}} P_{a,b}\log[P(a|b)/P_a^{\,0}] \\ &= \sum_X P_{X,X}\log[P(X|X)/P_X^{\,0}] + \sum\sum_{X<Y} P_{X,Y}\log\{[P(X|Y)/P_X^{\,0}][P(Y|X)/P_Y^{\,0}]\} \\ &\equiv \{\sum_X N^{ion}(X)\} + \{\sum\sum_{X<Y} N^{ion}(X,Y)\} \equiv N_{oc}^{\,ion}(M) + N_{tc}^{\,ion}(M). \end{aligned} \qquad (8.7.2)$$

Hence, the resulting overall *one*-centre entropy/information index of Xth AIM,

$$N(X) = N^{cov}(X) + N^{ion}(X) = -P_{X,X}\log P_X^{\,0}. \qquad (8.7.3)$$

Of interest is also the total *two*-centre entropy/information index for the specified pair (X,Y) of bonded atoms:

$$N(X,Y) = N^{cov}(X,Y) + N^{ion}(X,Y) = -P_{X,Y}\log(P_X^{\,0}P_Y^{\,0}) \equiv -P_{X,Y}\log P_{X,Y}^{\,0}, \qquad (8.7.4)$$

where $P_{X,Y}^{\,0} \equiv P_{X,Y} - \Delta P_{X,Y}$ denotes the promolecular off-diagonal joint-probability for the (independent) free-atoms X^0 and Y^0, which is shifted by $\Delta P_{X,Y}$ relative to the corresponding molecular (correlated) value, a small correction relative to both $P_{X,Y}^{\,0}$ and $P_{X,Y}$. Hence, by expanding to first-order in $x_{X,Y} = (\Delta P_{X,Y}/P_{X,Y})$ the logarithm

$$\log P_{X,Y}^{\,0} = \log[P_{X,Y}(1 - x_{X,Y})] = \log P_{X,Y} + \log(1 - x_{X,Y}) \cong \log P_{X,Y} - (\Delta P_{X,Y}/P_{X,Y}),$$

gives the approximate value of the total two-centre bond index,

$$N(X, Y) \cong -P_{X,Y}\log P_{X,Y} + \Delta P_{X,Y} \equiv \mathcal{H}(X, Y) + \Delta P_{X,Y}, \qquad (8.7.5)$$

expressed as the sum of the Shannon diatomic term $\mathcal{H}(X,Y)$ and the molecular displacement $\Delta P_{X,Y}$ in the inter-atomic two-electron probability relative to the corresponding promolecular value.

In the 2-AO model of the ground (singlet) state of $M = A$—B (Section 7.5) for the output (nolecular) probabilities $\mathbf{P} = (P, Q = 1 - P)$ and input (promolecular) probabilities $\mathbf{P}^0 = (½, ½)$ these bond indices (in bits) are given by the following functions of atomic probabilities:

$$\begin{aligned} &N^{cov}(A) = -P^2\log_2 P, && N^{cov}(B) = -Q^2\log_2 Q, && N^{cov}(A, B) = -PQ\log_2(PQ); \\ &N^{ion}(A) = P^2\log_2(2P), && N^{ion}(B) = Q^2\log_2(2Q), && N^{ion}(A, B) = PQ\log_2(4PQ); \\ &N(A) = P^2, && N(B) = Q^2, && N(A, B) = 2PQ. \end{aligned} \qquad (8.7.6)$$

Thus, for the electron configuration $P = Q = ½$ of the pure-covalent bond one predicts (in bits),

$$N^{cov}(A) = N^{cov}(B) = N(A) = N(B) = ¼, \qquad N^{cov}(A, B) = N(A, B) = ½,$$
$$N^{ion}(A) = N^{ion}(B) = N^{ion}(A, B) = 0, \tag{8.7.7}$$

while the molecular configuration $P = 1$, representing the lone-pair on the acidic atom A, i.e., the ion pair [A^-B^+] relative to the electron-sharing AP-reference [A^0B^0], gives:

$$N^{cov}(A) = N^{cov}(B) = N^{cov}(A, B) = N^{ion}(B) = N^{ion}(A, B) = N(B) = N(A, B) = 0,$$
$$N^{ion}(A) = N(A) = 1. \tag{8.7.8}$$

For the single, purely-covalent bond we have thus obtained the entropic description, which strongly resembles that resulting from the two-electron *Difference Approach* of the MO theory (Nalewajski and Mrozek, 1994; Nalewajski, 2004b). The latter uses the displacements of the molecular *pair*-diagonal elements of the two-electron density matrix in AO representation, $\mathbf{\Gamma} = \{\Gamma_{\mu,\nu}\}$, relative to the corresponding SAL/promolecule values $\mathbf{\Gamma}^0 = \{\Gamma_{\mu,\nu}{}^0\}$,

$$\Delta\mathbf{\Gamma} = \{\Delta\Gamma_{\mu,\nu} = \Gamma_{\mu,\nu} - \Gamma_{\mu,\nu}{}^0 = \Delta\Gamma_{\mu,\nu}{}^{(1)} + \Delta\Gamma_{\mu,\nu}{}^{(2)}\} \equiv \Delta\mathbf{\Gamma}^{(1)} + \Delta\mathbf{\Gamma}^{(2)}, \tag{8.7.9}$$

where $\Delta\mathbf{\Gamma}^{(1)}$ and $\Delta\mathbf{\Gamma}^{(2)}$ denote contributions linear and quadratic in the corresponding displacements of the CBO matrix elements $\boldsymbol{\gamma}$ or $\{\boldsymbol{\gamma}^\sigma\}$ [Eqs. (8.6.9b) and (8.6.10)]:

$$\Delta\boldsymbol{\gamma} = \Delta\boldsymbol{\gamma} - \Delta\boldsymbol{\gamma}^0 = \{\Delta\gamma_{a,b} = \gamma_{a,b} - \gamma_{a,b}{}^0\} \qquad \text{or}$$
$$\Delta\boldsymbol{\gamma}^\sigma = \Delta\boldsymbol{\gamma}^\sigma - \Delta\boldsymbol{\gamma}^{\sigma,0} = \{\gamma_{a,b}{}^\sigma = \gamma_{a,b}{}^\sigma - \gamma_{a,b}{}^{\sigma,0}\}, \qquad \sigma = \alpha, \beta, \tag{8.7.10}$$

with the diagonal displacements representing shifts in the AO electron populations:

$$\Delta\gamma_{a,a} \equiv \Delta q_a = \gamma_{a,a} - \gamma_{a,a}{}^0 \quad \text{or} \quad \Delta\gamma_{a,a}{}^\sigma \equiv \Delta q_a{}^\sigma = \gamma_{a,a}{}^\sigma - \gamma_{a,a}{}^{\sigma,0}, \quad \sigma = \alpha, \beta. \tag{8.7.11}$$

The Difference-Approach indices are derived from the quadratic displacements: $-½\Delta\mathbf{\Gamma}^{(2)}$. The quadratic forms of shifts in orbital (atomic) electron populations give rise to *ionic*-contributions, while those involving the shifts in the *off*-diagonal CBO matrix elements are classified as *covalent*-terms. They represent *quadratic* forms of the CBO-displacements of Eq. (8.7.10). In the UHF approximation the *one*-centre (*oc*) and *two*-centre (*tc*) bond-indices of this MO-approach are defined as follows (Nalewajski and Mrozek, 1994; Nalewajski, 2004b):

one-centre:

$$\mathfrak{I}_{oc}{}^{cov}(A) = \sum\sum_{(a<a')\in A} \sum_{\sigma=\alpha,\beta} (\Delta\gamma^\sigma_{a,a'})^2,$$

$$\mathfrak{I}_{oc}{}^{ion}(A) = ½[\sum_{a\in A}\sum_{\sigma=\alpha,\beta} (\Delta q_a{}^\sigma)^2 - (\Delta N_A)^2]; \quad \Delta N_X = \sum_{x\in X}\Delta q_x = \sum_{x\in X}\sum_{\sigma=\alpha,\beta} \Delta q_x{}^\sigma,$$

two-centre:

$$\mathfrak{I}_{tc}^{cov}(A, B) = \sum_{a\in A}\sum_{b\in B}\sum_{\sigma=\alpha,\beta}(\Delta\gamma_{a,b}^{\sigma})^2,$$

$$\mathfrak{I}_{tc}^{ion}(A, B) = -\sum_{a\in A}\sum_{b\in B}\Delta q_a\,\Delta q_b = -\Delta N_A\,\Delta N_B. \tag{8.7.12}$$

Therefore, for the electron configuration $P = Q = ½$ of the purely-covalent bond in the 2-AO model the difference approach also ascribes a "half"-bond to the two-centre Wiberg-type covalency,

$$\mathfrak{I}_{tc}^{cov}(A, B) = ½, \tag{8.7.13}$$

with the remaining half-bond being attributed to the sum of the *one*-centre "ionicities" of two atoms (Nalewajski and Mrozek, 1994; Nalewajski, 2004b):

$$\mathfrak{I}_{oc}^{ion} = \mathfrak{I}_A^{ion}(A) + \mathfrak{I}_B^{ion}(B) = ½. \tag{8.7.14}$$

Clearly, the one-centre "ionicity" $\mathfrak{I}_A^{ion}(A)$ of the difference-approach describes the same effect of an effective atom-promotion in the molecular environment as does the one-centre entropy-covalency $N^{cov}(X)$ of the communication theory. Similarly, in the *ion-pair* cases, $P = (0, 1)$ both approaches correctly predict the vanishing covalent terms and a "single" ionic bond. Notice, however, that the IT bond-multiplicity ascribes 1 bit of the mutual-information measure to the *one*-centre ionic term of the atom, on which the lone electron pair is located, while the difference approach identifies this single ionic bond as due to the two-centre ionicity of the molecular ion-pair. One also observes that in the 2-AO model the sum of two-centre IT contributions in Eq. (8.7.6) gives rise to a half of the Wiberg (diatomic) covalency (see Fig. 7.1) and it equals to the sum of all covalent contributions of the two-electron difference approach:

$$N_{tc} = N^{cov}(A, B) + N^{ion}(A, B) = 2PQ$$
$$= ½\mathfrak{I}_{A,B}^{W}(P) = \mathfrak{I}_{tc}^{cov}(A, B) + \mathfrak{I}_{oc}^{cov}(A) + \mathfrak{I}_{oc}^{cov}(B) \equiv \mathcal{I}^{cov}(A, B), \tag{8.7.15}$$

8.8. BOND-ENTROPY CONCEPT

A direct IT measure (in bits) of the diatomic bond-order A–B is also provided by the *bond-entropy* concept (Nalewajski, 2000c, 2004b). It is defined in terms of the diatomic blocks of the molecular (pair-diagonal) two-electron density matrix in atomic resolution, $\mathbf{\Gamma} = \{\mathbf{\Gamma}_{X,Y}\}$:

$$H_{A,B} = H_{A,B}(\mathbf{\Gamma}_{A,B}) = -2\sum_{a\in A}\sum_{b\in B}\Gamma_{a,b}\log_2\Gamma_{a,b} \equiv \sum_{a\in A}\sum_{b\in B}H_{a,b}, \tag{8.8.1}$$

where indices a and b stand for the AO (basis functions) centered on atoms A and B, respectively, and $\mathbf{\Gamma}_{A,B} = \{\Gamma_{a\in A,\, b\in B}\}$. This entropic bond-order has been numerically validated for simple orbital models and π-electron systems in the Hückel approximation, where $\mathbf{\Gamma}_{A,B}$ contains a single element $\Gamma_{a,b}$. One observes that in all these models the bond-entropy of the diatomic fragment (A^0, B^0) in the promolecule M^0, $H_{A,B}{}^0 = H_{A,B}(\Gamma_{A,B}{}^0 = 1) = 0$, with $\Gamma_{A,B}{}^0$ denoting the free (non-bonded) atoms A^0 and B^0 in AP, each contributing a single AO and a single electron to the π-electron system. Therefore, for all these systems $\Delta H_{A,B} \equiv H_{A,B} - H_{A,B}{}^0 = H_{A,B}$, so that bond entropies automatically vanish in both the promolecular/SAL references.

This observation suggests that for general molecular systems, for which $H_{A,B}{}^0 \neq 0$, the bond-entropy index should be based either upon the bond-entropy displacement $\Delta H_{A,B}$ or the entropy-deficiency of Kullback and Leibler (1951) of Section 3.3, relative to the promolecular pair-distribution:

$$\Delta S_{A,B}[\mathbf{\Gamma}_{A,B} \,|\, \mathbf{\Gamma}_{A,B}{}^0] = 2\sum_{a\in A}\sum_{b\in B}\Gamma_{a,b}\log_2[\Gamma_{a,b}/\Gamma_{a,b}{}^0]$$
$$= 2\,N(N-1)\sum_{a\in A}\sum_{b\in B}P_{a,b}\log_2[P_{a,b}/P_{a,b}{}^0] \geq 0. \qquad (8.8.2)$$

The zero value marks the identical (isoelectronic) pair-distributions: $\mathbf{\Gamma}_{A,B} = \mathbf{\Gamma}_{A,B}{}^0$ or $\mathbf{P}_{A,B} = \mathbf{\Gamma}_{A,B}/[N(N-1)] = \{P_{a\in A,b\in B}\} = \mathbf{P}_{A,B}{}^0 = \mathbf{\Gamma}_{A,B}{}^0/[N(N-1)] = \{P_{a\in A,b\in B}{}^0\}$. This *cross-entropy* quantity provides the "information distance" (in bits) between the molecular density matrix $\mathbf{\Gamma}_{A,B} = N(N-1)\mathbf{P}_{A,B}$ and the reference promolecular matrix $\mathbf{\Gamma}_{A,B}{}^0 = N(N-1)\mathbf{P}_{A,B}{}^0$, proportional to that between the associated blocks $\mathbf{P}_{A,B}$ and $\mathbf{P}_{A,B}{}^0$ of the joint two-electron probabilities in the AO-resolution. The related divergence (symmetrized) measure of Kullback (1959),

$$\Delta S_{A,B}[\mathbf{\Gamma}_{A,B}, \mathbf{\Gamma}_{A,B}{}^0] = \Delta S_{A,B}[\mathbf{\Gamma}_{A,B} \,|\, \mathbf{\Gamma}_{A,B}{}^0] + \Delta S_{A,B}[\mathbf{\Gamma}_{A,B}{}^0 \,|\, \mathbf{\Gamma}_{A,B}]$$
$$= \sum_{a\in A}\sum_{b\in B}\Delta\Gamma_{a,b}\log_2[\Gamma_{a,b}/\Gamma_{a,b}{}^0] \geq 0, \qquad (8.8.3)$$

could also be used as an alternative direct information descriptor of the A—B bond, which also vanishes in the AP/SAL limits.

To summarize, all these *relative* IT bond-orders vanish identically in the atomic promolecule. Also, due to the specific structure of $\mathbf{\Gamma}_{A,B}{}^0$ in all orbital models considered in this chapter,

$$\Delta S_{A,B}[\mathbf{\Gamma}_{A,B} \,|\, \mathbf{\Gamma}_{A,B}{}^0] = \Delta S_{A,B}[\mathbf{\Gamma}_{A,B}, \mathbf{\Gamma}_{A,B}{}^0] = \Delta H_{A,B} = H_{A,B}. \qquad (8.8.4)$$

For example, the 2-AO model of Section 7.5 gives $\mathbf{\Gamma}_{\mathrm{A,B}} = \Gamma_{\mathrm{A,B}} = 2PQ$ and generates the bond-entropy function shown in Fig. 7.1:

$$H_{\mathrm{A,B}}(P) = -\,4PQ\log_2(2PQ). \qquad (8.8.5)$$

It is seen to give rise to the vanishing index for the two lone-pair configurations and 1 bit of information-measure for $P = Q = ½$, e.g., for the π-bond in ethylene. However, a reference to Fig. 7.1 also shows that the bond-entropy exhibits the local maxima of slightly above 1 bit of information in the regions of the MO polarization (probability) parameter $P \cong (¼, ¾)$.

The π bond-entropies (in bits) for alternative diatomics in the benzene ring and butadiene chain, predicted from the Hückel theory, compare favorably with the corresponding Wiberg indices $\{\mathfrak{I}_{i,j}{}^W\}$ reported in Section 8.6:

benzene:

$$H_{i,i+1} = 0.56\ (\mathfrak{I}_{i,i+1}{}^W = 0.44), \quad H_{i,i+2} = \mathfrak{I}_{i,i+2}{}^W = 0, \quad H_{i,i+3} = 0.16\ (\mathfrak{I}^W{}_{i,i+3} = 0.11);$$

butadiene:

$$H_{1,2} = 0.88\ (\mathfrak{I}_{1,2}{}^W = 0.80), \quad H_{1,3} = \mathfrak{I}_{1,3}{}^W = 0, \quad H_{2,3} = H_{1,4} = 0.27\ (\mathfrak{I}_{2,3}{}^W = \mathfrak{I}_{1,4}{}^W = 0.27).$$

These predictions show that the bond-entropies are generally in good agreement with the corresponding Wiberg-covalency indices. In particular, they give rise to a "chemical" covalency value of about "half" π-bond between the neighboring carbon atoms in benzene, when they are in the mutual *ortho*-positions. It also correctly predicts a much lower π-bond entropy between two carbons in the mutual *para*-arrangement, and a vanishing information index between two carbons in the mutual *meta*-location. In butadiene the two sets of π-bond covalency indices predict the peripheral bonds (1-2 or 3-4) to be much stronger than the central (2-3) bond and that between the terminal carbons (1-4); they both give rise to a vanishing index for the second-neighbor interactions (1-3 or 2-4).

8.9. REDUCED CHANNELS OF MOLECULAR FRAGMENTS

8.9.1. Input and Output Reductions of Molecular Channels

It is of interest to examine how the level of a resolution of the electron distributions in a molecule affects the information descriptors of the chemical bonds inside and between molecular fragments. Specific schemes of the atomic condensation of molecular electron densities represent in fact *reductions* of the *fine-grained* molecular communication channel (in the *local* resolution). Indeed, by integrating all distributions over the local events of constituent atoms one refers to the atomic units as whole building-blocks of molecules. They define the discrete, AIM-resolved probability distributions, which determine the molecular communication channel. This condensed picture of the one-electron distribution reduces all the *intra-atom* (local) events to a single event of finding an electron on specified atom. It generates the IT bond descriptors which take into account only the *external* communications, *between* constituent atoms, and completely disregard the *internal* communications within atoms. We have encountered another example of the probability fragment-reduction in Section 8.4, at the IFA stage of combining the separated fragment entropies into the global entrop/information indices of the molecule as a whole.

Clearly, the AIM-condensation of local electron probabilities in molecular systems can be pursued still further, throgh the molecular fragment resolution of *Subsystems-in-a-Molecule* to the ultimate trivial case of the *Molecule-in-a-Molecule* description, which contains only a single (sure) molecular event of finding electrons of M in M, which gives rise to the identically vanishing IT bond indices. The greater the reduction level of the canonical AIM-resolved probability distributions and the associated molecular communication channel, the greater the loss of communication links between AIM, and hence more chemical bonds between AIM escapes counting in terms of the entropy/information indices.

It should be realized that the fragment reduction of the molecular joint two-electron probabilities in atomic resolution, $\mathbf{P} = \{P_{i,j}\}$, or the associated conditional probabilities $\mathbf{P}(\mathbf{B}\,|\,\mathbf{A}) = \{P(j|i)\}$, can be performed in the input and/or output of the molecular channel. These reductions of the AIM-resolved electron probabilities involve relevant summations over the reduced-subsystem probabilities, which generate the condensed probabilities for the group of AIM defining the reduced subsystems in question. For example, the AIM *one*-electron probabilities $\boldsymbol{P} = \{P_i\}$, in the input or output of the molecular information system, can be condensed into the subsystem (group) probabilities (see Section 8.4): $\boldsymbol{P}^G = \{P_L = \sum_{l\in L} P_l\}$.

A summation of the AIM joint probabilities over atomic events of a specific *input*-fragment K gives the *input*-condensed two-electron probability vector $\boldsymbol{P}_K{}^G = \{P_{K,j} = \sum_{k\in K} P_{k,j}\}$. It determines the corresponding conditional probabilities in this reduction scheme, of all AIM outputs, given the condensed input-fragment K:

$$\boldsymbol{P}(\mathbf{B}|K) \equiv \boldsymbol{P}(\boldsymbol{j}\,|K) = \{P(j|K) = P_{K,j}/P_K\}. \tag{8.9.1}$$

The condensed conditional probabilities for the reduced *output*-fragment L, of finding an electron on L, when another electron has already been located on atom i in the molecule, are similarly derived from the output-condensed two-electron probability vector $\boldsymbol{P}_L{}^G = \{P_{i,L} = \sum_{l\in L} P_{i,l}\}$:

$$\boldsymbol{P}(L\,|\,\mathbf{A}) \equiv \boldsymbol{P}(L|\boldsymbol{i}) = \{P(L|i) = P_{i,L}/P_i = \sum_{l\in L} P(l|i)\}. \tag{8.9.2}$$

As indicated in the preceding equation the reduction of $\mathbf{P}(\mathbf{B}\,|\,\mathbf{A})$ into $\boldsymbol{P}(L|\mathbf{A})$ involves the summation over columns $\boldsymbol{l}\in L$ of the full molecular conditional probability matrix in atomic resolution $\boldsymbol{P}(\mathbf{B}|\mathbf{A}) \equiv \boldsymbol{P}(\boldsymbol{j}|\boldsymbol{i})$.

Finally, a simultaneous reduction of the AIM two-electron probabilities in the input and output of the molecular channel requires the double summation over the constituent atoms of the input (K) and output (L) fragments: $\mathbf{P}^G \equiv \boldsymbol{P}(\boldsymbol{GG}') = \{P_{K,L} = \sum_{k\in K} \sum_{l\in L} P_{k,l}\}$. It defines the associated conditional probabilities in the molecular fragment resolution:

$$\mathbf{P}^G(\boldsymbol{G}'|\boldsymbol{G}) = \{P(L|K) = P_{K,L}/P_K\}. \tag{8.9.3}$$

This fragment reduction implies that each of its constituent AIM is now described by the subsystem-average AIM probabilities and communications. Therefore, the above fully-condensed approach misses the fine complexity of the inter-AIM interactions between such condensed groups of AIM, which should influence both internal and external bond indices of molecular fragments. Instead it uses the average communications between atoms, averaged over the subsystem constituent atoms. Therefore, the global IT-descriptors of chemical bonds from such *average*-channel mises part of the exact entropy/information contributions generated by the AIM-resolved molecular channels. In some cases, however, e.g., in determining the external contributions from the SFA/IFA combination-rule (see Table 8.1), this channel-reduction does not result in any loss of information contained in the original AIM-resolved channel.

In what follows we shall again adopt the notation of Section 8.4, in which the bold fragment/molecule symbol implies the AIM-unreduced collection of its constituent atoms, thus recognizing each atom as a separate input- and/or output-unit of the molecular system and the associated information channel.

We again recall that the input resolution of the molecular channel characterizes the bond contributors (origins), while the output reduction level specifies the actual bonds generated by the input units. Therefore, by using the unreduced (AIM-resolved) input and the fragment-reduced output of the molecular channel one recognizes the detailed atomic origins of the chemical bonds accounted for by the output reduction scheme, i.e., all the *external* bonds, between the output-reduced fragments. In other words, the entropy/information quantities generated by such an output-reduced channel include all but the internal (intra-fragment) contributions to the global bond index of the whole system. One can thus extract the missing internal part by subtracting these external information measures from the corresponding global index, which contains all contributions.

Therefore, by appropriately selecting the fragment partition of a given molecular system, one can determine the subsystem IT-indices of chemical bonds involving any part or parts of the molecule, both internal and external, which are required in diverse *chemical* interpretations of the known (experimental or theoretical) distributions of electrons in a molecule. In fact, one can judiciously "tailor" the reduction scheme to extract the very bond-indices, which directly address a specific chemical problem at hand.

We shall now demonstrate that, by selecting the complementary output-partition into the fragment of interest, $\boldsymbol{L}$, and the reminder of the molecule, $\boldsymbol{R}_L \neq \boldsymbol{L}$, $\boldsymbol{M} = (\boldsymbol{L}, \boldsymbol{R}_L)$ and a subsequent reduction $\boldsymbol{L} \rightarrow L$ of the output-fragment AIM into a single unit, one can directly determine the IT-measure of the sum of both the fragment *external*-bonds $\boldsymbol{L}$— $\boldsymbol{R}_L$ with its molecular environment, and bonds $\boldsymbol{R}_L$ — $\boldsymbol{R}_L$ inside $\boldsymbol{R}_L$. It then allows an extraction of the exact *intra-**L*** bond measures. In order to limit the internal bond estimate to the fragment $\boldsymbol{L}$ only, one has to limit the *output* reduction exclusively to atoms $\boldsymbol{L}$, $\boldsymbol{L} \rightarrow L$, keeping the molecular environment of L

unreduced, in full atomic resolution $\boldsymbol{R}_L$, $\boldsymbol{M}^r(L) \equiv (L, \boldsymbol{R}_L)$, thus explicitly counting all bond contributions due to communications inside $\boldsymbol{R}_L$, with the molecular input kept unreduced: $\boldsymbol{i} = \mathbf{A} \equiv \boldsymbol{M}$ or $\boldsymbol{i}^0 = \mathbf{A}^0 \equiv \boldsymbol{M}^0$. The relevant bond indices,

$$S(\boldsymbol{M}^r(L)\,|\,\boldsymbol{M}) = -\sum_{m\in M}[P_{m,L}\log P(L|m) + \sum_{n\notin L} P_{m,n}\log P(n|m)] \equiv S_{ext}(\boldsymbol{L})$$
$$\equiv S(\boldsymbol{L}\text{–}\boldsymbol{R}_L) + S(\boldsymbol{R}_L\text{–}\boldsymbol{R}_L),$$

$$I(\boldsymbol{M}^0{:}\boldsymbol{M}^r(L)) = \sum_{m\in M}\{P_{m,L}\log[P(m|L)/P_m{}^0] + \sum_{n\notin L} P_{m,n}\log[P(m|n)/P_m{}^0]\} \equiv I_{ext}(\boldsymbol{L}),$$
$$\equiv I(\boldsymbol{L}\text{–}\boldsymbol{R}_L) + I(\boldsymbol{R}_L\text{–}\boldsymbol{R}_L),$$

$$N(\boldsymbol{M};\boldsymbol{M}^r(L)) = S(\boldsymbol{M}^r(L)|\boldsymbol{M}) + I(\boldsymbol{M}^0 : \boldsymbol{M}^r(L)) \equiv N_{ext}(\boldsymbol{L})$$
$$\equiv N(\boldsymbol{L}\text{–}\boldsymbol{R}_L) + N(\boldsymbol{R}_L\text{–}\boldsymbol{R}_L), \qquad (8.9.4a)$$

contain all but the *intra-**L*** contributions, i.e., they reflect the external bonds $\boldsymbol{L}$–$\boldsymbol{R}_L$ between atoms in $\boldsymbol{L}$ and the molecular remainder, as well as chemical bonds $\boldsymbol{R}_L$–$\boldsymbol{R}_L$ inside $\boldsymbol{R}_L$.

This allows an extraction of the fragment *internal* entropy/information bond descriptors, by subtracting the IT-indices of Eq. (8.9.4a) from the corresponding global bond descriptors:

$$S_{int}(\boldsymbol{L}) \equiv S(\boldsymbol{M}\,|\,\boldsymbol{M}) - S(\boldsymbol{M}^r(L)\,|\,\boldsymbol{M}), \quad I_{int}(\boldsymbol{L}) \equiv I(\boldsymbol{M}^0 : \boldsymbol{M}) - I(\boldsymbol{M}^0; \boldsymbol{M}^r(L)),$$
$$N_{int}(\boldsymbol{L}) \equiv N(\boldsymbol{M};\boldsymbol{M}) - N(\boldsymbol{M};\boldsymbol{M}^r(L)). \qquad (8.9.5)$$

In order to separate the intra-$\boldsymbol{R}_L$ bonds from all bonds external to $\boldsymbol{L}$, one has to simultaneously reduce both $\boldsymbol{L}$ into L and $\boldsymbol{R}_L$ into R_L in the molecular output, since then the intra-$\boldsymbol{R}_L$ bonds are no longer counted in the modified indices:

$$S(L, R_L\,|\,\boldsymbol{M}) \equiv S(\boldsymbol{L}\text{–}\boldsymbol{R}_L), \quad I(\boldsymbol{M}^0 {:} L, R_L) \equiv I(\boldsymbol{L}\text{–}\boldsymbol{R}_L), \quad N(\boldsymbol{M}; L, R_L) \equiv N(\boldsymbol{L}\text{–}\boldsymbol{R}_L). \quad (8.9.4b)$$

Finally, by subtracting these inter-fragment bond indices from the external quantities of Eq. (8.9.4a) one determines the IT bond multiplicities inside the chemical environment of L in M:

$$S(\boldsymbol{R}_L\text{–}\boldsymbol{R}_L) = S_{ext}(\boldsymbol{L}) - S(\boldsymbol{L}\text{–}\boldsymbol{R}_L), \qquad I(\boldsymbol{R}_L\text{–}\boldsymbol{R}_L) = I_{ext}(\boldsymbol{L}) - I(\boldsymbol{L}\text{–}\boldsymbol{R}_L),$$
$$N(\boldsymbol{R}_L\text{–}\boldsymbol{R}_L) = N_{ext}(\boldsymbol{L}) - N(\boldsymbol{L}\text{–}\boldsymbol{R}_L) = S(\boldsymbol{R}_L\text{–}\boldsymbol{R}_L) + I(\boldsymbol{R}_L\text{–}\boldsymbol{R}_L). \qquad (8.9.4c)$$

Consider next the *double* reduction scheme, when the above single-fragment reduction is simultaneously performed in both the input and output of the molecular communication system: $K = L$ and $\boldsymbol{R}_K = \boldsymbol{R}_L$. Only the "communications" between $\boldsymbol{M}^r(L)$ in the input and output then remain, with the fragment L acting as a whole (reduced) unit at both ends of the communication system. Since in the input of such a reduced molecular channel the collection of atoms $\boldsymbol{L}$ is considered as a single input L, only the *approximate* bond-indices can be extracted from this reduced information system. Indeed, the input reduction hides the true AIM origins of the chemical bonds counted in the channel output, since in the associated IT indices the AIM probabilities of $\boldsymbol{L}$ are now regarded as the fragment *average* atomic probabilities. Therefore, such a doubly reduced, complementary channel can be used to generate the information-theoretic descriptors of the sum of *external* chemical bonds L—$\boldsymbol{R}_L$ of L in $\boldsymbol{M}$, between the whole (reduced) L and all remaining atoms $\boldsymbol{R}_L$, and the internal bonds in the fragment molecular environment, $\boldsymbol{R}_L$ — $\boldsymbol{R}_L$:

$$S(L\!-\!\boldsymbol{R}_L) + S(\boldsymbol{R}_L\!-\!\boldsymbol{R}_L) \equiv S^{ext}(L) \equiv S(\boldsymbol{M}^r(L)\,|\,\boldsymbol{M}^r(L)) = -\,P_{L,L}\log P(L|L)$$
$$-\textstyle\sum_{n\notin L} P_{L,n}\log P(n|L) - \sum_{m\notin L} P_{m,L}\log P(L|m) - \sum_{n\notin L}\sum_{m\notin L} P_{m,n}\log P(n|m),$$

$$I(L\!-\!\boldsymbol{R}_L) + I(\boldsymbol{R}_L\!-\!\boldsymbol{R}_L) \equiv I^{ext}(L) \equiv I(\boldsymbol{M}^{r,0}(L){:}\boldsymbol{M}^r(L))$$
$$= P_{L,L}\log[P(L|L)/P_L^{\,0}] + \textstyle\sum_{n\notin L}P_{L,n}\log[P(L|n)/P_L^{\,0}]$$
$$+ \textstyle\sum_{m\notin L}P_{m,L}\log[P(m|L)/P_m^{\,0}] + \sum_{n\notin L}\sum_{m\notin L}P_{m,n}\log[P(m|n)/P_m^{\,0}],$$

$$N(L\!-\!\boldsymbol{R}_L) + N(\boldsymbol{R}_L\!-\!\boldsymbol{R}_L) \equiv N^{ext}(L) \equiv N(\boldsymbol{M}^r(L);\,\boldsymbol{M}^r(L)) = S^{ext}(L) + I^{ext}(L). \qquad (8.9.6)$$

As we have already remarked above, the approximate character of these external bond measures, identified here by the upper *ext* label, is due to the fact that they involve the reduced L in the input (bond-origin) part of the communication system, thus recognizing the L-averaged atoms $\boldsymbol{L}^{av}$, with equalized AIM probabilities, instead of the real bonded (or free) atoms, with generally non-equalized probabilities, as the source of the L-external communications in the molecule. In this channel one simultaneously "counts" the L—$\boldsymbol{R}_L$ and $\boldsymbol{R}_L$–$\boldsymbol{R}_L$ bonds, as specified by the output-reduction scheme generated by the (L, $\boldsymbol{R}_L$) input sources defined by the input-reduction scheme. It should be observed that the exact external bond quantities of Eq. (8.9.4a) have recognized each bonded atoms in the molecule or the free-atom of the promolecule as independent input (bond-generator) unit.

Let us examine, how closely the sum of the exact internal [Eq. (8.9.5)] and approximate external [Eq. (8.9.6)] quantities reproduce the corresponding global IT-measures of chemical bonds in the system as a whole:

$$S(L) = S_{int}(L) + S^{ext}(L) = S(\boldsymbol{M}|\boldsymbol{M}) + [S(\boldsymbol{M}^r(L)|\boldsymbol{M}^r(L)) - S(\boldsymbol{M}^r(L)|\boldsymbol{M})]$$
$$\equiv S(\boldsymbol{M}|\boldsymbol{M}) + \Delta S(L),$$

$$I(L) = I_{int}(L) + I^{ext}(L) = I(\boldsymbol{M}^0{:}\boldsymbol{M}) + [I(\boldsymbol{M}^{r,0}(L){:}\boldsymbol{M}^r(L)) - I(\boldsymbol{M}^0{:}\boldsymbol{M}^r(L))]$$
$$\equiv I(\boldsymbol{M}^0{:}\boldsymbol{M}) + \Delta I(L),$$

$$N(L) = N_{int}(L) + N^{ext}(L) = N(\boldsymbol{M};\,\boldsymbol{M}) + [N(\boldsymbol{M}^r(L);\,\boldsymbol{M}^r(L)) - N(\boldsymbol{M};\,\boldsymbol{M}^r(L))]$$
$$\equiv N(\boldsymbol{M};\,\boldsymbol{M}) + \Delta N(L). \qquad (8.9.7)$$

The deviations $\{\Delta S(L), \Delta I(L), \Delta N(L)\}$ defined in the foregoing equation should be expected to be small compared to the global index itself, since all internal and external communications for a given complementary division have been *effectively* exhausted in the underlying reduced molecular channels, be it in an approximate manner at the external L–$\boldsymbol{R}_L$ stage. Let us find out, under what conditions this is the case. A straightforward transformation of these entropy/information displacements gives:

$$\Delta S(L) = \Sigma_{m\in L} P_{m,L} \log [P(L|m)/P(L|L)] + \Sigma_{m\in L} \Sigma_{n\notin L} P_{m,n} \log [P(n|m)/P(n|L)],$$

$$\Delta I(L) = \Sigma_{m\in L} P_{m,L} \log [P_{L,L}P_m{}^0/(P_{m,L}P_L{}^0)] + \Sigma_{m\in L}\Sigma_{n\notin L}P_{m,n}\log [P_{L,n}P_m{}^0/(P_{m,n}P_L{}^0)],$$

$$\Delta N(L) = \Sigma_{m\in L}P_{m,M}\log (P_m/P_m{}^0) + P_{L,M}\log (P_L/P_L{}^0). \qquad (8.9.8)$$

The $\Delta N(L)$ displacement exactly vanishes for the identical molecular and promolecular input probabilities in the AIM resolution, $\boldsymbol{P} = \boldsymbol{P}^0$, e.g., for π-systems of the alternant hydrocarbons in the Hückel approximation, which also imply the identical condensed probabilities: $P_L = P_L{}^0$. Therefore, the approximate combination rule for the total IT bond-order $N(L)$ in Eq. (8.9.7) in general should be quite accurate, since the AIM electron probabilities differ only slightly from their promolecular analogs. It also directly follows from Eq. (8.9.8) that for $\boldsymbol{P} = \boldsymbol{P}^0$ one indeed obtains $\Delta I(L) = -\Delta S(L)$, and thus the vanishing deviation of the global entropy/information bond index $\Delta N(L)$.

However, the approximate combination rules in Eq. (8.9.7) for the global entropy-covalency $S(L)$ and information-ionicity $I(L)$ should be less accurate. One further predicts that the deviations $\Delta I(L)$ and $\Delta S(L)$ vanish identically for

$$P(L|m) = P(L|L),\ m\in L, \qquad \text{and} \qquad P(n|m) = P(n|L),\ m\in L,\ n\notin L. \qquad (8.9.9)$$

These conditions should be approximately satisfied by the molecular conditional probabilities, when there are small differences between the AIM probabilities in L, e.g., between the carbon atoms in the benzene ring or in butadiene chain, which we shall consider further in this section. Indeed, the atom averaging within the reduced fragment of the molecule then has only minor effect on individual atoms in L.

8.9.2. Reduced π Channels in Butadiene

The results of the reduced-channel analysis carried out for the internal and external π-bonds of alternative diatomics in the carbon chain of butadiene are reported in Table 8.6. The internal bond measures of Eq. (8.9.5) predict equal half-bit internal-bond for all pairs of carbons in the π chain, and hence 1.5 bit of the remaining chemical interactions involving the remaining carbon atoms in the molecule. This gives rise to a conserved "double" (two-bit) overall π bond in the system as a whole.

Although the peripheral bond covalency is predicted to be slightly higher that that for other pair of AIM, the bond components for all diatomics in a molecule are almost equalized in the communication theory description, contrary to the MO indices and bond-entropy description, which exhibit large variations for different choices of the diatomic fragment (see Section 8.8). The highest covalent component for the (1, 2) pair is only qualitatively consistent with the MO prediction and the entropy/information result of Scheme 8.2. All internal and external IT bond descriptors of diatomic fragments in butadiene are seen to predict almost purely entropy-covalent interactions. The strongest internal information-ionic part is detected for the middle (2-3) bond, while the terminal (1-2) and (3-4) bonds exhibit the highest external IT ionic contribution.

Table 8.6. A comparison of the external and internal IT-indices of π bonds involving diatomics in the butadiene carbon chain (the Hückel approximation).

Diatomic fragment	Bond indices	Internal bonds	External bonds		Overall indices	Global indices
			Eq. (8.9.6)	Eq. (8.9.4a)		
		a	b	c	$a+b$	$a+c$
(1,2), (3,4)	S	0.498	1.447	1.446	1.945	1.944
	I	0.002	0.053	0.054	0.055	0.056
	N	0.500	1.500	1.500	2.000	2.000
(1,3), (2,4)	S	0.472	1.480	1.472	1.952	1.944
	I	0.028	0.020	0.028	0.048	0.056
	N	0.500	1.500	1.500	2.000	2.000
(2,3), (1,4)	S	0.473	1.483	1.471	1.956	1.944
	I	0.027	0.017	0.029	0.044	0.056
	N	0.500	1.500	1.500	2.000	2.000

It follows from a comparison between columns b and c of Table 8.6 that the approximate measures of the external π bonds of column b reproduce semi-quantitatively the exact data reported in column c. The estimate of the overall indices from the reduced internal fragment descriptors and approximate external IT bond quantities is seen to reproduce in Table 8.6 the total global IT bond-order exactly. The resulting conditional-entropy and mutual-information components agree only semi-quantitatively with the exact values reported in the last column. While for the entropy-covalency index S the agreement is quite good, the results for the information-ionic index I are seen to be less satisfactory.

8.9.3. General Combination Rules for Reduced Channels

Let us combine the molecular *output* in butadiene (see Scheme 8.10a), which we again select as an illustrative example, into the two reduced complementary fragments $\boldsymbol{M}^r(L, R_L) \equiv \boldsymbol{M}^r(L, L') = \{(1,2), (3,4)\}$, where the brackets enclose the carbon atoms to be condensed. By preserving the full (unreduced) atomic input $\boldsymbol{M}$ in this output-reduced channel, one fully accounts for the atomic origins of the external L-L' bonds, which are counted in this reduced communication system. The corresponding exact external bond indices (in bits) are:

$$S(\boldsymbol{M}^r(L, L') \mid \boldsymbol{M}) \equiv S(L, L' \mid \boldsymbol{M}) \equiv S_{ext}(\boldsymbol{L}, \boldsymbol{L}') = -\sum_{i\in M}\sum_{G=L, L'} P_{i,G} \log(P_{i,G}/P_i) = 0.948,$$

$$I(\boldsymbol{M}^0:\boldsymbol{M}^r(L, L')) \equiv I(\boldsymbol{M}^0:L, L') \equiv I_{ext}(\boldsymbol{L}, \boldsymbol{L}') = \sum_{i\in M}\sum_{G=L, L'} P_{i,G} \log[P_{i,G}/(P_i^0 P_G)] = 0.052,$$

$$N(\boldsymbol{M};\boldsymbol{M}^r(L, L')) \equiv N(\boldsymbol{M};L, L') \equiv N_{ext}(\boldsymbol{L}, \boldsymbol{L}') = S(\boldsymbol{M}^r(L, L') \mid \boldsymbol{M}) + I(\boldsymbol{M}:\boldsymbol{M}^r(L, L')) = 1.000. \quad (8.9.10)$$

Here, $P_{i,G} = \sum_{g\in G} P_{i,g}$ is the joint probability of simultaneously detecting one electron on *i*th AIM and another electron on fragment G, and the condensed one-electron probability $P_G = \sum_{g\in G} P_g$.

These indices take into account all but the intra-subsystem communications (chemical interactions). In other words, they measure only the external L–L' bonds. Equation (8.9.10) defines the exact external complements of the corresponding intra-subsystem quantities [Eq. (8.9.5)] reported in Table 8.6:

$$\begin{aligned} S_{int}(\boldsymbol{L}) &= S(\boldsymbol{M} \mid \boldsymbol{M}) - S(L, \boldsymbol{L}' \mid \boldsymbol{M}) \\ &= -\sum_{i\in M}\sum_{j\in M} P_{i,j} \log(P_{i,j}/P_i) + \sum_{i\in M} P_{i,L} \log(P_{i,L}/P_i) \\ &\quad + \sum_{i\in M}\sum_{l\in L'} P_{i,l} \log(P_{i,l}/P_i) \\ &\equiv S(\boldsymbol{M} \mid \boldsymbol{M}) - S(L \mid \boldsymbol{M}) - S(\boldsymbol{L}' \mid \boldsymbol{M}), \end{aligned} \quad (8.9.11)$$

$$\begin{aligned} I_{int}(\boldsymbol{L}) &= I(\boldsymbol{M}^0:\boldsymbol{M}) - I(\boldsymbol{M}^0:L, \boldsymbol{L}') \\ &= \sum_{i\in M}\sum_{j\in M} P_{i,j} \log[P_{i,j}/(P_i^0 P_j)] - \sum_{i\in M} P_{i,L} \log[P_{i,L}/(P_i^0 P_L)] \\ &\quad + \sum_{i\in M}\sum_{l\in L'} P_{i,l} \log[P_{i,l}/(P_i^0 P_l)] \\ &\equiv I(\boldsymbol{M}^0:\boldsymbol{M}) - I(\boldsymbol{M}^0:L) - I(\boldsymbol{M}^0:\boldsymbol{L}'). \end{aligned} \quad (8.9.12)$$

The sum of these internal and external terms indeed reproduces the corresponding global quantity, e.g.,

$$\begin{aligned} S_{int}(\boldsymbol{L}) &+ S_{int}(\boldsymbol{L}') + S_{ext}(\boldsymbol{L}, \boldsymbol{L}') \\ &= 2S(\boldsymbol{M} \mid \boldsymbol{M}) - [S(\boldsymbol{L}' \mid \boldsymbol{M}) + S(\boldsymbol{L} \mid \boldsymbol{M})] + S(L, L' \mid \boldsymbol{M}) - [S(L \mid \boldsymbol{M}) + S(L' \mid \boldsymbol{M})] \\ &= S(\boldsymbol{M} \mid \boldsymbol{M}), \end{aligned} \quad (8.9.13)$$

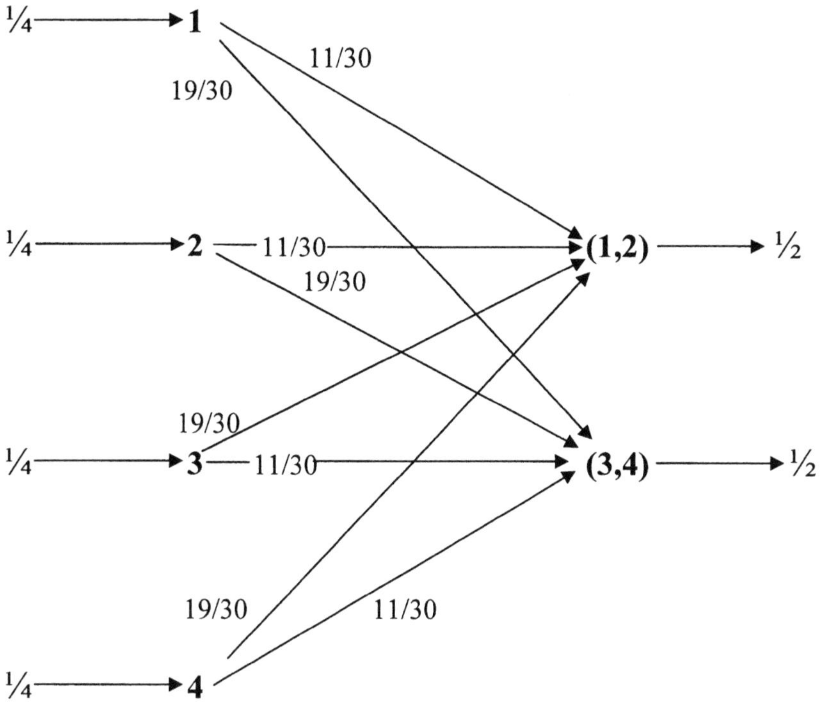

a)

$S(L, L' \mid \boldsymbol{M}) = 0.948, \qquad I(M^0: L, L') = 0.052, \qquad N(M^0: L, L') = 1.000$

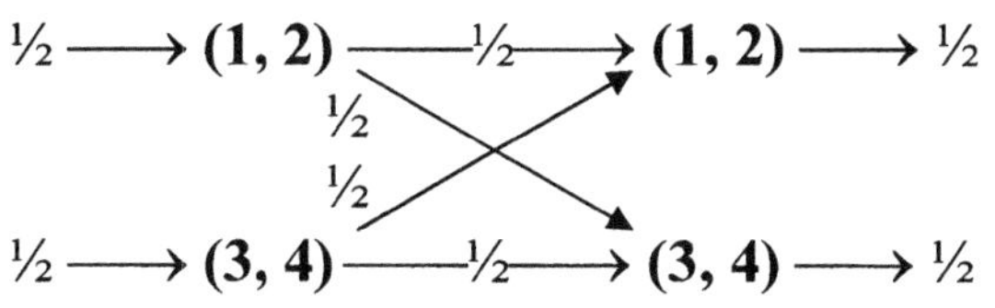

b)

$S(L, L' \mid L, L') = 1, \qquad I(L^0, L'^0: L, L') = 0, \qquad N(L^0, L'^0: L, L') = 1$

Scheme 8.10. Reduction of the molecular channel of butadiene (Scheme 7.11) into two complementary diatomics $\boldsymbol{G} = \{(1, 2), (3, 4)\} \equiv (L, L')$: *output*–only reduction (Panel *a*) and *input-and-output* reduction (Panel b). The reduced channel entropy/information data (in bits) are also reported.

since

$$S(\boldsymbol{M}\,|\,\boldsymbol{M}) = S(\boldsymbol{L}'\,|\,\boldsymbol{M}) + S(\boldsymbol{L}\,|\,\boldsymbol{M}) \quad \text{and} \quad S(L, L'\,|\,\boldsymbol{M}) = S(L\,|\,\boldsymbol{M}) + S(L'\,|\,\boldsymbol{M}). \qquad (8.9.14)$$

The same exhaustive character can be demonstrated for the internal and external mutual information quantities:

$$\begin{aligned} I_{int}(\boldsymbol{L}) + I_{int}(\boldsymbol{L}') + I_{ext}(\boldsymbol{L}, \boldsymbol{L}') \\ &= 2I(\boldsymbol{M}^0{:}\boldsymbol{M}) - [I(\boldsymbol{M}^0{:}\boldsymbol{L}) + I(\boldsymbol{M}^0{:}\boldsymbol{L}')] + I(\boldsymbol{M}^0{:}L, L') - [I(\boldsymbol{M}^0{:}L) + I(\boldsymbol{M}^0{:}L')] \\ &= I(\boldsymbol{M}^0{:}\boldsymbol{M}), \end{aligned} \qquad (8.9.15)$$

since also

$$I(\boldsymbol{M}^0{:}\boldsymbol{M}) = I(\boldsymbol{M}^0{:}\boldsymbol{L}) + I(\boldsymbol{M}^0{:}\boldsymbol{L}') \quad \text{and} \quad I(\boldsymbol{M}^0{:}L, L') = I(\boldsymbol{M}^0{:}L) + I(\boldsymbol{M}^0{:}L'). \qquad (8.9.16)$$

Because these combination rules are satisfied for the components of the total bond indices, they must be also satisfied for the total indices themselves:

$$N_{int}(\boldsymbol{L}) + N_{int}(\boldsymbol{L}') + N_{ext}(\boldsymbol{L}, \boldsymbol{L}') = N(\boldsymbol{M}; \boldsymbol{M}) = S(\boldsymbol{M}\,|\,\boldsymbol{M}) + I(\boldsymbol{M}^0{:}\boldsymbol{M}). \qquad (8.9.17)$$

Supplementing the external bond indices of Eq. (8.9.10), which are reported in Scheme 8.10*a*, with the internal contributions of both fragments from Table 8.6, (equal by symmetry), indeed reproduces exactly all global indices listed in the last column of the table: 0.948 + 2×0.498 = 1.944, 0.052 + 2×0.002 = 0.056, and 1.000 + 2×0.500 = 2.000.

As we have already observed in the preceding section, by additionally performing the same complementary reduction in the input of the molecular channel (Scheme 8.10*b*), for $K = L$ and $K' = L'$, one obtains the approximate estimates of the fragment-fragment (external) interactions amounting to a "single" (1 bit), purely covalent L–L' bond:

$$\begin{aligned} &S(L, L'\,|\,L, L') \equiv S^{ext}(L, L') = 1.000, \\ &I_{ext}(L, L'{:}L, L') \equiv I_{ext}(L, L') = 0.000, \qquad N(L, L'; L, L') = 1.000. \end{aligned} \qquad (8.9.18)$$

It recognizes the two *reduced*, complementary fragments L and L' in the input of the channel shown in Scheme 8.10b as the source of the L–L' chemical interactions counted in the output of this channel. As also seen in Panel *b* of Scheme 8.10 the numerical values of these IT bond indices approximate quite well the exact external indices for this partitioning, which are reported in Panel *a*.

Of interest in chemistry also are the atomic bond indices, which characterize all bond contributions between a given AIM i and its *reduced* molecular remainder $R(i)$. Examples of such entropy/information quantities have already been provided by the atomic partial *column* and *row* channels introduced in Section 8.5 (see also Appendix F). The corresponding approximate measures of such *external* interactions of a specified carbon atom of the π-electron system in butadiene with its complementary (reduced) molecular environment are identical in the Hückel approximation for all AIM positions in the carbon chain:

$$S(i, R(i) \,|\, i, R(i)) = 0.802, \quad I(i^0, R^0(i^0): i, R(i)) = 0.009, \quad N(i, R(i); i, R(i)) = 0.811.$$

They predict that each carbon participates in about 0.8 bond with all remaining atoms in the π system. These interactions are found to be almost purely IT-covalent.

8.9.4. Illustrative Results for Benzene

A similar prediction is valid for benzene: for all carbon atoms in the π-system ring

$$S(i, R(i) \,|\, i, R(i)) = 0.645, \quad I(i^0, R^0(i^0): i, R(i)) = 0.005, \quad N(i, R(i); i, R(i)) = 0.650.$$

Therefore, this equalized measure of the atomic *external* IT bond-order, which approximates an overall participation of a given carbon atom in all π bonds with the remaining carbon atoms, is distinctly higher in butadiene, in comparison to that in benzene. The same conclusion has been reached in Section 8.5 from examining the corresponding entropy/information data of the partial atomic channels in both these molecules. This result again reflects the fact that in the symmetric hexagon structure of the benzene ring, due to the carbon-carbon σ bonds, the system compromises the π bond-order, which otherwise could reach higher values, should the bond alternation be allowed (Shaik 1989; and Hiberty, 1991, 1995, 2004; Jug and Köster, 1990).

Let us now examine the remaining entropy/information data for the reduced fragments of the benzene ring, which involve a pair of carbon atoms defining the reduced diatomic L, and the four remaining atoms of the ring defining the reduced complementary (tetra-atomic) system L'. When the AIM-resolved fragments define the input and/or output units of the molecular communication system we shall again use the bold symbol notation: $\boldsymbol{L}$ and/or $\boldsymbol{L}'$.

Table 8.7 collects the relevant entropy/information data (in bits) for the reduced channels of interest, while the resulting internal and external bond indices (in bits), reflecting the intra-L and L–L' bonds, are listed in Table 8.8, where the relevant combination rules have also been tested. It should be recalled that the output reductions (L, $\boldsymbol{L}'$) and ($\boldsymbol{L}$, L'), with the full AIM resolved input $\boldsymbol{M}$ (or $\boldsymbol{M}^0$), which are the subject of the first two rows of each group of entries in Table 8.7, contain all but internal bond contributions of the reduced fragment. Therefore, by subtracting these exact external bond components from the corresponding global index one generates the fragment internal descriptors, which are listed in columns (a, b) of Table 8.8.

Table 8.7. The entropy/information characteristics of the reduced channels for a diatomic (L) and the complementary tetra-atomic (L') subsystems in the benzene ring $M(L, L')$ (the Hückel approximation).

Entropy/Information	L:	$(i, i+1)$	$(i, i+2)$	$(i, i+3)$
$S(L, \mathbf{L'}\|\mathbf{M})$		2.221	2.226	2.226
$S(\mathbf{L}, L'\|\mathbf{M})$		1.237	1.239	1.239
$S(L, L'\|\mathbf{M})$		0.908	0.914	0.915
$S(L, L'\| L, L')$		0.908	0.916	0.915
$I(\mathbf{M}^0 : L, \mathbf{L'})$		0.030	0.026	0.025
$I(\mathbf{M}^0 : \mathbf{L}, L')$		0.015	0.013	0.013
$I(\mathbf{M}^0 : L, L')$		0.011	0.004	0.003
$I(L^0, L'^0 : L, L')$		0.010	0.002	0.003
$N(\mathbf{M}; L, \mathbf{L'})$		2.252	2.252	2.252
$N(\mathbf{M}; \mathbf{L}, L')$		1.252	1.252	1.252
$N(\mathbf{M}; L, L')$		0.918	0.918	0.918
$N(L, L'; L, L')$		0.918	0.918	0.918

TABLE 8.8. Same as in Table 8.6 for benzene.

L	Information/ entropy	Internal in L a	Internal in L' b	External L–L': Exact c	Approx. d	Overall $e=a+b+c$	Global f
$i, i+1$	S	0.330	1.314	0.908	0.908	2.551	2.551
	I	0.004	0.019	0.011	0.010	0.034	0.034
	N	0.333	1.333	0.918	0.918	2.585	2.585
$i, i+2$	S	0.325	1.312	0.914	0.916	2.551	2.551
	I	0.008	0.021	0.004	0.002	0.034	0.034
	N	0.333	1.333	0.918	0.918	2.585	2.585
$i, i+3$	S	0.325	1.312	0.915	0.915	2.551	2.551
	I	0.009	0.021	0.003	0.003	0.034	0.034
	N	0.333	1.333	0.918	0.918	2.585	2.585

Again, the complementary (L, L') reduction in the output of the molecular channel (third rows of each group in Table 8.7), for the atom-resolved inputs $\boldsymbol{M}$ or $\boldsymbol{M}^0$, gives rise to the exact external L-L' bond contributions, listed in Column *c* of Table 8.8. Finally, a double complementary-reduction, in the input and output of the molecular communication network (fourth row of each group in Table 8.7), generates approximate inter-subsystem IT bond indices, which are collected in the Column *d* of Table 8.8.

A reference to column *e* in Table 8.8 reveals that combining the complementary internal and exact external bond indices indeed recovers the corresponding global descriptors for the system as a whole, which are listed in Column *f*. The approximate external bond indices of column *d* reproduce almost perfectly the exact external contributions of column *c*. The total fragment indices are the same for any choice of the diatomic fragment, with only the covalent/ionic components exhibiting minor variations, when the identity of the diatomic fragment changes from the *ortho* through *meta* to *para* mutual position of two carbons in the ring-fragment L. This picture is thus dramatically different from that emerging from the MO indices, e.g., the Wiberg index, which predicts for these three pairs of carbon atoms dramatically different covalent bond-orders: 0.44, 0.00, and 0.11, respectively.

8.10. CONCLUSION

We have explored in this chapter a variety of IT-descriptors of chemical bonds involving molecular fragments. Some of them have originated from the additive decomposition of the related global quantities, while others reflect the information bond indices of specific partial communication channels of molecular subsystems. Both the SFA, generating the entropy/information bond-indices of the disconnected molecular fragments, and a more realistic treatment using the partial channels of the mutually connected parts of the molecule, open relative to their surrounding AIM, have been shown to allow for a quantitative combination of the subsystem descriptors into the corresponding global quantities, characterizing the system as a whole.

The capabilities of alternative reductions of the molecular information channel in atomic resolution, by combining the constituent AIM of the reduced fragment into a single unit, have been investigated. This approach offers a variety of the IT diagnostic tools for exploring the bonding patterns in molecular subsystems and provides a systematic scheme to extract the entropy/information indices of the fragment internal and external bonds. They reflect the *intra-* and *inter*-fragment chemical interactions, respectively, e.g., chemical bonds between the specified molecular subsystem and its molecular environment. The reduction scheme can be targeted to answer specific chemical problems. In particular, the fragment complementary reduction has been shown to provide a convenient framework for formulating general combination rules for the entropy/information bond descriptors.

It has been demonstrated that the bonding pattern emerging from the IT descriptors of both atomic and diatomic fragments in a molecule is quite different from that generated by the bond-multiplicity measures from the molecular orbital theory. In general, the IT indices, derived from the electron probability distributions, strongly emphasize the information *equilibrium* of the system ground-state electronic structure. This is manifested by a remarkable equalization of various atomic and diatomic entropy/information indices, irrespectively of the fragment position in the molecule. This is in contrast to the related MO description, which gives rise to a much stronger differentiation between alternative sites and bonds. Often, as in the case of the π bonds C_1–C_3 in butadiene and C_i–C_{i+2} in benzene, the MO orbital approach unrealistically predicts exactly vanishing bond indices, which appears to be the artifact of the adopted bond-order measure. We have also demonstrated that the bond-entropy concept, designed for indexing the localized bonds between a given pair of bonded atoms, generates the bonding patterns, which strongly resemble the corresponding MO description.

We have also explored the additive atomic and diatomic contributions to the entropy/information indices of chemical bonds. They were shown to provide in the 2-AO model of the chemical bond a perspective on variations in the covalent and ionic bond components similar to that emerging from the quadratic bond-descriptors of the two-electron difference-approach formulated within the MO theory.

This molecular fragment development should be also useful in exploring the reactant fragments of the bimolecular reactive system. This is the main purpose of the next chapter, which will apply some of the concepts developed in this and previous chapters to characterize reactive systems.

9. REACTIVE SYSTEMS

In the Donor-Acceptor reactive systems the *Fukui function* descriptors and the *Charge-Transfer Affinities* (CTA) of the Hirshfeld reactants are introduced, combining the derivative properties of the molecular energy and entropy-deficiency functionals. The stockholder subsystems are shown to correspond to the equilibrium amount of CT, marking the vanishing affinities ("forces") for the flow of electrons between the basic and acidic reactants. The 3-AO model of a symmetric transition-state complex of the atom-exchange reactions in the atom-diatom reactive collisions is introduced and applied to further illustrate and test the use of alternative approaches developed in the preceding chapter in generating the bond descriptors of specified groups of AIM. The entropy deficiencies of IT are used to measure the information-distance between the electron densities of the transition-state complex in collinear atom-exchange reactions and the relevant "promolecular" densities obtained from the electron distributions of the reaction substrates and products. These quantities are used to probe relative similarities between the electronic structures of the transition-state complex and those of the reactant and product subsystems of the forward and reverse reactions involving the hydrogen molecule and halogen atom, thus providing entropic measures of the "closeness" between the compared electron distributions. This gives rise to a quantitative formulation of the qualitative Hammond postulate of the reactivity theory. The electronic and geometric components of such information "distances" are separated and discussed.

9.1. CHARGE AFFINITIES OF STOCKHOLDER REACTANTS

Consider the $M_b \rightarrow M_a$ CT, from the *basic* (*b*) reactant M_b to its *acidic* (*a*) partner M_a in the donor-acceptor (DA) reactive system $M_R = M_a$---$M_b \equiv (M_a \,\vert\, M_b)$, for the fixed external potential due to nuclei of the constituent atoms of both reactants: $v(\boldsymbol{r}) = v_a(\boldsymbol{r}) + v_b(\boldsymbol{r})$. Here the vertical broken line symbolizes the freedom of these bonded, *mutually*-open reactants to exchange electrons. For greater simplicity we further assume that the geometries of the isolated reactants M_a^0 and M_b^0 are held frozen in the DA complex M_R, so that there exists the uniquely specified *promolecular* reactive system reference, $M_R^0 \equiv (M_a^0 | M_b^0)$, consisting of the free ("frozen"–internal-geometry) reactants shifted to their mutual orientation and separation in M_R; here, the vertical solid line separating the non-bonded reactants, implies that they are *mutually closed*, unable to exchange electrons and to form chemical bonds between them.

For the electron-transfer processes in the *externally*-closed M_R, for which $N_a + N_b = N_R = const.$, the overall electron populations on reactants, $\boldsymbol{N}_R = (N_a, N_b)$, resulting from the integration of the subsystem densities $\boldsymbol{\rho}_R = (\rho_a, \rho_b)$, $N_\alpha = \int \rho_\alpha(\boldsymbol{r})\, d\boldsymbol{r}$, $\alpha = a, b$, determine the current *amount* of the *internal* (inter-reactant) CT in M_R:

$$N_{CT} = N_a - N_a^0 = N_b^0 - N_b > 0. \tag{9.1.1}$$

Here the row-vector $\boldsymbol{N}_R^0 = (N_a^0, N_b^0)$ groups the reference electron populations of the separated (free) reactants of the promolecular reactive system. This quantity represents the independent "reaction-coordinate" for the *intra*-M_R (*inter*-reactant) displacement of the system electronic structure. It should be observed that for the equilibrium geometries of the isolated reactants, fixing the external potentials of the separated-fragments, the electron densities of infinitely-separated reactants are functions of their overall electron populations, $\boldsymbol{\rho}_R^0 = \{\rho_\alpha^0(N_\alpha^0)\}$, since N_α^0 and $v_\alpha(\boldsymbol{r})$ uniquely identify the electronic Hamiltonian of the separate reactant α.

The equilibrium densities $\boldsymbol{\rho}_R^* = \boldsymbol{\rho}_R^*(\boldsymbol{N}_R^0)$ of the polarized (promoted) reactants in $M_R^* \equiv (M_a^* | M_b^*)$, for $\boldsymbol{N}_R^* = \boldsymbol{N}_R^0$, can be obtained from the relevant minimum entropy-deficiency principle (see Sections 5.3 and 5.4). It should be realized, that for the externally closed M_R, i.e., for the fixed value of $N_R = N_a^* + N_b^* = N_a^0 + N_b^0 = N_R^0$, only one global Lagrange multiplier, say for the closure-constraint of M_a, $N_a^* = N_a^0$, is required to simultaneously enforce the specified numbers of electrons on both reactants. Indeed, the requirement of an exhaustive division of the overall density, that the optimum subsystem densities reproduce at each point in space the given density of M_R as a whole, $\sum_{\alpha=a,b} \rho_\alpha^*(\boldsymbol{r}) = \rho_R(\boldsymbol{r})$, as is the case in the Hirshfeld division scheme for bonded atoms, already enforces the global constraints of the specified number of electrons in the system as a whole:

$$\int \rho_R(\boldsymbol{r})\, d\boldsymbol{r} = N_R = N_a^* + N_b^* = N_R^0. \tag{9.1.2}$$

Hence, when the fullfilment of the system constraint is enforced by the local subsidiary condition it is sufficient to satisfy the *global* closure-requirement on one reactant, say $\int \rho_a^*(\boldsymbol{r})\, d\boldsymbol{r} = N_a^* = N_a^0$, to guarantee the closure-constraint of the other reactant: $\int \rho_b^*(\boldsymbol{r})\, d\boldsymbol{r} = N_b^* = N_b^0$.

In order to enforce the local constraint of the exhaustive division, one has to use the appropriate Lagrange-multiplier function $\lambda(\boldsymbol{r})$ in the following variational principle for the minimum of the entropy-deficiency in the reactant resolution:

$$\delta\mathcal{L}[\boldsymbol{\rho}_R^*|N_a^0, \boldsymbol{\rho}_R^0, \rho_R] \equiv \delta\{\Delta S[\boldsymbol{\rho}_R^*|\boldsymbol{\rho}_R^0] - \int\lambda(\boldsymbol{r})[\rho_a^*(\boldsymbol{r}) + \rho_b^*(\boldsymbol{r})]\, d\boldsymbol{r} - \lambda_a \int \rho_a^*(\boldsymbol{r})\, d\boldsymbol{r}\} = 0. \tag{9.1.3}$$

Above, the auxiliary information-distance functional $\mathcal{L}[\boldsymbol{\rho}_R^*|N_a^0, \boldsymbol{\rho}_R^0, \rho_R]$ depends upon the two "variable" functions $\boldsymbol{\rho}_R^*$ and involves the references $\boldsymbol{\rho}_R^0$ and constraint terms corresponding to the fixed "parameters" (N_a^0, ρ_R). The first term in this functional represents the information "penalty" for the current subsystem densities deviating from their free-reactant shapes.

This entropy deficiency between the polarized densities of subsystems $\boldsymbol{\rho}_R^*$ and the reference densities $\boldsymbol{\rho}_R^0$ of the separated reactants, of the same ("frozen") internal geometry as in M_R,

$$\Delta S[\boldsymbol{\rho}_R \,|\, \boldsymbol{\rho}_R^0] = \sum_{s=a,b} \int \rho_s(\boldsymbol{r}) \ln[\rho_s(\boldsymbol{r})/\rho_s^0(\boldsymbol{r})]\, d\boldsymbol{r}, \tag{9.1.4}$$

calls for the maximum similarity between the equilibrium subsystem densities $\boldsymbol{\rho}_R^*$ and the corresponding electron distributions in the free reactants. The second term in Eq. (9.1.3) enforces the local constraint of the conserved density of M_R as a whole, while the last term guarantees the prescribed overall number of electrons in the internally polarized M_a^*. As we have argued above, these two subsidiary conditions automatically guarantee the complementary number of electrons in M_b^*.

The resulting Euler equations for the optimum densities $\boldsymbol{\rho}_R^* = (\rho_a^*, \rho_b^*)$ of the polarized-reactants are:

$$\ln[\rho_a^*(\boldsymbol{r})/\rho_a^0(\boldsymbol{r})] + 1 - \lambda(\boldsymbol{r}) - \lambda_a \equiv \ln\{\rho_a^*(\boldsymbol{r})/[\rho_a^0(\boldsymbol{r})C_a(\boldsymbol{r})]\} = 0,$$

$$\ln[\rho_b^*(\boldsymbol{r})/\rho_b^0(\boldsymbol{r})] + 1 - \lambda(\boldsymbol{r}) \equiv \ln\{\rho_b^*(\boldsymbol{r})/[\rho_b^0(\boldsymbol{r})C_b(\boldsymbol{r})]\} = 0, \tag{9.1.5a}$$

where $C_b(\boldsymbol{r}) \equiv \exp[\lambda(\boldsymbol{r}) - 1]$ and $C_a(\boldsymbol{r}) = \exp(\lambda_a)C_b(\boldsymbol{r}) \equiv CC_b(\boldsymbol{r})$. Hence,

$$\rho_a^*(\boldsymbol{r}) = \rho_a^0(\boldsymbol{r})C_a(\boldsymbol{r}) \quad \text{and} \quad \rho_b^*(\boldsymbol{r}) = \rho_b^0(\boldsymbol{r})C_b(\boldsymbol{r}). \tag{9.1.5b}$$

The local constraint then gives the following expression for $C_b(\boldsymbol{r})$ in terms of C,

$$C_b(\boldsymbol{r}) = \rho_R(\boldsymbol{r})/[C\rho_a^0(\boldsymbol{r}) + \rho_b^0(\boldsymbol{r})], \tag{9.1.6}$$

while the global constraint gives the integral equation for determining C:

$$N_a^* = \int \rho_a^*(\boldsymbol{r})\, d\boldsymbol{r} = C \int \rho_a^0(\boldsymbol{r})\, \rho_R(\boldsymbol{r}) / [C\rho_a^0(\boldsymbol{r}) + \rho_b^0(\boldsymbol{r})]\, d\boldsymbol{r} = N_a^0. \qquad (9.1.7)$$

It can be easily verified that one recovers the stockholder subsystem solutions for $\lambda_a = 0$ $(C = 1)$, when the subsystem global constraint is absent.

Clearly, one could enforce in this way any admissible value of the average electron populations of subsystems, $N_a \neq N_a^0$ and the associated value of $N_b = N - N_a \neq N_b^0$. Therefore, the entropy-deficiency principle of Eq. (9.1.3) allows one to determine the equilibrium reactant densities, which exactly reproduce the density of the reactive system as a whole, thus conserving the overall number of electrons in the whole reactive complex, and carry the prescribed average electron populations on both reactants.

This variational principle thus establishes the unique dependence of the subsystem densities on their average numbers of electrons in M_R, $\boldsymbol{\rho}_R = \boldsymbol{\rho}_R(\boldsymbol{N}_R)$. This feature allows one to define the subsystem *Fukui-Function* (FF) descriptors [Eq. (2.3.1)] (see, e.g., Nalewajski, 1995b, 1997b, 1999, 2002d,e, 2003a; Nalewajski *et al.*, 1996; Nalewajski and Korchowiec, 1997) for such complementary subsystems of the DA complex, which define the 2×2 square matrix (Nalewajski and Świtka, 2002):

$$\mathbf{f}_R(\boldsymbol{r}) \equiv \left(\frac{\partial}{\partial \boldsymbol{N_R}} \left(\frac{\partial \tilde{E}_v[\boldsymbol{\rho_R}]}{\partial \boldsymbol{v_R}(\boldsymbol{r})} \right)_{\rho_R} \right)_v = \left(\frac{\partial \boldsymbol{\rho_R}(\boldsymbol{r})}{\partial \boldsymbol{N_R}} \right)_v \equiv \left\{ f_{s,s'}(\boldsymbol{r}) = \left(\frac{\partial \rho_{s'}(\boldsymbol{r})}{\partial N_s} \right)_{v, s \neq s'} \right\}, \qquad (9.1.8)$$

where we have used the Hellmann-Feynman theorem for subsystems [Eq. (D.2.9)], the electronic energy density functional in the subsystem resolution $\tilde{E}_v[\boldsymbol{\rho_R}] \equiv E_v[\boldsymbol{\rho}_R] = E[\boldsymbol{N}_R, \boldsymbol{v}_R] \equiv E_v[\rho]$, $\boldsymbol{v}_R(\boldsymbol{r}) = v(\boldsymbol{r})\boldsymbol{1}$, and the row unit vector $\boldsymbol{1} = (1, 1)$ (see Appendix D). These FF components satisfy the usual normalizations (Nalewajski *et al.*, 1996; Nalewajski and Korchowiec, 1997) for the *independent* electron populations of both reactants:

$$\int f_{s,s'}(\boldsymbol{r})\, d\boldsymbol{r} = \partial N_{s'} / \partial N_s = \delta_{s,s'}. \qquad (9.1.9)$$

Here, the partial differentiation with respect to the subsystem electron population implies the fixed number of electrons in the complementary subsystem, as reflected by the off-diagonal ($s \neq s'$) normalization condition. The diagonal ($s = s'$) relation expresses the normalization of the diagonal FF components to the single electron inflow to (or outflow from) the reactant in question.

Next, let us define the entropy-deficiency conjugates $\boldsymbol{\Phi}_R = \{\Phi_a, \Phi_b\}$ of the subsystem electron populations $\boldsymbol{N}_R = (N_a, N_b)$. Using the functional chain-rule gives:

$$\Phi_s = \frac{\partial \Delta S[\boldsymbol{\rho_R}(\boldsymbol{N_R}) | \boldsymbol{\rho_R^0}]}{\partial N_s} = \sum_{s'} \int \left(\frac{\partial \Delta S[\boldsymbol{\rho_R}(\boldsymbol{N_R}) | \boldsymbol{\rho_R^0}]}{\partial \rho_{s'}(\boldsymbol{r})} \right) \left(\frac{\partial \rho_{s'}(\boldsymbol{r})}{\partial N_s} \right) d\boldsymbol{r} = \sum_{s'} \int S_{s'}(\boldsymbol{r}) f_{s',s}(\boldsymbol{r})\, d\boldsymbol{r},$$

$$s = a, b, \qquad (9.1.10a)$$

or in the matrix form:

$$\boldsymbol{\Phi}_R = \int \boldsymbol{S}_R(\boldsymbol{r})\, \mathbf{f}_R(\boldsymbol{r})^{\mathrm{T}}\, d\boldsymbol{r}. \tag{9.1.10b}$$

Here the row vector $\boldsymbol{S}_R(\boldsymbol{r})$ groups the entropy-deficiency conjugates of the reactant densities:

$$\boldsymbol{S}_R(\boldsymbol{r}) = \partial\Delta S[\boldsymbol{\rho}_R(\boldsymbol{N}_R) \,|\, \boldsymbol{\rho}_R^{\,0}]/\partial\boldsymbol{\rho}_R(\boldsymbol{r}) = \{\ln[\rho_a(\boldsymbol{r})/\rho_a^{\,0}(\boldsymbol{r})] + 1,\ \ln[\rho_b(\boldsymbol{r})/\rho_b^{\,0}(\boldsymbol{r})] + 1\}, \tag{9.1.11}$$

and the FF-matrix of the reactive system is defined by Eqs. (9.1.8) and (D.3.2b).

These FF-type quantities in the entropy-deficiency representation, called the reactant *CT-Affinities* (CTA), represent the generalized derivative ("force") indices of reactants measuring the "intensities" associated with the "extensive" state-variables of electron population $\boldsymbol{N}_R$. They combine the entropy $[\boldsymbol{S}_R(\boldsymbol{r})]$ and energy $[\mathbf{f}_R(\boldsymbol{r})]$ derivatives, thus providing more adequate CT-indices, which recognize the information-distance dependence of charge sensitivities of the reactant subsystems. Indeed, since the unique definition of molecular fragments themselves requires the entropy-deficiency representation, a presence of the cross-entropy factors in CTA should indeed be expected. As we have argued in Section 6.5, the equilibrium condition in the *energy*-representation, represented by the equalization of the subsystem chemical potentials (electronegativities), is insufficient to define the equilibrium partitioning of the electron density of the whole molecular system into the densities of reactants. For their unique, unbiased definition the minimum principle of the promolecule-referenced entropy deficiency is required. The generalized FF indices $\boldsymbol{\Phi}_R$ reflect the effect of these information-theoretic origins of molecular subsystems upon the charge affinities of reactants in the relative-entropy representation.

A reference to Eq. (9.1.2) shows that the electron populations $\boldsymbol{N}_R = (N_a = N_a^{\,0} + N_{CT},\ N_b = N_b^{\,0} - N_{CT}) \equiv \boldsymbol{N}_R(N_{CT})$ are linear functions of the amount of CT, N_{CT}, which represents the *independent* variable measuring the progress of the internal $M_b \rightarrow M_a$ CT. Therefore, of interest in the theory of chemical reactivity (see, e.g., Nalewajski *et al.*, 1996; Nalewajski and Korchowiec, 1997) is also the derivative of the global entropy deficiency with respect to this electronic "reaction coordinate":

$$\Phi_R^{CT} = \frac{\partial\Delta S[\boldsymbol{\rho}_R(\boldsymbol{N}_R(N_{CT}))|\boldsymbol{\rho}_R^0]}{\partial N_{CT}} = \sum_\alpha \left(\frac{\partial\Delta S[\boldsymbol{\rho}_R(\boldsymbol{N}_R(N_{CT}))|\boldsymbol{\rho}_R^0]}{\partial N_\alpha}\right)\left(\frac{\partial N_\alpha}{\partial N_{CT}}\right) = \Phi_a - \Phi_b$$

$$= \int\{[f_{a,a}(\boldsymbol{r}) - f_{b,a}(\boldsymbol{r})]\, S_a(\boldsymbol{r}) + [f_{a,b}(\boldsymbol{r}) - f_{b,b}(\boldsymbol{r})]\, S_b(\boldsymbol{r})\}\, d\boldsymbol{r}$$

$$\equiv \int\{f_a^{\,CT}(\boldsymbol{r})\, S_a(\boldsymbol{r}) + f_b^{\,CT}(\boldsymbol{r})\, S_b(\boldsymbol{r})\}\, d\boldsymbol{r}, \tag{9.1.12}$$

where the reactant *in situ* FF of reactants, $\boldsymbol{f}_R^{\,CT}(\boldsymbol{r}) = [f_a^{\,CT}(\boldsymbol{r}),\ f_b^{\,CT}(\boldsymbol{r})]$, represent the subsystem internal CT-derivatives:

$$f_R^{CT}(\boldsymbol{r}) = [\partial \boldsymbol{\rho}_R(\boldsymbol{r})/\partial N_{CT}]_v. \qquad (9.1.13)$$

As indicated in the first line of Eq. (9.1.12), the CTA of the DA reactive system, Φ_R^{CT}, representing the *in situ* FF-type derivative in the entropy-deficiency representation, is measured by the difference between the CT-affinities of the acidic and basic reactants.

It should be recalled that the usual definition of the FF [Eq. (2.3.1)] of Parr and Yang (1984) refers to the system as a whole. It measures the response of the equilibrium density of the whole, externally-open system M_R,

$$f_R(\boldsymbol{r}) = [\partial \rho_R(\boldsymbol{r})/\partial N_R]_v, \qquad (9.1.14)$$

reflecting the overall density displacement per inflow of a single electron to M_R from an external electron reservoir. This charge sensitivity, representing an important reactivity index of molecular systems, can be expressed in terms of the FF-matrix elements of Eq. (9.1.8):

$$f_R(\boldsymbol{r}) = \sum_\alpha [\partial \rho_\alpha(\boldsymbol{r})/\partial N_R]_v \equiv \sum_\alpha f_\alpha^{\ R}(\boldsymbol{r})$$
$$= \sum_\alpha \sum_{\alpha'} (\partial N_{\alpha'}/\partial N_R)_v\, [\partial \rho_\alpha(\boldsymbol{r})/\partial N_{\alpha'}]_v \equiv \sum_\alpha \sum_{\alpha'} F_{\alpha'} f_{\alpha',\,\alpha}. \qquad (9.1.15)$$

where the *condensed* FF of reactants,

$$\boldsymbol{F}_R = \{F_\alpha = (\partial N_{\alpha'}/\partial N_R)_v\}, \qquad (9.1.16)$$

measure the electron-population responses of subsystems per unit displacement in the global number of electrons in M_R.

In the ρ_R-constrained ("vertical") CT-problem, one determines the equilibrium entropy deficiency partition of the fixed $\rho_R(\boldsymbol{r})$ among the two reactant subsystems, when the barrier preventing the inter-reactant CT is lifted. This implies $\lambda_a = 0$ in the MED principle (9.1.3), which then gives rise to the Hirshfeld reactant densities, $\boldsymbol{\rho}_R^{\ H}(\boldsymbol{r}) = [\rho_a^{\ H}(\boldsymbol{r}),\ \rho_b^{\ H}(\boldsymbol{r})]$, and the corresponding electron populations of both subsystems, $\boldsymbol{N}_R^{\ H} = \{N_\alpha^{\ H} = \int \rho_\alpha^{\ H}\, d\boldsymbol{r}\}$. For this equilibrium division, $\rho_R(\boldsymbol{r}) = \rho_a^{\ H}(\boldsymbol{r}) + \rho_b^{\ H}(\boldsymbol{r})$, the subsystem entropic intensities of Eq. (9.1.11) are equalized at the global value describing the DA-complex as a whole:

$$S_a^{\ H}(\boldsymbol{r}) = S_b^{\ H}(\boldsymbol{r}) = S_R(\boldsymbol{r}) = \ln[\rho_R(\boldsymbol{r})/\rho_R^{\ 0}(\boldsymbol{r})] + 1 \equiv \ln w_R(\boldsymbol{r}) + 1. \qquad (9.1.17)$$

Since $\sum_{\alpha'} f_{\alpha,\,\alpha'}(\boldsymbol{r}) = [\partial \rho_R(\boldsymbol{r})/\partial N_\alpha]_v$, the preceding equation gives rise to the vanishing subsystem CTA (9.1.10a) for the ρ_R-constrained ("vertical") CT between reactants in M_R (identified by the subscript ρ),

$$(\Phi_s^{\ H})_\rho = \int S_R(\boldsymbol{r}) \sum_{\alpha'} f_{\alpha,\alpha'}(\boldsymbol{r})\, d\boldsymbol{r} = \int S_R(\boldsymbol{r})\, [\partial \rho_R(\boldsymbol{r})/\partial N_\alpha]_{v,\rho}\, d\boldsymbol{r} = 0, \quad \alpha = a, b, \qquad (9.1.18)$$

thus confirming the equilibrium character of the Hirshfeld reactant densities. Moreover, since the CTA of M_R as a whole [Eq. (9.1.12)] is the difference between the reactant CTA's, one also finds that

$$(\Phi_R^{CT}[\boldsymbol{\rho}_R^H])_\rho = (\Phi_s^H)_\rho - (\Phi_s^H)_\rho = 0. \tag{9.1.19}$$

To summarize, the Hirshfeld reactants do indeed correspond to the equilibrium amount of the *vertical* CT in DA reactive systems. We have explicitly demonstrated that the internal CTA's ("forces"), which combine the entropy deficiency intensities and the FF's or reactants, identically vanish for the Hirshfeld subsystems, which conserve the overall density of the whole reactive system. This confirms the equilibrium character of the subsystem electron distributions of both reactants.

9.2. SIMPLE ORBITAL MODEL OF A SYMMETRIC TRANSITION-STATE COMPLEX

9.2.1. Three-Orbital Model

The typical electronic structure calculations do not allow for a control over the non-equilibrium electronic distributions for the fixed positions of the nuclei (Born-Oppenheimer approximation). Only the optimum (equilibrium) density of electrons can be uniquely determined in this way. The simple orbital models, e.g., the 2-AO model of Section 7.5, give a unique opportunity to test variations in the bond-orders for such displaced (non-equilibrium) electronic configurations, which automatically satisfy the Pauli exclusion-principle.

The 3-AO model (Nalewajski and Mrozek, 1994; Nalewajski *et al.*, 1994; Nalewajski and Jug, 2002) describes three electrons contributed by the three constituent atoms of a symmetric *Transition-State* (TS) complex in the atom-exchange reaction,

$$A_1\text{–}B + A_2 \;\rightarrow\; [A_1\text{—}B\text{—}A_2]^{\ddagger} \;\rightarrow\; A_1 + B\text{–}A_2, \tag{9.2.1}$$

e.g., in the H_2 + H reactive system. The system electrons occupy the two lowest MO obtained by combining the three (orthonormal) AO centered on respective atoms: $\boldsymbol{X}(\boldsymbol{r}) = \{a(\boldsymbol{r}), b(\boldsymbol{r}), c(\boldsymbol{r})\}$.

The relevant AP reference assumes one electron on each atom/orbital with the alternanting spin orientations on neighboring atoms: $A_1^0(\uparrow) + B^0(\downarrow) + A_2^0(\uparrow)$. This promolecule $M^0 = (A_1^0|B^0|A_2^0)$ exhibits the highest degree of spin-pairing between the adjacent atoms and thus - the highest spin-distribution similarity to that of the molecular ground-state of the TS-complex.

The model electron configuration is controlled by the charge of the middle atom B, $q = q_B$, and its spin polarization $\Sigma = q_B^\alpha - q_B^\beta$. For a given value of Σ the allowed values of q are in the range $|\Sigma| \leq q \leq 2 - |\Sigma|$ and the overall spin polarization of AP, $N^{\alpha,0} - N^{\beta,0} = 1$, is preserved in the bond-breaking–bond-forming process.

Let us now examine representative distributions of electronic spins in this model, for the *uniform* distribution of electrons among three constituent atoms, i.e., for $q = 1$. The molecular spin distribution for ($q = 1$, $\Sigma = -1$) in the molecular TS $M = (A¦B¦C)$, consisting of the mutually open (bonded) AIM, is shown in Scheme 9.1*a*. It is identical with that corresponding to the AP electron configuration, of the mutually closed (non-bonded) free-atoms in $M^0 = (A^0|B^0|C^0)$. It should be emphasized, however, that these two species exhibit dramatically different networks of conditional probabilities (compare the communication channels shown in Panels *a* and *d* of Scheme 9.2). The other extreme configuration ($q = 1$, $\Sigma = 1$), has been previously linked to the *Atom-Diatom Limit* (ADL), $B^0 + A_1–A_2$ (Nalewajski and Jug, 2002). However, such a dissociation limit calls for the communication network in which the middle atom is closed (exhibiting no "communication" links) relative to the pair of the mutually-open peripheral atoms: M^*(ADL) = $(A¦C|B)$. This is not the case in Scheme 9.2*b*, which corresponds to the spin distribution 9.1*b*.

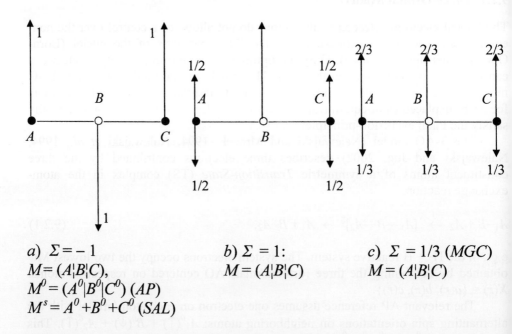

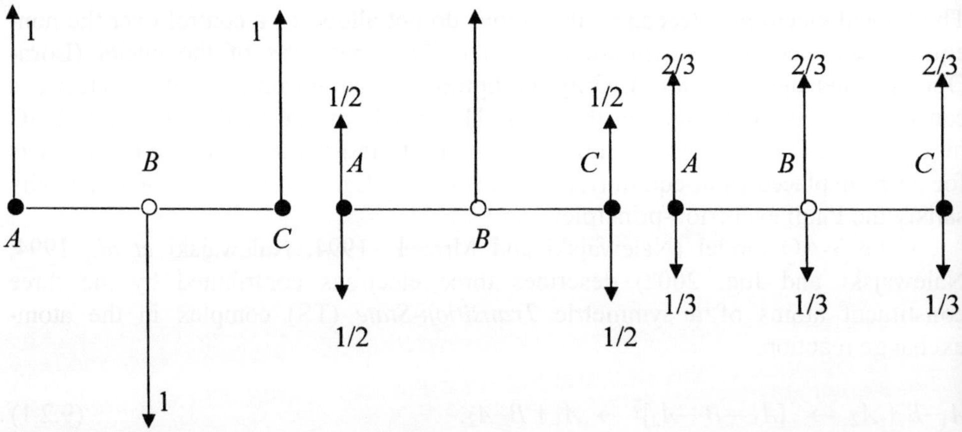

a) $\Sigma = -1$
$M = (A¦B¦C)$,
$M^0 = (A^0|B^0|C^0)$ (*AP*)
$M^s = A^0 + B^0 + C^0$ (*SAL*)

b) $\Sigma = 1$:
$M = (A¦B¦C)$

c) $\Sigma = 1/3$ (*MGC*)
$M = (A¦B¦C)$

Scheme 9.1. The molecular AIM populations of the spin-up (↑) and spin-down (↓) electrons along the *Maximum-Covalency Path* (MCP) ($q = 1$) in the 3-AO model of the symmetric TS for selected values of the spin polarization parameter: $\Sigma = \{-1$ (Panel *a*), 1 (Panel *b*), 1/3 (Panel *c*)$\}$. As indicated in the diagram the Panel *a* distribution also represents the spin distribution of the *Atomic Promolecule* (AP) and the *Separated Atom Limit* (SAL). The Panel *c* generates the system *Maximum Global Covalency* (MGC) structure.

9.2.2. Global Bond Descriptors from MO Theory

It follows from the extended basis set UHF calculations for H_3 that the TS-configuration exhibits almost uniform distribution of electrons among constituent AIM, $q = 1.086$, and a positive, fractional spin polarization $\Sigma = 0.135$ (a.u.). Clearly, in the TS complex all pairs of AIM exhibit fractional bond multiplicities. They give rise to the overall bond-multiplicity index from the two-electron difference approach (Nalewajski and Mrozek, 1994; Nalewajski *et al.*, 1994): $\mathcal{I}^{\ddagger,UHF} = 1.208$. It has been obtained by combining all covalent and ionic contributions of Eq. (8.7.12), which roughly preserve the initial overall bond-order in the atom-diatom limit H + H_2: $\mathcal{I}^{ADL,UHF} = 1.046$. The corresponding predictions from the 3-AO model, $\mathcal{I}^{\ddagger} = 1 + \Sigma = 1.135$ and $\mathcal{I}^{ADL} = 1$ (Nalewajski and Mrozek, 1994), compare favorably with the UHF estimate.

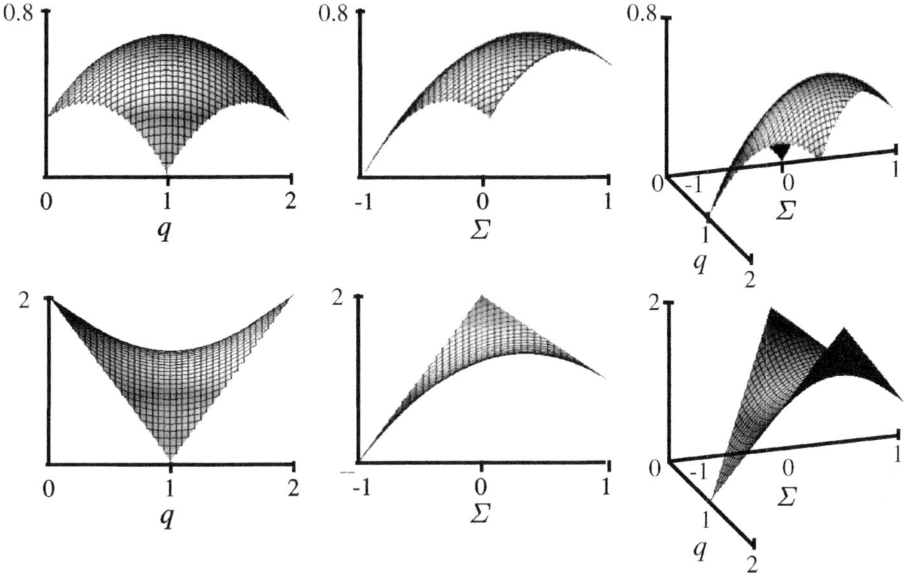

Figure 9.1. The bond-order indices from the MO theory for the 3-AO model of the symmetric TS: Wiberg's covalent component $\mathcal{I}_{tc}(q, \Sigma)$ [Eq. (9.2.2)] (upper panels) and the index $\mathcal{M}(q, \Sigma)$ [Eq. (9.2.7)] (lower panels).

It is of interest to examine the crucial *inter-atomic* contributions to the global bond indices, characterizing the system as a whole. In the upper panels of Fig. 9.1 we have displayed a half of the overall inter-atomic covalent index of Wiberg (1968), given by the sum of squares of the upper-diatomic elements of the spin-resolved CBO matrix elements $\{\gamma_{X,Y}^{\sigma}\}$, with $\gamma_{X,Y} = \sum_{\sigma=\alpha,\beta} \gamma_{X,Y}^{\sigma}$, which represents the overall two-centre covalency in the difference-approach [Eq. (8.7.12)]:

$$\tfrac{1}{2}\mathfrak{I}^W(q,\Sigma) = \sum\sum_{X<Y}\sum_{\sigma=\alpha,\beta}[\gamma_{X,Y}{}^{\sigma}(q,\Sigma)]^2 \equiv \mathfrak{I}_{tc}{}^{cov}(q,\Sigma)$$
$$= \sum_{\sigma=\alpha,\beta}\{[(\gamma_{A,B}{}^{\sigma}(q,\Sigma)]^2 + [\gamma_{A,C}{}^{\sigma}(q,\Sigma)]^2 + [\gamma_{B,C}{}^{\sigma}(q,\Sigma)]^2\}$$
$$= (1/8)[(1+\Sigma)(5-3\Sigma) - 3(1-q)^2]. \qquad (9.2.2)$$

It follows from the three upper panels of this figure that for a given value of the spin polarization parameter the maximum covalency is always achieved for the uniform AIM electron populations $q_{max} = 1$, with the maximum global covalency (MGC) being reached in the TS-complex of Scheme 9.1*c* (see also Fig. 9.1):

$$\mathfrak{I}_{tc}{}^{cov}(q_{max}, \Sigma_{max}) = \tfrac{1}{2}\mathfrak{I}^W(q_{max}, \Sigma_{max}) = \mathfrak{I}_{tc}{}^{cov}(1, 1/3) = 2/3. \qquad (9.2.3)$$

As we have already remarked in Section 8.7, the two-electron indices of the difference-approach to the chemical bond multiplicities [see Eq. (8.7.12)] are derived from the displacements of the molecular two-electron density matrix $\mathbf{\Gamma}$ in the AO resolution, relative to the SAL reference matrix $\mathbf{\Gamma}^0$: $\Delta\mathbf{\Gamma} = \mathbf{\Gamma} - \mathbf{\Gamma}^0 = \Delta\mathbf{\Gamma}^{(1)} + \Delta\mathbf{\Gamma}^{(2)}$. The so called *quadratic* indices (Nalewajski and Mrozek, 1994; Nalewajski *et al.*, 1994, 1996, 1997} are based upon the *second*-order displacements $\Delta\mathbf{\Gamma}^{(2)}$. They are given by quadratic functions of displacements in the relevant CBO matrix elements, $\{\Delta\gamma^{\sigma} = \gamma^{\sigma} - \gamma^{\sigma,0}\}$, in the spirit of the Wiberg index of Eq. (8.6.9). Another set of the two-electron indices (Nalewajski *et al.*, 1993; Nalewajski, 2004b) has been defined in terms of the *overall* displacements $\Delta\mathbf{\Gamma}$, which also contain the first-order shifts $\Delta\mathbf{\Gamma}^{(1)}$.

The Wiberg-type index of Eq. (9.2.2) is an example of the *quadratic* two-electron bond multiplicity. The quadratic functions of displacements in the orbital electron populations are classified as *ionic*, while the quadratic forms of displacements in the *off*-diagonal CBO matrix elements determine the *covalent* bond-order contributions [see Eq. (8.7.12)]. The products of displacements on the same atom contribute to the *one-center* (*oc*, intra-atom) quantity, while the products of displacements on different atoms determine the *two-center* (*tc*, inter-atom) bond components.

In simple orbital models, where each atom contributes a single valence orbital to form the chemical bonds, the one-center covalent terms identically vanish, so that $\mathfrak{I}^{cov} = \mathfrak{I}_{tc}{}^{cov}$. However, the ionic component in general exhibits the non-vanishing *one*-center and *two*-center contributions: $\mathfrak{I}^{ion} = \mathfrak{I}_{oc}{}^{ion} + \mathfrak{I}_{tc}{}^{ion}$.

The second-order displacements $\Delta\mathbf{\Gamma}^{(2)}$ in the 3-AO model give rise to the following expressions for the overall bond components (Nalewajski and Mozek, 1994; Nalewajski *et al.*, 1994; Nalewajski, 2004b):

$$\mathfrak{I}^{cov}(q,\Sigma) = \mathfrak{I}_{tc}{}^{cov}(q,\Sigma),$$
$$\mathfrak{I}^{ion}(q,\Sigma) = \mathfrak{I}_{oc}{}^{ion}(q,\Sigma) + \mathfrak{I}_{tc}{}^{ion}(q,\Sigma) = (3/8)\,[(1+\Sigma)^2 + (1-q)^2]. \qquad (9.2.4)$$

The preceding total ionicity contains the *one*-center and *two*-center ionicities:

$$\mathfrak{I}_{oc}^{ion}(q, \Sigma) = (3/8)(q + \Sigma)(\Sigma - q + 2) \quad \text{and} \quad \mathfrak{I}_{tc}^{ion}(q, \Sigma) = (3/4)(1 - q)^2. \qquad (9.2.5)$$

These contributions predict a linear dependence of the *global* bond multiplicity of the two-electron difference-approach on the spin polarization parameter (Nalewajski and Mrozek, 1994):

$$\mathfrak{I}(q, \Sigma) = \mathfrak{I}^{ion}(q, \Sigma) + \mathfrak{I}^{cov}(q, \Sigma) = 1 + \Sigma \equiv \mathfrak{I}(\Sigma). \qquad (9.2.6)$$

This global index $\mathfrak{I}(\Sigma)$ correctly predicts the vanishing bond-order for the Promolecule/SAL configuration, for $\Sigma = -1$, but fails in the ADL, A_1–A_2 + B, for $\Sigma_{ADL} = q_{ADL} = 1$, when $\mathfrak{I}(\Sigma = 1) = 2$. The latter index is composed of $\mathfrak{I}^{cov}(1, 1) = 1/2$ and $\mathfrak{I}_{oc}^{ion}(1, 1) = 3/2$, with $\mathfrak{I}_{tc}^{ion}(1, 1) = 0$.

Another bond-multiplicity index recently proposed within the two-electron difference approach (Nalewajski, 2004b),

$$\begin{aligned}\mathcal{M}(q, \Sigma) &= \mathrm{tr}\, \Delta\mathbf{\Gamma}(q, \Sigma) = (1/4)[(1 + \Sigma)(5 - 3\Sigma) + 3(1 - q)^2] \\ &= 2[\mathfrak{I}^{cov}(q, \Sigma) + \mathfrak{I}_{tc}^{ion}(q, \Sigma)] = 2[\mathfrak{I}(q, \Sigma) - \mathfrak{I}_{oc}^{ion}(q, \Sigma)],\end{aligned} \qquad (9.2.7)$$

is shown in the lower panels of Fig. 9.1. As indicated in the preceding equation, it measures the sum of the global covalent and two-center ionic bond components, and exhibits the saddle-point at the TS configuration ($q = 1$, $\Sigma = 1/3$), along the *Maximum Covalency Path* (MCP), representing the Σ-constrained extrema (see Fig. 9.1):

$$\{\min_q \mathcal{M}(q, \Sigma) \ \text{or} \ \max_q \mathfrak{I}^{W}(q, \Sigma)\} \ \Rightarrow q_{extr.} = 1. \qquad (9.2.8)$$

It should be also observed that the profile of $\mathcal{M}(q, \Sigma)$ along the MCP,

$$\mathcal{M}(1, \Sigma) = (1/4)(1 + \Sigma)(5 - 3\Sigma) \equiv \mathcal{M}(\Sigma) = 2\mathfrak{I}^{cov}(q, \Sigma), \qquad (9.2.9)$$

provides a more realistic representation of a dependence of the overall chemical bond-order in the 3-AO model on the system spin-polarization parameter: $\mathcal{M}(-1) = 0$, in the SAL, $\mathcal{M}(1) = 1$ in the ADL, and $\mathcal{M}(1/3) = 4/3$, for the MGC structure of Scheme 9.1*c*, when all atomic spin populations are equalized throughout the system:

$$N_X^{\alpha} = 2/3 \qquad \text{and} \qquad N_X^{\beta} = 1/3, \qquad X = A, B, C.$$

The MGC value represents a slightly increased bond multiplicity in the TS configuration, relative to the single chemical bond in the ADL. These predictions are in good agreement with both the intuitive chemical expectations and the UHF predictions for H_3 system, which were mentioned at the beginning of this section.

9.2.3. Overall Bond Multiplicities from Communication Theory

After this short overview of the bond-order measures from the MO theory let us now summarize the corresponding IT bond indices from the communication theory (Nalewajski, 2000, 2004b; Nalewajski and Jug, 2002) for the three representative molecular electron configurations of Scheme 9.1 along the MCP. We shall also examine the bond descriptors of the AP and SAL reference states. All these chemical species correspond to the uniform distributions of the AIM electron populations: q = 1. The corresponding communication channels of the three *molecular* systems of Scheme 9.1, all exhibiting the identical input and output entropies, $S(\mathbf{A}^0) = S(\mathbf{B}^0) = S(\mathbf{A}) = S(\mathbf{B}) = \log_2 3 = 1.585$ bits, are shown in Schemes 9.2*a-c*. In Panel *d* of this scheme the communication channels corresponding to the non-bonded constituent atoms in M^0 (for $\boldsymbol{P}(\mathbf{A}^0) = \{P_i^0 = 1/3\}$) or M^s (for $\boldsymbol{P}(\mathbf{A}^s) = \{P_i^s = 1\}$) are displayed, representing a collection of the mutually closed (disconnected), free atoms.

It should be emphasized that the non-zero *off*-diagonal conditional probabilities of the *molecular* configuration of Panel *a* in Scheme 9.2, for $\Sigma = -1$, characterize the mutually-*open* AIM in M, allowed to delocalize their electronic spins onto remaining atoms. They give rise to the same resultant spin distribution as that characterizing the collection of the mutually closed (free) atoms in Panel *d* of the scheme, in which electrons are forbidden to spread throughout the triatomic system.

The corresponding values (in bits) of the *two*-electron entropies $S(\mathbf{AB})$, $S(\mathbf{A}^0\mathbf{B}^0)$ or $S(\mathbf{A}^s\mathbf{B}^s)$, the conditional entropies measuring the global entropy-covalencies, $S(\mathbf{B}\,|\,\mathbf{A}) = N^{cov}(1, \Sigma)$, $S(\mathbf{B}^0\,|\,\mathbf{A}^0) = N_0^{cov}(1, \Sigma = -1) = 0$ or $S(\mathbf{B}^s\,|\,\mathbf{A}^s) = N_s^{cov}(1, \Sigma = -1) = 0$, and the mutual informations measuring the global entropy-ionicities, $I(\mathbf{A}^0{:}\mathbf{B}) = N^{ion}(1, \Sigma)$, $I(\mathbf{A}^0{:}\mathbf{B}^0) = N_0^{ion}(1, \Sigma = -1) = S(\mathbf{A}^0)$, or $I(\mathbf{A}^s{:}\mathbf{B}^s) = N_s^{ion}(1, \Sigma = -1) = 0$, are also reported in Scheme 9.2. A reference to this scheme indicates that the overall entropic bond-order measure is preserved in all molecular and AP configurations:

$$N(1, \Sigma) = N^{cov}(1, \Sigma) + N^{ion}(1, \Sigma) = N^0(1, \Sigma = -1) = N_0^{ion}(1, \Sigma = -1)$$
$$= S(\mathbf{A}) = S(\mathbf{A}^0) = \log_2 3 = 1.585. \qquad (9.2.10)$$

This represents an increased overall bond multiplicity level, in comparison to the bond entropy index of 1 bit in the diatomic model of the preceding section, which provides the IT bond order for the ADL. It also follows from Scheme 9.2*d* that the overall IT bond-order in the SAL, $N^s(q = 1, \Sigma = -1) = 0$, while the MGC molecular configuration exhibits almost purely covalent bonds in the 3-AO model of TS.

The MGC entropy/information indices are almost identical with those predicted for the separate allyl-fragment of the butadiene π-system (Scheme 8.4B) and the π-channel for the ground-state of allyl radical itself, shown in Scheme 7.10. Indeed, the three communication channels have much in common, exhibiting almost identical input and output probabilities and similar conditional-probability networks, thus giving rise to similar entropy/information characteristics of the system chemical bonds.

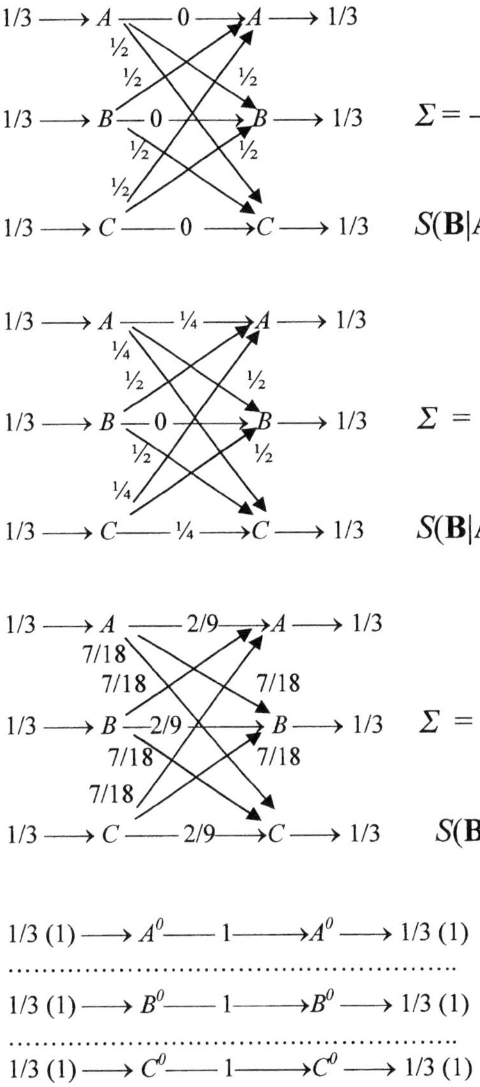

$\Sigma = -1{:}\ M = (A|B|C)$ ***a)***

$S(\mathbf{AB}) = 2.585$

$S(\mathbf{B}|\mathbf{A}) = 1.000 \quad I(\mathbf{A}{:}\mathbf{B}) = 0.585$

$\Sigma = 1{:}\ M = (A|B|C)$ ***b)***

$S(\mathbf{AB}) = 2.918$

$S(\mathbf{B}|\mathbf{A}) = 1.333 \quad I(\mathbf{A}{:}\mathbf{B}) = 0.252$

$\Sigma = 1/3{:}\ M = (A|B|C)$ (MGC) ***c)***

$S(\mathbf{AB}) = 3.127$

$S(\mathbf{B}|\mathbf{A}) = 1.542 \quad I(\mathbf{A}{:}\mathbf{B}) = 0.043$

$\Sigma = -1{:}$ ***d)***

$M^0 = (A^0|B^0|C^0)$ (AP): $S(\mathbf{A}^0\mathbf{B}^0) = 1.585, \quad S(\mathbf{B}^0|\mathbf{A}^0) = 0, \quad I(\mathbf{A}^0{:}\mathbf{B}^0) = 1.585$

$M^s = A^0 + B^0 + C^0$(SAL): $S(\mathbf{A}^s\mathbf{B}^s) = S(\mathbf{B}^s|\mathbf{A}^s) = I(\mathbf{A}^s{:}\mathbf{B}^s) = 0$

Scheme 9.2. The molecular communication channels for the mutually bonded (open) AIM for representative MCP ($q = 1$) structures $\Sigma = \{-1(a), 1(b)$, and $1/3(c)\}$ in the 3-AO model of the symmetric TS (see Scheme 9.1). Panel *d* shows the AP/SAL channels representing the mutually non-bonded (closed) free-atoms, giving rise to the three (disconnected, deterministic, noiseless) atomic channels. For each channel the two-electron and conditional entropies, as well as the mutual-information bond descriptors (in bits) are also reported.

The $(q = 1, \Sigma = 1)$ case of Panel *b* is seen to exhibit 4/3 bits of the covalent entropy and around ¼ bit of bond ionicity. This prediction clearly shows that such a spin distribution for a collection of three mutually open AIM, to which the communication channel of Scheme 9.2*b* applies, does not in fact describe the $B^0 + AC$ dissociation limit, for which the diatomic index equals exactly 1 bit of information (see Section 7.5).

As expected, the entropy/information indices of bond-components in the three $\Sigma = -1$ configurations of Panels *a* and *d* in Scheme 9.2, representing the mutually *open* (bonded) AIM and *closed* (free) atoms in AP and SAL, respectively, which give rise to the same spin-distribution of Scheme 9.1*a*, are quite different (compare the conditional entropies in Panels *a* and *d* of Scheme 9.2). The disconnected channels of the infinitely-separated atoms in the SAL generate the vanishing entropy covalency and information ionicity, while for the free-atoms in the AP the whole (conserved) molecular bond index $N = 1.585$ bits is attributed exclusively to the system information-ionicity. Finally, the bonded atoms in the "molecular" TS-complex generate for this spin polarization 1 bit of the entropy bond-covalency and roughly a half bit of the IT-ionicity.

The "molecular" predictions reported in Scheme 9.2 agree qualitatively with a similar bond-strengthening effect in TS complex predicted by the two-electron difference-approach (Nalewajski and Mrozek, 1994; Nalewajski *et al.*, 1994, Nalewajski, 2004b). The competition between the bond components is also well reflected by the entropy/information data displayed in Scheme 9.2, with the maximum value of the mutual-information measure of the bond IT-ionicity being observed in the minimum covalency "molecular" and AP structures, for $\Sigma = -1$. Accordingly, the minimum, almost zero, information-ionicity is found in the MGC structure, where all atoms have equalized their spin-populations, thus generating the highest level of the entropy covalency reflecting the highest uncertainty in the electron distribution. Finally, for $\Sigma = 1$, when the spin equalization takes place only between the "terminal" atoms *A* and *C*, intermediate levels of the entropy-covalency and the complementary information-ionicity of all chemical bonds are observed in the 3-AO model. The vanishing bond indices in the SAL communication system also conform to the intuitive expectations.

Therefore, as already observed in the 2-AO model of Section 7.5, the overall IT bond index remains conserved, when the spin polarization changes for the identical overall electron distributions among AIM: $q_A = q_B = q_C = 1$. Only the covalent/ionic bond composition reflects changes in the two-electron configuration defining the molecular communication system. The additional degrees-of-freedom for the spin delocalization in the molecular MGC structure of Scheme 9.2*c*, relative to the $\Sigma = -1$ molecular system of Scheme 9.2*a*, which exhibits 3 vanishing diagonal conditional probabilities, are seen to give rise to higher information-noise measure of the bond covalency and lower imutual-information measure of bond ionicity in the MGC configuration.

A reference to Panel *b* of the same scheme also shows that in the $\Sigma = 1$ channel, which exhibits only one vanishing diagonal conditional probability, the probability scattering (communication noise) is only partially restricted relative to the unconstrained MGC network, thus generating relatively less entropy-covalency (more information-ionicity). Accordingly, since the $\Sigma = 1$ channel is less restrictive in comparison to the $\Sigma = -1$ molecular channel, it is seen to generate relatively higher entropy-covalency (lower information-ionicity).

One also observes in Scheme 9.2*a* that the "molecular" $\Sigma = -1$ TS gives rise to the non-vanishing entropy covalency. Only the truly non-bonded free-atoms in the SAL (Scheme 9.2*d*) are seen to give rise to the vanishing bond components in this atomic dissociation limit, in qualitative agreement with the two-electron MO indices of Eqs. (9.2.4)-(9.2.7) and the Wiberg index of Eq. (9.2.2).

In the remaining part of this section we shall examine alternative IT descriptors of the reactant-subsystems in the 3-AO model of the symmetric TS-complex, which have been developed in the preceding chapter. In particular, the three main strategies involving the fragment separate, partial, and reduced communication channels, respectively, will be used to generate for the IT descriptors of the TS complex, includinmg the weakened internal bond of the diatomic reactant and the newly formed external bond between this reactant and the approaching atom.

9.2.4. Separate Diatomic Channels

Consider the three representative electron configurations of Scheme 9.1 of the 3-AO model of the symmetric TS complex for the uniform AIM electron populations ($q = 1$) in the system input and output. The renormalized channels for the (A, B)–fragment are shown in Scheme 9.3, while the corresponding channels for the (A,C)–fragment are displayed in Scheme 9.4. The associated overall (internal) bond indices (in bits) of such separate diatomic subsystems,

$$N_{X\text{-}Y}{}^r = S_{X\text{-}Y}(\mathbf{B}^r \,|\, \mathbf{A}^r) + I_{X\text{-}Y}(\mathbf{A}^r{:}\mathbf{B}^r), \tag{9.2.11}$$

are also reported together with their entropy-covalency $S_{X\text{-}Y}(\mathbf{B}^r \,|\, \mathbf{A}^r)$ and information-ionicity $I_{X\text{-}Y}(\mathbf{A}^r{:}\mathbf{B}^r)$ contributions of Section 8.2.

A reference to Schemes 9.3 and 9.4 shows that the predicted overall entropic bond-orders $N_{A\text{-}B}{}^r$ ($= N_{B\text{-}C}{}^r$) and $N_{A\text{-}C}{}^r$ remain roughly preserved at about 1 bit level, while the bond composition is seen to be strongly affected by both the spin polarization and the selection of the diatomic fragment. For the system MGC structure, for $\Sigma = ⅓$ (Panels *c* of the two schemes) the internal *A-B* and *A-C* bonds in these two separate diatomics are seen to be identical and almost purely covalent. For the $\Sigma = 1$ configuration, shown in Panels *b* of these diagrams, the internal *A-B* bond is predicted to be approximately half-covalent and half-ionic, while the *A-C* bond remains purely covalent. Finally, for the AP-value $\Sigma = -1$ of the "molecular" spin polarization (Panels *a*), both the *A-B* and *A-C* bonds are diagnosed as purely ionic, as indeed reflected by the spin separation shown in Scheme 9.1*a*.

$$M^*(A, B) = (A_!^!B|C)$$

$a)$ $\Sigma = -1$: $S_{A-B}(\mathbf{B}^r \mid \mathbf{A}^r) = 0$

$I_{A-B}(\mathbf{A}^r{:}\mathbf{B}^r) = 1$ $N_{A-B}{}^r = 1$

$b)$ $\Sigma = 1$: $S_{A-B}(\mathbf{B}^r \mid \mathbf{A}^r) = 0.551$

$I_{A-B}(\mathbf{A}^r{:}\mathbf{B}^r) = 0.420$ $N_{A-B}{}^r = 0.971$

$c)$ $\Sigma = \tfrac{1}{3}$: $S_{A-B}(\mathbf{B}^r \mid \mathbf{A}^r) = 0.946$

$I_{A-B}(\mathbf{A}^r{:}\mathbf{B}^r) = 0.054$ $N_{A-B}{}^r = 1$

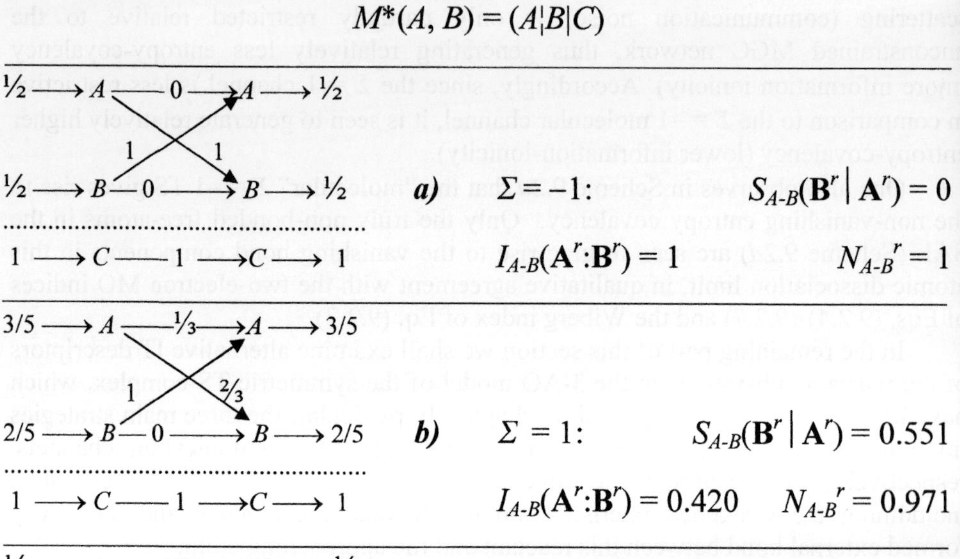

Scheme 9.3. The renormalized communication systems of the separate (A,B)–fragment in $M^*(A, B) = (A_!^!B|C)$ in the 3-AO model of the symmetric TS complex for $q = 1$ and $\Sigma = \{-1$ (a), 1 (b), $\tfrac{1}{3}$ $(c)\}$.

9.2.5. Partial Channels

In Table 9.1 we have listed the IT indices of the atomic partial channels in the 3-AO model of the symmetric TS-complex for the representative electron configurations of Scheme 9.1. These "sub-channels" were constructed from the corresponding molecular communication systems shown in Scheme 9.2. Moreover, in the partial *row*-channels the *molecular* output probabilities, including the contributions due to the missing input atoms, were kept unchanged (see also Scheme 8.8) in accordance with the definition of Eq. (8.5.5). It ensures the additivity of the information-ionicity contributions, which sum up to the corresponding global quantity, describing the molecule as a whole.

$$M^*(A, C) = (A\dot{|}C|B)$$

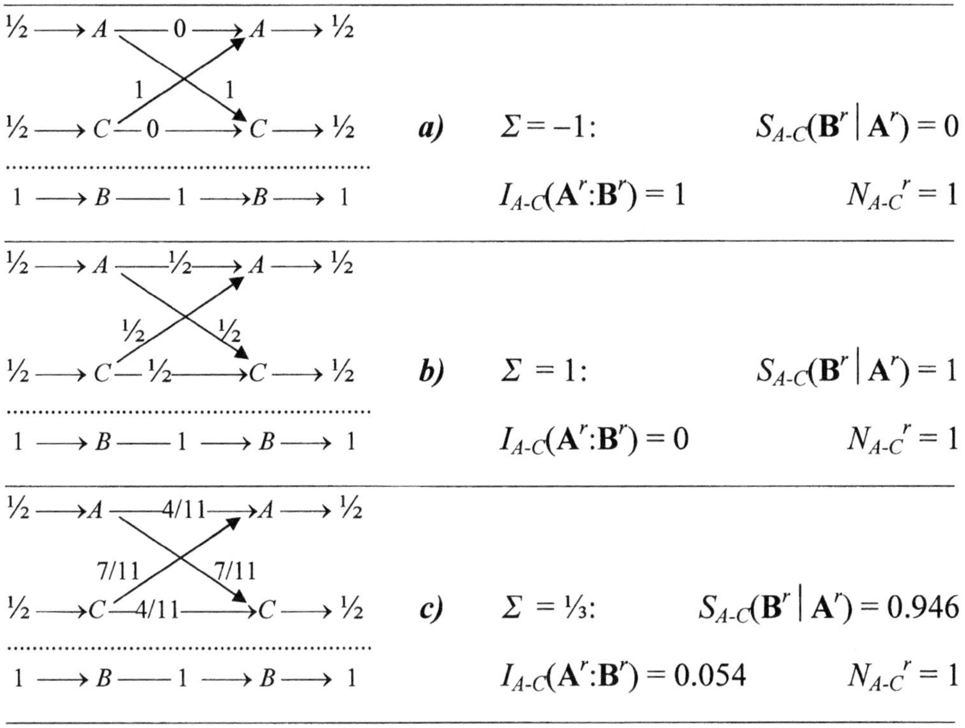

Scheme 9.4. The renormalized communication systems of the separate (A,C)–fragment in $M^*(A, C) = (A\dot{|}C|B)$ in the 3-AO model of the symmetric TS complex for $q = 1$ and $\Sigma = \{-1\ (a), 1(b), ⅓\ (c)\}$.

Table 9.1. A comparison of the entropy/information indices (in bits) of the *atomic* partial *row*- and *column*-channels in the 3-AO model of TS complex for the three electron configurations of Scheme 9.1.

Configuration	Atom	row-channels			column-channels		
	X	S_X^r	I_X^r	N_X^r	S_X^c	I_X^c	N_X^c
$\Sigma = -1$	A, B, C	⅓	0.195	0.528	⅓	0.195	0.528
$\Sigma = 1$	A, C	½	0.028	0.528	½	0.028	0.528
	B	⅓	0.195	0.528	⅓	0.195	0.528
$\Sigma = ⅓$	A, B, C	0.514	0.014	0.528	0.514	0.014	0.528

As we have already argued in Section 8.5 and in Appendix F, the IT-descriptors of the atomic *row*-channel reflect the AIM contributions $N_X^r = I_X^r + S_X^r$ to all bonds in the molecular system and their covalent (S_X^r) and ionic (I_X^r) components. Similarly, the *column*-channel indices $N_X^c = S_X^c + I_X^c$, S_X^c and I_X^c, provide the IT-measures of the associated atomic valence-numbers. A reference to Table 9.1 shows that the two sets of atomic indices in the 3-AO model are identical for the assumed equal atomic electron populations and the three selected values of the spin-polarization parameter, with the overall atomic index [see Scheme 9.2 and Eq. (8.2.10)]:

$$N_X^c(1, \Sigma) = N_X^r(1, \Sigma) = N(1, \Sigma)/3 = 0.528 \text{ bits.} \qquad (9.2.12)$$

Therefore, the overall atomic IT-valences in the 3-AO model of TS complex are equal to their contributions to the system chemical bonds.

It should be observed that in the MGC configuration, for $\Sigma = ⅓$, each atom is engaged in almost exclusively covalent interactions with the remaining atoms, while a relatively high ionic indices are detected for all atoms in the $\Sigma = -1$ molecular state and for the middle atom B for the $\Sigma = 1$ electron configuration. It should be observed that these ionicities cannot be given a simple *one*-electron interpretation since the atomic charges of all three AIM vanish in all three electron distributions of Scheme 9.1. These information-ionicities reflect the IT two-electron quantities, measuring the information flow in the corresponding atomic row-channels and reflecting a degree of independence between the input- and output-probabilities in the molecular channel.

A reference to the atomic bond contributions in Table 9.1 shows that all atoms exhibit different IT bond-composition, for the conserved overall measure of about half a bit. The highest covalent bond-contribution or valence-number, about 1 bit per atom, is again observed in the MGC configuration, with the lowest atomic entropy/covalency index being observed for all atoms in the $\Sigma = -1$ configuration and for the middle atom in the $\Sigma = 1$ state.

In the row channels the reduced values of IT-indices, per unit normalization of the fragment input probabilities, are reported. The actual values of these indices in the "molecular" system are obtained by multiplying the reduced data by the condensed probability of the input diatomic fragment $\boldsymbol{K}$ in the 3-AO model for $q = 1$: $P_K = ⅔$. Additional insight is gained by partitioning the overall indices of these partial channels into their respective *internal* and *external* contributions [see Eqs. (8.6.7) and (8.6.8)], which are also reported in Schemes 9.5 and 9.6.

It follows from Scheme 9.5 that the output fragment index $N(\boldsymbol{M}; \boldsymbol{L})$ for $\boldsymbol{L} = \{(A, B), (B, C), (A, C)\}$, is preserved: $N(\boldsymbol{M}; \boldsymbol{L}) = ⅔ \log_2 3 = 1.057$ bits. Therefore, a given pair of atoms in the model is involved in a roughly "single" (1 bit) bond. The information-theoretic composition of these interactions, however, is changing substantially, when the electronic spins are allowed to delocalize throughout the system in the $\Sigma = (⅓, 1)$ electron configurations, in comparison to the $\Sigma = -1$ structure, where this effect is not allowed.

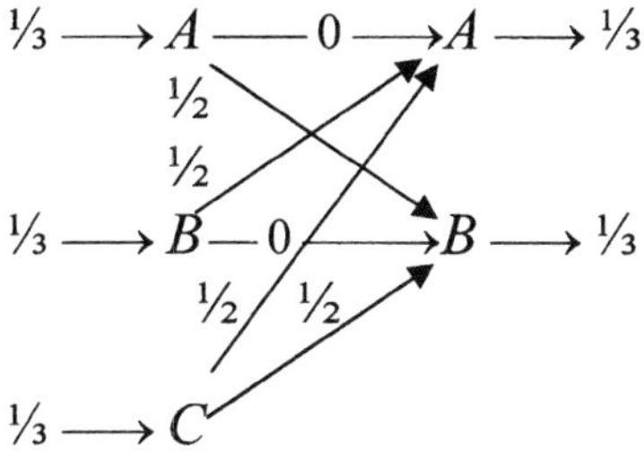

a) $\Sigma = -1$: $\quad \boldsymbol{L} = (A, B), (B, C), (A, C)$

$N(\boldsymbol{M}; \boldsymbol{L}) = 1.057, \quad S(\boldsymbol{L}|\boldsymbol{M}) = ⅔, \quad I(\boldsymbol{M}^0:\boldsymbol{L}) = 0.390$

$N^{int}(\boldsymbol{L}) = 0.528, \quad S^{int}(\boldsymbol{L}) = ⅓, \quad I^{int}(\boldsymbol{L}) = 0.195$

$N^{ext}(\boldsymbol{L}) = 0.528, \quad S^{ext}(\boldsymbol{L}) = ⅓, \quad I^{ext}(\boldsymbol{L}) = 0.195$

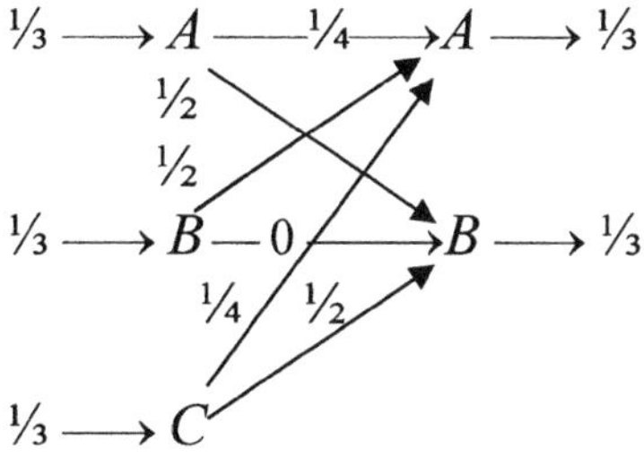

b) $\Sigma = 1$: $\quad \boldsymbol{L} = (A, B), (B, C)$

$N(\boldsymbol{M}; \boldsymbol{L}) = 1.057, \quad S(\boldsymbol{L}|\boldsymbol{M}) = 0.833, \quad I(\boldsymbol{M}^0:\boldsymbol{L}) = 0.223$

$N^{int}(\boldsymbol{L}) = 0.660, \quad S^{int}(\boldsymbol{L}) = ½, \quad I^{int}(\boldsymbol{L}) = 0.160$

$N^{ext}(\boldsymbol{L}) = 0.396, \quad S^{ext}(\boldsymbol{L}) = ⅓, \quad I^{ext}(\boldsymbol{L}) = 0.063$

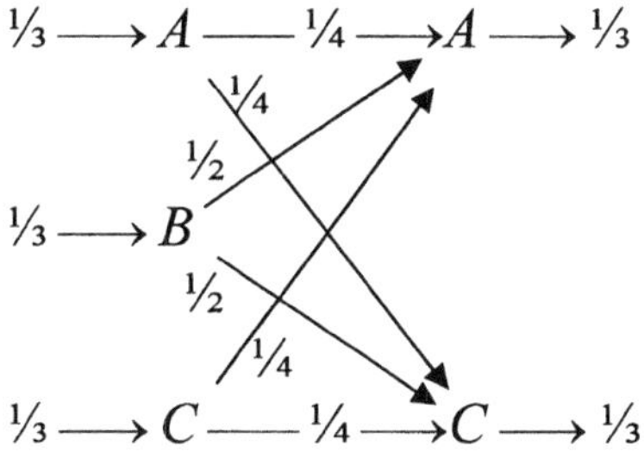

$\boldsymbol{L} = (A, C)$

$N(\boldsymbol{M}; \boldsymbol{L}) = 1.057, \quad S(\boldsymbol{L}|\boldsymbol{M}) = 1, \quad I(\boldsymbol{M}^0:\boldsymbol{L}) = 0.057$

$N^{int}(\boldsymbol{L}) = 0.528, \quad S^{int}(\boldsymbol{L}) = ⅔, \quad I^{int}(\boldsymbol{L}) = -0.138$

$N^{ext}(\boldsymbol{L}) = 0.528, \quad S^{ext}(\boldsymbol{L}) = ⅓, \quad I^{ext}(\boldsymbol{L}) = 0.195$

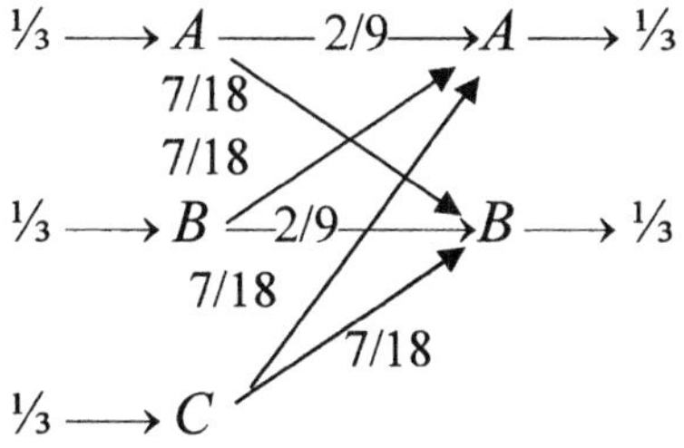

c) $\Sigma = ⅓$: $\quad \boldsymbol{L} = (A, B), (B, C), (A, C)$

$N(\boldsymbol{M}; \boldsymbol{L}) = 1.057, \quad S(\boldsymbol{L}|\boldsymbol{M}) = 1.028, \quad I(\boldsymbol{M}^0:\boldsymbol{L}) = 0.029$

$N^{int}(\boldsymbol{L}) = 0.646, \quad S^{int}(\boldsymbol{L}) = 0.675, \quad I^{int}(\boldsymbol{L}) = -0.029$

$N^{ext}(\boldsymbol{L}) = 0.411, \quad S^{ext}(\boldsymbol{L}) = 0.353, \quad I^{ext}(\boldsymbol{L}) = 0.058$

Scheme 9.5. The partial *column*-channels for alternative diatomic output-fragments in the 3-AO model of TS complex, and the related entropy/information indices for the three electron configurations of Scheme 9.1. For the definition of the internal and external bond contributions see Eq. (8.6.7).

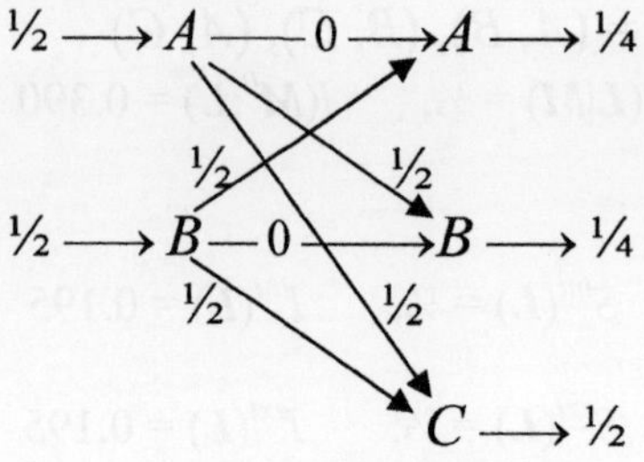

a) $\Sigma = -1$: $\boldsymbol{K} = (A, B), (B, C), (A, C)$

$N^r(\boldsymbol{K}; \boldsymbol{M}) = 3/2$, $S^r(\boldsymbol{M}|\boldsymbol{K}) = 1$, $I^r(\boldsymbol{K}^0{:}\boldsymbol{M}) = ½$

$N^{r,int}(\boldsymbol{K}) = 1$, $S^{r,int}(\boldsymbol{K}) = ½$, $I^{r,int}(\boldsymbol{K}) = ½$

$N^{r,ext}(\boldsymbol{K}) = ½$, $S^{r,ext}(\boldsymbol{K}) = ½$, $I^{r,ext}(\boldsymbol{K}) = 0$

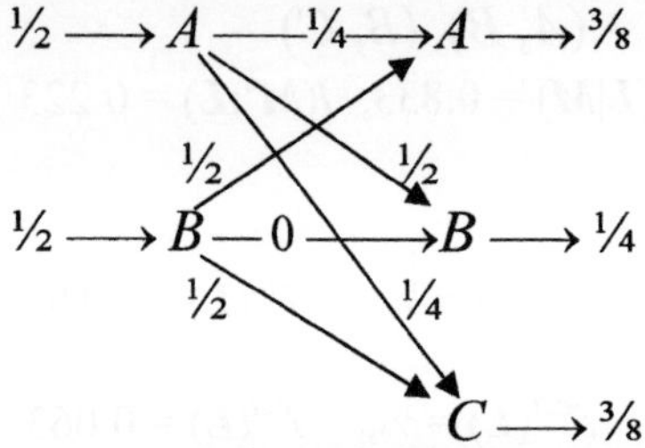

b) $\Sigma = 1$: $\boldsymbol{K} = (A, B), (B, C)$

$N^r(\boldsymbol{K}; \boldsymbol{M}) = 1.561$, $S^r(\boldsymbol{M}|\boldsymbol{K}) = 5/4$, $I^r(\boldsymbol{K}^0{:}\boldsymbol{M}) = 0.311$

$N^{r,int}(\boldsymbol{K}) = 1.031$, $S^{r,int}(\boldsymbol{K}) = ¾$, $I^{r,int}(\boldsymbol{K}) = 0.281$

$N^{r,ext}(\boldsymbol{K}) = 0.531$, $S^{r,ext}(\boldsymbol{K}) = ½$, $I^{r,ext}(\boldsymbol{K}) = 0.031$

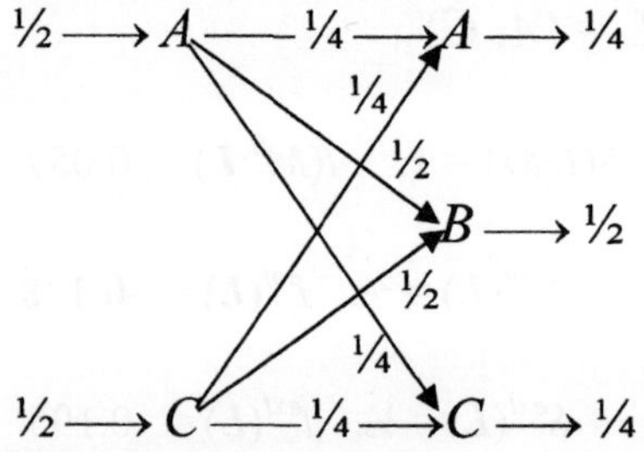

$\boldsymbol{K} = (A, C)$

$N^r(\boldsymbol{K}; \boldsymbol{M}) = 3/2$, $S^r(\boldsymbol{M}|\boldsymbol{K}) = 3/2$, $I^r(\boldsymbol{K}^0{:}\boldsymbol{M}) = 0$

$N^{r,int}(\boldsymbol{K}) = 1$, $S^{r,int}(\boldsymbol{K}) = 1$, $I^{r,int}(\boldsymbol{K}) = 0$

$N^{r,ext}(\boldsymbol{K}) = ½$, $S^{r,ext}(\boldsymbol{K}) = ½$, $I^{r,ext}(\boldsymbol{K}) = 0$

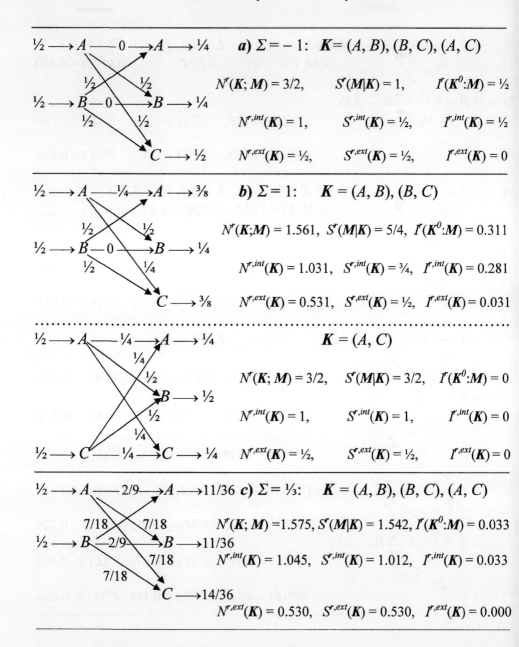

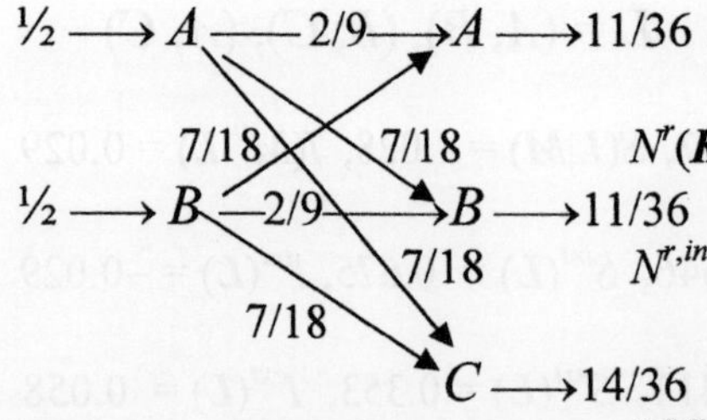

c) $\Sigma = ⅓$: $\boldsymbol{K} = (A, B), (B, C), (A, C)$

$N^r(\boldsymbol{K}; \boldsymbol{M}) = 1.575$, $S^r(\boldsymbol{M}|\boldsymbol{K}) = 1.542$, $I^r(\boldsymbol{K}^0{:}\boldsymbol{M}) = 0.033$

$N^{r,int}(\boldsymbol{K}) = 1.045$, $S^{r,int}(\boldsymbol{K}) = 1.012$, $I^{r,int}(\boldsymbol{K}) = 0.033$

$N^{r,ext}(\boldsymbol{K}) = 0.530$, $S^{r,ext}(\boldsymbol{K}) = 0.530$, $I^{r,ext}(\boldsymbol{K}) = 0.000$

Scheme 9.6. The partial row-channels for diatomic input fragments in the 3-AO model of TS complex and the associated "reduced" IT indices, per unit condensed probability of the input fragment, for the three electron configurations of Scheme 9.1. The absolute ("molecular") values of bond indices are obtained by multiplying these reduced results by the diatomic fragment probability in TS complex, $P_K = ⅔$. The internal and external IT contributions are defined in Eq. (8.6.8).

Consider first $\boldsymbol{L} = (A, C)$ fragment, composed of the terminal atoms A_1 and A_2 of this model TS complex. For the $\Sigma = ⅓$ configuration the (A, C) unit is predicted to be involved in almost purely covalent interactions, with 97% of the entropy-covalency $S(\boldsymbol{M}|\boldsymbol{L})$ contribution in the total index of roughly 1 bit, while for $\Sigma = -1$ the fragment information-ionicity $I(\boldsymbol{M}^0{:}\boldsymbol{L})$ already amounts to about 37% of the total index. The partially delocalized $\Sigma = 1$ structure exhibits an intermediate level of about 5% of the ionic contribution. A similar trend is observed in the entropy/information indices for the output fragments $\boldsymbol{L} = \{(A, B), (B, C)\}$, which are equivalent by symmetry: 37% contribution from $I(\boldsymbol{M}^0{:}\boldsymbol{L})$ for $\Sigma = -1$ becomes gradually lowered with the increasing spin delocalization in the system: 12% for $\Sigma = 1$ and 3% for $\Sigma = ⅓$.

The predicted conservation of the overall index of the diatomic column channels in 3-AO model reflects the bond-order conservation in all electron configurations: the overall information bond multiplicity of the molecular fragment, due to both the internal and external chemical interactions, is preserved at about one bit level. The spin distribution in the molecular system, for $\Sigma = -1$, corresponds to the equal (single electron) spin-up populations on peripheral atoms $A_1 = A$ and $A_2 = C$, with no spin-delocalization being allowed between each of these terminal atoms and the middle atom B, on which the spin-down electron is located.

It follows from Scheme 9.5*a* that only for $\Sigma = -1$ the total IT bond indices of the partial column channel are equally divided between the respective internal and external (environmental) contributions. Let us examine first the internal conditional entropies reported in Scheme 9.5. This internal bond-covalency index increases with a growing equalization of electronic spin populations on the constituent AIM, from Panel *a* to Panel *c*, with the highest level being observed for the totally equalized spin populations of the last panel. All three configurations exhibit fractional diatomic covalencies for all atomic pairs, so that none of them represents the dissociation limit, in which some bonds are truly "broken". Such almost separated fragments of the TS-complex would require the *disconnected* communication channels, with no probability scattering between the *separated* subsystems.

The observed trends in the intra-bond covalency reflect some chemical expectations. For example, in the molecular configuration of the first panel all bonds exhibit only ⅓ of the covalency expected for a single, purely covalent bond. In the second panel, previously ascribed to the $AC + B$ dissociation limit (Nalewajski and Jug, 2002), the fractional A-C bond is indeed diagnosed as internally the most covalent one, as expected from the equalized spins within the AC fragment for this molecular configuration (see Scheme 9.1*b*). It should be realized, however, that a transition from the $\Sigma = -1$ to $\Sigma = 1$ configuration marks a delocalization of both the spin-up and spin-down electrons in the model, with one spin-up electron replacing the spin-down electron on atom B. In fact, in the $\Sigma = 1$ structure one detects a growing fractional covalency of both A-B and A-C bonds, which reaches the maximum level in the equivalent-bond configuration for $\Sigma = ⅓$.

The external covalency index for a diatomic output-fragment $\boldsymbol{L}$ remains roughly preserved at the $S^{ext}(\boldsymbol{L}) \approx ⅓$ level in all three electron configurations considered. For the "molecular" electron configuration of the first panel, not to be confused with the AP (M^0) and SAL (M^S) systems, which give rise to the same values of the electron configuration parameters $q = 1$ and $\Sigma = -1$ but differ dramatically in the respective patterns of the communication connections, the internal and external indices are the same so that both atoms in $\boldsymbol{L}$ are equally involved in covalent interactions between themselves and with the third atom. In the remaining molecular configurations the internal covalency exceeds the external one. For example, in Panel *b*, the *A-C* bond exhibits twice as large internal entropy index relative to the external contribution. Similar proportions are detected for all bonds in the MGC configuration of Panel *c*.

The largest ionicity index, observed for $\Sigma = -1$ (Panel *a*) configuration, is seen to be equally distributed between the internal and external contributions. For $\Sigma = 1$ the external mutual-information index of the *A-B* output fragment is substantially diminished, in contrast to the *A-C* subsystem, for which it remains unchanged relative to the $\Sigma = -1$ value. Of interest also are the negative values of the internal information-ionicity index for the *A-C* fragment in the $\Sigma = 1$ configuration, and that for all atomic pairs in the last panel. A reference to these communication channels reveals that in the former case all (equal) contributions to $I^{int}(L)$ are negative, since $P(b \mid a)/P_b = ¾$, where $(a, b) = A, B$ (Panel *b*), while in the latter case there should be a substantial cancellation of two negative diagonal contributions due to $P(X \mid X)/P_X^0 = ⅔$, for $X = A, B$, and two positive off-diagonal terms due to $P(A \mid B)/P_A^0 = P(B \mid A)/P_B^0 = 7/6$ (Panel *c*).

As we have already argued in Chapter 7, a negative value of the mutual information between the input and output events implies that the occurrence of events in one set makes the non-occurrence of events in the other set more likely. We can therefore conclude that in the two communication systems, which exhibit a negative information-ionicity, locating an electron on $\boldsymbol{L}$ in the channel input makes the event of finding an electron outside $\boldsymbol{L}$ in the output more probable. Thus, a negative internal ionicity index may be expected, when a large portion of electrons on $\boldsymbol{L}$ are transferred to the third atom in the molecule. This factor is similar in all four channels shown in Scheme 9.5, as reflected by the fragment conditional probability $P(L \mid L) = \sum_{a \in \boldsymbol{l}} \sum_{b \in \boldsymbol{l}} P_{a,b}$ that the electron originating from L in the input of the partial channel will stay in L in the molecular output: ⅓ (Panel *a*), 5/12, ⅓ (Panel *b*), and 11/27 (Panel *c*). The other factor conducive for reaching a negative value of the internal ionicity index is how evenly the input probability of a given atom l in L distributes itself between the fragment constituent AIM $\boldsymbol{l}$ in the output. The more equalized the conditional probabilities $\boldsymbol{P}(\boldsymbol{L} \mid \boldsymbol{l})$, the more likelihood of the negative internal ionicity index. A reference to the partial channels of Scheme 9.5 shows that this is indeed a decisive factor, which distinguishes the first two channels, characterized by a low degree of the intra-fragment conditional-probability equalization, from the remaining two channels, where these probabilities are practically equalized, thus giving rise to a relatively high level of the internal fragment covalency $S^{int}(\boldsymbol{L})$ measuring the fragment average communication "noise".

One concludes from these results that the output fragment entropy/information descriptors of the corresponding partial (column) channels differentiate between various diatomic fragments to much lesser degree than do the orbital index of Wiberg and the related two-electron indices from the difference-approach in the MO theory. This should be expected, since the IT quantities are derived from the AIM electron probabilities, which reflect the overall atom involvement in *all* chemical bonds in a molecule, which should be comparable for all pairs of atoms in the model system examined in this section. However, the observed variations in the bond composition in the 3-AO model are in a general agreement with the chemical intuition. For example, the highest bond covalency for $\Sigma = ⅓$, exactly 1 bit of the A–C conditional entropy index (a "single" electron-sharing bond), and a weaker covalent component of the A–B (B–C) bond for $\Sigma = 1$, all agree with intuitive expectations based upon the spin-equalization levels observed in Scheme 9.1.

Examples of the *row*-channels for alternative diatomic input-fragments $\boldsymbol{K}$ in the 3-AO model of TS complex are reported in Scheme 9.6 together with the associated entropy/information indices. Following the division of the bond indices of the output fragment (column) channels, the total "reduced" descriptors $S^r(\boldsymbol{M}|\boldsymbol{K})$, $I^r(\boldsymbol{K}^0\!:\!\boldsymbol{M})$ and $N^r(\boldsymbol{K};\ \boldsymbol{M})$, per unit overall probability within the input-fragment $\boldsymbol{K}$, have been partitioned into the *internal* and *external* components, combining the contributions due to the *intra*-fragment conditional probability scattering and that involving the subsystem third AIM, respectively [see Eq.(8.6.8)].

The IT indices of Scheme 9.6 provide the "reduced" *valence*-numbers of diatomic fragments in the triatomic TS complex (see Appendix F). The total indices have been subsequently decomposed into the internal and external parts, using Eq. (8.6.8). The "absolute" values, obtained by multiplying the "reduced" quantities of Scheme 9.6 by the condensed probability $P_K = P_K^{\,0} = ⅔$, are compared in Table 9.2 with the corresponding information characteristics of the partial column-channels, of the diatomic output-fragments (from Scheme 9.5).

A reference to the *column*-channel part of Table 9.2 indicates that roughly a single-electron of any diatomic fragment L is engaged in forming the chemical bonds in the model TS, while the *row*-channel part implies that an equalized, practically a single IT-measure of all chemical bonds in TS is generated by any of its diatomic fragments. It also follows from this table that the numerical values of the entropy/information indices reflecting, the bond-contributions and effective-valencies of diatomic fragments in the 3-AO model, respectively are almost identical. This implies that the fragment effective IT-bond contributions to the chemical interactions in the molecule, reflected by the row-channel data, are practically identical with the corresponding effective valence-numbers of this fragment provided by the respective column-channel data. Therefore, the associated partial column- and row-channels IT descriptors give rise to a consistent description of the overall involvement of a given diatomic subsystem in all chemical bonds in this model TS complex.

Table 9.2. A comparison of the entropy/information indices (in bits) of the partial column- and row-channels of diatomic fragments in the 3-AO model of TS complex for the three electron configurations of Scheme 9.1.

Partial channels:		row-channels			column-channels		
Configuration	Fragment	$N(\boldsymbol{K}; \boldsymbol{M})$	$S(\boldsymbol{M} \mid \boldsymbol{K})$	$I(\boldsymbol{K}^0{:}\boldsymbol{M})$	$N(\boldsymbol{M}; \boldsymbol{L})$	$S(\boldsymbol{L} \mid \boldsymbol{M})$	$I(\boldsymbol{M}^0{:}\boldsymbol{L})$
$\Sigma = -1$	(A, B)	1.00	⅔	⅓	1.06	⅔	0.39
$\Sigma = 1$	(A, B)	1.04	0.83	0.21	1.06	0.83	0.22
	(A, C)	1.00	1.00	0.00	1.06	1.00	0.06
$\Sigma = ⅓$	(A, B)	1.05	1.03	0.02	1.06	1.03	0.03

The diatomic *column*-channels of Scheme 9.5 measure the overall IT-valence number and its covalent and ionic components of the given *output*-fragment $\boldsymbol{L} = (l_1, l_2)$, while the diatomic *row*-channels of Scheme 9.6 generate the IT indices reflecting the contribution of the *input*-fragment atoms $\boldsymbol{K} = (k_1, k_2)$ to all bonds in the 3-AO model of TS complex. Therefore, a reference to Table 9.1 indicates that each pair of atoms $\boldsymbol{K}$ gives rise to approximately conserved, single IT bond-order measure in the system, with differing ionic/covalent composition. The other message from this table is that each pair of atoms in the output diatomic $\boldsymbol{L}$ gives rise to roughly a single bond in TS complex, with the strongest, dominating covalent part predicted for $\Sigma = ⅓$ and weakest covalent part found for $\Sigma = -1$ electron configurations. The largest differentiation of the covalent components, of bonds between peripheral (A, C) and neighboring (A, B) or (B, C) atoms, is observed for $\Sigma = 1$.

9.2.6. Reduced Channels

As an illustration of a flexibility of the channel-reduction approach of Section 8.9 let us now apply it to the diatomic fragments in the 3-AO model of TS complex. The *output*–reduced channels corresponding to alternative selections of a diatomic subsystems $\boldsymbol{L} = (l_1, l_2)$ of this triatomic model are shown in Scheme 9.7, where the corresponding *internal* IT bond descriptors [Eq. (8.9.5)] have also been reported. The corresponding *input-and-output*–reduced channels and the associated (approximate) external bond indices they generate [Eq. (8.9.6)] are the subject of Scheme 9.8.

In Table 9.3 we have compared the internal, external and overall bond descriptors to check how these complementary bond measures combine into the global indices of the system as a whole. It follows from the table that the overall intra-fragment bond indices and the approximate, complementary quantities characterizing the information-coupling between the diatomic fragment and the third atom of the model combine into the exact overall global IT bond-order. In most cases the global bond covalent/ionic composition is also reproduced to a remarkably high accuracy. Clearly, the external indices generated by Scheme 9.7,

$$S(\boldsymbol{M}^r(L)|\boldsymbol{M}) = S_{ext}(\boldsymbol{L}), \quad I(\boldsymbol{M}^0{:}\boldsymbol{M}^r(L)) = I_{ext}(\boldsymbol{L}), \quad N(\boldsymbol{M};\boldsymbol{M}^r(L)) = S_{ext}(\boldsymbol{L}) + I_{ext}(\boldsymbol{L}), \quad (9.2.13)$$

where $\boldsymbol{M}^r(L) = (L, Z)$, $\boldsymbol{L} = (l_1, l_2)$ and $Z \notin \boldsymbol{L}$, combine exactly with the internal indices reported in this scheme into the corresponding global index.

Table 9.3. A comparison of the *internal* and *external* bond indices in the 3-AO model of TS complex for the three electron configurations of Scheme 9.1, obtained from the reduced channels of Schemes 9.7 and 9.8, respectively. The last two columns test the performance of the approximate combination rules of Eq. (8.9.7).

Σ	Diatomic fragment	Bond component	Internal indices	External indices	Overall indices	Global indices
−1	*AB, BC, AC*	*S*	0.333	0.667	1.000	1.000
		I	0.333	0.252	0.585	0.585
		N	0.667	0.918	1.585	1.585
+1	*AB, BC*	*S*	0.459	0.907	1.366	1.333
		I	0.208	0.011	0.219	0.252
		N	0.667	0.918	1.585	1.585
	AC	*S*	0.667	0.667	1.333	1.333
		I	0.000	0.252	0.252	0.252
		N	0.667	0.918	1.585	1.585
+ ⅓	*AB, BC, AC*	*S*	0.645	0.897	1.542	1.542
		I	0.022	0.021	0.043	0.043
		N	0.667	0.918	1.585	1.585

The conditional-entropy index $S(\boldsymbol{L}|\boldsymbol{M})$ of the partial (column) channel in Scheme 9.5 for the output fragment $\boldsymbol{L} = (l_1, l_2)$ (see also Table 9.2) describes the intra-fragment covalent bond in TS, as does the internal entropy-covalency index $S_{int}(\boldsymbol{L})$ from the *output*-reduced channel of this fragment in Scheme 9.7 (see also Table 9.3). However, as seen in these schemes and tables, the internal covalency indices have slightly different numerical values. Indeed the relation between these internal bond descriptors reads:

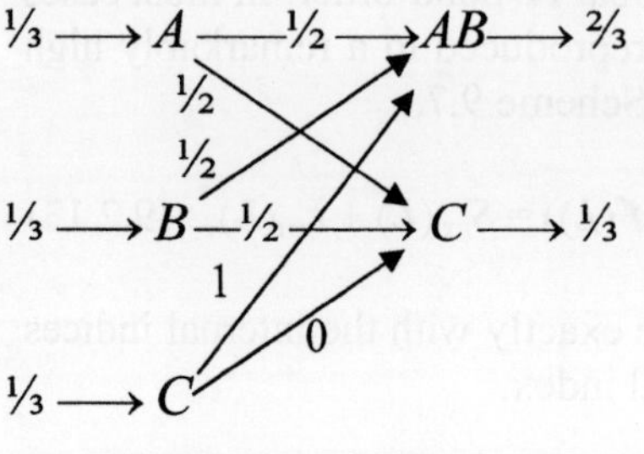

a) $\Sigma = -1$: $M^r = (L, R_L)$, $L = AB, BC, AC$

$N(M;M^r) = 0.918$, $S(M^r|M) = ⅔$, $I(M^0:M^r) = 0.252$

$N_{int}(L) = ⅔$, $S_{int}(L) = ⅓$, $I_{int}(L) = ⅓$

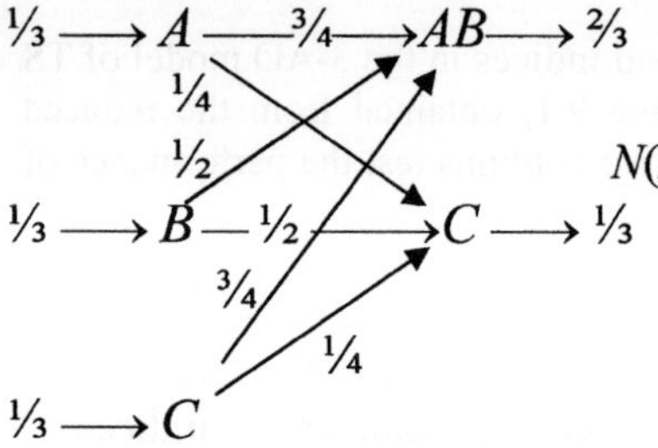

b) $\Sigma = 1$: $M^r = (L, R_L)$, $L = AB, BC$

$N(M;M^r) = 0.918$, $S(M^r|M) = 0.874$, $I(M^0:M^r) = 0.044$

$N_{int}(L) = ⅔$, $S_{int}(L) = 0.459$, $I_{int}(L) = 0.208$

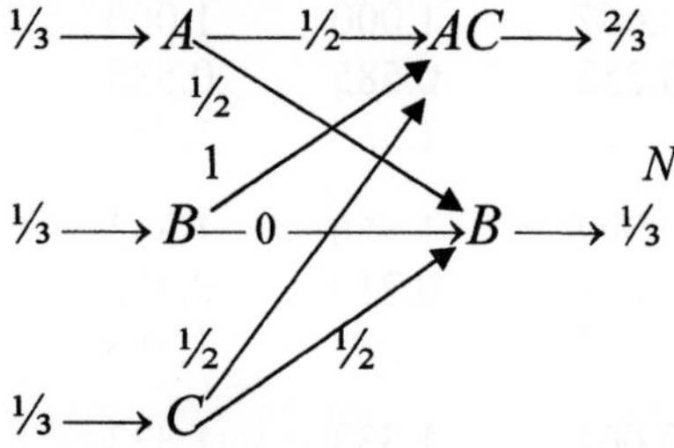

$L = AC$

$N(M;M^r) = 0.918$, $S(M^r|M) = ⅔$, $I(M^0:M^r) = 0.252$

$N_{int}(L) = ⅔$, $S_{int}(L) = ⅔$, $I_{int}(L) = 0$

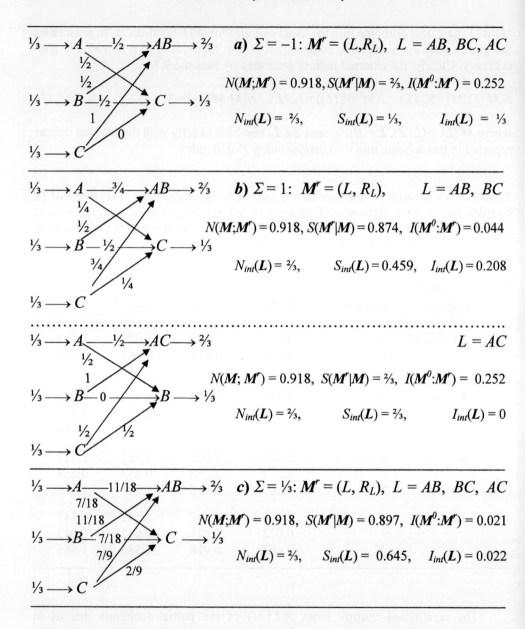

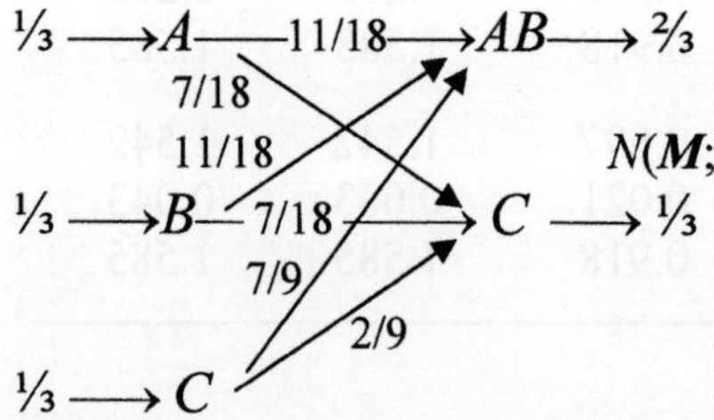

c) $\Sigma = ⅓$: $M^r = (L, R_L)$, $L = AB, BC, AC$

$N(M;M^r) = 0.918$, $S(M^r|M) = 0.897$, $I(M^0:M^r) = 0.021$

$N_{int}(L) = ⅔$, $S_{int}(L) = 0.645$, $I_{int}(L) = 0.022$

Scheme 9.7. The *output*-reduced channels for alternative *output*-diatomics $L = AB, BC, AC$ in the 3-AO model of TS complex, for the three electron configurations of Scheme 9.1. The reduced channel entropy/information descriptors and the resulting internal indices of L (in bits) are also reported.

$$S_{int}(\boldsymbol{L}) = S(\boldsymbol{M}\,|\,\boldsymbol{M}) - S(L, Z\,|\,\boldsymbol{M}) = S(\boldsymbol{L}\,|\,\boldsymbol{M}) + S(Z\,|\,\boldsymbol{M}) - S(L, Z\,|\,\boldsymbol{M}), \tag{9.2.14}$$

where $\boldsymbol{M} = (\boldsymbol{L}, Z)$, the fragment external covalency $S(L, Z\,|\,\boldsymbol{M}) = S_{ext}(\boldsymbol{L})$ is listed in Scheme 9.7, and the atomic partial entropy $S(Z\,|\,\boldsymbol{M})$ is available in Table 9.1.

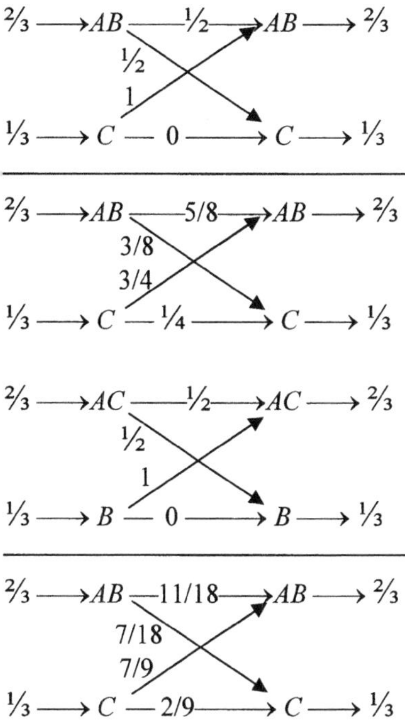

a) $\Sigma = -1$: $\boldsymbol{M}^r = (L, R_L)$, $K = L = AB, BC, AC$

$N^{ext}(L) = 0.918$, $S^{ext}(L) = ⅔$, $I^{ext}(L) = 0.252$

b) $\Sigma = 1$: $\boldsymbol{M}^r = (L, R_L)$, $K = L = AB,\ BC$

$N^{ext}(L) = 0.918$, $S^{ext}(L) = 0.907$, $I^{ext}(L) = 0.011$

$K = L = AC$

$N^{ext}(L) = 0.918$, $S^{ext}(L) = ⅔$, $I^{ext}(L) = 0.252$

a) $\Sigma = -1$: $\boldsymbol{M}^r = (L, R_L)$, $K = L = AB,\ BC,\ AC$

$N^{ext}(L) = 0.918$, $S^{ext}(L) = 0.897$, $I^{ext}(L) = 0.021$

Scheme 9.8. The *input-and-output*–reduced channels of alternative diatomics $K = L = \{AB, BC, AC\}$ in the 3-AO model for the three electron configurations of Scheme 9.1. The reduced channel entropy/information descriptors (in bits), representing the approximate external indices of Eq. (8.9.6) are also reported.

9.3. INFORMATION-DISTANCE APPROACH TO HAMMOND POSTULATE

The original, *qualitative* Hammond (1955) postulate and its more recent *quantitative* formulations (Marcus, 1968; Agmon and Levine, 1977, 1979; Miller, 1978; Nalewajski, 1980, Ciosłowski, 1991; Nalewajski *et al.*, 1994) all emphasize a relative resemblance of the TS complex to substrates (products) of the exoergic (endoergic) reactions. Therefore, the activation barrier appears as *"early"* in the exothermic reactions, with the substrates being only slightly modified in the TS complex, both electronically and geometrically. In the endothermic bond-forming-bond-breaking processes, when the barrier is *"late"* along the reaction coordinate, the activated complex resembles more the reaction products.

Various criteria of similarity between the electronic and geometrical structures of the TS-complex and the dissociation products of the *forward* and *reverse* reactions, respectively, e.g., energies, electron distributions, bond-orders, etc., have been used in the past to probe the electronic and geometrical aspects of the Hammond postulate. As we have demonstrated in Chapters 4 and 5, the cross-entropy (entropy-deficiency, missing information) of the molecular electron density, relative to the appropriate reference-distribution (Kullback and Leibler, 1951; Kullback, 1959; see also Section 3.3), can be used as the direct, quantitative IT criterion of molecular similarity. Therefore, the information-distances can be used as alternative tools to quantify the Hammond postulate (Nalewajski and Broniatowska, 2003).

It is of interest to separate the purely electronic component, called the *electronic–distance*, from the overall *electronic/geometric–distance* measure of an entropic degree of closeness of the two compared molecular structures. The Kullback-Leibler (1951) measure, $N\Delta S[p_1|p_2] = \Delta S[\rho_1|\rho_2] \geq 0$, of the information distance between two normalized probability distributions $p_1(\boldsymbol{r})$ and $p_2(\boldsymbol{r})$, representing the current and reference *shape*-factors of the corresponding isoelectronic densities $\rho_1(\boldsymbol{r}) = Np_1(\boldsymbol{r})$ and $\rho_2(\boldsymbol{r}) = Np_2(\boldsymbol{r})$, respectively, is defined by Eq. (3.3.2), with the zero value of this functional being reached for the identical distributions. The lower the value of $\Delta S[p_1|p_2]$ (or $\Delta S[\rho_1|\rho_2]$) the more $p_1(\boldsymbol{r})$ [$\rho_1(\boldsymbol{r})$] resembles $p_2(\boldsymbol{r})$ [$\rho_2(\boldsymbol{r})$]. In other words, the lower the entropy-deficiency value the "shorter" is the information "distance" between the two compared densities. The symmetrized (*divergence*) measure of the information distance (Kullback, 1959), $N\Delta S[p_1, p_2] = \Delta S[\rho_1, \rho_2] \geq 0$, is given by the sum of two directed divergencies of Eq. (3.3.7). It exhibits a non-negative density: $\Delta d(\boldsymbol{r}) \equiv N\Delta p(\boldsymbol{r})\ I[p_1(\boldsymbol{r})/p_2(\boldsymbol{r})] = \Delta\rho(\boldsymbol{r})\ I[\rho_1(\boldsymbol{r})/\rho_2(\boldsymbol{r})] \geq 0$, which vanishes only for the identical probability distributions (electron densities): $N\Delta p(\boldsymbol{r}) = \Delta\rho(\boldsymbol{r}) = 0$.

These information distances for a single-component distribution can be straightforwardly generalized to cover the multi-component (vector) distributions [see, e.g., Eqs. (5.3.4) and (5.3.7)]. In particular, the Hirshfeld (1977), "stockholder" AIM pieces $\boldsymbol{p}^H(\boldsymbol{r})$ [or $\boldsymbol{\rho}^H(\boldsymbol{r})$] [Eq. (5.2.1)] of the molecular electron distributions $p(\boldsymbol{r})$ [or $\rho(\boldsymbol{r})$], where shown in Section 5.3 to minimize these information-distances relative to the corresponding distributions of free-atoms $\boldsymbol{p}^0(\boldsymbol{r}) = \{p_X^0(\boldsymbol{r})\}$ [or $\boldsymbol{\rho}^0(\boldsymbol{r}) = \{\rho_X^0(\boldsymbol{r})\}$], which give rise to the overall promolecular distribution $p^0(\boldsymbol{r})$ [or $\rho^0(\boldsymbol{r})$], by shifting the "frozen" atomic densities to their actual positions $\mathbf{R} = \{\boldsymbol{R}_X\}$ in a molecule.

Obviously, the same missing information (entropy deficiency) concepts can be used to characterize similarities between larger molecular fragments, consisting of several atoms. In reference to the Hammond postulate we seek measures of the information distance which reflect relative similarities between the electron distribution in the TS complex (‡), identified by the nuclear configuration $\mathbf{R}^{\ddagger}$, and the relevant "promolecule" distributions obtained from the densities of separated reactants (or products). For example, in the collinear atom-exchange reaction between the hydrogen molecule and the halogen atom X = F, Cl, Br, I,

$$\text{H—H} + X \rightarrow [\text{H --- H --- }X]^{\ddagger} \rightarrow \text{H} + \text{H—}X, \qquad (9.3.1)$$

the electron density of the transition-state complex $\rho^{\ddagger}(\boldsymbol{r})$ should be compared with the overall densities of reactants, $\rho_{reactants}{}^{0}(\boldsymbol{r}) = \rho_{\text{HH}}(\boldsymbol{r}) + \rho_{X}(\boldsymbol{r})$, or products, $\rho_{products}{}^{0}(\boldsymbol{r}) = \rho_{\text{H}}(\boldsymbol{r}) + \rho_{\text{H}X}(\boldsymbol{r})$, defining the respective "promolecular" reactive systems.

These reference densities may include densities of the separated (free) subsystems, i.e., those of the free atom A, $\rho_{\text{A}}{}^{0}(\boldsymbol{r})$, and the diatomic B—C for its *equilibrium* inter-nuclear distance $R_{BC}{}^{0}$, $\rho_{BC}{}^{0}(\boldsymbol{r}; R_{BC}{}^{0})$, shifted to a relative separation giving rise to the maximum overlap with $\rho^{\ddagger}(\boldsymbol{r})$. In the present illustrative application to the chemical reaction (9.3.1) this relative atom-diatom separation has been chosen in such a way that the distance between the center-of-mass (CM) of the separate diatomic and remaining atom is equal to the corresponding distance in the TS complex. By assumption, in this "promolecular" reference the free-atom assumes the same position as its bonded analog in the TS complex. In what follows such a linear mutual alignment of the free diatomic and the corresponding fragment of the transition-state complex will be identified by the CM label (see also Fig. 9.2).

The entropy deficiencies $\Delta S[\rho^{\ddagger}(CM)\,|\,\rho_{\alpha}{}^{0}(CM)] = \Delta S_{\alpha}(CM)$ and $\Delta S[\rho^{\ddagger}(CM), \rho_{\alpha}{}^{0}(CM)] \equiv \Delta D_{\alpha}(CM)$, $\alpha = reactants, products$, will used as the *overall* IT probes of the relative closeness of the TS electronic structure to reactants or products, respectively. These quantities reflect the sum of the *nuclear* (geometrical) contribution, due to the diatomic bond elongation in the TS complex, and the *electronic* component, due to a change in the electronic structure in $[\text{ABC}]^{\ddagger}$ relative to the "promolecular" atom-diatom system $A + (BC)^{\ddagger}$ of the same geometry as in TS.

The electronic missing information can be extracted by the entropy deficiency between $\rho^{\ddagger}(r)$ and the modified "promolecule" density $\rho_{\alpha}{}^{0}(\mathbf{R}^{\ddagger})$, generated by the sum of the free-atom density $\rho_{A}{}^{0}(\boldsymbol{r})$ and the ground-state density $\rho_{BC}{}^{0}(\boldsymbol{r}; \boldsymbol{R}^{\ddagger})$ of the *activated* (elongated) diatomic $(\text{BC})^{\ddagger}$, for its inter-nuclear distance $R_{BC}{}^{\ddagger}$ equal to that in TS complex. In this purely-electronic index the corresponding atoms of the "promolecule" and TS complex assume exactly the same positions $\text{R}^{\ddagger}$, so that only changes in the electronic structure contribute to the information-distances $\Delta S[\rho^{\ddagger}(\mathbf{R}^{\ddagger})\,|\,\rho_{\alpha}{}^{0}(\mathbf{R}^{\ddagger})] = \Delta S_{\alpha}(\mathbf{R}^{\ddagger})$, and $\Delta S[\rho^{\ddagger}(\mathbf{R}^{\ddagger}), \rho_{\alpha}{}^{0}(\mathbf{R}^{\ddagger})] = \Delta D_{\alpha}(\mathbf{R}^{\ddagger})$, $\alpha = reactants, products$.

The internuclear distances $R_{\text{HH}}{}^{\ddagger}$ and $R_{\text{H}X}{}^{\ddagger}$ of Dunning (1984) have been assumed for the four TS complexes in reactions (9.3.1) for X = F, Cl, Br, I. (see Table 9.4). The electron densities for the transition-state complexes, reactants and products have been obtained from the LDA DFT calculations using the extended (DZVP) basis set.

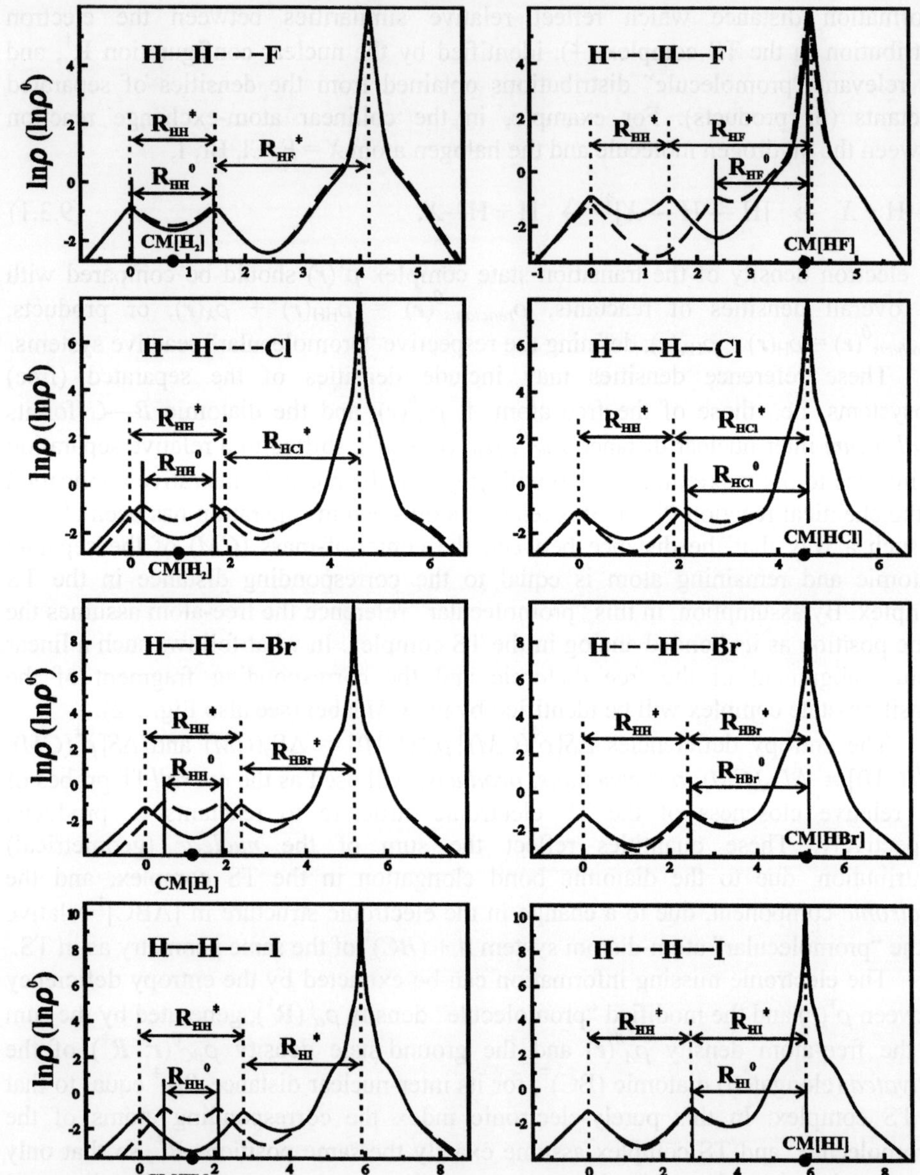

Fig. 9.2. Logarithmic plots of the electron densities in the transition-state complexes, $\rho^{\ddagger}(\mathbf{R}^{\ddagger})$, compared against the promolecule densities ρ_{α}^{0}(CM), for α = *reactants* (left column) and *products* (right column). The corresponding plots for the $\rho_{\alpha}^{0}(\mathbf{R}^{\ddagger})$ densities are practically indistinguishable from those for the $\rho^{\ddagger}(\mathbf{R}^{\ddagger})$ densities.

Table 9.4. Summary of the internuclear distances $\{R_{AB}^{\ddagger}\}$, net charges $\{q_X^H \equiv q^H(X)\}$ (a.u.) of the Hirshfeld atoms, and bond multiplicities N from the Difference (^{D}) and Mayer (^{M}) MO approaches, in the collinear TS-complexes $[H\text{---}H\text{---}X]^{\ddagger}$ of the atom exchange reactions (9.3.1): X = F, Cl, Br, I.

Property	$H_{(1)}$ -- $H_{(2)}$ – F	$H_{(1)}$ -- $H_{(2)}$ -- Cl	$H_{(1)}$ -- $H_{(2)}$ -- Br	$H_{(1)}$ -- $H_{(2)}$ – I
$R_{HH}^{\ddagger}$ (Å)	0.790	1.010	1.260	1.460
$R_{HX}^{\ddagger}$(Å)	1.410	1.430	1.490	1.670
$N^D(H_{(1)}–H_{(2)})^{\ddagger}$	0.575	0.318	0.189	0.130
$N^M(H_{(1)}–H_{(2)})^{\ddagger}$	0.668	0.413	0.260	0.184
$N^D(H_{(2)}–X)^{\ddagger}$	0.562	0.848	1.088	1.164
$N^M(H_{(2)}–X)^{\ddagger}$	0.317	0.570	0.709	0.788
$N^D(H_{(1)}–X)^{\ddagger}$	0.370	0.297	0.236	0.189
$N^M(H_{(1)}–X)^{\ddagger}$	0.182	0.142	0.116	0.113
$N^D(H\text{-}H\text{-}X)^{\ddagger}$	1.506	1.463	1.512	1.483
$N_{cov}^D(H\text{-}H\text{-}X)^{\ddagger}$	1.109	1.129	1.069	1.169
$N^M(H\text{-}H\text{-}X)^{\ddagger}$	1.167	1.125	1.085	1.085
$q^H(H_{(1)})^{\ddagger}$	0.11	0.10	0.09	0.07
$q^H(H_{(2)})^{\ddagger}$	0.06	0.05	0.05	0.04
$q^H(X)^{\ddagger}$	-0.17	-0.15	-0.14	-0.11

The Kohn-Sham determinants have then been used to generate the bond-orders of the TS-complex, $N^D(A–B)^{\ddagger}$ and $N^M(A–B)^{\ddagger}$, obtained using the difference (D) (Nalewajski *et al.*, 1997) and Mayer (M) (1983, 1985) approaches, respectively, which are also reported in Table 9.4. These bond multiplicities have then been used to obtain the overall bond indices $N^D(H\text{---}H\text{---}X)^{\ddagger}$ and $N^M(H\text{---}H\text{---}X)^{\ddagger}$, for the system as a whole, by summing over the contributions from two bonds between neighboring atoms and that corresponding to a pair of peripheral atoms in the collinear structure. The overall covalent bond-orders from the difference approach, $N_{cov}^D(H\text{---}H\text{---}X)^{\ddagger}$ are also reported in the table together with the net-charges $\{q_X^H\}$of the Hirshfeld atoms. The corresponding results for the equilibrium ground-state structures $\{R_{XY}\}$ of the diatomic reactant and product molecules are reported in Table

9.5. Finally, in Table 9.6, the overall information distances are examined, which test the quantitative performance of the Hammond postulate in IT.

Table 9.5. The same as in Table 9.4 for the equilibrium geometries of the ground-state reactant and product diatomics.

Property	HH	HF	HCl	HBr	HI
$R_{HH(X)}$ (Å)	0.767	0.940	1.298	1.449	1.652
$N^D(HX)$	1.010	1.268	1.330	1.416	1.452
$N_{cov}{}^D(HX)$	0.742	0.917	0.994	0.963	1.102
$N^M(HX)$	1.000	0.748	0.901	0.904	0.935
$q^H(H)$	0.00	0.24	0.14	0.12	0.09
$q^H(X)$	0.00	-0.24	-0.14	-0.12	-0.09

A comparison between the internuclear distances in Tables 9.4 and 9.5 directly reveals the *geometrical* aspect of the Hammond postulate: one observes a growing elongation of the H–H bond in the TS complex in the series X = F, Cl, Br and I. This is in accordance with the observed trends in corresponding reaction energies, reflected by the difference between the experimental dissociation energy (D_e, kcal/mol) of the broken H–H bond (109.4) and the newly formed H–X bond: H–F (140.2) - exothermic reaction, and {H–Cl (107.2), H–Br (93.8), H–I (81.1)} - endothermic reactions (Dunning, 1984). The same conclusion follows from a comparison of the HX structural data listed in Tables 9.4 and 9.5.

Table 9.6. Overall information distances between the TS-complex ("molecular") and substrate/product ("promolecular") densities for the atom exchange reactions (9.3.1).

	$H_2 + F \rightarrow$	$H + HF$	$H_2 + Cl \rightarrow$	$H + Cl$	$H_2 + Br \rightarrow$	$H + Br$	$H_2 + I \rightarrow$	$H + HI$
$\Delta S(\mathbf{R}^\ddagger)$	0.052	0.085	0.059	0.018	0.084	0.006	0.092	0.004
$\Delta D(\mathbf{R}^\ddagger)$	0.104	0.168	0.123	0.038	0.172	0.013	0.192	0.008
$\Delta S(CM)$	0.052	0.548	0.147	0.072	0.401	0.013	0.661	0.005
$\Delta D(CM)$	0.105	1.057	0.306	0.142	0.848	0.027	1.400	0.011

In Table 9.4 a given type of the overall bond-order measure in TS complex, calculated from the MO theory using either Mayer's (*M*) or Difference (*D*) bond-multiplicities, is seen to be practically identical in the four TS complexes considered. The difference approach generates the overall *covalent* component, of about a single bond multiplicity, which is similar to that predicted by the Mayer scheme. However, the overall (*ionic+covalent*) difference index of the difference approach predicts the TS bond-multiplicity by about half-bond higher, which remains roughly conserved in all four reactions. The $N(\text{H–H})^{\ddagger}$ and the $N(\text{H–}X)^{\ddagger}$ diatomic bond-orders in Table 9.4, which are complementary in the Hammond postulate context, do indeed reflect a relative closeness to the *initial* (reactant) or *final* (product) bond-multiplicities, respectively, which are reported in Table 9.5.

In Table 9.5 the total diatomic bond-orders from the difference approach are seen to increase from about a single-bond measure in H_2 to around one-and-a-half bond in HI. The growing covalent component reflects an increased participation of the virtual atomic orbitals on a heavy (soft) halogen atom. A reference to the net charges of "stockholder" atoms reported in Table 9.4 shows that in TS-comoplex the amount of electronic charge accepted by the approaching halogen from the hydrogen molecule reflects decreasing halogen electronegativity in this series. One also finds that the (positive) charge on the terminal hydrogen is roughly twice as large as that on the middle hydrogen. The corresponding entries in Table 9.5 also reflect the decreasing CT in H*X*, when the size of halogen increases (electronegativity decreases).

In Table 9.6 the entropy-deficiencies ΔS and ΔD are listed for the purely-electronic ($\mathbf{R}^{\ddagger}$) and the electronic-geometric (*CM*) approaches, respectively. These missing-information indices are indeed seen to be in agreement with the qualitative Hammond Postulate. The displacements with respect to the reactants or products, respectively, are found to be relatively more emphasized in the *CM*-scheme. The same conclusion follows from Fig. 9.2. For example, the *CM*-promolecular density constructed from the isolated *substrates* H_2 and F differs less from the transition-state electron distribution, than the density constructed from the isolated *products*: H and HF. The opposite trend is detected for the other extreme case of $X = \text{I}$: the transition-state density is now almost identical with the promolecule density derived from the *products* H and HI, while exhibiting substantial differences relative to the promolecular electron density constructed using the densities of the free reactants H_2 and I. The differences between electron distributions of Fig. 9.2 are well reflected by the integral information-distance quantities reported in Table 9.6.

9.4. CONCLUSION

In this chapter we have used various IT concepts and techniques developed in the preceding chapters in an exploration of reactants and transition-state complexes. Since the molecular subsystems can be objectively defined only in the entropy-deficiency representation, the information-distance element of the molecular fragment definition must affect the descriptors of molecular subsystems of reactive systems, e.g., their reactivity indices. We have examined in this chapter an example of such entropic charge-sensitivities of reactants, called charge affinities. These FF-like entropic quantities represent the entropy-deficiency conjugates of the fragment electron populations. They combine the subsystem energetic derivative (Fukui function) and the entropic derivative measuring the information-conjugate of the subsystem electron density. Such mixed energy/entropy derivative quantities can be expected to generate a more complete description of reactants in the Donor-Acceptor reactive systems. We have illustrated their applicability to the vertical, density-partitioning problems of the molecular electronic structure by demonstrating that the charge affinities of the stockholder subsystems vanish identically, thus confirming their entropy-equilibrium character.

A simple 3-AO model of the TS-complex in the atom-diatom reactive system has been introduced and applied in a theoretical study of displacements in the overall and diatomic bond-orders in TS, relative to both the atom-diatom and separated-atom limits. The bond-order measures from the MO theory have been compared with those generated by the communication theory of the chemical bond. Specific strategies for molecular subsystems, which have been developed in Chapter 8, have been applied to this model TS to study the influence of the spin polarization on subsystem bond-multiplicities and their ionic/covalent composition for equal distribution of electrons among the three constituent atoms.

Finally, the information-distance concept of IT has been applied as quantitative measure of the closeness (similarity) of the TS electronic and geometric structures to those of reactants and products of the atom exchange reaction in the H_2— halogen systems. This analysis has been undertaken to test the qualitative Hammond postulate of reactivity theory. The electronic and nuclear components of these resemblance-measures from IT have been separated. This illustrative application further confirms the versatility of the IT diagnostic tools in a chemical interpretation of the electronic structure of molecules and reactive systems.

10. ELEMENTS OF THE INFORMATION–DISTANCE THERMODYNAMICS

Thermodynamic analogies in the IT-description of the stationary (equilibrium) and instantaneous (non-equilibrium) electron distributions in molecules and their constituent fragments are explored. The "horizontal" displacements in the molecular electronic-structure, from one ground-state density to another, and "vertical" flows between subsystems, which preserve the molecular electron density, are examined. The coupling between the energy and entropy-deficiency descriptors of the electron densities in moleculess is examined for both the equilibrium (ground-state) and non-equilibrium (variational) distributions. The Lagrange multipliers of the constrained variational principles in both the energy and missing-information representations are identified. They are related to the information "temperature" parameters of the thermodynamic-like interpretation. It is demonstrated that the global information "temperature", which characterizes the *equilibrium* electron distribution, is represented by the system chemical potential. The *non-equilibrium* electron densities are shown to require the local (non-equalized) information temperature, measured by the local chemical potential of such trial electron distributions. The illustrative case of the hydrogen-like atom is examined in a more detail. The alternative *vertical* descriptions corresponding to alternative choices of the independent state-parameters of molecular fragments will be explored. The thermodynamic-like affinities ("forces") and fluxes ("responses") will be established. They determine the associated local entropy-deficiency "source". The vertical IT-thermodynamic description of molecular subsystems gives rise to the equilibrium ("stockholder") molecular fragments, which correspond to the minimum of the subsystem entropy-deficiency relative to the promolecular reference. The Hirshfeld subsystems-in-molecules, previously viewed as static pieces of the molecular density, are interpreded as *dynamical* entities with the distribution of local fluctuations in the unconstrained state-parameters being related in the thermodynamic-like fashion to the local value of the relevant Legendre-transform of the missing-information density. The associated "reciprocity" relations are derived for the vertical processes in molecular systems. They represent the IT-analogs of the familiar Onsager relations of the ordinary irreversible thermodynamics.

10.1. INTRODUCTION

We have demonstrated in Section 5.3 that the classical *vertical* problem of partitioning a given molecular electron density into pieces representing molecular fragments, e.g., AIM, can be given an unbiased IT-description, much in the spirit of the phenomenological approach of ordinary thermodynamics, in terms of the bonded fragments, which retain as much as possible the information contained in the free atoms defining the promolecule. One needs the information-entropy representation to define the equilibrium state of molecular subsystems, since such energetical quantities as the system overall electronic energy, chemical potential, etc., are fixed by the conserved molecular density for its all admissible divisions into molecular fragments. We have also argued in Section 6.1 that the Hirshfeld partitioning of the electron density appears as the "equilibrium" entropy-deficiency scheme satisfying the inter-subsystem equalization rules of several information-distance densities.

As we have also mentioned before, the Hirshfeld fragments represent the entropy-deficiency stable, equilibrium pieces of the molecular electronic density. This implies that any hypothetical deviation from the Hirshfeld division will increase the missing-information of the molecular fragments, relative to the lowest value reached in this "stockholder" division, which corresponds to the relative entropy between the molecular and "promolecular" probability densities. The energy equilibrium criterion, the chemical potential equalization principle, is satisfied by any arbitrary set of the mutually open subsystems. Therefore, it does not discriminate between alternative divisions of a given molecular density. Hence, it cannot be used as the criterion for distinguishing the optimum partition of the electronic density from a multitude of alternative divisions. The information-theoretic justification of the Hirshfeld scheme emphasizes the need for an independent, *entropy/information* criterion, complementary to the familiar *energy* criterion of the subsystem chemical potential (electronegativity) equalization. Only the combined energy and information-entropy representations provide an adequate theory of the electronic structure of molecular subsystems which covers both the horizontal and vertical density displacements. These two descriptions have to be united in the global description of the molecular ground-state since the Hohenberg-Kohn theorems of DFT state that the molecular electron density specifies uniquely all physical/chemical properties of the system as a whole.

Therefore, only the combined energy-representation of the molecular quantum mechanics and the entropy representation of IT provide the *complete*, fully unbiased treatment of molecular fragments. Such an objective description is vital for the chemical understanding of molecular processes. This double-representation perspective resembles that used in the ordinary thermodynamics (Callen, 1962), in which the equilibrium states can be determined from the equivalent *minimum-energy* or the *maximum-entropy* principles (see Section 2.1). In the thermodynamic-like description of molecular systems IT establishes the entropy representation, complementary to the familiar energy representation emphasized in most of the theories of the electronic structure (Nalewajski and Parr, 2001; Nalewajski, 2002c,

2003a,b, 2004a, 2005d). The entropy-deficiency approach to molecular systems complements earlier local "thermodynamical" DFT treatments (Ghosh *et al.*, 1984; Ghosh and Berkowitz, 1985; Nagy and Parr, 1994).

The fundamental relations of the phenomenological thermodynamics in both the energy and entropy representations simultaneously involve the system internal energy and its entropy. Indeed, the coupling between the energetical and entropic aspects of physical processes constitutes the essence of the thermodynamical approach. The corresponding variational principles, of the minimum energy for constant entropy or the maximum entropy for constant energy, respectively, require the relevant *temperature*-related Lagrange multipliers, which enforce the constraints involved. It has been recently shown that a similar *coupled* "thermodynamic" description is possible within the IT-treatment of the electron distributions in molecules (Nalewajski and Parr, 2001; Nalewajski, 2005d). Prior applications of IT to molecular systems were limited mainly to extracting the thermodynamic-like interpretation of the known ground-state molecular densities of electrons. In such a *decoupled* approach the IT-analysis was applied to the equilibrium molecular densities, which were previously obtained from the minimum principle of the system electronic energy, subject to the relevant constraints of either the wave-function normalization (Schrödinger variational principle of the wave-function theory) or the electron density normalization (Hohenberg-Kohn principle od DFT). Such treatment misses the information "temperature" descriptors of molecules, which measure a degree of the coupling between the information/entropy quantities of the electron gas and its energy. In this chapter we shall investigate such "temperature" related concepts, which are involved in the *coupled*, thermodynamic-like variational principles. Both the equilibrium (ground-state) and non-equilibrium (trial) distributions of electrons will be examined and the illustrative example of the hydrogen-like atom will be examined.

In chemistry one is interested not only in the equilibrium distributions of electrons in molecules and their constituent fragments, but also in processes characterized by *rates*. The conceptual structure of such a phenomenological *"dynamical"* description of the subsystem density fluctuations, due to instantaneous flows of electrons in a molecule, should parallel that of the ordinary irreversible thermodynamics (Callen, 1960). It calls for two types of the conjugate quantities: *"affinities"*, to describe thermodynamic *"forces"* that drive irreversible processes, and *"fluxes"* describing the system *responses* to the applied forces. In this chapter an overview of such a thermodynamic-like perspective on both the stationary and instantaneous states of molecules and their constituent fragments will be given.

10.2. "HORIZONTAL" PROCESSES IN MOLECULES

The ultimate goal of a thermodynamic-like description of molecular systems and their fragments is to determine both the *"horizontal"* and *"vertical"* displacements of the electronic structure (see Section 1.6). The former, which we shall examine in this section, represent transitions from one molecular ground-state density to another: $\rho_1[v_1] \rightarrow \rho_2[v_2]$, where $\rho[v]$ stands for the ground-state density for the external potential v.

In order to relate the cross-entropy $\Delta S[\rho'|\rho^0]$ for the trial (variational) density ρ', relative to the reference distribution ρ^0, to the system ground-state energy constraint, $\mathcal{F}_0[\rho'] \equiv E_v[\rho'] = E_v[\rho] \equiv \mathcal{F}_0^0$, and possibly to other physical constraints $\{\mathcal{F}_k[\rho'] = \mathcal{F}_k[\rho] \equiv \mathcal{F}_k^0,\ k > 0\}$, one uses the EPI principle of Eq. (3.8.1a):

$$\delta\{\Delta S[\rho'|\rho^0] - \Sigma_k \lambda_k \mathcal{F}_k[\rho']\} = 0, \qquad (10.2.1)$$

where λ_k is the Lagrange multiplier associated with k-th constraint. An example of such an application of the information theory in determining the exchange-correlation part of the effective one-body potential of Kohn and Sham (1965) has recently been reported by Parr and Wang (1997). This procedure constitutes the information-theoretic variant of the Zhao–Morrison–Parr procedure (Zhao and Parr, 1992, 1993; Zhao, Morrison and Parr, 1994). The entropy term in the foregoing equation represents a "device", which allows one to assimilate in the optimum density ρ the physical information contained in the constraints, in reference to the promolecule density ρ^0, in the most unbiased manner possible.

10.2.1. "Thermodynamic" Principle in Energy Representation

Let us again consider the variational principle for the minimum of the electronic energy (see Section 1.5), which determines the ground-state density $\rho(\boldsymbol{r}) = \rho[N, v]$ for the molecular electronic Hamiltonian $\hat{H}(N, v)$ defined by the overall (integer) number of electrons $N[\rho] \equiv \int \rho(\boldsymbol{r})\, d\boldsymbol{r} = N$ and the external potential $v(\boldsymbol{r})$ due to the system nuclei in their fixed positions (Born-Oppenheimer approximation). In what follows we shall build up upon the basic DFT variational equation, which summarizes the second Hohenberg-Kohn theorem, and develop the equivalent (coupled) "thermodynamic"-like principles, which simultaneously involve the electronic energy and the Kullback-Leibler entropy-deficiency (information-penalty) term (Nalewajski, 2005d).

We recall that the Hohenberg-Kohn (1964) principle of Eq. (1.5.20) involves the constrained minimization of the system energy $E_v[\rho'] = \int \rho'(\boldsymbol{r})\, v(\boldsymbol{r})\, d\boldsymbol{r} + F[\rho']$ for the trial (variational) density ρ':

$$\delta\{E_v[\rho'] - \mu[N, v]\, N[\rho']\} = \delta E_v[\rho'] - \mu[\rho]\, \delta N[\rho'] = 0 \qquad \Rightarrow \rho. \qquad (10.2.2)$$

The universal (v-independent) $F[\rho']$ part of the energy functional combines the electronic kinetic and repulsion energies for the trial density ρ', and the Lagrange multiplier $\mu[N, v] = (\partial E[N, v]/\partial N)_v = \mu[\rho]$, which enforces in the optimum solutions the correct number of electrons, $N[\rho] = N$, represents the system chemical potential [Eq. (1.5.21)], which is equalized throughout the space:

$$\mu[\rho] = \frac{\delta E_v[\rho]}{\delta\rho(\boldsymbol{r})} = v(\boldsymbol{r}) + \frac{\delta F[\rho]}{\delta\rho(\boldsymbol{r})} = \mu(\boldsymbol{r}). \tag{10.2.3}$$

In a more compact form of Eq. (1.5.22) this Euler equation involves the relative external potential, $u(\boldsymbol{r}) \equiv v(\boldsymbol{r}) - \mu = -\delta F[\rho]/\delta\rho(\boldsymbol{r}) = u[\rho; \boldsymbol{r}]$, the unique functional of the ground-state density. As we have also emphasized in Section 1.5, this DFT Euler equation is exactly equivalent to the associated Schrödinger principle (1.5.7), with respect to the trial wave-functions of N electrons, $\Psi(N)$, for the minimum expectation value of the electronic energy subject to the constraint of the wave-function normalization, $\mathsf{N}\,[\Psi] \equiv \langle\Psi|\Psi\rangle = 1$, enforced by the energy Lagrange multiplier $E[N, v] = (\partial E_v/\partial \mathsf{N}\,])_v$:

$$\delta\{E_v[\Psi(N)] - E[N, v]\,\langle\Psi(N)|\Psi(N)\rangle\}$$
$$\equiv \delta E_v[\Psi(N)] - E[N, v]\,\delta\mathsf{N}\,[\Psi(N)] = 0 \qquad \Rightarrow \qquad \Psi[N, v], \tag{10.2.4}$$

One then defines the deviation of the current (variational) energy from the exact ground-state energy level for a given trial density ρ',

$$\Delta E_v[\rho'|\rho] = E_v[\rho'] - E_v[\rho], \tag{10.2.5}$$

so that $\delta\Delta E_v[\rho'|\rho] = \delta E_v[\rho']$ and $\Delta E_v[\rho\,|\rho] = 0$.

Clearly, the subsidiary condition of the required normalization of the optimum electron density in the DFT variational principle of Eq. (10.2.2) can be automatically satisfied, when the variational procedure guarantees that the optimum density $\rho^{opt.} = \rho[N, v]$. For example, this can be done *directly*, by imposing the *local* constraint $\rho'(\boldsymbol{r}) = \rho(\boldsymbol{r})$ multiplied by the corresponding local Lagrange multiplier $\lambda(\boldsymbol{r})$ in a trivial Euler-Lagrange problem:

$$\delta\{E_v[\rho] - \int \lambda(\boldsymbol{r})\rho(\boldsymbol{r})\,d\boldsymbol{r}\} = 0, \qquad \Rightarrow \qquad \rho(\boldsymbol{r}). \tag{10.2.6}$$

One then identifies $\lambda(\boldsymbol{r})$ as the local chemical potential of Eq. (10.2.3),

$$\lambda(\boldsymbol{r}) = \frac{\delta E_v[\rho]}{\delta\rho(\boldsymbol{r})} = \mu(\boldsymbol{r}) = \mu[\rho], \tag{10.2.7}$$

so that that the minimum-energy principles of Eqs. (10.2.2) and (10.2.6) become identical.

Alternatively, this density constraint can be imposed *indirectly*, in a thermodynamic-like manner, by using an appropriate information-penalty term, e.g., the cross-entropy in the current density ρ' relative to ρ, representing the information distance between the two compared electron distributions. It introduces the entropic "penalty" for the trial density deviating from the true ground-state density ρ, and combines the system electronic energy and its average missing-information functional. The most natural candidate for such a penalty function is the *directed-divergence* (entropy-deficiency) functional of Kullback and Leibler (see Section 3.3),

$$\Delta S[\rho'|\rho] = \int \rho'(\boldsymbol{r}) \ln[\rho'(\boldsymbol{r})/\rho(\boldsymbol{r})]\, d\boldsymbol{r} \equiv \int \Delta s(\boldsymbol{r})\, d\boldsymbol{r}, \qquad (10.2.8)$$

which identically vanishes for $\rho'(\boldsymbol{r}) = \rho(\boldsymbol{r})$.

The $\Delta S[\rho' \mid \rho]$ "intensive" conjugate of the "extensive" variable ρ is then defined by the functional derivative

$$\left.\frac{\delta \Delta S[\rho'|\rho]}{\delta \rho(\boldsymbol{r})}\right|_{\rho'=\rho} = \{\ln\left[\frac{\rho'(\boldsymbol{r})}{\rho(\boldsymbol{r})}\right] + 1\}_{\rho'=\rho} \equiv g'(\boldsymbol{r})|_{\rho'=\rho} \equiv g(\boldsymbol{r}) = 1, \qquad (10.2.9)$$

which is seen to be equalized throughout the physical space for $\rho'(\boldsymbol{r}) = \rho(\boldsymbol{r})$.

Hence, by enforcing the overall entropy deficiency constraint, $\Delta S[\rho' \mid \rho] = 0$, which is satisfied only for $\rho'(\boldsymbol{r}) = \rho(\boldsymbol{r})$, multiplied by the appropriate *global* Lagrange multiplier τ in the constrained minimum energy principle,

$$\delta\{\Delta E_v[\rho'|\rho] - \tau\, \Delta S[\rho'|\rho]\} = 0, \qquad (10.2.10)$$

one indirectly satisfies at the solution point the density normalization constraint $N' \equiv N[\rho'] = N[\rho] = N$. Therefore, the energy-minimum-principle of Eq. (10.2.2) and the foregoing "thermodynamic" principle, of the minimum electronic energy for constant (vanishing) entropy-deficiency, are equivalent having both the system ground-state density as their solutions.

This coupled ("thermodynamic") principle identifies τ as the system global information "temperature":

$$\tau = \frac{\partial \Delta E_v}{\partial \Delta S}. \qquad (10.2.11)$$

It measures the linear response of the system electronic energy,

$$\delta \Delta E_v[\delta\rho] = \int \frac{\delta E_v[\rho]}{\delta \rho(\boldsymbol{r})}\, \delta\rho(\boldsymbol{r})\, d\boldsymbol{r} = \mu \delta N, \qquad (10.2.12)$$

where $\delta\rho(\boldsymbol{r}) = \delta[\rho'(\boldsymbol{r}) - \rho(\boldsymbol{r})] \equiv \delta\Delta\rho(\boldsymbol{r})$ and $\delta N = \int \delta\rho(\boldsymbol{r}) d\boldsymbol{r}$, per unit displacement in the system global entropy deficiency:

$$\delta \Delta S[\delta\rho] = \int \left.\frac{\delta \Delta S[\rho'|\rho]}{\delta \rho'(\boldsymbol{r})}\right|_{\rho'=\rho} \delta\rho(\boldsymbol{r})\, d\boldsymbol{r} = \delta N, \qquad (10.2.13)$$

which amounts to the displacement in the system number of electrons. Hence the physical interpretation of the global information-temperature as the system chemical potential:

$$\tau = \lim_{\delta\rho\to 0} \frac{\delta \Delta E_v[\delta\rho]}{\delta \Delta S[\delta\rho]} = \mu\,(\delta N/\delta N) = \mu. \qquad (10.2.14)$$

This somewhat surprising identification is a direct consequence of Eqs. (10.2.9) and (10.2.13). The latter explicitly shows that the linear variations of the system entropy-deficiency and its overall number of electrons are identical.

Alternatively, the density constraint $\rho'(\boldsymbol{r}) = \rho(\boldsymbol{r})$ can be imposed through the *local* "thermodynamic" constraint, by fixing the the entropy-deficiency density of [Eq. (10.2.8)], $\Delta s(\boldsymbol{r}) = 0$, using the local Lagrange multiplier $\tau(\boldsymbol{r})$:

$$\delta\{\Delta E_v[\rho'|\rho] - \int \tau(\boldsymbol{r})\,\Delta s(\boldsymbol{r})\,d\boldsymbol{r}\} = 0 \qquad \Rightarrow \qquad \rho(\boldsymbol{r}). \qquad (10.2.15)$$

The latter is now defined by the functional derivative

$$\tau(\boldsymbol{r}) = \left.\frac{\delta \Delta E_v}{\delta \Delta s(\boldsymbol{r})}\right|_{\rho'=\rho}, \qquad (10.2.16)$$

which can thus be interpreted as the *local* information temperature. The associated Euler equation [see Eqs. (10.2.3) and (10.2.9)] then gives:

$$\tau(\boldsymbol{r}) = \frac{\delta E_v[\rho]}{\delta\rho(\boldsymbol{r})} \Bigg/ \left.\frac{d\Delta s(\boldsymbol{r})}{d\rho'(\boldsymbol{r})}\right|_{\rho'=\rho} = \mu(\boldsymbol{r}) = \mu = \tau. \qquad (10.2.17)$$

Hence, this alternative, local formulation also identifies the equilibrium local chemical potential for the ground-state density, $\mu(\boldsymbol{r}) = \mu$, as the local (equalized) value of the information temperature: $\tau(\boldsymbol{r}) = \tau$.

We therefore conclude that the information temperature in "thermodynamic" variational principles of Eqs. (10.2.10) and (10.2.15) is given by the system chemical potential of Eq. (10.2.3). Indeed, in the Hohenberg-Kohn [Eq. (10.2.2)] and global "thermodynamic" [Eq. (10.2.15)] variational principles in the electronic-energy representation this Lagrange multiplier plays a similar role, of enforcing the correct density normalization: directly in the DFT energy minimum principle, and indirectly in the coupled "thermodynamic" principle, through the entropy-deficiency penalty from IT. The DFT variational principle is thus interpreted as being equivalent to the "thermodynamic" principle of the IT approach. In the latter the electronic chemical potential plays the role of the information "temperature", which enforces the ground-state entropy-deficiency constraint. This global temperature parameter has also been shown to represent the equalized value of the local information temperature, $\tau(\boldsymbol{r}) = \mu(\boldsymbol{r})$, characterizing the specified infinitesimal volume element of the ground-state (*equilibrium*) distribution of electrons. In what follows we shall demonstrate that a *non*-equalized (local) information temperature is required in the "thermodynamic" principle determining the *non*-equilibrium electron distribution ρ': $\Delta\rho = \rho' - \rho \neq 0$.

10.2.2. Illustrative Example: Hydrogen-Like Atom

Consider the one-electron atom described by the electron density (a.u.) originating from the trial atomic orbital $1s(\boldsymbol{r}; \alpha)$,

$$\rho'(\boldsymbol{r}) = \rho(\boldsymbol{r}; \alpha) = (\alpha^3/\pi)\exp(-2\alpha r) = |1s(\boldsymbol{r}; \alpha)|^2 \equiv \rho(\alpha), \tag{10.2.18}$$

where the exponent α is the variational parameter. The exact ground-state density is reached for $\alpha = Z$ (the nuclear charge in a.u.), $\rho(\boldsymbol{r}) = \rho(\boldsymbol{r}; Z)$, when

$$E_v[\rho] = -Z^2/2 = \mu = \mu(\boldsymbol{r}). \tag{10.2.19}$$

The density functional for the electronic energy of this hydrogen-like system is known exactly. It consists of the Fisher information (kinetic energy) term, reminiscent of von Weizsäcker's (1935) non-homogeneity correction, and the trivial functional for the nuclear attraction energy:

$$E_v[\rho'] = \frac{1}{8}\int \frac{|\nabla\rho'|^2}{\rho'}d\boldsymbol{r} - Z\int \frac{\rho'}{r}d\boldsymbol{r}\,. \tag{10.2.20}$$

Hence, for $\rho'(\boldsymbol{r}) = \rho(\alpha)$,

$$E_v[\rho(\alpha)] = \tfrac{1}{2}\,\alpha^2 - \alpha Z. \tag{10.2.21}$$

The missing information functional of Eq. (10.2.8) for this trial density reads:

$$\Delta S[\rho(\alpha)|\rho] = 3[\ln(\alpha/Z) + (Z/\alpha) - 1)]. \tag{10.2.22}$$

It gives rise to the associated entropic intensity of Eq. (10.2.9),

$$g(\boldsymbol{r}; \alpha) = 3\ln(\alpha/Z) + 2(Z - \alpha) + 1, \tag{10.2.23}$$

which determines the (local) kernel

$$\left.\frac{\delta\Delta s(\boldsymbol{r}')}{\delta\rho(\boldsymbol{r})}\right|_{\rho(\alpha)} = \left.\frac{d\Delta s(\boldsymbol{r})}{d\rho(\boldsymbol{r})}\right|_{\rho(\alpha)} \delta(\boldsymbol{r} - \boldsymbol{r}') = g(\boldsymbol{r}; \alpha)\,\delta(\boldsymbol{r} - \boldsymbol{r}'). \tag{10.2.24}$$

It should be also observed that $g(\boldsymbol{r}; Z) = 1$, in accordance with Eq. (10.2.9). The functional derivative of the density functional for the system energy,

$$\mu(\boldsymbol{r};\alpha) = \left.\frac{\delta E_v[\rho']}{\delta\rho'(\boldsymbol{r})}\right|_{\rho(\alpha)} = -Z/r + (1/8)\,\{(\nabla\rho/\rho)^2 - 2\,(\nabla^2\rho)/\rho\}|_{\rho(\alpha)}$$

$$= -\tfrac{1}{2}\,\alpha^2 + (\alpha - Z)/r, \qquad (10.2.25)$$

then correctly predicts [Eq. 10.2.19)]

$$\mu(\boldsymbol{r};Z) = -Z^2/2 = \mu = \mu(\boldsymbol{r}). \qquad (10.2.26)$$

Also, by the density chain-rule,

$$\tau(\boldsymbol{r}) = \left.\frac{\delta\Delta E_v}{\delta\Delta s(\boldsymbol{r})}\right|_{\rho'=\rho} = \int\left(\frac{\delta E_v[\rho]}{\delta\rho(\boldsymbol{r}')}\right)\Big/\left(\frac{\delta\Delta s(\boldsymbol{r}')}{\delta\rho(\boldsymbol{r})}\right)d\boldsymbol{r}' = \mu(\boldsymbol{r})/g(\boldsymbol{r}) = \mu(\boldsymbol{r}) = \mu. \qquad (10.2.27)$$

Therefore, when the equilibrium (ground-state) distribution is reached, the local information-temperature (chemical potential) in the hydrogen-like atom equalizes throughout the space at the global chemical potential value, equal to the exact 1*s* orbital energy.

10.2.3. "Thermodynamic" Principle for Non-Equilibrium Density

By the Hohenberg-Kohn theorem of DFT the ground-state density for the external potential v_1, $\rho[v_1]$, represents the non-equilibrium distribution of electrons for another external potential $v_2 \neq v_1 + const$: $\rho[v_1] \neq \rho[v_2]$. The relevant ground-state Euler equations for v_1 and v_2, are satisfied by the ground-state densities ρ_1 and ρ_2, respectively,

$$\mu_1 = \tau_1 = v_1(\boldsymbol{r}) + \frac{\delta F[\rho_1]}{\delta\rho_1(\boldsymbol{r})} \equiv \mu_1(\boldsymbol{r}) \quad \text{and} \quad \mu_2 = \tau_2 = v_2(\boldsymbol{r}) + \frac{\delta F[\rho_2]}{\delta\rho_2(\boldsymbol{r})} \equiv \mu_2(\boldsymbol{r}). \qquad (10.2.28)$$

Clearly, the density ρ_1 does not satisfy the minimum principle,

$$\delta\{\Delta E_{v_2}[\rho'|\rho_1] - \tau_2\Delta S[\rho'|\rho_1]\} = 0, \qquad (10.2.29)$$

since the vanishing functional derivative in the preceding equation, for $\rho' = \rho_1$ would then require

$$\mu_2[\rho_1;\boldsymbol{r}] \equiv \left.\frac{\delta\Delta E_{v_2}[\rho'|\rho_1]}{\delta\rho'(\boldsymbol{r})}\right|_{\rho'=\rho_1} = \frac{\delta E_{v_2}[\rho_1]}{\delta\rho_1(\boldsymbol{r})} = v_2(\boldsymbol{r}) + \frac{\delta F[\rho_1]}{\delta\rho_1(\boldsymbol{r})} \equiv \tau_2, \qquad (10.2.30)$$

which contradicts the Euler equations (10.2.28). Indeed, the quantity $\mu_2[\rho_1; \boldsymbol{r}]$ in the preceding equation marks the *non*-equilibrium (*non*-equalized) chemical potential for the external potential v_2, corresponding to the trial density $\rho_1 \neq \rho[v_2]$.

Such a non-equilibrium (with respect to v_2) distribution of electrons calls for the *local* entropy constraint,

$$\Delta s[\rho'(\boldsymbol{r})|\rho_1(\boldsymbol{r})] \equiv \Delta s_1(\boldsymbol{r}) = \rho'(\boldsymbol{r}) \ln[\rho'(\boldsymbol{r})/\rho_1(\boldsymbol{r})] = 0, \qquad (10.2.31)$$

in the modified "thermodynamic" principle

$$\delta\{ \Delta E_{v_2}[\rho'|\rho_1] - \int \tau_2(\boldsymbol{r})\, \Delta s_1(\boldsymbol{r})\, d\boldsymbol{r}\} = 0. \qquad (10.2.32)$$

It involves the *non*-equalized information-temperature, $\tau_2(\boldsymbol{r})$, of the molecular system identified by v_2:

$$\tau_2(\boldsymbol{r}) = \left.\frac{\delta \Delta E_{v_2}[\rho'|\rho_1]}{\delta \Delta s_1(\boldsymbol{r})}\right|_{\rho'=\rho_1} = \mu_2[\rho_1; \boldsymbol{r}] = v_2(\boldsymbol{r}) - v_1(\boldsymbol{r}) + \mu_1 \equiv \mu_1 - \Delta v(\boldsymbol{r})$$

$$= \mu_2 - \frac{\delta F[\rho_2]}{\delta \rho_2(\boldsymbol{r})} + \frac{\delta F[\rho_1]}{\delta \rho_1(\boldsymbol{r})} \equiv \mu_2 - \Delta \frac{\delta F[\rho]}{\delta \rho(\boldsymbol{r})}. \qquad (10.2.33)$$

To summarize, the non-equilibrium densities for a given external potential are characterized by the non-equalized information-temperature (chemical-potential) descriptors associated with the "thermodynamic" principle in the electronic energy representation, which couples the electronic energy with the local entropy-deficiency constraint.

10.2.4. Variational Principles in Entropy-Deficiency Representation

In this section we shall briefly summarize the equivalent principles of the minimum entropy-deficiency, which involve the relevant energy constraints. In the equilibrium (ground-state) case one searches for the minimum of the entropy deficiency $\Delta S[\rho'|\rho]$ subject to the constraint of the vanishing energy displacement $\Delta E_v[\rho'|\rho] = 0$:

$$\delta\, \{\Delta S[\rho'\, |\rho] - \omega\, \Delta E_v[\rho'|\rho]\} = 0. \qquad (10.2.34)$$

The resulting Euler equation,

$$\ln[\rho'(\boldsymbol{r})/\rho(\boldsymbol{r})]_{\rho'=\rho} + 1 = 1 = \omega\mu, \qquad (10.2.35)$$

then identifies the Lagrange multiplier ω as the inverse global information temperature:

$$\omega = \left(\frac{\partial \Delta S}{\partial \Delta E_v}\right)_{\rho'=\rho} = \left(\frac{\partial \Delta S}{\partial E_v}\right)_{\rho'=\rho} = 1/\mu = 1/\tau, \qquad (10.2.36)$$

in close analogy to the ordinary thermodynamics.

Consider next the electron distribution of the preceding section $\rho_1 = \rho[v_1]$, which represents non-equilibrium density for the external potential v_2. For the entropic variational principle to give ρ_1, the local energy constraint is required for fixing the deviation of the local value of the energy density:

$$\Delta\varepsilon_{v_2}(\boldsymbol{r}) \equiv \varepsilon_{v_2}[\rho';\boldsymbol{r}] - \varepsilon_{v_2}[\rho_1;\boldsymbol{r}] = 0, \qquad (10.2.37)$$

where:

$$E_{v_2}[\rho'] \equiv \int \varepsilon_{v_2}[\rho';\boldsymbol{r}]\, d\boldsymbol{r} \quad \text{and} \quad E_{v_2}[\rho_1] \equiv \int \varepsilon_{v_2}[\rho_1;\boldsymbol{r}]\, d\boldsymbol{r}. \qquad (10.2.38)$$

It should be observed that in terms of these energy densities the non-equalized chemical potential of Eqs. (10.2.30) and (10.2.33) is then given by the local functional derivative

$$\mu_2[\rho_1;\boldsymbol{r}] = \frac{\delta E_{v_2}[\rho_1]}{\delta \rho_1(\boldsymbol{r})} = \frac{\delta \varepsilon_{v_2}[\rho_1;\boldsymbol{r}]}{\delta \rho_1(\boldsymbol{r})}. \qquad (10.2.39)$$

The minimum entropy-deficiency principle including the local energy constraint (10.2.37) then reads:

$$\delta\,\{\Delta S[\rho'|\rho_1] - \int \omega_2(\boldsymbol{r})\, \Delta\varepsilon_{v_2}(\boldsymbol{r})\, d\boldsymbol{r}\} = 0, \qquad (10.2.40)$$

where the Lagrange-multiplier function

$$\omega_2(\boldsymbol{r}) = \left.\frac{\delta \Delta S[\rho'|\rho_1]}{\delta \Delta\varepsilon_{v_2}(\boldsymbol{r})}\right|_{\rho'=\rho_1}. \qquad (10.2.41)$$

It then follows from the associated Euler equation that

$$\left.\ln\frac{\rho'(\boldsymbol{r})}{\rho_1(\boldsymbol{r})}\right|_{\rho'=\rho_1} + 1 = 1 = \omega_2(\boldsymbol{r})\, \mu_2[\rho_1;\boldsymbol{r}] \quad \text{or} \quad \omega_2(\boldsymbol{r}) = 1/\mu_2[\rho_1;\boldsymbol{r}] = 1/\tau_2(\boldsymbol{r}). \qquad (10.2.42)$$

again in a close analogy to the phenomenological thermodynamics.

Therefore, in the entropy-deficiency principle of (10.2.40) the inverse of the local information-temperature (local chemical potential), provides the Lagrange multiplier for the constraint of the fixed energy density for the non-equilibrium electron density. It equalizes at the inverse of the global chemical potential, when the optimum, ground-state density for a given external potential is reached. This equalized inverse information temperature, valid for the system as a whole, is responsible for enforcing the global ground-state energy value in the entropy-deficiency principle of Eq. (10.2.34). As in ordinary thermodynamics these entropic variational principles are exactly equivalent to the complementary energetical principles discussed in Sections 10.2.1 and 10.2.3.

10.3. VERTICAL PROCESSES IN MOLECULAR FRAGMENTS

The molecular system can be formally considered as being externally-closed or open. The former case corresponds to the fixed (integer) overall number of electrons in the system as a whole, $N = \int \rho(\boldsymbol{r})\, d\boldsymbol{r}$, e.g., due to the fixed electron density ρ in the *vertical* processes involving molecular fragments. The latter case represents a system in contact with the macroscopic reservoir of electrons, thus exhibiting different values of the instantaneous overall numbers of electrons, $N^i = \int \rho^i(\boldsymbol{r}) d\boldsymbol{r}$, due to the fluctuations in instantaneous densities $\{\rho^i(\boldsymbol{r})\}$, at the constant value of the electronic chemical potential, equal to that of the external reservoir. Clearly, any selection of molecular fragments can be similarly externally opened with respect to their separate reservoirs (see Appendix D), thus corresponding to the fixed, reservoir levels of the electron chemical potentials, while the remaining density components are being kept externally closed, preserving their overall (integer) numbers of electrons.

In the open molecular system in the grand-canonical ensemble, characterized by its *average* (in general fractional) number of electrons, the *instantaneous* density $\rho^i(\boldsymbol{r})$ at a given location in space fluctuates due to the electron transfers to/from the reservoir. Similarly, the instantaneous densities of molecular subsystems in vertical processes, $\boldsymbol{\rho}^i(\boldsymbol{r}) = \{\rho_k^{\,i}(\boldsymbol{r})\}$, which give rise to the fixed molecular electron density $\rho(\boldsymbol{r}) = \sum_k \rho_k^{\,i}(\boldsymbol{r})$, and the corresponding instantaneous electron populations on AIM, $\boldsymbol{N}^i = \int \boldsymbol{\rho}^i(\boldsymbol{r})\, d\boldsymbol{r} = \{N_k^{\,i}\}$, also fluctuate while preserving the overall number of electrons $N^i = \sum_k N_k^{\,i} = N$:

$$\delta\rho^i(\boldsymbol{r}) = \sum_k \delta\rho_k^{\,i}(\boldsymbol{r}) = 0, \qquad \delta N_k^{\,i} = \int \delta\rho^i(\boldsymbol{r})\, d\boldsymbol{r} = 0. \tag{10.3.1}$$

Alternatively, one can treat a small portion of a molecule as a *local* "subsystem", with the molecular remainder then representing the complementary *microscopic* "reservoir", which is infinitely large relative to the local subsystem itself. The latter, chosen mentally as being of constant (infinitesimal) volume, undergoes intrinsic fluctuations in the electron density and exhibits the associated fluctuations in the energy density.

The IT-optimum, Hirshfeld AIM were shown to exhibit several attractive features discussed in detail in Chapters 5 and 6, which make these pieces of the molecular electron density unique and useful concepts for a "thermodynamic" interpretation of the vertical equilibria between the mutually-open atomic fragments of the molecule, when the overall density of the system as a whole is "frozen". In particular, only for the Hirshfeld partitioning the local values of several measures of the entropy-deficiency density of bonded atoms, relative to the free-atom reference, e.g., subsystem surprisals, equalize at the corresponding global values reflecting the molecular missing-information relative to the promolecule. As we have also argued before, this inter-subsystem equalization of the "intensive" information-distance parameters of molecular fragments can be regarded as a local criterion of the vertical equilibrium, thus providing a "thermodynamic" complement of the energetic principle of the chemical-potential (electronegativity) equalization. While the latter fails to identify the equilibrium partitioning of the molecular density, the alternative entropy-deficiency criteria uniquely select the Hirshfeld division as the equilibrium partition. More specifically, one reaches these equilibrium pieces of the molecular density, when the local intensive information distances of all the mutually-open subsystems equalize at the corresponding local value of the whole system. Thus, in IT the "stockholder" subsystems represent the *equilibrium* bonded fragments of a molecule, which resemble the corresponding free fragments to the maximum degree possible, by exhibiting the lowest average missing-information relative to the non-bonded (free) fragments defining the promolecule.

Following the ordinary thermodynamics (e.g., Callen, 1962) one can summarize such a vertical thermodynamic-like, phenomenological approach to molecular subsystems (Nalewajski and Parr, 2001; Nalewajski, 2002c, 2003b, 2004a) in terms of the following three basic postulates:

Postulate I. Among all possible partitions of the molecular ground-state density ρ into densities $\boldsymbol{\rho} = \{\rho_\alpha\}$ of molecular fragments (bonded-atoms), for the fixed "promolecule" (free-atom) reference, there exists the *equilibrium* division into the "stockholder" fragments that is characterized completely by ρ and by the reference densities alone.

Postulate II. There exists a functional $\Delta S[\boldsymbol{\rho}|\boldsymbol{\rho}^0]$, called the *entropy deficiency* (missing information) of the "extensive" subsystem parameters $\{\rho_\alpha\}$ of a composite (molecular) system, defined for all equilibrium divisions of ρ and having the following property: the values assumed by $\{\rho_\alpha\}$ in the absence of the internal constraints are those that minimize $\Delta S[\boldsymbol{\rho}|\boldsymbol{\rho}^0]$ over the manifold of the constrained equilibrium states, defined by the subsystem *effective* external potentials $\{v_\alpha^{eff}(\boldsymbol{r}) = v_\alpha^{eff}[\rho; \boldsymbol{r}]\}$ [Eq. (1.6.7)].

Postulate III. The entropy-deficiency of a composite system and its density are *additive* over the constituent components.

It has been recently demonstrated (Nalewajski, 2002c, 2003b, 2004a) that IT also provides tools for the local phenomenological description of the overall density conserving (vertical) fluctuations and flows of electrons involving the molecular fragments. It too closely follows the thermodynamic theory of irreversible processes (see, e.g., Callen, 1962). In what follows we shall give an overview of such an approach. A short summary of the standard thermodynamic description of such phenomena (Callen, 1962) is provided in Appendix G.

10.3.1. Equilibrium Distributions of Molecular Subsystems: "Thermodynamic" Interpretation

State Parameters

We shall now examine fluctuations in the instantaneous densities $\boldsymbol{\rho}^i(\boldsymbol{r})$ and the associated electron populations $\boldsymbol{N}^i = \int \boldsymbol{\rho}^i(\boldsymbol{r})d\boldsymbol{r} = \{N_k^i\}$ of molecular fragments, e.g., AIM, reactants, functional groups, etc., relative to the Hirshfeld (equilibrium) densities of Section 5.2, $\boldsymbol{\rho}^H(\boldsymbol{r}) = \{\rho_k^H(\boldsymbol{r})\}$, and their electron populations, $\boldsymbol{N}^H = \int \boldsymbol{\rho}^H(\boldsymbol{r})d\boldsymbol{r} = \{N_k^H\}$, respectively. As we have already shown in Section 5.3, the *"stockholder"* pieces $\boldsymbol{\rho}^H$ of a given molecular density ρ represent the entropy-deficiency *equilibrium* subsystems, which satisfy the relevant minimum principle of the system missing-information in subsystem resolution {Eq. (5.3.5)}. These Hirshfeld density contributions are defined by Eq. (5.2.1),

$$\boldsymbol{\rho}^H(\boldsymbol{r}) = \boldsymbol{\rho}^0(\boldsymbol{r})\,[\rho(\boldsymbol{r})/\rho^0(\boldsymbol{r})] \equiv \boldsymbol{\rho}^0(\boldsymbol{r})\,w(\boldsymbol{r}) = \rho(\boldsymbol{r})\,[\boldsymbol{\rho}^0(\boldsymbol{r})/\rho^0(\boldsymbol{r})] \equiv \rho(\boldsymbol{r})\,\boldsymbol{d}^H(\boldsymbol{r}),$$
$$\rho(\boldsymbol{r}) = \textstyle\sum_k \rho_k^H(\boldsymbol{r}), \qquad \rho^0(\boldsymbol{r}) = \textstyle\sum_k \rho_k^0(\boldsymbol{r}), \qquad (10.3.2)$$

either through the unbiased (molecular) *enhancement* $w(\boldsymbol{r})$ of the free-subsystem densities $\boldsymbol{\rho}^0(\boldsymbol{r}) = \{\rho_k^0(\boldsymbol{r})\}$, which is common to all constituent fragments, or as the promolecular *shares* (conditional probabilities) $\boldsymbol{d}^H(\boldsymbol{r}) = \{d_k^H(\boldsymbol{r}) = d_k^0(\boldsymbol{r})\}$ in the molecular density $\rho(\boldsymbol{r})$. Here $\rho^0(\mathbf{r})$ is the electron density of the "promolecule", consisting of the free subsystem densities shifted to their actual positions in the molecule, which also serves as the reference in the density difference function extracting effects due to bond formation: $\Delta\rho(\boldsymbol{r}) = \rho(\boldsymbol{r}) - \rho^0(\boldsymbol{r})$.

It has already been demonstrated in Section 5.3 that the stockholder densities of molecular subsystems have a solid basis in the Information Theory. They were shown to minimize the *instantaneous* (i) entropy deficiency (missing information) functional of Kullback and Leibler [Eq. (5.3.5)] exhibited by the trial densities $\boldsymbol{\rho}^i(\boldsymbol{r})$ relative to the fixed reference densities $\boldsymbol{\rho}^0$ of free (separate) fragments,

$$\Delta S^i[\boldsymbol{\rho}^i\,|\,\boldsymbol{\rho}^0] = \textstyle\sum_k\int \rho_k^i(\boldsymbol{r})\,\ln[\rho_k^i(\boldsymbol{r})/\rho_k^0(\boldsymbol{r})]\,d\boldsymbol{r}$$
$$\equiv \textstyle\sum_k\int \Delta s_k^i(\boldsymbol{r})\,d\boldsymbol{r} \equiv \int \Delta s^i(\boldsymbol{r})\,d\boldsymbol{r} \equiv \sum_k \Delta S_k^i[\boldsymbol{\rho}^i\,|\,\boldsymbol{\rho}^0], \qquad (10.3.3)$$

subject to the local constraint of the exhaustive partitioning of the given molecular electron density, $\rho(\boldsymbol{r}) = \sum_k \rho_k^i(\boldsymbol{r})$:

$$\min_i\{\Delta S^i[\boldsymbol{\rho}^i|\boldsymbol{\rho}^0] - \int F(\boldsymbol{r})\,[\textstyle\sum_k \rho_k^i(\boldsymbol{r}) - \rho(\boldsymbol{r})]\,d\boldsymbol{r}\}$$
$$= \Delta S[\boldsymbol{\rho}^H|\boldsymbol{\rho}^0] = \textstyle\sum_k \int \rho_k^H(\boldsymbol{r}) \ln[\rho_k^H(\boldsymbol{r})/\rho_k^0(\boldsymbol{r})]\,d\boldsymbol{r} \equiv \sum_k \int \Delta s_k^H(\boldsymbol{r})\,d\boldsymbol{r}$$
$$= \int \rho(\boldsymbol{r}) \ln w(\boldsymbol{r})\,d\boldsymbol{r} = \Delta S[\rho|\rho^0] \equiv \int \Delta s(\boldsymbol{r})\,d\boldsymbol{r}. \qquad (10.3.4)$$

Here the Lagrange multiplier function $F(\boldsymbol{r})$ represents the inter-subsystem equalized local "intensity" (entropic "force"),

$$F(\boldsymbol{r}) = \frac{\delta\Delta S[\rho|\rho^0]}{\delta\rho(\boldsymbol{r})} = 1 + \ln w(\boldsymbol{r}) = \frac{\partial\Delta s(\boldsymbol{r})}{\partial\rho(\boldsymbol{r})} = \frac{\partial\Delta s_k^H(\boldsymbol{r})}{\partial\rho_k^H(\boldsymbol{r})} \equiv F_k^H(\boldsymbol{r}),\ \ k = 1, 2, \ldots, m. \qquad (10.3.5)$$

We have explicitly indicated in the preceding equation that $F(\boldsymbol{r})$ is the entropy-deficiency *"intensive"* conjugate of the *"extensive"* density "variables" of both the molecule as a whole and its constituent fragments:

$$\boldsymbol{F}^H(\boldsymbol{r}) = \frac{\partial\Delta S[\boldsymbol{\rho}^H|\boldsymbol{\rho}^0]}{\partial\boldsymbol{\rho}^H(\boldsymbol{r})} = \frac{\partial\Delta s(\boldsymbol{r})}{\partial\boldsymbol{\rho}^H(\boldsymbol{r})} = F(\boldsymbol{r})\boldsymbol{1}, \qquad (10.3.6)$$

where the unit row-vector $\boldsymbol{1} = (1, 1, \ldots)$. In the last equation the functional derivative with respect to the subsystem density is the partial one, calculated for the fixed densities of the remaining subsystems.

It also follows from Eq. (10.3.5) that the shape of these entropy-deficiency intensities reflects the *surprisal* function of the overall distribution of electrons in a molecule, relative to that in the promolecular reference:

$$I[\rho(\boldsymbol{r})\,|\,\rho^0(\boldsymbol{r})] = \ln w(\boldsymbol{r}) \equiv I(\boldsymbol{r}). \qquad (10.3.7a)$$

The subsystem surprisals are equalized in the Hirshfeld partitioning at the corresponding value of the above global surprisal function (see Section 6.1):

$$I[\rho_k^H(\boldsymbol{r})\,|\,\rho_k^0(\boldsymbol{r})] = \ln\frac{\rho_k^H(\boldsymbol{r})}{\rho_k^0(\boldsymbol{r})} \equiv I_k^H(\boldsymbol{r}) = I(\boldsymbol{r}), \qquad k = 1, 2, \ldots, m. \qquad (10.3.7b)$$

Following the standard thermodynamic approach described in Appendix G, one similarly introduces the *instantaneous* entropy-deficiency intensities, conjugates of the instantaneous subsystem densities $\boldsymbol{\rho}^i$, defined as the corresponding partial functional derivatives of the instantaneous entropy deficiency of Eq. (10.3.3):

$$\boldsymbol{F}^i(\boldsymbol{r}) = \frac{\partial\Delta S^i[\boldsymbol{\rho}^i|\boldsymbol{\rho}^0]}{\partial\boldsymbol{\rho}^i(\boldsymbol{r})} = \frac{\partial\Delta s^i(\boldsymbol{r})}{\partial\boldsymbol{\rho}^i(\boldsymbol{r})}$$
$$= \{F_k^i(\boldsymbol{r}) = \frac{\partial\Delta s_k^i(\boldsymbol{r})}{\partial\rho_k^i(\boldsymbol{r})} = 1 + \ln\frac{\rho_k^i(\boldsymbol{r})}{\rho_k^0(\boldsymbol{r})} = 1 + I[\rho_k^i(\boldsymbol{r})\,|\,\rho_k^0(\boldsymbol{r})] \equiv 1 + I_k^i(\boldsymbol{r})\}. \qquad (10.3.8)$$

They exhibit the inter-subsystem *non-equalized* deviations relative to the Hirshfeld (equilibrium) values of Eq. (10.3.6):

$$\delta \boldsymbol{F}^i(\boldsymbol{r}) = \boldsymbol{F}^i(\boldsymbol{r}) - \boldsymbol{F}^H(\boldsymbol{r}) = \{\delta F_k^{\,i}(\boldsymbol{r}) = \ln\frac{\rho_k^i(\boldsymbol{r})}{\rho_k^H(\boldsymbol{r})} = I_k^{\,i}(\boldsymbol{r}) - I_k^{\,H}(\boldsymbol{r}) \equiv \delta I_k^{\,i}(\boldsymbol{r}) \neq 0\}. \qquad (10.3.9)$$

In what follows we shall use these displacements to define the instantaneous "thermodynamic" *forces* triggering the vertical flows of electrons involving molecular subsystems.

Expressing the instantaneous density in terms of the equilibrium distribution, $\rho_k^{\,i}(\boldsymbol{r}) = \rho_k^{\,H}(\boldsymbol{r}) + \delta\rho_k^{\,i}(\boldsymbol{r})$, gives the following first-order expansion of the force $\delta F_k^{\,i}(\boldsymbol{r})$:

$$\delta F_k^{\,i}(\boldsymbol{r}) = \ln\frac{\rho_k^i(\boldsymbol{r})}{\rho_k^H(\boldsymbol{r})} = \ln[1+\frac{\delta\rho_k^i(\boldsymbol{r})}{\rho_k^H(\boldsymbol{r})}] \cong \frac{\delta\rho_k^i(\boldsymbol{r})}{\rho_k^H(\boldsymbol{r})} \equiv \delta^{(1)}F_k^i(\boldsymbol{r}), \qquad (10.3.10a)$$

linear in the component relative density displacement $|\delta\rho_k^{\,i}(\boldsymbol{r})/\rho_k^{\,H}(\boldsymbol{r})| \cong 0$. In this linear-response approximation the shift in the component density is thus proportional to the component thermodynamic force:

$$\delta\rho_k^{\,i}(\boldsymbol{r}) \cong \rho_k^{\,H}(\boldsymbol{r})\delta^{(1)}F_k^{\,i}(\boldsymbol{r}) \equiv \delta^{(1)}\rho_k^{\,i}(\boldsymbol{r}) \approx \rho_k^{\,H}(\boldsymbol{r})\,\delta F_k^{\,i}(\boldsymbol{r}). \qquad (10.3.10b)$$

The instantaneous displacement of the information-distance corresponding to a given ρ-conserving fluctuation $\delta\boldsymbol{\rho}^i$ in the subsystem densities, when $\sum_k \delta\rho_k^{\,i}(\boldsymbol{r}) = \delta\rho^i(\boldsymbol{r}) = 0$, then reads:

$$\begin{aligned}\delta\Delta S^i[\delta\boldsymbol{\rho}^i] &\equiv \Delta S^i[\boldsymbol{\rho}^i\,|\,\boldsymbol{\rho}^0] - \Delta S[\boldsymbol{\rho}^H\,|\,\boldsymbol{\rho}^0]\\ &= \textstyle\sum_k \int \rho_k^{\,i}(\boldsymbol{r})\, I_k^{\,i}(\boldsymbol{r})\, d\boldsymbol{r} - \sum_k \int [\rho_k^{\,i}(\boldsymbol{r}) - \delta\rho_k^{\,i}(\boldsymbol{r})] I_k^{\,H}(\boldsymbol{r})\, d\boldsymbol{r}\\ &= \textstyle\int [\sum_k \delta\rho_k^{\,i}(\boldsymbol{r})]\, I(\boldsymbol{r})\, d\boldsymbol{r} + \sum_k \int \rho_k^{\,i}(\boldsymbol{r})\, [I_k^{\,i}(\boldsymbol{r}) - I_k^{\,H}(\boldsymbol{r})]\, d\boldsymbol{r}\\ &= \textstyle\int \delta\rho(\boldsymbol{r})\, I(\boldsymbol{r})\, d\boldsymbol{r} + \sum_k \int \rho_k^{\,i}(\boldsymbol{r})\, \delta F_k^{\,i}(\boldsymbol{r})\, d\boldsymbol{r} = \sum_k \int \rho_k^{\,i}(\boldsymbol{r})\, \delta F_k^{\,i}(\boldsymbol{r})\, d\boldsymbol{r}\\ &\equiv \textstyle\sum_k \int \delta\Delta s_k^{\,i}(\boldsymbol{r})\, d\boldsymbol{r} \equiv \int \delta\Delta s^i(\boldsymbol{r})\, d\boldsymbol{r} \equiv \sum_k \delta\Delta S_k^{\,i}[\delta\rho_k^{\,i}].\end{aligned} \qquad (10.3.11)$$

It thus follows from the preceding equation that the subsystem density and its force represent the (partial) $\delta\Delta S^i$-conjugates of each other:

$$\rho_k^{\,i}(\boldsymbol{r}) = \left(\frac{\partial\delta\Delta S^i}{\partial \boldsymbol{F}^i(\boldsymbol{r})}\right)_{\boldsymbol{\rho}^i} = \left(\frac{\partial\delta\Delta s^i(\boldsymbol{r})}{\partial \boldsymbol{F}^i(\boldsymbol{r})}\right)_{\boldsymbol{\rho}^i} = \left\{\left(\frac{\partial\delta\Delta s_k^i(\boldsymbol{r})}{\partial F_k^i(\boldsymbol{r})}\right)_{\rho_k^i}\right\}, \qquad (10.3.12)$$

$$\delta F_k^{\,i}(\boldsymbol{r}) = \left(\frac{\partial\delta\Delta S^i}{\partial \boldsymbol{\rho}^i(\boldsymbol{r})}\right)_{\delta\boldsymbol{F}^i} = \left(\frac{\partial\delta\Delta s^i(\boldsymbol{r})}{\partial \boldsymbol{\rho}^i(\boldsymbol{r})}\right)_{\delta\boldsymbol{F}^i} = \left\{\left(\frac{\partial\delta\Delta s_k^i(\boldsymbol{r})}{\partial \rho_k^i(\boldsymbol{r})}\right)_{\delta F_k^i}\right\}. \qquad (10.3.13)$$

Using Eq. (10.3.10b) to express $\delta\Delta S^i$ of Eq. (10.3.11) in terms of the density shifts alone gives the quadratic function

$$\delta^{(1)}\Delta S^i[\delta\boldsymbol{\rho}^i] = \Sigma_k \int \rho_k^i(\boldsymbol{r})\, \delta^{(1)}F_k^i(\boldsymbol{r})\, d\boldsymbol{r} = \Sigma_k \int [\rho_k^H(\boldsymbol{r}) + \delta\rho_k^i(\boldsymbol{r})] \frac{\delta\rho_k^i(\boldsymbol{r})}{\rho_k^H(\boldsymbol{r})}\, d\boldsymbol{r}$$

$$= \int \delta\rho^i(\boldsymbol{r})\, d\boldsymbol{r} + \Sigma_k \int \frac{[\delta\rho_k^i(\boldsymbol{r})]^2}{\rho_k^H(\boldsymbol{r})}\, d\boldsymbol{r} = \Sigma_k \int \frac{[\delta\rho_k^i(\boldsymbol{r})]^2}{\rho_k^H(\boldsymbol{r})}\, d\boldsymbol{r}. \qquad (10.3.14)$$

It also follows from Eq. (10.3.10b) that this displacement in the entropy deficiency can be alternatively written as the associated quadratic function of forces:

$$\delta^{(1)}\Delta S^i[\delta\boldsymbol{\rho}^i[\delta^{(1)}\boldsymbol{F}^i]] \equiv \delta^{(1)}\Delta S^i[\delta^{(1)}\boldsymbol{F}^i] = \Sigma_k \int \rho_k^H(\boldsymbol{r})[\delta^{(1)}F_k^i(\boldsymbol{r})]^2\, d\boldsymbol{r}. \qquad (10.3.15)$$

These expressions establish the entropy-conjugation relations for the vertical processes in the linear response approximation:

$$\delta\rho_k^i(\boldsymbol{r}) \cong \delta^{(1)}\rho_k^i(\boldsymbol{r}) = \frac{1}{2}\left(\frac{\partial\delta^{(1)}\Delta S^i}{\partial\delta F_k^i(\boldsymbol{r})}\right)_\rho, \quad \delta F_k^i(\boldsymbol{r}) \cong \delta^{(1)}F_k^i(\boldsymbol{r}) = \frac{1}{2}\left(\frac{\partial\delta^{(1)}\Delta S^i}{\partial\rho_k^i(\boldsymbol{r})}\right)_\rho. \qquad (10.3.16)$$

Thermodynamic Principles

It follows from the Eqs. (10.3.2) and (10.3.4) that the equilibrium densities of subsystems $\boldsymbol{\rho}^H$ are uniquely determined by the system overall density and the atomic/promolecule references. The ground-state density ρ of a given molecular system can be in principle determined from the Hohenberg-Kohn variational principle in the energy representation [Eq. (10.2.2)], which we have interpreted in Section 10.2.1 as the thermodynamic-like rule [Eq. (10.2.10)], with the system global chemical potential μ measuring its information "temperature", the Lagrange multiplier enforcing the global entropy-deficiency constraint: $\mu = \tau$. A similar "thermodynamic" interpretation can be attributed to the equilibrium partitioning $\boldsymbol{\rho}^H$, of the molecular electron density ρ between the mutually closed, but externally open molecular subsystems (see Appendix D). We also recall that this general, constrained-equilibrium case also covers the global equilibrium, with all components (Hirshfeld densities) coupled to the same electron reservoir, when the subsystem chemical potentials are equalized at the same global, molecular level: $\mu_k = \mu$, $k = 1, 2, \ldots, m$. Throughout this section we drop the upper subscript i in the *trial* (variational) densities $\boldsymbol{\rho}$ of subsystems, which will be reserved for the *instantaneous* densities during spontaneous fluctuations in the electronic structure of molecular fragments.

Consider the variational principle in the energy representation [Eq. (D.1.8)] for determining the equilibrium densities of the mutually-closed, externally-open subsystems. By assumption they are coupled to *separate* electron reservoirs, which in general exhibit different levels of the electronic chemical potential. It should be observed that the global-equilibrium (Hirshfeld) subsystems correspond to the composite system, in which all molecular fragments are coupled to the same (molecular) electron reservoir, thus equalizing their chemical potentials:

$$\{\mu_k[\boldsymbol{\rho}^H[\rho]] \equiv \mu_k = \mu[\rho] \equiv \mu\}. \tag{10.3.17}$$

The subsystem chemical potentials, equal to those of their reservoirs, enforce the associated electron populations of molecular fragments, $\{N_k^H[\rho_k^H[\rho]] \equiv N_k^H\}$, in the Euler-Lagrange principle:

$$\delta\{E_v[\boldsymbol{\rho}] - \sum_k \mu_k(N_k[\rho_k] - N_k^H)\} = \delta E_v[\boldsymbol{\rho}] - \mu\, \delta N[\rho] = 0, \tag{10.3.18}$$

where the energy density functional $E_v[\boldsymbol{\rho}] = E_v[\rho]$. It follows from the foregoing equation that in the vertical density partitioning problems, for which $\delta\rho = \delta N = 0$, this variational rule is automatically satisfied by any ρ-conserving partition, since the ground-state density minimizes the energy. Indeed, as we have already emphasized in Section 6.2, the equilibrium criterion (10.3.17) of the energy representation is not sufficient to identify the equilibrium densities of subsystems. One has to supplement the variational principle (10.3.18) with additional cross-entropy (penalty) terms, in order to distinguish the stockholder components from a multitude of other admissible pieces of the molecular density. For example, adding the constraint of the vanishing entropy-deficiency displacement,

$$\delta S[\delta\boldsymbol{\rho}] \equiv \Delta S[\boldsymbol{\rho} \,|\, \boldsymbol{\rho}^0] - \Delta S[\boldsymbol{\rho}^H \,|\, \boldsymbol{\rho}^0], \tag{10.3.19}$$

where $\delta\boldsymbol{\rho} = \boldsymbol{\rho} - \boldsymbol{\rho}^H$, in the modified variational rule

$$\delta E_v[\boldsymbol{\rho}] - \mu\, \delta N[\rho] - \xi\, \delta S[\delta\boldsymbol{\rho}] = 0, \tag{10.3.20a}$$

gives the Euler equation

$$\left.\frac{\partial E_v[\boldsymbol{\rho}]}{\partial \rho_k(\boldsymbol{r})}\right|_{\boldsymbol{\rho}^H[\rho]} - \mu - \xi \left.\frac{\partial \delta S[\delta\boldsymbol{\rho}]}{\partial \rho_k(\boldsymbol{r})}\right|_{\boldsymbol{\rho}^H[\rho]} = \mu_k - \mu - \xi\, \delta F_k^H(\boldsymbol{r}) = -\,\xi\, \delta F_k^H(\boldsymbol{r}) = 0. \tag{10.3.20b}$$

This equation is satisfied only by the Hirshfeld components of the molecular density, for which the entropic forces $\{\delta F_k^H(\boldsymbol{r}) = 0\}$.

Alternatively, the thermodynamic-like principles (see Section 10.2) can be used to uniquely identify the equilibrium electron distributions of molecular fragments. In the energy representation they involve the minimization of the system energy coupled to the relevant missing-information constraints enforced by the corresponding Lagrange multipliers measuring the associated information temperatures $\{\tau_k\}$ or τ. For example, the equilibrium subsystem densities can be assured either by the vanishing entropy deficiencies of each molecular fragment:

$$\Delta S_k[\rho_k \,|\, \rho_k^H] = 0, \qquad k = 1, 2, ..., m, \tag{10.3.21a}$$

or by the global cross-entropy condition:

$$\Delta S[\boldsymbol{\rho} \,|\, \boldsymbol{\rho}^0] = \Sigma_k \, \Delta S_k[\rho_k \,|\, \rho_k^H] = 0. \tag{10.3.21b}$$

The corresponding IT-"thermodynamic" variational principles of the energy representation in the molecular subsystem resolution then read:

$$\delta\{E_v[\boldsymbol{\rho}] - \Sigma_k \, \tau_k \, \Delta S_k[\rho_k \,|\, \rho_k^H]\} = 0, \qquad \tau_k = \left.\frac{\partial E_v}{\partial \Delta S_k}\right|_{\boldsymbol{\rho}^H[\rho]}; \tag{10.3.22a}$$

$$\delta\{E_v[\boldsymbol{\rho}] - \tau \, \Delta S[\boldsymbol{\rho} \,|\, \boldsymbol{\rho}^0] = 0, \qquad \tau = \left.\frac{\partial E_v}{\partial \Delta S}\right|_{\boldsymbol{\rho}^H[\rho]}. \tag{10.3.22b}$$

The associated Euler equations identify the IT-temperatures as the corresponding chemical potentials:

$$\left.\frac{\partial E_v[\boldsymbol{\rho}]}{\partial \rho_k(\boldsymbol{r})}\right|_{\boldsymbol{\rho}^H[\rho]} = \mu_k = \mu = \tau_k \left.\frac{\partial \Delta S_k[\rho_k | \rho_k^H]}{\partial \rho_k(\boldsymbol{r})}\right|_{\boldsymbol{\rho}^H[\rho]} = \tau_k, \tag{10.3.23a}$$

$$\left.\frac{\partial E_v[\boldsymbol{\rho}]}{\partial \rho_k(\boldsymbol{r})}\right|_{\boldsymbol{\rho}^H[\rho]} = \mu_k = \mu = \tau \left.\frac{\partial \Delta S[\boldsymbol{\rho} | \boldsymbol{\rho}^H]}{\partial \rho_k(\boldsymbol{r})}\right|_{\boldsymbol{\rho}^H[\rho]} = \tau. \tag{10.3.23b}$$

Therefore, in the molecular (global) ground-state equilibrium the information temperatures of the stockholder components of the system electron density are measured by the fragment chemical potentials. They are all equalized throughout the whole space at the global chemical potential level.

In the non-equilibrium case, $\rho' \neq \rho[N, v]$ (see Section 10.2.3), the thermodynamic-like principles in the subsystem resolution, which recover the Hirshfeld densities $\boldsymbol{\rho}^H[\rho'; \boldsymbol{r}] = \{\rho_k^H(\rho'; \boldsymbol{r})\}$, must involve *local* IT-temperatures to enforce the constraints of the vanishing cross-entropy densities [see Eq. (10.2.32)],

$$\Delta s_k(\rho_k(\boldsymbol{r}) \,|\, \rho_k^H(\rho'; \boldsymbol{r})) \equiv \Delta s_k^H(\rho'; \boldsymbol{r}) = \rho_k(\boldsymbol{r}) \ln[\rho_k(\boldsymbol{r})/\rho_k^H(\rho'; \boldsymbol{r})] = 0, \tag{10.3.24}$$

$$\delta\{E_v[\boldsymbol{\rho}] - \Sigma_k \!\int\! \tau_k(\rho'; \boldsymbol{r}) \, \Delta s_k^H(\rho'; \boldsymbol{r}) \, d\boldsymbol{r}\} = 0, \qquad \tau_k(\rho'; \boldsymbol{r}) = \left.\frac{\partial E_v[\boldsymbol{\rho}]}{\partial \Delta s_k(\rho'; \boldsymbol{r})}\right|_{\boldsymbol{\rho}^H[\rho']}. \tag{10.3.25}$$

The associated Euler equation,

$$\left.\frac{\partial E_v[\rho]}{\partial \rho_k(\boldsymbol{r})}\right|_{\rho^H[\rho']} \equiv \mu_k(\rho'; \boldsymbol{r}) = \tau_k(\rho'; \boldsymbol{r}) \left.\frac{\partial \Delta s_k^H(\rho';\boldsymbol{r})}{\partial \rho_k(\boldsymbol{r})}\right|_{\rho^H[\rho']} = \tau_k(\rho'; \boldsymbol{r}), \qquad (10.3.26)$$

then identifies the local IT-temperature of the molecular subsystem as its local chemical potential $\mu_k(\rho'; \boldsymbol{r})$. Therefore, reaching the molecular vertical equilibrium can be interpreted as a process of finding the partition of the ground-state density for which the local information temperatures (chemical potentials) of subsystems equalize throughout the physical space at the global temperature (chemical potential) levels.

The equivalent (vertical) variational principles in the entropy-deficiency representation (see Section 10.2.4) involve the minimization of the entropy-deficiency functional in the subsystem resolution, subject to the appropriate energy constraints enforced by the inverses of the information temperatures (chemical potentials) acting as the Lagrange multipliers. However, in order to make the subsystem energy constraints unique, one has to invoke the multi-component version of the non-interacting KS system (Kohn and Sham, 1965; see also Appendix C), involving the non-interacting subsystems (Nalewajski, 2001), containing the non-interacting electrons with the same electron densities (chemical potentials) as the interacting subsystems in the real molecule. In this limit each electronic component experiences the same effective external potential $v_{KS}(\boldsymbol{r})$ as the overall molecular density [Eq. (C.17)], so that the electronic energy of a composite molecular system is exactly given by the sum of subsystem energies:

$$E_v^{KS}[\rho] = T_s[\rho] + \int v_{KS}(\boldsymbol{r})\,\rho(\boldsymbol{r})\,d\boldsymbol{r}$$
$$= \sum_k \{T_s[\rho_k] + \int v_{KS}(\boldsymbol{r})\,\rho_k(\boldsymbol{r})\,d\boldsymbol{r}\} \equiv \sum_k E_v^{KS}[\rho_k] \equiv E_v^{KS}[\rho], \qquad (10.3.27)$$

where $T_s[\rho] = \sum_k T_s[\rho_k]$ denotes the density functional for the non-interacting kinetic energy [Eq. (C.9)]. The relevant multi-component generalization (Capitani *et al.*, 1982) of the Hohenberg-Kohn (1964) theorems, ensures that the set of subsystem densities of the non-degenerate molecular state, carries the full information about the composite system, thus providing a unique mapping from the subsystem densities to the energies of both the hypothetical (non-interacting) KS system and the real (interacting) system of molecular fragments.

Thus the constraints of the vanishing displacements of the electronic energies of the non-interacting components relative to the corresponding Hirshfeld values,

$$E_v^{KS}[\rho_k] - E_v^{KS}[\rho_k^H[\rho]] = \Delta E_v^{KS}[\rho_k] \equiv \Delta E_k^{KS} = 0, \qquad k = 1, 2, \ldots, m, \qquad (10.3.28)$$

uniquely identify the Hirshfeld densities $\rho_k^H[\rho]$ derived from the molecular ground-state density ρ. Using these constraints in the minimum principle of $\Delta S_k[\rho_k | \rho_k^H]$ gives the thermodynamic principle for kth component in the entropy-deficiency representation:

$$\delta\{\Delta S_k[\rho_k\,|\,\rho_k^H] - \omega_k\,\Delta E_v^{KS}[\rho_k]\} = 0, \qquad \omega_k = \left.\frac{\partial \Delta S_k}{\partial E_k^{KS}}\right|_{\rho^H[\rho]} \equiv \tau_k^{KS}, \qquad (10.3.29)$$

where τ_k^{KS} denotes the information temperature of kth component of the non-interacting system. The corresponding Euler equation then identifies the Lagrange multiplier ω_k as the inverse of the subsystem chemical potential (IT-temperature) of the non-interacting system:

$$\left.\frac{\delta \Delta S_k[\rho_k|\rho_k^H]}{\delta \rho_k(\boldsymbol{r})}\right|_{\rho^H[\rho]} = 1 = \omega_k \left.\frac{\delta E_v[\boldsymbol{\rho}]}{\delta \rho_k(\boldsymbol{r})}\right|_{\rho^H[\rho]} = \mu_k^{KS} \quad \text{or} \quad \omega_k = \tau_k^{KS} = 1/\mu_k^{KS}. \qquad (10.3.30)$$

Finally, since by assumption the KS and real composite systems combine the same set of subsystem densities, they exhibit the same chemical potentials:

$$\mu_k^{KS}[\rho] = \mu_k[\rho] = \tau_k[\rho] = \mu = \tau. \qquad (10.3.31)$$

To summarize, the equivalent *vertical* thermodynamic principles in the entropy-deficiency representation, which determine the stockholder pieces of the molecular ground-state density, involve the inverses of the (equalized) subsystem chemical potentials (information temperatures) as Lagrange multipliers enforcing the energy constraints of Eq. (10.3.28). This is in close analogy to both the *horizontal* development of Section 10.2 and the ordinary thermodynamics (Callen, 1962).

10.3.2. Instantaneous Processes

Legendre Transforms of Entropy Deficiency

We have demonstrated that within the information-theoretic "thermodynamics" of molecular systems and their fragments (Nalewajski, 2002c, 2003b, 2004a) the principle of the minimum entropy-deficiency replaces the familiar maximum entropy principle of the ordinary thermodynamics. Thus, in the IT approach the maximum principle of the generalized Massieu function of Eq. (G.6) that the equilibrium values of the unconstrained parameters of a system in contact with reservoirs characterized by their intensive parameters $\boldsymbol{f}^r$ maximize $\tilde{S}[\boldsymbol{f}^r]$ at constant intensities $\boldsymbol{f}^i = \boldsymbol{f}^r$, is replaced by an appropriate extremum (minimum) principle of the Legendre transform of the missing-information functional [Eq. (10.3.4)]. Let us examine the auxiliary functional which is minimized in the information principle for the optimum (Hirshfeld) densities $\boldsymbol{\rho}^i = \boldsymbol{\rho}^H[\rho]$. It indeed represents the Legendre transform of the system entropy deficiency $\Delta S[\rho\,|\,\rho^0] = \Delta S[\boldsymbol{\rho}^H[\rho]\,|\,\boldsymbol{\rho}^0]$:

$$\Delta S[\rho\,|\,\rho^0] - \int F(\boldsymbol{r})\,\rho(\boldsymbol{r})\,d\boldsymbol{r} = \Delta S[\rho\,|\,\rho^0] - \int \frac{\delta \Delta S[\rho|\rho^0]}{\delta \rho(\boldsymbol{r})}\rho(\boldsymbol{r})\,d\boldsymbol{r} \equiv \Delta\tilde{S}[F|\rho^0]$$
$$= \Delta S[\boldsymbol{\rho}^H\,|\,\rho^0] - \int \boldsymbol{F}^H(\boldsymbol{r})\,\boldsymbol{\rho}^H(\boldsymbol{r})^{\mathrm{T}}\,d\boldsymbol{r}$$
$$= \Delta S[\boldsymbol{\rho}^H\,|\,\rho^0] - \int \frac{\delta \Delta S[\boldsymbol{\rho}^H|\rho^0]}{\delta \boldsymbol{\rho}^H(\boldsymbol{r})}\boldsymbol{\rho}^H(\boldsymbol{r})^{\mathrm{T}}\,d\boldsymbol{r} \equiv \Delta\tilde{S}[\boldsymbol{F}^H|\rho^0]$$
$$= -\int \Sigma_k \rho_k{}^H(\boldsymbol{r})\,d\boldsymbol{r} = -\int \rho(\boldsymbol{r})\,d\boldsymbol{r} = -N, \qquad (10.3.32)$$

where we have used Eq. (10.3.5). In the first line of the preceding equation the "extensive" variable ρ of the cross-entropy $\Delta S[\rho\,|\,\rho^0]$ has been replaced in the modified list of the system equilibrium state-parameters of the Legendre transform $\Delta\tilde{S}[F|\rho^0]$ by the density-conjugate "intensity" F. Accordingly, in the subsystem-resolved functional $\Delta\tilde{S}[\boldsymbol{F}^H|\rho^0]$ the equilibrium (stockholder) densities $\boldsymbol{\rho}^H$ of molecular fragments have been substituted by their entropy-deficiency conjugates $\boldsymbol{F}^H$ in the list of the system vertical state-variables. The instantaneous intensities of Eq. (10.3.8) similarly define the corresponding instantaneous Legendre transform of the entropy deficiency:

$$\Delta\tilde{S}^i[\boldsymbol{\rho}^i, \boldsymbol{F}^i\,|\,\rho^0] = \Delta S^i[\boldsymbol{\rho}^i\,|\,\rho^0] - \Sigma_k \int F_k^i(\boldsymbol{r})\,\rho_k^i(\boldsymbol{r})\,d\boldsymbol{r} \equiv \int \Delta\tilde{s}^i[\boldsymbol{\rho}^i(\boldsymbol{r}), \boldsymbol{F}^i(\boldsymbol{r}); \rho^0(\boldsymbol{r})]\,d\boldsymbol{r}$$
$$\equiv \Sigma_k\, \Delta\tilde{S}_k^i[\rho_k^i, F_k^i\,|\rho_k{}^0]. \qquad (10.3.33)$$

The equilibrium values of the "intensive" state-variables for the Hirshfeld subsystems, $\boldsymbol{F}^i(\boldsymbol{r}) = \boldsymbol{F}^H(\boldsymbol{r}) = F(\boldsymbol{r})\boldsymbol{1}$, can be thus interpreted as resulting from coupling the molecular fragments to a common Hirshfeld "reservoir", characterized by the entropic intensity distribution $F(\boldsymbol{r})$ related to the global molecular surprisal function $I(\boldsymbol{r})$. This observation stresses the global equilibrium character of the stockholder partition with respect to the hypothetical intra- and inter-subsystem flows of electrons. The subsystem analog of Eq. (G.6) now reads:

$$\Delta\tilde{S}[\boldsymbol{F}^H|\rho^0] = \Delta S[\boldsymbol{\rho}^H(\boldsymbol{F}^H)\,|\,\rho^0] - \Sigma_k \int F_k{}^H(\boldsymbol{r})\,\rho_k{}^H(\boldsymbol{r})\,d\boldsymbol{r} = \min_i \Delta\tilde{S}^i[\boldsymbol{\rho}^i, \boldsymbol{F}^H\,|\rho^0]. \quad (10.3.34)$$

The minimum entropy deficiency principle of Eq. (10.3.4) now becomes:

$$\min_i\{\Delta S^i[\boldsymbol{\rho}^i\,|\,\rho^0] - \Sigma_k \int F_k{}^H(\boldsymbol{r})\,\rho_k^i(\boldsymbol{r})\,d\boldsymbol{r}\} \equiv \min_i\, \Delta\tilde{S}^i[\boldsymbol{\rho}^i, \boldsymbol{F}^H\,|\rho^0]$$
$$\equiv \min_i \int \Delta\tilde{s}^i[\boldsymbol{\rho}^i(\boldsymbol{r}), \boldsymbol{F}^H(\boldsymbol{r}); \rho^0(\boldsymbol{r})]\,d\boldsymbol{r} = \Delta S[\boldsymbol{\rho}^H(\boldsymbol{F}^H)\,|\,\rho^0] - \Sigma_k \int F_k{}^H(\boldsymbol{r})\,\rho_k{}^H(\boldsymbol{r})\,d\boldsymbol{r}$$
$$= \Delta\tilde{S}[\boldsymbol{F}^H|\rho^0] = -N. \qquad (10.3.35)$$

In other words, the equilibrium subsystem densities $\{\langle\rho_k^i(\boldsymbol{r})\rangle = \rho_k{}^H(\boldsymbol{r})\}$ of molecular subsystems in contact with the common Hirshfeld "reservoir", which exhibits the entropy deficiency intensity $F(\boldsymbol{r})$, minimize the instantaneous Massieu function $\Delta\mathfrak{S}^i[\boldsymbol{\rho}^i, \boldsymbol{F}^H|\boldsymbol{\rho}^0]$ at the equilibrium value of Eq. (10.3.32). This is precisely the minimum principle yielding the generalized Massieu function of Eq. (10.3.34), the Legendre transform of the equilibrium entropy deficiency, in which the subsystem densities have been replaced by the local, inter-subsystem equalized Hirshfeld intensities in the list of the subsystem state-parameters. Due to the additive character of this equilibrium Legendre transform, its overall value can be naturally decomposed into the corresponding Hirshfeld subsystem contributions $\{\Delta\mathfrak{S}_k[F_k^H|\rho_k^0]\}$, equal to the negative values of the corresponding electron-populations of the Hirshfeld fragments $\{N_k^H\}$:

$$\Delta\mathfrak{S}[\boldsymbol{F}^H|\boldsymbol{\rho}^0] \equiv \sum_k \Delta\mathfrak{S}_k[F_k^H|\rho_k^0] = -\sum_k \int \rho_k^H(\boldsymbol{r})\, d\boldsymbol{r} \equiv -\sum_k N_k^H = -N. \tag{10.3.36}$$

In this "thermodynamic" interpretation the state of each fragment can be thus considered independently, with only the molecular intensity of the Hirshfeld reservoir reflecting its bonded (open) character relative to the remaining fragments.

Local Description

The above *global* development can be given an equivalent *local* interpretation by taking all quantities per unit volume and using the integrand $\Delta s^i(\boldsymbol{r}) = \sum_k \Delta s_k^i(\boldsymbol{r})$ or its subsystem components $\{\Delta s_k^i(\boldsymbol{r})\}$ as the local measures of the missing information (entropy deficiency). Their equilibrium value are reached for the Hirshfeld densities of molecular fragments, when $\langle\Delta s^i(\boldsymbol{r})\rangle = \Delta s(\boldsymbol{r}) \equiv \sum_k \Delta s_k(\boldsymbol{r})$, and the corresponding intensive-conjugates of the subsystem electron densities are the inter-subsystem equalized at the overall intensity value, giving rise to the corresponding instantaneous and equilibrium Massieu function densities:

$$\Delta s^i(\boldsymbol{r}) - \sum_k F_k^i(\boldsymbol{r})\, \rho_k^i(\boldsymbol{r}) \equiv \Delta\mathfrak{s}^i[\boldsymbol{F}^i(\boldsymbol{r})] \equiv \sum_k \Delta\mathfrak{s}_k^i[F_k^i(\boldsymbol{r})] = -\sum_k \rho_k^i(\boldsymbol{r}) \equiv -\rho^i(\boldsymbol{r}), \tag{10.3.37}$$

$$\begin{aligned}\Delta s(\boldsymbol{r}) - \sum_k F_k^H(\boldsymbol{r})\rho_k^H(\boldsymbol{r}) &\equiv \Delta\mathfrak{s}[\boldsymbol{F}^H(\boldsymbol{r})] = \Delta s(\boldsymbol{r}) - F(\boldsymbol{r})\rho(\boldsymbol{r})\\ &\equiv \Delta\mathfrak{s}[F(\boldsymbol{r})] \equiv \sum_k \Delta\mathfrak{s}_k[F_k^H(\boldsymbol{r})] = \min_i \Delta\mathfrak{s}^i[\rho^i(\boldsymbol{r}), \boldsymbol{F}^H(\boldsymbol{r}); \boldsymbol{\rho}^0(\boldsymbol{r})]\\ &= -\sum_k \rho_k^H(\boldsymbol{r}) = -\rho(\boldsymbol{r}).\end{aligned} \tag{10.3.38}$$

Hence, the instantaneous (vertical) divisions $\boldsymbol{\rho}^i$ of the given molecular density $\rho(\boldsymbol{r})$, which conserve the overall electron density, $\rho^i(\boldsymbol{r}) = \sum_k \rho_k^i(\boldsymbol{r}) = \rho(\boldsymbol{r}) = \sum_k \rho_k^H(\boldsymbol{r})$, preserve the molecular instantaneous entropy-deficiency density: $\Delta\mathfrak{s}[F(\boldsymbol{r})] = \Delta\mathfrak{s}^i[\boldsymbol{F}^i(\boldsymbol{r})] = -\rho(\boldsymbol{r})$. Only the subsystem local information distances, $\{\Delta\mathfrak{s}_k^i[F_k^i(\boldsymbol{r})] = -\rho_k^i(\boldsymbol{r})\}$ discriminate between alternative (vertical) partitions of the "frozen" molecular density.

To summarize, the subsystem contributions to the local Legendre transforms are additive. The equilibrium function $\Delta\mathfrak{F}[F^H(r)]$ is "normalized" to the negative value of the molecular density, with the negative densities of the stockholder subsystems $-\rho_k^H(r) = -\rho_k^H[\rho; r]$ measuring the subsystem contributions $\Delta\mathfrak{F}_k[F_k^H(r)]$. It also follows from the preceding two equations that the negative densities are the entropy-conjugates of the corresponding information intensities:

$$\frac{\partial\Delta\tilde{\mathfrak{s}}[F(r)]}{\partial F(r)} = -\rho(r), \quad \frac{\partial\Delta\tilde{\mathfrak{s}}[F^H(r)]}{\partial F^H(r)} = -\rho^H(r), \quad \frac{\partial\Delta\tilde{\mathfrak{s}}_k^i[F_k^i(r)]}{\partial F_k^i(r)} = -\rho_k^i(r), \quad (10.3.39)$$

where the partial differentiation implies that all remaining independent variables of the Legendre-transform density are held fixed.

To simplify notation, in what follows we shall omit the specification of the given position in space, r, which identifies a particular local subsystem under consideration: $\rho(r) = \rho$, $\rho_k^H[\rho; r] = \rho_k^H[\rho] = \rho_k^H$, $\Delta s^i(r) = \Delta s^i$, $\Delta\mathfrak{F}_k[F_k^H(r)] = \Delta\mathfrak{F}_k[F_k^H]$.

Gaussian Distribution Function

The local information-distance analog of the exponent in the thermodynamic distribution function of Eq. (G.4) reads:

$$-\kappa\{\Delta\tilde{\mathfrak{s}}^i[\rho^i, F^H; \rho^0] - \Delta\mathfrak{F}[F^H]\} = -\kappa[(\Delta s^i - \Delta s) - \Sigma_k F_k^H(\rho_k^i - \rho_k^H)]$$
$$\equiv -\kappa(\delta\Delta s^i - F\delta\rho^i) = -\kappa\,\Sigma_k(\delta\Delta s_k^i - F_k^H\,\delta\rho_k^i), \quad (10.3.40)$$

where the factor $\kappa = \rho^{-1}$ has been chosen to make the exponent dimensionless. Hence, the local IT distribution function becomes:

$$W(\rho^i) = \omega\exp[-\rho^{-1}(\delta\Delta s^i - F\delta\rho^i)]$$
$$= \Pi_k\,\omega_k\exp[-\rho^{-1}(\delta\Delta s_k^i - F_k^H\,\delta\rho_k^i)] \equiv \Pi_k\,W_k(\rho_k^i), \quad (10.3.41)$$

where $\{W_k(\rho_k^i)\}$ are the (mutually uncorrelated) subsystem distributions, $\omega = \Pi_k\omega_k$ is the overall normalization constant, with the subsystem normalization factors $\{\omega_k\}$ such that

$$\int W_k(\rho_k^i)\,d\rho_k^i = 1, \qquad k = 1, 2, \ldots, m. \quad (10.3.42)$$

The Gaussian distribution of the Einstein method (see Appendix G) can be obtained by expanding the instantaneous entropy-deficiency density Δs^i around the equilibrium value $\Delta s \equiv \Delta s^i|_0$, for the vanishing displacements of subsystem densities, $\delta\rho^i = \{\delta\rho_k^i\} \equiv \rho^i - \rho^H = 0$, in powers of the current instantaneous displacements $\delta\rho^i$,

$$\delta\Delta s^i = \Delta s^i - \Delta s = \sum_k F_k^H\, \delta\rho_k^i + \tfrac{1}{2}\sum_k\sum_l\, \delta\rho_k^i\, s_{k,l}^H\, \delta\rho_l^i + \ldots, \qquad (10.3.43)$$

where:

$$s_{k,l}^H = \frac{\partial^2 \Delta s}{\partial\rho_k^H \partial\rho_l^H} = \frac{\partial}{\partial\rho_k^H}\ln\frac{\rho_l^H}{\rho_l^0} = \frac{\delta_{k,l}}{\rho_k^H} = \frac{\rho^0\delta_{k,l}}{\rho\rho_k^0}. \qquad (10.3.44)$$

Therefore, neglecting the higher-order terms in Eq. (10.3.43) gives:

$$\delta\Delta s^i \cong F\delta\rho^i + \tfrac{1}{2}\sum_k[\rho^0/(\rho\rho_k^0)]\,(\delta\rho_k^i)^2 = F\delta\rho^i + \tfrac{1}{2}\sum_k(\delta\rho_k^i)^2/(\rho\, d_k^H), \qquad (10.3.45)$$

where $d_k^H = \rho_k^0/\rho^0$ denotes the subsystem *share*-factor in the Hirshfeld partition [see Eqs. (5.2.1) and (10.3.2)]. Finally, Eqs. (10.3.41) and (10.3.45) define the following Gaussian distribution in terms of the *relative* density displacements $\delta y^i = \{\delta y_k^i \equiv \delta\rho_k^i/\rho\}$:

$$W(\delta y^i) \cong \prod_k \varpi_k \exp[-(2d_k^H)^{-1}(\delta y_k^i)^2] \equiv \prod_k W_k^{\mathrm{G}}(\delta y_k^i) \equiv \prod_k W_k^{\mathrm{G}}(\delta\rho_k^i). \qquad (10.3.46)$$

where ϖ_k stands for the appropriately modified normalization constant.

A comparison between the subsystem distribution function $W_k^{\mathrm{G}}(\delta y_k^i)$ and the standard normal distribution of Eq. (3.2.6) identifies the local subsystem variance $D_k^H = (\sigma_k^H)^2 = \rho^2 d_k^H$ and hence the local dispersion of ρ_k^i around ρ_k^H,

$$\sigma_k^H \equiv \sigma[\delta\rho_k^i] = \rho(d_k^H)^{1/2}, \qquad (10.3.47)$$

the *relative* dispersion

$$\sigma[\delta y_k^i] \equiv \sigma[\delta\rho_k^i]/\rho = (d_k^H)^{1/2}, \qquad (10.3.48)$$

and the subsystem normalization constant $\varpi_k = 1/[(2\pi)^{1/2}\sigma_k^H]$.

Therefore, this Gaussian distribution predicts the local dispersion of the subsystem density to be proportional to the local value of the molecular density and the square root of the subsystem share-factor. In other words, the largest local dispersion is exhibited by the AIM which contributes the most to the local value of molecular density. It also follows from Eq. (6.6.25) that the local FF (softness) of the Hirshfeld subsystem is proportional to the corresponding global quantity, of the molecule as a whole, and the subsystem share factor d_k^H. We can therefore conclude that the local FF/softness of the molecular fragment in fact reflects the local variance of the subsystem density and thus the magnitude of its density fluctuations.

The second moments of the Hirshfeld subsystem densities can be also expressed in the form of Eq. (G.1.9). One first calculates the derivative

$$\frac{\partial W}{\partial F_k^H} = W \frac{\delta \rho_k^i}{\rho}. \tag{10.3.49}$$

Hence, the "diagonal" second moment can be written as

$$(\sigma_k^H)^2 = \langle(\delta\rho_k^{\,i})^2\rangle = \int (\delta\rho_k^{\,i})^2\, W\, d\boldsymbol{p}^i = \rho \int \delta\rho_k^{\,i} \frac{\partial W}{\partial F_k^H}\, d\boldsymbol{p}^i$$

$$= \rho\, \{ \frac{\partial}{\partial F_k^H} [\int \delta\rho_k^{\,i}\, W\, d\boldsymbol{p}^i] - \int W \frac{\partial \delta\rho_k^i}{\partial F_k^H}\, d\boldsymbol{p}^i \}$$

$$= \rho\, \frac{\partial \langle \delta\rho_k^i \rangle}{\partial F_k^H} + \rho\, \frac{\partial \rho_k^H}{\partial F_k^H} = \rho\, \frac{\partial \rho_k^H}{\partial F_k^H} = \rho\, \rho_k^H = \rho^2 d_k^H. \tag{10.3.50}$$

Above we have observed that $\langle\delta\rho_k^{\,i}\rangle = 0$ and recognized the derivative of $\delta\rho_k^{\,i} = \rho_k^{\,i} - \rho_k^H(\boldsymbol{F}^H)$ using Eqs. (10.3.10b) and (10.3.39):

$$\frac{\partial \delta\rho_k^i}{\partial F_k^H} = -\frac{\partial \rho_k^H}{\partial F_k^H} = \frac{\partial^2 \Delta\tilde{s}[\boldsymbol{F}^H]}{\partial F_k^H \partial F_k^H} \equiv S_{k,k} = -\rho_k^H = -\rho d_k^H. \tag{10.3.51}$$

The corresponding expression for the off-diagonal second moment reads:

$$\langle \delta\rho_k^{\,i}\, \delta\rho_l^{\,i} \rangle = \rho \frac{\partial \rho_k^H}{\partial F_l^H} = \rho \frac{\partial \rho_l^H}{\partial F_k^H} = -\rho\, \frac{\partial^2 \Delta\tilde{s}[\boldsymbol{F}^H]}{\partial F_k^H \partial F_l^H} \equiv -\rho\, S_{k,l} = 0, \tag{10.3.52}$$

since the distribution of Eq. (10.3.46) predicts that the fluctuation $\delta\rho_k^{\,i}$ is equally likely to be accompanied by the fluctuations $|\,\delta\rho_l^{\,i}\,|$ and $-|\,\delta\rho_l^{\,i}\,|$, since $\delta\rho_k^{\,i}$ and $\delta\rho_l^{\,i}$ are uncorrelated.

It should be emphasized that the *partial* derivatives in Eqs. (10.3.49)-(10.3.52) are calculated for the fixed (inter-subsystem equalized) Hirshfeld intensities of the remaining subsystems. The partial differentiation with respect to F_k^H thus corresponds to a "thermodynamic" description, in which the k-th subsystem is coupled to the separately controlled local reservoir, characterized by intensity F_k^H. Thus, e.g., in the derivative $\partial\rho_k^H/\partial F_k^H$ one monitors the response in ρ_k^H per unit displacement in δF_k^H, with the remaining local density components being coupled to the common *reservoir* of the initial equilibrium: $F_l^H = F[\rho]$, $l \neq k$.

To further simplify notation, in what follows we shall drop the upper index i indicating the instantaneous quantity: $\delta\rho_k^{\,i} = \delta\rho_k$, $\delta\Delta s^i = \delta\Delta s$, etc.

Affinities, Fluxes and Reciprocity Relations

Following the conceptual structure of the phenomenological irreversible thermodynamics (see, e.g., Callen, 1960) we require two types of quantities: *affinities*, to describe "thermodynamic" *forces* that drive an irreversible process, and *fluxes*, measuring *responses* to these forces. The fluxes vanish when the affinities vanish, and nonzero affinities lead to nonzero fluxes. The rates of irreversible processes are characterized by the relationship between fluxes and affinities. As in ordinary thermodynamics specific definitions of such quantities depend on the set of the system independent state-parameters. However, in the IT-approach to the electronic structure of molecules specific definitions of fluxes and affinities, and hence also that of the associated local *source* of the system entropy-deficiency, must also depend upon the adopted measure of the information-distance (Nalewajski, 2002c, 2003b, 2004a; see also Appendix G). Here, the identification of the local affinities for both the molecular system and its constituent fragments will be carried out by considering the rate of the local production of the Kullback-Leibler missing information in a continuous system (Nalewajski, 2003b). The vertical, ρ-conserving processes are assumed.

As in the ordinary irreversible thermodynamics one defines the entropy deficiency in a non-equilibrium system by postulating that the functional dependence of the cross-entropy density on the instantaneous state-parameters is taken to be identical to the dependence in the Hirshfeld equilibrium. Also by analogy to the irreversible thermodynamics we introduce the *vector* subsystem affinities $\{\boldsymbol{F}_k = (\mathcal{F}_{k\alpha}, \alpha = x, y, z), k = 1, 2, \ldots, m\} \equiv \{\mathcal{F}_t\}$, representing the IT-"thermodynamic" forces of subsystems that drive the "vertical" processes in the subsystem-resolved, *vertical* electronic structure of molecules. They are defined as gradients of the local IT-intensities for the Kullback-Leibler measure of the missing information. For the thermodynamic forces to identically vanish at the equilibrium (Hirshfeld) partitioning of a given (fixed) molecular density we define the affinities relative to the Hirshfeld equilibrium, as gradients of displacements of the subsystem intensities, $\delta\boldsymbol{F} \equiv \boldsymbol{F} - \boldsymbol{F}^H$ [Eqs. (10.3.9) and (10.3.13)]:

$$\boldsymbol{F}_k = \nabla\delta F_k = \nabla\ln(\rho_k/\rho_k^H) = (\nabla\rho_k)/\rho_k - (\nabla\rho_k^H)/\rho_k^H. \qquad (10.3.53a)$$

The latter represent the entropy-conjugates of the fragment densities $\boldsymbol{\rho}$ and their displacements $\delta\boldsymbol{\rho} \equiv \boldsymbol{\rho} - \boldsymbol{\rho}^H$ [Eqs. (10.3.12) and (10.3.16)]:

$$\delta\boldsymbol{F} = \frac{\partial\Delta S[\boldsymbol{\rho}|\boldsymbol{\rho}^0]}{\partial\boldsymbol{\rho}} - \frac{\partial\Delta S[\boldsymbol{\rho}^H|\boldsymbol{\rho}^0]}{\partial\boldsymbol{\rho}^H} = \left\{\ln\frac{\rho_k}{\rho_k^H} \equiv \delta I_k\right\} = \frac{\partial\Delta S[\boldsymbol{\rho}|\boldsymbol{\rho}^H]}{\partial\delta\boldsymbol{\rho}} - 1 = \frac{\partial\delta\Delta S[\boldsymbol{\rho}|\boldsymbol{\rho}^H]}{\partial\delta\boldsymbol{\rho}},$$

$$\frac{\partial\delta\Delta S[\boldsymbol{\rho}|\boldsymbol{\rho}^H]}{\partial\delta\boldsymbol{F}} = \boldsymbol{\rho}. \qquad (10.3.53b)$$

The responses to these forces are measured by the local subsystem fluxes characterized by the rates of change of the subsystem densities $\boldsymbol{\rho}$ and their sum ρ (see Appendix G). As in the irreversible thermodynamics of continuous systems they are defined as components of the *vector* current densities of electrons: $\{\boldsymbol{J}_k = (J_{k\alpha}, \alpha = x, y, z), k = 1, 2, \ldots, m\} \equiv \{\boldsymbol{J}_r\}$ and $\boldsymbol{J} = \sum_k \boldsymbol{J}_k$. The magnitude and direction of each vector respectively reflect the amount and direction of the corresponding electron flow across the unit area in unit time. The instantaneous local current density $\boldsymbol{J}_s = \boldsymbol{J}_s[\boldsymbol{\rho}]$ of the entropy deficiency transported through unit area per unit time is then given by the combination of electron flows of the subsystem components,

$$\boldsymbol{J}_s = \sum_k \delta F_k \, \boldsymbol{J}_k = \sum_k \ln(\rho_k / \rho_k^H)] \, \boldsymbol{J}_k, \tag{10.3.54}$$

in accordance with the differential for the displacement $\delta\Delta s(\boldsymbol{r})$ in the density of the entropy-deficiency $\delta\Delta S[\boldsymbol{\rho}\,|\,\boldsymbol{\rho}^H] \equiv \int \delta\Delta s(\boldsymbol{r})\, d\boldsymbol{r}$ [see Eq. (10.3.11)]:

$$d\delta\Delta s[\boldsymbol{\rho}] = \sum_k \delta F_k \, d\rho_k = \sum_k \delta F_k \, d\delta\rho_k. \tag{10.3.55}$$

We now seek the associated expression for the rate of production of the entropy-deficiency density:

$$\delta\sigma = \frac{d\delta\Delta s}{dt} = \frac{d\Delta s}{dt}. \tag{10.3.56}$$

It represents the net effect of the rate of a local increase of the missing information per unit volume, measured by the partial derivative $\partial(\delta\Delta s)/\partial t$, and the entropy-deficiency leaving the region, measured by the divergence term $\nabla \bullet \boldsymbol{J}_s$, in accordance with the *continuity equation* (see Appendix G) for the missing information in vertical processes:

$$\frac{d\delta\Delta s}{dt} = \frac{\partial\delta\Delta s}{\partial t} + \nabla \bullet \boldsymbol{J}_s. \tag{10.3.57}$$

In order to ensure that the instantaneous densities $\{\tilde{\rho}_k\} = \tilde{\boldsymbol{\rho}}$ reproduce the known (equilibrium) electron populations of the Hirshfeld fragments $\{\rho_k^H\} = \boldsymbol{\rho}^H[\rho]$, $\boldsymbol{N} = \{N_k\} = \boldsymbol{N}^H = \{N_k^H\}$, one appropriately renormalizes the "trial" densities:

$$\rho_k = \frac{N_k^H}{\int \tilde{\rho}_k(\boldsymbol{r}')d\boldsymbol{r}'}\tilde{\rho}_k. \tag{10.3.58}$$

Since this normalization fixes the fragment electron populations the relevant continuity equations for $\{\rho_k\} = \boldsymbol{\rho}$ are those for the mutually-closed components [Eqs. (G.2.3) and (G.2.8)]. The density parameters of the closed molecule and its now effectively closed subsystems can be neither produced nor destroyed, so that the equation of continuity for the displacement $\delta\rho_k = \rho_k - \rho_k^H$ of the kth subsystem electron density and that for the overall density become:

$$0 = \frac{\partial \delta\rho_k}{\partial t} + \nabla \bullet \boldsymbol{J}_k, \qquad k = 1, 2, ..., m; \qquad\qquad 0 = \frac{\partial \delta\rho}{\partial t} + \nabla \bullet \boldsymbol{J}. \qquad (10.3.59)$$

Therefore, using the continuity equations (10.3.57) and (10.3.59) gives the expression for the local entropy-deficiency production relative to the Hirshfeld level as combination of products of forces and fluxes:

$$\delta\sigma = \frac{d\delta\Delta s}{dt} = \frac{\partial \delta\Delta s}{\partial t} + \nabla \bullet \boldsymbol{J}_s = \sum_k \left(\frac{\partial \delta\Delta s}{\partial \rho_k}\right)\left(\frac{\partial \rho_k}{\partial t}\right) + \sum_k (\nabla \delta F_k) \bullet \boldsymbol{J}_k + \sum_k \delta F_k \nabla \bullet \boldsymbol{J}_k$$
$$= \sum_k \delta F_k (-\nabla \bullet \boldsymbol{J}_k) + \sum_k \boldsymbol{\mathsf{F}}_k \bullet \boldsymbol{J}_k + \sum_k \delta F_k \nabla \bullet \boldsymbol{J}_k = \sum_k \boldsymbol{\mathsf{F}}_k \bullet \boldsymbol{J}_k. \qquad (10.3.60)$$

For the Markoffian system (with no dynamical "memory"), to which we restrict this analysis, each local flux depends only on *all* forces (relative intensities) $\delta\boldsymbol{F}$ and affinities $\{\boldsymbol{\mathsf{F}}_k\} = \{\mathcal{F}_t\}$, $J_r = J_r(\{\mathcal{F}_t\},\{\delta F_k\})$, and it vanishes, by definition, when all affinities vanish. It should be observed that the Hirshfeld AIM keep a "memory" about the free atoms, but this *static* "memory" about the bonded fragment origins is automatically embedded in the entropy deficiency functional through the reference densities.

Expanding J_r in powers of affinities and neglecting the third- and higher-order terms expresses the linear effect of affinities on the rth flux,

$$J_r = \Sigma_t\, \mathcal{F}_t\, L_{t,r}, \qquad (10.3.61)$$

in terms of the *kinetic coefficients* $\mathbf{L}(\delta\boldsymbol{F}_k) = \{L_{t,r}(\delta\boldsymbol{F}_k)\}$,

$$\mathbf{L} \equiv \{\mathbf{L}_{k,l} \equiv \left(\frac{\partial \boldsymbol{J}_l}{\partial \boldsymbol{\mathsf{F}}_k}\right)_H \equiv \{L_{t,r} = (\partial J_r/\partial \mathcal{F}_t)_H\}, \qquad (10.3.62)$$

calculated for the equilibrium (Hirshfeld) subsystems, for which all affinities and forces identically vanish: $\delta\boldsymbol{F}^H = \mathbf{0}$ and $\{\boldsymbol{\mathsf{F}}_k = \nabla(\delta F_k^H) = \mathbf{0}\}$. Here, the coefficient $L_{t,r}$ measures the local linear effect of the tth affinity on the rth flux. Combining Eqs. (10.3.60) and (10.3.61) gives the for the relative entropy deficiency source the quadratic function of affinities:

$$\delta\sigma = \Sigma_r\, \Sigma_t\, \mathcal{F}_t\, L_{t,r}\, \mathcal{F}_r. \qquad (10.3.63)$$

In the absence of an externally applied magnetic field the following *reciprocity theorem* of Onsager (Callen, 1962) couples the local molecular subsystems:

$$L_{t,r} = L_{r,t}. \tag{10.3.64}$$

That is, in the linear (Markoff) process involving a network of electron flows between local molecular fragments the linear effect of the *t*th affinity on the *r*th flux is the same as that of the *r*th affinity on the local *t*th flux. This symmetry law reflects the Maxwell cross-differentiation relation. More specifically, from Eq. (10.3.60) one finds $J_r = [\partial(\delta\sigma)/\partial\mathcal{F}_r]_H$ and hence

$$L_{t,r} = (\partial J_r/\partial\mathcal{F}_t)_H = ½\,[\partial^2(\delta\sigma)\,/\partial\mathcal{F}_t\,\partial\mathcal{F}_r]_H$$
$$= ½\,[\partial^2(\delta\sigma)\,/\partial\mathcal{F}_r\,\partial\mathcal{F}_t]_H = (\partial J_t/\partial\mathcal{F}_r)_H = L_{r,t}. \tag{10.3.65}$$

These reciprocity relations can be also justified through fluctuations, using the *time symmetry* of physical laws (Callen, 1962). Consider again the spontaneous fluctuations $\delta\boldsymbol{\rho}$ of the local subsystem densities $\boldsymbol{\rho}$ around the equilibrium Hirshfeld values $\boldsymbol{\rho}^H$. The correlation moment $\langle\delta\rho_k\,\delta\rho_l(\tau)\rangle$, where $\delta\rho_l$ is observed a time τ after $\delta\rho_k$, must remain unchanged if we replace τ by $-\tau$, since only the time interval between the two observations is significant:

$$\langle\delta\rho_k\,\delta\rho_l(\tau)\rangle = \langle\delta\rho_k(-\tau)\,\delta\rho_l\rangle = \langle\delta\rho_k(\tau)\,\delta\rho_l\rangle. \tag{10.3.66}$$

Subtracting $\langle\delta\rho_k\,\delta\rho_l\rangle$ from each side of the foregoing equation and dividing by τ gives in the limit $\tau\to 0$ the associated relation involving time derivatives of the instantaneous densities of subsystems:

$$\langle\delta\rho_k\,\delta\left(\frac{d\rho_l}{dt}\right)\rangle = \langle\delta\left(\frac{d\rho_k}{dt}\right)\delta\rho_l\rangle. \tag{10.3.67}$$

Let us now modify the definition of affinities and fluxes. In what follows we shall adopt the subsystem entropic intensities $\{\delta F_k\}$ as *direct* measures of the system affinities. One then appropriately modifies the definitions of the local fluxes $\{ J_k = \delta(d\rho_k/dt) = d(\delta\rho_k)/dt\}$, since ρ_k^H in $\delta\rho_k = \rho_k - \rho_k^H$ is fixed by the molecular density ρ. This gives rise to the modified set of kinetic coefficients $\{\mathcal{L}_{l,k}\}$, which differs from $\mathbf{L}$ [Eq. (10.3.62)]. These conjugate forces and responses also define the modified rate of the local entropy production [from Eq. (10.3.55)], again given by products of conjugated forces and fluxes:

$$\delta\sigma = d(\delta\Delta s)/dt = \textstyle\sum_k \delta F_k\,J_k\,. \tag{10.3.68}$$

Hence, assuming the linear dynamical process for a decay of the fluctuation $\delta(d\rho_k/dt)$ through the relevant phenomenological equations

$$\delta\left(\frac{d\rho_k}{dt}\right) = \Sigma_l\, \delta F_l\, \mathcal{L}_{l,k}, \qquad k = 1, 2, \ldots, m, \tag{10.3.69a}$$

$$\mathcal{L}_{l,k} = \frac{\partial \delta\left(\frac{d\rho_k}{dt}\right)}{\partial \delta F_l} = \frac{\partial J_k}{\partial \delta F_l}, \tag{11.3.69b}$$

gives:

$$\Sigma_r\, \mathcal{L}_{r,l} \langle \delta\rho_k\, \delta F_r \rangle = \Sigma_r\, \mathcal{L}_{r,k} \langle \delta F_r\, \delta\rho_l \rangle. \tag{10.3.70}$$

Next, from the Hirshfeld distribution function of Eq. (10.3.41) one obtains [see also Eq. (10.3.49)],

$$-\rho \frac{\partial W}{\partial \delta\rho_k} = W \delta F_k\,. \tag{10.3.71}$$

Hence, by a straightforward integration by parts of the correlation moment in Eq. (10.3.70),

$$\langle \delta\rho_k\, \delta F_r \rangle = \int \delta\rho_k\, W \delta F_r\, d\boldsymbol{\rho} = -\rho \int \delta\rho_k \frac{\partial W}{\partial \delta\rho_r}\, d\boldsymbol{\rho} = \rho \int W \frac{\partial \delta\rho_k}{\partial \delta\rho_r}\, d\boldsymbol{\rho} = \rho\, \delta_{r,k} \int W\, d\boldsymbol{\rho} = \rho\, \delta_{r,k}. \tag{10.3.72}$$

Finally, inserting the preceding equation into Eq. (10.3.70) gives the subsystem Onsager theorem in the absence of magnetic fields: $\mathcal{L}_{k,l} = \mathcal{L}_{l,k}$.

10.4. CONCLUSION

In this chapter we have examined the "thermodynamic"-like principles for both the "horizontal" and "vertical" density displacements in molecular systems. The corresponding variational rules, which couple the system density functionals for the electronic energy and entropy deficiency of Kullback and Leibler, have been explored in both the energy and information-entropy representations. This analysis has identified the system/subsystem chemical potentials as the corresponding information temperatures appearing as Lagrange multipliers enforcing the energetic or entropic constraints in such coupled thermodynamic-like principles determining the optimum densities for the system as a whole or its constituent fragments.

In Information Theory the free-subsystem referenced variational principle of the entropy deficiency, subject to the auxiliary conditions of the exhaustive partitioning of the molecular density at each point in space, leads to the "stockholder" electron densities of molecular fragments, which thus mark the entropy-stable, *vertical* equilibrium state in the subsystem resolution. In this chapter we have developed the local "instantaneous" approach to fluctuations in subsystem densities conserving the given molecular density, in which the stockholder subsystems appear as equilibrium fragments. In this development the Hirshfeld subsystems, previously regarded as static entities, can be viewed as the averages of the instantaneous (dynamic) entities, with the distribution of local fluctuations being related to the relevant missing-nformation density in the thermodynamic-like fashion. In this hydrodynamic (non-equilibrium) description the subsystem "affinity" (force) and "flux" (response) quantities, which determine the local cross-entropy production (source), have been established in close analogy to the phenomenological irreversible thermodynamics. For the linear dynamical processes, in the Markoffian system approximation, they imply the local reciprocity rules, the molecular analogs of the familiar Onsager relations. These relations reflect basic symmetries between the linear effects of the subsystem affinities on fluxes. This phenomenological development should facilitate a better understanding of a hierarchy of the coupled flows of electrons between molecular fragments.

This analysis has demonstrated that the IT-utlook on the electronic structure of molecular systems introduces the vital entropy-representation, which is crucial for a "thermodynamic" interpretation of sub-molecular processes. The experience of the ordinary thermodynamics can thus be applied to uncover the role played by the information contained in the electron distributions in a variety of processes involving molecules, AIM, reactants, etc. This extends the range of applications of the information-theoretic treatment of the sub-molecular processes to the realm of non-equilibrium states of subsystems, e.g., the electronic structure reorganizations in a chemical reaction. The present analysis also establishes a convenient theoretical framework for the *"dynamical"* indexing of the non-equilibrium reactivity phenomena. With this development we have identified the basic conceptual ingredients of the IT-hydrodynamic approach to the instantaneous internal density fluctuations in molecules which are always present in the open molecular subsystems. These fluctuations are the key ingredients of many chemical concepts, e.g., the chemical softness and Fukui function quantities.

Finally, one should emphasize that the "vertical", sub-molecular reality of the subsystem resolution, which is so important for the *"language"* and understanding in chemistry, cannot be validated by a direct experiment, since it cannot be formulated as the unique quantum-mechanical "observable". The bonded atoms and chemical bonds have to be ultimately classified as Kantian noumenons. However, as we have argued throughout this book, their partial understanding and indirect probes are available from several different perspectives. The close analogy between the IT phenomenological description of molecules and their fragments and the ordinary thermodynamics further validates these chemical concepts. It introduces the thermodynamic-like causality into relations between perturbations and responses of molecular subsystems and hence brings more consistency into the theory of chemistry.

APPENDICES

APPENDIX A: Functional Derivatives

The functional derivatives and the Taylor expansion they define are introduced. The localized displacements of the argument function are defined using the singular delta function of Dirac, the representations and properties of which are briefly summarized. The rules of functional differentiation are outlined, including the chain-rule transformations of derivatives.

For simplicity, let us assume the functional of a single function $f(x)$ depending upon of a continuous coordinate x,

$$F[f] = \int L\,[x, f(x), f'(x), ...]\, dx\,. \tag{A.1}$$

It is defined by the functional density $L\,[x, f(x), f'(x), ...]$, which in a general case may depend on the current value of x, the argument function itself, $f(x)$, and its derivatives: $f'(x) = df(x)/dx$, etc. The functional thus attributes to the argument function f the scalar F. An important problem, which we shall often encounter throughout this book, is to find the variations of the functional, due to a small modification of the argument function, $\delta f(x) = \varepsilon\, h(x)$, where ε is a small parameter and $h(x)$ stands for the displacement function or the perturbation.

Such an overall shift of the functional argument can be alternatively viewed as a superposition of local manipulations of $f(x)$, $\{\delta f(x') = h(x')\delta(x' - x)\}$,

$$\delta f(x) = \varepsilon\, h(x) = \int \varepsilon\, h(x')\delta(x'-x)\,dx' \equiv \int \delta f(x')\delta(x'-x)\,dx'\,, \tag{A.2}$$

where $\delta(x' - x)$ stands for the Dirac delta function centered at $x' = x$. The latter can be envisaged, as the limiting form of the ordinary Gaussian (normal) distribution of the probability theory in the limit of vanishing variance:

$$\delta(x'-x) = \lim_{\sigma \to 0} \frac{1}{\sqrt{2\pi\sigma^2}} \exp\left(-\frac{(x'-x)^2}{2\sigma^2}\right). \tag{A.3}$$

It thus represents an infinitely sharp, normalized displacement, $\int \delta(x' - x)\, dx' = 1$, which is localized at $x' = x$ and exhibits the following unique property:

$$\int f(x')\delta(x' - x)\, dx' = f(x) \tag{A.4a}$$

Selected additional equalities satisfied by this singular function are:

$$\delta(x) = \delta(-x), \quad x\delta(x) = 0, \quad \delta(ax) = |a|^{-1}\delta(x), \quad f(x')\delta(x' - x) = f(x)\delta(x' - x),$$
$$\int \delta(x' - x)\, \delta(x - x'')\, dx = \delta(x' - x''), \quad \delta(x^2 - a^2) = (2|a|)^{-1}[\delta(x - a) + \delta(x + a)]. \tag{A.4b}$$

Of interest also are the related properties of the derivative of the Dirac delta, $\delta'(x) \equiv d\delta(x)/dx$,

$$\int f(x)\delta'(x)\, dx = -f'(0), \qquad x\delta'(x) = -\delta(x). \tag{A.4c}$$

This function can be viewed as the *continuous* generalization of the familiar (*discrete*) Kronecker's delta:

$$\delta_{i,j} = \{0, \text{ for } i \neq j;\ 1, \text{ for } i = j\}, \qquad \sum_j \delta_{i,j} = 1, \qquad \sum_j \delta_{i,j} f_j = f_i . \tag{A.5}$$

Another analytical representation of the singular Dirac's δ-function originates from the Fourier transform relations, e.g., between the wave-function in the position and momentum representations of quantum mechanics,

$$\Phi(k) = \frac{1}{\sqrt{2\pi}} \int \exp(-ikx) f(x)\, dx \quad \text{and} \quad f(x) = \frac{1}{\sqrt{2\pi}} \int \exp(ik'x)\Phi(k')\, dk', \qquad i = \sqrt{-1}. \tag{A.6}$$

Substituting the second, inverse transformation into the first one gives

$$\Phi(k) = \frac{1}{2\pi} \int \Phi(k') \{ \int \exp[ix(k'-k)]dx \} dk' \tag{A.7}$$

and hence from Eq. (A.4a)

$$\delta(k'-k) = \frac{1}{2\pi} \int \exp[ix(k'-k)] dx . \tag{A.8}$$

The local perspective (A.2) on varying the argument function of $F[f]$ introduces the concept of the functional derivatives, which define the consecutive terms in the functional Taylor-Volterra expansion,

$$\delta F[f] = \int \frac{\delta F}{\delta f(x)} \delta f(x) dx + \frac{1}{2} \iint \delta f(x) \frac{\delta^2 F}{\delta f(x) \delta f(x')} \delta f(x') dx\, dx' + \ldots \equiv \delta^{(1)} F[f] + \delta^{(2)} F[f] + \ldots, \tag{A.9}$$

where $\delta f(x)$ is a local variation of the argument function f. Therefore, the first functional derivative

$$\frac{\delta F}{\delta f(x)} = \lim_{\varepsilon \to 0} \frac{F[f(x') + \varepsilon\delta(x'-x)] - F[f(x')]}{\varepsilon} \equiv g[f;x], \tag{A.10}$$

transforms the local displacements of the argument function into the first differential $\delta^{(1)}F[f]$. The first term in Eq. (A.9) can thus be viewed as the continuous generalization of the differential of a function of several variables: $d^{(1)}f(x_1, x_2, \ldots) = \sum_i (\partial f/\partial x_i)\, dx_i$.

One can interpret in a similar way the second functional derivative

$$\frac{\delta^2 F}{\delta f(x) \delta f(x')} = \frac{\delta g[f;x]}{\delta f(x')} = \lim_{\varepsilon \to 0} \frac{g[f(x'') + \varepsilon\delta(x''-x');\, x] - g[f;x]}{\varepsilon}, \tag{A.11}$$

as determining the continuous transformation of the two-point displacements of the argument function, $\delta f(x)\delta f(x')$, into the second differential $\delta^{(2)}F[f]$, which parallels the corresponding discrete expression: $d^{(2)}f(x_1, x_2, \ldots) = \frac{1}{2} \sum_i \sum_j (\partial^2 f/\partial x_i \partial x_j)\, dx_i\, dx_j$.

The rules of the functional differentiation thus represent the continuous generalization of those characterizing the differentiation of functions, e.g.:

$$\frac{\delta}{\delta f(x)}\{aF[f]+bG[f]\}=a\frac{\delta F}{\delta f(x)}+b\frac{\delta G}{\delta f(x)},\qquad \frac{\delta}{\delta f(x)}\{F[f]G[f]\}=G\frac{\delta F}{\delta f(x)}+F\frac{\delta G}{\delta f(x)}. \tag{A.12}$$

The *chain-rule* transformation of functional derivatives also holds. Consider the *composite* functional $F[f]=F[f[g]]\equiv\overline{F}[g]$. Substituting the first differential of $f(x) = f[g; x]$,

$$\delta^{(1)}f[g;x]=\int\frac{\delta f(x)}{\delta g(x')}\delta g(x')\,dx', \tag{A.13}$$

into $\delta^{(1)}F[f]$ of Eq. (A.9) gives:

$$\delta^{(1)}\overline{F}[g]=\int\frac{\delta\overline{F}}{\delta g(x')}\delta g(x')\,dx'=\int\frac{\delta F}{\delta f(x)}\Big[\int\frac{\delta f(x)}{\delta g(x')}\delta g(x')\,dx'\Big]dx. \tag{A.14}$$

Hence,

$$\frac{\delta\overline{F}}{\delta g(x')}=\int\frac{\delta F}{\delta f(x)}\frac{\delta f(x)}{\delta g(x')}\,dx. \tag{A.15}$$

One similarly derives the chain-rules for *implicit* functionals. If functional $F[f, g]$ is held constant, the variations of the two argument functions are not independent, since the relation $F[f, g]$ = *const.* implies a functional relation between them, e.g., $g = g[f]_F$. The vanishing first differential

$$\delta^{(1)}F[f,g]=\int\left[\left(\frac{\partial F}{\partial f(x)}\right)_g[\delta f(x)]_F+\left(\frac{\partial F}{\partial g(x)}\right)_f[\delta g(x)]_F\right]dx=0,\qquad\text{or}$$

$$\int\left(\frac{\partial F}{\partial f(x)}\right)_g[\delta f(x)]_F\,dx=-\int\left(\frac{\partial F}{\partial g(x')}\right)_f[\delta g(x')]_F\,dx', \tag{A.16}$$

is determined by the *partial* functional derivatives, a natural, continuous extension of the ordinary partial derivatives of a function of several variables, e.g.,

$$\left(\frac{\partial F}{\partial f(x)}\right)_g=\lim_{\varepsilon\to 0}\frac{F[f(x')+\varepsilon\delta(x'-x),\,g]-F[f,g]}{\varepsilon}. \tag{A.17}$$

Finally, differentiating both sides of Eq. (A.16) with respect to one of the argument functions for constant F then gives the following implicit chain rules:

$$\left(\frac{\partial F}{\partial f(x)}\right)_g=-\int\left(\frac{\partial F}{\partial g(x')}\right)_f\left(\frac{\partial g(x')}{\partial f(x)}\right)_F dx',\quad \left(\frac{\partial F}{\partial g(x')}\right)_f=-\int\left(\frac{\partial F}{\partial f(x)}\right)_g\left(\frac{\partial f(x)}{\partial g(x')}\right)_F dx. \tag{A.18}$$

These relations parallel familiar manipulations of derivatives of the classical thermodynamics.

For the fixed value of the composite functional $F[f[u], g[u]] = \widetilde{F}[u]$ = *const.* one similarly finds

$$\left(\frac{\partial g(x')}{\partial f(x)}\right)_{\tilde{F}} = \int\left(\frac{\partial g(x')}{\partial u(x'')}\right)_{\tilde{F}}\left(\frac{\partial u(x'')}{\partial f(x)}\right)_{\tilde{F}} dx'', \quad \left(\frac{\partial f(x)}{\partial g(x')}\right)_{\tilde{F}} = \int\left(\frac{\partial f(x)}{\partial u(x'')}\right)_{\tilde{F}}\left(\frac{\partial u(x'')}{\partial g(x')}\right)_{\tilde{F}} dx''. \quad (A.19)$$

Let us further assume that functions $f(x)$ and $g(x)$ are unique functionals of each other, $f(x) = f[g; x]$ and $g(x') = g[f; x']$. Substitution of Eq. (A.13) into

$$\delta^{(1)} g[f; x''] = \int \frac{\delta g(x'')}{\delta f(x)} \delta f(x) dx, \quad (A.20)$$

then gives:

$$\delta^{(1)} g[f; x''] = \int \frac{\delta g(x'')}{\delta f(x)} \delta f(x) dx = \iint \frac{\delta g(x'')}{\delta f(x)} \frac{\delta f(x)}{\delta g(x')} \delta g(x') dx' dx. \quad (A.21)$$

This equation identifies the Dirac delta function as the functional derivative of the function at one point with respect to its value at another point, as also implied by Eq. (A.2):

$$\int \frac{\delta g(x'')}{\delta f(x)} \frac{\delta f(x)}{\delta g(x')} dx = \frac{\delta g(x'')}{\delta g(x')} = \delta(x''-x'), \quad (A.22)$$

where we have applied the functional chain rule. Equation (A.22) also defines the inverse functional derivatives

$$\frac{\delta g(x')}{\delta f(x)} = \left(\frac{\delta f(x)}{\delta g(x')}\right)^{-1}. \quad (A.23)$$

Let us assume the functional (A.1) in the typical form including the dependence of its density on the argument function itself and its first n derivatives: $f^{(i)}(x) = d^i f(x)/dx^i$, $i = 1, 2, \ldots, n$:

$$\mathsf{L}(x) = \mathsf{L}(x, f(x), f^{(1)}(x), f^{(2)}(x), \ldots, f^{(n)}(x)). \quad (A.24)$$

The functional derivative of $F[f]$ is then given by the following general expression:

$$\frac{\delta F}{\delta f(x)} = \frac{\partial \mathsf{L}(x)}{\partial f(x)} - \frac{d}{dx}\left(\frac{\partial \mathsf{L}(x)}{\partial f^{(1)}(x)}\right) + \frac{d^2}{dx^2}\left(\frac{\partial \mathsf{L}(x)}{\partial f^{(2)}(x)}\right) - \ldots + (-1)^n \frac{d^n}{dx^n}\left(\frac{\partial \mathsf{L}(x)}{\partial f^{(n)}(x)}\right). \quad (A.25)$$

The first term in r.h.s. of the preceding equation defines the so called *variational* derivative. It determines the functional derivative of the local functionals, the densities of which depend solely upon the argument function itself.

This development can be straightforwardly generalized into the multi-component functionals of functions in 3 dimensions, which depend on the position vector in physical space $\boldsymbol{r} = (x, y, z)$. For example, the Dirac delta function $\delta(\boldsymbol{r}' - \boldsymbol{r})$ becomes the product of the one-dimensional delta functions of its coordinates, $\delta(\boldsymbol{r}' - \boldsymbol{r}) = \delta(x' - x)\, \delta(y' - y)\, \delta(z' - z)$, and satisfies the properties [see Eq. (A.4a)]:

$$\int \delta(\boldsymbol{r}' - \boldsymbol{r})\, d\boldsymbol{r}' = 1, \qquad \int f(\boldsymbol{r}')\delta(\boldsymbol{r}' - \boldsymbol{r})\, d\boldsymbol{r}' = f(\boldsymbol{r}), \qquad \text{etc.} \tag{A.26}$$

The corresponding Fourier-transform representations [compare Eq.(A.8)] are:

$$\delta(\boldsymbol{k}'-\boldsymbol{k}) = \frac{1}{(2\pi)^3}\int \exp[i\boldsymbol{r}(\boldsymbol{k}'-\boldsymbol{k})]d\boldsymbol{r} \quad \text{and} \quad \delta(\boldsymbol{r}'-\boldsymbol{r}) = \frac{1}{(2\pi)^3}\int \exp[i\boldsymbol{k}(\boldsymbol{r}'-\boldsymbol{r})]d\boldsymbol{k}\,. \tag{A.27}$$

Equation (A.25) can be extended into the $f = f(\boldsymbol{r})$ case by replacing the operator d/dx by its 3-dimensional analog, the gradient $\nabla \equiv d/d\boldsymbol{r}$. For example, for $L(\boldsymbol{r}) = L(\boldsymbol{r}, f(\boldsymbol{r}), |\nabla f(\boldsymbol{r})|)$ the functional derivative of $F[f]$ is given by the expression:

$$\frac{\delta F}{\delta f(\boldsymbol{r})} = \frac{\partial L(\boldsymbol{r})}{\partial f(\boldsymbol{r})} - \nabla\left(\frac{\partial L(\boldsymbol{r})}{\partial |\nabla f(\boldsymbol{r})|}\right). \tag{A.28}$$

APPENDIX B: Geometric Interpretation of Density Displacements and Charge Sensitivities

The "geometric" interpretation of the electronic density displacements in the Hilbert space is given and the associated projection-operator partitioning of the hardness and softness operators (kernels) is developed. The eigenvectors $|\boldsymbol{\alpha}\rangle = \{|\alpha\rangle\}$ of the hardness operator define the complete (identity) projector $\hat{\mathrm{P}} = \sum_\alpha |\alpha\rangle\langle\alpha| = 1$ for general displacements of the electron density, including the *charge-transfer* (*CT*) component, while the eigenvectors $|\boldsymbol{i}\rangle = \{|i\rangle\}$ of the linear-response operator determine the *polarizational* *P*-projector, $\hat{\mathrm{P}}_P = \sum_i |i\rangle\langle i|$. Their difference thus defines the complementary *CT*-projector: $\hat{\mathrm{P}}_{CT} = 1 - \hat{\mathrm{P}}_P$. The complete vector space for density displacements can be also spanned by supplementing the *P*-modes with the homogeneous *CT*-mode. These subspaces separate the integral (normalization) and local aspects of density shifts in molecular systems. The (*P*,*CT*)-resolution of the Fukui Function, as well as the softness and hardness operators(kernel) allows one to separate the polarizational and charge-transfer components of these quantities and to quantify the mutual coupling between these conventional degrees-of-freedom of molecular and/or reactive systems.

B.1. Hilbert Space of Independent Density Displacement Modes

The shift $\Delta\rho(\boldsymbol{r}) \equiv \langle \boldsymbol{r}|\Delta\rho\rangle$ in the equilibrium distribution of electrons in a molecule has a transparent "geometrical" interpretation in the Hilbert space spanned by the independent *Density Displacement Modes* (DDM) (Nalewajski 2002e, 2005g) in which it is represented by the vector $|\Delta\rho\rangle$. This vector space, *complete* for expanding a general $|\Delta\rho\rangle$, consists of the *Polarizational* (*P*) and *Charge-Transfer* (*CT*) subspaces, defined by the complementary projection operators $\hat{\mathrm{P}}_P$ and $\hat{\mathrm{P}}_{CT}$, respectively,

$$\hat{\mathrm{P}} = \hat{\mathrm{P}}_{CT} + \hat{\mathrm{P}}_P = 1. \tag{B.1.1}$$

The *P*-subspace is spanned by the orthonormal *P*-modes $|\boldsymbol{i}\rangle = \{|i\rangle\}$, $\langle j|i\rangle = \delta_{i,j}$, or their position representations $\boldsymbol{b}(\boldsymbol{r}) = \langle \boldsymbol{r}|\boldsymbol{i}\rangle = \{b_i(\boldsymbol{r}) = \langle \boldsymbol{r}|i\rangle\}$, which conserve the overall number of electrons N in the system as a whole:

$$\int b_i(\boldsymbol{r})\, d\boldsymbol{r} = 0, \qquad i = 1, 2, \ldots \tag{B.1.2}$$

These polarization functions are the eigenvectors of the density *Linear Response* (LR) kernel (Nalewajski, 1988, 1993, 1995a,b, 1997a; Nalewajski *et. al.*, 1988, 1996; Nalewajski and Korchowiec, 1997),

$$\beta(\boldsymbol{r},\boldsymbol{r}') = \left(\frac{\partial\rho(\boldsymbol{r}')}{\partial v(\boldsymbol{r})}\right)_N \equiv \langle \boldsymbol{r}|\hat{\beta}|\boldsymbol{r}'\rangle, \tag{B.1.3}$$

where v denotes the external potential due to the nuclei, $N = \int\rho(\boldsymbol{r})\, d\boldsymbol{r}$ is the system overall number of electrons, and $\hat{\beta}$ stands for the underlying operator in the Hilbert space:

$$\int\beta(\boldsymbol{r},\boldsymbol{r}')b_i(\boldsymbol{r}')d\boldsymbol{r}' = \beta_i\, b_i(\boldsymbol{r}) \quad \text{or} \quad \hat{\beta}|\, i\rangle = \beta_i|\, i\rangle \quad \text{and} \quad \langle i|\hat{\beta} = \beta_i\langle i|,$$
$$\langle i\,|\, j\rangle = \int b_i^*(\boldsymbol{r})b_j(\boldsymbol{r})d\boldsymbol{r} = \delta_{i,j}, \qquad i,j = 1, 2, \ldots \tag{B.1.4}$$

In terms of the eigensolutions $|\boldsymbol{i}\rangle$ and $\boldsymbol{\beta} = \{\beta_i\}$ the density LR operator is given by its spectral representation

$$\hat{\beta} = \Sigma_i |i\rangle\, \beta_i\, \langle i| \qquad \text{or} \qquad \beta(\boldsymbol{r},\boldsymbol{r}') = \langle \boldsymbol{r}|\hat{\beta}\,|\boldsymbol{r}'\rangle = \sum_i b_i(\boldsymbol{r})\,\beta_i\, b_i^*(\boldsymbol{r}'). \tag{B.1.5}$$

The associated projection operator

$$\hat{\mathrm{P}}_P = \sum_i |\, i\rangle\langle i\,| \quad \text{or} \quad \langle \boldsymbol{r}|\,\hat{\mathrm{P}}_P\,|\boldsymbol{r}'\rangle \equiv \gamma(\boldsymbol{r},\boldsymbol{r}') = \sum_i \langle \boldsymbol{r}|\, i\,\rangle\langle\, i\,|\boldsymbol{r}'\rangle = \sum_i b_i(\boldsymbol{r})\, b_i^*(\boldsymbol{r}'), \tag{B.1.6}$$

separates the *P*-component, $|\Delta\rho\rangle_P$, of a general density displacement vector $|\Delta\rho\rangle$:

$$\hat{\mathrm{P}}_P|\Delta\rho\rangle \equiv |\Delta\rho\rangle_P = \sum_i |\, i\,\rangle\langle\, i\,|\Delta\rho\rangle \equiv \sum_i |\, i\,\rangle\,\Delta\rho_i\,, \quad \Delta\rho_i = \int b_i^*(\boldsymbol{r})\,\Delta\rho(\boldsymbol{r})\, d\boldsymbol{r}, \qquad \text{or}$$
$$\Delta\rho_P(\boldsymbol{r}) = \langle \boldsymbol{r}|\Delta\rho\rangle_P = \int\gamma(\boldsymbol{r},\boldsymbol{r}')\,\Delta\rho(\boldsymbol{r}')\,d\boldsymbol{r}' = \sum_i b_i(\boldsymbol{r})\,\Delta\rho_i\,. \tag{B.1.7}$$

Next, let us express a general displacement of the system density, $\rho(\boldsymbol{r}) \equiv N\, p(\boldsymbol{r})$, in terms of the associated changes in its normalization, ΔN, and the shape" (probability) factor, $\Delta p(\boldsymbol{r})$:

$$|\Delta\rho\rangle = N\,|\Delta p\rangle + |p\rangle\,\Delta N \equiv |\Delta\rho\rangle_N + |\Delta\rho\rangle_{CT} \quad \text{or} \quad \Delta\rho(\boldsymbol{r}) = N\,\Delta p(\boldsymbol{r}) + p(\boldsymbol{r})\,\Delta N \equiv \Delta\rho_N(\boldsymbol{r}) + \Delta\rho_{CT}(\boldsymbol{r}), \tag{B.1.8a}$$

where $\int p(\boldsymbol{r})\, d\boldsymbol{r} = 1$, $N = \int \rho(\boldsymbol{r})\, d\boldsymbol{r} \equiv N[\rho]$, and hence: $\int \Delta p(\boldsymbol{r})\, d\boldsymbol{r} = 0$, $\int \Delta\rho(\boldsymbol{r})\, d\boldsymbol{r} = \Delta N$. It follows from the preceding equation that the closed system polarization part $\Delta\rho_N(\boldsymbol{r})$, which integrates to zero, is purely polarizational: $\hat{\mathrm{P}}_P\,|\Delta\rho\rangle_N = |\Delta\rho\rangle_N$ (see Fig. B.1). The ΔN normalized CT–component $\Delta\rho_{CT}(\boldsymbol{r})$, however, must contain the nonvanishing P- and CT-displacements, since the added (or removed) ΔN electrons are distributed in accordance with the initial probability factor. Indeed, this density change can be viewed as a combination of the homogeneous density displacement $\Delta\rho_{CT}{}^{hom}$, such that $\Delta\rho_{CT}{}^{hom}\ V = \Delta N \neq 0$, where V denotes the (finite) molecular volume, which does not bias any local volume element in the molecule, and the remaining, polarizational part $\Delta\rho_{CT}{}^{inhom}(\boldsymbol{r})$, which integrates to zero:

$$\Delta\rho_{CT}(\boldsymbol{r}) = \Delta\rho_{CT}^{hom}(\boldsymbol{r}) + \Delta\rho_{CT}^{inhom}(\boldsymbol{r}). \tag{B.1.8b}$$

These two components, shown in Fig. B.1, are represented by the corresponding vectors in the molecular Hilbert space:

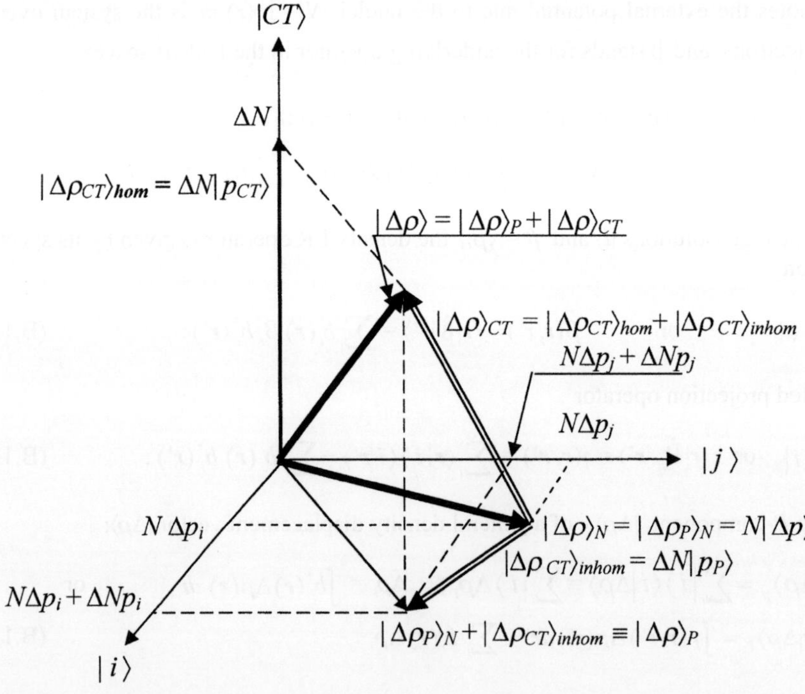

Figure B1. The density displacement vectors $|\Delta\rho\rangle_N$ and $|\Delta\rho\rangle$, of the closed and open molecular systems, respectively, and their partitions into the P- and CT-components in the Hilbert space spanned by the two polarizational modes $\{|i\rangle, |j\rangle\}$ and the homogeneous CT-mode $|CT\rangle$.

$$\hat{\mathrm{P}}_P\,|\Delta\rho\rangle_{CT} = \Delta N\ \hat{\mathrm{P}}_P\,|p\rangle \equiv \Delta N\ |p_P\rangle = |\Delta\rho_{CT}\rangle_{inhom}\,,$$

$$|\Delta\rho\rangle_{CT} = \Delta N\ \hat{\mathrm{P}}_{CT}\,|p\rangle \equiv \Delta N\ |p_{CT}\rangle = |\Delta\rho_{CT}\rangle_{hom}\,. \tag{B.1.8c}$$

The sum of the closed-system polarization and the inhomogeneous *CT*-component then defines the overall polarization in the open molecular system:

$$|\Delta\rho\rangle_P = \hat{\mathrm{P}}_P\,|\Delta\rho\rangle = |\Delta\rho\rangle_N + |\Delta\rho_{CT}\rangle_{inhom} \ \text{ or } \ \Delta\rho_P(\boldsymbol{r}) = \langle\boldsymbol{r}\,|\,\hat{\mathrm{P}}_P\,|\Delta\rho\rangle = \Delta\rho_N(\boldsymbol{r}) + \Delta\rho_{CT}^{inhom}(\boldsymbol{r})\,. \tag{B.1.8d}$$

In the density displacement Hilbert space the chemical softness and hardness operators, $\hat{\sigma}$ and $\hat{\eta}$, respectively, can be similarly defined in terms of their common (orthonormal) eigenvectors $|\boldsymbol{\alpha}\rangle = \{|\alpha\rangle\}$, $\langle\alpha|\beta\rangle = \delta_{\alpha,\beta}$, or their position representations $\boldsymbol{c}(\boldsymbol{r}) = \langle\boldsymbol{r}|\boldsymbol{\alpha}\rangle = \{c_\alpha(\boldsymbol{r}) = \langle\boldsymbol{r}|\alpha\rangle\}$, and the corresponding eigenvalues, $\boldsymbol{\eta} = \{\eta_\alpha\}$ and $\boldsymbol{\sigma} = \{\sigma_\alpha = 1/\eta_\alpha\}$:

$$\hat{\eta}\,|\alpha\rangle = \eta_\alpha\,|\alpha\rangle, \qquad \hat{\sigma}\,|\alpha\rangle = \sigma_\alpha\,|\alpha\rangle, \qquad \alpha, \beta = 1, 2, \ldots, \tag{B.1.9}$$

or in the electron position representation

$$\int\langle\boldsymbol{r}|\hat{\eta}|\boldsymbol{r}'\rangle\langle\boldsymbol{r}'|\alpha\rangle\,d\boldsymbol{r}' \equiv \int\eta(\boldsymbol{r},\boldsymbol{r}')\,c_\alpha(\boldsymbol{r}')\,d\boldsymbol{r}' = \eta_\alpha\,c_\alpha(\boldsymbol{r})\,,$$

$$\int\langle\boldsymbol{r}|\hat{\sigma}|\boldsymbol{r}'\rangle\langle\boldsymbol{r}'|\alpha\rangle\,d\boldsymbol{r}' \equiv \int\sigma(\boldsymbol{r},\boldsymbol{r}')\,c_\alpha(\boldsymbol{r}')\,d\boldsymbol{r}' = \sigma_\alpha\,c_\alpha(\boldsymbol{r})\,,$$

$$\langle\alpha|\beta\rangle = \int\langle\alpha|\boldsymbol{r}\rangle\langle\boldsymbol{r}|\beta\rangle \equiv \int c_\alpha^*(\boldsymbol{r})\,c_\beta(\boldsymbol{r})\,d\boldsymbol{r} = \delta_{\alpha,\beta}\,. \tag{B.1.10}$$

Here, the eigenvalues $\boldsymbol{\eta}$ and $\boldsymbol{\sigma}$ represent the system *principal* hardnesses and softnesses, respectively (see, e.g., Nalewajski *et al.*, 1996; Nalewajski and Korchowiec, 1996). The resulting *spectral resolutions* of these two operators then read:

$$\hat{\eta} = \Sigma_\alpha\,|\alpha\rangle\,\eta_\alpha\,\langle\alpha|, \qquad \text{or} \qquad \eta(\boldsymbol{r},\boldsymbol{r}') = \langle\boldsymbol{r}|\,\hat{\eta}\,|\boldsymbol{r}'\rangle = \sum_\alpha c_\alpha(\boldsymbol{r})\,\eta_\alpha\,c_\alpha^*(\boldsymbol{r}')\,; \tag{B.1.11}$$

$$\hat{\sigma} = \Sigma_\alpha\,|\alpha\rangle\,\sigma_\alpha\,\langle\alpha|, \qquad \text{or} \qquad \sigma(\boldsymbol{r},\boldsymbol{r}') = \langle\boldsymbol{r}|\,\hat{\sigma}\,|\boldsymbol{r}'\rangle = \sum_\alpha c_\alpha(\boldsymbol{r})\,\sigma_\alpha\,c_\alpha^*(\boldsymbol{r}')\,. \tag{B.1.12}$$

These independent, collective density displacement modes, called the *Density Normal Modes* (DNM) (see, e.g., Nalewajski *et al.*, 1996; Nalewajski and Korchowiec, 1996) span the *complete* vector space, capable of describing general displacements of the electron density, including both the *CT*- and *P*-components:

$$\hat{\mathrm{P}}\,|\Delta\rho\rangle \equiv \Sigma_\alpha\,|\alpha\rangle\,\langle\alpha|\Delta\rho\rangle \equiv \Sigma_\alpha\,|\alpha\rangle\,\Delta\rho_\alpha = |\Delta\rho\rangle, \qquad \hat{\mathrm{P}} \equiv \Sigma_\alpha\,|\alpha\rangle\,\langle\alpha| = 1, \qquad \langle\boldsymbol{r}|\,\hat{\mathrm{P}}\,|\boldsymbol{r}'\rangle = \delta(\boldsymbol{r}'-\boldsymbol{r}),$$

$$\int\langle\boldsymbol{r}|\,\hat{\mathrm{P}}\,|\boldsymbol{r}'\rangle\,\langle\boldsymbol{r}'|\Delta\rho\rangle\,d\boldsymbol{r}' = \int\delta(\boldsymbol{r}'-\boldsymbol{r})\Delta\rho(\boldsymbol{r}')d\boldsymbol{r}' = \Delta\rho(\boldsymbol{r}) = \Sigma_\alpha\,c_\alpha(\boldsymbol{r})\Delta\rho_\alpha, \qquad \Delta\rho_\alpha = \int c_\alpha^*(\boldsymbol{r})\,\Delta\rho(\boldsymbol{r})\,d\boldsymbol{r}\,. \tag{B.1.13}$$

Therefore, the *CT*-projector is given by the difference

$$\hat{\mathrm{P}}_{CT} = \hat{\mathrm{P}} - \hat{\mathrm{P}}_P = \Sigma_\alpha\,|\alpha\rangle\,\langle\alpha| - \Sigma_i\,|i\rangle\,\langle i| = 1 - \hat{\mathrm{P}}_P\,. \tag{B.1.14}$$

It projects out the homogeneous CT component [see Eq. (B.1.8c)] of a general density displacement

$$\hat{P}_{CT}|\Delta\rho\rangle \equiv |\Delta\rho_{CT}\rangle_{hom} = |\Delta\rho\rangle - |\Delta\rho\rangle_P = |p_{CT}\rangle\, \Delta N \qquad \text{or}$$

$$\Delta\rho_{CT}^{hom}(\boldsymbol{r}) = \langle \boldsymbol{r}|\Delta\rho_{CT}\rangle_{hom} = \Delta\rho(\boldsymbol{r}) - \Delta\rho_P(\boldsymbol{r}) = p_{CT}\,\Delta N\,, \tag{B.1.15}$$

where $p_{CT} = 1/V = \langle \boldsymbol{r}|p_{CT}\rangle$.

In the "geometric", vector analysis of density displacements (Nalewajski. 2002e, 2005g) the complete vector space capable of representing any displacement of the molecular electron density was constructed by supplementing the closed system *P*-modes of Eq. (B.1.4) with the *homogeneous* *CT*-mode, capable of representing changes in the density normalization. For this additional mode $a(\boldsymbol{r}) = \langle \boldsymbol{r}|CT\rangle$ to be orthogonal to all polarizational functions $\boldsymbol{b}(\boldsymbol{r})$ [see Eq. (B.1.2)] it has to be constant throughout space: $a(\boldsymbol{r}) \equiv V^{-1/2} = const.$ Alternatively, to avoid problems with an infinite molecular volume, one can select V locally, as the volume $V \equiv V_s(\boldsymbol{r}) = 4\pi r_s(\boldsymbol{r})^3/3$ of the Wigner-Seitz sphere of the homogeneous electron gas, with the density dependent radius $r_s(\boldsymbol{r}) = \rho^{-1}(\boldsymbol{r})$, so that $V_s(\boldsymbol{r})$ by definition contains a single electron, $V_s(\boldsymbol{r})\rho(\boldsymbol{r}) = 1$. In this way one can interpret the constant (renormalized) *CT*-mode $|\boldsymbol{CT}\rangle$ as

$$\bar{a}(\boldsymbol{r}) = \langle \boldsymbol{r}|\boldsymbol{CT}\rangle \equiv V_s(\boldsymbol{r})\rho(\boldsymbol{r}) = 1. \tag{B.1.16}$$

These vectors explicitly define the corresponding CT-projectors:

$$\bar{\hat{P}}^{CT} = |\boldsymbol{CT}\rangle\langle \boldsymbol{CT}|, \quad \langle \boldsymbol{r}|\bar{\hat{P}}^{CT}|\boldsymbol{r}'\rangle = 1; \qquad \hat{P}^{CT} = |CT\rangle\langle CT|, \qquad \langle \boldsymbol{r}|\hat{P}_{CT}|\boldsymbol{r}'\rangle = V^{-1}. \tag{B.1.17}$$

The *P*-modes then reflect the *local* aspect of shifts in the electron density, while the *CT*-mode corresponds to the homogeneous (normalization) facet of the electron distribution in the molecule. Since the two homogeneous *CT*-projectors are renormalized versions of each other, in what follows we shall explicitly use the $\hat{P}^{CT}$ operator, which makes reference to the molecular volume. When $\hat{P} = \hat{P}^{CT} + \hat{P}_P = 1$ the overall projection onto the whole (*CT*, *P*)-vector space gives:

$$\langle \boldsymbol{r}|\hat{P}|\boldsymbol{r}'\rangle = \delta(\boldsymbol{r}'-\boldsymbol{r}) = V^{-1} + \gamma(\boldsymbol{r}, \boldsymbol{r}'). \tag{B.1.18}$$

Let us again examine the result of the *CT*-projection of $|\Delta\rho\rangle$:

$$\begin{aligned}\langle \boldsymbol{r}|\hat{P}^{CT}|\Delta\rho\rangle &= \Delta N\,\langle \boldsymbol{r}|\hat{P}^{CT}|p\rangle = \Delta N\,\langle \boldsymbol{r}|p_{CT}\rangle \equiv \Delta N\; p_{CT}(\boldsymbol{r})\\ &= \int \langle \boldsymbol{r}|\hat{P}_{CT}|\boldsymbol{r}'\rangle\,\langle \boldsymbol{r}'|\Delta\rho\rangle\, d\boldsymbol{r}' = V^{-1}\int \Delta\rho(\boldsymbol{r}')\, d\boldsymbol{r}' = \Delta N/V \equiv \Delta\rho_{CT}^{hom}\,,\end{aligned} \tag{B.1.19}$$

Thus the additional *CT*-mode indeed describes the homogeneous *CT*-displacement of the system electronic density (see Fig. B1), which does not bias any location in space.

The *P*-projection of the density displacement vector similarly gives [Eqs. (B.1.7) and (B.1.8d)]:

$$\begin{aligned}\langle \boldsymbol{r}|\,\hat{\mathrm{P}}_P\,|\Delta\rho\rangle &= N\langle \boldsymbol{r}\,|\,\hat{\mathrm{P}}_P\,|\Delta p\rangle + \Delta N\langle \boldsymbol{r}\,|\,\hat{\mathrm{P}}_P\,|p\rangle \\ &\equiv N\langle \boldsymbol{r}\,|\Delta p\rangle + \Delta N\langle \boldsymbol{r}\,|\,p_P\rangle \equiv N\,\Delta p(\boldsymbol{r}) + \Delta N\;p_P(\boldsymbol{r}) \\ &= \int \langle \boldsymbol{r}\,|\,\hat{\mathrm{P}}_P\,|\boldsymbol{r}'\rangle\,\langle \boldsymbol{r}'|\Delta\rho\rangle = \int[\delta(\boldsymbol{r}'-\boldsymbol{r}) - V^{-1}]\Delta\rho(\boldsymbol{r}')\,d\boldsymbol{r}' \equiv \;\Delta\rho_P(\boldsymbol{r}) \\ &= \Delta\rho(\boldsymbol{r}) - \Delta N/V = N\Delta p(\boldsymbol{r}) + \Delta N\,[p(\boldsymbol{r}) - V^{-1}] \equiv \Delta\rho_N(\boldsymbol{r}) + \Delta\rho_{CT}^{inhom}(\boldsymbol{r})\,. \end{aligned} \qquad \text{(B.1.20)}$$

Rewriting $\Delta\rho_P(\boldsymbol{r})$ in terms of the P-modes $\boldsymbol{b}(\boldsymbol{r})$ also gives (see Fig. B1):

$$\begin{aligned}\langle \boldsymbol{r}|\,\hat{\mathrm{P}}_P\,|\Delta\rho\rangle &= \Sigma_i\, b_i(\boldsymbol{r})\,[\int b_i^*(\boldsymbol{r}')\,\Delta\rho(\boldsymbol{r}')\,d\boldsymbol{r}'] \equiv \Sigma_i\, b_i(\boldsymbol{r})\,\Delta\rho_i \\ &= \Sigma_i\, b_i(\boldsymbol{r})\,\{N\,[\int b_i^*(\boldsymbol{r}')\,\Delta p(\boldsymbol{r}')\,d\boldsymbol{r}'] + \Delta N\,[\int b_i^*(\boldsymbol{r}')\,p(\boldsymbol{r}')\,d\boldsymbol{r}']\} \\ &\equiv \Sigma_i\, b_i(\boldsymbol{r})\,(N\,\Delta p_i + \Delta N\,p_i). \end{aligned} \qquad \text{(B.1.21)}$$

Therefore, the *homogeneous* CT-component, $\Delta\rho_{CT}^{hom}(\boldsymbol{r}) \equiv \langle \boldsymbol{r}|\Delta\rho_{CT}\rangle_{hom}$, of the open system displacement in the electron density indeed corresponds to the uniform distribution of ΔN electrons in the whole molecular volume, while its overall P-component $\Delta\rho_P(\boldsymbol{r}) \equiv \langle \boldsymbol{r}|\Delta\rho\rangle_P$ now contains the closed-system polarization density, $\Delta\rho_N(\boldsymbol{r}) \equiv N\,\Delta p(\boldsymbol{r}) \equiv \langle \boldsymbol{r}|\Delta\rho_P\rangle_N$, and the *inhomogeneous* part of the CT-component, $\Delta\rho_{CT}^{inhom}(\boldsymbol{r}) \equiv \langle \boldsymbol{r}|\Delta\rho_{CT}\rangle_{inhom}$ (see Fig. B1). As also shown in the figure the two CT-related vectors define the overall density displacement vector:

$$|\Delta\rho\rangle_{CT} = |\Delta\rho_{CT}\rangle_{hom} + |\Delta\rho_{CT}\rangle_{inhom} = \Delta N\,|p\rangle. \qquad \text{(B.1.22)}$$

The above inhomogeneous CT-component should not be confused with the real *CT-induced polarization*, which represents the *second-order* effect involving the product of ΔN and $|\Delta p\rangle$.

As indicated in Eq. (B.1.8c) the above partitioning of the density CT-displacement into its homogeneous and inhomogeneous components results from the division of the probability (shape-factor) vector:

$$|p\rangle = (\,\hat{\mathrm{P}}_P + \hat{\mathrm{P}}_{CT}\,)|p\rangle \equiv |p_P\rangle + |p_{CT}\rangle. \qquad \text{(B.1.23)}$$

It then follows from Eqs. (B.1.19) and (B.1.20) that

$$\langle \boldsymbol{r}\,|\,p_{CT}\rangle = p_{CT}(\boldsymbol{r}) = V^{-1} \qquad \text{and} \qquad \langle \boldsymbol{r}\,|\,p_P\rangle = p_P(\boldsymbol{r}) = p(\boldsymbol{r}) - V^{-1}. \qquad \text{(B.1.24)}$$

The resultant polarization vector of the open molecular system [Eqs. (B.1.8d), (B.1.20), and (B.1.21)] thus includes the inhomogeneous ΔN-dependent term, representing the pseudo-polarization relative to the homogeneous background of the CT-mode, which results from distributing the ΔN electrons among local volume elements in proportion to the local shape factor $p(\boldsymbol{r})$. This extra displacement is explicitly shown in Fig. B1. It follows from Eq. (B.1.21) that the closed system term $\langle \boldsymbol{r}|\Delta\rho\rangle_N = N\,\Sigma_i\, b_i(\boldsymbol{r})\,\Delta p_i = N\,\Delta p(\boldsymbol{r})$, while the inhomogeneous part of the CT-displacement of the electron density $\langle \boldsymbol{r}|\Delta\rho_{CT}\rangle_{inhom} = \Delta N\;p_P(\boldsymbol{r}) = \Delta N\,\Sigma_i\, b_i(\boldsymbol{r})\,p_i$.

B.2. Density-Potential Relations

Let us summarize in the vector form the relations linking the shifts $|\Delta\rho\rangle$ in the system electron density $\Delta\rho(\boldsymbol{r})$ and the corresponding vectors representing changes in the external potential, $\langle\boldsymbol{r}|\Delta v\rangle = \Delta v(\boldsymbol{r})$, and the relative potential, $\langle\boldsymbol{r}|\Delta u\rangle = \Delta u(\boldsymbol{r}) = \Delta v(\boldsymbol{r}) - \Delta\mu(\boldsymbol{r})$, where $\Delta\mu(\boldsymbol{r}) = \langle\boldsymbol{r}|\Delta\mu\rangle$ stands for the displacement in the system local chemical potential of electrons. The latter is equalized for the ground-state (equilibrium) density at the global chemical potential level $\mu[N, v]$ (Parr *et al.*, 1978, Parr and Yang, 1989; Nalewajski *et al.*, 1996; Nalewajski and Korchowiec, 1997),

$$\mu(\boldsymbol{r}) = \frac{\delta E_v[\rho]}{\delta\rho(\boldsymbol{r})} = v(r) + \frac{\delta F[\rho]}{\delta\rho(\boldsymbol{r})} = \mu[N,v] = \left(\frac{\partial E[N,v]}{\partial N}\right)_v \text{ or } u(\boldsymbol{r}) \equiv v(\boldsymbol{r}) - \mu = -\frac{\delta F[\rho]}{\delta\rho(\boldsymbol{r})}, \quad \text{(B.2.1a)}$$

where, the universal (v-independent) part $F[\rho] = T_e[\rho] + V_{ee}[\rho]$ of the density functional for the system electronic energy,

$$E_v[\rho] = \int \rho(\boldsymbol{r}) v(\boldsymbol{r}) d\boldsymbol{r} + F[\rho], \quad \text{(B.2.2)}$$

generates the sum of the electronic kinetic (T_e) and repulsion (V_{ee}) energies. The chemical potential (electronegativity) equalization of Eq. (25a) is implied by the Hohenberg-Kohn (1964) Euler equation derived from the N-constrained minimum principle for the system electronic energy:

$$\delta\{E_v[\rho] - \mu[N, v]\, N[\rho]\} = 0. \quad \text{(B.2.1b)}$$

Equation (B.2.1a) implies the following expression for the linear response in the relative external potential:

$$\Delta u(\boldsymbol{r}) \equiv \Delta v(\boldsymbol{r}) - \Delta\mu = -\Delta\left(\frac{\delta F[\rho]}{\delta\rho(\boldsymbol{r})}\right) = -\int \Delta\rho(\boldsymbol{r}')\frac{\delta^2 F[\rho]}{\delta\rho(\boldsymbol{r}')\delta\rho(\boldsymbol{r})} d\boldsymbol{r}' \equiv -\int \Delta\rho(\boldsymbol{r}')\eta(\boldsymbol{r}', \boldsymbol{r})\, d\boldsymbol{r}', \quad \text{(B.2.3)}$$

where the *chemical hardness kernel*,

$$\eta(\boldsymbol{r}', \boldsymbol{r}) = \langle\boldsymbol{r}'|\hat{\eta}|\boldsymbol{r}\rangle = -\frac{\delta u(\boldsymbol{r})}{\delta\rho(\boldsymbol{r}')} \equiv -\frac{\langle\delta u|\boldsymbol{r}\rangle}{\langle\boldsymbol{r}'|\delta\rho\rangle}, \quad \text{(B.2.4)}$$

represents the chemical *hardness operator* $\hat{\eta} = -\frac{\langle\delta u|}{|\delta\rho\rangle}$. Equation (B.2.3) can be written in the equivalent vector form:

$$|\Delta u\rangle = |\Delta v\rangle - |\Delta\mu\rangle = -\hat{\eta}\,|\Delta\rho\rangle. \quad \text{(B.2.5)}$$

The inverse relation,

$$|\Delta\rho\rangle = -\hat{\sigma}\,|\Delta u\rangle, \quad \text{(B.2.6)}$$

is obtained by acting on both sides of Eq. (B.2.5) with the chemical *softness operator* $\hat{\sigma} = \hat{\eta}^{-1} = -\frac{\langle\delta\rho|}{|\delta u\rangle}$, which gives rise to the *softness kernel*:

$$\sigma(\boldsymbol{r}', \boldsymbol{r}) = \eta^{-1}(\boldsymbol{r}', \boldsymbol{r}) = \langle \boldsymbol{r}' | \hat{\sigma} | \boldsymbol{r} \rangle = -\frac{\delta\rho(\boldsymbol{r})}{\delta u(\boldsymbol{r}')} \equiv -\frac{\langle\delta\rho|\boldsymbol{r}\rangle}{\langle\boldsymbol{r}'|\delta u\rangle}. \tag{B.2.7}$$

These inverse kernels satisfy the functional reciprocity relation:

$$\int\sigma(\boldsymbol{r},\boldsymbol{r}'')\,\eta(\boldsymbol{r}'',\boldsymbol{r}')\,d\boldsymbol{r}'' = \int\frac{\delta\rho(\boldsymbol{r}'')}{\delta u(\boldsymbol{r})}\frac{\delta u(\boldsymbol{r}')}{\delta\rho(\boldsymbol{r}'')}\,d\boldsymbol{r}'' = \frac{\delta u(\boldsymbol{r}')}{\delta u(\boldsymbol{r})} = \delta(\boldsymbol{r}'-\boldsymbol{r}). \tag{B.2.8}$$

In the externally *closed* molecular system, for which the density displacement $\Delta\rho_N(\boldsymbol{r}) = \langle\boldsymbol{r}|\Delta\rho\rangle_N = N\,\Delta p(\boldsymbol{r})$ [see Eq. (B.1.3)],

$$_N\langle\Delta\rho| = N\,\langle\Delta p| = \langle\Delta v|\,\hat{\beta} = \langle\Delta v|\,\frac{_N\langle\delta\rho|}{|\delta v\rangle}. \tag{B.2.9}$$

In the position representation this operator equation in the DDM Hilbert space assumes the familiar form:

$$\Delta\rho_N(\boldsymbol{r}) = {}_N\langle\Delta\rho|\boldsymbol{r}\rangle = N\,\Delta p(\boldsymbol{r}) = \int\langle\Delta v|\boldsymbol{r}'\rangle\,\frac{_N\langle\delta\rho|\boldsymbol{r}\rangle}{\langle\boldsymbol{r}'|\delta v\rangle}\,d\boldsymbol{r}' = \int\Delta v(\boldsymbol{r}')\,\beta(\boldsymbol{r}', \boldsymbol{r})\,d\boldsymbol{r}'. \tag{B.2.10}$$

The corresponding inverse relations are:

$$\langle\Delta v| = {}_N\langle\Delta\rho|\,\hat{\beta}^{-1} = N\,\langle\Delta p|\,\frac{\langle\delta v|}{|\delta\rho\rangle_N}, \tag{B.2.11}$$

or in the position representation:

$$\langle\Delta v|\boldsymbol{r}\rangle = \Delta v(\boldsymbol{r}) = N\int\langle\Delta p|\boldsymbol{r}'\rangle\,\frac{\langle\delta v|\boldsymbol{r}\rangle}{\langle\boldsymbol{r}'|\delta\rho\rangle_N}\,d\boldsymbol{r}' = N\int\Delta p(\boldsymbol{r}')\,\beta^{-1}(\boldsymbol{r}', \boldsymbol{r})\,d\boldsymbol{r}'. \tag{B.2.12}$$

The so called *internal* hardness operator (Nalewajski, 2002e, 2005g),

$$\hat{\eta}^{int} \equiv -\,\hat{\beta}^{-1} = -\,\Sigma_i\,|i\rangle\,\beta_i^{-1}\,\langle i| = -\frac{\langle\delta v|}{|\delta\rho\rangle_N}, \tag{B.2.13}$$

defines the internal hardness kernel of the closed molecular system:

$$\eta^{int}(\boldsymbol{r}, \boldsymbol{r}') = -\frac{\langle\delta v|\boldsymbol{r}'\rangle}{\langle\boldsymbol{r}|\delta\rho\rangle_N} = -\,\beta^{-1}(\boldsymbol{r},\boldsymbol{r}') = -\left(\frac{\partial v(\boldsymbol{r}')}{\partial\rho(\boldsymbol{r})}\right)_N \equiv -\,\langle\boldsymbol{r}|\,\hat{\beta}^{-1}\,|\boldsymbol{r}'\rangle. \tag{B.2.14}$$

The negative density LR kernel of Eq. (B.1.3) similarly defines the *internal* softness kernel of the closed molecular system, the position representation of the corresponding operator

$$\hat{\sigma}^{int} \equiv -\hat{\beta} = -\Sigma_i |i\rangle\, \beta_i \langle i| = -\frac{{}_N\langle\delta\rho|}{|\delta v\rangle}, \tag{B.2.15}$$

$$\sigma^{int}(\boldsymbol{r},\boldsymbol{r}') = -\frac{{}_N\langle\delta\rho|\boldsymbol{r}'\rangle}{\langle\boldsymbol{r}|\delta v\rangle} = -\beta(\boldsymbol{r},\boldsymbol{r}') = -\left(\frac{\partial\rho(\boldsymbol{r}')}{\partial v(\boldsymbol{r})}\right)_N \equiv -\langle\boldsymbol{r}|\hat{\beta}|\boldsymbol{r}'\rangle. \tag{B.2.16}$$

The operator equations linking the density and the external potential vectors of the *open* molecular system are given in Eqs. (B.2.5) and (B.2.6). The linear *responses* in the system relative potential due to a given *perturbation* in the electron distribution are defined by the hardness operator/kernel:

$$\langle\Delta u| = \langle\Delta v| - \langle\Delta\mu| = -\langle\Delta\rho|\,\hat{\eta} \quad = -N\langle\Delta p|\,\hat{\eta} - \Delta N\langle p|\,\hat{\eta}, \tag{B.2.17}$$

$$\begin{aligned}\langle\Delta u|\boldsymbol{r}\rangle &= \Delta u(\boldsymbol{r}) = \Delta v(\boldsymbol{r}) - \Delta\mu = [\Delta u(\boldsymbol{r})]_\mu + [\Delta u(\boldsymbol{r})]_v = -\int\langle\Delta\rho|\boldsymbol{r}'\rangle\langle\boldsymbol{r}'|\,\hat{\eta}\,|\boldsymbol{r}\rangle\, d\boldsymbol{r}' = -\int\Delta\rho(\boldsymbol{r}')\,\eta(\boldsymbol{r}',\boldsymbol{r})\,d\boldsymbol{r}' \\ &= -\int[N\,\Delta p(\boldsymbol{r}') + \Delta N\,p(\boldsymbol{r}')]\,\eta(\boldsymbol{r}',\boldsymbol{r})\,d\boldsymbol{r}' = -\int\{[\Delta\rho(\boldsymbol{r}')]_N + [\Delta\rho(\boldsymbol{r}')]_p\}\,\eta(\boldsymbol{r}',\boldsymbol{r})\,d\boldsymbol{r}'.\end{aligned} \tag{B.2.18}$$

The corresponding inverse relations linking perturbations in the relative potential with the equilibrium linear responses of the system electron density are determined by the system softness operator/kernel:

$$\langle\Delta\rho| = N\langle\Delta p| - \Delta N\langle p| = -\langle\Delta u|\,\hat{\sigma} = -\langle\Delta v|\,\hat{\sigma} + \langle\Delta\mu|\,\hat{\sigma}, \tag{B.2.19}$$

$$\begin{aligned}\langle\Delta\rho|\boldsymbol{r}\rangle &= \Delta\rho(\boldsymbol{r}) = N\,\Delta p(\boldsymbol{r}) + \Delta N\,p(\boldsymbol{r}) = [\Delta\rho(\boldsymbol{r})]_N + [\Delta\rho(\boldsymbol{r})]_p = -\int\langle\Delta u|\boldsymbol{r}'\rangle\langle\boldsymbol{r}'|\,\hat{\sigma}\,|\boldsymbol{r}\rangle\,d\boldsymbol{r}' \\ &= -\int\Delta u(\boldsymbol{r}')\,\sigma(\boldsymbol{r}',\boldsymbol{r})\,d\boldsymbol{r}' = \int[\Delta\mu - \Delta v(\boldsymbol{r}')]\sigma(\boldsymbol{r}',\boldsymbol{r})\,d\boldsymbol{r}' \\ &= -\int\{[\Delta u(\boldsymbol{r}')]_\mu + [\Delta u(\boldsymbol{r}')]_v\}\,\sigma(\boldsymbol{r}',\boldsymbol{r})\,d\boldsymbol{r}'.\end{aligned} \tag{B.2.20}$$

It follows from Eq. (B.2.18) that the hardness kernel of Eq. (B.2.4) can be alternatively expressed by the following *partial* derivatives involving the external and chemical potentials:

$$\begin{aligned}\eta(\boldsymbol{r}',\boldsymbol{r}) &= -\left(\frac{\partial u(\boldsymbol{r})}{\partial\rho(\boldsymbol{r}')}\right)_N = -\left(\frac{\partial v(\boldsymbol{r})}{\partial\rho(\boldsymbol{r}')}\right)_N + \left(\frac{\partial\mu}{\partial\rho(\boldsymbol{r}')}\right)_N = -\frac{\delta u(\boldsymbol{r})}{N\delta p(\boldsymbol{r}')} \\ &= -\left(\frac{\partial u(\boldsymbol{r})}{\partial\rho(\boldsymbol{r}')}\right)_p = -\left(\frac{\partial v(\boldsymbol{r})}{\partial\rho(\boldsymbol{r}')}\right)_p + \left(\frac{\partial\mu}{\partial\rho(\boldsymbol{r}')}\right)_p = -\frac{\delta u(\boldsymbol{r})}{p(\boldsymbol{r}')\delta N}.\end{aligned} \tag{B.2.21a}$$

Let us first examine the *internal* derivatives, for the fixed *N*, in the first row of the preceding equation. The first partial derivative represents the internal hardness kernel of Eq. (B.2.14), given by the negative inverse of the density LR kernel. It represents the external potential response due to the local change in the density resulting from the displacements in the density shape factor, for its fixed normalization. Therefore, the other derivative, with respect to the closed system electronic density, corresponding to the shift in the system chemical potential due to such an internal density displacement, is given by the difference

$$\left(\frac{\partial \mu}{\partial \rho(\boldsymbol{r}')}\right)_N = \eta(\boldsymbol{r}', \boldsymbol{r}) - \eta^{int}(\boldsymbol{r}', \boldsymbol{r}) \quad \text{or} \quad \left(\frac{\partial \mu}{\partial p(\boldsymbol{r}')}\right)_N = N[\eta(\boldsymbol{r}', \boldsymbol{r}) - \eta^{int}(\boldsymbol{r}', \boldsymbol{r})]. \tag{B.2.21b}$$

The first partial derivative of the second row in Eq. (B.2.21a), measuring the external potential response per unit change in the overall number of electrons, for the fixed density probability factor, results from the chain-rule for implicit functionals:

$$-\left(\frac{\partial v(\boldsymbol{r})}{\partial \rho(\boldsymbol{r}')}\right)_p = -\left(\frac{\partial v(\boldsymbol{r})}{p(\boldsymbol{r}')\partial N}\right)_p = \int \left(\frac{\partial \rho(\boldsymbol{r}'')}{p(\boldsymbol{r}')\partial N}\right)_v \left(\frac{\partial v(\boldsymbol{r})}{\partial \rho(\boldsymbol{r}'')}\right)_N d\boldsymbol{r}''$$
$$= p(\boldsymbol{r}')^{-1} \int f(\boldsymbol{r}'')\beta^{-1}(\boldsymbol{r}'', \boldsymbol{r}) d\boldsymbol{r}'' \equiv \frac{\beta^{-1}(\boldsymbol{r})}{p(\boldsymbol{r}')}, \tag{B.2.21c}$$

where the electronic *Fukui Function* (FF) (Parr and Yang, 1984)

$$f(\boldsymbol{r}) = \left(\frac{\partial \rho(\boldsymbol{r})}{\partial N}\right)_v = \left(\frac{\partial \mu}{\partial v(\boldsymbol{r})}\right)_N, \tag{B.2.22}$$

and $\beta^{-1}(\boldsymbol{r})$ stands for the internal local hardness, which is not equalized. Thus, from Eq. (B.2.21a),

$$\left(\frac{\partial \mu}{\partial \rho(\boldsymbol{r}')}\right)_p = \left(\frac{\partial \mu}{p(\boldsymbol{r}')\partial N}\right)_p = \eta(\boldsymbol{r}', \boldsymbol{r}) - \frac{\beta^{-1}(\boldsymbol{r})}{p(\boldsymbol{r}')}. \tag{B.2.21d}$$

A similar division can be carried out for the softness kernel. It can be expressed using the functional chain-rule transformation in terms of the closed system density response, and the extra term present in the open molecular systems, which involves the electronic FF and the system global softness (Parr and Pearson 1983; Parr and Yang, 1989; Nalewajski and Korchowiec 1997). For this purpose we express the ground-state density as functional of N and v, $\rho = \rho[N, v]$:

$$\sigma(\boldsymbol{r}, \boldsymbol{r}') = \eta(\boldsymbol{r}', \boldsymbol{r})^{-1} = -\frac{\delta \rho(\boldsymbol{r}')}{\delta u(\boldsymbol{r})} = -\left(\frac{\partial \rho(\boldsymbol{r}')}{\partial v(\boldsymbol{r})}\right)_\mu = -\left(\frac{\partial \rho(\boldsymbol{r}')}{\partial v(\boldsymbol{r})}\right)_N - \left(\frac{\partial N}{\partial v(\boldsymbol{r})}\right)_\mu \left(\frac{\partial \rho(\boldsymbol{r}')}{\partial N}\right)_v$$
$$= -\beta(\boldsymbol{r}, \boldsymbol{r}') + s(\boldsymbol{r}) f(\boldsymbol{r}') = -\beta(\boldsymbol{r}, \boldsymbol{r}') + f(\boldsymbol{r})\, S f(\boldsymbol{r}'). \tag{B.2.23}$$

Here the *local softness*

$$s(\boldsymbol{r}) = \left(\frac{\partial \rho(\boldsymbol{r})}{\partial \mu}\right)_v = \left(\frac{\partial \rho(\boldsymbol{r})}{\partial N}\right)_v \left(\frac{\partial N}{\partial \mu}\right)_v = f(\boldsymbol{r})\, S = -\left(\frac{\partial N}{\partial v(\boldsymbol{r})}\right)_\mu = \int \sigma(\boldsymbol{r}, \boldsymbol{r}')\, d\boldsymbol{r}', \tag{B.2.24}$$

and the *global softness*

$$S = \left(\frac{\partial N}{\partial \mu}\right)_v = \iint \frac{\delta \rho(\boldsymbol{r}')}{\delta u(\boldsymbol{r})} \left(\frac{\partial u(\boldsymbol{r})}{\partial \mu}\right)_v d\boldsymbol{r}\, d\boldsymbol{r}' = \iint \sigma(\boldsymbol{r}, \boldsymbol{r}')\, d\boldsymbol{r}\, d\boldsymbol{r}' = \int \left(\frac{\partial \rho(\boldsymbol{r})}{\partial \mu}\right)_v d\boldsymbol{r} = \int s(\boldsymbol{r})\, d\boldsymbol{r}\,. \tag{B.2.25}$$

It also follows from Eq. (B.2.20) that

$$\sigma(\boldsymbol{r}', \boldsymbol{r}) = -\left(\frac{\partial\rho(\boldsymbol{r})}{\partial u(\boldsymbol{r}')}\right)_{\mu} = -\left(\frac{\partial\rho(\boldsymbol{r})}{\partial v(\boldsymbol{r}')}\right)_{\mu} = -\left(\frac{\partial\rho(\boldsymbol{r})}{\partial u(\boldsymbol{r}')}\right)_{v} = -\left(\frac{\partial\rho(\boldsymbol{r})}{\partial \mu(\boldsymbol{r}')}\right)_{v}, \tag{B.2.26}$$

where the non-equilibrium (non-equalized) local displacement of the system chemical potential $[\delta\mu(\boldsymbol{r})]_v = \int [\delta\rho(\boldsymbol{r}')]_v\, \eta(\boldsymbol{r}', \boldsymbol{r})\, d\boldsymbol{r}'$. Using Eqs. (B.1.8a) and (B.2.24) then gives:

$$\sigma(\boldsymbol{r}',\boldsymbol{r}) = -N\left(\frac{\partial p(\boldsymbol{r})}{\partial v(\boldsymbol{r}')}\right)_{\mu} - p(\boldsymbol{r})s(\boldsymbol{r}') = N\left(\frac{\partial p(\boldsymbol{r})}{\partial \mu(\boldsymbol{r}')}\right)_{v} - p(\boldsymbol{r})s(\boldsymbol{r}'). \tag{B.2.27}$$

Finally, combining Eqs. (B.2.11) and (B.2.17) gives an explicit expression for the chemical potential displacement:

$$\langle\Delta\mu| = \langle\Delta v| - \langle\Delta u| = \langle\Delta\rho|(\hat{\mathrm{P}}_P\, \hat{\beta}^{-1} + \hat{\eta}) = {}_P\langle\Delta\rho|(\hat{\beta}^{-1} + \hat{\eta}) + {}_{CT}\langle\Delta\rho|\,\hat{\eta} \tag{B.2.28}$$

or

$$\Delta\mu(\boldsymbol{r}) = \Delta\mu = \int \Delta\rho_N(\boldsymbol{r}')[\beta^{-1}(\boldsymbol{r}', \boldsymbol{r}) + \eta(\boldsymbol{r}', \boldsymbol{r})]\, d\boldsymbol{r}' + \int \Delta\rho_{CT}(\boldsymbol{r}')\eta(\boldsymbol{r}', \boldsymbol{r})\, d\boldsymbol{r}'$$

$$= \int \Delta\rho(\boldsymbol{r}')[\int \gamma(\boldsymbol{r}', \boldsymbol{r}'')\beta^{-1}(\boldsymbol{r}'', \boldsymbol{r})\, d\boldsymbol{r}'' + \eta(\boldsymbol{r}', \boldsymbol{r})]\, d\boldsymbol{r}' \equiv \int \Delta\rho(\boldsymbol{r}')\left(\frac{\delta\mu(\boldsymbol{r})}{\delta\rho(\boldsymbol{r}')}\right) d\boldsymbol{r}', \tag{B.2.29}$$

where the position representation $\gamma(\boldsymbol{r}, \boldsymbol{r}')$ of the P-projector has been defined in Eqs. (B.1.6) and (B.1.18). One also identifies the first term in Eq. (B.2.20) [see also Eqs. (B.2.1) and (B.2.2)] as

$$\langle\Delta v| = \langle\Delta\rho|\,\hat{\mathrm{P}}_P\,\hat{\beta}^{-1} = {}_P\langle\Delta\rho|\hat{\beta}^{-1} \quad \text{or}$$

$$\Delta v(\boldsymbol{r}) = \int [\int \Delta\rho(\boldsymbol{r}')\,\gamma(\boldsymbol{r}', \boldsymbol{r}'')\, d\boldsymbol{r}']\beta^{-1}(\boldsymbol{r}'', \boldsymbol{r})\, d\boldsymbol{r}'' \equiv \int \Delta\rho_P(\boldsymbol{r}'')\left(\frac{\delta v(\boldsymbol{r})}{\delta\rho(\boldsymbol{r}'')}\right)_N d\boldsymbol{r}''. \tag{B.2.30}$$

In Eq. (B.2.21a) we have effectively partitioned the hardness kernel into the additive components representing the linear responses in the system chemical and external potentials:

$$\eta(\boldsymbol{r}', \boldsymbol{r}) = -\frac{\delta u(\boldsymbol{r})}{\delta\rho(\boldsymbol{r}')} = -\frac{\langle\delta u|\boldsymbol{r}\rangle}{\langle\boldsymbol{r}'|\delta\rho\rangle} = \left(\frac{\delta\mu(\boldsymbol{r})}{\delta\rho(\boldsymbol{r}')}\right) - \left(\frac{\delta v(\boldsymbol{r})}{\delta\rho(\boldsymbol{r}')}\right) = \frac{\langle\delta\mu|\boldsymbol{r}\rangle}{\langle\boldsymbol{r}'|\delta\rho\rangle} - \frac{\langle\delta v|\boldsymbol{r}\rangle}{\langle\boldsymbol{r}'|\delta\rho\rangle}. \tag{B.2.31}$$

Equation (B.2.23) identifies the corresponding inverse functional derivatives

$$\sigma(\boldsymbol{r}',\boldsymbol{r}) = -\frac{\delta\rho(\boldsymbol{r})}{\delta u(\boldsymbol{r}')} = -\frac{\langle\delta\rho|\boldsymbol{r}\rangle}{\langle\boldsymbol{r}'|\delta u\rangle} = \left(\frac{\delta\rho(\boldsymbol{r})}{\delta\mu(\boldsymbol{r}')}\right)_v = \left(\frac{\langle\delta\rho|\boldsymbol{r}\rangle}{\langle\boldsymbol{r}'|\delta\mu\rangle}\right)_v = -\left(\frac{\delta\rho(\boldsymbol{r})}{\delta v(\boldsymbol{r}')}\right)_{\mu} = -\left(\frac{\langle\delta\rho|\boldsymbol{r}\rangle}{\langle\boldsymbol{r}'|\delta v\rangle}\right)_{\mu}. \tag{B.2.32}$$

In Section B.4 we shall use the (P,CT)-projections to extract from the hardness and softness operators/kernels the specific components reflecting the pure P or CT effects, and those reflecting the coupling between these two degrees-of-freedom of electronic densities in molecular systems.

B.3. Geometric Decomposition of the Fukui Function

Equation (B.2.23) can be also interpreted as the position representation of the underlying operator equation in the molecular Hilbert space for density displacements

$$\hat{\sigma} = -\hat{\beta} + |f\rangle S\langle f| \equiv -\hat{\beta} + S\hat{\mathrm{P}}_f, \tag{B.3.1}$$

where $|f\rangle$ represents the electronic FF [Eq. (B.2.22)] in the density-displacement vector space: $f(\boldsymbol{r}) = \langle \boldsymbol{r}|f\rangle$, which defines the associated FF projector $\hat{\mathrm{P}}_f = |f\rangle\langle f|$. The complementary ($P$, CT)-projections of the density displacements in the *open* molecular systems then give rise to the corresponding geometric interpretation of the FF vector:

$$|f\rangle = \left(\frac{|\partial\rho\rangle}{\partial N}\right)_v = N\left(\frac{|\partial p\rangle}{\partial N}\right)_v + |p\rangle \equiv |f\rangle_N + |f\rangle_{CT} = |f\rangle_N + (|p_{CT}\rangle + |p_P\rangle) \equiv |f\rangle_N + [\,|f_{CT}\rangle_{hom} + |f_{CT}\rangle_{inhom}]$$

$$= (|f\rangle_N + |p_P\rangle) + |p_{CT}\rangle = (|f\rangle_N + |f_{CT}\rangle_{inhom}) + |f_{CT}\rangle_{hom} \equiv |f\rangle_P + |f_{CT}\rangle_{hom}, \tag{B.3.2}$$

where the *internal*, closed system component,

$$|f\rangle_N = \left(\frac{\partial|\rho\rangle_N}{\partial N}\right)_v = \left(\frac{N\partial|p\rangle}{\partial N}\right)_v, \tag{B.3.3}$$

is due to a change in the shape factor of the electron distribution, and the *external CT*-component,

$$|f\rangle_{CT} = \left(\frac{\partial|\rho\rangle_{CT}}{\partial N}\right)_v = |p\rangle, \tag{B.3.4}$$

contains both the *inhomogeneous* (polarizational) component,

$$|f_{CT}\rangle_{inhom} = \left(\frac{\partial(\hat{\mathrm{P}}_P|\rho\rangle_{CT})}{\partial N}\right)_v = |p_P\rangle, \tag{B.3.5}$$

and the homogeneous part

$$|f_{CT}\rangle_{hom} = \left(\frac{\partial(\hat{\mathrm{P}}_{CT}|\rho\rangle_{CT})}{\partial N}\right)_v = |p_{CT}\rangle = 1/V, \tag{B.3.6}$$

which reflects the shift in the density normalization due to the a unit change in the system overall number of electrons.

Therefore, as in the case of the density displacement, the identity projection of the FF vector,

$$|f\rangle = (\hat{\mathrm{P}}_{CT} + \hat{\mathrm{P}}_P)|f\rangle \equiv [|f\rangle_N + |f_{CT}\rangle_{inhom}] + |f_{CT}\rangle_{hom} = |f\rangle_P + |f_{CT}\rangle_{hom}, \tag{B.3.7}$$

separates the overall *intra-system*, *P*-component of an open molecule, $|f\rangle_P = |f\rangle_N + |p_P\rangle$, from the homogeneous external component, $|p_{CT}\rangle = 1/V$ [Eq. (B.1.15)]. They give rise to the associated contributions to the FF in the position representation:

$$f(\boldsymbol{r}) = \langle \boldsymbol{r}|f\rangle = \langle \boldsymbol{r}|f\rangle_P + \langle \boldsymbol{r}|f_{CT}\rangle_{hom} = f_P(\boldsymbol{r}) + 1/V, \tag{B.3.8}$$

$$f_P(\boldsymbol{r}) = \langle \boldsymbol{r}|f\rangle_P = [f(\boldsymbol{r})]_N + p_P(\boldsymbol{r}) = N\left(\frac{\partial p(\boldsymbol{r})}{\partial N}\right)_v + \int \gamma(\boldsymbol{r},\boldsymbol{r}')\,p(\boldsymbol{r}')\,d\boldsymbol{r}'. \tag{B.3.9}$$

B.4. Geometric Decomposition of Density-Potential Kernels

The density partitioning of Eq. (B.1.8) and the related geometric (*P*,*CT*)-projections provide additional tools for interpreting energy changes due to shifts in the electron density. Consider, as an illustrative example, the second differential of the universal density functional $F[\rho] \equiv F[Np]$ equal to that of $E_v[\rho] = E[N, v]$ [see Eqs. (B.2.1a), (B.2.2) and (B.2.18)] :

$$\Delta^{(2)}E_v[\rho] = \Delta^{(2)}F[\rho] = \frac{1}{2}\iint \Delta\rho(\boldsymbol{r})\,\eta(\boldsymbol{r},\boldsymbol{r}')\,\Delta\rho(\boldsymbol{r}')\,d\boldsymbol{r}\,d\boldsymbol{r}' = -\frac{1}{2}\int \Delta u(\boldsymbol{r}')\,\Delta\rho(\boldsymbol{r}')\,d\boldsymbol{r}'. \tag{B.4.1}$$

It should be emphasized that this energy change also represents the sum of the first two differentials of the grand-potential

$$\Omega[u] = E[N,v] - \left(\frac{\partial E[N,v]}{\partial N}\right)_v N = E[N,v] - N\mu = \int u(\boldsymbol{r})\rho(\boldsymbol{r})\,d\boldsymbol{r} + F[\rho] = \Omega_u[\rho[u]] \tag{B.4.2}$$

which provides the "thermodynamic" potential for the open molecular systems in contact with the external electron reservoir [Eq. (B.2.1b)]. Indeed, by the Euler equation (B.2.1a) the first differential of $\Omega_u[\rho]$ vanishes at the equilibrium distribution of electrons and hence

$$\Delta^{(1+2)}\,\Omega_u[\rho] = \Delta^{(2)}\Omega_u[\rho] = \Delta^{(2)}F[\rho]. \tag{B.4.3}$$

One should also observe that the Euler equation (B.2.1a) implies the reciprocal functional dependences between the ground-state density and the relative potential: $\rho = \rho[u]$ and $u = u[\rho]$. This further implies that $F[\rho] = F[\rho[u]] \equiv \widetilde{F}[u]$. Using Eqs. (B.2.6) and (B.4.1) then gives

$$\Delta^{(2)}\widetilde{F}[u] = \frac{1}{2}\iint \Delta u(\boldsymbol{r})\,\sigma(\boldsymbol{r},\boldsymbol{r}')\,\Delta u(\boldsymbol{r}')\,d\boldsymbol{r}\,d\boldsymbol{r}'. \tag{B.4.4}$$

Using Eq. (B.1.8a) allows one to separate the closed-system and external *CT*-effects in this second-order energy:

$$\Delta^{(2)}F[\rho] = \Delta^{(2)}F[Np] = \frac{1}{2}\left[\overline{\eta}(\Delta N)^2 + 2N\,\Delta N\int \overline{\eta}(\boldsymbol{r})\Delta p(\boldsymbol{r})\,d\boldsymbol{r} + N^2\iint \Delta p(\boldsymbol{r})\eta(\boldsymbol{r},\boldsymbol{r}')\Delta p(\boldsymbol{r}')\,d\boldsymbol{r}\,d\boldsymbol{r}'\right], \tag{B.4.5}$$

where:

$$\left(\frac{\partial^2 F[Np]}{\partial N^2}\right)_p = \iint \left(\frac{\partial \rho(\boldsymbol{r})}{\partial N}\right)_p \left(\frac{\partial^2 F[\rho]}{\partial \rho(\boldsymbol{r})\,\partial \rho(\boldsymbol{r}')}\right)\left(\frac{\partial \rho(\boldsymbol{r}')}{\partial N}\right)_p d\boldsymbol{r}\, d\boldsymbol{r}'$$
$$= \iint p(\boldsymbol{r})\,\eta(\boldsymbol{r},\boldsymbol{r}')\,p(\boldsymbol{r}')d\boldsymbol{r}\, d\boldsymbol{r}' \equiv \bar{\eta}, \tag{B.4.6}$$

$$-\left(\frac{\partial u(\boldsymbol{r})}{\partial N}\right)_p \equiv \bar{\eta}(\boldsymbol{r}) = -\int \frac{\delta u(\boldsymbol{r})}{\delta \rho(\boldsymbol{r}')}\left(\frac{\partial \rho(\boldsymbol{r}')}{\partial N}\right)_p d\boldsymbol{r}' = \int p(\boldsymbol{r}')\eta(\boldsymbol{r}',\boldsymbol{r})d\boldsymbol{r}'. \tag{B.4.7}$$

This resolution separates the first contribution, due to ΔN for constant p, which involves the *average hardness* $\bar{\eta}$, from the last term, due to Δp for constant N, and the two coupling terms containing the *average local hardness* $\bar{\eta}(\boldsymbol{r})$, which is not equalized throughout the space.

Alternatively, the geometric (P,CT)-projections can be used to resolve, for interpretative purposes, the second-order energy of Eq. (B.4.1). The identity projection $\hat{\mathrm{P}} = \hat{\mathrm{P}}_{CT} + \hat{\mathrm{P}}_P = 1$ placed between the density displacements and the hardness operator of the Hilbert space interpretation of $\Delta^{(2)}E_v[\rho] = \Delta^{(2)}F[\rho]$,

$$\Delta^{(2)}F[\rho] = -\frac{1}{2}\langle \Delta u | \Delta\rho\rangle = \frac{1}{2}\langle \Delta\rho | \hat{\mathrm{P}}\,\hat{\eta}\,\hat{\mathrm{P}} | \Delta\rho\rangle, \tag{B.4.8}$$

can be attributed to either the density displacements,

$$\Delta^{(2)}F[\rho] = \frac{1}{2}(\langle \Delta\rho|\hat{\mathrm{P}})\hat{\eta}\,(\hat{\mathrm{P}}|\Delta\rho\rangle) = -\frac{1}{2}\langle \Delta u|(\hat{\mathrm{P}}|\Delta\rho\rangle) = -\frac{1}{2}(\langle \Delta u|\hat{\mathrm{P}})\,|\Delta\rho\rangle, \tag{B.4.9}$$

or to the hardness operator itself,

$$\Delta^{(2)}F[\rho] = \frac{1}{2}\langle \Delta\rho|(\hat{\mathrm{P}}\,\hat{\eta}\,\hat{\mathrm{P}})|\Delta\rho\rangle . \tag{B.4.10}$$

The former gives rise to the associated partitioning the density displacements, while the latter amounts to dividing the hardness operator into the four matrix elements

$$\hat{\eta} = \sum_X \sum_Y \hat{\mathrm{P}}_X \hat{\eta}\, \hat{\mathrm{P}}_Y = \sum_X \sum_Y \hat{\eta}_{X,Y}, \qquad (\hat{\mathrm{P}}_X, \hat{\mathrm{P}}_Y) \in \{\hat{\mathrm{P}}_{CT}, \hat{\mathrm{P}}_P\}, \tag{B.4.11}$$

and the associated division of the hardness kernel:

$$\eta(\boldsymbol{r},\boldsymbol{r}') = \langle \boldsymbol{r}|\hat{\eta}|\boldsymbol{r}'\rangle \equiv \sum_X\sum_Y \langle \boldsymbol{r}|\hat{\eta}_{X,Y}|\boldsymbol{r}'\rangle = \sum_X\sum_Y \eta_{X,Y}(\boldsymbol{r},\boldsymbol{r}'), \qquad (X, Y) \in \{P, CT\}. \tag{B.4.12}$$

This operator partitioning resolves the differential of Eq. (B.4.1) into the (P,CT)-resolved contributions:

$$\Delta^{(2)}F[\rho] = \frac{1}{2}\iint \langle \Delta\rho|\boldsymbol{r}\rangle\, \langle \boldsymbol{r}|(\hat{\mathrm{P}}\,\hat{\eta}\,\hat{\mathrm{P}})|\boldsymbol{r}\rangle\, \langle \boldsymbol{r}'|\Delta\rho\rangle\, d\boldsymbol{r}\, d\boldsymbol{r}' \equiv \frac{1}{2}\sum_X\sum_Y \iint \Delta\rho(\boldsymbol{r})\,\eta_{X,Y}(\boldsymbol{r},\boldsymbol{r}')\,\Delta\rho(\boldsymbol{r}')d\boldsymbol{r}\, d\boldsymbol{r}'. \tag{B.4.13}$$

A similar projection of the softness operator partitions it into the four matrix components,

$$\hat{\sigma} = \sum_X \sum_Y \hat{P}_X \hat{\sigma} \hat{P}_Y = \sum_X \sum_Y \hat{\sigma}_{X,Y}, \qquad (X, Y) \in \{P, CT\}. \tag{B.4.14}$$

In the position representation this operator projection gives rise to the corresponding division of the softness kernel:

$$\sigma(\boldsymbol{r}, \boldsymbol{r}') = \langle \boldsymbol{r}|\hat{\sigma}|\boldsymbol{r}'\rangle \equiv \sum_X \sum_Y \langle \boldsymbol{r}|\hat{\sigma}_{X,Y}|\boldsymbol{r}'\rangle = \sum_X \sum_Y \sigma_{X,Y}(\boldsymbol{r}, \boldsymbol{r}'). \tag{B.4.15}$$

It also provides the associated partitioning of the differential of Eq. (B.4.4):

$$\Delta^{(2)} \tilde{F}[u] = \frac{1}{2} \iint \langle \Delta u|\boldsymbol{r}\rangle \langle \boldsymbol{r}|(\hat{P} \hat{\sigma} \hat{P})|\boldsymbol{r}'\rangle \langle \boldsymbol{r}'|\Delta u\rangle \, d\boldsymbol{r} \, d\boldsymbol{r}' \equiv \frac{1}{2} \sum_X \sum_Y \iint \Delta u(\boldsymbol{r}) \, \sigma_{X,Y}(\boldsymbol{r}, \boldsymbol{r}') \Delta u(\boldsymbol{r}') d\boldsymbol{r} \, d\boldsymbol{r}'. \tag{B.4.16}$$

The (*P*,*CT*)-resolved kernels:

$$\boldsymbol{\eta}(\boldsymbol{r}, \boldsymbol{r}') \equiv \begin{bmatrix} \eta_{CT,CT}(\boldsymbol{r}, \boldsymbol{r}') & \eta_{CT,P}(\boldsymbol{r}, \boldsymbol{r}') \\ \eta_{P,CT}(\boldsymbol{r}, \boldsymbol{r}') & \eta_{P,P}(\boldsymbol{r}, \boldsymbol{r}') \end{bmatrix}, \qquad \boldsymbol{\sigma}(\boldsymbol{r}, \boldsymbol{r}') \equiv \begin{bmatrix} \sigma_{CT,CT}(\boldsymbol{r}, \boldsymbol{r}') & \sigma_{CT,P}(\boldsymbol{r}, \boldsymbol{r}') \\ \sigma_{P,CT}(\boldsymbol{r}, \boldsymbol{r}') & \sigma_{P,P}(\boldsymbol{r}, \boldsymbol{r}') \end{bmatrix}, \tag{B.4.17}$$

jointly denoted as $\mathbf{x}(\boldsymbol{r}, \boldsymbol{r}') = \{x_{X,Y}(\boldsymbol{r}, \boldsymbol{r}')\}$, $x = (\eta, \sigma)$, can be directly expressed in terms of the eigenfunctions $\boldsymbol{b}(\boldsymbol{r})$ of Eq. (B.1.4). For the diagonal *P*-components one finds:

$$x_{P,P}(\boldsymbol{r}, \boldsymbol{r}') = \langle \boldsymbol{r}|\hat{P}_P \, \hat{x} \, \hat{P}_P|\boldsymbol{r}'\rangle = \sum_i \sum_j \langle \boldsymbol{r}|i\rangle \langle i|\hat{x}|j\rangle \langle j|\boldsymbol{r}'\rangle = \sum_i \sum_j b_i(\boldsymbol{r}) \, x_{i,j} \, b_j^*(\boldsymbol{r}'), \tag{B.4.18}$$

where $x_{i,j} = \iint b_i^*(\boldsymbol{r}) x(\boldsymbol{r}, \boldsymbol{r}') b_j(\boldsymbol{r}') d\boldsymbol{r} \, d\boldsymbol{r}'$. For the two off-diagonal terms one obtains:

$$\begin{aligned} x_{CT,P}(\boldsymbol{r}, \boldsymbol{r}') &= \langle \boldsymbol{r}|(1 - \hat{P}_P) \, \hat{x} \, \hat{P}_P|\boldsymbol{r}'\rangle = \langle \boldsymbol{r}| \, \hat{x} \, \hat{P}_P|\boldsymbol{r}'\rangle - x_{P,P}(\boldsymbol{r}, \boldsymbol{r}') \\ &= \sum_j \langle \boldsymbol{r}| \, \hat{x} \, |j\rangle \langle j|\boldsymbol{r}'\rangle - x_{P,P}(\boldsymbol{r}, \boldsymbol{r}') = \sum_j x_j(\boldsymbol{r}) \, b_j^*(\boldsymbol{r}') - x_{P,P}(\boldsymbol{r}, \boldsymbol{r}'), \end{aligned} \tag{B.4.19}$$

where $x_j(\boldsymbol{r}) = \int x(\boldsymbol{r}, \boldsymbol{r}') b_j(\boldsymbol{r}') d\boldsymbol{r}'$, and

$$\begin{aligned} x_{P,CT}(\boldsymbol{r}, \boldsymbol{r}') &= \langle \boldsymbol{r}|\hat{P}_P \, \hat{x} (1 - \hat{P}_P)|\boldsymbol{r}'\rangle = \langle \boldsymbol{r}| \, \hat{P}_P \hat{x} \, |\boldsymbol{r}'\rangle - x_{P,P}(\boldsymbol{r}, \boldsymbol{r}') \\ &= \sum_i \langle \boldsymbol{r}|i\rangle \langle i|\hat{x}|\boldsymbol{r}'\rangle - x_{P,P}(\boldsymbol{r}, \boldsymbol{r}') = \sum_i b_i(\boldsymbol{r}) x_i^*(\boldsymbol{r}') - x_{P,P}(\boldsymbol{r}, \boldsymbol{r}'), \end{aligned} \tag{B.4.20}$$

with $x_i^*(\boldsymbol{r}') = \int b_i^*(\boldsymbol{r}) \, x(\boldsymbol{r}, \boldsymbol{r}') \, d\boldsymbol{r}$. Finally, the diagonal *CT*-kernel is obtained by subtracting the above three partial kernels from the total kernel:

$$x_{CT,CT}(\boldsymbol{r}, \boldsymbol{r}') = x(\boldsymbol{r}, \boldsymbol{r}') - x_{CT,P}(\boldsymbol{r}, \boldsymbol{r}') - x_{P,CT}(\boldsymbol{r}, \boldsymbol{r}') - x_{P,P}(\boldsymbol{r}, \boldsymbol{r}'). \tag{B.4.21}$$

This illustrative geometric partitioning of the softness and hardness kernels should facilitate a better assessment of the relative roles played by the charge-transfer and polarization effects in molecular processes, and an evaluation of the strength of their mutual coupling. It also provides a powerful tool for dividing the associated changes in the electronic energy, which is vital for such a diagnosis and ultimately for an understanding of the physical and chemical implications these components have in diverse density rearrangements in molecules.

APPENDIX C: The Kohn-Sham Method

The modern Density Functional Theory, succeeding the Thomas-Fermi model, determines the density and energy of the *interacting* N-electron system by solving the Self-Consistent-Field equations for the optimum Spin-Orbitals defining the Kohn-Sham determinant of the associated *non-interacting* system, which by hypothesis gives rise to the same electron distribution as the interacting system of interest. In this way the complex many-body effects in the atomic, molecular and solid state systems are represented by the effective local potentials, which in principle include the Fermi (exchange) and Coulomb electron correlations. The density functional for the exchange-correlation energy $E_{xc}[\rho]$ is introduced and the Kohn-Sham equations are derived. The relevant energy expressions are summarized and the concept of the correlation holes is outlined.

It follows from the celebrated *Hohenberg-Kohn* (1964) (HK) theorems (see Section 1.5), which constitute the formal basis of the modern *Density Functional Theory* (DFT) that the (non-degenerate) ground-state density $\rho = \rho[N, v]$ of the molecule containing N electrons moving in the external field $v(\boldsymbol{r})$, due to the system nuclei in their fixed positions $\mathbf{R} = \{\boldsymbol{R}_\alpha\}$ (Born-Oppenheimer approximation), uniquely determines the ground-state wave-function $\Psi[N, v] = \Psi[\rho] \equiv \Psi(\{x_i\}) \equiv \Psi(\mathbf{x})$, the external potential $v = v[\rho]$, and the system electronic energy,

$$E_v[\rho] \equiv \int v(\boldsymbol{r})\rho(\boldsymbol{r})\, d\boldsymbol{r} + F[\rho] = E[N, v]. \qquad \text{(C.1)}$$

Here the first, external potential term determines the attraction energy between electrons and nuclei and the universal (v-independent) functional $F[\rho] = T[\rho] + V_{ee}[\rho]$ generates the sum of the expectation values of the electronic kinetic and repulsion energies. Above, the electronic wave-function arguments $\mathbf{x} = \{\boldsymbol{x}_i\}$ group the position coordinates $\{\boldsymbol{r}_i\}$ and spin variables $\{\sigma_i\}$ of all N electrons: $\mathbf{x} = \{\boldsymbol{x}_i = (\boldsymbol{r}_i, \sigma_i)\}$.

Clearly, since ρ uniquely identifies the system electronic Hamiltonian, $\hat{\mathrm{H}}(N[\rho], v[\rho]) = \hat{\mathrm{H}}[\rho]$, the ground-state electron density in principle uniquely determines other physical properties as well, e.g., the system chemical potential:

$$\mu[\rho] = \frac{\delta E_v[\rho]}{\delta\rho(\boldsymbol{r})} = \left(\frac{\partial E[N, v]}{\partial N}\right)_v = \mu[N, v]. \qquad \text{(C.2)}$$

The second HK theorem provides the variational principle for determining the ground-state density $\rho[N, v]$ and energy $E[N, v]$ by solving the Euler-Lagrange equation for the fixed external potential v:

$$\delta\{E_v[\rho'] - \mu[N, v]\, N[\rho']\}\Big|_{\rho[N,v]} = 0 \qquad \Rightarrow \qquad v(\boldsymbol{r}) + \frac{\delta F[\rho]}{\delta\rho(\boldsymbol{r})} = \mu = \frac{\delta E_v[\rho]}{\delta\rho(\boldsymbol{r})}, \tag{C.3}$$

with the system chemical potential $\mu[N, v]$ enforcing the subsidiary condition of the fixed number of electrons: $N[\rho] \equiv \int \rho(\boldsymbol{r})\, d\boldsymbol{r} = N$. This constraint term is not required, when the trial density $\rho' = \rho_N$ automatically conserves, by construction, the specified number of electrons:

$$\frac{\delta E_v[\rho_N]}{\delta\rho_N(\boldsymbol{r})}\Bigg|_{\rho[N,v]} = 0\,. \tag{C.4}$$

Most DFT calculations for molecular systems employ, after Kohn and Sham (1965) (KS), the exact orbital dependent functional for the kinetic energy of the *non-interacting* system described by the sum of the *separable* (*s*), effective one-electron Hamiltonians (a.u.):

$$\hat{\mathrm{H}}_s(N) = \sum_{i=1}^{N} \hat{\mathrm{H}}_{KS}(\boldsymbol{r}_i), \qquad \hat{\mathrm{H}}_{KS}(\boldsymbol{r}) = -\tfrac{1}{2}\nabla^2 + v_{KS}(\boldsymbol{r}). \tag{C.5}$$

The link with the *real* system of *N interacting* electrons is realized through the requirement that this hypothetical, non-interacting system gives rise to the ground-state density of the interacting system of interest: $\rho = \rho[N, v]$.

Therefore, by the first HK theorem, the effective one-body potential $v_{KS}(\boldsymbol{r})$ is uniquely determined by the electron density, $v_{KS}(\boldsymbol{r}) = v_{KS}[\rho; \boldsymbol{r}]$, and so is the exact ground-state wave-function of the non-interacting system:

$$\Psi_s[N, v] = \Psi_s[\rho] = \det\{\psi_n\} \equiv |\,\psi_1\ \psi_2 \ldots \psi_N\,|\,. \tag{C.6}$$

It is exactly given by the KS determinant constructed from the N orthonormal, singly occupied molecular *Spin-Orbitals* (SO) $\psi(\boldsymbol{x}_i) = \{\psi_n(\boldsymbol{x}_i) = \varphi_n(\boldsymbol{r}_i)\,\chi_n(\sigma_i)\}$, which correspond to the lowest eigenvalues $\{\varepsilon_n\}$ of the effective KS Hamiltonian $\hat{\mathrm{H}}_{KS}(\boldsymbol{r})$. These one-electron Schrödinger equations, called the *KS equations*, determine the spatial parts of SO, called *Molecular Orbitals* (MO), $\boldsymbol{\varphi}(\boldsymbol{r}_i) = \{\varphi_n(\boldsymbol{r}_i)\}$:

$$\hat{\mathrm{H}}_{KS}(\boldsymbol{r})\,\varphi_n(\boldsymbol{r}) = \varepsilon_n\,\varphi_n(\boldsymbol{r}), \qquad n = 1, 2, \ldots, N. \tag{C.7}$$

The KS eigenvalues $\boldsymbol{\varepsilon} = \{\varepsilon_n\}$ represent the orbital energies of the non-interacting system and $\boldsymbol{\chi}(\sigma) = \{\chi_n(\sigma)\}$ groups the spin functions of occupied SO, which depend upon the discrete spin variable σ representing the two admissible spin orientations of an electron: $\sigma = \{\uparrow, \downarrow\}$.

Since the one-body Hamiltonian $\hat{\mathrm{H}}_{KS}(\boldsymbol{r})$ of the above effective eigenvalue problem is uniquely specified by the ground-state electron density ρ, its solutions are also functionals of this equilibrium distribution of electrons: $\boldsymbol{\varphi} = \boldsymbol{\varphi}[\rho]$ and $\boldsymbol{\varepsilon} = \boldsymbol{\varepsilon}[\rho]$. The electron density is straightforwardly generated by the sum of orbital densities $\boldsymbol{\rho} = \{\rho_n\}$,

$$\rho(\boldsymbol{r}) = \sum_n |\varphi_n(\boldsymbol{r})|^2 \equiv \sum_n \rho_n(\boldsymbol{r})$$
$$= \sum_{\sigma=\uparrow,\downarrow} \sum_n |\varphi_{n\sigma}(\boldsymbol{r})|^2 \equiv \sum_{\sigma=\uparrow,\downarrow} \sum_n \rho_{n\sigma}(\boldsymbol{r}) \equiv \sum_{\sigma=\uparrow,\downarrow} \rho_\sigma(\boldsymbol{r}), \quad (C.8)$$

where $\{\rho_\sigma(\boldsymbol{r})\}$ stand for the system spin-densities, and the exact kinetic energy of the non-interacting system is similarly given by the sum of the occupied orbital expectation values:

$$T_s[\rho] = -\ \tfrac{1}{2} \sum_n \langle \varphi_n[\rho] | \nabla^2 | \varphi_n[\rho] \rangle. \quad (C.9)$$

In the KS theory the density functionals for the electronic energies of the non-interacting and real molecular systems, respectively, are given by the following expressions:

$$E_s[\rho] = T_s[\rho] + \int \rho(\boldsymbol{r}) v_{KS}(\boldsymbol{r})\, d\boldsymbol{r} = \sum_n \varepsilon_n[\rho], \quad (C.10)$$
$$E_v[\rho] = T_s[\rho] + \int \rho(\boldsymbol{r}) v(\boldsymbol{r})\, d\boldsymbol{r} + V_{ee}^{class}[\rho] + E_{xc}[\rho], \quad (C.11)$$

where the classical (Hartree) energy of the Coulomb repulsion between electrons

$$V_{ee}^{class}[\rho] = \frac{1}{2} \iint \frac{\rho(\boldsymbol{r})\rho(\boldsymbol{r}')}{|\boldsymbol{r} - \boldsymbol{r}'|}\, d\boldsymbol{r}\, d\boldsymbol{r}' \quad (C.12)$$

and $E_{xc}[\rho]$ stands for the density functional for the KS *exchange-correlation energy*:

$$E_{xc}[\rho] = F[\rho] - T_s[\rho] - V_{ee}^{class}[\rho]. \quad (C.13)$$

The exact form of this functional, which contains all electron correlation contributions, is not known exactly, but its reliable approximations, explicitly density and/or orbital dependent representations, e.g., the sophisticated density-gradient-dependent functionals, give excellent results for atoms, molecules and solid-state systems.

It should be stressed that the kinetic energy of interacting electrons includes both the non-interacting (s) and correlation (c) contributions:

$$T[\rho] = T_s[\rho] + T_c[\rho]. \quad (C.14)$$

Therefore, the total electron repulsion energy of the interacting system is given by the sum of the Hartree and exchange-correlation terms minus the correlation kinetic energy:

$$V_{ee}[\rho] = V_{ee}^{class}[\rho] + (E_{xc}[\rho] - T_c[\rho]) = V_{ee}^{class}[\rho] + V_{ee}^{corr}[\rho], \quad (C.15)$$

where $V_{ee}^{corr}[\rho]$ denotes the correlation part of the electron repulsion energy.

Since, by assumption, the densities of the interacting and non-interacting systems are identical, the chemical potential of Eq. (C.2) is equal to the functional derivative of $E_s[\rho]$:

$$\mu[\rho] = \frac{\delta E_v[\rho]}{\delta\rho(\boldsymbol{r})} = \frac{\delta T_s[\rho]}{\delta\rho(\boldsymbol{r})} + v(\boldsymbol{r}) + \frac{\delta V_{ee}^{class}[\rho]}{\delta\rho(\boldsymbol{r})} + \frac{\delta E_{xc}[\rho]}{\delta\rho(\boldsymbol{r})} = \frac{\delta E_s[\rho]}{\delta\rho(\boldsymbol{r})} = \frac{\delta T_s[\rho]}{\delta\rho(\boldsymbol{r})} + v_{KS}(\boldsymbol{r}). \quad (C.16)$$

Hence, the effective one-body potential of the hypothetical non-interacting system includes the external potential due to the nuclei, $v(\boldsymbol{r})$, corrected by the two electronic terms: the classical *Hartree potential* $v_H(\boldsymbol{r})$, originating from $V_{ee}^{class}[\rho]$, and the *exchange-correlation potential* $v_{xc}(\boldsymbol{r})$, resulting from $E_{xc}[\rho]$:

$$v_{KS}(\boldsymbol{r}) = v(\boldsymbol{r}) + \int \frac{\rho(\boldsymbol{r}')}{|\boldsymbol{r}-\boldsymbol{r}'|} d\boldsymbol{r}' + \frac{\delta E_{xc}[\rho]}{\delta\rho(\boldsymbol{r})} \equiv v(\boldsymbol{r}) + v_H(\boldsymbol{r}) + v_{xc}(\boldsymbol{r}). \tag{C.17}$$

The KS equations (C.7) have to be solved iteratively, since the effective potential is density dependent. Given the explicit functional $E_{xc}[\rho]$, one calculates from the last equation the effective one-body potential of the KS non-interacting system for the current variational density, generated by the initial KS orbitals, and by solving KS equations determines the next, better approximation to KS orbitals, etc. The resulting scheme is easy to solve numerically.

The KS equations (C.7), which determine the optimum MO of the hypothetical non-interacting system and the electron density and energy of the interacting system, can be straightforwardly derived from the variational principle for the system electronic energy subject to subsidiary conditions of the MO orthogonality and normalization, $\{\langle\varphi_m|\varphi_n\rangle = \delta_{m,n}\}$,

$$\delta\,\{E_v[\rho] - \textstyle\sum_n \sum_m \theta_{m,n}\langle\varphi_m|\varphi_n\rangle\} = 0. \tag{C.18}$$

where $\boldsymbol{\theta} = \{\theta_{m,n}\}$ stands for the matrix of the associated Lagrange multipliers. It should be realized that in the *canonical representation,* which defines the KS MO, this matrix becomes diagonal: $\theta_{m,n} = \varepsilon_n\delta_{m,n}$.

The key functional $E_{xc}[\rho]$ can be formally expressed in terms of the coupling-constant (λ) averaged *xc-hole*,

$$h_{xc}(\boldsymbol{r}'|\boldsymbol{r}) = \int_0^1 h_{xc}^{\lambda}(\boldsymbol{r}'|\boldsymbol{r})\,d\lambda, \tag{C.19}$$

which corrects the independent-particle conditional density $\rho^{ind}(\boldsymbol{r}'|\boldsymbol{r}) = \rho(\boldsymbol{r}')$, of finding an electron at $\boldsymbol{r}'$, when another electron is known to has already been found at $\boldsymbol{r}$, into the associated conditional density $\rho(\boldsymbol{r}'|\boldsymbol{r})$ of the real, correlated system:

$$\rho(\boldsymbol{r}'|\boldsymbol{r}) = \rho_2(\boldsymbol{r}, \boldsymbol{r}')/\rho(\boldsymbol{r}) = \rho(\boldsymbol{r}') + h_{xc}(\boldsymbol{r}'|\boldsymbol{r}), \tag{C.20}$$

where the electron pair-density

$$\rho_2(\boldsymbol{r}, \boldsymbol{r}') = N(N-1)\int\ldots\int |\Psi(\mathbf{x})|^2 \delta(\boldsymbol{r}_1 - \boldsymbol{r})\,\delta(\boldsymbol{r}_2 - \boldsymbol{r}')\,d\mathbf{x} \equiv \langle\Psi|\hat{\rho}_2(\boldsymbol{r}, \boldsymbol{r}')|\Psi\rangle. \tag{C.21}$$

It should be observed that in the definition of the correlation hole $h_{xc}(\boldsymbol{r}'|\boldsymbol{r})$ the roles of the two electrons are not symmetrical, with the *reference*-electron position coordinates $\boldsymbol{r}$ representing *parameters* and those of the *dependent*-electron, $\boldsymbol{r}'$, denoting *variables* of the conditional distribution. In the familiar *pair correlation function*, however,

$$g(\boldsymbol{r}, \boldsymbol{r}') = \rho_2(\boldsymbol{r}, \boldsymbol{r}')/[\rho(\boldsymbol{r})\,\rho(\boldsymbol{r}')] = h_{xc}(\boldsymbol{r}'|\boldsymbol{r})/\rho(\boldsymbol{r}), \tag{C.22}$$

the two electrons are treated in a symmetrical way. Expressing $h_{xc}(\boldsymbol{r}'|\boldsymbol{r})$ in terms of $g(\boldsymbol{r}, \boldsymbol{r}')$ gives:

$$h_{xc}(\boldsymbol{r}'|\boldsymbol{r}) = \rho(\boldsymbol{r}')[\, g(\boldsymbol{r}, \boldsymbol{r}') - 1]. \tag{C.23}$$

The coupling constant λ in Eq. (C.19) controls the electron repulsion in the *scaled* electronic Hamiltonian,

$$\hat{\mathrm{H}}^{\lambda}(N) = \hat{\mathrm{T}}(N) + \sum_{i=1}^{N} v^{\lambda}(\boldsymbol{r}_i) + \lambda \hat{\mathrm{V}}_{\mathrm{ee}}(N), \qquad \hat{\mathrm{H}}^{\lambda}(N)\Psi^{\lambda}(N) = E^{\lambda}(N)\Psi^{\lambda}(N), \tag{C.24}$$

with the modified external potential $v^{\lambda}(\boldsymbol{r})$, which makes the ground-state electron density$\rho^{\lambda}(\boldsymbol{r})$ identical to that common to both the non-interacting and interacting systems: $\rho^{\lambda}(\boldsymbol{r}) = \rho(\boldsymbol{r})$. In this *adiabatic connection* $v^{\lambda=0}(\boldsymbol{r}) = v_{KS}(\boldsymbol{r})$, $v^{\lambda=1}(\boldsymbol{r}) = v(\boldsymbol{r})$, and

$$\rho^{\lambda}(\boldsymbol{r}'|\boldsymbol{r}) = \rho_2^{\lambda}(\boldsymbol{r}, \boldsymbol{r}')/\rho(\boldsymbol{r}) = \rho(\boldsymbol{r}') + h_{xc}^{\lambda}(\boldsymbol{r}'|\boldsymbol{r}), \tag{C.25}$$

where the scaled electron pair-density: $\rho_2^{\lambda}(\boldsymbol{r}, \boldsymbol{r}') = \langle \Psi^{\lambda} | \hat{\rho}_2(\boldsymbol{r}, \boldsymbol{r}') | \Psi^{\lambda} \rangle$.

It then directly follows from the Hellmann-Feynman theorem that $E_{xc}[\rho]$ represents the classical Coulomb interaction between the electron density and the average hole of Eq. (C.19):

$$E_{xc}[\rho] = \frac{1}{2} \int\!\!\int \rho(\boldsymbol{r}) \frac{h_{xc}(\boldsymbol{r}'|\boldsymbol{r})}{|\boldsymbol{r}' - \boldsymbol{r}|} d\boldsymbol{r}'\, d\boldsymbol{r}. \tag{C.26}$$

It should be realized that the hole anisotropy does not affect the exchange-correlation energy, which can be expressed as the functional of the *spherically averaged* hole,

$$h_{xc}(u|\boldsymbol{r}) = \int h_{xc}(\boldsymbol{r} + \boldsymbol{u}|\boldsymbol{r}) \frac{d\Omega_{\boldsymbol{u}}}{4\pi}, \tag{C.27}$$

$$E_{xc}[\rho] = 2\pi \int\!\!\int u\, h_{xc}(u|\boldsymbol{r})\, du\, d\boldsymbol{r}, \tag{C.28}$$

where $\boldsymbol{u} = \boldsymbol{r}' - \boldsymbol{r} = (u, \theta, \phi) = (u, \Omega_u)$.

The exact *exchange energy*, corresponding to $\lambda = 0$, is given by the following functional of the KS MO:

$$E_x[\rho] = -\frac{1}{2} \sum_{\sigma=\uparrow,\downarrow} \sum_{m,n=1}^{N_\sigma} \int\!\!\int |\boldsymbol{r}' - \boldsymbol{r}|^{-1} \varphi_{n\sigma}(\boldsymbol{r})\varphi^*_{m\sigma}(\boldsymbol{r})\, \varphi_{m\sigma}(\boldsymbol{r}')\varphi^*_{n\sigma}(\boldsymbol{r}')\, d\boldsymbol{r}'\, d\boldsymbol{r} \equiv \frac{1}{2} \int\!\!\int \rho(\boldsymbol{r}) \frac{h_x(\boldsymbol{r}'|\boldsymbol{r})}{|\boldsymbol{r}' - \boldsymbol{r}|} d\boldsymbol{r}'\, d\boldsymbol{r}, \tag{C.29}$$

where $N_\sigma = \int \rho_\sigma(\boldsymbol{r})\, d\boldsymbol{r}$ is the overall number of electrons of the spin variety σ. Thus, the *exchange* (*Fermi*) *hole* of the preceding equation is given by the following expression in terms of KS orbitals:

$$h_x(\boldsymbol{r}'|\boldsymbol{r}) = -\frac{1}{\rho(\boldsymbol{r})} \sum_{\sigma=\uparrow,\downarrow} \sum_{m,n=1}^{N_\sigma} \varphi_{n\sigma}(\boldsymbol{r})\varphi^*_{m\sigma}(\boldsymbol{r})\, \varphi_{m\sigma}(\boldsymbol{r}')\varphi^*_{n\sigma}(\boldsymbol{r}'). \tag{C.30}$$

One can also separate the average *Coulomb hole*,

$$h_c(\boldsymbol{r}'|\boldsymbol{r}) = h_{xc}(\boldsymbol{r}'|\boldsymbol{r}) - h_x(\boldsymbol{r}'|\boldsymbol{r}), \tag{C.31}$$

which defines the (Coulomb) *correlation energy*:

$$E_c[\rho] = \frac{1}{2}\int\int \rho(\boldsymbol{r})\frac{h_c(\boldsymbol{r}'|\boldsymbol{r})}{|\boldsymbol{r}'-\boldsymbol{r}|}d\boldsymbol{r}'\,d\boldsymbol{r} = E_{xc}[\rho] - E_x[\rho]. \tag{C.32}$$

Using the approximate, explicitly orbital-dependent functionals for the Coulomb correlation energy established in Quantum Chemistry, e.g., those derived in the *Møller-Plessett* (1934) (MP) theory, which uses the *Perturbation Theory* (PT) to determine the *Configuration Interaction* (CI) coefficients, one can also write down an explicit expression for the Coulomb hole in terms of MO.

The physically meaningful correlation holes must satisfy the following sum rules:

$$\int h_x(\boldsymbol{r}'|\boldsymbol{r})\,d\boldsymbol{r}' = \int h_{xc}(\boldsymbol{r}'|\boldsymbol{r})\,d\boldsymbol{r}' = -1, \qquad \int h_c(\boldsymbol{r}'|\boldsymbol{r})\,d\boldsymbol{r}' = 0. \tag{C.33}$$

APPENDIX D: Constrained Equilibria in Molecular Subsystems

The constrained (internal) equilibria in the *mutually* closed molecular subsystems are examined, when they are *externally* open or closed, relative to the independent electron reservoirs. Using the Euler equations for the equilibrium electron densities of molecular fragments the alternative sets of the subsystem state-parameters are identified and the corresponding "thermodynamic" potentials, defined by the relevant Legendre transforms of the system electronic energy, are introduced. Their derivative properties are investigated, with a special emphasis placed upon the molecular charge sensitivities in the subsystem resolution, including the chemical *hardness*, *softness* and *Fukui function* quantities of molecular fragments. This development constitutes the subsystem extension of the global approach described in Chapter 2. In the subsystem resolution a separation of the *additive* and *non-additive* contributions to the density functional for the universal part of the electronic energy provides a basis for an exact definition of the *effective* external potentials of molecular subsystems. They include the molecular potential due to the system nuclei, which is modified by the embedding potential due to the subsystem non-additive interactions. The additive and non-additive contributions to the hardness and softness kernels of molecular fragments are examined, and the transformations between the equilibrium displacements of the subsystem electron densities and potentials are summarized.

D.1. State-Parameters in the Subsystem Description

The subsystem chemical potentials $\boldsymbol{\mu} = \{\mu_\alpha\}$ are defined by the *partial* functional derivatives of Eqs. (6.5.7a), (6.5.12), and (6.5.15):

$$\mu_\alpha(\boldsymbol{r}) = \{\partial E_v[\boldsymbol{\rho}]/\partial \rho_\alpha(\boldsymbol{r})\}_{v,\beta\neq\alpha} = \mu_\alpha = \mu\,, \qquad \alpha = 1, 2, \ldots, m. \tag{D.1.1}$$

Here the row vector of the subsystem densities $\boldsymbol{\rho} = (\rho_\alpha, \rho_\beta, \ldots)$ gives rise to the overall density $\rho = \sum_\gamma \rho_\gamma = \boldsymbol{1}\boldsymbol{\rho}^{\mathrm{T}}$, where the unit row vector $\boldsymbol{1} = (1, 1, \ldots, 1)$, and the row vector of the subsystem external potentials $\boldsymbol{v}(\boldsymbol{r}) = v(\boldsymbol{r})\boldsymbol{1}$. They are calculated for the fixed external potential $v(\boldsymbol{r})$, due to all nuclei in the molecule, and the "frozen" embedding densities $\{\rho_{\beta\neq\alpha}(\boldsymbol{r})\}$ of all remaining subsystems. They are equalized at the molecular chemical potential level μ, $\boldsymbol{\mu} = \mu\boldsymbol{1}$, when these fragments are mutually *opened*. This is the case in the *global* equilibrium state considered in Chapter 2. We shall denote such mutually-open condition of subsystems by the vertical *broken* lines in the symbolic representation of the composite molecular system in the *global* (*g*) (inter-subsystem) *equilibrium* of the ground-state of an externally open molecule: $M_g = (\alpha \,\vdots\, \beta \,\vdots\, \gamma \,\vdots \ldots)$ (see, e.g., Nalewajski and Korchowiec, 1977; Nalewajski *et al.*, 1996; Nalewajski, 1993, 2002d, 2003a).

This equalization of the subsystem chemical potentials, $\boldsymbol{\mu} = (\mu_\alpha, \mu_\beta, \mu_\gamma, \ldots) = \mu\boldsymbol{1}$, can be attributed to a common external reservoir ($\mathfrak{R}$) of electrons, exhibiting the chemical potential

$$\mu^{\mathfrak{R}} = \mu_\alpha = \mu\,, \qquad \alpha = 1, 2, \ldots, m, \tag{D.1.2}$$

to which all molecular subsystems are coupled in the hypothetical combined system $\mathfrak{M}[M_g] \equiv (\mathfrak{R} \,\vdots\, \alpha \,\vdots\, \beta \,\vdots\, \gamma \,\vdots \ldots)$.

In the subsystem resolution one also considers the *constrained* (intra-subsystem) *equilibrium* states, when all subsystems are mutually *closed*, which is symbolized by the vertical *solid* lines in the symbolic representation of a collection of the mutually closed AIM of $M_c = (\alpha \mid \beta \mid \ldots)$. In order to probe the externally-open subsystem characteristics in M_c, when each molecular fragment is characterized by the intra-subsystem equalized, generally different level of its chemical potential,

$$\mu_\alpha(\boldsymbol{r}) = \mu_\alpha \neq \mu_\beta(\boldsymbol{r}) = \mu_\beta \neq \ldots \neq \mu\,, \tag{D.1.3}$$

one envisages separate electron reservoirs $\{\mathfrak{R}_\alpha\}$ for each subsystem, characterized by the independently controlled chemical potentials $\{\mu_\alpha^{\mathfrak{R}}\}$, in the combined system $\mathfrak{M}[M_c] \equiv (\mathfrak{R}_\alpha \,\vdots\, \alpha \mid \mathfrak{R}_\beta \,\vdots\, \beta \mid \ldots)$. Notice that only for the global equilibrium $\boldsymbol{\mu} = \mu\boldsymbol{1}$, when $\mathfrak{R}_\alpha = \mathfrak{R}_\beta = \ldots = \mathfrak{R}$. Similarly, when only a single subsystem or a subset of molecular fragments is considered externally open, while the remaining subsystems are externally closed, one envisages a coupling of a specified selection of subsystems to their corresponding reservoirs, e.g., in $\mathfrak{M}[M_\alpha] \equiv (\mathfrak{R}_\alpha \,\vdots\, \alpha \mid \beta \mid \gamma \mid \ldots)$. The equilibrium state of the subsystem α in contact with $\mathfrak{R}_\alpha$ is characterized by the equalization of this subsystem chemical potential and that of its reservoir:

$$\mu_\alpha^{\mathfrak{R}} = \mu_\alpha\,. \tag{D.1.4}$$

Therefore, the hypothetical, independent displacements of the subsystem chemical potentials reflect those of the corresponding subsystem reservoirs: $d\mu_\alpha^{\mathfrak{R}} = d\mu_\alpha$.

In such a subsystem resolution one defines the row vector of the AIM relative external potentials:

$$\boldsymbol{u}(\boldsymbol{r}) \equiv -\,\partial \widetilde{F}[\boldsymbol{\rho}]/\partial \boldsymbol{\rho}(\boldsymbol{r}) = v(\boldsymbol{r})\boldsymbol{1} - \boldsymbol{\mu} \equiv \boldsymbol{v}(\boldsymbol{r}) - \boldsymbol{\mu}$$

$$= \{u_\alpha(\boldsymbol{r}) \equiv v(\boldsymbol{r}) - \mu_\alpha = -\,\partial \widetilde{F}[\boldsymbol{\rho}]/\partial \rho_\alpha(\boldsymbol{r}) = -\,\partial F^a[\boldsymbol{\rho}]/\partial \rho_\alpha(\boldsymbol{r}) - \partial F^n[\boldsymbol{\rho}]/\partial \rho_\alpha(\boldsymbol{r})\}, \qquad \text{(D.1.5)}$$

They are defined by the negative partial derivatives of the functional $\widetilde{F}[\boldsymbol{\rho}]$, which can be partitioned into the additive (a) and non-additive (n) parts [Eq. (1.6.8)]:

$$\widetilde{F}[\boldsymbol{\rho}] = F[\Sigma_\gamma \rho_\gamma] = F^n[\boldsymbol{\rho}] + \Sigma_\gamma F[\rho_\gamma] \equiv F^n[\boldsymbol{\rho}] + F^a[\boldsymbol{\rho}]. \qquad \text{(D.1.6)}$$

It provides the universal (v-independent) part of the energy density functional $E_v[\boldsymbol{\rho}]$ in the subsystem resolution:

$$E_v[\boldsymbol{\rho}] = \int \boldsymbol{v}(\boldsymbol{r})\, \boldsymbol{\rho}(\boldsymbol{r})^{\mathrm{T}}\, d\boldsymbol{r} + \widetilde{F}[\boldsymbol{\rho}]. \qquad \text{(D.1.7)}$$

This directly follows from the Euler equations determining the optimum (open) subsystem densities in the internal equilibrium state [Eqs. (6.5.7a-c)]:

$$\delta\{E_v[\boldsymbol{\rho}] - \Sigma_\gamma\, \mu_\gamma N[\rho_\gamma]\} = 0, \qquad \text{or}$$

$$\{\partial E_v[\boldsymbol{\rho}]/\partial \rho_\alpha(\boldsymbol{r})\}_{\beta\neq\alpha} \equiv \mu_\alpha(\boldsymbol{r}) = v(\boldsymbol{r}) + \{\partial F[\boldsymbol{\rho}]/\partial \rho_\alpha(\boldsymbol{r})\}_{\beta\neq\alpha} = \mu_\alpha, \qquad \alpha = 1, 2, \ldots, m. \qquad \text{(D.1.8)}$$

D.2. Legendre-Transformed Representations

It follows from the foregoing equation that the equilibrium densities of the open subsystems are unique functionals of the molecular external potential and the reservoir chemical potentials, $\boldsymbol{\rho} = \boldsymbol{\rho}[\boldsymbol{u}[\boldsymbol{\mu}, v]] = \boldsymbol{\rho}[\boldsymbol{\mu}, v]$, and so is the associated row vector of the average (fractional) numbers of electrons: $\boldsymbol{N}[\boldsymbol{\rho}] = \int \boldsymbol{\rho}[\boldsymbol{\mu}, v; \boldsymbol{r}]\, d\boldsymbol{r}$.

Similarly, when all subsystems are both mutually and externally closed in $M \equiv (\alpha \,|\, \beta \,|\, \ldots)$, the external potential and the numbers of electrons in subsystems uniquely determine the equilibrium densities of molecular fragments, $\boldsymbol{\rho} = \boldsymbol{\rho}[\boldsymbol{N}, v]$, the system energy, $E_v[\boldsymbol{\rho}] = E_v[\boldsymbol{\rho}[\boldsymbol{N}, \boldsymbol{v}]] = E[\boldsymbol{N}, v]$, and all its physical properties, e.g., the subsystem chemical potentials $\boldsymbol{\mu} = \boldsymbol{\mu}[\boldsymbol{N}, v]$.

Following the global equilibrium development of Chapter 2, one defines the "thermodynamic" potentials corresponding to the four alternative sets of the subsystem state-parameters, which define the corresponding constrained-equilibrium states,

$$\{\boldsymbol{N}, v\}, \qquad \{\boldsymbol{\mu}, v\} = \{\boldsymbol{u}\}, \qquad \{\boldsymbol{N}, \boldsymbol{\rho}\} = \{\boldsymbol{\rho}\}, \qquad \{\boldsymbol{\mu}, \boldsymbol{\rho}\}, \qquad \text{(D.2.1)}$$

as the relevant Legendre transforms of the electronic energy $E[\boldsymbol{N}, v]$:

$$\Omega[\boldsymbol{\mu}, \boldsymbol{v}] = E - \boldsymbol{N}\,(\partial E/\partial \boldsymbol{N})_{\boldsymbol{v}}^{\mathrm{T}} = E[\boldsymbol{\rho}[\boldsymbol{u}]] - \boldsymbol{N}[\boldsymbol{\rho}[\boldsymbol{u}]]\,\boldsymbol{\mu}[\boldsymbol{\rho}[\boldsymbol{u}]]^{\mathrm{T}} = \Omega[\boldsymbol{u}], \tag{D.2.2}$$

$$\widetilde{F}\,[\boldsymbol{\rho}] = E - \int \boldsymbol{v}(\boldsymbol{r})\{\partial E/\partial \boldsymbol{v}(\boldsymbol{r})\}_{\boldsymbol{N}}^{\mathrm{T}}\, d\boldsymbol{r} = E - \int \boldsymbol{v}(\boldsymbol{r})\,\boldsymbol{\rho}(\boldsymbol{r})^{\mathrm{T}} d\boldsymbol{r}\,, \tag{D.2.3}$$

$$R[\boldsymbol{\mu}, \boldsymbol{\rho}] = E - \boldsymbol{N}(\partial E/\partial \boldsymbol{N})_{\boldsymbol{v}}^{\mathrm{T}} - \int \boldsymbol{v}(\boldsymbol{r})\{\partial E/\partial \boldsymbol{v}(\boldsymbol{r})\}_{\boldsymbol{N}}^{\mathrm{T}} d\boldsymbol{r} = F[\boldsymbol{\rho}] - \boldsymbol{N}[\boldsymbol{\rho}]\,\boldsymbol{\mu}[\boldsymbol{\rho}]^{\mathrm{T}}, \tag{D.2.4}$$

Clearly, in this subsystem resolution one could also consider all intermediate specifications of the molecular constrained equilibria, when only a part of the subsystems remains externally open (characterized by the fixed chemical potentials of a common or separate reservoirs) with the remaining, complementary set of subsystems being closed (characterized by the fixed subsystem numbers of electrons). We would like to observe that in the theory of chemical reactivity these partially opened situations do indeed arise, e.g., in the surface reactions, when one adsorbate reactant is opened (chemisorbed) and the other closed (physisorbed) with respect to the catalyst surface, with the latter acting as the external electron reservoir.

The corresponding differentials of the system electronic energy and its Legendre transforms read:

$$dE[\boldsymbol{N}, \boldsymbol{v}] = \boldsymbol{\mu}[\boldsymbol{N}, \boldsymbol{v}]\, d\boldsymbol{N}^{\mathrm{T}} + \int \boldsymbol{\rho}[\boldsymbol{N}, \boldsymbol{v}; \boldsymbol{r}]\, d\boldsymbol{v}(\boldsymbol{r})^{\mathrm{T}}\, d\boldsymbol{r}. \tag{D.2.5}$$

$$d\Omega[\boldsymbol{\mu}, \boldsymbol{v}] = -\boldsymbol{N}\, d\boldsymbol{\mu}^{\mathrm{T}} + \int \boldsymbol{\rho}[\boldsymbol{N}, \boldsymbol{v}; \boldsymbol{r}]\, d\boldsymbol{v}(\boldsymbol{r})^{\mathrm{T}}\, d\boldsymbol{r} = \int \boldsymbol{\rho}[\boldsymbol{u}; \boldsymbol{r}]\, d\boldsymbol{u}(\boldsymbol{r})^{\mathrm{T}}\, d\boldsymbol{r} = d\Omega[\boldsymbol{u}], \tag{D.2.6}$$

$$d\widetilde{F}\,[\boldsymbol{\rho}] = \boldsymbol{\mu}\, d\boldsymbol{N}^{\mathrm{T}} - \int \boldsymbol{v}(\boldsymbol{r})\, d\boldsymbol{\rho}(\boldsymbol{r})^{\mathrm{T}}\, d\boldsymbol{r} = -\int \boldsymbol{u}[\boldsymbol{\rho}; \boldsymbol{r}]\, d\boldsymbol{\rho}(\boldsymbol{r})^{\mathrm{T}}\, d\boldsymbol{r}\,, \tag{D.2.7}$$

$$dR[\boldsymbol{\mu}, \boldsymbol{\rho}] = -\boldsymbol{N}\, d\boldsymbol{\mu}^{\mathrm{T}} - \int \boldsymbol{v}(\boldsymbol{r})\, d\boldsymbol{\rho}(\boldsymbol{r})^{\mathrm{T}}\, d\boldsymbol{r}. \tag{D.2.8}$$

It should be observed that the displacements $d\boldsymbol{v}(\boldsymbol{r})$ also allow for the independent changes of the external potential of each subsystem, due to the appropriate external sources. The above differential expressions identify the parameters conjugate to those defining the representation under consideration:

$$\boldsymbol{\mu} = (\partial E[\boldsymbol{N}, \boldsymbol{v}]/\partial \boldsymbol{N})_{\boldsymbol{v}}, \qquad \boldsymbol{\rho}(\boldsymbol{r}) = \{\partial E[\boldsymbol{N}, \boldsymbol{v}]/\partial \boldsymbol{v}(\boldsymbol{r})\}_{\boldsymbol{N}}; \tag{D.2.9}$$

$$\boldsymbol{N} = -(\partial \Omega[\boldsymbol{\mu}, \boldsymbol{v}]/\partial \boldsymbol{\mu})_{\boldsymbol{v}}, \quad \boldsymbol{\rho}(\boldsymbol{r}) = \{\partial \Omega[\boldsymbol{\mu}, \boldsymbol{v}]/\partial \boldsymbol{v}(\boldsymbol{r})\}_{\boldsymbol{\mu}} = \delta\Omega[\boldsymbol{u}]/\delta \boldsymbol{u}(\boldsymbol{r})\,; \tag{D.2.10}$$

$$-\boldsymbol{u}(\boldsymbol{r}) = \boldsymbol{\mu} - \boldsymbol{v}(\boldsymbol{r}) = \delta\widetilde{F}\,[\boldsymbol{\rho}]/\delta \boldsymbol{\rho}(\boldsymbol{r}); \tag{D.2.11}$$

$$-\boldsymbol{N} = (\partial R[\boldsymbol{\mu}, \boldsymbol{\rho}]/\partial \boldsymbol{\mu})_{\boldsymbol{\rho}}, \qquad -\boldsymbol{v}(\boldsymbol{r}) = \{\partial R[\boldsymbol{\mu}, \boldsymbol{\rho}]/\partial \boldsymbol{\rho}(\boldsymbol{r})\}_{\boldsymbol{\mu}}\,. \tag{D.2.12}$$

D.3. Charge Sensitivities

The corresponding second-order Taylor expansion of the molecular electronic energy in powers of displacements of the canonical state-parameters, $[d\boldsymbol{N}, d\boldsymbol{v}(\boldsymbol{r})]$, involves the relevant principal derivatives of the energy representation in the subsystem resolution:

$$\begin{aligned}\Delta^{(1+2)}E[\boldsymbol{N}, \boldsymbol{v}] = {} & (\partial E[\boldsymbol{N}, \boldsymbol{v}]/\partial \boldsymbol{N})_{\boldsymbol{v}}\, d\boldsymbol{N}^{\mathrm{T}} + \int \{\partial E[\boldsymbol{N}, \boldsymbol{v}]/\partial \boldsymbol{v}(\boldsymbol{r})\}_{\boldsymbol{N}}\, d\boldsymbol{v}(\boldsymbol{r})^{\mathrm{T}}\, d\boldsymbol{r} \\ & + \tfrac{1}{2}\{d\boldsymbol{N}(\partial^2 E[\boldsymbol{N}, \boldsymbol{v}]/\partial \boldsymbol{N}\,\partial \boldsymbol{N})_{\boldsymbol{v}}\, d\boldsymbol{N}^{\mathrm{T}} + 2\, d\boldsymbol{N}\,[\partial/\partial \boldsymbol{N}\!\int \{\partial E[\boldsymbol{N}, \boldsymbol{v}]/\partial \boldsymbol{v}(\boldsymbol{r})\}_{\boldsymbol{N}}]_{\boldsymbol{v}}\, d\boldsymbol{v}(\boldsymbol{r})^{\mathrm{T}}\, d\boldsymbol{r}\} \\ & + \iint d\boldsymbol{v}(\boldsymbol{r})\, \{\partial^2 E[\boldsymbol{N}, \boldsymbol{v}]/\partial \boldsymbol{v}(\boldsymbol{r})\, \partial \boldsymbol{v}(\boldsymbol{r}')\}_{\boldsymbol{N}}\, d\boldsymbol{v}(\boldsymbol{r}')^{\mathrm{T}}\, d\boldsymbol{r}\, d\boldsymbol{r}'\} \end{aligned} \tag{D.3.1a}$$

$$\begin{aligned} \equiv {} & \boldsymbol{\mu}\, d\boldsymbol{N}^{\mathrm{T}} + \int \boldsymbol{\rho}(\boldsymbol{r})\, d\boldsymbol{v}(\boldsymbol{r})^{\mathrm{T}}\, d\boldsymbol{r} \\ & + \tfrac{1}{2}\{d\boldsymbol{N}\,\mathbf{H}\, d\boldsymbol{N}^{\mathrm{T}} + 2\, d\boldsymbol{N}\!\int \mathbf{f}(\boldsymbol{r})\, d\boldsymbol{v}(\boldsymbol{r})^{\mathrm{T}}\, d\boldsymbol{r} + \iint d\boldsymbol{v}(\boldsymbol{r})\, \mathbf{B}(\boldsymbol{r}, \boldsymbol{r}')\, d\boldsymbol{v}(\boldsymbol{r}')^{\mathrm{T}}\, d\boldsymbol{r}\, d\boldsymbol{r}'\}. \end{aligned} \tag{D.3.1b}$$

The second differential in the preceding equation is determined by the following matrices of the principal charge sensitivities in the subsystem resolution:

i) the *hardness* matrix, $\mathbf{H} = (\partial\boldsymbol{\mu}/\partial\boldsymbol{N})_v = \{H_{\alpha,\beta} = (\partial\mu_\beta/\partial N_\alpha)_v\}$; (D.3.2a)

ii) the *Fukui function* matrix,

$$\mathbf{f}(\boldsymbol{r}) = [\partial\boldsymbol{\rho}(\boldsymbol{r})/\partial\boldsymbol{N}]_v = [\partial\boldsymbol{\mu}/\partial\boldsymbol{v}(\boldsymbol{r})]_N^{\mathrm{T}} = \{f_{\alpha,\beta}(\boldsymbol{r}) = [\partial\rho_\beta(\boldsymbol{r})/\partial N_\alpha]_v = [\partial\mu_\alpha/\partial v_\beta(\boldsymbol{r})]_N\}; \quad \text{(D.3.2b)}$$

iii) the *density linear-response* matrix,

$$\mathbf{B}(\boldsymbol{r}, \boldsymbol{r}') = [\partial\boldsymbol{\rho}(\boldsymbol{r}')/\partial\boldsymbol{v}(\boldsymbol{r})]_N = \{B_{\alpha,\beta}(\boldsymbol{r}, \boldsymbol{r}') = [\partial\rho_\beta(\boldsymbol{r}')/\partial v_\alpha(\boldsymbol{r})]_N\}. \quad \text{(D.3.2c)}$$

One also defines the corresponding matrices of the *softness* quantities in the subsystem resolution:

i) the *softness* matrix,

$$\mathbf{S} = (\partial\boldsymbol{N}/\partial\boldsymbol{\mu})_v = \mathbf{H}^{-1} = \{S_{\alpha,\beta} = (\partial N_\beta/\partial\mu_\alpha)_v\} = \int \mathbf{s}(\boldsymbol{r})\, d\boldsymbol{r}; \quad \text{(D.3.4a)}$$

ii) the *local softness* matrix,

$$\mathbf{s}(\boldsymbol{r}) = -\partial\boldsymbol{N}/\partial\boldsymbol{u}(\boldsymbol{r}) = [\partial\boldsymbol{\rho}(\boldsymbol{r})/\partial\boldsymbol{\mu}]_v = \{s_{\alpha,\beta}(\boldsymbol{r}) = -\,\partial N_\beta/\partial u_\alpha(\boldsymbol{r}) = [\partial\rho_\beta(\boldsymbol{r})/\partial\mu_\alpha]_v\}; \quad \text{(D.3.4b)}$$

iii) the *softness kernel* matrix,

$$\boldsymbol{\sigma}(\boldsymbol{r}, \boldsymbol{r}') = -\partial\boldsymbol{\rho}(\boldsymbol{r}')/\partial\boldsymbol{u}(\boldsymbol{r}) = -[\partial\boldsymbol{\rho}(\boldsymbol{r}')/\partial\boldsymbol{v}(\boldsymbol{r})]_\mu = \{\sigma_{\alpha,\beta}(\boldsymbol{r}, \boldsymbol{r}') = -[\partial\rho_\beta(\boldsymbol{r}')/\partial v_\alpha(\boldsymbol{r})]_\mu\}. \quad \text{(D.3.4c)}$$

A straightforward chain rule transformation gives the following expression for $\mathbf{s}(\boldsymbol{r})$ in terms of $\boldsymbol{\sigma}(\boldsymbol{r}, \boldsymbol{r}')$:

$$\mathbf{s}(\boldsymbol{r}) = -\int [\partial\boldsymbol{u}(\boldsymbol{r}')/\partial\boldsymbol{\mu}]_v\, [\partial\boldsymbol{\rho}(\boldsymbol{r})/\partial\boldsymbol{u}(\boldsymbol{r}')]\, d\boldsymbol{r}' = \int \mathbf{I}\, \boldsymbol{\sigma}(\boldsymbol{r}, \boldsymbol{r}')\, d\boldsymbol{r}', \quad \text{(D.3.4d)}$$

where the identity matrix $\mathbf{I} = \{\delta_{\alpha,\beta}\}$. A similar transformation gives the expression for the Fukui function matrix in terms of the local softnesses:

$$\mathbf{f}(\boldsymbol{r}) = \int [\partial\boldsymbol{u}(\boldsymbol{r}')/\partial\boldsymbol{N}]_v\, [\partial\boldsymbol{\rho}(\boldsymbol{r})/\partial\boldsymbol{u}(\boldsymbol{r}')]\, d\boldsymbol{r}' = [\partial\boldsymbol{\mu}/\partial\boldsymbol{N}]_v \int \boldsymbol{\sigma}(\boldsymbol{r}, \boldsymbol{r}')\, d\boldsymbol{r}'$$
$$= \mathbf{H}\, \mathbf{s}(\boldsymbol{r}) = \mathbf{S}^{-1}\mathbf{s}(\boldsymbol{r}). \quad \text{(D.3.4e)}$$

Multiplying both sides of the last equation from the left by $\mathbf{S} = \mathbf{H}^{-1}$ also gives:

$$\mathbf{s}(\boldsymbol{r}) = \mathbf{S}\, \mathbf{f}(\boldsymbol{r}). \quad \text{(D.3.4f)}$$

The matrix of hardness kernels in subsystem resolution is the inverse matrix of the softness kernels:

$$\boldsymbol{\eta}(\boldsymbol{r}, \boldsymbol{r}') = \{\partial^2 E_v[\boldsymbol{\rho}]/\partial\boldsymbol{\rho}(\boldsymbol{r})\,\partial\boldsymbol{\rho}(\boldsymbol{r}')\}_v = \partial^2 \widetilde{F}\,[\boldsymbol{\rho}]/\partial\boldsymbol{\rho}(\boldsymbol{r})\partial\boldsymbol{\rho}(\boldsymbol{r}') = -\,\partial\boldsymbol{u}(\boldsymbol{r}')/\partial\boldsymbol{\rho}(\boldsymbol{r}) = -\,[\partial\boldsymbol{v}(\boldsymbol{r}')/\partial\boldsymbol{\rho}(\boldsymbol{r})]_\mu$$
$$= \{\eta_{\alpha,\beta}(\boldsymbol{r}, \boldsymbol{r}') = -\partial u_\beta(\boldsymbol{r}')/\partial\rho_\alpha(\boldsymbol{r}) = -\,[\partial v_\beta(\boldsymbol{r}')/\partial\rho_\alpha(\boldsymbol{r})]_\mu\} = \boldsymbol{\sigma}^{-1}(\boldsymbol{r}', \boldsymbol{r}). \quad \text{(D.3.5)}$$

These two sets of kernels satisfy the following reciprocity relation:

$$\int \boldsymbol{\eta}(\boldsymbol{r}', \boldsymbol{r})\, \boldsymbol{\sigma}(\boldsymbol{r}, \boldsymbol{r}'')\, d\boldsymbol{r} = \{\textstyle\sum_\gamma \int \eta_{\alpha,\gamma}(\boldsymbol{r}', \boldsymbol{r})\sigma_{\gamma,\beta}(\boldsymbol{r}, \boldsymbol{r}'')d\boldsymbol{r} = \partial\rho_\beta(\boldsymbol{r}'')/\partial\rho_\alpha(\boldsymbol{r}') = \delta_{\alpha,\beta}\,\delta(\boldsymbol{r}'-\boldsymbol{r}'')\}, \quad \text{(D.3.6)}$$

since in the subsystem resolution the subsystem densities are the independent function "variables".

The hardness matrix of Eq. (D.3.2a) can be expressed in terms of the hardness kernel matrix using the double chain-rule transformation:

$$\mathbf{H} = (\partial\boldsymbol{\mu}/\partial\boldsymbol{N})_v = \iint [\partial\boldsymbol{\rho}(\boldsymbol{r})/\partial\boldsymbol{N}]_v\, [\partial^2 E_v[\boldsymbol{\rho}]/\partial\boldsymbol{\rho}(\boldsymbol{r})\,\partial\boldsymbol{\rho}(\boldsymbol{r}')]_v\, [\partial\boldsymbol{\rho}(\boldsymbol{r}')/\partial\boldsymbol{N}]_v^{\mathrm{T}}\, d\boldsymbol{r}\, d\boldsymbol{r}'$$
$$= \iint \mathbf{f}(\boldsymbol{r})\, \boldsymbol{\eta}(\boldsymbol{r}, \boldsymbol{r}')\, \mathbf{f}(\boldsymbol{r}')^{\mathrm{T}}\, d\boldsymbol{r}\, d\boldsymbol{r}'. \qquad \text{(D.3.7)}$$

The subsystem resolved analog of the local *hardness equalization* principle of Eq. (2.3.37) now reads:

$$\boldsymbol{\eta}(\boldsymbol{r}) = [\partial\boldsymbol{\mu}(\boldsymbol{r})/\partial\boldsymbol{N}]_v = \int [\partial^2 E_v[\boldsymbol{\rho}]/\partial\boldsymbol{\rho}(\boldsymbol{r})\partial\boldsymbol{\rho}(\boldsymbol{r}')]_v\, [\partial\boldsymbol{\rho}(\boldsymbol{r}')/\partial\boldsymbol{N}]_v^{\mathrm{T}}\, d\boldsymbol{r}' = \int \boldsymbol{\eta}(\boldsymbol{r}, \boldsymbol{r}')\, \mathbf{f}(\boldsymbol{r}')^{\mathrm{T}}\, d\boldsymbol{r}'$$
$$= [\partial\boldsymbol{\mu}/\partial\boldsymbol{N}]_v = \mathbf{H}, \qquad \text{(D.3.8)}$$

where we have used the intra-subsystem chemical potential equalization, $\boldsymbol{\mu}(\boldsymbol{r}) = \mu\boldsymbol{1}$, which identifies the intra-fragment equilibrium state [see Eq. (D.1.8)].

The differential of the subsystem densities of the externally closed subsystems, $\boldsymbol{\rho} = \boldsymbol{\rho}[\boldsymbol{N},\boldsymbol{v}; \boldsymbol{r}]$ reads:

$$d\boldsymbol{\rho}(\boldsymbol{r}) = d\boldsymbol{\rho}[\boldsymbol{N},\boldsymbol{v}; \boldsymbol{r}] = d\boldsymbol{N}\, \mathbf{f}(\boldsymbol{r}) + \int d\boldsymbol{v}(\boldsymbol{r}')\, \mathbf{B}(\boldsymbol{r}', \boldsymbol{r})\, d\boldsymbol{r}'. \qquad \text{(D.3.9)}$$

It can be alternatively expressed as $d\boldsymbol{\rho}(\boldsymbol{r}) = d\boldsymbol{\rho}[\boldsymbol{u}; \boldsymbol{r}]$:

$$d\boldsymbol{\rho}(\boldsymbol{r}) = \int [d\boldsymbol{\mu} - d\boldsymbol{v}(\boldsymbol{r}')]\, \boldsymbol{\sigma}(\boldsymbol{r}', \boldsymbol{r})\, d\boldsymbol{r}'. \qquad \text{(D.3.10)}$$

where

$$d\boldsymbol{\mu} = d\boldsymbol{\mu}[\boldsymbol{N},\boldsymbol{v}] = d\boldsymbol{N}\, \mathbf{H} + \int d\boldsymbol{v}(\boldsymbol{r})\, \mathbf{f}(\boldsymbol{r})^{\mathrm{T}}\, d\boldsymbol{r}. \qquad \text{(D.3.11)}$$

Using Eqs. (D.3.10) and (B.3.11) gives equivalent expression for $d\boldsymbol{\rho}[\boldsymbol{N},\boldsymbol{v}; \boldsymbol{r}]$ in terms of the charge sensitivities for the externally open molecular subsystems:

$$d\boldsymbol{\rho}[\boldsymbol{N},\boldsymbol{v}; \boldsymbol{r}] = \int [d\boldsymbol{N}\, \mathbf{H} + \int d\boldsymbol{v}(\boldsymbol{r}'')\, \mathbf{f}(\boldsymbol{r}'')^{\mathrm{T}}\, d\boldsymbol{r}'' - d\boldsymbol{v}(\boldsymbol{r}')]\, \boldsymbol{\sigma}(\boldsymbol{r}', \boldsymbol{r})\, d\boldsymbol{r}'$$
$$= d\boldsymbol{N}\, [\mathbf{H}\, \mathbf{s}(\boldsymbol{r})] + \int d\boldsymbol{v}(\boldsymbol{r}')\, [\mathbf{f}(\boldsymbol{r}')^{\mathrm{T}}\mathbf{s}(\boldsymbol{r}) - \boldsymbol{\sigma}(\boldsymbol{r}', \boldsymbol{r})]\, d\boldsymbol{r}'. \qquad \text{(D.3.12)}$$

A comparison between Eqs. (D.3.9) and (D.3.12) finally gives Eq. (D.3.4.e) and a generalization of the Berkowitz-Parr (1988) relation [see Eqs. (2.3.31) and (B.2.23)] in the subsystem resolution:

$$\mathbf{B}(\boldsymbol{r}', \boldsymbol{r}) = \mathbf{f}(\boldsymbol{r}')^{\mathrm{T}}\, \mathbf{s}(\boldsymbol{r}) - \boldsymbol{\sigma}(\boldsymbol{r}', \boldsymbol{r}) = \mathbf{f}(\boldsymbol{r}')^{\mathrm{T}}\, \mathbf{S}\, \mathbf{f}(\boldsymbol{r}) - \boldsymbol{\sigma}(\boldsymbol{r}', \boldsymbol{r}). \qquad \text{(D.3.14)}$$

D.4. Additive and Non-Additive Components of Hardness and Softness Kernels

It follows from the Euler equation for subsystems [Eqs. (6.5.7a-c)] that by adding to both sides of Eq. (D.1.5) the subsystem embedding (*non-additive*) potential of Eqs. (1.6.7) and (6.5.7a), defined by the partial functional derivative $v_\alpha^e(\boldsymbol{r}) \equiv \{\partial F^n[\boldsymbol{\rho}]/\partial\rho_\alpha(\boldsymbol{r})\}_{\beta\neq\alpha}$, one obtains their alternative form

$$w_\alpha(\boldsymbol{r}) = v_\alpha^{eff}(\boldsymbol{r}) - \mu_\alpha = -\{\partial F^a[\rho]/\partial\rho_\alpha(\boldsymbol{r})\}_{\beta\neq\alpha} = -\delta F[\rho_\alpha]/\delta\rho_\alpha(\boldsymbol{r}) \equiv -v_\alpha^a(\boldsymbol{r}), \tag{D.4.1a}$$

or in the matrix form,

$$\boldsymbol{w}(\boldsymbol{r}) = \boldsymbol{v}^{eff}(\boldsymbol{r}) - \boldsymbol{\mu} = -\partial F^a[\boldsymbol{\rho}]/\partial\boldsymbol{\rho}(\boldsymbol{r}) \equiv -\boldsymbol{v}^a(\boldsymbol{r}), \tag{D.4.1b}$$

where the row vectors $\boldsymbol{w}(\boldsymbol{r}) = \{w_\alpha(\boldsymbol{r})\}$ and $\boldsymbol{v}^{eff}(\boldsymbol{r}) = \{v_\alpha^{eff}(\boldsymbol{r})\}$ groups the subsystem effective external potentials,

$$\boldsymbol{v}^{eff}(\boldsymbol{r}) \equiv \boldsymbol{v}(\boldsymbol{r}) + \partial F^n[\boldsymbol{\rho}]/\partial\boldsymbol{\rho}(\boldsymbol{r}) = \{v_\alpha^{eff}(\boldsymbol{r}) = v(\boldsymbol{r}) + \{\partial F^n[\boldsymbol{\rho}]/\partial\rho_\alpha(\boldsymbol{r})\}_{\beta\neq\alpha} \equiv v(\boldsymbol{r}) + v_\alpha^e(\boldsymbol{r})\}, \tag{D.4.1c}$$

and $\boldsymbol{v}^a(\boldsymbol{r}) = \{v_\alpha^a(\boldsymbol{r})\}$ stands for a row vector of the subsystem *additive* potentials, due to the additive part $F^a[\boldsymbol{\rho}]$ of $F[\boldsymbol{\rho}]$ [Eq. (D.1.6)].

These subsystem equilibrium conditions in the form of the corresponding global condition, with the relative effective external potential $w_\alpha(\boldsymbol{r})$ of subsystem α replacing the global relative external potential $u(\boldsymbol{r})$ [see Eq. (1.5.22)].

It should also be noticed that by adding to both sides of Eq. (D.1.5) the subsystem additive potential of Eqs. (D.4.1a,b) one obtains yet another form of the Euler equations in the subsystem resolution:

$$z_\alpha(\boldsymbol{r}) \equiv v(\boldsymbol{r}) + v_\alpha^a(\boldsymbol{r}) - \mu_\alpha = -v_\alpha^e(\boldsymbol{r}), \qquad \alpha = 1, 2, \ldots, m, \tag{D.4.2a}$$

or in the matrix form:

$$\boldsymbol{z}(\boldsymbol{r}) \equiv \boldsymbol{v}(\boldsymbol{r}) + \boldsymbol{v}^a(\boldsymbol{r}) - \boldsymbol{\mu} = -\partial F^n[\boldsymbol{\rho}]/\partial\boldsymbol{\rho}(\boldsymbol{r}) = -\boldsymbol{v}^e(\boldsymbol{r}). \tag{D.4.2b}$$

The partitioning of Eq. (D.1.6) provides the related division of the hardness kernels into the corresponding subsystem *additive* and *non-additive* components:

$$\boldsymbol{\eta}(\boldsymbol{r}, \boldsymbol{r}') = \partial^2 F^a[\boldsymbol{\rho}]/\partial\boldsymbol{\rho}(\boldsymbol{r})\,\partial\boldsymbol{\rho}(\boldsymbol{r}') + \partial^2 F^n[\boldsymbol{\rho}]/\partial\boldsymbol{\rho}(\boldsymbol{r})\,\partial\boldsymbol{\rho}(\boldsymbol{r}') \equiv \boldsymbol{\eta}^a(\boldsymbol{r}, \boldsymbol{r}') + \boldsymbol{\eta}^n(\boldsymbol{r}, \boldsymbol{r}'). \tag{D.4.3}$$

It also follows from the definition of the additive functional, $F^a[\boldsymbol{\rho}] = \Sigma_\gamma F[\rho_\gamma]$ that the matrix of the additive hardness kernels is the subsystem-diagonal:

$$\eta_{\alpha,\beta}^a(\boldsymbol{r}, \boldsymbol{r}') = \{\partial^2 F[\rho_\alpha]/\partial\rho_\alpha(\boldsymbol{r})\,\partial\rho_\alpha(\boldsymbol{r}')\}\,\delta_{\alpha,\beta} = \eta_{\alpha,\alpha}^a(\boldsymbol{r}, \boldsymbol{r}')\delta_{\alpha,\beta}. \tag{D.4.4}$$

It follows from Eq. (D.3.5) that the matrix of the subsystem hardness kernels transforms a given displacement (perturbation) of the subsystem densities, $d\boldsymbol{\rho}$ into the conjugate responses` in the negative relative external potentials of subsystems:

$$-d\boldsymbol{u}(\boldsymbol{r}) = \int d\boldsymbol{\rho}(\boldsymbol{r}')\,\boldsymbol{\eta}(\boldsymbol{r}', \boldsymbol{r})\,d\boldsymbol{r}'. \tag{D.4.5}$$

The two components of $\boldsymbol{\eta}(\boldsymbol{r}', \boldsymbol{r})$ in Eq. (D.4.3) similarly transform the hypothetical shifts of the subsystem densities into the corresponding responses in the conjugate potentials:

$$-d\boldsymbol{w}(\boldsymbol{r}) = \int d\boldsymbol{\rho}(\boldsymbol{r}')\,\boldsymbol{\eta}^a(\boldsymbol{r}', \boldsymbol{r})\,d\boldsymbol{r}' = d\boldsymbol{v}^a(\boldsymbol{r}), \quad -d\boldsymbol{z}(\boldsymbol{r}) = \int d\boldsymbol{\rho}(\boldsymbol{r}')\,\boldsymbol{\eta}^n(\boldsymbol{r}', \boldsymbol{r})\,d\boldsymbol{r}' = d\boldsymbol{v}^e(\boldsymbol{r}), \tag{D.4.6}$$

since

$$\eta^a(\mathbf{r}',\mathbf{r}) = -\partial w(\mathbf{r})/\partial \rho(\mathbf{r}') = \partial v^a(\mathbf{r})/\partial \rho(\mathbf{r}') \quad \text{and} \quad \eta^n(\mathbf{r}',\mathbf{r}) = -\partial z(\mathbf{r})/\partial \rho(\mathbf{r}') = \partial v^e(\mathbf{r})/\partial \rho(\mathbf{r}'). \quad \text{(D.4.7)}$$

These two matrices of the hardness kernels define the corresponding inverses, which represent the associated matrices of the softness kernels:

$$\sigma^a(\mathbf{r},\mathbf{r}') = -\,\partial \rho(\mathbf{r}')/\partial w(\mathbf{r}) = \eta^a(\mathbf{r},\mathbf{r}')^{-1}, \qquad \sigma^n(\mathbf{r},\mathbf{r}') = -\,\partial \rho(\mathbf{r}')/\partial z(\mathbf{r}) = \eta^n(\mathbf{r},\mathbf{r}')^{-1}. \quad \text{(D.4.8)}$$

Clearly, these two components do not add up to the matrix of the overall softness kernels: $\sigma^a(\mathbf{r},\mathbf{r}') + \sigma^n(\mathbf{r},\mathbf{r}') \neq \sigma(\mathbf{r},\mathbf{r}')$. The matrices of the softness kernels of Eqs. (D.3.4c) and (D.4.8) transform the displacements in the relevant potentials into the conjugate linear responses of the subsystem densities:

$$d\rho(\mathbf{r}) = -\int du(\mathbf{r}')\,\sigma(\mathbf{r}',\mathbf{r})\,d\mathbf{r}' = -\int dw(\mathbf{r}')\,\sigma^a(\mathbf{r}',\mathbf{r})\,d\mathbf{r}' = -\int dz(\mathbf{r}')\,\sigma^n(\mathbf{r}',\mathbf{r})\,d\mathbf{r}'. \quad \text{(D.4.9)}$$

APPENDIX E: The Molecular Channels: Elaboration

The *singly*- and *doubly*-excited configurations of the 2-AO model of *two*-electron system (see Section 8.5) are examined within the communication theory approach to chemical bonds. First the ground-state entropy/information bond indices are generated for the *non-bonding* (singly-excited) singlet and triplet states of the model. This analysis shows that these ground-state entropy/information concepts identify an intermediate MO polarization which gives rise to the single entropy-covalent bond. The subsequent application to the doubly-excited (*anti-bonding*) configuration fails to recognize its repulsive character. This failure is due to fact that the relative phases of AO in MO, and those of the elementary VB-structures in the configuration wave-function, respectively, are being lost in the corresponding *one*- and *two*-electron probabilities in atomic resolution. The latter, being generated from the quantum-mechanical *superposition principle*, determine the model communication channels, which are shown to be identical for both the bonding (ground-state) and anti-bonding (doubly-excited) singlet-states of the model. A resolution of this difficulty is proposed through the partition of the overall molecular channel into the partial *"forward"* and *"reverse"* channels, which correspond to the "positive" and "negative" phases of the wave-function components in AO representation. The average triplet spin-channel, of the singly-excited electron configuration of the 2-AO model, is examined and the entropy/information descriptors of the local *Hirshfeld-channel* in the AIM resolution are explored.

E.1. "Phase" Problem in Excited Configurations of 2-AO Model

In order to gain a better understanding of the information-theoretic description of the chemical bonds, it is of interest to examine the *non*-bonding and *anti*-bonding excited configurations of two-electrons in the 2-AO model of Section 8.5, when they occupy, partially or completely, the *anti*-bonding MO of Eq. (8.5.5). The new problem one then encounters is that of *phases* of the LCAO MO coefficients and those of the elementary VB-structures in the configuration two-electron wave-function, which are lost, when the *one*- or *two*-electron probabilities are generated using the *superposition principle* of quantum mechanics. In order to identify the problem and to ultimately resolve the difficulty it creates, we shall first examine the *singly*- and *doubly*-excited configurations of the 2-AO model using the entropy/ information concepts developed in the ground-state approach of Chapter 8, which describe the *resultant* (phase-independent) communication channel of a molecule.

E.1.1. Singly-Excited Configurations

Consider the singly-excited state of two electrons in the 2-AO model, with one electron occupying the *bonding* MO, $\varphi_b(\boldsymbol{r}) = Ca(\boldsymbol{r}) + Db(\boldsymbol{r})$, and the other electron being placed on the *anti-bonding* MO, $\varphi_a(\boldsymbol{r}) = -Da(\boldsymbol{r}) + Cb(\boldsymbol{r})$, where $C^2 + D^2 \equiv P + Q = 1$ [see Eq. (8.5.5)]. The relevant singlet wave-function $\Psi_S^e(1,2) \equiv \Phi_S^e(1,2)\,\Xi_S(1,2)$ is given by the product of the spatial function of two electrons, $\Phi_S^e(1, 2)$, *symmetric* with respect to permutation of two electrons, and the *anti-symmetric* spin-factor, $\Xi_S(1,2)$ [Eq. (8.2.22a)]. Accordingly, the triplet wave-functions $\{\Psi_T(1, 2; M_S) \equiv \Phi_T(1, 2)\ \Xi_T(1, 2; M_S)\}$, $M_S \in \{1, 0, -1\}$, are defined as products of the *anti-symmetric* spatial function $\Phi_T(1,2)$ and the *symmetric* spin part, $\Xi_T(1, 2; M_S)$. The triplet spin-components for the specified projection M_S of the resultant spin of two electrons are given by the following products of the two spin-states $\alpha(i)$ and $\beta(i)$ of an electron:

$$\Xi_T(1, 2; M_S) = \{\alpha(1)\alpha(2),\ \ 2^{-1/2}[\alpha(1)\beta(2) + \beta(1)\alpha(2)],\ \ \beta(\sigma_1)\beta(\sigma_2)\}. \tag{E.1.1}$$

The orbital part of the singly-excited (e) singlet state of the model reads:

$$\begin{aligned}\Phi_S^e(1, 2) &= 2^{-1/2}\,[\varphi_b(1)\varphi_a(2) + \varphi_a(1)\varphi_b(2)] = |\,\varphi_b\ \varphi_a\,| + 2^{1/2}\varphi_a(1)\varphi_b(2)\\ &= 2^{1/2}CD\,[b(1)\,b(2) - a(1)\,a(2)] + 2^{-1/2}\,(C^2 - D^2)[\,a(1)\,b(2) + b(1)\,a(2)]\\ &= (2PQ)^{1/2}\,[S_{b,b}(1, 2) - S_{a,a}(1, 2)] + 2^{-1/2}\,(P - Q)\,[S_{a,b}(1, 2) + S_{b,a}(1, 2)]\\ &\equiv 2(PQ)^{1/2}\,\overline{\Phi}_{ion}(1,2) + (P - Q)\,\Phi_{cov}(1,2),\end{aligned} \tag{E.1.2}$$

where $\{S_{m,n}(1, 2)\}$ denote the elementary VB-structures of Section 8.2.3. It is seen to be defined by the combination of the four elementary VB-structures, with both positive and negative coefficients. We have introduced in the foregoing equation the modified VB-ionic wave-function component, $\overline{\Phi}_{ion}(1,2)$, defined by the difference between the elementary VB-ionic structures of Section 8.2.3.

The triplet orbital function, to complement one of the spin-functions of Eq. (E.1.1), is given by the spatial Slater determinant $|\,\varphi_b\ \varphi_a\,|$, i.e., the difference of the elementary *VB*-covalent structures, which defines the modified VB-covalent wave-function component $\overline{\Phi}_{cov}(1,2)$ (see Section 8.2.3):

$$\Phi_T(1, 2) = 2^{-1/2} [\varphi_b(1)\, \varphi_a(2) - \varphi_b(2)\, \varphi_a(1)] = |\, \varphi_b\, \varphi_a \,| = 2^{-1/2}[a(1)\, b(2) - b(1)\, a(2)] = |\, a\, b \,|$$
$$= 2^{-1/2} [S_{a,b}(1, 2) - S_{b,a}(1, 2)] \equiv \overline{\Phi}_{cov}(1,2)\,. \qquad \text{(E.1.3)}$$

One easily verifies, using the superposition principle of quantum mechanics, that the *anti*-symmetric combinations $\{\,\overline{\Phi}_{cov}(1,2), \overline{\Phi}_{ion}(1,2)\,\}$ give rise to the same communication channels as their symmetric analogs $\{\,\Phi_{cov}(1,2), \Phi_{ion}(1,2)\,\}$ (see Scheme 8.4), since the signs of coefficients before the elementary VB-structures do not affect their conditional probabilities in the state under consideration.

It should be realized that the two singly-occupied MO in the excited triplet electron configuration of 2-AO model are physically *equivalent* to the original AO. This is because these two sets of one-electron functions, both singly-occupied, are related through the orthogonal transformation defined by the LCAO coefficients, thus defining the same spatial Slater determinants: $|\,\varphi_b\,\varphi_a\,| = |\,ab\,|$. This confirms the orbitally *non*-bonded (n) character of this singly-excited triplet configuration of the model. An independent confirmation of this statement is the vanishing off-diagonal CBO matrix element in this state, $\gamma_{a,b}^{\ e} = CD - DC = 0$, while the diagonal elements, representing the AO electron populations, $\gamma_{a,a}^{\ e} = q_a^{\ e} = \gamma_{b,b}^{\ e} = q_b^{\ S,e} = C^2 + D^2 = 1$, are identical with those in the electron-sharing ($x = ½$) *Atomic Promolecule* (AP). Therefore, the CBO matrix $\boldsymbol{\gamma}^n = \{\delta_{m,n}\}$ is identical with that describing the two free atoms of the electron-sharing AP reference $\mathbf{A}^0(x = ½)$: $\boldsymbol{\gamma}^e = \boldsymbol{\gamma}^0(x = ½)$. In other words, the orbital structure in this singly-excited configuration exactly reflects a collection of two *non*-bonded atoms of the AP reference.

A reference to Eq. (E.1.2) shows that this equivalence is no longer the case in the symmetric spatial wave-function of two-electrons in the excited singlet state, since it is not given by the spatial Slater determinant. Therefore, this state is not equivalent to a collection of two free-atoms, thus representing a presence of some chemical bond in the 2-AO model. In what follows we shall examine this difference between the singly-excited singlet and triplet configurations in a more detail. The issue of the bond covalency in the apparently *orbitally* non-bonded singlet state is particularly intriguing and requires a special attention.

It should be also emphasized that the entropy/information bond descriptors have no direct implications for the bond-energy. They are more closely related to the chemist notion of the bond multiplicity, which reflects a "discrete" representation of a specific condition of the electron spin-pairing pattern in a molecule. Thus, a "single" bond in this chemical terminology is still classified as being single irrespectively of the actual value of the bond-energy.

Triplet Orbital-Channel

As we have observed above, the orbitally non-bonding character of the triplet configuration is directly reflected by Eq. (E.1.3). It follows from this spatial wave-function that the condensed probabilities of simultaneously finding the two electrons on specified AIM/AO are identical with those corresponding to the electron-sharing promolecular input reference: $\mathbf{P}^T = \{P_{X,Y}^{\ T} = ½(1 - \delta_{X,Y})\} = \mathbf{P}^0(x = ½)$. They generate the condensed one-electron probability vector $\boldsymbol{P}^T = (½, ½) = \boldsymbol{P}^0(x = ½)$. These AO resolved probabilities give rise to the conditional probability matrix $\mathbf{P}_T(\mathbf{B}|\mathbf{A}) = \{P_T(Y|X) = 1 - \delta_{X,Y}\}$, generating the symmetric (noiseless, deterministic) binary channel of Scheme E.1, which is identical with that in Scheme 8.4 corresponding to the *VB-covalent* wave-function $\Phi_{cov}(1, 2)$:

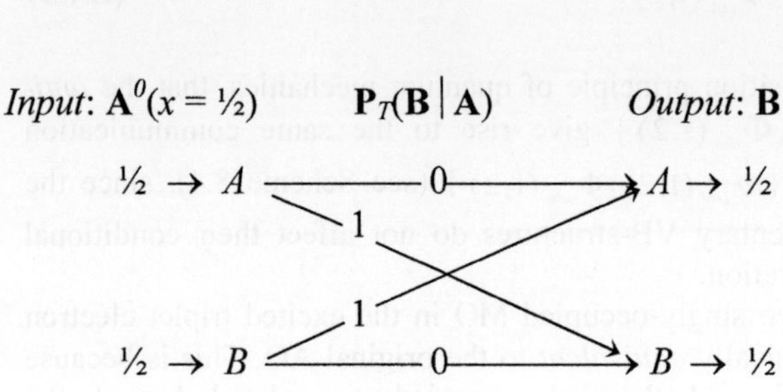

Scheme E.1. The symmetric orbital channel of the singly-excited, orbitally *non*-bonding triplet state of two electrons in the 2-AO model.

The one- and two-electron entropies (in bits) of this orbital triplet orbital channel are: $S_T(\mathbf{A}^0) = S_T(\mathbf{A}) = S_T(\mathbf{B}) = S_T(\mathbf{AB}) = 1$. Since the conditional probabilities in such a deterministic system are either 1 or 0, they generate the average conditional entropy $S_T(\mathbf{B}|\mathbf{A}) = 0 \equiv N_T^{cov}$, measuring the exactly vanishing covalent component of this orbitally non-bonding state. The associated mutual-information quantity, measuring the triplet orbital-ionicity, is equal to the input entropy: $I_T(\mathbf{A}^0:\mathbf{B}) \equiv N_T^{ion} = S_T(\mathbf{A}) - S_T(\mathbf{B}|\mathbf{A}) = S_T(\mathbf{A}) = 1$. It reflects the exchange-correlation in the spatial part of the triplet wave-function. Indeed, one gains the full information about the output probabilities, given the input probabilities, so there is no loss of the information in the channel.

Therefore, the orbitally non-bonding triplet state of two electrons in the 2-AO model is indeed characterized by the vanishing orbital-covalency and 1 bit of the molecular orbital-ionicity. These complementary descriptors define the vector entropy/information "signature", which is identical with that characterizing the stationary promolecular channel of Scheme 8.7/III for the electron-sharing input probabilities ($x = ½$): $\aleph_T^{orb} = (N_T^{cov}, N_T^{ion}) = (0, 1) \equiv \aleph_0^{orb}(x = ½)$.

Average Triplet Spin-Channel

Three components of the triplet spin-function, $\{\Xi_T(\sigma_1, \sigma_2; M_S) \equiv \Xi_T(1, 2; M_S),\ M_S = 1, 0, -1\}$, generate the corresponding *two*-electron spin probability matrices:

$$\boldsymbol{\pi}_2^T(M_S) = \{\pi_2^T(\sigma, \sigma'; M_S) = |\Xi_T(\sigma, \sigma'; M_S)|^2\},$$

$$\boldsymbol{\pi}_2^T(1) = \begin{bmatrix} 1 & 0 \\ 0 & 0 \end{bmatrix}, \qquad \boldsymbol{\pi}_2^T(0) = \begin{bmatrix} 0 & 1/2 \\ 1/2 & 0 \end{bmatrix}, \qquad \boldsymbol{\pi}_2^T(-1) = \begin{bmatrix} 0 & 0 \\ 0 & 1 \end{bmatrix}. \tag{E.1.4}$$

Therefore, the $\boldsymbol{\pi}_2^T(0)$ matrix, identical with that for the singlet spin-channel, $\boldsymbol{\pi}_2^T(0) = \boldsymbol{\pi}_2^S$ [Eq. (8.5.31)], also determines the deterministic, noiseless spin channel of Scheme 8.9. The elementary spin-channels for the other triplet components, isomorphic with the orbital-channels for the elementary VB-ionic structures $S_{a,a}(1, 2)$ and $S_{b,b}(1, 2)$ in Scheme 8.3, which give rise to the same two-electron probability matrix: $\boldsymbol{\pi}_2^T(1) = \mathbf{P}_{a,a}$ and $\boldsymbol{\pi}_2^T(-1) = \mathbf{P}_{b,b}$, are shown in Scheme E.2.

Of interest also is the *effective* spin-channel for the triplet state, which takes into account all resultant-spin orientations. Using the ensemble approach, with equal probabilities for each orientation of the resultant spin of two electrons, $\{p^{ens}(M_S) = ⅓\}$, gives rise to the *average* two-electron spin probabilities:

$$\boldsymbol{\pi}_2^T = \frac{1}{3}\sum_{M_S}\boldsymbol{\pi}_2^T(M_S) = \begin{bmatrix} 1/3 & 1/6 \\ 1/6 & 1/3 \end{bmatrix}. \qquad \text{(E.1.5)}$$

The summation over one spin-index gives the correct (unbiased) one-electron spin distribution: $\boldsymbol{\pi}_1^T = \boldsymbol{\pi}_1^S = (½, ½)$. These probabilities determine the corresponding ensemble average conditional probabilities:

$$\boldsymbol{\pi}_T(\Sigma_\mathbf{B} | \Sigma_\mathbf{A}) = \begin{bmatrix} 2/3 & 1/3 \\ 1/3 & 2/3 \end{bmatrix}, \qquad \text{(E.1.6)}$$

which define the effective, symmetric spin-channel for the triplet state shown in Scheme E.3.

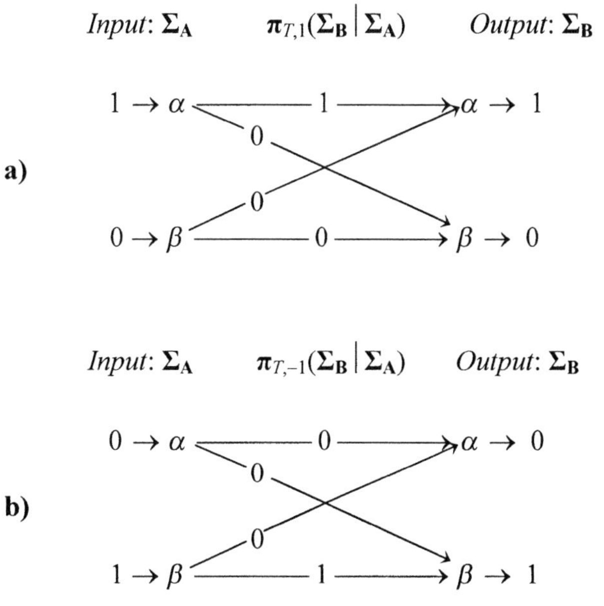

Scheme E.2. Examples of the partial triplet spin-channels for $M_S = 1$ (Panel a) and $M_S = -1$ (Panel b).

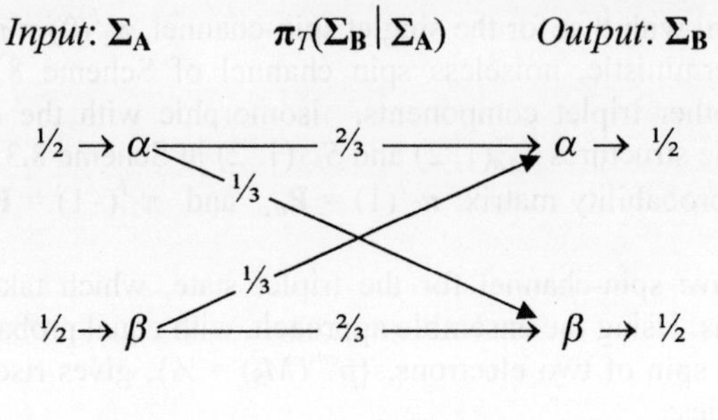

Scheme E.3. The effective spin-channel for the triplet state.

The relevant entropies (in bits) for this *effective* triplet channel are: $S(\boldsymbol{\pi}_1{}^T) \equiv S_T(\boldsymbol{\Sigma}_\mathbf{A}) \equiv S_T(\boldsymbol{\Sigma}_\mathbf{B}) = 1$, $S(\boldsymbol{\pi}_2{}^T) \equiv S_T(\boldsymbol{\Sigma}_\mathbf{A}\boldsymbol{\Sigma}_\mathbf{B}) = \log_2 3 + ⅓ = 1.92$. Hence, the spin bond indices for the ensemble averaged triplet channel are: $\mathcal{V}_T{}^{cov} \equiv S_T(\boldsymbol{\Sigma}_\mathbf{B} | \boldsymbol{\Sigma}_\mathbf{A}) = S(\boldsymbol{\pi}_2{}^T) - S_T(\boldsymbol{\pi}_1{}^T) = 0.92$, $\mathcal{V}_T{}^{ion} \equiv I_T(\boldsymbol{\Sigma}_\mathbf{A}{:}\boldsymbol{\Sigma}_\mathbf{B}) = 2S(\boldsymbol{\pi}_1{}^T) - S(\boldsymbol{\pi}_2{}^T) = 0.08$. Thus, they contain altogether 1 bit of information, mainly in the spin-covalent component. One observes a significant lowering of the spin-ionic (correlation) contribution, relative to the $\mathcal{V}_T{}^{ion}(M_S = 0) = 1$ bit value, in accordance with an intuitive expectation of a lower spin correlation in the statistical mixture of spin states for $M_S = 1, 0, -1$, relative to that in the pure $M_S = 0$ state.

Singlet State

It follows from the spatial wave-function of Eq. (E.1.2) that the two-electron probabilities in atomic resolution define the following matrix:

$$\mathbf{P}_S^e(P) = \begin{bmatrix} R(P) & 1/2 - R(P) \\ 1/2 - R(P) & R(P) \end{bmatrix}, \qquad R(P) = 2P(1-P) = 2PQ. \tag{E.1.7}$$

They give rise to the associated molecular one-electron electron probabilities, which are identical with those characterizing the electron-sharing *AP* reference: $\boldsymbol{P}_S{}^e(P) = (½, ½) = \boldsymbol{P}^0(x = ½)$. These probabilities define a network of the communication links of the associated molecular channel, defined by the conditional probabilities,

$$\mathbf{P}_S^e(\mathbf{B}(P)|\mathbf{A}(P)) = \mathbf{P}_S^e(\mathbf{B}(P)|\mathbf{A}^0(x = \frac{1}{2})) = \begin{bmatrix} 2R(P) & 1 - 2R(P) \\ 1 - 2R(P) & 2R(P) \end{bmatrix} \equiv \begin{bmatrix} 1-\omega & \omega \\ \omega & 1-\omega \end{bmatrix}. \tag{E.1.8}$$

Therefore, as we have explicitly indicated in the preceding equation, this excited singlet-state defines the *symmetric binary channel* (SBC) of Section 3.7 (see Scheme 3.2), the entropy/information descriptors of which have been fully discussed in Example 3.13. Hence, the crossover probability $\omega(P)$, which explicitly depends upon the MO-polarization, probability parameter P, fully determines the molecular communication system. A reference to Scheme 3.2 indicates that in this case the molecular output explicitly depends on the input parameter x. Therefore, the input and output probability distributions exhibit a generally non-vanishing values of the mutual-information quantity of Eq. (3.7.5), $I_S^e(\mathbf{A}^0(x):\mathbf{B}(P)) = H(z[x, \omega(P)]) - H(\omega(P))$, which measures the bond information-ionicity in this molecular channel. It reaches the maximum (capacity) level of Eq. (3.7.6), $C_S^e(\omega(P)) = 1 - H(\omega(P))$, for the electron-sharing promolecular input: $x = ½$. The corresponding entropy-covalence index is given by the conditional entropy of Eq. (3.7.3): $S_S^e(\mathbf{B}(P)|\mathbf{A}(P)) = H(\omega(P))$.

Let us now explore the entropy/information descriptors of specific cases of this molecular channel for the electron-sharing AP, when $x = ½$. We first observe that the *extreme* (*e*) values of $\omega(P^e) = (0, 1)$, for which $H(\omega(P^e)) = 0$ (see Fig. 3.2), are defined by equations:

$$1 - 4P^e(1 - P^e) = 0 \qquad \text{or} \qquad 1 - 4P^e(1 - P^e) = 1. \tag{E.1.9}$$

The former identifies the perfectly *delocalized* MO, for $P^e = ½$, e.g., in H_2, as giving rise to the covalently non-bonded singly-excited singlet state, in accordance with the orbitally non-bonded character of this configuration. The latter similarly classifies the singly excited singlet state originating from the perfectly *localized* MO solutions, $P^e = (1, 0)$, identical with the original AO. Indeed, in the latter case the singly occupied bonding and anti-bonding MO amount to the promolecular electron configuration, so that there should be no bond-covalency present in the system. It should be also observed that these zero entropy-covalency solutions correspond to 1 bit of the orbital information-ionicity, which distinguishes the promolecular channel (Scheme 8.7/III) from its SAL analog (Scheme 8.7/IV): $\mathbb{N}_S^{e,orb}(P^e) = (N_S^{e,cov}(P^e), N_S^{e,ion}(P^e)) = (0, 1) = \mathbb{N}_0^{orb}(x = ½)$. For the non-polarized, perfectly delocalized MO, when $P^e = Q^e = ½$ and hence $\omega(P^e) = 0$, represented by the spatially *symmetric* (*s*) φ_b and *anti-symmetric* φ_a, e.g., in H_2, this excited singlet configuration generates the communication channel shown in Scheme 8.4 for the $\Phi_{ion}(1, 2)$ wave-function, which indeed have been previously identified as representing the covalently non-bonding state exhibiting 1 bit of the information-ionicity.

Consider now the hypothetical *bonding* (*b*) MO polarizations $\{P^b\}$ in this excited singlet state, defined by the equation $\omega(P^b) = ½$, or

$$P^b(1 - P^b) = 1/8, \tag{E.1.10}$$

which give rise to the maximum $H(\omega(P^b)) = 1$ bit of the entropy-covalency and hence the vanishing information-ionicity $C_S^e(\omega(P^b)) = 1 - H(\omega(P^b)) = 0$. There are two solutions of the foregoing equation, symmetrically placed relative to the perfect delocalization value of the MO polarization parameter $P = ½$:

$$P_1^b = ½(1 - 2^{-1/2}) = 0.15 \qquad \text{or} \qquad (C_1^b = 0.38,\ D_1^b = 0.92),$$
$$P_2^b = 1 - P_1^b = ½(1 + 2^{-1/2}) = 0.85 \qquad \text{or} \qquad (C_2^b = 0.92,\ D_2^b = 0.38). \tag{E.1.11}$$

Hence, for $P = P_1^b$, $\varphi_b \approx b$ and $\varphi_a \approx a$, while $P = P_2^b$ gives the complementary polarized, partially delocalized shapes of MO, $\varphi_b \approx a$ and $\varphi_a \approx b$, for which one detects full 1 bit of entropy-covalency and zero information-ionicity in the singly-excited singlet state of the 2-AO model. This distinguishes the singlet configuration from the triplet one, with the latter representing the perfectly non-covalent bond irrespectively of the current value of the MO polarization parameter P. This agrees with our previous conjecture that the excited singlet configuration of the 2-AO model, which is not equivalent to the separate AO of the free-atoms, should exhibit some electron delocalization, i.e., $0 < P < 1$, and non-zero covalent bond component.

Therefore, the two-electron IT approach using the ground-state–like entropy/information bond indices, generated by the *resultant* communication system for the excited (singlet) state of the model, identifies an intermediate degree of the bond polarization, representing a pair of the *localized* MO, strongly resembling the original AO, which gives rise to the full, purely covalent bond index amounting to 1 bit of the entropy-covalency. In the (singlet) ground-state such a diagnosis of the chemical bond was possible only for the perfectly delocalized MO, $P = Q = ½$, which were shown to give rise to the covalently non-bonded state of two electrons in the singly-excited configuration.

E.1.2. Doubly-Excited Configuration

Consider next the *anti*-bonding (*a*), doubly-excited singlet state of the 2-AO model, when the anti-bonding MO, φ_a is occupied by the two spin-paired electrons:

$$\Psi_S^a(1, 2) = |\varphi_a\alpha \ \varphi_a\beta| \equiv \Phi_S^a(1, 2)\ \Xi_S(1, 2),$$

$$\begin{aligned}\Phi_S^a(1, 2) &= \varphi_a(1)\varphi_a(2) = D^2a(1)a(2) + C^2 b(1)b(2) - CD[a(1)b(2) + b(1)a(2)]\\ &= QS_{a,a}(1, 2) + PS_{b,b}(1, 2) - (PQ)^{1/2}\,[S_{a,b}(1, 2) + S_{b,a}(1, 2)]. \qquad (E.1.12)\end{aligned}$$

It is seen to be defined by the anti-bonding combination of the weighted average of the diagonal (ionic) VB-structures and the covalent part $\Phi_{cov}(1,2)$ grouping the off-diagonal elementary VB-structures, with the *negative* coefficient before the latter wave-function component. In this doubly-excited state the corresponding *one*- and *two*-electron probabilities in the AO resolution,

$$\boldsymbol{P}_S^a = [Q, P], \qquad \mathbf{P}_S^a = \begin{bmatrix} Q^2 & PQ \\ PQ & P^2 \end{bmatrix}, \qquad (E.1.13)$$

and the resulting conditional probability matrix,

$$\mathbf{P}_S^a(\mathbf{B}\,|\,\mathbf{A}) = \begin{bmatrix} Q & P \\ Q & P \end{bmatrix}, \qquad (E.1.14)$$

give rise to the communication channel of Scheme E.4.

This permuted information channel, relative to the ground-state channel of Scheme 8.7, generates the same entropy-information bond descriptors: $N_S^{a,cov}(P) = H(P)$, $N_S^{a,ion}(x, P) = H(x) - H(P)$, as does the bonding-state, when φ_b is doubly occupied (see Section 8.5.1). Therefore, the IT bond indices designed for the molecular ground-state channels are unable to recognize the anti-bonding character of the doubly-excited configuration for the symmetric case of H_2 and ethylene, when $P = Q = x = ½$, predicting the same amount of 1 bit of the entropy covalency as in the bonding, ground-state configuration.

Input: $\mathbf{A}[\mathbf{A}^0(x = \frac{1}{2})]$ $\mathbf{P}_S^a(\mathbf{B}\,|\,\mathbf{A})$ *Output:* $\mathbf{B}$

$Q\,(\frac{1}{2}) \rightarrow A$ ———————— Q ————————→ $A \rightarrow Q$
          P
                     $\times$
          Q
$P\,(\frac{1}{2}) \rightarrow B$ ———————— P ————————→ $B \rightarrow P$

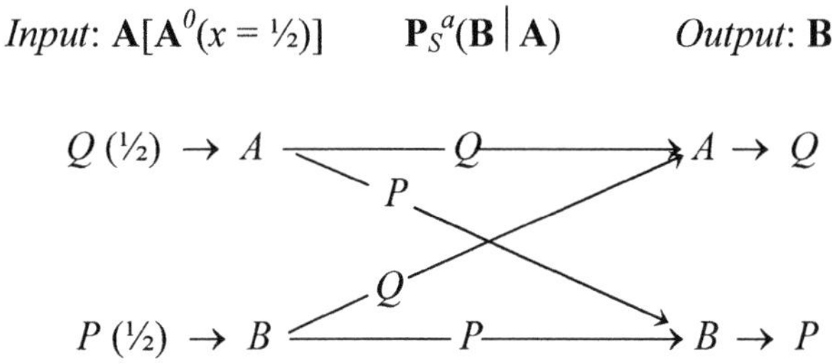

Scheme E.4. The orbital channel of the doubly-excited, *anti*-bonding singlet state of two electrons in the 2-AO model, for both molecular and promolecular ($x = \frac{1}{2}$) inputs.

The conditional-entropy (average noise) measure of the bond-covalency from the overall (resultant) input of this communication channel properly "counts" the spin-paired electrons in this excited state but it fails to reflect the implications of the phases (nodes) of the anti-bonding MO. This is due to the fact that the negative sign of the covalent wave-function component in Eq. (E.1.12) disappear in the probability matrices (E.1.13) and (E.1.14), thus having no effect on the entropy/information indices.

In the next section we shall remedy this difficulty by partitioning the overall inputs of the molecular communication system into the *"directed"* (partial) inputs, which result from the wave-function components in the AO representation exhibiting the *positive* and *negative* coefficients, respectively. They give rise to the *"forward"* and *"backward"* probability inputs of the overall molecular channel, which remains fixed, when manipulating the input data. These phase-sensitive, partial molecular inputs define the *"forward"* and *"reverse"* (*"backward"*) flows of the information through the molecular communication network.

Defining the *effective* IT indices of the chemical bond in excited configurations as differences between the entropy/information descriptors of the "forward" and "backward" inputs of the molecular channel introduces the element of the bond multiplicity *reduction* due to the actual excitations from the bonding to anti-bonding MO. We shall demonstrate below that this phase-resolved reappraisal of molecular channels gives rise to a better agreement with chemical intuition. It also reproduces the ground-state indices of the 2-AO model of Section 8.5. The information-theoretic indices generated in this approach can thus be regarded as the *generalized* entropic descriptors of the chemical bond.

E.1.3. The Partial "Forward" and "Reverse" Flows of Entropy/Information

We shall illustrate this generalized approach by applying it to the doubly-excited, anti-bonding configuration $\Phi_S^a(1, 2)$ of the preceding section. Let us separate in the two-electron probabilities $\mathbf{P}_S^a$, and hence also in their partial sums defining the associated one-electron probabilities P_S^a, the terms which originate from the VB-ionic (diagonal) structures in Eq. (E.1.12), exhibiting the *positive* coefficients in $\Phi_S^a(1, 2)$, from those originating from VB-covalent (off-diagonal) structures, exhibiting the *negative* coefficients in this doubly-excited configuration of the 2-AO model. They respectively define the partial (directed) probability inputs of the phase-dependent communication approach, $\vec{\boldsymbol{P}}_S^a$ and $\overleftarrow{\boldsymbol{P}}_S^a$ i.e., the forward (→) and backward (←) one-electron probability distributions in the AO resolution:

$$\vec{\mathbf{P}}_S^a = \begin{bmatrix} Q^2 & 0 \\ 0 & P^2 \end{bmatrix}, \quad \vec{\boldsymbol{P}}_S^a = (Q^2, P^2); \qquad \overleftarrow{\mathbf{P}}_S^a = \begin{bmatrix} 0 & PQ \\ PQ & 0 \end{bmatrix}, \quad \overleftarrow{\boldsymbol{P}}_S^a = (PQ, PQ). \qquad \text{(E.1.15)}$$

The abobe partial probability vectors constitute the phase–resolved inputs of the fixed molecular channel of Eq. (E.1.14): $\vec{\mathbf{P}}_S^a(\vec{\mathbf{B}}|\vec{\mathbf{A}}) = \overleftarrow{\mathbf{P}}_S^a(\overleftarrow{\mathbf{B}}|\overleftarrow{\mathbf{A}}) = \mathbf{P}_S^a(\mathbf{B}\,|\,\mathbf{A})$. The directed flows of information in this molecular channel are shown in Scheme E.5. By convention, the *"forward"* (*fwd*) flow is from the left to right, and *"backward"* (*bwd*) flow is in the reverse direction.

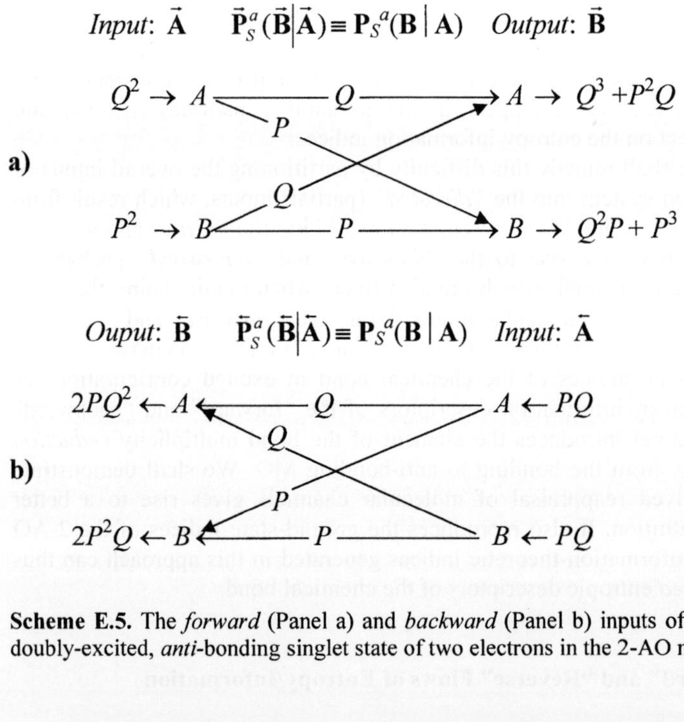

Scheme E.5. The *forward* (Panel a) and *backward* (Panel b) inputs of the molecular channel of the doubly-excited, *anti*-bonding singlet state of two electrons in the 2-AO model.

The conditional entropies of these directed flows of the electron probability are:

$$S_S^{a,fwd}(\mathbf{B}|\mathbf{A}) = (Q^2 + P^2)\, H(P), \qquad S_S^{a,bwd}(\mathbf{B}|\mathbf{A}) = 2PQ\, H(P). \qquad \text{(E.1.16)}$$

Their sum recovers the previous erroneous prediction of the entropy-covalency $H(P)$ of the preceding section. In order to introduce its reduction due to the MO excitations, from bonding to anti-bonding orbitals, we propose the difference between the forward and backward flows of the entropy/information entropies defined in the preceding equation as a *generalized* measure of the bond entropy-covalency:

$$\mathcal{N}_S^{a,cov}(P) = S_S^{a,fwd}(\mathbf{B}|\mathbf{A}) - S_S^{a,bwd}(\mathbf{B}|\mathbf{A}) = [P - Q(P)]^2\, H(P). \qquad \text{(E.1.17)}$$

This phase-sensitive measure of the bond entropy-covalency also reproduces the ground-state result of Eq. (8.5.12), since in this state [see Eq.(8.5.7a)] all input contributions belong to the *forward* set, with the *backward* set of inputs being empty.

The new measure of the bond entropy-covalency is seen to properly predict the vanishing covalency for the extreme MO polarization $P = (1, 0)$, which gives rise to the zero electron delocalization in the model, with the two MO being identical with the original AO. Also, for the maximum delocalization case of $P = Q = ½$ one correctly predicts the vanishing entropy-covalency in the anti-bonding configuration $\Phi_S^a(1, 2)$. This *minimum-covalency* result is in full agreement with both the chemical intuition and the elementary MO description of the chemical bond.

It is also of interest to examine the maxima of the modified covalency measure of Eq. (E.1.17). The extremum condition,

$$\frac{\partial \mathcal{N}_S^{a,cov}(P)}{\partial P} = 0, \qquad \text{(E.1.18)}$$

identifies the minimum value of $\mathcal{N}_S^{a,cov}(P_{min}) = 0$, at $P_{min} = ½$, and two maxima $\mathcal{N}_S^{a,cov}(P_{max}) = 0.31$, at $P_{max} = (0.12, 0.88)$. Therefore, the intermediate MO polarization $P_{max} = 0.12$, for which $C = 0.35$ and $D = 0.94$, so that MO strongly resemble the original AO, gives rise to the modified measure of the entropy-covalency amounting to about (⅓)-bond.

Now let us examine the mutual-information descriptors of the partial channels of Scheme E.5, which measure the amount of information flowing in the forward and reverse directions respectively. As we have already observed in Sections 3.7 and 8.5, they also reflect a degree of the mutual dependence between the partial inputs and outputs of these model communication systems. The relevant mutual-information quantities read:

$$I_S^{a,fwd}(\mathbf{A}{:}\mathbf{B}) = -(Q^2 + P^2)\log(Q^2 + P^2) = H(2PQ) + 2PQ\log(2PQ),$$

$$I_S^{a,bwd}(\mathbf{A}{:}\mathbf{B}) = -2PQ\log(2PQ) = H(Q^2 + P^2) + (Q^2 + P^2)\log(Q^2 + P^2). \qquad \text{(E.1.19)}$$

They give rise to the phase-dependent, generalized information-ionicity descriptor of the chemical bond in this excited configuration:

$$\mathcal{N}_S^{a,ion}(P) = I_S^{a,fwd}(\mathbf{A}{:}\mathbf{B}) - I_S^{a,bwd}(\mathbf{A}{:}\mathbf{B}) = H(2PQ) + 4PQ\log(2PQ). \qquad \text{(E.1.20)}$$

This *molecular* information-ionicity identically vanishes for both the original AO, when $P = (0, 1)$, and for the perfectly delocalized MO marking the minimum of the generalized entropy covalency: $P = Q = P_{min} = ½$, e.g., in H_2 or the π-bond in ethylene. This is in perfect agreement with the chemical and MO intuitions for the anti-bonding electron configuration $\Phi_S^a(1, 2)$. The maximum covalency, intermediate MO polarizations for $P = P_{max} = (0.12, 0.88)$ give the negative $\mathcal{N}_S^{a,ion}(P_{max}) = -0.20$ bits, and the generalized overall bond-order measure:

$$\mathcal{N}_S^{a}(P_{max}) \equiv \mathcal{N}_S^{a,cov}(P_{max}) + \mathcal{N}_S^{a,ion}(P_{max}) = 0.10 \text{ bits}. \qquad \text{(E.1.21)}$$

E.1.4. Singly-Excited Configurations: Reappraisal

Let us now examine the generalized entropy/information descriptors in the singly excited configurations $\Phi_S^e(1, 2)$ and $\Phi_T(1, 2)$ [Eqs. (E.1.2) and (E.1.3)]. The forward and reverse flows in the triplet case are shown in Scheme E.6 (see also Scheme E.1).

A reference to diagrams in Scheme E.6 indicates that all generalized bond indices identically vanish in the whole range of the admissible MO polarizations $0 \le P \le 1$, again in full accord with both the chemical intuition and the standard MO description.

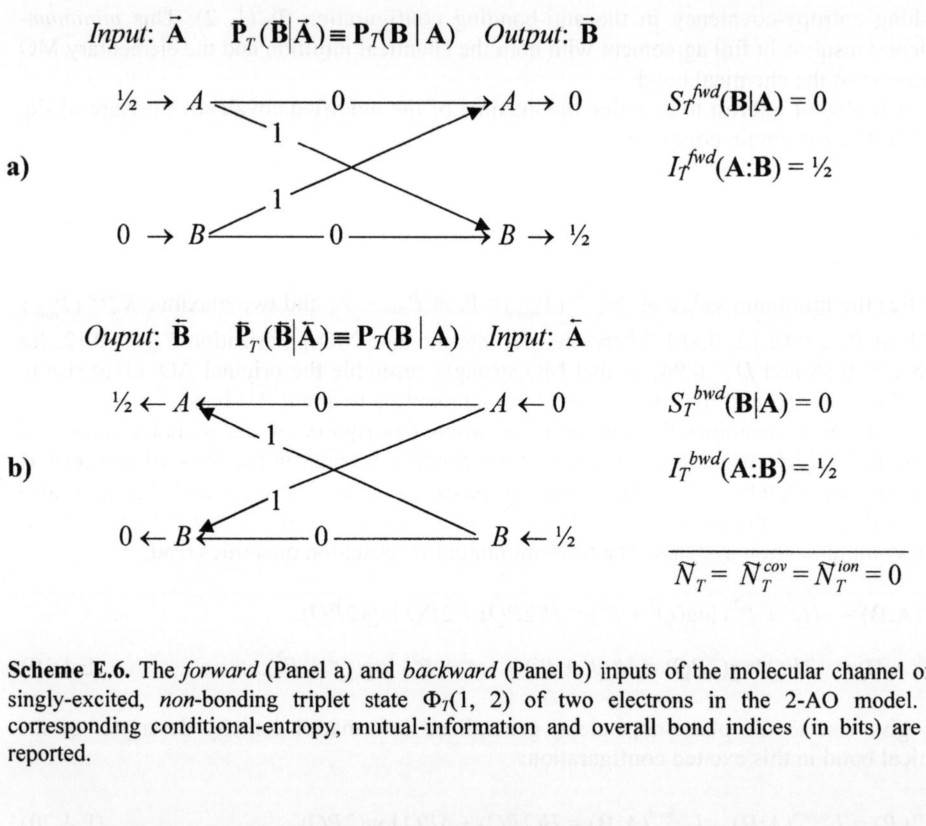

Scheme E.6. The *forward* (Panel a) and *backward* (Panel b) inputs of the molecular channel of the singly-excited, *non*-bonding triplet state $\Phi_T(1, 2)$ of two electrons in the 2-AO model. The corresponding conditional-entropy, mutual-information and overall bond indices (in bits) are also reported.

The orbital channel of the singlet *non-bonded* configuration $\Phi_S^e(1, 2)$ [Eq. (E.1.8) defines the SBC in terms of the conditional probability $\omega(P) = 1 - 4P(1-P)$. In Scheme E.7 we have partitioned the molecular inputs into the corresponding *forward*- and *backward*-contributions, in accordance with the phases of the wave-function components in Eq. (E.1.2). For definiteness the $P \ge ½$ MO polarization has been assumed, thus attributing a relative *acidic* character to atom *A* and a relative basic character to atom *B* in the diatomic molecule. In these diagrams the expressions for the corresponding conditional-entropy and mutual-information descriptors of the partial channels are also reported.

The generalized entropy-covalency index [see Eq. (E.1.17)],

$$\tilde{N}_S^{e,cov}(\omega(P)) = S_S^{e,fwd}(\mathbf{B}|\mathbf{A}) - S_S^{e,bwd}(\mathbf{B}|\mathbf{A}) = \omega\, H(\omega), \tag{E.1.22}$$

vanishes for $\omega = (0, 1)$. The former value implies the maximum electron delocalization in the bonding MO, $P = Q = \frac{1}{2}$, while the latter value corresponds in the ground-state to the lone-pair on A, $P = 1$, i.e., zero electron delocalization. These predictions for the non-bonded excited state are in perfect agreement with the chemical intuition and with the bonding diagnosis based upon the phase-independent indices (Section E.1.1).

Input: $\vec{\mathbf{A}}$ $\vec{\mathbf{P}}_S^e(\vec{\mathbf{B}}|\vec{\mathbf{A}}) \equiv \mathbf{P}_S^e(\mathbf{B}|\mathbf{A})$ *Output:* $\vec{\mathbf{B}}$

a)

$$\frac{1}{2}\omega \to A \xrightarrow{\quad 1-\omega \quad} A \to \omega(1 - \omega/2)$$

ω

ω

$$\frac{1}{2} \to B \xrightarrow{\quad 1-\omega \quad} B \to \frac{1}{2}[1 + \omega(\omega - 1)]$$

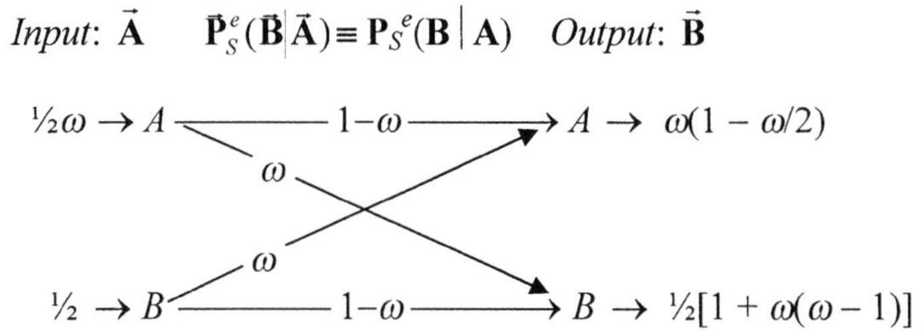

$$S_S^{e,fwd}(\mathbf{B}|\mathbf{A}) = \frac{1}{2}(1+\omega)H(\omega);$$

$$I_S^{e,fwd}(\mathbf{A}:\mathbf{B}) = \frac{(1-\omega)\omega}{2}\log\frac{2(1-\omega)}{(2-\omega)\omega} + \frac{\omega^2}{2}\log\frac{2\omega}{(\omega^2 - \omega + 1)}$$

$$+ \frac{(1-\omega)}{2}\log\frac{2(1-\omega)}{(\omega^2 - \omega + 1)} + \frac{\omega}{2}\log\frac{2}{(2-\omega)};$$

Ouput: $\vec{\mathbf{B}}$ $\vec{\mathbf{P}}_S^e(\vec{\mathbf{B}}|\vec{\mathbf{A}}) \equiv \mathbf{P}_S^e(\mathbf{B}|\mathbf{A})$ *Input:* $\vec{\mathbf{A}}$

b)

$$\frac{1}{2}(1-\omega)^2 \leftarrow A \xleftarrow{\quad 1-\omega \quad} A \leftarrow \frac{1}{2}(1-\omega)$$

ω

ω

$$\frac{1}{2}(1-\omega)\omega \leftarrow B \xleftarrow{\quad 1-\omega \quad} B \leftarrow 0$$

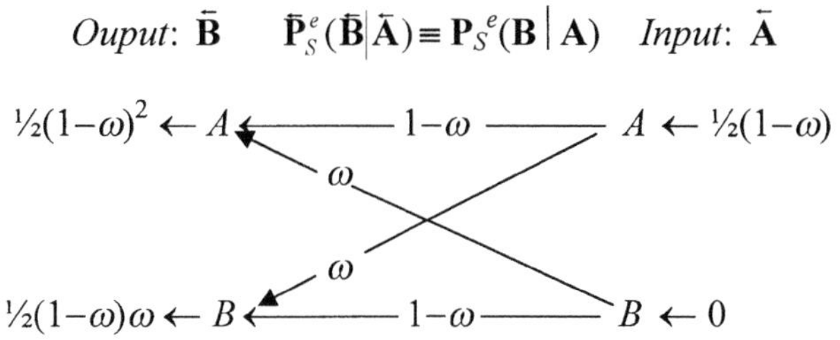

$$S_S^{e,bwd}(\mathbf{B}|\mathbf{A}) = \frac{1}{2}(1-\omega)H(\omega); \qquad I_S^{e,bwd}(\mathbf{A}:\mathbf{B}) = (1-\omega)^2[1 - \log(1-\omega)];$$

Scheme E.7. The *forward* (Panel a) and *backward* (Panel b) inputs of the molecular channel of the singly-excited, *non*-bonding singlet state $\Phi_S^e(1, 2)$ of two electrons in the 2-AO model. The expressions for the conditional-entropy and mutual-information bond indices are displayed below each directed channel.

However, this phase-dependent description also identifies some entropy-covalency for the $\tilde{N}_S^{e,cov}$ - maximum configuration at $\omega_{max} = 0.70$, $\tilde{N}_S^{e,cov}(\omega_{max}) = 0.62$ bits, for $P_{max} = 0.92$, i.e. $C = 0.96$. This "marginal" entropy-covalency reflects the symmetrizing influence of the (anti-symmetric) singlet spin function exerted in this excited state upon the spatial wave-function. As a result, there is some degree of the communication "noise" left in this orbitally non-bonding electron configuration, due to the symmetrized MO product in Eq. (E.1.2). This prediction parallels a similar result reported in Section E.1.1, which was obtained using the ground-state (phase-independent) bond indices, of the full 1 bit entropy-covalency for $P_{max} = 0.85$ ($C = 0.92$).

Next, let us examine the phase-dependent information-ionicity expressions reported in Scheme E.7. These partial-channel expressions give rise to the generalized ionicity index

$$\tilde{N}_S^{e,ion}(\omega(P)) = I_S^{e,fwd}(\mathbf{A{:}B}) - I_S^{e,bwd}(\mathbf{A{:}B})$$
$$= ½[-2\omega^2 + 5\omega - 1 + (1-\omega)(3-\omega)\log(1-\omega) + \omega(2\omega-1)\log\omega$$
$$-\omega(2-\omega)\log(2-\omega) - (\omega^2 - \omega + 1)\log(\omega^2 - \omega + 1)]. \qquad (E.1.23)$$

For the perfectly delocalized (non-polarized) bonding MO, when $\omega = 0$ ($P = ½$), this phase-dependent index gives $\tilde{N}_S^{e,ion}(0) = -½$ bit, while the non-delocalized $\omega = P = 1$ channel predicts $\tilde{N}_S^{e,ion}(1) = 1$ bit. The maximum entropy-covalency value $\omega_{max} = 0.7$ gives rise to the marginal information-ionicity $\tilde{N}_S^{e,ion}(0.7) = 0.06$ bits and hence the overall index $\tilde{N}_S^{e}(0.7) = \tilde{N}_S^{e,cov}(0.7) + \tilde{N}_S^{e,ion}(0.7) = 0.68$ bits.

With this development we have extended the range of applicability of the entropy/information descriptors of the chemical bond to the excited configurations by the introduction of the *directed* information channels, which reflect the relative phases of the wave-function components in AO representation. An illustrative application to the 2-AO model has demonstrated that the generalized IT indices of the bond covalency and ionicity, which are generated by these phase-dependent inputs of the molecular channel, are in general agreement with the familiar MO description and chemical intuition. In the ground-state of the model the generalized bond indices reproduce the previous, phase-independent entropy/information bond-orders. The latter where shown to be only partly successful, when applied to the excited configurations.

E.2. Local Hirshfeld Channel

In the fine-grained, local resolution the communication system approach to chemical bonds would require a complicated integration over the *two*-point electron probability distributions. Its generalization to entropies relating several probability schemes in IT would require integration over simultaneous probabilities of many electrons, which would severely limit the applicability of such an information-theoretic approach in the local resolution. In this section we shall briefly outline the elements of the *local* treatment of the entropy/information descriptors of the bonded (Hirshfeld) AIM, (Nalewajski 2004c,e; Nalewajski and Broniatowska, 2005a), relative to a the free constituent atoms of the promolecular reference. Due to the local character of the one-electron Hirshfeld division scheme (see Section 5.2) this approach requires only the integration over the *one*-electron probability densities. It provides the molecularly (or promolecularly) averaged entropies of the local probability networks in the molecule, in which the atom allocation signals are locally propagated between infinitesimal volume elements of all constituent atoms in the molecule.

We first observe that the Hirshfeld division of the molecular electron density into atomic components [Eq. (5.2.1)], $\rho(\boldsymbol{r}) = Np(\boldsymbol{r}) = \sum_X \rho_X^H(\boldsymbol{r})$, where $\rho_X^H(\boldsymbol{r}) = Np_X^H(\boldsymbol{r})$, is *local* in character. Indeed, at a given point in space each AIM participates in the local value of the overall electron density $\rho(\boldsymbol{r})$ in proportion to the promolecular conditional probability (*share -factor*) of Eq. (5.2.4),

$$\pi^0(X|\boldsymbol{r}) = p_X^0(\boldsymbol{r})/p^0(\boldsymbol{r}) = \rho_X^0(\boldsymbol{r})/\rho^0(\boldsymbol{r}) = \rho_X^H(\boldsymbol{r})/\rho(\boldsymbol{r}) = p_X^H(\boldsymbol{r})/p(\boldsymbol{r}) = \pi^H(X|\boldsymbol{r}) \equiv \pi(X|\boldsymbol{r}) = d_X^H(\boldsymbol{r}),$$
$$\sum_X \pi(X|\boldsymbol{r}) = 1, \qquad \text{(E.2.1)}$$

representing the subsystem relative contribution to the promolecular density $\rho^0(\boldsymbol{r}) = \sum_X \rho_X^0(\boldsymbol{r}) = Np^0(\boldsymbol{r})$. This equation summarizes the "stockholder" principle of Hirshfeld (1977) that each atom participates in the overall molecular *"profit"* $\rho(\boldsymbol{r})$ in proportion to its "share" in the promolecular *"investment"* $\rho^0(\boldsymbol{r})$.

In the conditional probability $\pi(X|\boldsymbol{r})$ the AIM label X represents a *variable*, while the vector $\boldsymbol{r}$ plays the role of a continuous electron-position *parameter*, as indeed reflected by the normalization condition in the foregoing equation. This probability density corresponds to a given local, infinitesimal volume element of atom X. Our aim now is to determine the local network of the electron probability density and the related electron density flow chart, connecting different atoms at the specified electron location in space. In Section 1.6 we have classified such displacements in the distribution of electrons as the *vertical* flows, carried out for a given (fixed) value of the overall (molecular) electron density $\rho(\boldsymbol{r})$, or its shape factor $p(\boldsymbol{r})$. We combine the subsystem densities for a given location in space into the row vectors: molecular,

$$\boldsymbol{\rho}^H(\boldsymbol{r}) \equiv \{\rho_X^H(\boldsymbol{r})\} = N\{p_X^H(\boldsymbol{r})\} \equiv N\,\boldsymbol{p}^H(\boldsymbol{r}), \qquad \text{(E.2.2a)}$$

and promolecular,

$$\boldsymbol{\rho}^0(\boldsymbol{r}) \equiv \{\rho_X^0(\boldsymbol{r})\} = \{N\,p_X^0(\boldsymbol{r})\} \equiv N\,\boldsymbol{p}^0(\boldsymbol{r}). \qquad \text{(E.2.2b)}$$

We also introduce in atomic resolution the unit *row*-vector $\boldsymbol{1}$ = (1, 1, ...), so that the corresponding closure relations in the matrix notation now read:

$$\rho(\boldsymbol{r}) = \boldsymbol{\rho}^H(\boldsymbol{r})\,\boldsymbol{1}^{\mathrm{T}} = N\,\boldsymbol{p}^H(\boldsymbol{r})\,\boldsymbol{1}^{\mathrm{T}}, \qquad \rho^0(\boldsymbol{r}) = \boldsymbol{\rho}^0(\boldsymbol{r})\boldsymbol{1}^{\mathrm{T}} = N\,\boldsymbol{p}^0(\boldsymbol{r})\,\boldsymbol{1}^{\mathrm{T}}. \qquad \text{(E.2.3)}$$

As we have already observed in Eq. (5.2.8), the Hirshfeld partition of the molecular electron distribution can be also interpreted as the universal (subsystem independent, unbiased) local enhancement of the free-atom distribution (Nalewajski and Parr, 2001; Nalewajski *et al.*, 2002):

$$p_X^H(\boldsymbol{r}) = p_X^0(\boldsymbol{r})\,[p(\boldsymbol{r})/p^0(\boldsymbol{r})] \equiv p_X^0(\boldsymbol{r})\,w(\boldsymbol{r}) \quad \text{or} \quad \boldsymbol{p}^H(\boldsymbol{r}) = w(\boldsymbol{r})\,\boldsymbol{p}^0(\boldsymbol{r}). \qquad \text{(E.2.4)}$$

By partitioning the molecular probability density, the shape-factor of the electron density, in the numerator of this local enhancement-factor,

$$w(\boldsymbol{r}) = \sum_X [p_X^H(\boldsymbol{r})/p^0(\boldsymbol{r})] \equiv \sum_X w_X^H(\boldsymbol{r}), \qquad \text{(E.2.5)}$$

one can express this overall enhancement as the sum of the AIM enhancements $\boldsymbol{w}(\boldsymbol{r}) = \{w_X^H(\boldsymbol{r})\}$ (a row vector):

$$w_X^H(\boldsymbol{r}) = \pi(X|\boldsymbol{r})\, w(\boldsymbol{r}) \quad \text{or} \quad \boldsymbol{w}(\boldsymbol{r}) = w(\boldsymbol{r})\, \boldsymbol{\pi}(X|\boldsymbol{r}), \quad w(\boldsymbol{r}) = \boldsymbol{w}(\boldsymbol{r})\boldsymbol{1}^{\mathrm{T}}, \quad \boldsymbol{\pi}(X|\boldsymbol{r})\, \boldsymbol{1}^{\mathrm{T}} = 1, \qquad \text{(E.2.6)}$$

where the row vector $\boldsymbol{\pi}(X|\boldsymbol{r}) = \{\pi(X|\boldsymbol{r})\}$ groups the conditional probabilities of the stockholder atoms.

Therefore, the Hirshfeld conditional probabilities, equal to the promolecular share-factors, partition the overall local enhancement factor into the associated subsystem contributions. For simplicity, in what follows we shall explicitly refer to a diatomic case: $X = A, B$. Rewriting the molecular closure relation of equation (E.2.3) and using Eqs. (E.2.4) and (E.2.6) finally gives the following expression for the molecular probability density $p(\boldsymbol{r})$ in terms of the local, additive contributions from the system constituent atoms, grouped in the square matrix $\mathbf{W}(\boldsymbol{r}) = \{W_{X,Y}(\boldsymbol{r})\}$:

$$p(\boldsymbol{r}) = \boldsymbol{1}\, \boldsymbol{p}^H(\boldsymbol{r})^{\mathrm{T}} = \boldsymbol{1}\, \boldsymbol{p}^0(\boldsymbol{r})^{\mathrm{T}} w(\boldsymbol{r}) = \boldsymbol{1}[\boldsymbol{p}^0(\boldsymbol{r})^{\mathrm{T}} \boldsymbol{w}(\boldsymbol{r})]\boldsymbol{1}^{\mathrm{T}} \equiv \boldsymbol{1}\mathbf{W}(\boldsymbol{r})\boldsymbol{1}^{\mathrm{T}}. \qquad \text{(E.2.7)}$$

Therefore, the matrix-element contributions of this *double* AIM resolution,

$$W_{X,Y}(\boldsymbol{r}) = p_X^0(\boldsymbol{r})\, w_Y^H(\boldsymbol{r}) = [p_X^0(\boldsymbol{r})/p^0(\boldsymbol{r})]\, p_Y^H(\boldsymbol{r}) = \pi(X|\boldsymbol{r})\, p_Y^H(\boldsymbol{r}), \qquad \text{(E.2.8)}$$

reproduce exactly the local value of the molecular probability density.

Dividing next $\mathbf{W}(\boldsymbol{r})$ by $p(\boldsymbol{r})$ defines the renormalized matrix,

$$\mathbf{V}(\boldsymbol{r}) = \mathbf{W}(\boldsymbol{r})/p(\boldsymbol{r}) = \{V_{X,Y}(\boldsymbol{r}) = W_{X,Y}(\boldsymbol{r})/p(\boldsymbol{r})\},$$

$$V_{X,Y}(\boldsymbol{r}) = \pi(X|\boldsymbol{r})[p_Y^H(\boldsymbol{r})/p(\boldsymbol{r})] = \pi(X|\boldsymbol{r})\pi(Y|\boldsymbol{r}), \quad \textstyle\sum_X\sum_Y V_{X,Y}(\boldsymbol{r}) = [\sum_X \pi(X|\boldsymbol{r})][\sum_Y \pi(Y|\boldsymbol{r})] = 1, \quad \text{(E.2.9)}$$

where the last line also expresses the *overall* normalization condition satisfied by this probability matrix. The two *partial* normalizations of $\mathbf{V}(\boldsymbol{r})$ similarly give:

$$\textstyle\sum_X V_{X,Y}(\boldsymbol{r}) = [\sum_X \pi(X|\boldsymbol{r})]\, \pi(Y|\boldsymbol{r}) = \pi(Y|\boldsymbol{r}), \qquad \sum_Y V_{X,Y}(\boldsymbol{r}) = \pi(X|\boldsymbol{r})[\sum_Y \pi(Y|\boldsymbol{r})] = \pi(X|\boldsymbol{r}). \qquad \text{(E.2.10)}$$

This matrix provides the local "product" (joint) probability matrix in atomic resolution,

$$\mathbf{P}[\mathbf{A}^0(\boldsymbol{r})\, \mathbf{B}^H(\boldsymbol{r})] = \mathbf{V}(\boldsymbol{r}) = \boldsymbol{P}(X|\boldsymbol{r}) \otimes \boldsymbol{P}(Y|\boldsymbol{r}), \qquad \text{(E.2.11)}$$

linking the local volume element of the promolecular (free-atom) *inputs*, $\mathbf{A}^0(\boldsymbol{r}) = \{A^0(\boldsymbol{r}), B^0(\boldsymbol{r})\}$, with the same local volume element of the molecular (Hirshfeld AIM) *outputs*, $\mathbf{B}^H(\boldsymbol{r}) = \{A^H(\boldsymbol{r}), B^H(\boldsymbol{r})\}$ (see Scheme E.8).

The local Hirshfeld conditional probabilities can thus be regarded as the *input* and *output* probabilities of the *local* Hirshfeld "communication" system shown in Scheme E.8:

$$\boldsymbol{P}[\mathbf{A}^0(\boldsymbol{r})] = \boldsymbol{P}[\mathbf{B}(\boldsymbol{r})] = \boldsymbol{\pi}(X|\boldsymbol{r}). \qquad \text{(E.2.12)}$$

As also shown in this scheme, a network for a propagation of the local AIM allocation signals is then determined by the Hirshfeld conditional probabilities themselves:

$$\mathbf{P}[\mathbf{B}^H(\boldsymbol{r})|\mathbf{A}^H(\boldsymbol{r})] = \{P[Y^H(\boldsymbol{r})|X^H(\boldsymbol{r})] = V_{XY}(\boldsymbol{r})/\pi(X|\boldsymbol{r}) = \pi(Y|\boldsymbol{r}). \qquad \text{(E.2.13)}$$

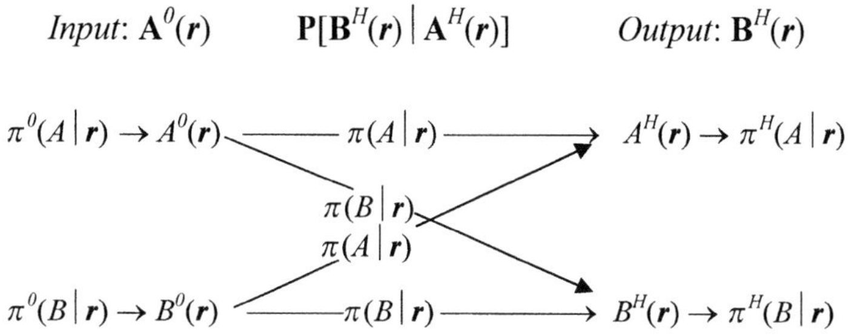

Scheme E.8. The local ("vertical") Hirshfeld communication channel for a diatomic molecule *A*—*B*.

These conditional probabilities, which completely define the local Hirshfeld channel of Scheme E.8, generate the identical input and output entropies, $S[\mathbf{A}^0(\boldsymbol{r})] = S[\mathbf{B}^H(\boldsymbol{r})] = S[\pi(X \mid \boldsymbol{r})]$, and the "product" entropy $S[\mathbf{A}^H(\boldsymbol{r})\mathbf{B}^H(\boldsymbol{r})] = 2S[\pi(X \mid \boldsymbol{r})]$. Hence, they give rise to the channel conditional entropy $S[\mathbf{B}^H(\boldsymbol{r}) \mid \mathbf{A}^H(\boldsymbol{r})] = S[\pi(X \mid \boldsymbol{r})]$ and to the vanishing local mutual-information quantity, $I[\mathbf{A}^0(\boldsymbol{r}){:}\mathbf{B}^H(\boldsymbol{r})] = 0$, reflecting the independent input and output probabilities of this local communication network.

The local entropies of the Hirshfeld channel can be averaged over space, using either the molecular, $p(\boldsymbol{r})$, or promolecular, $p^0(\boldsymbol{r})$, probability distributions as local "weights". Hence, for alternative local probability densities,

$$\boldsymbol{P}(\boldsymbol{r}) = \{\boldsymbol{p}^H(\mathbf{A}^0(\boldsymbol{r})), \boldsymbol{p}^H(\mathbf{B}^H(\boldsymbol{r})), \mathbf{P}[\mathbf{A}^H(\boldsymbol{r})\mathbf{B}^H(\boldsymbol{r})], \mathbf{P}[\mathbf{B}^H(\boldsymbol{r}) \mid \mathbf{A}^H(\boldsymbol{r})], \text{etc.}\},$$

one defines the corresponding two types of the average entropies:

$$\langle S(\boldsymbol{P})\rangle = \int p(\boldsymbol{r})\, S[\boldsymbol{P}(\boldsymbol{r})]\, d\boldsymbol{r} \qquad \text{and} \qquad \langle S(\boldsymbol{P})\rangle_0 = \int p^0(\boldsymbol{r})\, S[\boldsymbol{P}(\boldsymbol{r})]\, d\boldsymbol{r}, \qquad \text{(E.2.14)}$$

with the identically vanishing average mutual-information measures. For example, the average conditional entropies,

$$\langle S(\mathbf{B}^H \mid \mathbf{A}^H)\rangle = \int p(\boldsymbol{r})\, S[\mathbf{B}^H(\boldsymbol{r}) \mid \mathbf{A}^H(\boldsymbol{r})]\, d\boldsymbol{r} = \int p(\boldsymbol{r})\, S[\pi(X \mid \boldsymbol{r})]\, d\boldsymbol{r} \qquad \text{or}$$

$$\langle S(\mathbf{B}^H \mid \mathbf{A}^H)\rangle_0 = \int p^0(\boldsymbol{r})\, S[\mathbf{B}^H(\boldsymbol{r}) \mid \mathbf{A}^H(\boldsymbol{r})]\, d\boldsymbol{r} = \int p^0(\boldsymbol{r})\, S[\pi(X \mid \boldsymbol{r})]\, d\boldsymbol{r}, \qquad \text{(E.2.15)}$$

measuring the average noise in the local channel of Scheme E.8, can be regarded as alternative measures of the local (vertical) covalency of the Hirshfeld AIM.

Consider next the average input and output entropies of the local Hirshfeld channel. It is natural to use the *molecular* weights to calculate the average (molecular) output entropy, and the promolecular weights, to determine the average (promolecular) input entropy:

$$\langle S(\mathbf{B}^H)\rangle = \langle S(\mathbf{B}^H\,|\,\mathbf{A}^H)\rangle = -\int p(\boldsymbol{r})\,[\Sigma_Y\,\pi(Y|\,\boldsymbol{r})\,\log\pi(Y|\,\boldsymbol{r})]\,d\boldsymbol{r} = -\int \Sigma_Y\,p_Y{}^H(\boldsymbol{r})\,\log\,[p_Y{}^H(\boldsymbol{r})/p(\boldsymbol{r})]\,d\boldsymbol{r}$$
$$= -\,\Sigma_Y \int p_Y{}^H(\boldsymbol{r})\,\log p_Y{}^H(\boldsymbol{r})\,d\boldsymbol{r} + \int p(\boldsymbol{r})\,\log p(\boldsymbol{r})\,d\boldsymbol{r} = -\,\{S[p] - \Sigma_Y\,S[p_Y{}^H]\} \equiv -\,S^{nadd}[\boldsymbol{p}^H], \quad (E.2.16)$$

$$\langle S(\mathbf{A}^0)\rangle_0 = -\int p^0(\boldsymbol{r})\,[\Sigma_X\,\pi(X|\,\boldsymbol{r})\,\log\pi(X|\,\boldsymbol{r})]\,d\boldsymbol{r} = -\int \Sigma_X\,p_X{}^0(\boldsymbol{r})\,\log\,[p_X{}^0(\boldsymbol{r})/p^0(\boldsymbol{r})]\,d\boldsymbol{r}$$
$$= -\{S[p^0] - \Sigma_X\,S[p_X{}^0]\} \equiv -\,S^{nadd}[\boldsymbol{p}^0]. \quad (E.2.17)$$

As indicated in the preceding two equations, the average output and input entropies of the Hirshfeld information channel measure the negative *non-additive* entropy contributions of the overall molecular and promolecular probability distributions, respectively. This accords with the intuitive expectation that the bond covalent (electron delocalisation) component, reflecting the information noise in the molecular channel due to the inter-atomic communications, should be related to the non-additive part of the information entropy of the bonded atoms.

It is also of interest to explore the *displacement* in the entropy non-additivity, due to the electron delocalisation accompanying the bond formation. It is measured by the difference between the two non-additive entropies of Eqs. (E.2.16) and (E.2.17):

$$\Delta S^{nadd}[\boldsymbol{p}^H; \boldsymbol{p}^0] \equiv S^{nadd}[\boldsymbol{p}^H] - S^{nadd}[\boldsymbol{p}^0] = \langle S(\mathbf{A}^0)\rangle_0 - \langle S(\mathbf{B}^H)\rangle. \quad (E.2.18)$$

This entropy difference reflects the information quantity associated with the effect of the quantum-mechanical "mixing" of electrons in the molecule, thus providing an additional index of the bond entropy-covalency acquired by the stockholder atoms in the molecule, relative to the free-atoms of the promolecule.

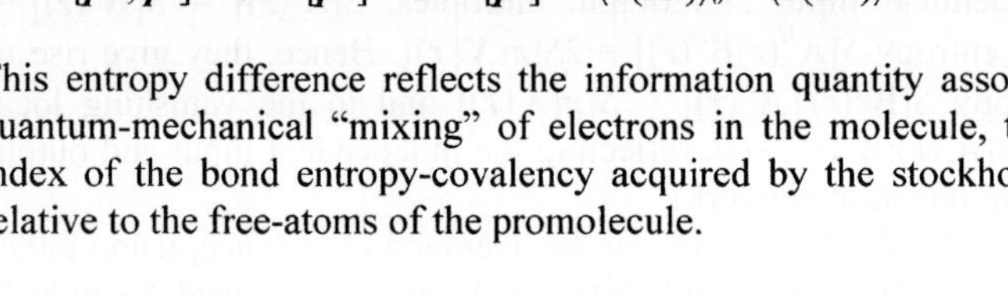

Scheme E.9. The Hirshfeld local ("vertical") flow diagram for the atomic densities in a diatomic molecule *A*—*B*.

One also observes that equations (E.2.8) and (E.2.9) imply the associated flow diagram for atomic densities, which is shown in Scheme E.9. In this flow-chart for the local (vertical) Hirshfeld communication channel the electron densities of free-atoms define the input, while the stockholder atomic pieces of the molecular electron density determine the system output. This density channel for the given location in space shows that the Hirshfeld conditional probabilities and the molecular enhancement function describe in full detail all the vertical rearrangements of the electron density in molecular systems, from the non-bonded atoms of the promolecule to the final (equilibrium) densities of the stockholder AIM.

APPENDIX F: Atomic Resolution of Bond Descriptors in Two-Orbital Model

The *probability*-resolution scheme for partitioning the molecular bond-multiplicities and/or valence-numbers into atomic contributions, suggested by the *partial* atomic channels of the *Communication Theory* of the chemical bond (Section 9.5), is established and applied to the 2-AO model of Section 8.5. The plots of the entropy/information bond indicators of the atomic partial *row*- and *column*-channels of Section 9.5 (Schemes 9.8 and 9.9) for the 2-AO model are generated in the whole range of the electron probability parameter $0 \leq P \leq 1$, which measures a degree of the bonding-MO polarization. These plots are compared with the related MO indices for the model. The corresponding plots of the atomic bond/valence numbers in both these approaches agree semi-quantitatively with one another. The *row*-channel predictions reflect the atomic contributions to the overall bond-multiplicity and its covalent and ionic components. In both IT and MO theoretical frameworks they giving rise to a conserved "single" bond-order measure for the total molecular index. The molecular *column*-channels generate the atomic *valence* numbers, which reflect upon the *promoted* states of bonded atoms. It offers a unique perspective on a degree of the atom involvement in the covalent and ionic interactions with its molecular environment. The overall atomic valence-number in the 2-AO model is found to be equal to the probability of finding an electron on the atom under consideration.

In what follows we shall examine in a more detail the IT descriptors of the chemical bond, which result from the *partial* atomic channels of Schemes 9.8 [*row*(*r*)-channels] and 9.9 [*column*(*c*)-channels] for the ground-state electron configuration in the 2-AO model of Section 8.5. The corresponding molecular information system is shown in Scheme 8.7. We assume the electron-sharing promolecular reference, $x = y = ½$, in which each atom contributes a single electron to form the chemical bond: $\boldsymbol{P}^0 = (P_A^{\ 0}, P_B^{\ 0}) = (x, y) = (½, ½)$. The entropy-covalency (S_X), information-ionicity (I_X), and the overall bond indices ($N_X = I_X + S_X$) of two constituent atoms, $X = A$, B, will be examined for the whole range of the parameter $P = C^2 = 1 - Q$, which measures the probability of finding an electron on atom A: $0 \leq P \leq 1$. Here Q stands for the complementary condensed one-electron probability of atom B. Parameter P uniquely specifies the molecular probabilities $\boldsymbol{P} = (P, Q)$, the MO polarization [Eq. (8.5.5)], orbital populations on both atoms, $\boldsymbol{q} = (q_a, q_b) = 2\boldsymbol{P}$, and the ground-state CBO matrix γ^S in AO resolution, with the diagonal elements equal to the orbital electron populations, $\{\gamma_{m,m}^{\ S} = q_m\}$, and the two off-diagonal elements $\gamma_{a,b}^{\ S} = \gamma_{b,a}^{\ S} = 2(PQ)^{½}$.

F.1. Row-Channels

The *row*-channel for atom X reflects communications between this input atom X (or X^0) and all bonded atoms of the molecular output. It is determined by the Xth row of the molecular conditional probability matrix **P**(**B**|**A**), of the molecular output **B** given the molecular input **A**. It generates the contributions of atom X to the chemical bond A—B.

For the assumed promolecular probabilities $\boldsymbol{P}^0 = (P^0 = x = ½, Q^0 = y = ½)$ in Scheme 9.8 the IT bond indices (in bits) are given by the following functions of the molecular probability parameter P:

$$S_A{}^r(P) \equiv S(\mathbf{B}\,|\,A) = PH(P), \qquad I_A{}^r(P) \equiv I(A^0:\mathbf{B}) = P(\log_2 P + 1),$$
$$N_A{}^r(P) \equiv N(A;\mathbf{B}) = PH(P) + P(\log_2 P + 1);$$

$$S_B{}^r(P) \equiv S(\mathbf{B}\,|\,B) = QH(P), \qquad I_B{}^r(P) \equiv I(B^0:\mathbf{B}) = Q(\log_2 Q + 1),$$
$$N_A{}^r(P) \equiv N(A;\mathbf{B}) = QH(P) + Q(\log_2 Q + 1), \qquad \text{(F.1.1)}$$

where $H(P)$ stands for the binary entropy function of Fig. 3.2.

These atomic bond components are shown in Panel a of Fig. F.1. They reproduce the overall entropy-covalency of the whole diatomic system,

$$S^r(P) = S_A{}^r(P) + S_B{}^r(P) = H(P) = S(\mathbf{B}\,|\,\mathbf{A}) \equiv S(P), \qquad \text{(F.1.2)}$$

and the molecular information-ionicity,

$$I^r(P) = I_A{}^r(P) + I_B{}^r(P) = 1 - H(P) = I(\mathbf{A}^0:\mathbf{B}) \equiv I(P). \qquad \text{(F.1.3)}$$

Together they give rise to the conserved 1-bit of the IT total index of a single chemical bond in the 2-AO model (see Figs. 8.1 and F1):

$$N^r(P) = S^r(P) + I^r(P) = S(\boldsymbol{P}^0) \equiv N(P) = 1 \text{ bit}. \qquad \text{(F.1.4)}$$

Figure F1a reflects a competition between two atoms for the bonding, shared electrons, with the maximum $S_A{}^r(P_A{}^{max})$ $[S_B{}^r(P_B{}^{max})]$ being observed for $P_A{}^{max} > ½$ $(P_B{}^{max} < ½)$. It follows from this atomic perspective that the maximum *molecular* entropy-covalency $H(P)$, for $P = Q = ½$, marks the "compromise" values of the corresponding atomic contributions $S_A{}^r(½) = S_B{}^r(½) = ½$ bit, for which the overall covalency exactly exhausts the total IT bond-order $H(½) = N^r(P) = 1$ bit, thus marking the purely covalent bond: $I^r(½) = 0$.

The same competition between the two AIM can be detected in the atomic ionic indices, which reach the highest values $I_A{}^r(1) = I_B{}^r(0) = 1$ bit when the two electrons occupy the AO on the same atom. Indeed, these extreme MO polarizations mark the lone-electron pair configurations, i.e., the ion-pairs A^-B^+ or A^+B^- bonded by the single, purely ionic bond, for which all covalent contributions in the model exactly vanish.

It should be observed that the atomic ionic index $I_B{}^r(P)$ reaches the minimum (negative) value at slightly higher value of P compared to $P_A{}^{max}$, which also gives rise to the negative minimum feature of $N_B{}^r(P)$ in this region, where the covalency index of the other atom, $S_A{}^r(P)$, reaches the maximum value. The negative mutual-information value reflects the donor-acceptor (coordination) interaction $B\to A$ in this region of the MO polarization parameter P: $½ < P < 1$. A similar, complementary behavior is observed in $I_A{}^r(P)$ and the associated overall index $N_A{}^r(P)$, with both exhibiting negative-valued minima for $0 < P < ½$, where $S_B{}^r(P)$ reaches the maximum value.

Figure F1. A comparison between the atomic bond contributions from the Communication (Panel *a*) and MO (Panel *b*) theories. The entropy-covalency S_X^r, information-ionicity I_X^r, and overall atomic bond index $N_X^r = S_X^r + I_X^r$ (in bits) in Panel *a* describe the atomic *row*-channels of Scheme 8.8, while the corresponding MO indices $\Im_X^{cov}$, $\Im_X^{ion}$, $\Im_X = \Im_X^{cov} + \Im_X^{ion}$ in Panel *b* represent the atomic MO covalencies, ionicities and overall bond indices. The covalent contributions are from the probability-partitioning of the Wiberg index $\Im_{A,B}^W = 4PQ$, while the atomic ionicities represent the probability weighted charge displacements: $\Im_X^{ion} = P_X \Delta q_X = 2P_X \Delta P_X$. The overall bond-orders in Panel *a*: $S^r(P) = S_A^r(P) + S_B^r(P) = H(P)$, $I^r(P) = I_A^r(P) + I_B^r(P) = 1 - H(P)$, and the conserved total bond-multiplicity, $N = S^r(P) + I^r(P) = 1$ bit, are also plotted. In Panel *b* the overall covalency $\Im^{cov}(P) = \Im_A^{cov}(P) + \Im_B^{cov}(P) = \Im_{A,B}^W(P) = 4PQ$, and the conserved total bond-order in the singlet (*S*) ground-state, $N_S = \Im^{cov}(P) + [\Im_A^{ion}(P) + \Im_B^{ion}(P)] \equiv \Im^{cov}(P) + \Im^{ion}(P) = 1$, are also shown.

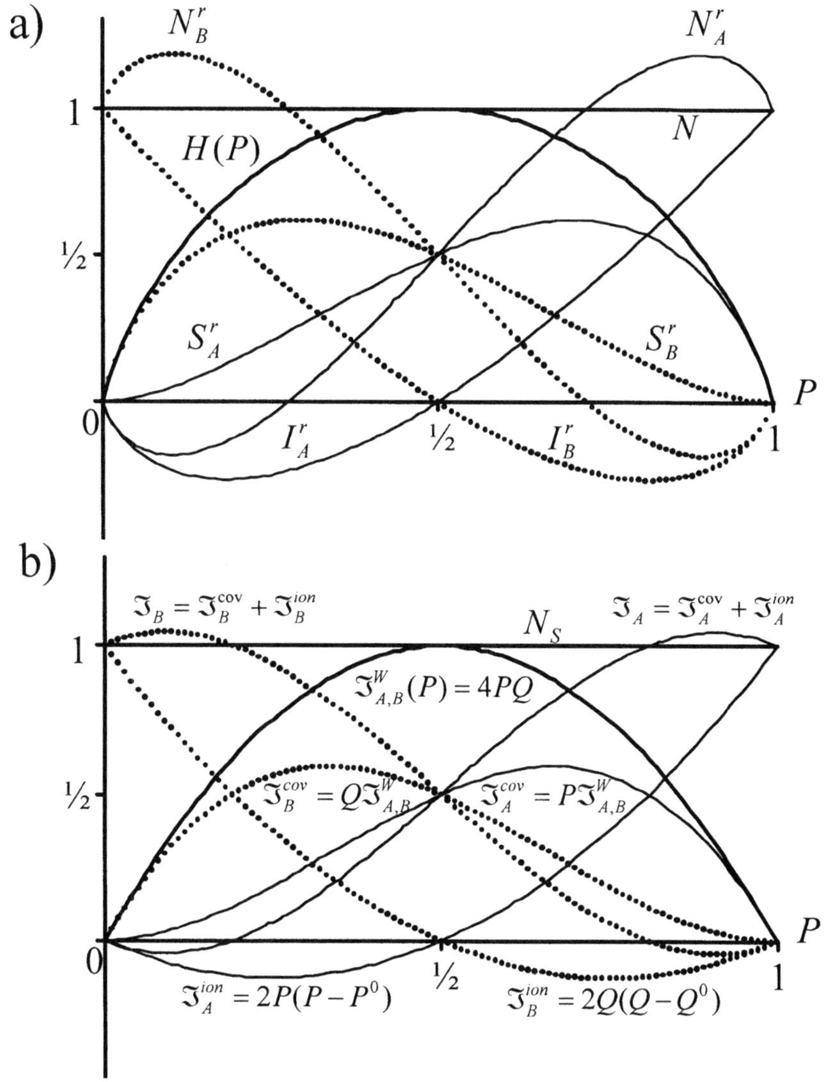

To summarize, we observe that the atomic partial *row*-channels generate an attractive perspective on atomic contributions to IT bond-orders. In fact, they generate the exhaustive *probability*-partitioning scheme, for dividing the diatomic chemical-bond descriptors into atomic contributions. The diagram of Fig. F1*a* also generates a framework for estimating relative roles played by specific bond components for different MO polarizations. These plots display directly a competition between the two atoms for a share in both the overall molecular bond-multiplicity and its ionic and covalent components. As we shall demonstrate in Section F3 the atomic bond indices resulting from the molecular bond-multiplicities obtained from the MO theory by using the same *probability*-partitioning rule give rise to a similar representation of the interplay between atomic bond contributions for changing MO polarization P of the model chemical bond.

F.2. Column-Channels

Let us examine next the *column*-channels of Scheme 9.9. Each of them consists of two atomic inputs and a single atomic output. As such they count no inter-atomic bonds, but rather, their IT indices describe the "promoted" (valence) state of the output atom in the molecule, in terms of the entropy/information *valence*-numbers, which reflect the effective number of electrons on the atom in question, which are active in the bond formation process.

For the assumed promolecular reference $\boldsymbol{P}^0 = (\frac{1}{2}, \frac{1}{2})$ the entropy/information quantities (*valence*-numbers) listed in Scheme 9.9 are given by the following functions of the probability parameter P:

$$S_A^{\,c}(P) \equiv S(A\,|\,\mathbf{A}) = PH(P) = PS(P), \qquad I_A^{\,c}(P) \equiv I(\mathbf{A}^0 : A) = P[1 - H(P)] = PI(P),$$
$$N_A^{\,c}(P) \equiv N(\mathbf{A}; A) = P = PN(P);$$

$$S_B^{\,c}(P) \equiv S(B\,|\,\mathbf{A}) = QH(P) = QS(P), \qquad I_B^{\,c}(P) \equiv I(\mathbf{A}^0 : B) = Q[1 - H(P)] = QI(P),$$
$$N_B^{\,c}(P) \equiv N(\mathbf{A}; B) = Q = QN(P). \tag{F.2.1}$$

The plots of these functions are displayed in Fig. F2*a*. Again, the covalent-valencies of both atoms reproduce the conserved overall entropy-covalency in the model,

$$S^c(P) = S_A^{\,c}(P) + S_B^{\,c}(P) = H(P) = S(\mathbf{B}\,|\,\mathbf{A}) = S(P), \tag{F.2.2}$$

while the ionic-valencies reproduce the system information-ionicity,

$$I^c(P) = I_A^{\,c}(P) + I_B^{\,c}(P) = 1 - H(P) = I(\mathbf{A}^0{:}\mathbf{B}) = I(P). \tag{F.2.3}$$

Together they give rise to the conserved total valence of both atoms in the 2-AO model:

$$N^c(P) = S^c(P) + I^c(P) = N(P) = 1. \tag{F.2.4}$$

The valence-functions of Eq. (F.2.1) demonstrate that the molecular IT bond multiplicity and its covalent (conditional-entropy) and ionic (mutual-information) bond-order measures, are consistently divided between the constituent AIM in accordance with the atomic probabilities $\boldsymbol{P} = (P, Q)$. This probability-division principle generates the overall atomic valence-number equal to the atom electron probability in the molecule:

$$N_X^{\,c}(P_X) = P_X, \qquad X = A, B. \tag{F.2.5}$$

Figure F2. A comparison between the atomic *valence*-numbers from the Communication (Panel *a*) and MO (Panel *b*) theories. The covalent (S_X^c), ionic (I_X^c), and overall, $N_X^c = S_X^c + I_X^c$ atomic valencies (in bits) in Panel *a* describe the atomic *column*-channels of Scheme 8.9, while the corresponding MO valencies V_X^{cov}, V_X^{ion}, $V_X = V_X^{cov} + V_X^{ion}$ in Panel *b* are from the probability-partitioning of the diatomic covalency number, $V^{cov}(P) = V_A^{cov}(P) + V_B^{cov}(P) = \Im_{A,B}^{W}(P)$, and the diatomic ionicity $V^{ion}(P) = \Im_{A,B}^{ion}(P)$. The overall valencies in Panel *a*, $S^c(P) = S_A^c(P) + S_B^c(P) = H(P)$, $I^c(P) = I_A^c(P) + I_B^c(P) = 1 - H(P)$, and the conserved total valency, $N = S^c(P) + I^c(P) = 1$, are also displayed. In Panel *b* the overall MO covalent number, $V^{cov}(P) = V_A^{cov}(P) + V_B^{cov}(P) = \Im_{A,B}^{W}(P)$, and the conserved total valence of the singlet (S) ground state, $N_S = V^{cov}(P) + \Im_{A,B}^{ion}(P) = 1$, are also plotted.

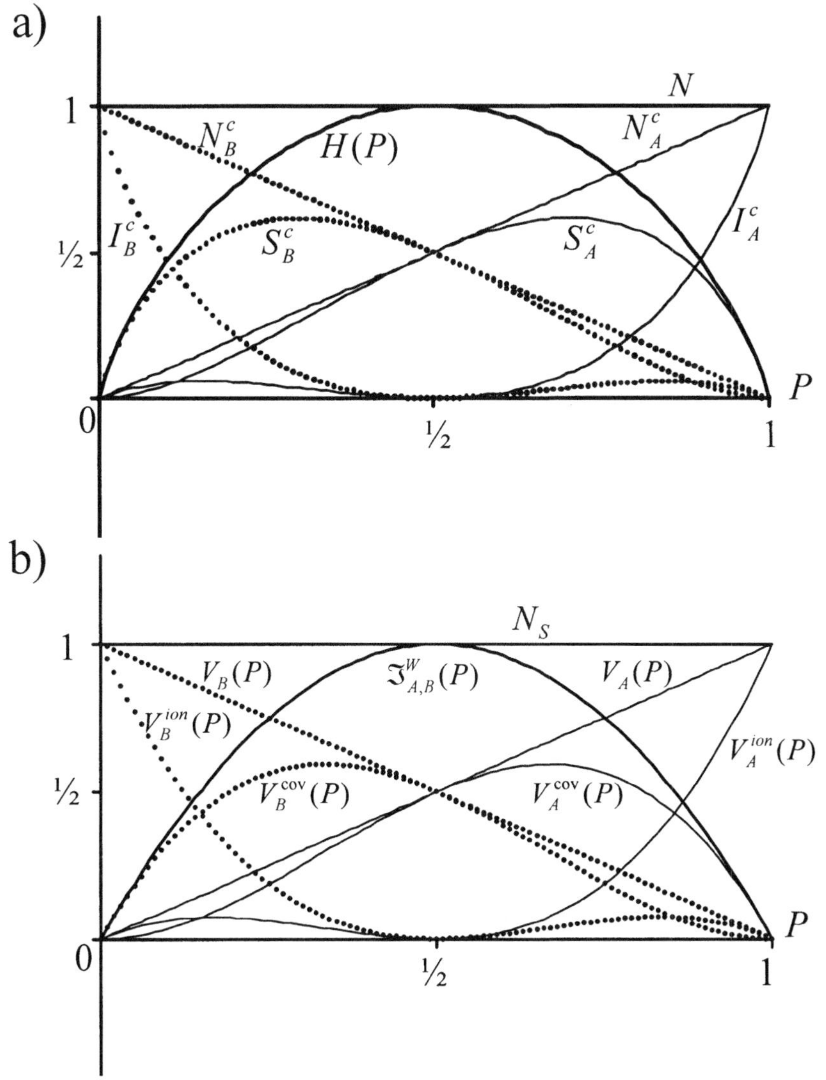

In order to reproduce the conventional, intuitive valence number $N^{val} = 1$ for a single *active* electron on each atom in the symmetric combination of the two AO into the bonding MO, $P = Q = ½$, when both atoms are equally engaged in forming the single covalent bond of the shared spin-paired electrons of the model, e.g., the σ bond in H_2 or the π bond in ethylene, one has to express these probability-valencies in terms of the AIM/orbital electron populations $\boldsymbol{q} = \{q_X = 2P_X\}$. Indeed, the *probability* measures of the overall AIM valencies from the column-channels [Eq. (F.2.5)], $N_A{}^c(½) = N_B{}^c(½) = ½$, then give rise to the intuitive value of a single bond-active electron on each atom:

$$\tilde{N}_A^c(½) \equiv 2N_A{}^c(½) = \tilde{N}_B^c(½) \equiv 2N_A{}^c(½) = 1. \qquad \text{(F.2.6)}$$

It agrees with the conventional, chemical measure of the atomic valence indicator in such a model, purely-covalent bond.

The limiting lone-pair values of $\{\tilde{N}_A^c(0) = 0,\ \tilde{N}_B^c(0) = 2\}$ and $\{\tilde{N}_A^c(1) = 2,\ \tilde{N}_B^c(0) = 0\}$, also correctly reflect the actual electron populations of AO in these extreme-polarization configurations.

A comparison between atomic indices from the partial column- and row-channels in the 2-AO model also reveals that the atomic covalent contributions of Panel *a* in Figs. F1 and F2 are identical. This is not the case for the ionic contributions, with the ionic-valence term being *non*-negative. This is appropriate for the quantity reflecting the bonded atom effective valence counts the effective number of its bond-active electrons.

We can thus conclude that the atomic partial channels provide a consistent framework for extracting both the AIM contributions to the system chemical bonds (from row-channels) and their effective valence indices (from column-channels) in the molecule. These atomic bond indices from IT result from the probability-partition of the molecular bond indices and their entropy-covalency and information-ionicity components. We shall demonstrate in the next section that the same division rule applied to the molecular bond-multiplicities from the MO theory generates the atomic diagrams, which semi-quantitatively agree with the corresponding IT plots.

F. 3. Atomic Bond-Indices and Valence-Numbers from MO Theory

In order to get a better understanding of the molecular and atomic IT bond indices reported in Figs F1*a* and F2*a* we now aim at providing the MO analogs of these quantities, obtained from the relevant molecular bond-multiplicities, the physical meaning of which is fairly well understood. When partitioning these molecular bond/valence descriptors into the complementary atomic indices we shall adopt the same probability-partitioning principle, which has been established in the preceding Section, by using the partial communication channels of the 2-AO model.

We first compare the IT bond-order contributions of Fig. 1*a* with the corresponding MO predictions for the 2-AO model. The Wiberg (1968) index,

$$\mathfrak{I}_{A,B}{}^W(P) = (\gamma_{a,b})^2 = 4PQ, \qquad \text{(F.3.1)}$$

provides a realistic representation of the model covalent component of the chemical bond, while the two-centre ionicity of the two-electron difference approach (Nalewajski and Mrozek, 1994; Nalewajski, 2004b),

$$\mathfrak{I}_{tc}^{ion}(A, B) = -\Delta q_a \Delta q_b = (1 - 2P)(2Q - 1) \equiv \mathfrak{I}_{A,B}^{ion}(P) = 1 - 4PQ, \tag{F.3.2}$$

generates the complementary ionic component. Together these terms give rise to the conserved total bond index in the ground (singlet) state of the model (see Panels *b* of Figs. F1 and F2):

$$\mathfrak{I}_{A,B}^{ion}(P) + \mathfrak{I}_{A,B}^{W}(P) = N_S(P) = 1. \tag{F.3.3}$$

In order to obtain the atomic contributions to these complementary diatomic bond-orders we adopt the *probability*-division principle, which was established by the atomic partial row-channels. Hence, the quadratic index (Nalewajski and Mrozek, 1994) of the ionic bond-component of atom *X*, to conform the molecular indices of Eqs. (F.3.1) and (F.3.2), must be proportional to the AIM condensed one-electron probability P_X, of finding an electron on this atom in the molecule, and to the molecular shift in the atom electron population $q_X = 2P_X$ relative to the promolecular value $N_X^0 = 2P_X^0 = 1$ (see Fig. F1*b*):

$$\mathfrak{I}_A^{ion}(P) = 2P_A(P_A - P_A^0) = 2P(P - P^0) = P(N_A - N_A^0),$$
$$\mathfrak{I}_A^{ion}(P) = 2P_A(P_A - P_A^0) = 2P(P - P^0) = P(N_A - N_A^0). \tag{F.3.4}$$

A related partition of the Wiberg covalent term gives the following atomic MO covalencies shown in Fig. F1*b*:

$$\mathfrak{I}_A^{cov}(P) = P\mathfrak{I}_{A,B}^{W}(P) = 4P^2Q, \quad \mathfrak{I}_B^{cov}(P) = Q\mathfrak{I}_{A,B}^{W}(P) = 4PQ^2. \tag{F.3.5}$$

The atomic bond components of Eqs. (F.3.4) and (F.3.5) define the corresponding overall MO indices of the AIM bond contributions in the 2-AO model,

$$\mathfrak{I}_A(P) = \mathfrak{I}_A^{cov}(P) + \mathfrak{I}_A^{ion}(P), \qquad \mathfrak{I}_B(P) = \mathfrak{I}_B^{cov}(P) + \mathfrak{I}_B^{ion}(P), \tag{F.3.6}$$

which are also displayed in Fig. F1*b*.

A reference to the corresponding plots of Panels *a* and *b* in Fig. F1 confirms a *semi*-quantitative agreement between the Communication- and MO-estimates of the atomic bond-order contributions and their covalent and ionic components, thus validating the IT predictions from the partial row-channels of constituent atoms.

We can thus conclude that the communication-theory perspective on the mutual interplay between the two atoms in the process of forming the chemical bond parallels that resulting from the MO theory. Both treatments emphasize the fact that the purely covalent chemical bond, for the symmetrical combination of two atomic orbitals, when $P = Q = ½$, maximize the *sum* of the atomic bond-orders, i.e., the resultant molecular bond-multiplicity, while each atom alone maximizes its bond contribution for the non-symmetrical combination of two orbitals, $P_A^{max} > ½$ and $P_B^{max} < ½$, marking the polarized bonding-MO.

We turn next to a comparison between the IT and MO atomic *valence*-numbers shown in Fig. F2. We apply the probability-division rule in partitioning the diatomic valence-numbers of Eqs. (F.3.1) and (F.3.2), $V^{cov}(P) \equiv \mathfrak{I}_{A,B}^{W}(P)$ and $V^{ion}(P) \equiv \mathfrak{I}_{A,B}^{ion}(P)$, with each atom participating in the diatomic valence-number proportionally to its condensed one-electron probability. This gives the following expressions for the overall atomic valencies and their ionic and covalent components:

$$V_A^{cov}(P) = P\mathfrak{I}_{A,B}^{W}(P), \quad V_A^{ion}(P) = P\mathfrak{I}_{A,B}^{ion}(P), \quad V_A(P) = \mathfrak{I}_A^{cov} + \mathfrak{I}_A^{ion} = PN_S(P) = P,$$

$$V_B^{cov}(P) = Q\mathfrak{I}_{A,B}^{W}(P), \quad V_B^{ion}(P) = Q\mathfrak{I}_{A,B}^{ion}(P), \quad V_B(P) = \mathfrak{I}_B^{cov} + \mathfrak{I}_B^{ion} = QN_S(P) = Q. \quad \text{(F.3.7)}$$

These atomic quantities are shown in the atomic-valence diagram of Fig. F2*b*.

Again an inspection of the related plots in both panels of Fig. F2 shows a remarkable agreement between the two perspectives on the AIM valence-descriptors of their promoted state in the molecule.

We have thus established, using the partial *row*-channels of constituent atoms in the two-orbital model of the chemical bond, the *probability*-partitioning principle for extracting the atomic contributions to molecular bond-multiplicities. The same rule, following from the atomic *column*-channels, was shown to divide the molecular valency descriptors into the corresponding atomic valence-numbers. The same principle applied to the bond-orders from the MO theory gives a similar perspective of the competition between AIM in the bond formation process, and between their covalent and ionic bond components. In fact, the agreement between predictions from the Communication Theory approach of IT and the MO treatment of the atomic bond contributions and their effective valencies in the 2-AO model was shown to be semi-quantitative.

Similar probability-division schemes are quite common in molecular physics. To mention just one example, one recalls the familiar expression for the exchange-correlation energy as functional of the correlation-hole [Eq. (C.26)]. It also involves a separation of the average two-electron quantity into formally separarable one-electron distributions of the marginal electron density of one electron, and the conditional probability distribution representing the correlation correction to the conditional probability density of the other electron. Similar ideas have been used in the σ-π separation approximations in aromatic systems (Jug and Köster, 1990), and in a search for the density functional of the electronic kinetic energy (Sears, 1980; Nalewajski, 2003f).

APPENDIX G: Elements of Thermodynamic Description of Instantaneous Processes in Continuous Systems

The distribution function of the ordinary thermodynamics for the fluctuating state-variables is introduced as function of the relevant Legendre transform of the system instantaneous entropy, which reflects the constraints and the reservoirs involved. The continuity equations of the hydrodynamic-like approach to molecular density components are summarized.

G.1. Distributions of Instantaneous State-Variables

In ordinary thermodynamics (Callen, 1962) the *instantaneous* (i) *extensive* parameters $\{x_\alpha^{\,i}\}$ of the system in contact with the *reservoirs* (r) appropriate to a given set of the free parameters $\boldsymbol{x}^i \equiv (x_0^{\,i}, x_1^{\,i}, \ldots, x_s^{\,i})$ undergo continual fluctuations due to transfers between the system and the reservoirs. Their equilibrium values $\{x_\alpha\}$, predicted by the maximum-entropy principle of Section 2.1, are the *average* values $\{\langle x_\alpha^{\,i}\rangle \equiv x_\alpha\}$, denoted by the $\langle\rangle$ brackets, so that the average value of the deviation $\delta x_\alpha^{\,i} \equiv x_\alpha^{\,i} - x_\alpha$ vanishes:

$$\langle \delta x_\alpha^{\,i}\rangle \equiv \langle x_\alpha^{\,i} - x_\alpha\rangle = \langle x_\alpha^{\,i}\rangle - x_\alpha = 0, \qquad \alpha = 0, 1, 2, \ldots, s. \tag{G.1.1}$$

The reservoirs are assumed to be restrictive with respect to the remaining extensive variables $\boldsymbol{x}^c \equiv (x_{s+1}, \ldots, x_t)$, which are kept constant.

In the thermodynamic theory of fluctuations the *average* (or the most probable) *values* of state-parameters are determined by the *statistical distribution function* $W(\boldsymbol{x})$ of all extensive parameters $\boldsymbol{X}^i \equiv (\boldsymbol{x}^i, \boldsymbol{x}^c)$, which determines the probability

$$W(\boldsymbol{X}^i)\, dx_0^{\,i}\, dx_1^{\,i} \ldots dx_s^{\,i} \equiv W(\boldsymbol{X}^i)\, d\boldsymbol{x}^i = P(d\boldsymbol{x}^i) \tag{G.1.2}$$

that simultaneously $x_0^{\,i}$ will be found in the interval $dx_0^{\,i}$, $x_1^{\,i}$ will be found in the range $dx_1^{\,i}$, …, and $x_s^{\,i}$ will be found in $dx_s^{\,i}$. It depends on the *instantaneous entropy*, a function of the system extensive parameters, $S^i(\boldsymbol{X}^i) = S^i(\boldsymbol{x}^i, \boldsymbol{x}^c)$, which assumes the equilibrium value $S(\boldsymbol{x})$, when the instantaneous extensive parameters are given their average, equilibrium values $\boldsymbol{x} = \langle \boldsymbol{X}^i\rangle = (\langle \boldsymbol{x}^i\rangle, \boldsymbol{x}^c)$:

$$S(\boldsymbol{x}) \equiv \langle S^i(\boldsymbol{X}^i)\rangle = S^i(\boldsymbol{x}). \tag{G.1.3}$$

A separate postulate specifies $W(\boldsymbol{X}^i)$ as the exponential function of a displacement in the relevant *Legendre transform* (see Section 2.2) of $S^i(\boldsymbol{x})$, the generalized *Massieu function* appropriate for the reservoirs involved. The probability $P(d\boldsymbol{x}^i)$ for a system in contact with the reservoirs corresponding to the set $\boldsymbol{x}^i$ is given by

$$W(\boldsymbol{X}^i) = \Omega \exp(k_B^{-1}\{\delta[S^i(\boldsymbol{X}^i) - \Sigma_\alpha f_\alpha x_\alpha^{\,i}]\}) = \Omega \exp(k_B^{-1}[\delta S^i(\boldsymbol{X}^i) - \Sigma_\alpha f_\alpha\, \delta x_\alpha^{\,i}]). \tag{G.1.4}$$

Here Ω is the normalizing constant such that $\int W(\boldsymbol{X}^i)\, d\boldsymbol{x}^i = 1$, k_B is Boltzmann's constant, $\delta S^i(\boldsymbol{X}^i) = S^i(\boldsymbol{X}^i) - S(\boldsymbol{x})$, the sum is over fluctuating variables $\{x_\alpha^{\,i}\} \in \boldsymbol{x}^i$, $\boldsymbol{f} \equiv \{f_\alpha\} = \{f_\alpha^{\,r}\} \equiv \boldsymbol{f}^r$, $\alpha = 0, 1, 2, \ldots, s$, are their conjugate *intensities* in the entropy representation, equalized at the corresponding *reservoir* values $\boldsymbol{f}^r = \partial S^r(\boldsymbol{x}^r)/\partial \boldsymbol{x}^r$, where $S^r(\boldsymbol{x}^r)$ is the entropy function of the reservoirs, with $\boldsymbol{x}^r$ grouping their extensive state-parameters.

The displacement function in the exponent of the distribution function of Eq. (G.1.4),

$$\begin{aligned}\delta[S^i(\boldsymbol{X}^i) - \Sigma_\alpha f_\alpha x_\alpha^{\,i}] &= [S^i(\boldsymbol{X}^i) - \Sigma_\alpha f_\alpha^{\,r} x_\alpha^{\,i}] - [S(\boldsymbol{x}) - \Sigma_\alpha f_\alpha^{\,r} x_\alpha] \\ &= [S^i(\boldsymbol{X}^i) - S(\boldsymbol{x})] - \Sigma_\alpha f_\alpha^{\,r}(x_\alpha^{\,i} - x_\alpha) = \delta S^i(\boldsymbol{X}^i) - \Sigma_\alpha f_\alpha\, \delta x_\alpha^{\,i},\end{aligned} \tag{G.1.5}$$

is defined with respect to the *maximum* value of the instantaneous quantity $S^i(\boldsymbol{X}^i) - \Sigma_\alpha f_\alpha x_\alpha^{\,i}$, over the whole ranges of the fluctuating variables:

$$\tilde{S}(\boldsymbol{f}^r, \boldsymbol{x}^c) = \max{}_i[S^i(\boldsymbol{x}^i, \boldsymbol{x}^c) - \Sigma_\alpha f_\alpha^{\,r} x_\alpha^{\,i}] = S(\boldsymbol{x}) - \Sigma_\alpha f_\alpha^{\,r} x_\alpha. \tag{G.1.6}$$

Therefore, the quantity defined in the preceding equation is no longer a function of the fluctuating extensive parameters $\boldsymbol{x}^i$, being instead solely determined by the corresponding intensive parameters of the reservoirs, $\boldsymbol{f}^r$, and the system constrained extensive parameters, $\boldsymbol{x}^c$. It represents the Legendre transform of the system equilibrium entropy, since the most probable, equilibrium values of $\boldsymbol{x}^i$, which maximize W (a monotonically increasing function of the exponent $\delta[S^i(X^i) - \Sigma_\alpha f_\alpha x_\alpha{}^i]$), are determined by the condition

$$S^i(x^i, x^c) - \Sigma_\alpha f_\alpha{}^r x_\alpha{}^i = \text{maximum}. \qquad \text{(G.1.7)}$$

Hence, the maximum of Eq. (G.1.6) occurs, when the fluctuating variables have their *equilibrium values*. The instantaneous entropy then reaches the equilibrium value of Eq. (G.1.3). Equation (G.1.6) then defines the Legendre transform of the system equilibrium entropy, $\mathcal{S}(\boldsymbol{f}^r, \boldsymbol{x}^c) \equiv L^S_{x^i}[\boldsymbol{f}^r] \equiv \mathcal{S}[\boldsymbol{f}^r]$, corresponding to the replacement of the fluctuating extensive variables $\boldsymbol{x}^i$ by their entropy intensive conjugates $\boldsymbol{f} = \boldsymbol{f}^r$ in the list of the system state-parameters.

One also defines the Legendre transform of the *instantaneous* entropy, the instantaneous Massieu function,

$$\mathcal{S}^i(\boldsymbol{f}^i, \boldsymbol{x}^c) = S^i(\boldsymbol{x}^i, \boldsymbol{x}^c) - \Sigma_\alpha f_\alpha{}^i x_\alpha{}^i \equiv \mathcal{S}^i[\boldsymbol{f}^i], \qquad \text{(G.1.8)}$$

where the entropic *instantaneous intensive parameters* are defined as the corresponding partial derivatives of the instantaneous entropy: $\boldsymbol{f}^i = \partial S^i(\boldsymbol{x}^i, \boldsymbol{x}^c)/\partial \boldsymbol{x}^i$. Therefore, the *equilibrium condition* of Eq. (G.1.7) can be rephrased as follows: the equilibrium values of the unconstrained extensive parameters of a system in contact with reservoirs maximize $\mathcal{S}^i[\boldsymbol{f}^i]$ at constant $\boldsymbol{f}^i = \boldsymbol{f}^r$.

If the distribution function is given, all the moments can be computed by integration. A simple, approximate analytical representation for the distribution function, the $(s+1)$-dimensional Gaussian distribution of Einstein, is obtained by expanding the exponent in Eq. (G.1.4) to the second-order and neglecting higher-order terms. It gives the correct second-moments, but predicts inexact higher moments.

It follows from the form of the distribution function of Eq. (G.1.4) that the thermodynamic *fluctuation* (second) *moment* is given by the expression:

$$\langle \delta x_\alpha{}^i \delta x_\beta{}^i \rangle = \int \delta x_\alpha{}^i \delta x_\beta{}^i\, W(\mathrm{x})\, d\mathrm{x}^i = -k_B (\partial x_\alpha / \partial f_\beta)_{c(\beta)} = -k_B (\partial x_\beta / \partial f_\alpha)_{c(\alpha)}, \qquad \text{(G.1.9)}$$

where the constraints $c(\beta) = (f_0, \ldots, f_{\beta-1}, f_{\beta+1}, \ldots f_s, \boldsymbol{x}^c) \equiv (\boldsymbol{f}_\beta{}', \boldsymbol{x}^c)$ and $c(\alpha) = (\boldsymbol{f}_\alpha{}', \boldsymbol{x}^c)$. In particular, the diagonal second moment, for $\alpha = \beta$, represents the *mean square deviation* $\langle(\delta x_\alpha{}^i)^2\rangle = \langle(x_\alpha{}^i)^2\rangle - x_\alpha{}^2$, which measures the magnitude of fluctuations of the variable $x_\alpha{}^i$, with the ratio $\langle(\delta x_\alpha{}^i)^2\rangle^{1/2}/x_\alpha$ characterizing the "*sharpness*" of the αth section of the distribution function. The off-diagonal second moment of Eq. (G.1.9), for $\alpha \neq \beta$, similarly reflects the *correlation* between the fluctuations in $x_\alpha{}^i$ and $x_\beta{}^i$.

G.2. Elements of Hydrodynamic Description

In order to describe the vertical flows of the electron gas, between the constituent molecular fragments, e.g. AIM, we use the state-variables that are local in space and time, e.g., the component electron densities $\boldsymbol{\rho}(\boldsymbol{r}, t) = \{\rho_k(\boldsymbol{r}, t)\}$ and the overall density they give rise to, $\rho(\boldsymbol{r},$

$t) = \sum_k \rho_k(\boldsymbol{r}, t)$. The basic conservation laws, which reflect this assumed "continuous" distribution of electrons, are called the *equations of continuity*. As an illustration let us first consider the conservation of the overall number of electrons in the externally closed molecular system, when $\int \rho(\boldsymbol{r}, t)\, d\boldsymbol{r} = N(t) = N^0 = const.$ One envisages an arbitrary closed (fixed) volume V inside the molecular electron fluid, with the total number of electrons contained in V at time t given by the volume integral $N_V(t) = \int_V \rho(\boldsymbol{r}, t)\, d\boldsymbol{r}$. Because electrons are neither produced nor destroyed in the closed molecular system, $N_V(t)$ can vary with time only as a result of the electron flow across the closed boundary surface S that defines V. At a given point in V one defines the *vector flux density* which counts the number of electrons per unit area and unit time,

$$\boldsymbol{J}(\boldsymbol{r}, t) = \rho(\boldsymbol{r}, t)\, \boldsymbol{v}(\boldsymbol{r}, t), \qquad \text{(G.2.1)}$$

where $\boldsymbol{v}(\boldsymbol{r}, t)$ is the local fluid velocity. This vector points in the direction of the local flow of electrons. The total number of electrons leaving V in unit time must be equal to the surface integral of $\boldsymbol{J}(\boldsymbol{r}, t)$ over S:

$$\frac{dN_V(t)}{dt} = \int \frac{\partial \rho(\boldsymbol{r},t)}{\partial t} d\boldsymbol{r} = -\int_S \boldsymbol{J}(\boldsymbol{r},t) \bullet d\boldsymbol{s} = -\int_V \nabla \bullet \boldsymbol{J}(\boldsymbol{r},t)\, d\boldsymbol{r}, \qquad \text{(G.2.2)}$$

where a vector $d\boldsymbol{s}$ corresponding to an infinitesimal section of S, ds, is perpendicular to this section pointing outward from V, with $|\, d\boldsymbol{s}\, | = ds$, and the minus sign reflects the convention in assigning the direction of $d\boldsymbol{s}$. In the last part of the preceding equation we have converted the surface integral to a volume integral using the Gauss divergence theorem. Because V is an arbitrary volume the foregoing equation must be also satisfied for an infinitesimal volume around the specified location in space thus giving the associated equation of continuity:

$$\frac{\partial \rho(\boldsymbol{r},t)}{\partial t} = -\nabla \bullet \boldsymbol{J}(\boldsymbol{r},t)\,. \qquad \text{(G.2.3)}$$

Therefore, for the fixed molecular density corresponding to the electron fluid at rest, when the local velocity of electrons in the system as a whole vanishes at all locations, $\boldsymbol{v}(\boldsymbol{r}, t) = 0$ and hence $\boldsymbol{J}(\boldsymbol{r}, t) = 0$, this equation simply reads:

$$\frac{\partial \rho(\boldsymbol{r},t)}{\partial t} = 0. \qquad \text{(G.2.4)}$$

Equation (G.2.3) can be expressed in the alternative form, in terms of the *total* derivative

$$\frac{d\rho(\boldsymbol{r},t)}{dt} = \frac{\partial \rho(\boldsymbol{r},t)}{\partial t} + \boldsymbol{v} \bullet \nabla \rho(\boldsymbol{r},t), \qquad \text{(G.2.5)}$$

$$\frac{d\rho(\boldsymbol{r},t)}{dt} + \rho(\boldsymbol{r},t) \nabla \bullet \boldsymbol{v} = 0\,. \qquad \text{(G.2.6)}$$

The total derivative $d\rho/dt$ represents the time rate of change of the density of an volume element of the electron fluid as it moves in space, while the partial derivative $\partial\rho/\partial t$ measures the rate of change of the density of the electron fluid at the fixed point in space, inside the infinitesimal volume element at rest. Thus, for the molecular electron fluid at rest: $d\rho/dt$ $\partial\rho/\partial t = 0$.

Next let us consider the continuity equations expressing the conservation of the number of electrons in the subsystem components of the fixed *molecular* density, again describing the molecular electron fluid at rest: $\rho(\boldsymbol{r}, t) = \sum_k \rho_k(\boldsymbol{r}, t) \equiv \rho(\boldsymbol{r})$. The rate of change of $\rho(\boldsymbol{r}, t)$ is still given by Eq. (G.2.3) provided that $\boldsymbol{v}\rho$ measures the flux of the molecular centre of mass: $\boldsymbol{v}\rho = \sum_k \rho_k \boldsymbol{v}_k = 0$, where $\boldsymbol{v}_k(\boldsymbol{r}, t)$ is the velocity of the component k. We now introduce the flux of this component relative to that carried along the motion of the centre of mass,

$$\boldsymbol{J}_k(\boldsymbol{r}, t) = \rho_k(\boldsymbol{r}, t)[\boldsymbol{v}_k(\boldsymbol{r}, t) - \boldsymbol{v}(\boldsymbol{r}, t)] = \rho_k(\boldsymbol{r}, t)\, \boldsymbol{v}_k(\boldsymbol{r}, t), \tag{G.2.7}$$

in terms of which the continuity equation expressing the conserved number of electrons in the *closed* component k, $\int \rho_k(\boldsymbol{r}, t)\, d\boldsymbol{r} = N_k(t) = N_k^{\,0} = const.$, assumes the form of the global continuity equation (G.2.3) or (G.2.6):

$$\frac{\partial \rho_k(\boldsymbol{r},t)}{\partial t} = -\nabla \bullet \boldsymbol{J}_k(\boldsymbol{r},t) \quad \text{or} \quad \frac{d\rho_k(\boldsymbol{r},t)}{dt} + \rho_k(\boldsymbol{r},t)\nabla \bullet \boldsymbol{v}_k = 0\,. \tag{G.2.8}$$

It can be further recast as the "diffusion" equation using Fick's phenomenological relation expressing the component flux density as proportional to the gradient of its density,

$$\boldsymbol{J}_k(\boldsymbol{r}, t) = -D_k \nabla \rho_k(\boldsymbol{r}, t), \tag{G.2.9}$$

where the negative sign before the (positive) diffusion coefficient D_k, measuring the mean square displacement of electrons in kth subsystem, expresses the observation that spontaneous diffusion flow occurs from region of high density to a region of low density. The resulting Fick's law for component k finally reads:

$$\frac{\partial \rho_k(\boldsymbol{r},t)}{\partial t} = D_k \nabla^2 \rho_k(\boldsymbol{r},t)\,. \tag{G.2.10}$$

REFERENCES

Abramson, N. (1963): *Information Theory and Coding*. McGraw-Hill, New York.

Acharya, P. K., Bartolotti L. J., Sears, S. B., and Parr, R. G. (1980): An Atomic Kinetic Energy Functional with Full Weizsäacker Correction. *Proc. Natl. Acad. Sci. USA* **77**, 6978-6984.

Agmon, N. and Levine R. D. (1977): Energy, Entropy and the Reaction Coordinate: Thermodynamic-like Relations in Chemical Kinetics. *Chem. Phys. Lett.* **52**, 197-201.

Agmon, N. and Levine R.D. (1979): Empirical Triatomic Potential Energy Surfaces Defined over Orthogonal Bond-Order Coordinates. *J. Chem. Phys.* **71**, 3034-3041.

Albright, T.A., Burdett, J.K., and Whangbo, M.-H. (1985): *Orbital Interactions in Chemistry*. Wiley-Interscience, New York.

Alonso, J.A. and Balbás, L.C. (1993): Hardness of Metallic Clusters. In *Hardness: Structure and Bonding* **80**. Sen, K., Ed. Springer-Verlag, Berlin, pp. 229-257.

Ash, R. B. (1965): *Information Theory*. Interscience, New York.

Aslangul, C.; Constanciel, R.; Daudel, R.; Kottis, P. (1972): Aspects of the Localizability of Electrons in Atoms and Molecules: Loge Theory and Related Methods. *Adv. Quantum Chem.* **6**, 94-141.

Ayers, P.W. (2000a): Density per Particle as a Descriptor of Coulomb Systems. *Proc. Natl. Acad. Sci. USA* **97**, 1959-1964.

Ayers, P.W. (2000b): Atoms in Molecules, an Axiomatic Approach: I. Maximum Transferability. *J. Chem. Phys.* **113**, 10886-10898.

Ayers, P.W. (2001b): Methods for Computing Chemical Reactivity Indices. *Theoret. Chem. Acc.* **106**, 271-279.

Ayers, P.W. and Levy M. (2000): Perspective on "Density Functional Approach to the Frontier-Electron Theory of Chemical Reactivity " [Parr, R.G. and Yang, W. (1984). *J. Am. Chem. Soc.* **106:** 4049-4050]. *Theor. Chem. Acc.* **103**, 353-360.

Ayers, P.W. and Parr, R.G. (2000): Variational Principles for Site Selectivity in Chemical Reactivity: the Fukui Function and Chemical Hardness Revisited. *J. Am. Chem. Soc.* **122**; 2010-2018.

Ayers, P.W. and Parr, R.G. (2001): Variational Principles for Describing Chemical Reactions: Reactivity Indices Based on the External Potential. *J. Am. Chem. Soc.* **123**, 2007-2017.

Bader, R.F. (1960): An Interpretation of Potential Interaction Constants in Terms of Low-Lying Excited States. *Mol. Phys.* **3**, 137-151.

Bader, R.F. (1968): Relaxation of the Molecular Charge Distribution and the Vibrational Force Constant. *J. Chem. Phys.* **49**, 1666-1675.

Bader, R. F. W. (1990): *Atoms in Molecules*. Oxford University Press, New York, and references therein.

Baekelandt, B.G., Mortier, W.J., and Schoonheydt, R.A. (1993): The EEM Approach to Chemical Hardness in Molecules and Solids: Fundamentals and Applications. In *Hardness*; *Structure and Bonding* **80**, Sen, K.D., Ed. Springer-Verlag, Berlin pp. 187-227.

Baekelandt, B. G., Janssens, G. O. A., Toufar, H., Mortier W. J., Schoonheydt, R. A., and Nalewajski, R. F. (1995): Mapping between Electron Population and Vibrational Modes within the Charge Sensitivity Analysis. *J. Phys. Chem.* **99**, 9784-9794.

Becke, A. D. and Edgecombe, K. E. (1990): A Simple Measure of Electron Localization in Atomic and Molecular Systems. *J. Chem. Phys.* **92**, 5397-5403.

Berkowitz, M. and Parr, R.G. (1988): Molecular Hardness and Softness, Local Hardness and Softness, Hardness and Softness Kernels, and Relations among these Quantities. *J. Chem. Phys.* **88**, 2554-2557.

Bernstein, R. B. (1982): *Chemical Dynamics via Molecular Beam and Laser Techniques*. Oxford, Clarendon.

Białynicki-Birula, I. and Mycielski J. (1976): Nonlinear Wave Mechanics. *Ann. Phys.* **100**, 62-94.

Bochicchio, R., Ponec, R., Lain, L., and Torre, A. (1998): On the Physical Meaning of Bond Indices from the Population Analysis of Higher Order Densities. *J. Phys. Chem. A* **102**, 7176-7180.

Brillouin, L. (1956): *Science and Information Theory*. Academic Press, New York.

Broniatowska, E. (2005): *Ph. D. Thesis*. Jagiellonian University, Cracow.

Callen, H. B. (1962): *Thermodynamics: an Introduction to the Physical Theories of Equilibrium Thermostatics and Irreversible Thermodynamics.* Wiley, New York.

Capitani, J. F., Nalewajski, R. F. and Parr, R. G. (1982): Non-Born-Oppenheimer Density Functional Theory of Molecular Systems. *J. Chem. Phys.* **76**, 568-573.

Carr, R. and Parrinello, M. (1985): Unified Approach for Molecular Dynamics and Density-Functional Theory. *Phys. Rev. Lett.* **55**, 2471-2474.

Cedillo, A., Chattaraj, P.K., and Parr, R.G. (2000): Atoms-in-Molecules Partitioning of a Molecular Density. *Int. J. Quantum Chem.* **77**, 403-407.

Chandra, A. K., Michalak, A., Nguyen, M. T., and Nalewajski R. F. (1998): Regional Matching of Atomic Softnesses in Chemical Reactions: A Two-Reactant Charge Sensitivity Study. *J. Phys. Chem. A* **102**, 10182-10188.

Chapman, N.B. and Shorter, J. Eds. (1972): *Advances in Linear Free Energy Relationships.* Plenum, New York.

Chattaraj, P.K. and Parr, R.G. (1993): Density Functional Theory of Chemical Hardness. In *Chemical Hardness*; *Structure and Bonding* **80**, Sen, K.D., Ed. Springer-Verlag, Berlin, pp. 11-25.

Ciosłowski, J. (1991): Quantifying the Hammond Postulate: Intramolecular Proton Transfer in Substituted Hydrogen Catecholate Anions. *J. Am. Chem. Soc.* **113**, 6756-6761.

Ciosłowski, J. and Mixon, S. T.(1993): Electronegativities in Situ, Bond Hardnesses, and Charge-Transfer Components of Bond Energies from the Topological Theory of Atoms in Molecules. *J. Am. Chem. Soc.* **115**, 1084-1088.

Cohen, M.H. (1996): Strengthening the Foundations of Chemical Reactivity Theory. In *Density Functional Theory IV: Theory of Chemical Reactivity*; *Topics in Current Chemistry* **183**, Nalewajski, R.F., Ed. Springer-Verlag Berlin, pp. 143-184.

Cohen, M.H., Ganguglia-Pirovano, M.V., and Kurdnovský, J. (1994): Electronic and Nuclear Chemical Reactivity. *J. Chem. Phys.* **101**, 8988-8997.

Cohen, M.H., Ganguglia-Pirovano, M.V., and Kurdnovský, J. (1995): Reactivity Kernels, the Normal Modes of Chemical Reactivity, and the Hardness and Softness Spectra. *J. Chem. Phys.* **103**, 3543-3551.

Colonna, F. and Savin, A. (1999): Correlation Energies for some Two- and Four-electron Systems along the Adiabatic Connection in Density Functional Theory. *J. Chem. Phys.*.**110**: 2828-2835.

Cooper, D. L., Ed. (2002): *Valence Bond Theory*. Elsevier, Amsterdam.

Coulson, C.A. and Longuet-Higgins, H.C. (1947a): The Electronic Structure of Conjugated Systems I: General Theory. *Proc. Roy. Soc. A* (London) **191**, 39-60.

Coulson, C.A. and Longuet-Higgins, H.C. (1947b): The Electronic Structure of Conjugated Systems II: Unsaturated Hydrocarbons and Their Hetero-Derivatives. *Proc. Roy. Soc. A* (London) **192**, 16-32.

Daudel, R. (1969): *The Fundamentals of Theoretical Chemistry*. Pergamon, Oxford.

Daudel, R. (1974): *The Quantum Theory of the Chemical Bond*. D. Reidel, Dordrecht.

Decius, J.C. (1963): Compliance Matrix and Molecular Vibrations. *J. Chem. Phys.* **38**, 241-248.

Dewar, M. J. S. (1969): *Molecular Orbital Theory of Organic Chemistry*. McGraw-Hill, New York.

Dewar, M. J. S. and Dougherty, R. C. (1975): *The PMO Theory of Organic Chemistry*. Plenum, New York.

Dreizler, R.M. and Gross, E.K.U. (1990): *Density Functional Theory: An Approach to the Quantum Many-Body Problem*. Springer-Verlag, Berlin.

Dunning, T. H., Jr. (1984): Theoretical Studies of the Energetics of the Abstraction and Exchange Reactions in H + HX, with X = F-I. *J. Phys. Chem.* **88**, 2469-2477.

Epiotis, N. D. (1978): *Theory of Organic Reactions*. Springer-Verlag, Berlin.

Epstein, S.T., Hurley, A.C., Wyatt, R.E., and Parr, R.G. (1967): Integrated and Integral Hellmann-Feynman Formulas. *J. Chem. Phys.* **47**, 1275-1286.

Esquivel, R. O., Rodriquez, A. L., Sagar, R. P., Hõ, M., and Smith, V. H., Jr. (1996): Physical Interpretation of Information Entropy: Numerical Evidence of the Collins Conjecture. *Phys. Rev. A* **54**, 259-265.

Fisher, R. A. (1922): On the Mathematical Foundations of Theoretical Statistics. *Phil. Trans. R. Soc. A* (London). **222**, 309-368.

Fisher, R. A. (1925): Theory of Statistical Estimation. *Proc. Cambridge Phil. Soc.* **22**, 700-725.

Fisher, R. A. (1959): *Statistical Methods and Scientific Inference*, 2nd edn. Oliver and Boyd, London.

Frieden, B. R. (2000): *Physics from the Fisher Information - A Unification*. Cambridge University Press, Cambridge.

Fujimoto, H. and Fukui, K. (1974): Intermolecular Interactions and Chemical Reactivity. In *Chemical Reactivity and Reaction Paths*, Klopman, G., Ed. Wiley-Interscience, New York, pp. 23-54.

Fukui, K. (1975): *Theory of Orientation and Stereoselection*. Springer-Verlag, Berlin.

Fukui, K. (1987): Role of Frontier Orbitals in Chemical Reactions. *Science* **218**, 747-754.

Gadre, S. R. (1984): Information Entropy and Thomas-Fermi Theory. *Phys. Rev. A* **30**, 620-621.

Gadre, S. R. (2002): Information Theoretical Approaches to Quantum Chemistry. In *Reviews of Modern Quantum Chemistry: A Celebration of the Contributions of Robert G. Parr*, K. D. Sen, Ed. Vol. I. World Scientific, Singapore, pp. 108-147.

Gadre, S. R. and Shirsat, R. N. (2000): *Electrostatics of Atoms and Molecules*. Universities Press, Hyderabad.

Gadre, S. R. and Bendale, R. D. (1985): Maximization of Atomic Information-Entropy Sum in Configuration and Momentum Space. *Int. J. Quantum. Chem.* **28**, 311-314.

Gadre, S. R., Bendale R. D., and Gejii, S. P. (1985): Analysis of Atomic Electron Momentum Space Densities: Use of Information Entropies in Coordinate and Momentum Space. *Chem. Phys. Lett.* **117**, 138-142.

Gadre, S. R. and Sears, S. B. (1979): An Application of Information Theory to Compton Profiles. *J. Chem. Phys.* **71**, 4321-4323.

Gadre, S.R., Sears, S.B., Chakravorty, S.J., Bendale, R.D. (1985): Some Novel Characteristics of Atomic Information Entropies. *Phys. Rev. A* **32**, 2602-2606.
Gázquez, J. L. (1993): Hardness and Softness in Density Functional Theory. In *Hardness*; *Structure and Bonding* **80**, Sen, K.D., Ed. Springer-Verlag, Berlin, pp. 27-43.
Gázquez, J. L. and Parr, R. G. (1978): Two-Parameter Statistical Model for Atoms. *J. Chem. Phys.* **68**, 2323-2326.
Gázquez, J. L., Vela, A., and Galvàn (1987): Fukui Function, Electronegativity and Hardness in the Kohn-Sham Theory. In *Electronegativity*; *Structure and Bonding* **66**, Sen, K.D. and Jørgensen, C.K., Eds. Springer-Verlag, Berlin, pp. 79-97.
Geerlings, P., De Proft, F., and Langenaeker, W. (2003): Conceptual Density Functional Theory. *Chem. Rev.* **103**, 1793-1873.
Ghosh, S.K., Berkowitz, M., and Parr, R.G. (1984): Transcription of Ground-State Density Functional Theory into a Local Thermodynamics. *Proc. Natl. Acad. Sci. USA* **81**, 8028-8031.
Ghosh, S.K. and Berkowitz, M. (1985): A Classical Fluid-Like Approach to the Density Functional Formalism of Many-Electron Systems. *J. Chem. Phys.* **83**, 2976-2983.
Gilchrist, T.L. and Storr, R.C. (1972): *Organic Reactions and Orbital Symmetry*. Cambridge University, Cambridge.
Gill, G. B. and Willis, M. R. (1974): *Pericyclic Reactions*. Chapman and Hall, New York.
Goddard III, W. A. and Harding, L. B. (1978): The Description of Chemical Bonding from Ab-Initio Calculations. *Annu. Rev. Phys. Chem.* **29**, 363-396.
Gopinathan, M. S. and Jug., K. (1983): Valency. I. A Quantum Chemical Definition and Properties. *Theoret. Chim. Acta* (Berl.) **63**, 497-510; Valency. II. Applications to Molecules with First-Row Atoms. *Theoret. Chim. Acta* (Berl.) **63**, 511-527.
Grabo, T., Kreibich, T., Kurth, S., and Gross, E. K. U. (1998): Orbital Functionals in Density Functional Theory: The Optimized Effective Potential Method. In *Strong Coulomb Correlations in Electronic Structure: Beyond the Local Density Approximation*, Anisimov, V. I., Ed. Gordon & Breach, Tokyo, pp. 1-98.
Gutmann, V. (1978): *The Donor-Acceptor Approach to Molecular Interactions*. Plenum, New York.
Gyftopoulos, E. P. and Hatsopoulos, G. N. (1965): Quantum-Thermodynamic Definition of Electronegativity. *Proc. Natl. Acad. Sci. USA* **60**, 786-793.
Halevi, E. A. (1992): *Orbital Symmetry and Reaction Mechanism – The Orbital Correspondence in Maximum Symmetry View*. Springer Verlag, Berlin.
Hammett, L. P. (1935): Some Relations between Reaction Rates and Equilibrium Constants. *Chem. Rev.* **17**, 125-136.
Hammett, L. P. (1937): The Effect of Structure upon the Reactions of Organic Compounds. Benzene Derivatives. *J. Am. Chem. Soc.* **59**, 96-103.
Hammond, G. S. (1955): A Correlation of Reaction Rates. *J. Am. Chem. Soc.* **77**, 334-338.
Hartley, R. V. L. (1928): Transmission of Information. *Bell System Tech. J.* 7:535-563.
Heitler, W. and London, F. (1927): Wechselwirkung Neutraler Atome und Homöopolare Bindung nach der Quantenmechanik *Z. Physik* **44**, 455-472; for an English translation see: Hettema, H. (2000): *Quantum Chemistry Classic Scientific Paper*. World Scientific, Singapore.
Hirshfeld, F. L. (1977): Bonded-Atom Fragments for Describing Molecular Charge Densities. *Theoret. Chim. Acta* (Berl.) **44**, 129-138.

Hõ, M., Sagar, R. B., Schmider, H., Weaver, D. F., and Smith, V. H. Jr. (1995): Measures of Distance for Atomic Charge and Momentum Densities and Their Relationship to Physical Properties. *Int. J. Quantum Chem.* **53**, 627-633.

Hohenberg, P. and Kohn, W. (1964): Inhomogeneous Electron Gas. *Phys. Rev.* **136**B, 864-871.

Huber, P. J. (1981): *Robust Statistics*. Wiley, New York.

Iczkowski, R. P. and Margrave, J. L. (1961): Electronegativity. *J. Am. Chem. Soc.* **83**, 3547-3551.

Johnson, C. D. (1973): *The Hammett Equation.* Cambridge University, London.

Johnson, C. D. (1975): Linear Free Energy Relationships and the Reactivity-Selectivity Principle. *Chem. Rev.* **75**, 755-765.

Jones, L. H. and Ryan, R. R. (1970): Interaction Coordinates and Compliance Constants. *J. Chem. Phys.* **52**, 2003-2004.

Jones, R. A. Y. (1979): *Physical and Mechanistic Organic Chemistry.* Cambridge University, Cambridge.

Jaynes, E. T. (1957a): Information Theory and Statistical Mechanics. *Phys. Rev.* **106**, 620-630; (1957b): Information Theory and Statistical Mechanics. II. *Phys. Rev.* **108**, 171-190.

Jaynes, E. T. (1985), in *Maximum Entropy and Bayesian Methods in Inverse Problems*, Smith, C. R. and Grandy, W. T., Eds. Reidel, Dordrecht.

Jug, K. and Gopinathan, M. S. (1990). Valence in Molecular Orbital Theory. In *Theoretical Models of Chemical Bonding*, Maksić, Z. B., Ed., Vol. II. Springer-Verlag, Heidelberg, p. 77.

Jug, K. and Köster, A. M. (1990): Influence of σ and π Electrons on Aromaticity. *J. Am. Chem. Soc.* **112**, 6772-6777.

Kato, T. (1957): On the Eigenfunctions of Many-Particle Systems in Quantum Mechanics. *Commun. Pure Appl. Math.* **10**, 151-177.

Khinchin, A. I. (1957): *Mathematical Foundations of the Information Theory*, Dover, New York.

Klopman, G. (1968): Chemical Reactivity and the Concept of Charge and Frontier-Controlled Reactions. *J. Am. Chem. Soc.* **90**, 223-234.

Klopman, G., Ed. (1974a): *Chemical Reactivity and Reaction Paths.* Wiley-Interscience, New York.

Klopman, G. (1974b): The Generalized Perturbational Theory of Chemical Reactivity and Its Applications. In *Chemical Reactivity and Reaction Paths*. Klopman, G., Ed. Wiley-Interscience, New York, pp. 55-165.

Kohn, W. and Sham, L. J. (1965): Self-Consistent Equations Including Exchange and Correlation Effects. *Phys. Rev.* **140**A, 1133-1138.

Korchowiec, J. and Uchimaru, T. (1998): The Charge Transfer Fukui Function: Extension of the Finite-Difference Approach to Reactive Systems. *J. Phys. Chem. A* **102**, 10167-10172.

Kraka, E. and Cremer, D. (1990): Chemical Implications of Local Features of the Electron Density Distribution. In *Theoretical Models of Chemical Bonding, Part 3: The Concept of the Chemical Bond.* Maksić Z.B., Ed. Springer-Verlag, Berlin, pp. 453-542.

Krieger, J. B., Li, Y., and Iafrate, G. J. (1995): Recent Developments in Kohn-Sham Theory for Orbital-Dependent Exchange-Correlation Energy Functionals. In *Density Functional Theory*, Gross E. K. U. and Dreizler R. M., Eds. Pleum Press, New York, pp. 191-216.

Kullback, S. and Leibler, R. A. (1951): On Information and Sufficiency. *Ann. Math. Stat.* 22, 79-86.

Kullback, S. (1959): *Information Theory and Statistics*. Wiley, New York.

Leeuwen, R. van, Gritsenko, O. V. and Baerends, E. J. (1996): Analysis and Modeling of Atomic and Molecular Kohn-Sham Potentials. In *Density Functional Theory I: Topics in Current Chemistry* Vol. **180**, Nalewajski, R. F., Ed., Springer-Verlag, Berlin, pp. 107-167.

Lewis, G. N. (1916): The Atom and the Molecule. *J. Am. Chem. Soc.* **38**, 762-785.

Levine, R.D. (1978): Information Theory Approach to Molecular Reaction Dynamics. *Annu. Rev. Phys. Chem.* **29**, 59-92.

Lieb, E. H. (1983): Density Functionals for Coulomb Systems. *Int. J. Quantum. Chem.* 24, 243-277.

London, F. (1928): Zur Quantentheorie der homöopolaren Valenzzahlen (On the Quantum Theory of Homo-Polar Valence Numbers). *Z. Phys.* **46**, 455-477.

Marcus, R. A. (1968): Theoretical Relations among Rate Constants, Barriers, and Broensted Slopes of Chemical Reactions. *J. Phys. Chem.* **72**, 891-899.

Marcus, R. A. (1969): Unusual Slopes of Free Energy Plots in Kinetics. *J. Am. Chem. Soc.* **91**, 7224-7225.

Marques, M. A. L. and Gross, E. K. U. (2004): Time Dependent Density Functional Theory. *Annu. Rev. Phys. Chem.* **55**, 427-455.

Mathai, A. M. and Rathie, P. M. (1975): *Basic Concepts in Information Theory and Statistics: Axiomatic Foundations and Applications*. Wiley: New York.

Mayer, I. (1983): Charge, Bond Order and Valence in the Ab Initio SCF Theory. *Chem. Phys. Lett.* **97**, 270-274; (1985): Bond Orders and Valences in the SCF Theory: a Comment. *Theoret. Chim. Acta* (Berl.) **67**, 315-322.

McWeeny, R. (1979): *Coulson's Valence*. Oxford University, Oxford.

Michalak, A., De Proft, F., Geerlings, P., and Nalewajski, R. F. (1999): Fukui Functions from the Relaxed Kohn-Sham Orbitals. *J. Phys. Chem. A.* **103**, 762-771.

Miller, A. R. (1978): A Theoretical Relation for the Position of the Energy Barrier between Initial and Final States of Chemical Reactions. *J. Am. Chem. Soc.* **100**, 1984-1992.

Morrison, R. C. and Parr, R. G. (1991): Approximate Density Matrices and Husimi Functions Using the Maximum Entropy Formulation with Constraints. *Int. J. Quantum Chem.* **39**, 823-837.

Morrison, R. C., Yang. W., Parr, R. G., and Lee, C. (1990): Approximate Density Matrices and Wigner Distribution Functions from Density, Kinetic Energy Density, and Idempotency Constraints. *Int. J. Quantum Chem.* **38**, 819-830.

Mortier, W. M. and Schoonheydt, R. A. Eds. (1997): *Developments in the Theory of Chemical Reactivity and Heterogeneous Catalysis*. Research Signpost, Trivandrum.

Møller, C. and Plessett, M. S. (1934): Note on an Approximation Treatment for Many-Electron Systems. *Phys. Rev.* **46**, 618-622.

Mrozek, J., Nalewajski, R. F. and Michalak, A. (1998). Exploring Bonding Patterns of Molecular Systems Using Quantum Mechanical Bond Multiplicities. *Polish J. Chem.* **72**, 1779-1791.

Mulliken, R. S. (1934): A New Electronegativity Scale: together with Data on Valence States and on Ionization Potentials and Electron Affinities. *J. Chem. Phys.* **2**, 782-793.

Murray, J. S. and Sen, K. Eds. (1996): *Molecular Electrostatic Potentials: Concepts and Applications.* Elsevier, Amsterdam.

Murrell, J. N., Carter, S., Farantos, S. C., Huxley, P., and Varandas, A. J. C. (1984): *Molecular Potential Energy Functions. Wiley, New York.*

Mycielski, J. and Białynicki-Birula, I. (1975): Uncertainty Relations for Information Entropy in Wave Mechanics. *Commun. Math. Phys.* **44**, 129-132.

Nagy, Á. and Parr, R. G. (1994): Density Functional Theory as Thermodynamics. *Proc. Indian Acad. Sci. (Chem. Sci.)* **106**, 217-227.

Nagy, Á. and Parr, R.G. (1996): Information Entropy as a Measure of the Quality of an Approximate Electronic Wave Function. *Int. J. Quantum Chem.* **58**, 323-327.

Nagy, Á. and Parr, R.G. (2000): Remarks on Density Functional Theory as a Thermodynamics. *J. Mol. Struct. (Theochem)* **501**, 101-106.

Nakatsuji, H. (1973): Electrostatic Force Theory for a Molecule and Interacting Molecules. I. Concept and Illustrative Applications. *J. Am. Chem. Soc.* **95**, 345-354.

Nakatsuji, H. (1974a). Common Natures of the Electron Cloud of the System Undergoing Change in Nuclear Configuration. *J. Am. Chem. Soc.* **96**, 24-30.

Nakatsuji, H. (1974b). Electron-Cloud Following and Preceding and the Shapes of Molecules. *J. Am. Chem. Soc.* **96**, 30-37.

Nalewajski, R. F. (1980): Virial Theorem Implications for the Minimum Energy Reaction Paths. *Chem. Phys.* **50**, 127-136.

Nalewajski, R. F. (1983): Reduction of Derivatives and Simple Applications of the Legendre Transformed Density Functional Theory. *J. Chem. Phys.* **78**, 6112-6120.

Nalewajski, R. F. (1984): Electrostatic Effects in Interactions between Hard (Soft) Acids and Bases. *J. Am. Chem. Soc.* **106**, 944-945.

Nalewajski, R.F. (1985): A Study of Electronegativity Equalization. *J. Phys. Chem.* **89**, 2831-2837.

Nalewajski, R. F. (1988): General Relations between Molecular Sensitivities and Their Physical Content. *Z. Naturforsch.* **43a**, 65-72.

Nalewajski, R. F. (1989): Recursive Combination Rules for Molecular Hardnesses and Electronegativities. *J. Chem. Phys.* **93**, 2658-2666.

Nalewajski, R. F. (1993): The Hardness Based Molecular Charge Sensitivities and Their Use in the Theory of Chemical Reactivity. *Structure and Bonding* **80**, 115-186.

Nalewajski, R. F. (1995a): Chemical Reactivity Concepts in Charge Sensitivity Analysis. *Int. J. Quantum Chem.* **56**, 453-476.

Nalewajski, R. F. (1995b): Charge Sensitivity Analysis as Diagnostic Tool for Predicting Trends in Chemical Reactivity. *Proceedings of the NATO ASI on Density Functional Theory*, Gross, E. K. U. and Dreizler, R. M., Eds. Plenum Press, New York, pp. 339-389.

Nalewajski, R.F. Ed. (1996a): *Density Functional Theory I-IV; Topics in Current Chemistry* **180-183**. Springer-Verlag, Berlin. .

Nalewajski, R.F. Ed. (1996b): Density Functional Theory IV: Theory of Chemical Reactivity; *Topics in Current Chemistry* **183**. Springer-Verlag, Berlin.

Nalewajski, R. F. (1997a): Charge Response Criteria of Chemical Reactivity: Fukui Function Indices and Populational Reference Frames Reflecting the Inter-Reactant Charge Coupling. *Int. J. Quantum Chem.* **61**, 181-196.

Nalewajski, R. F. (1997b): Consistent Two-Reactant Approach to Chemisorption Complexes in Charge Sensitivity Analysis. In *Developments in the Theory of Chemical Reactivity and Heterogeneous Catalysis*, Mortier, W. M. and Schoonheydt, R. A., Eds.. Research Signpost, Trivandrum, pp.135-196.

Nalewajski, R.F. (1998a): On the Chemical Potential/Electronegativity Equalization in Density Functional Theory. *Polish J. Chem.* **72**, 1763-1778.

Nalewajski, R. F. (1998b): Kohn-Sham Description of Equilibria and Charge Transfer in Reactive Systems. *Int. J. Quantum Chem.* **69**, 591-605.

Nalewajski, R. F. (1999): A Coupling between the Equilibrium State Variables of Open Molecular and Reactive Systems. *Phys. Chem. Chem. Phys.* **1**, 1037-1049.

Nalewajski, R. F. (2000a): Coupling Relations between Molecular Electronic and Geometrical Degrees of Freedom in Density Functional Theory and Charge Sensitivity Analysis. *Computers Chem.* **24**, 243-257.

Nalewajski, R. F. (2000b): Charge Sensitivities of the Externally Interacting Open Reactants. *Int. J. Quantum Chem.* **78**, 168-178 (2000).

Nalewajski, R. F. (2000c): Entropic Measures of Bond Multiplicity from the Information Theory. *J. Phys. Chem. A* **104**, 11940-11951.

Nalewajski, R. F. (2000d): Manifestations of the Maximum Complementarity Principle for Matching Atomic Softnesses in Model Chemisorption Systems. *Topics in Catal.* **11/12**, 469-485.

Nalewajski, R. F. (2001): Coupling Constant Integration Analysis of Density Functionals for Subsystems. *Adv. Quantum Chem.* **38**, 217-277.

Nalewajski, R. F. (2002a): Hirshfeld Analysis of Molecular Densities: Subsystem Probabilities and Charge Sensitivities. *Phys. Chem. Chem. Phys.* **4**, 1710-1721.

Nalewajski, R. F. (2002b): Applications of the Information Theory to Problems of Molecular Electronic Structure and Chemical Reactivity. *Int. J. Mol. Sci.* **3**, 237-259.

Nalewajski, R. F. (2002c): Electron Flows in Molecules: An Information-Theoretic Approach. *Acta Chim. Phys. Debr.* **34/35**, 131-145.

Nalewajski, R. F. (2002d): Charge Sensitivities of Molecules and Their Fragments. In *Reviews of Modern Quantum Chemistry: A Celebration of the Contributions of Robert G. Parr*, K. D. Sen, Ed. Vol. II. World Scientific, Singapore, pp. 1071-1105.

Nalewajski, R. F. (2002e): Geometric Separation of the Polarization and Charge Transfer Components of Charge Sensitivities of Open Molecular Systems. *Chem. Phys. Lett.* **353**, 143-153.

Nalewajski, R. F. (2002f): Studies of the Nonadditive Kinetic Energy Functional and the Coupling between Electronic and Geometrical Structures. In *Recent Advances in Density Functional Methods*, Barone, V., Bencini, A., and Fantucci, P., Eds. Part III. World Scientific, Singapore, pp. 257-277.

Nalewajski, R. F. (2003a): Electronic Structure and Chemical Reactivity: Density Functional and Information Theoretic Perspectives. *Adv. Quantum Chem.* **43**, 119-184.

Nalewajski, R. F. (2003b): Information Theoretic Approach to Fluctuations and Electron Flows between Molecular Fragments. *J. Phys. Chem. A.* **107**, 3792-3802.

Nalewajski, R. F. (2003c): Information Principles in the Theory of Electronic Structure, *Chem. Phys. Lett.* **372**, 28-34.

Nalewajski, R. F. (2003d): Information Principles in the Loge Theory. *Chem. Phys. Lett.* **375**, 196-203.

Nalewajski, R. F. (2003e): Aspects of the Kinetic Energy Non-Additivity in Molecular and Model Systems. *Mol. Phys.* **101**, 2369-2379.

Nalewajski, R. F. (2003f): Kinetic Energy as Fuctional of the Correlation Hole. *Chem. Phys. Lett.* **367**, 414-422.
Nalewajski, R. F. (2004a): Local Information-Distance Thermodynamics of Molecular Fragments. *Ann. Phys.* (Leipzig) **13**, 201-222.
Nalewajski, R. F. (2004b): Entropic and Difference Bond Multiplicities from the Two-Electron Probabilities in Orbital Resolution. *Chem. Phys. Lett.* **386**, 265-271.
Nalewajski, R. F. (2004c): Entropy Descriptors of the Chemical Bond in Information Theory: I. Basic Concepts and Relations. *Mol. Phys.* **102**, 531-546.
Nalewajski, R. F. (2004d): Entropy Descriptors of the Chemical Bond in Information Theory: II. Applications to Simple Orbital Models. *Mol. Phys.* **102**, 547-566.
Nalewajski, R. F. (2004e): Communication Theory Approach to the Chemical Bond. *Structural Chemistry* **15**, 395-407.
Nalewajski, R. F. (2005a): Reduced Communication Channels of Molecular Fragments and Their Entropy/Information Bond Indices. *Theoret. Chem. Acc.* **114**, 4-18.
Nalewajski, R. F. (2005b): Partial Communication Channels of Molecular Fragments and Their Entropy/Information Indices. *Mol. Phys.* **103**, 451-470.
Nalewajski, R. F. (2005c): Entropy/Information Bond Indices of Molecular Fragments. *J. Math. Chem.* **38**, 43-66.
Nalewajski, R. F. (2005d): Electronic Chemical Potential as Information Temperature. *Mol. Phys.*, in press.
Nalewajski, R. F. (2005e): Fukui Function as Correlation Hole. *Chem Phys. Lett.* **410**, 335-338.
Nalewajski, R. F. (2005f): A Comparison between the Valence-Bond and Communication Theories of the Chemical Bond in H_2. *Mol. Phys.*, in press.
Nalewajski, R. F. (2005g): Molecular Communication Channels of Model Excited Electron Configurations. *Mol. Phys.*, in press.
Nalewajski, R. F. (2005h): Atomic Resolution of Bond Descriptors in the Two-Orbital Model. *Mol. Phys.*, in press.
Nalewajski, R. F. (2005g): Geometric Interpretation of Density Displacements and Charge Sensitivities. *J. Chem. Sci.* **117**, 455-466.
Nalewajski, R.F. and Broniatowska (2003a): Entropy Displacement and Information Distance Analysis of Electron Distributions in Molecules and Their Hirshfeld Atoms. *J. Phys. Chem. A* **107**, 6270-6280.
Nalewajski, R.F. and Broniatowska (2003b): Information Distance Approach to Hammond Postulate. *Chem. Phys. Lett.* **376**, 33-39.
Nalewajski, R.F. and Broniatowska, E. (2005a): Entropy/Information Indices of the "Stockholder" Atoms-in-Molecules. *Int. J. Quantum Chem.* **101**, 349-362.
Nalewajski, R. F. and Broniatowska, E. (2005b): Atoms-in-Molecules from the Stockholder Partition of the Molecular Two-Electron Distribution. *Theor. Chem. Acc.*, in press.
Nalewajski, R. F., Formosinho, S. J., Varandas, A. J. C., and Mrozek, J. (1994): Quantum Mechanical Valence Study of a Bond Breaking-Bond Forming Process in Triatomic Systems. *Int. J. Quantum Chem.* **52**, 1153-1176.
Nalewajski, R.F. and Jug, K. (2002): Information Distance Analysis of Bond Multiplicities in Model Systems. In *Reviews of Modern Quantum Chemistry: A Celebration of the Contributions of Robert G. Parr*, K. D. Sen, Ed. Vol. I. World Scientific, Singapore, pp. 148-203.
Nalewajski, R. F. and Koniński, M. (1988): General Relations between Sensitivities of Atoms and Atoms-in-a-Molecule. *Acta Phys. Polon. A* **74**, 255-268.

Nalewajski, R. F. and Korchowiec, J. (1989): Chemical Reactivity and Charge Sensitivities of Reactants: Interaction Energy and Application to Protonation of Pyrrole and Cyclopentadiene. *Acta Phys. Polon. A***76**,747-788.

Nalewajski, R. F. and Korchowiec, J. (1997). *Charge Sensitivity Approach to Electronic Structure and Chemical Reactivity*. World-Scientific, Singapore.

Nalewajski, R. F., Korchowiec J., and Michalak, A. (1996): Reactivity Criteria in Charge Sensitivity Analysis. *Topics in Current Chemistry* **183**, 25-141.

Nalewajski, R. F. , Korchowiec, J. and Z. Zhou (1988): Molecular Hardness and Softness Parameters and Their Use in Chemistry. *Int. J. Quantum Chem. Symp.* **22**,349-366.

Nalewajski, R. F., Köster, A. M., and Escalante, S. (2005): Electron Localization Function as Information Measure, *J. Phys. Chem. A*, in press.

Nalewajski, R.F., Köster, A. M., Jug, K. (1993): Chemical Valence from the Two-Particle Density Matrix. *Theoret. Chim. Acta (Berl.)* **85**, 463-484.

Nalewajski, R. F. and Loska, R. (2001): Bonded Atoms in Sodium Chloride – the Information-Theoretic Approach. *Theor. Chem. Acc.* **105**, 374-382.

Nalewajski, R. F. and Mrozek, J. (1994): Modified Valence Indices from the Two-Particle Density Matrix. *Int. J. Quantum Chem.* **51**, 187-200.

Nalewajski, R. F. and Mrozek, J. (1996): Hartree-Fock Difference Approach to Chemical Valence: Three-Electron Indices in UHF Approximation. *Int. J. Quantum Chem.* **57**, 377-389.

Nalewajski, R. F., Mrozek, J. and Mazur, G. (1996): Quantum Mechanical Valence Indices from the One-Determinantal Difference Approach. *Can. J. Chem.* **100**, 1121-1130.

Nalewajski, R. F., Mrozek, J. and Michalak, A. (1997): Two-Electron Valence Indices from the Kohn-Sham Orbitals. *Int. J. Quantum Chem.* **61**, 589-601.

Nalewajski, R. F. and Parr, R. G. (1982): Legendre Transforms and Maxwell Relations in Density Functional Theory. *J. Chem. Phys.* **77**, 399-407. The extremum principle in Eqs. (69) and (70) of this paper is a maximum principle, not a minimum principle.

Nalewajski, R. F. and Parr, R. G. (2000): Information Theory, Atoms in Molecules, and Molecular Similarity. *Proc. Natl. Acad. Sci. USA* **97**, 8879-8882.

Nalewajski, R. F. and Parr, R. G. (2001): Information Theory Thermodynamics of Molecules and Their Hirshfeld Fragments. *J. Phys. Chem. A* **105**, 7391-7400.

Nalewajski, R.F.and Sikora, O. (2000): Electron-Following Mapping Transformations from the Electronegativity Equalization Principle. *J. Phys. Chem. A* **104**, 5638-5646.

Nalewajski, R. F., Świtka, E. (2002): Information Theoretic Approach to Molecular and Reactive Systems. *Phys. Chem. Chem. Phys.* **4**, 4952-4958.

Nalewajski, R. F., Świtka, E., and Michalak, A. (2002): Information Distance Analysis of Molecular Electron Densities. *Int. J. Quantum Chem.* **87**, 198-213.

Parr, R. G., Donnelly, R. A., Levy, M., and Palke, W. E. (1978): Electronegativity: the Density Functional Viewpoint. *J. Chem. Phys.* **69**, 4431-4439.

Parr, R. G. and Pearson, R. G. (1983). Absolute Hardness: Companion Parameter to Absolute Electronegativity. *J. Am. Chem. Soc.* **105**, 7512-7516.

Parr, R. G. and Yang, W. (1984): Density Functional Approach to the Frontier-Electron Theory of Chemical Reactivity. *J. Am. Chem. Soc.* **106**, 4049-4050.

Parr, R. G. and Yang, W. (1989): *Density-Functional Theory of Atoms and Molecules*. Oxford University Press, New York.

Parr, R. G. and Wang, Y. (1997): Kohn-Sham Method as a Free-Energy Minimization at Infinite Temperature. *Phys. Rev. A* **55**, 3226-3228.

Pearson, R. G. (1973): *Hard and Soft Acids and Bases.* Dowden, Hutchinson, and Ross, Stroudsburg.

Pearson, R. G. (1976): *Symmetry Rules for Chemical Reactions: Orbital Topology and Elementary Processes.* Wiley, New York.

Perdew, J. P. (1985): What do the Kohn-Sham Orbital Energies Mean? How do Atoms Dissociate? In *Density Functional Methods in Physics*, Dreizler, R. M. and da Providencia, J., Eds. Plenum, New York, pp. 265-308.

Perdew, J. P. and Levy, M. (1983): Physical Content of the Exact Kohn-Sham Orbital Energies: Band Gaps and Derivative Discontinuities. *Phys. Rev. Lett.* **51**, 1884-1887.

Perdew, J. P. and Levy, M. (1984): Density Functional Theory for Open Systems. In *Many-Body Phenomena at Surfaces*, Langreth, D. and Suhl, H., Eds. Academic Press, Orlando, pp. 71-89.

Perdew, J. P. and Norman, M. R. (1982): Electron Removal Energies in Kohn-Sham Density Functional Theory. *Phys. Rev. B* **26**, 5445-5450.

Perdew, J. P., Parr, R. G., Levy, M., and Balduz, J. L. (1982): Density Functional Theory for Fractional Particle Number: Derivative Discontinuities of the Energy. *Phys. Rev. Lett.* **49**, 1691-1694.

Pfeiffer, P. E. (1978): *Concepts of Probability Theory.* Dover, New York.

Politzer, P. and Parr, R.G. (1976): Separation of Core and Valence Regions in Atoms. *J. Chem. Phys.* **64**, 4634-4637.

Politzer, P. and Truhlar, D. Eds. (1981): *Chemical Applications of Atomic and Molecular Electrostatic Potentials.* Plenum, New York.

Ponec, R. and Strnad, M. (1994): Population Analysis of Pair Densities: A Link between Quantum Chemical and Classical Picture of Chemical Structure. *Int. J. Quantum Chem.* **50**, 43-53.

Ponec, R. and Uhlik, F. (1997): Electron Pairing and Chemical Bonds. On the Accuracy of the Electron Pair Model of Chemical Bond. *J. Mol. Struct.* (*Theochem*) **391**, 159-168.

Salem, L. (1968a): Intermolecular Orbital Theory of the Interaction between Conjugated Systems. I. General Theory. *J. Am. Chem. Soc.* **90**, 543-552.

Salem, L. (1968b): Intermolecular Orbital Theory of the Interaction between Conjugated Systems. II. Thermal and Photochemical Cycloadditions. *J. Am. Chem. Soc.* **90**, 553-566.

Salem, L. (1969): Orbital Interactions in Reaction Paths. *Chem. in Britain* **5**, 449-458.

Sanderson, R. T. (1951): An Interpretation of Bond Lengths and a Classification of Bonds. *Science* **114**, 670-672.

Sanderson, R. T. (1976): *Chemical Bonds and Bond Energy*, 2d Edn. Academic Press, New York.

Savin, A., Nesper, R., Wengert, S. and Fässler, T.F. (1997): ELF: The Electron Localization Function. *Angew. Chem. Int. Ed. Engl.* **36**, 1808-1832.

Sears, S. B. (1980). *Applications of Information Theory in Chemical Physics.* Ph.D Thesis, The University of North Carolina at Chapel Hill.

Sears, S. B., Parr, R. G., and Dinur, U. (1980): On the Quantum-Mechanical Kinetic Energy as a Measure of Information in a Distribution. *Israel J. Chem.* **19**, 165-173.

Sen, K. D. and Jørgensen, C. K. Eds. (1987): *Electronegativity*; *Structure and Bonding* **66**. Springer-Verlag, Berlin.

Sen, K. D. Ed. (1993): *Chemical Hardness*; *Structure and Bonding* **80**. Springer-Verlag, Berlin.

Senet., P. (1996) Non-linear Electronic Responses, Fukui Functions and Hardnesses as Functionals of the Ground-State Electronic Density. *J. Chem. Phys.* **105**, 6471-6489.

Senet., P. (1997): Kohn-Sham Orbital Formulation of the Chemical Electronic Responses, Including the Hardness. *J. Chem. Phys.* **107**, 2516-2524.

Shaik, S. (1989): A Qualitative Valence Bond Model for Organic Reactions. In *New Theoretical Concepts for Understanding Organic Reactions*, NATO ASI Series, Vol. C267. Bertran, J. and Czismadia, I. G., Eds..Kluwer Academic Publ., Dordrecht, pp. 165-217.

Shaik, S. and Hiberty, P. C. (1991): Curve Crossing Diagrams as General Models for Chemical Structure and Reactivity. In *Theoretical Models of Chemical Bonding*, Vol. 4, Maksić Z. B., Ed., pp. 269-322.

Shaik, S. and Hiberty, P. C. (1995): Valence Bond Mixing and Curve Crossing Diagrams in Chemical Reactivity and Bonding. *Adv. Quant. Chem.* **26**, 100.

Shaik, S. and Hiberty, P. C. (2004): Valence Bond Theory, Its History, Fundamentals, and Applications: A Primer. In *Reviews in Computational Chemistry*, Vol. 20, Lipkowitz, K. B., Larter, L., and Cundari, T. R., Eds., pp. 1-100.

Shannon, C. F. (1948): The Mathematical Theory of Communication, *Bell System Technol. J.* 27, 379-493, 623-656.

Shannon, C. E. and Weaver, W. (1949): *The Mathematical Theory of Communication*. University of Illinois, Urbana.

Silvi, B. and Savin A. (1994): Classification of Chemical Bonds Based on Topological Analysis of Electron Localization Functions. *Nature* **371**, 683-686.

Stone, A. J. (1978): Theories of Organic Reactions. In *Specialist Periodical Reports: Theoretical Chemistry, Vol. 3*, Dixon, R.N. and Thomson, C., Eds. Bartholomew Press, Dorking, pp. 39-69.

Swanson, B. I. (1976): Minimum Energy Coordinates. A Relationship between Molecular Vibrations and Reaction Coordinates. *J. Am. Chem. Soc.* **98**, 3067-3071.

Swanson, B. I. and Satija, S. K. (1977): Molecular Vibrations and Reaction Pathways. Minimum Energy Coordinates and Compliance Constants for some Tetrahedral and Octahedral Complexes. *J. Am. Chem. Soc.* **99**, 987-991.

Tisza, L. (1977): *Generalized Thermodynamics*. M.I.T. Press, Cambridge, (Massachusetts).

Wang, W. P. and Parr, R. G. (1977): Statistical Atomic Models with Piecewise Exponentially Decaying Electron Densities. *Phys. Rev. A* **16**, 891-902.

Weizsäcker, C. F. von (1935): Zur Theorie der Kernmassen. *Z. Phys.* **96**, 431- 458.

Wiberg, K. B. (1968): Application of the Pople-Santry-Segal CNDO Method to the Cyclopropylcarbinyl and Cyclobutyl Cation and to Bicyclobutane. *Tetrahedron*, **24**, 1083-1096.

Woodward, R. B. and Hoffmann, R. (1970): *The Conservation of Orbital Symmetry*. Academic Press, London.

Woodward, R. B. and Hoffmann, R. (1971): *The Conservation of Orbital Symmetry*. Verlag Chemie, Weinheim.

Yáñez, R. J., Angulo, J. C., and Dehesa, S. J. (1995): Information Entropies of Many-Electron Systems. *Int. J. Quantum Chem.* **56**, 489-498.

Yang, W. and Parr, R.G. (1985): Hardness, Softness, and the Fukui Function in the Electronic Theory of Metals and Catalysis. *Proc. Natl. Acad. Sci. USA* **82**, 6723-6726.

Yang, W., Parr, R.G., and Pucci, R. (1984): Electron Density, Kohn-Sham Frontier Orbitals, and Fukui Functions. *J. Chem. Phys.* **81**, 2862-2863.

Zhao, Q. and Parr, R. G.(1992): Quantities $T_s[n]$ and $T_c[n]$ in Density-Functional Theory. *Phys. Rev. A* **46**, 2337-2343; (1993): Constrained-Search Method to Determine Electronic Wave Functions from Electronic Densities. *J. Chem. Phys.* **98**, 543-548.

Zhao, Q., Morrison, R. C. and Parr, R. G. (1994): From Electron Densities to Kohn-Sham Kinetic Energies, Orbital Energies, Exchange Correlation Potentials, and Exchange-Correlation Energies. *Phys. Rev. A* **50**, 2138-2142.

Ziesche, P. (1995): Correlation Strength and Information Entropy. *Int. J. Quantum Chem.* **56**, 363-369.

INDEX

Printed and bound by CPI Group (UK) Ltd, Croydon, CR0 4YY
08/05/2025
01864933-0001